Advances in Potato Chemistry and Technology

Advances in Potato Chemistry and Technology

Second Edition

Editors

Jaspreet Singh

Lovedeep Kaur

Riddet Institute and Massey Institute of Food Science and Technology, Massey University, Palmerston North, New Zealand

AMSTERDAM • BOSTON • HEIDELBERG • LONDON
NEW YORK • OXFORD • PARIS • SAN DIEGO
SAN FRANCISCO • SINGAPORE • SYDNEY • TOKYO

Academic Press is an imprint of Elsevier

Academic Press is an imprint of Elsevier
125 London Wall, London EC2Y 5AS, UK
525 B Street, Suite 1800, San Diego, CA 92101-4495, USA
50 Hampshire Street, Cambridge, MA 02139, USA
The Boulevard, Langford Lane, Kidlington, Oxford OX5 1GB, UK

Notices
Knowledge and best practice in this field are constantly changing. As new research and experience broaden our understanding, changes in research methods, professional practices, or medical treatment may become necessary.

Practitioners and researchers must always rely on their own experience and knowledge in evaluating and using any information, methods, compounds, or experiments described herein. In using such information or methods they should be mindful of their own safety and the safety of others, including parties for whom they have a professional responsibility.

To the fullest extent of the law, neither the Publisher nor the authors, contributors, or editors, assume any liability for any injury and/or damage to persons or property as a matter of products liability, negligence or otherwise, or from any use or operation of any methods, products, instructions, or ideas contained in the material herein.

ISBN: 978-0-12-800002-1

British Library Cataloguing-in-Publication Data
A catalogue record for this book is available from the British Library

Library of Congress Cataloging-in-Publication Data
A catalog record for this book is available from the Library of Congress

For information on all Academic Press publications
visit our website at www.elsevier.com

Working together
to grow libraries in
developing countries

www.elsevier.com • www.bookaid.org

Publisher: Nikki Levy
Acquisition Editor: Nancy Maragioglio
Editorial Project Manager: Billie Jean Fernandez
Production Project Manager: Caroline Johnson
Designer: Matthew Limbert

Typeset by TNQ Books and Journals
www.tnq.co.in

Contents

Chapter 10: Postharvest Storage of Potatoes 283
Reena Grittle Pinhero and Rickey Y. Yada

Chapter 11: Organic Potatoes 315
Vaiva Bražinskienė and Kristina Gaivelytė

Chapter 19: Advanced Analytical Techniques for Quality Evaluation of Potato and Its Products 563

Carmen Jarén, Ainara López and Silvia Arazuri

Chapter 20: The Role of Potatoes in Biomedical/Pharmaceutical and Fermentation Applications 603

Shrikant A. Survase, Jaspreet Singh and Rekha S. Singhal

List of Contributors

María Dolores Álvarez Torres   Department of Characterization, Quality, and Safety, Institute of Food Science, Technology and Nutrition (ICTAN-CSIC), Madrid, Spain

Silvia Arazuri   Department of Agricultural Projects and Engineering, Universidad Pública de Navarra, Pamplona, Navarra, Spain

Cristina Barsan   Université de Toulouse, INP-ENSA Toulouse, Génomique et Biotechnologie des Fruits, Castanet-Tolosan, France; INRA, Génomique et Biotechnologie des Fruits, Chemin de Borde Rouge, Castanet-Tolosan, France; Led Academy, Toulon, France

Eric Bertoft   Department of Food Science and Nutrition, University of Minnesota, St Paul, MN, USA

Andreas Blennow   Department of Plant and Environmental Sciences, University of Copenhagen, Frederiksberg C, Denmark

Vaiva Bražinskienė   Faculty of Business and Technologies, Utena University of Applied Sciences, Utena, Lithuania

Fanny Buffetto   INRA, UR1268 Biopolymères, Interactions et Assemblages, Nantes, France

Mary E. Camire   School of Food and Agriculture, University of Maine, Orono, ME, USA

Isabelle Capron   INRA, UR1268 Biopolymères, Interactions et Assemblages, Nantes, France

Marie-Christine Ralet   INRA, UR1268 Biopolymères, Interactions et Assemblages, Nantes, France

Rosana Colussi   Riddet Institute and Massey Institute of Food Science and Technology, Massey University, Palmerston North, New Zealand; Departamento de Ciência e Tecnologia Agroindustrial, Universidade Federal de Pelotas, Pelotas, Brazil

Virginia Corrigan   The New Zealand Institute for Plant & Food Research Ltd, Palmerston North, New Zealand

Pablo Cortés   Department of Chemical Engineering and Bioprocesses, Pontificia Universidad Católica de Chile, Santiago, Chile

Stef de Haan   International Center for Tropical Agriculture (CIAT), Tu Liem, Hanoi, Vietnam

Bruno De Meulenaer   NutriFOODchem Unit, Department of Food Safety and Food Quality, Faculty of Bioscience Engineering, Ghent University, Ghent, Belgium

Mendel Friedman   United States Department of Agriculture, Agricultural Research Service, Western Regional Research Center, Albany, CA, USA

Kristina Gaivelytė   Department of Pharmacognosy, Lithuanian University of Health Science, Kaunas, Lithuania

Fabienne Guillon   INRA, UR1268 Biopolymères, Interactions et Assemblages, Nantes, France

Hanjo Hellmann   School of Biological Sciences, Washington State University, Pullman, WA, USA

Carmen Jarén   Department of Agricultural Projects and Engineering, Universidad Pública de Navarra, Pamplona, Navarra, Spain

Lachman Jaromír   Department of Chemistry, Faculty of Agrobiology, Food and Natural Resources, Czech University of Life Sciences, Prague, Czech Republic

Salwa Karboune   Department of Food Science and Agricultural Chemistry, McGill University, Ste-Anne de Bellevue, Quebec, Canada

Hamouz Karel   Department of Plant Production, Faculty of Agrobiology, Food and Natural Resources, Czech University of Life Sciences, Prague, Czech Republic

Lovedeep Kaur   Riddet Institute and Massey Institute of Food Science and Technology, Massey University, Palmerston North, New Zealand

Régis Kesteloot   Régis Kesteloot Conseil, Lambersart, France

Carol E. Levin   United States Department of Agriculture, Agricultural Research Service, Western Regional Research Center, Albany, CA, USA

Ainara López   Department of Agricultural Projects and Engineering, Universidad Pública de Navarra, Pamplona, Navarra, Spain

María Salomé Mariotti   Department of Chemical Engineering and Bioprocesses, Pontificia Universidad Católica de Chile, Santiago, Chile

Orsák Matyáš   Department of Chemistry, Faculty of Agrobiology, Food and Natural Resources, Czech University of Life Sciences, Prague, Czech Republic

Owen J. McCarthy   Massey Institute of Food Science and Technology, Massey University, Palmerston North, New Zealand

Marian McKenzie   The New Zealand Institute for Plant & Food Research Ltd, Palmerston North, New Zealand

Raquel Medeiros   NutriFOODchem Unit, Department of Food Safety and Food Quality, Faculty of Bioscience Engineering, Ghent University, Ghent, Belgium

Frédéric Mestdagh   NutriFOODchem Unit, Department of Food Safety and Food Quality, Faculty of Bioscience Engineering, Ghent University, Ghent, Belgium

Duroy A. Navarre   USDA-ARS, Washington State University, Prosser, WA, USA

Wenceslao Canet Parreño   Department of Characterization, Quality, and Safety, Institute of Food Science, Technology and Nutrition (ICTAN-CSIC), Madrid, Spain

Anna Patsioura   INRA, UMR 1145 Ingénierie Procédés Alimentaires, Group Interaction between Materials and Media in Contact, Massy, France; AgroParisTech, UMR 1145 Ingénierie Procédés Alimentaires, Massy, France

Franco Pedreschi   Department of Chemical Engineering and Bioprocesses, Pontificia Universidad Católica de Chile, Santiago, Chile

Reena Grittle Pinhero   Department of Food Science, University of Guelph, Guelph, Ontario, Canada

Mohamed Fawzy Ramadan   Agricultural Biochemistry Department, Faculty of Agriculture, Zagazig University, Zagazig, Egypt

M.A. Rao   Cornell University, Geneva, NY, USA

Flor Rodriguez   International Potato Center (CIP), La Molina, Lima, Peru

Roshani Shakya   USDA-ARS, Washington State University, Prosser, WA, USA

Jaspreet Singh   Riddet Institute and Massey Institute of Food Science and Technology, Massey University, Palmerston North, New Zealand

Rekha S. Singhal   Food Engineering and Technology Department, Institute of Chemical Technology, Mumbai, Maharashtra, India

Paul Smith   Cargill R&D Centre Europe, Vilvoorde, Belgium

Shrikant A. Survase   Food Engineering and Technology Department, Institute of Chemical Technology, Mumbai, Maharashtra, India

Gilles Trystram   AgroParisTech, UMR 1145 Ingénierie Procédés Alimentaires, Group Interaction between Materials and Media in Contact, Massy, France

Jean-Michaël Vauvre   INRA, UMR 1145 Ingénierie Procédés Alimentaires, Group Interaction between Materials and Media in Contact, Massy, France; AgroParisTech, UMR 1145 Ingénierie Procédés Alimentaires, Massy, France; McCain Alimentaire S.A.S., Parc d'entreprises de la Motte du Bois, Harnes, France

Olivier Vitrac   INRA, UMR 1145 Ingénierie Procédés Alimentaires, Group Interaction between Materials and Media in Contact, Massy, France; AgroParisTech, UMR 1145 Ingénierie Procédés Alimentaires, Massy, France

Amanda Waglay   Department of Food Science and Agricultural Chemistry, McGill University, Ste-Anne de Bellevue, Quebec, Canada

Rickey Y. Yada   Department of Food Science, University of Guelph, Guelph, Ontario, Canada; Food, Nutrition, and Health Program, Faculty of Land and Food Systems, University of British Columbia, Vancouver, British Columbia, Canada

Foreword

Advances in Potato Chemistry & Technology, Second Edition

Originating in the Andes, where there are over 4000 varieties of native potatoes, the humble potato has been cultivated since at least 5000 BC as a food source for humans.

Today, it is the fourth most important crop in the world after rice, wheat, and maize, with more than a billion people globally eating potatoes as part of their diet on a regular basis. Annual production exceeds 300 million metric tonnes in over 100 countries.

Globally, trends in potato growing and consumption are changing. Traditionally in the domain of Europe and the Americas, there has been a dramatic increase in potato production and consumption in Asia and Africa since 1990.

A global population increasing in both size and wealth is predicted to require the production of 70% more food than today by 2050. In this context, the potato's role in global nutrition will become increasingly critical. To ensure the most efficient use of increasingly scarce natural resources, we must ensure that we understand the functional components within a potato and how to apply them to both food and nonfood applications.

Since its publication in June 2009, the first edition of *Advances in Potato Chemistry and Technology* has been embraced as a, if not the only, comprehensive and convenient single reference of current developments in the field of potato chemistry.

With advances in the field developing rapidly, the second edition is a comprehensive review that keeps this book at the cutting edge by adding the latest information on:

- Potato flavor: the effects of cooking, texture, and metabolites
- Potato proteins: their extraction, nutritional, and health-promoting properties and their application in the biogeneration of peptides
- Colored potatoes: their chemistry and health-related attributes
- Acrylamide in potato products
- Microstructure, starch digestion, and the glycemic index of potatoes
- Composition differences in organically and conventionally grown potatoes
- Mechanisms of oil uptake in french fries

- Potato proteomics: a new approach for the potato processing industry
- Advanced analytical techniques for quality evaluation in potato and its products
- Novel food and nonfood uses of potato and its by-products
- Potatoes and human health
- Novel methods of potato starch modification
- The role of potatoes in global food security
- Lipids of transgenic potato cultivars

We at Potatoes NZ, Inc., are very proud to be associated with the authors and congratulate them on this very comprehensive extensive and cohesive update to what is already a celebrated contribution to the field.

I have no hesitation in recommending this book to scientists, researchers, academics, and graduate students working in the fields of food chemistry, agronomy, genetics, horticulture, and nutrition.

I'm sure that they will find it an indispensable companion as they seek to deepen their understanding of the potato at the molecular level and develop new applications that will ensure that the humble potato remains a cornerstone of human nutrition into the next millennium.

Champak Mehta
CEO, Potatoes NZ, Inc.
September 2015

Chemistry, Processing, and Nutritional Attributes of Potatoes—An Introduction

Jaspreet Singh, Lovedeep Kaur
Riddet Institute and Massey Institute of Food Science and Technology, Massey University, Palmerston North, New Zealand

The potato (*Solanum tuberosum* L.) has an annual world production exceeding 376 million metric tonnes (2013), with China being the top producer (FAOSTAT, 2015). Higher yield per unit area and nutritional value have led to an increase in potato production over past years compared with other tuber crops. In fact, the potato production from developing countries exceeds that of the developed countries (FAO, 2010). The potato plant, a perennial herb belonging to the family Solanaceae, bears white to purple flowers with yellow stamens, and some cultivars bear small green fruits, each containing up to 300 seeds. The potato tuber develops as an underground stem (swollen part of a subterranean rhizome or stolon) bearing auxiliary buds and scars of scale leaves and is rich in starch and storage proteins. Potatoes can be grown from the botanical seeds or propagated vegetatively by planting pieces of tubers. The eyes on the potato tuber surface, which are actually dormant buds, give rise to new shoots (sprouts) when grown under suitable conditions. A sprouted potato is not acceptable for consumption and processing. But optimum sprouting is a desired attribute when the tubers are used for propagation. The production potential of potatoes is quite high, as nearly 80% of the potato plant biomass constitutes economic yield (Osaki et al., 1996).

New cultivars of potatoes with better yield, disease resistance, and desirable end use are being developed with the help of breeding techniques. In the past many years, several potato cultivars with desired yield, dry matter, cooking texture (such as waxy, floury), flesh color, and disease resistance have been developed with the help of breeding. Following the rational development of genetic engineering, many genetically modified potatoes with very high amylose/amylopectin content, antioxidant levels, and tuber yield have also been developed. However, these transgenic varieties of potatoes are not permitted for food use in many countries because of the concerns related to consumer health and the environment. Until these genetically modified potatoes have been given proper clearance by the food authorities and acceptance by the consumers, they may have a good scope for their use in nonfood or other industrial applications.

Morphologically, a potato tuber is usually oval to round in shape, with white flesh and a pale brown skin, although variations in size, shape, and flesh/skin color are also frequently encountered, depending on the genetics of the cultivar. The color, size, and texture of potatoes are the main quality attributes assessed by the consumer for acceptability. Good-quality potatoes are considered to be relatively smooth, firm, and free from sprouts or any other disorders. In a potato tuber, about 20% is dry matter and the rest is water. The yield, the dry matter, and the composition of the dry matter vary among potato cultivars, soil type and temperature, location, cultural practices, maturity, postharvest storage conditions, and other factors (Burton, 1989). Starch is the major component of the dry matter, accounting for approximately 70% of the total solids. The major part of the fresh potato tuber comprises storage parenchyma in which the starch granules are stored as a reserve material. Potatoes are a rich source of high-value protein, essential vitamins, minerals, and trace elements. The average range of raw material composition of a potato tuber is as follows: starch (10–18%) having 22–30% amylose content, total sugars (1–7%), protein (1–2%), fiber (0.5%), lipids (0.1–0.5%), vitamin A (trace/100 g fresh weight, FW), vitamin C (30 mg/100 g FW), minerals (trace), and glycoalkaloids (1–3 mg/100 g FW). The average composition of a potato tuber is presented in Table 1.

In past years, potato breeding programs have targeted only the crop yield and disease resistance; therefore significant gaps exist in the knowledge of the nutrient range, processability, and health-related attributes of the new germplasm. The less well-known constituents of potato tuber are carotenoids and phenolics, which are potent antioxidants. Carotenoid content of potatoes ranges from 50 to 100 μg/100 g FW in white-fleshed cultivars to 2000 μg/100 g FW in deeply yellow- to orange-fleshed cultivars. Potatoes also contain phenolic compounds, predominantly chlorogenic acid, and up to 30 μg/100 g FW of flavonoids in white-fleshed potatoes and roughly

Table 1: Average Composition of a Potato Tuber, per 100 g, After Boiling in Skin and Peeling Before Consumption

Potato	
Water	77 g
Carbohydrates	20.13 g
Energy	87 kcal
Protein	1.87 g
Fat	0.1 g
Calcium	5 mg
Potassium	379 mg
Phosphorus	44 mg
Iron	0.31 mg
Niacin	1.44 mg
Thiamin	0.106 mg
Riboflavin	0.02 mg

USDA, National Nutrient Database.

$60\,\mu g/100\,g$ FW in red- and purple-fleshed potatoes. The total anthocyanin content of whole unpeeled red- and purple-fleshed potatoes may be around $40\,mg/100\,g$ FW (Brown, 2005). The colored potatoes, if processed in a way that does not destroy their anthocyanins and carotenoids, may help in lowering the incidence of several chronic diseases in humans.

Potato is generally processed through boiling, mashing, frying, etc., before consumption. The influence of the chemical composition of potatoes during processing is of significance to maintain the quality of processed potato products. As an example, the texture of potato crisps is dependent mainly on the starch content of the raw potato tubers. The nonstarch polysaccharides (cell wall) also play a crucial role in determining the quality of the crisps while contributing to the tuber fiber content, also. Potatoes with closely packed small and irregular parenchymatous cells have been observed to be relatively hard and cohesive. In contrast, potatoes with large, loosely packed cells are generally less hard. The cell wall characteristics of cooked potato also play an important role in the release of glucose during starch digestion in our body (Singh et al., 2013).

Starch is the major component of potato dry matter and consists of amylose and amylopectin. The structural characteristics and amylose-to-amylopectin ratio of potato starch vary among cultivars. The nutritional and processing quality of potatoes and potato products (frozen and dry) are greatly affected by their starch characteristics. Several chemical, physical, and enzymatic modifications are performed to improve the processing performance of potato starch. Most of these modifications are listed as generally recognized as safe by the safety authorities. Several modified potato starches with slow digestibility are being developed that may provide nutritional benefits for humans. These starches have the potential to be used for the treatment of certain medical conditions (e.g., glycogen storage disease and diabetes mellitus).

The total sugars in potato tuber range from 1 to $7\,g/kg$. The reducing sugars (glucose, fructose) are at the highest levels in young tubers and decrease considerably toward the end of the growing season. The starch-to-sugar conversion during postharvest storage also causes variation in the sugar content of tubers, which is an important consideration in the potato crisp industry. In the past few years, advanced analytical and instrumentation techniques have been introduced to evaluate the quality of the potato and its products. These techniques provide in-depth information about the structure and functionality of potato components, which helps to tailor potato products for desirable attributes.

Potato is an essential and safe source of energy and dietary fiber for children and pregnant women. The nutritional characteristics such as starch digestibility, glycemic index, and relative glycemic impact are important in human health. The microstructure of potatoes, whether natural or created during processing/storage, plays an important role during digestion of starch in the gastrointestinal tract and affects the glycemic index of potatoes. The relationships between the composition of potato tubers and its impact on the release of glucose in the blood have been studied by various researchers through in vitro and in vivo methods. The careful selection of suitable potato cultivars, processing techniques, and storage conditions can prove helpful in getting better nutritional benefits from potatoes. New formats of processed potato products suitable for the nutritional needs and taste of various population groups may help to further increase the consumption of potatoes. Apart from food use, potato products are being used for nonfood applications such as biodegradable packaging, fermentation, vaccines, and pharmaceuticals. New applications are being developed for the utilization of potato by-products and waste, which are otherwise an expensive waste management challenge.

References

Brown, C.R., 2005. Antioxidants in potato. American Journal of Potato Research 82, 163–172.

Burton, W.G., 1989. The Potato. John Wiley and Sons, Inc., New York.

FAOSTAT, 2015. Food and Agriculture Organization of the United Nations. http://www.faostat3.fao.org (accessed 17.09.15.).

FAO, 2010. Strengthening Potato Value Chains: Technical and Policy Options for Developing Countries. Food and Agricultural Organization of United Nations, Rome, Italy.

Osaki, M., Matsumoto, M., Shinano, T., Tadano, T., 1996. A root-shoot interaction hypothesis for high productivity of root crops. Soil Science and Plant Nutrition 42 (2), 289–301.

Singh, J., Kaur, L., Singh, H., 2013. Food microstructure and starch digestion. Advances in Food and Nutrition Research 70, 137–179.

CHAPTER 1

Potato Origin and Production

Stef de Haan[1], Flor Rodriguez[2]
[1]*International Center for Tropical Agriculture (CIAT), Tu Liem, Hanoi, Vietnam;* [2]*International Potato Center (CIP), La Molina, Lima, Peru*

1. Introduction

Western South America is the primary center of the origin and diversity of the potato crop and its wild relatives. Contemporary landrace gene pools occur from 45° south in Chile to 12° northern latitude in Colombia (Hawkes, 1990). Wild relatives of the potato (*Solanum* section *Petota*; Solanaceae) have a much wider distribution range and occur from northern Patagonia to the southern US and the western Atacama desert to eastern South America (Hijmans et al., 2002; Spooner et al., 2004). The genetic diversity of landraces and wild relatives has been and continues to be an extremely valuable source of variation for genetic enhancement, crop improvement, and understanding of chemical variability (Brown et al., 2007; Hanneman, 1989; Jansky et al., 2013; Osman et al., 1978; Väänänen, 2007). At the same time, ongoing evolution of potato diversity in farmers' hands is anticipated to allow for adaptation to climate change and continued food security in extreme agroecologies (Johns and Keen, 1986; Zimmerer, 2014).

The case of the potato in its center of origin is special because landraces are still widely grown by smallholder farmers in semitraditional and market-oriented production systems. Predictions regarding landrace loss, genetic erosion, and full-fledged extinction of genetic diversity (e.g., Fowler and Mooney, 1990; Hawkes, 1973; Ochoa, 1975) have not materialized in the Andes, because autonomous farming rationales and multiple innovations such as revaluation of local cuisines, farmers' markets, and biodiversity seed fairs have renewed local interest in conservation (de Haan et al., 2010b; Monteros, 2011).

Opportunities to screen and evaluate potato genetic resources to explore chemical profiles, determine concentration ranges of (anti)nutritional compounds, and test technological innovations are abundant. The extent of the potato gene pool, with its abundant landrace diversity and numerous wild relatives, offers a wide range of options for prospecting, prebreeding, and niche market development. At the same time, advancements in biotechnology, genomics, analytical techniques, and postharvest technologies open up many new possibilities for the enhanced use of genetic resources (Jo et al., 2014; Singh and Kaur, 2009; Slattery et al., 2010).

Advances in Potato Chemistry and Technology. http://dx.doi.org/10.1016/B978-0-12-800002-1.00001-7

2. Origin, Domestication, and Diversity

2.1 Biosystematics and Evolution

The evolutionary origin of the cultivated potato has not yet been conclusively unraveled, and geneticists, archaeobotanists, and taxonomists alike have explored different hypotheses for nearly 9 decades (Spooner et al., 2014). However, at a cultivated species level it is well documented that different species (*Solanum tuberosum*, *Solanum curtilobum*, *Solanum ajanhuiri*, and *Solanum juzepczukii*) and groups (*S. tuberosum* Chilotanum and Andigenum groups) are the result of unique evolutionary pathways and have different biogeographical distribution patterns.

Of all cultivated species, the *S. tuberosum* Chilotanum group ($2n=4x=48$) has contributed most to the founder effect of the European and North American gene pool and global crop improvement (Figure 1). According to van der Berg and Groendijk-Wilders (2014), over 99% of European extant modern cultivars possess Chilean cytoplasm. Yet, the level of intraspecific diversity within the *S. tuberosum* Chilotanum group is modest compared with the Andigenum group. Its contemporary distribution range is basically restricted to the Chiloe Island of south-central Chile (Contreras and Castro, 2008; Manzur, 2012). Two hypotheses are commonly put forth regarding the origin of Chilotanum landraces (Spooner et al., 2012). The first suggests that they originated independently in southern Chile, possibly involving the putative wild ancestor *Solanum maglia* (Dillehay, 1997; Ugent et al., 1987) or hybrids of *Solanum tarijense* (*Solanum berthaultii*) (Hosaka, 2003; Spooner et al., 2014). The second sustains an Andean origin with early introduction into Chile and a gradual adaptation of Andigenum landraces into long-day adapted Chilotanum landraces (Hawkes, 1990, 1999; Salaman, 1946; Simmonds, 1964, 1966).

The *S. tuberosum* Andigenum group as proposed in the taxonomic treatments of Ovchinnikova et al. (2011) and Spooner et al. (2005a, 2014) includes diploid, triploid, and tetraploid

Figure 1
Sample of mixed Chilotanum landraces from the island of Castro, Chile. *Photo: S. de Haan.*

subgroups previously considered separate species by Hawkes (1990) and Ochoa (1990, 1999). The biogeographical distribution of the group spans from northern Chile and Argentina to Colombia. Spooner et al. (2005a, 2014) proposed a single origin of Andigenum landraces from the *Solanum brevicaule* complex (*S. brevicaule* and *Solanum candolleanum*) involving hybridization, natural variation, polyploidization events, and anthropogenic selection. On the other hand, Hawkes (1990) and Ochoa (1990, 1999) previously proposed separate origins based on taxonomies that recognized multiple likely ancestral species such as *Solanum ambosinum*, *Solanum bukasovii*, *Solanum canasense*, *Solanum leptophyes*, and *Solanum sparsipilum* within what is now considered the *S. brevicaule* complex.

The *S. tuberosum* Andigenum group contains more intraspecific diversity than all other species and the *S. tuberosum* Chilotanum group combined (Figure 2). Based on ploidy, ethnobiological classification, altitudinal distribution, and hotspot concentration, several cultivar groups can be distinguished. A clear example of such a group is the Phureja cultivar group, which is vernacularly classified as Phureja (northwestern Bolivia) or Chaucha (Peru), which has differential agronomic characters (early bulking, lack of dormancy, and up to three cropping cycles a year) and is spatially separated by altitude in its distribution range from other cultivar groups within the *S. tuberosum* Andigenum group (Ghislain et al., 2006; Zimmerer, 1991a,b).

The three species, *S. ajanhuiri* ($2n=2x=24$), *S. juzepczukii* ($2n=3x=36$), and *S. curtilobum* ($2n=5x=60$), each evolved through unique pathways. The distribution of *S. ajanhuiri* is restricted to the Bolivian-Peruvian altiplano region around Lake Titicaca and evolved through hybridization events between the cultivated diploid *Solanum stenotomum* (*S. tuberosum* Andigenum group) and the wild *Solanum megistacrolobum* (*Solanum boliviense*) (Huamán et al., 1982; Johns, 1985; Johns and Keen, 1986). *Solanum juzepczukii* and *S. curtilobum* are so-called bitter species containing high levels of glycoalkaloids. They are commonly used by

Figure 2
Sample of mixed Andigenum landraces from central Peruvian Andes. *Photo: S. de Haan.*

Andean farmers for traditional freeze-drying (Figure 3). The biogeographical distribution range of both species spans from southern Bolivia to central Peru. *Solanum juzepczukii* has commonly been proposed to originate from hybridizations between *S. stenotomum* and tetraploid *Solanum acaule* (Hawkes, 1962; Schmiediche et al., 1980). Different possible origins for the pentaploid *S. curtilobum* have been proposed, involving the tetraploid Andigenum group × *S. juzepczukii* (Hawkes, 1962, 1990) or the triploid Andigenum group × *S. acaule* (Gavrilenko et al., 2013; Spooner et al., 2014).

2.2 Wild Relatives

The taxonomy of *Solanum* section *Petota* is complicated by sexual compatibility among species, interspecific hybridization, auto- and allopolyploidy, a mixture of sexual and asexual reproduction, possible species divergence, and phenotypic plasticity. The resulting complexity causes difficulty in defining and distinguishing species (Camadro et al., 2012; Huamán and Spooner, 2002; Knapp, 2008; Masuelli et al., 2009; Rodríguez et al., 2010; Spooner, 2009; Spooner and van den Berg, 1992). The potato has prezygotic and postzygotic hybridization barriers. *Solanum* species have been assigned endosperm balance numbers (EBNs) based on their ability to hybridize with each other (Hanneman, 1994; Johnston et al., 1980; Ortiz and Ehlenfeldt, 1992). The EBN is a strong postzygotic crossing barrier. Excluding other crossing barriers, successful hybridization is expected when male and female gametes have matching EBN values, regardless of ploidy. The other mechanism is unilateral incompatibility, the prezygotic hybridization barrier, in which pollen tube elongation is inhibited by stylar tissue but the reciprocal cross, when the self-compatible species is used as the female, is successful (Spooner et al., 2014).

The first modern comprehensive taxonomic treatment of the *Petota* section was provided by Hawkes (1956) and was followed by treatments prepared by Correll (1962),

Figure 3
Farmer from southern Andes of Peru preparing tubers from bitter potatoes for freeze-drying.
Photo: C. Fonseca.

Hawkes (1963, 1990), Bukasov (1978), Gorbatenko (2006), and Spooner et al. (2014). Until 1990, only morphology was used to define species. A wide range of molecular markers and deoxyribonucleic acid sequences were progressively combined with morphology to obtain better insights into species boundaries. Morphological studies throughout the range of the *Petota* section showed wide variation of character states within and overlap among closely related species (Spooner et al., 2014). A reclassification by Spooner et al. (2014) recognizes 107 wild species instead of the 228 previously proposed by Hawkes (1990). The latest taxonomic treatments use morphology and molecular markers to reinvestigate species boundaries, combined with the practical ability to distinguish species, following a phylogenetic species concept. The complete list of wild potato species recognized by Spooner et al. (2014) is listed in Table 1.

The degree of relatedness can also be classified based on Harlan and de Wet's (1971) gene pool concept. It is based on the degree of hybridization among species and recognizes three gene pools: primary (GP-1), secondary (GP-2), and tertiary (GP-3). The primary gene pool consists of biological species, and crossing within this gene pool is easy. Hybrids are vigorous, exhibit normal meiotic chromosome pairing, and possess total fertility. In the GP-2 pool are wild relatives crossable with some manipulations, and in GP-3 are those species that need radical techniques to allow gene transfer. Bradeen and Haynes (2011), Jansky et al. (2013), and Veilleux and De Jong (2007) have attempted to determine the three potato gene pools, but EBN and other pre- and postzygotic barriers to hybridization have made it difficult to apply the gene pool concept to potatoes (Chen et al., 2004; Jansky et al., 2013; Masuelli and Camadro, 1997). Spooner et al. (2014) proposed five crossability groups based on EBN and self-compatible/self-incompatible systems. They predicted most possible successful crosses and pointed out that whereas hybridization across groups is less likely to be successful than hybridization within groups, no barrier is complete.

2.3 Landraces

Intraspecific diversity of potato is high and the clonal nature of the crop contributes to consistent folk taxonomic classifications of cultivar groups and individual cultivars (Brush, 1980; de Haan et al., 2007; La Barre, 1947). The basic unit of management and consequent conservation for Andean and Chiloe island farmers is the landrace. It is commonly recognized and named. The International Potato Center's gene bank and most gene banks in the center of origin maintain clonal landraces as well as botanical seed. At the level of use, three landrace groups can be distinguished: (1) commercial or cosmopolitan floury landraces, (2) noncommercial floury landraces, and (3) bitter landraces.

Commercial or cosmopolitan landraces are a selected group of cultivars that enjoy market demand, consumer recognition, or relatively large crop areas (Brush, 2004). In the countries within the center of crop origin, these are among the most abundant landraces (Table 2). Some landraces such as the diploid Peruanita in Peru and Criollo Amarilla in Colombia are grown extensively and are widely available in regular markets and supermarkets alike. Others,

Table 1: *Solanum* Section *Petota* Wild Species with Three-Letter Standard Abbreviations, Countries of Occurrence, Ploidy (and EBN), and Gene Pool Assignment.

Wild Species[a]	Code[a]	Countries[a]	Ploidy[a] and (EBN)[a]	Gene Pool[b]
Solanum acaule Bitter	acl	Argentina, Bolivia, Peru	4x(2EBN), 6x	Primary
Solanum acroglossum Juz.	acg	Peru	2x(2EBN)	Secondary
Solanum acroscopicum Ochoa	acs	Peru	2x	Secondary
Solanum × *aemulans* Bitter and L. Wittmack	aem	Argentina	3x, 4x(2EBN)	
Solanum agrimonifolium Rydberg	agf	Guatemala, Honduras, Mexico	4x(2EBN)	Secondary
Solanum albicans (Ochoa) Ochoa	alb	Ecuador, Peru	6x(4EBN)	Secondary
Solanum albornozii Correll	abz	Ecuador	2x(2EBN)	Secondary
Solanum anamatophilum Ochoa	amp	Peru	2x(2EBN)	Tertiary
Solanum andreanum Baker	adr	Colombia, Ecuador	2x(2EBN),4x(4EBN)	Secondary
Solanum augustii Ochoa	agu	Peru	2x(1EBN)	Tertiary
Solanum ayacuchense Ochoa	ayc	Peru	2x(2EBN)	Secondary
Solanum berthaultii J.G. Hawkes	ber	Argentina, Bolivia	2x(2EBN), 3x	Primary
Solanum × *blanco-galdosii* Ochoa	blg	Peru	2x(2EBN)	
Solanum boliviense M.F. Dunal in DC	blv	Bolivia, Peru	2x(2EBN)	Secondary
Solanum bombycinum C.M. Ochoa	bmb	Bolivia	4x	Secondary
Solanum brevicaule bitter	brc	Argentina, Bolivia	2x(2EBN), 4x(4EBN), 6x(4EBN)	Primary
Solanum × *brucheri* D.S. Correll	bru	Argentina	3x	Secondary
Solanum buesii Vargas	bue	Peru	2x(2EBN)	Tertiary
Solanum bulbocastanum Dunal in Poiret	blb	Guatemala, Mexico	2x(1EBN), 3x	Secondary
Solanum burkartii Ochoa	brk	Peru	2x	Secondary
Solanum cajamarquense Ochoa	cjm	Peru	2x(1EBN)	Primary
Solanum candolleanum berthault	buk	Peru	2x(2EBN), 3x	Secondary
Solanum cantense Ochoa	cnt	Peru	2x(2EBN)	Tertiary
Solanum cardiophyllum Lindley	cph	Mexico	2x(1EBN), 3x	Secondary
Solanum chacoense bitter	chc	Argentina, Bolivia, Paraguay, Peru, Uruguay	2x(2EBN), 3x	
Solanum chilliasense Ochoa	chl	Ecuador	2x(2EBN)	Secondary
Solanum chiquidenum Ochoa	chq	Peru	2x(2EBN)	Secondary
Solanum chomatophilum bitter	chm	Peru	2x(2EBN)	Secondary
Solanum clarum D.S. Correll	clr	Guatemala, Mexico	2x	Secondary
Solanum colombianum Dunal	col	Colombia, Ecuador, Panama, Venezuela	4x(2EBN)	Secondary

Species	Code	Distribution	Ploidy (EBN)	Series
Solanum commersonii M.F. Dunal	cmm	Argentina, Brazil, Uruguay	2x(1EBN), 3x	Tertiary
Solanum contumazaense Ochoa	ctz	Peru	2x(2EBN)	Secondary
Solanum demissum Lindley	dms	Guatemala, Mexico	6x(4EBN)	Secondary
Solanum × doddsii D.S. Correll	dds	Bolivia	2x(2EBN)	
Solanum dolichocremastrum bitter	dcm	Peru	2x(1EBN)	Tertiary
Solanum × edinense P. Berthault	edn	Mexico	5x	
Solanum ehrenbergii (bitter) Rydberg	ehr	Mexico	2x(1EBN)	Tertiary
Solanum flahaultii bitter	flh	Colombia	4x	Secondary
Solanum gandarillasii H.M. Cárdenas	gnd	Bolivia	2x(2EBN)	Secondary
Solanum garcia-barrigae Ochoa	gab	Colombia	4x	Secondary
Solanum gracilifrons bitter	grc	Peru	2x	Secondary
Solanum guerreroense D.S. Correll	grr	Mexico	6x(4EBN)	Secondary
Solanum hastiforme Correll	hsf	Peru	2x(2EBN)	Secondary
Solanum hintonii D.S. Correll	hnt	Mexico	2x	Secondary
Solanum hjertingii J.G. Hawkes	hjt	Mexico	4x(2EBN)	Secondary
Solanum hougasii D.S. Correll	hou	Mexico	6x(4EBN)	Secondary
Solanum huancabambense Ochoa	hcb	Peru	2x(2EBN)	Secondary
Solanum humectophilum Ochoa	hmp	Peru	2x(1EBN)	Tertiary
Solanum hypacrarthrum bitter	hcr	Peru	2x(1EBN)	Tertiary
Solanum immite Dunal	imt	Peru	2x(1EBN), 3x	Tertiary
Solanum incasicum Ochoa	ins	Peru	2x(2EBN)	Secondary
Solanum infundibuliforme R.A. Philippi	Inf	Argentina, Bolivia	2x(2EBN)	Primary
Solanum iopetalum (bitter) J.G. Hawkes	iop	Mexico	6x(4EBN)	Secondary
Solanum jamesii J. Torrey	jam	Mexico, US	2x(1EBN)	Tertiary
Solanum kurtzianum bitter and L. Wittmack	ktz	Argentina	2x(2EBN)	Secondary
Solanum laxissimum bitter	lxs	Peru	2x(2EBN)	Secondary
Solanum lesteri J.G. Hawkes and Hjerting	les	Mexico	2x	Secondary
Solanum lignicaule Vargas	lgl	Peru	2x(1EBN)	Tertiary
Solanum limbaniense Ochoa	lmb	Peru	2x(2EBN)	Secondary
Solanum lobbianum bitter	lbb	Colombia	4x(2EBN)	Secondary
Solanum longiconicum bitter	lgc	Costa Rica, Panama	4x	Secondary
Solanum maglia D.F.L. von Schlechtendal	mag	Chile	2x, 3x	Secondary
Solanum malmeanum bitter		Argentina, Brazil, Paraguay, Uruguay	2x(1EBN), 3x	Tertiary
Solanum medians bitter	med	Peru	2x(2EBN), 3x	Secondary
Solanum × michoacanum (bitter) Rydberg	mch	Mexico	2x	Secondary

Continued

Table 1: Solanum Section Petota Wild Species with Three-Letter Standard Abbreviations, Countries of Occurrence, Ploidy (and EBN), and Gene Pool Assignment.—cont'd

Wild Species[a]	Code[a]	Countries[a]	Ploidy[a] and (EBN)[a]	Gene Pool[b]
Solanum microdontum bitter	mcd	Argentina, Bolivia	2x(2EBN), 3x	Secondary
Solanum minutifoliolum Correll	min	Ecuador	2x(1EBN)	Tertiary
Solanum mochiquense Ochoa	mcq	Peru	2x(1EBN)	Tertiary
Solanum morelliforme bitter and Muench	mrl	Guatemala, Mexico, Honduras	2x	Secondary
Solanum multiinterruptum bitter	mtp	Peru	2x(2EBN), 3x	Secondary
Solanum neocardenasii J.G. Hawkes and J.P. Hjerting	ncd	Bolivia	2x	Secondary
Solanum neorossii J.G. Hawkes and J.P. Hjerting	nrs	Argentina	2x	Secondary
Solanum neovavilovii Ochoa	nvv	Peru	2x(2EBN)	Secondary
Solanum × neoweberbaueri Wittm	nwb	Peru	3x	
Solanum nubicola Ochoa	nub	Peru	4x(2EBN)	Secondary
Solanum okadae J.G. Hawkes and J.P. Hjerting	oka	Bolivia	2x	Primary
Solanum olmosense Ochoa	olm	Ecuador, Peru	2x(2EBN)	Secondary
Solanum oxycarpum Schiede in D.F.L. von Schlechtendal	oxc	Mexico	4x(2EBN)	Secondary
Solanum paucissectum Ochoa	pcs	Peru	2x(2EBN)	Secondary
Solanum pillahuatense Vargas	pll	Peru	2x(2EBN)	Secondary
Solanum pinnatisectum Dunal	pnt	Mexico	2x(1EBN)	Tertiary
Solanum piurae bitter	pur	Peru	2x(2EBN)	Secondary
Solanum polyadenium Greenman	pld	Mexico	2x	Secondary
Solanum raphanifolium Cárdenas and Hawkes	rap	Peru	2x(2EBN)	Secondary
Solanum raquialatum Ochoa	raq	Peru	2x(1EBN)	Tertiary
Solanum × rechei J.G. Hawkes and J.P. Hjerting	rch	Argentina	2x, 3x	
Solanum rhomboideilanceolatum Ochoa	rhl	Peru	2x(2EBN)	Secondary
Solanum salasianum Ochoa	sls	Peru	2x	Secondary
Solanum × sambucinum Rydberg	smb	Mexico	2x	Tertiary
Solanum scabrifolium Ochoa	scb	Peru	2x	Secondary
Solanum schenckii bitter	snk	Mexico	6x(4EBN)	Tertiary
Solanum simplicissimum Ochoa	smp	Peru	2x(1EBN)	Secondary
Solanum sogarandinum Ochoa	sgr	Peru	2x(2EBN), 3x	Tertiary
Solanum stenophyllidium bitter	sph	Mexico	2x(1EBN)	Secondary
Solanum stipuloideum Rusby	stp	Bolivia	2x(1EBN)	Tertiary
Solanum stoloniferum D.F.L. von Schlechtendal	sto	Mexico	4x(2EBN)	Secondary
Solanum tarnii J.G. Hawkes and Hjerting	trn	Mexico	2x	Tertiary

Species		Country		
Solanum trifidum D.S. Correll	trf	Mexico	2x(1EBN)	Tertiary
Solanum trinitense Ochoa	trt	Peru	2x(1EBN)	Tertiary
Solanum ×vallis-mexici Juz	vll	Mexico	3x	Secondary
Solanum venturii J.G. Hawkes and J.P. Hjerting	vnt	Argentina	2x(2EBN)	Primary
Solanum vernei bitter and L. Wittmack	vrn	Argentina	2x(2EBN)	Secondary
Solanum verrucosum D.F.L. von Schlechtendal	ver	Mexico	2x(2EBN), 3x, 4x	Secondary
Solanum violaceimarmoratum bitter	vio	Bolivia, Peru	2x(2EBN)	Tertiary
Solanum wittmackii bitter	wtm	Peru	2x(1EBN)	

[a]Spooner et al. (2014).
[b]Castañeda-Álvarez et al. (2015).

Table 2: Contemporary Crop Area, Areal Proportion, and Diversity of Landraces in the Center of Origin and Diversity.

Country	Potato Area (ha)	Estimated Areal Proportion with Landraces (%)	Approximate Total In Situ Diversity of Landraces	Well-Known Commercial or Cosmopolitan Landraces	Well-Known Bitter Landraces
Argentina	69,500	<5%	50–70	Chicarera, Tuni, Perija, Negrita, Churqueña	–
Bolivia	192,277	70–80%	1000–1500	Alqa Imilla, Yana Imilla, Yuraq Imilla, Imilla Rosada, Waycha Paceña	Azul Luki, Wila Luki, Laran Luki, Qanqu Chuqipitu
Chile	49,576	<5%	300–400	Michuñe Roja, Michuñe Negra, Michuñe Blanca, Cabra, Murta, Clavela	–
Colombia	114,715	10–12%	180–240	Criolla Amarilla, Tuquerreña, Carriza, Argentina, Salentuna, Colombina, Bandera, Mambera, Ratona, Tornilla	–
Ecuador	47,302	<5%	350–450	Yema de huevo, Uvilla, Leona Blanca, Leona Negra, Coneja Negra, Coneja Blanca, Puña, Bolona, Jubaleña, Chaucha Amarilla	–
Peru	317,132	30–40%	2800–3300	Peruanita, Camotillo, Muru Huayro, Huayro Macho, Huamantanga, Amarilla Tumbay, Amarilla del Centro, Ccompis	Yuraq Siri, Yana Siri, Piñaza, Qanchillu, Locka
Venezuela	35,233	<5%	30–40	Arbolona Negra, Cucuba, Tocana, Concha Gruesa, Tiniruca, Guadalupe	–

Coca Morante (2002), Contreras and Castro (2008), FAO (2013), Iriarte et al. (2009), Monteros (2011), Monteros et al. (2010), Pumisacho and Sherwood (2002), Terrazas et al. (2008), and Ugarte and Iriarte (2000); expert opinion validation by Universidad Nacional de Colombia, CORPOICA, INIAP Ecuador, INIA Peru, INIAF Bolivia.

such as the tetraploid Mechuñe Roja in Chile or Tuni in Argentina, are available only in specialty restaurants and fairs. Since the early 2000s several value chain projects in Bolivia, Ecuador, and Peru have brought about shifts in the market demand of landraces that were previously noncommercial.

The bulk of landraces managed by smallholder farmers in the Andes and Chiloe islands are noncommercial and floury (nonbitter). This group consists of thousands of cultivars that are grown almost exclusively for home consumption. Farmers' reasons for cultivating this diversity consist of multiple complementary livelihood rationales including risk mitigation and yield stability, management options, preference traits, and superior quality in local cuisines, and prestige and cultural identity, among other factors (Brush, 2004; Zimmerer, 1996). Some regions can be considered contemporary landrace hotpots with high levels of diversity, such as Huancavelica (Peru) (CIP, 2006; de Haan et al., 2013), Paucartambo (Peru) (Pérez Baca, 1996; Zimmerer, 1996), northern La Paz (Bolivia) (Iriarte et al., 2009), and northern Potosí (Bolivia) (Terrazas et al., 2008). Yet, whereas other regions may harbor only modest levels of total diversity, this same diversity could consist of unique or endemic landraces.

Native bitter landraces are grown by smallholder farmers in central southern Peru and Bolivia. It concerns a group of landraces that are robust and commonly frost resistant (Christiansen, 1977; Condori et al., 2014). They are almost exclusively used to freeze tubers into *chuño*, *moraya*, or *tunta* (de Haan et al., 2010a). Intraspecific diversity of *S. juzepczukii* and *S. curtilobum* consists exclusively of bitter landraces. However, a common misconception is that intraspecific diversity of *S. ajanhuiri* also consists exclusively of bitter landraces and that tetraploid Andigenum landraces are always nonbitter. *Solanum ajanhuiri* landraces are commonly nonbitter whereas Andigenum landraces are sometimes bitter and destined for freeze-drying. The diversity of bitter landraces is modest compared with floury landraces, but between Peru and Bolivia they amount to at least 100 distinct cultivars.

3. Production in the Center of Origin

Cropping systems of the potato crop in the center of origin are extremely diverse. They can be differentiated according to their main orientation and agroecological setting. Orientation ranges from small-scale, nonmechanized family farming for household food security to medium- to large-scale mechanized production for the market, and to medium-scale mixed production systems that pursue both self-sufficiency and income generation. Macrolevel potato agroecologies in the region are the central dry Andes (Peru and Bolivia), northern wet Andes (Ecuador to Venezuela), and Southern Cone.

Small-scale family farming is predominant in the region in terms of the number of households that depend on this activity. In Peru alone, more than 600,000 households depend on potato production to support their livelihoods (Table 3). Small-scale family

Table 3: Number of Production Units, Average Yield, Production, and Consumption of Potato at Center of Origin.

Country	Number of Production Units (households/companies[a])	Average Yield (kg/ha)	Total Production (tons)	Consumption (kg/capita)
Argentina	3500	28,777	2,000,000	42
Bolivia	240,000	5768	1,108,994	43
Chile	59,265	23,379	1,159,022	51
Colombia	100,000	20,900	2,129,319	45
Ecuador	88,130	7313	345,922	32
Peru	600,000	14,413	4,570,673	68
Venezuela	12,500	17,694	623,399	25

[a]Private sector in Argentina and Chile.
Devaux et al. (2010), FAO (2013), INIA and Red LatinPapa (2012), and Scott (2011b); expert opinion validation by Universidad Nacional de Colombia, CORPOICA Colombia, INIA Venezuela, INTA, and McCain Argentina.

farming (0.1–2 ha/household) is often characterized by the use of animal traction or foot plows, field scattering, relatively high landrace diversity, limited use of external inputs, and the use of semitraditional practices such as zero-tillage cropping (de Haan and Juarez, 2010; Oswald et al., 2009). Medium- to large-scale mechanized production (5–80 ha/household) is frequently intensive and orientated toward the production of bred varieties or cosmopolitan landraces. Although the demand for processing varieties has been growing steadily, the bulk of commercial production is oriented toward the supply of ware potatoes (Scott, 2011a). Medium-scale mixed production systems (2–5 ha/household) that pursue both self-sufficiency and income generation are common in the Andes and southern Chile. Depending on the final destination of the production, either for home consumption or markets, the use of external inputs is often differential.

Potato production in the central dry Andes of Peru and Bolivia follows two cropping calendars, the big campaign or *qatun tarpuy* (November to May), which coincides with the rainy season, and the small or early campaign (*miska* or *maway*, June to December). Planting dates are generally the same for bred varieties and landraces, although the harvest of bred varieties tends to be earlier. Potatoes are produced under short-day conditions at altitudes up to 4300 m. Production in the northern wet Andes of Ecuador to Venezuela occurs up to 3300 m altitude under short-day conditions and follows several cropping calendars that are less defined compared with the dry Andes. In southern Chile, potatoes are grown during summer under long-day conditions with main planting for the *papa guarda* season starting in October and harvests lasting until April.

Seed tubers are predominantly supplied through farmer or informal seed systems (Thiele, 1999). The exception concerns the supply of certified tuber seeds of bred varieties in Chile and Argentina. Climate change and the consequent pressures of alterations in pest and diseases are a threat to seed quality and ware potato production alike. In particular, the altitudinal expansion of potato late blight (*Phytophthora infestans*) and potato tuber moth (*Phthorimaea operculella*) pose a threat to the potato in the high Andes (Giraldo et al., 2010; Kroschel et al., 2013).

4. Spread and Global Production Trends Outside the Center of Origin

4.1 Spread of a New World Crop

Botanists, historians, economists, anthropologists, and novelists alike have been interested in the fascinating story involving the spread of the potato from its center of origin to the rest of the world (Figure 4). At best, historical records of the early introduction of potatoes to Europe are sparse (Glendinning, 1983; Oliemans, 1988; Salaman, 1949). Initial fourteenth-century introductions have been shown to be of Andean origin; the introduction from Chile occurred as early as 1811, 34 years before the European late blight (*P. infestans*) epidemics (Ames and Spooner, 2008). The first written record of potatoes in Europe dates from 1567. The actual document reports a consignment of tubers from Gran Canary Island destined for Belgium

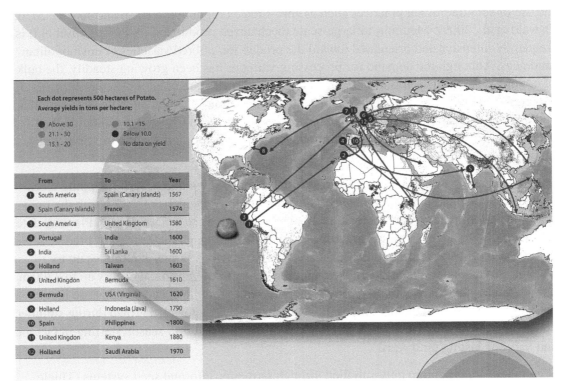

Figure 4
Chronological time series map with some of major events of global potato diffusion.

(Lobo-Cabrera, 1988; Ríos et al., 2007; Spooner et al., 2005b). Hawkes and Francisco-Ortega (1993) speculated that potatoes were introduced into the Canary Islands as early as 1562. Eleven years later, in 1573, potatoes were reported in continental Spain (Hamilton, 1934; Hawkes and Francisco-Ortega, 1992; Salaman, 1937). The second record of a shipment of potatoes from the Canary Islands destined for France is from 1574 (Hawkes and Francisco-Ortega, 1993; Ríos et al., 2007).

According to Salaman (1949) potatoes were introduced to England by Sir Francis Drake around 1586, to India by Portuguese sailors in the early seventeenth century, and around the same time to Sri Lanka by the British (Graves, 2001; Nunn and Qian, 2011). The British took potatoes to Bermuda around 1613 and from there to Virginia (United States) in 1621 (Graves, 2001). The potato reached Ireland in the late sixteenth century (Salaman, 1949). Unquestionably, the potato transformed European society because farmers could produce more food in less time and in small plots. This jump boosted population growth and enabled the Industrial Revolution (Reader, 2008). In 1603, potatoes were reportedly taken by Dutch settlers to the Penghu Islands in the Taiwan Strait. Around that same period, toward the end of the Ming dynasty, cultivation also quickly spread throughout China. It is generally accepted that potatoes were first introduced to New Zealand in the late eighteenth century by Captain James Cook and the French explorer Marion du Fresne (Harris and Niha, 1999). Potatoes soon became both a staple crop, with various cultivars receiving Maori names. Introduction of the potato into Africa occurred late; German and English settlers and missionaries introduced it into East Africa in the nineteenth century (Kiple and Ornelas, 2000).

4.2 Production Trends

With a total cropping area of about 20 million hectares globally, the potato is currently the fourth most important staple crop after rice, wheat, and maize. Since the early 1960s, growth in potato production has rapidly overtaken all other food crops in Africa and Asia. In 2005, for the first time, the combined potato production of Africa, Asia, and South America exceeded that of Europe and the United States (Birch et al., 2012; FAO, 2008). Nowadays, roughly half of the global potato cropping area is concentrated in Asia (96 million hectares).

China, India, and Russia are by far the largest producers in the world, with a national production output of 88, 45, and 30 million tons, respectively, registered in 2013, compared with 52, 19, and 9 million tons for the 28 member countries of the European Union, the United States, and the center of crop origin, respectively. Over 3 decades (1981–2011), the total potato cropping area in the Americas and Oceania has remained more or less the stable. However, during that period, Africa and Asia have seen staggering growth, with 300% and 237% increases in total area. In contract, the cropping area in Europe halved during that period.

The global average potato yield is 18.9 metric tons per hectare (t/ha). The yield gap is particularly high in Africa, where average yields are 14.2 t/ha, compared with 18.3, 21.1, and

25.9 t/ha for Asia, Europe, and the Americas, respectively (FOASTAT, 2013). International and national breeding programs in low- to middle-income countries are focusing on bridging the yield gap, solving classical constraints such as viruses (Potato Virus Y, Potato Leafroll Virus) and late blight (*P. infestans*) through resistance breeding, as well new challenges. The latter increasingly include an emphasis on abiotic stress (heat, salt, and drought tolerance), quality aspects (micronutrient density and taste profiles), and early bulking and storability. Of course, improved and sustainable management options will have to accompany genetic gains to bridge the yield gap.

A clear trend in selected Asian, African, and Latin American countries concerns the increased investment of governments in potato research, development, and capacity building. Unfortunately, this positive trend is not observed in all low- and middle-income countries where potato is an important staple. Therefore, the divide between strong and weak potato research programs within the same continents remains notably sharp. Yet, potato research in China, India, Rwanda, Brazil, Ecuador, and Chile, among other countries, has benefitted tremendously from consistent government support and targeted research agendas in past decades. This has resulted in the development of nationally adapted expertise and the release of a diverse portfolio of technologies that have been nationally adopted, including bred varieties and seed potato technologies.

In Europe and the United States, more potatoes are processed to meet the rising demand from the fast food, snack, and convenience food industries. Food and Agriculture Organization of the United Nations data reveal that from 2003 to 2007, average shares of exports in total production quantities reached around 7% for the world as a whole (FAO, 2013). The major drivers behind this development include growing urban populations, rising incomes, and the diversification of diets and lifestyles that leave less time for preparing fresh tubers for consumption (FAO, 2008). Walker et al. (1999) highlight that per capita consumption of potato tends to steadily decline as per capita income rises. Yet, the global demand for fries continues to grow and exports from the top 10 exporting countries increased by 71.5% between 2000 and 2010. Demand in Africa, Asia, and South America also continues to expand but internal supply is relatively low, thus fueling imports from Europe, Oceania, and the United States.

International trade is dominated by french fries, starch, and seed potatoes. Developing countries are still net importers in the international potato trade. However, multinational and national processing companies have been expanding their operations and processing capacity in Asia and Africa. Thus, it can be foreseen that the internal supply of potato for processing and convenience food industries will increase in countries where the crop is currently a staple food (Bradshaw and Ramsay, 2009; Kirkman, 2007). China has been the world's largest potato producer since 1993 and has accounted for more than 80% of the increase in global potato production from 1990 to 2002 (Wang and Zhang, 2004). The growth in China's potato industry has yet not changed global markets and the country remains a modest exporter. China's annual exports of fresh potatoes has held stable at 300,000–350,000 metric tons since 2005, with Malaysia, Vietnam, and Russia being the most important importing countries.

Despite its importance as a staple food and in combating hunger and poverty, the potato has been largely neglected in agricultural development policies for food crops. This may be partly related to the image of the potato as a fast food contributing to rising levels of obesity in high-income countries. In this sense, an important new trend relates to the increased effort of the global potato community, including industry, to highlight the multiple health benefits of the potato when it is not fried. The "Potato goodness unearthed" campaign by the US Potato Board and the "Potatoes are healthy" message by HZPC in Holland are two examples of this important effort. Behind it is a growing body of research that provides an evidence base for the positive role that potatoes have in providing nonfat-derived energy, vitamin C, micronutrients, carotenoids, and antioxidants (Burgos et al., 2007, 2009a,b; Graham et al., 2007; King and Slavin, 2013; Nesterenko and Sink, 2003; Singh and Kaur, 2009; Vinson et al., 2012).

5. Conservation in the Center of Origin
5.1 Ex Situ Conservation

Ex situ conservation involves conserving components of biological diversity outside their natural habitat (UNCED, 1992) in botanical gardens, gene banks, or captive breeding programs. Ex situ conservation targets collections of cultivars of important food crops in gene banks, including old landraces and modern varieties, interspecific hybrids, breeding clones, and crop wild relatives (GCDT, 2006). Germplasm conserved ex situ is ideally demanded by users who need ready access for genetic enhancement, crop improvement, varietal selection, or rehabilitation of in situ reserves (Cohen et al., 1991).

In the center of the origin of the potato, at least eight important potato collections can be identified (Table 4). Together, these maintain approximately 19,777 accessions (GCDT, 2006; personal communication with curators). Potato genetic resources are largely conserved as botanical seeds (wild species) or vegetatively as tubers and in vitro plantlets (cultivated species and breeding stocks). Landraces form the largest group of accessions, followed by wild species. There is expected to be a high level of duplication among gene banks in the center of origin.

The International Potato Center (CIP) (Peru) is the custodian of the world's largest in vitro field and true seed collection of potato, with approximately 4461 landrace accessions from all cultivated species and 2520 accessions of wild relatives of cultivated potato (Castañeda-Álvarez et al., 2015). The genetic diversity conserved at CIP, and related information about each accession, are globally available through the Multilateral System designed by the International Treaty on Plant Genetic Resources for Food and Agriculture.

Table 4: Composition and Size of Potato Collections in Latin America.

Collection/Country	aCIP, Peru	aINTA, Argentina	bCORPOICA, Colombia	cINIAF, Bolivia	aUACH, Chile	aINIAP, Ecuador	aINIA, Peru	dINIA, Chile
Wild species, no. species	151	30	10	35	6	43	0	0
Wild species, total accessions	2363	1460	118	500	183	275	0	0
Landraces, total accessions	4461	551	947	1400	331	222	310	388
Cultivars (old/new)	314	0	20	7	83	14	20	540
Other materialse	3170	0	0	300	1500	0	300	
Total accessions	10,308	2011	1085	2207	2097	511	630	928

aGCDT (2006).
bCORPOICA (personal communication).
cINIAF Bolivia (personal communication).
dINIA Chile (personal communication).
eBreeding lines, hybrids, etc.

Important potato germplasm collections in Latin America are maintained at national gene banks. These include the Instituto Nacional de Tecnología Agropecuaria (INTA) (Argentina), Corporacion Colombiana de Investigacion Agropecuaria (CORPOICA) (Colombia), Universidad Nacional de Colombia (UNC) (Colombia), Instituto Nacional de Innovación Agropecuaria y Forestal (INIAF) (Bolivia), Instituto de Investigaciones Agropecuarias (INIA) (Chile), Universidad Austral de Chile (UACH) (Chile), Instituto Nacional de Investigaciones Agropecuarias (INIAP) (Ecuador), Centro Nacional de Recursos Genéticos–Instituto Nacional de Investigaciones Forestales, Agrícolas y Pecuarias (CNRG-INIFAB) (Mexico), Instituto Nacional de Innovación Agraria (INIA) (Peru), Instituto Nacional de Investigación Agropecuaria (INIA) (Uruguay), and Centro Nacional de Pesquisa de Recursos Genéticos e Biotecnologia (CENARGEN) at the Empresa Brasileira de Pesquisa Agropecuária (EMBRAPA) (Brazil).

Most accessions maintained at national gene banks are of native origin (FAO, 2010). This means that they are often uniquely representative of certain species or cultivar groups. For example, this is the case of the collection of the *S. tuberosum* Chilotanum group in Chile (INIA and UACH), *S. ajanhuiri* in Bolivia (INIAF), and the Phureja Group in Colombia (UNC). Unfortunately, not all of these unique collections are globally available. There are also numerous ex situ collections of potato in smaller gene banks in the region. For example, the AGROECO project in Peru (2011–2014) collected 3501 landraces from all potato-producing districts of the Cusco region. These were contributed by 233 smallholder farmers; 509 unique landraces were identified after morphological, molecular, ethnobotanical, and agronomic studies. The complete collection is maintained at the Universidad Nacional Agraria La Molina (Lima, Peru) and Universidad Nacional de San Antonio Abad (Cusco, Peru).

The International Potato Center (CIP) and the International Center for Tropical Agriculture (CIAT) have been involved in efforts to identify gaps in global collections of crop wild relatives. An assessment of the current ex situ conservation status of the closest relatives of the cultivated potato was conducted (Castañeda-Álvarez et al., 2015). A total of 49,164 records for 73 species of wild relatives of potato were compiled (75.8% with coordinates), corresponding to 11,100 germplasm accessions and 37,251 reference sightings. Thirty-two species (44%) were targeted as high-priority for further collecting owing to significant gaps in ex situ collections available to the international research community. Importantly, four species require urgent attention because no active germplasm accessions are currently available in any global collections: *S. ayacuchense*, *S. neovavilovii*, *S. olmosense*, and *S. salasianum*. In addition, CIP is currently conducting a genetic gap analysis for the cultivated *S. tuberosum* Chilotanum group. It involves the principal collections of the following gene bank: CIP (Peru, 129 accessions), INIA (Chile, 335), Vavilov Institute (Russia, 60), and the US Department of Agriculture (12) with the aim of identifying gaps and redundancy among collections at the landrace and genetic levels.

5.2 On-Farm Conservation

In the strictest sense, on-farm conservation of potato landraces is a historical and autonomous farmer-driven phenomenon (Brush, 2000). Despite severe shocks, negative trends, and overall societal change, farmers in the Andes and Chiloe islands have continued to grow and conserve impressive landrace diversity. Farmer-driven conservation is a dynamic management strategy resulting in shifts in landrace portfolios based on farmer demand and adaptation potential (Bretting and Duvick, 1997). Thus, its outcomes are not static or absolute, as would be expected of conservation in gene banks.

Since the early 1990s, there has been an increase in projects and interventions that aim to support potato farmers in their effort to conserve landrace diversity on-farm (Scott, 2011c). Most of these initiatives act on the notion that potato diversity is becoming lost. Yet, scientific evidence of landrace loss and genetic erosion is scarce, even though several studies have explored the phenomenon and possible underlying causes (Brush et al., 1992; de Haan et al., 2013; Dueñas et al., 1992; Winters et al., 2006; Zimmerer, 1991a). On-farm conservation interventions themselves differ hugely in their approach and are frequently based on contrasting views of development ranging from value chain promotion (Meinzen-Dick et al., 2009) to so-called cultural reaffirmation (Apffel-Marglin, 1998). Development and nongovernmental organizations have adopted a diversity of interventions including park models, community seed banks, youth engagement, and integrated pest management. An approach that has been particularly successful involves biodiversity seed fairs (Tapia, 2000; Tapia and Rosas, 1993).

Since the early 2000s, there have been multiple initiatives throughout the Andes to register contemporary levels of potato landrace diversity in catalogs at the community (Cosio Cuentas, 2006; Gutiérrez and Valencia, 2010; Hancco et al., 2008), provincial (CIP, 2006; González, 2013), or national level (Monteros et al., 2010). This information provides an initial although as yet incomplete baseline for future monitoring of landrace conservation dynamics. However, besides documenting the absolute landrace diversity, it is necessary to register their relative abundance, their spatial distribution, associated collective knowledge, as well as farmer-perceived disincentives and threats to conservation. The International Potato Center (CIP) has started to set up a network of complementary observatories for systematic baseline documentation and future monitoring so that scientists can start to draw evidence-based lessons about shifts in the conservation status of landraces.

5.3 Ongoing Evolution

Insights into the domestication and evolution of the cultivated potato have advanced significantly based on genetic, taxonomic, and archeological evidence. However, with a few notable exceptions (Johns, 1989; Johns and Keen, 1986), little is known about how farmers actually might have facilitated this process, how much was spontaneous (natural selection) or human-mediated (anthropogenic selection), and whether farmer-driven evolution of the potato is something from the past or is an ongoing process. Understanding evolutionary adaptation of crops requires understanding their human ecology and the selective pressures farmers exert when they manipulate landraces that grow in close proximity to compatible wild relatives (McKey et al., 2012).

Ongoing evolution of crop genetic resources in their center of origin and diversity, where they are exposed to a dynamic state of management, environmental stress, and proximity to crop wild relatives, among other forces of natural and human selection, is commonly considered an essential contribution of dynamic on-farm conservation (Bretting and Duvick, 1997; Maxted et al., 1997). At least four pathways can contribute to ongoing farmer-mediated evolution of the potato. First is geneflow among landraces and compatible wild relatives and the eventual incorporation of new genotypes into farmer landrace stocks. The second involves the collection of selected semiwild potatoes brought into cultivation. The third pathway concerns mutations leading to intraclonal variation. The fourth involves Darwinist selection based on exposure of landrace mixtures to stressors and other selection pressures resulting in a "survival of the fittest" of genotypes best adapted to new conditions.

High levels of landrace diversity can hardly be explained without considering the possibility of conscious use of botanical seed by pre-Columbian farmers. There are a few reports of the putative use of sexual seed as a traditional practice (IBPGR, 1991; Quiros et al., 1992). However, this is no longer common practice in the Andes or Chiloe islands. Alternatively, geneflow through open pollination and spontaneous sexual reproduction, consequent germination of viable hybrid seed, emergence and establishment of a weedy volunteer, and incorporation of seed tubers into farmer-managed landrace stocks may represent another hypothetical yet little-documented pathway (Johns and Keen, 1986; Johns et al., 1987; Lawson, 1983; Rabinowitz et al., 1990). The pathway of geneflow up to the germination of botanical seed has indeed been well documented (Celis et al., 2004; Scurrah et al., 2008), but the final stages of successful establishment of a hybrid volunteer and subsequent farmer uptake are hypothetical.

The *Araq* potato is a folk taxon with considerable intraspecific diversity characterized by its weediness or wild state, which is collected and consumed by Andean farmers (de Haan et al., 2007). These semiwild potatoes are widely spread throughout the Andean center of potato origin from Bolivia to Ecuador, where they are known by a variety of different vernacular names: *Araq Papa* (central and southern Peru), *Chayka Papa* (central Peru), *Papa Curao* (central and northern Peru), *Lelekkoya* (Bolivia), and *Semillu* (Bolivia). These tetraploid wild potatoes belong to the *S. tuberosum* Andigenum group and are characterized by long stolons and big tubers with thick tuber skin (Ochoa, 1990, 1999). Collection of these semiwild potatoes to bring them into cultivation has been observed and is likely to have a role in ongoing evolution through a dynamic cultivated–wild germplasm renewal system.

The occurrence of chimera or so-called sports of common European and North American potato varieties is well documented (Lawrence et al., 1994; Ramulu et al., 1984; van Harten, 1998). The phenomenon has not been described for Andean landraces but is highly likely to occur and have a role in morphological diversification, particularly taking into account the evolutionary longevity of landraces compared with bred varieties. Furthermore, high levels of ultraviolet radiation at high altitudes may result in higher rates of mutations (Blumthaler et al., 1997).

Farmers throughout the Andes grow diverse potato landraces in mixtures. These are commonly called *chaqru* in the Quechua language and literally offer a "basket of options" to smallholder farmers as environmental or socioeconomic conditions change. Climate changes, such as occurring during the Little Ice Age (1300–1850) or under the current scenario of global warming, without doubt exert selection pressure on landrace populations. This pressure may occur directly though climatic shifts or indirectly through altered levels of biotic stress. Little is known about the impact of Darwinist selection on intraspecific diversity. Baselines are generally unavailable and timeline series are difficult to reconstruct. However, changes in genetic composition, notably the frequencies of alleles and genes, within the landrace population through successive generations can be expected to occur. Such changes may also involve altered spatial distribution patterns.

6. Uses of Potato in the Center of Origin
6.1 Traditional Uses

The characterization of traditional uses, including regional variants of processes and transformations, and their impact on the biochemical composition of end products represent an as-yet underexplored field of research that may lead to discovery as well as a better understanding of food security implications. Potato has been a staple crop and essential component of western Latin American food systems for thousands of years (Coe, 1994). The domestication and diversification of the crop have coevolved with multiple uses. Food-related uses are abundant and follow different primary processes including boiling, steaming, freeze-drying, drying, and fermentation (Towle, 1961). These uses alter the biochemical composition of the potato and frequently enhance storability to ensure year-round food availability. In addition, the potato is used as medicine and animal fodder and for festive or ritual purposes (CIP, 2006; Graves, 2001; Iriarte et al., 2009).

One early traditional use that is still commonly practiced by Andean farmers around Lake Titicaca involves geophagy (Browman and Gundersen, 1996). Potatoes are boiled and consumed with a clay dip called *ch'ago*. The clay has a detoxification function and neutralizes the human intake of potentially harmful glycoalkaloids (Johns, 1986). The transition of geophagy from a general response to toxin-related stress to a more specialized detoxification technique such as freeze-drying can be interpreted as an important step that has facilitated expanded resource exploitation and domestication (Johns, 1989, 1990).

Freeze-drying, drying, and fermentation are processing techniques that facilitate long-term storage (Werge, 1979). There are many variants of traditional freeze-drying techniques; the main ones involve processing white *chuño* (*tunta*, *moraya*) and black *chuño* (Woolfe, 1987). The elaboration of white *chuño* involves tending, freezing, treading, washing, and drying of tubers. White *chuño* is always washed or soaked, in part to remove glycolalkaloids (Johns, 1990). On the other hand, processing of black *chuño* does not involve washing or soaking and its preparation is generally simpler compared with white *chuño* (Mamani, 1981). Farmers take advantage of severe frost at night alternated with high daytime levels of solar radiation and low levels of relative humidity during the dry season to process either type of *chuño* (Sattaur, 1988). Freeze-drying results in a decrease in the zinc, potassium, phosphorus, and magnesium content of both types of *chuño*, whereas the iron and calcium content values of white *chuño* tend to increase after processing (de Haan et al., 2010a, 2013).

Processing of *papa seca* (dried potato) is common in the relatively dry western Andes. On the other hand, anaerobic fermentation of *tocosh* is practiced exclusively by smallholder farmers from the comparatively wet eastern and northern Andes of Peru. *Tocosh* is used as a food and a natural medicine believed to contain penicillin. The biochemical composition of both *tocosh* and *papa seca* has yet to be researched. All traditional preparations of potato in the Andes and Chiloe island basically involve boiling and steaming. Potatoes consumed with *ch'ago*, and *chuño*, *papa seca*, and *tocosh*, are always boiled before consumption. Another traditional and widely practiced technique to prepare potatoes involves the preparation of earth-ovens. Numerous regional variants exist, including the *pachamanka* in central Peru, *huatia* in southern Peru to northern Bolivia, and *curanto* in southern Chile (Brush et al., 1980; Olivas Weston, 2001).

6.2 Modern Uses

Today, most Latin American breeding programs are working on durable resistance to late blight, virus resistance, heat and drought tolerance, quality traits, and adaptation to low–external input agriculture. To accomplish this goal, the use of wild relatives of the cultivated potato and landraces has intensified. For example, in 2000, INIA-Uruguay initiated a prebreeding program for late blight resistance using crosses involving *S. bulbocastanum*, *S. microdontum*, and *S. circaeifolium*. Selections have been made in collaboration with other programs in Latin America. Furthermore, in collaboration with the University of the Republic (UdelaR), INIA developed $4x$ *S. tuberosum* hybrids from crosses with *Solanum commersonii*—an endemic wild species in Uruguay—and *Solanum chacoense*, with heat tolerance and bacterial wilt resistance, using Phureja as a bridge with the aim of transferring resistance to bacterial wilt to cultivated germplasm and broadening the genetic base of the crop and incrementing resistance to abiotic stresses (Galván et al., 2007; González et al., 2013; Narancio et al., 2013).

EMBRAPA in Brazil has also developed prebreeding lines for bacterial wilt resistance from populations derived from the wild species *S. sparsipilum*, *S. chacoense*, *S. microdontum*, and cultivated Phureja. Parental clones and derived progenies with bacterial wilt resistance and good agronomical traits have been selected under warm conditions. Similarly, CIP has developed 4x *S. tuberosum* hybrids from crosses with wild species from the Piurana group that possess novel genes for late blight resistance and developed initial crosses with sources of resistance to bacterial wilt. The tuberosum background of these hybrids involves breeding lines with heat and/or drought tolerance.

Taking advantage of its collection of Phureja landraces, the breeding program of the National University of Colombia (UNC) has developed modern diploid Phureja varieties with enhanced dormancy. The university has also begun a diploid potato breeding program focused on nutritional traits. Initially they determined the content of nutritional compounds (sugars, proteins, minerals, starch, moisture, fat, dietary fiber, carotenoids, vitamin C, and other antioxidants) and antinutrient compounds (glycoalkyloids) present in their Phureja core collection (Duarte-Delgado et al., 2015; Peña et al., 2015). Variability in the content of macronutrients, micronutrients, and functional compounds has allowed for the selection of parental materials for breeding (Duarte et al., 2013; Piñeros et al., 2013).

High levels and appropriate combinations of nutritional compounds are an important objective of global crop improvement. Relatively high levels of micronutrient contents have been documented in the diploid cultivar groups Phureja, Stenotomum, and Goniocalyx (Burgos et al., 2009a,b; Paget et al., 2014). These early domesticates possess good agronomical attributes, and along the history of potato breeding, they have been shown to form heterotic tetraploid hybrids in crosses with tetraploid potatoes by means of heterozygous unreduced gametes. This diploid landrace germplasm has been selected at CIP as the first gene pool for the development of nutritious heterotic hybrids and were subjected to three cycles of recurrent selection to increase concentrations of iron and zinc through inter- and intragroup hybridization.

From a broad base of landraces and wild potatoes, CIP's two most advanced tetraploid breeding populations were subjected to divergent selection for adaptive trait complexes required for sustainable production in tropical highland (population B) and subtropical lowland (LTVR population) agroecologies (Bonierbale et al., 2003; Trognitz et al., 2001). Significant positive heterosis for marketable tuber yield has already been obtained by interpopulation hybridization of these two populations. The new hybrid population exceeded the mean of both parental populations under favorable environments but not under a warm environment less favorable for potato growth. More heterotic families were identified under favorable environments, but the few heterotic families identified under the unfavorable warm environment were also heterotic under the favorable one. This indicated that for full expression of heterosis, it is important for the two gene pools to show adaptation to the environment for which new hybrids are intended. For this reason, adaptation traits such as heat and salinity tolerance, drought tolerance, and a long critical photoperiod are traits selected for gene pool enhancement and heterosis breeding at CIP (Bonierbale et al., 2008; Khan et al., 2014).

7. Trend for Sustainable Conservation and Use

Understanding the impact of climate change on the population ecology, genetics, and reproductive dynamics of in situ populations of the wild relatives of the cultivated potato will be an essential research topic for research programs in Latin America. It has been predicted that several species could disappear in coming decades (Jarvis et al., 2008). Monitoring the conservation status of vulnerable wild relatives and implementation of mitigation strategies where needed will need to be a top priority (Cadima, 2014). Ecology and conservation genetics could speed their development through the use of next-generation sequencing to investigate more precisely organisms in their natural habitat and to evaluate the effect of the disturbance made by humans (Narum et al., 2013). It will also be essential to make progress with gap filling to ensure that ex situ collections fully represent the gene pool of wild relatives (Castañeda-Álvarez et al., 2015).

Similarly, the on-farm conservation of landraces will need to evolve to a more harmonized framework. Essentially, the use of standard procedures to measure the conservation status and robustly monitor the dynamics of landrace pools in selected hotspots will be a future trend for potatoes in the Andes and Chiloe islands. It will also be important to document the impact of different on-farm conservation strategies so that interventions can become increasingly evidence-based, and to target those sites where local stakeholders demand external support. Structural comparisons of ex situ collections and on-farm landrace populations based on next-generation markers are anticipated to make cultivar-level gap analysis more effective. They will also inform where gaps in ex situ collections exist.

Advances in genomics, proteomics, metabolomics, and bioinformatics will enable breeders to produce new generations of advanced germplasm more intelligently for emerging priorities or special needs. Now that high-throughput resequencing, new genotyping technologies, and reference genomes are easy to access, it is expected that the characterization of nucleotide differences and indels in thousands of individual plant species and genotypes will increase exponentially (Dereeper et al., 2011). This rapid development will need the parallel development of bioinformatics tools for efficient data management, analysis, and genetic resource documentation.

Wider use of the potatoes' wild relatives will Without doubt have an essential role in innovative breeding approaches (Jansky et al., 2013). Breeders globally and in Latin America are developing new varieties with climate-proof traits, including heat, drought, and salt tolerance, and with early bulking as an escape mechanism. Initial screening efforts for abiotic stress tolerance frequently involve landraces. Genetic modification techniques could be applied for the production of molecules with industrial interest through metabolic engineering to transform the potato in a biofactory for novel molecules with potential use in the pharmaceutical, cosmetics, or agrochemical industry (Feingold et al., 2010). Selected studies have also reported novel beneficial roles of the potato in human health, such as anticancer, antihypertension, anti-inflammatory, and antimicrobial properties, among others (Piñeros et al., 2014). More in-depth research on potential novel uses or underexplored properties, such as factors contributing to tuber flavor, texture, and sensory and physicochemical properties (Ezekiel et al., 2013; Seefeldt et al., 2011), will shape the future of potato research and use of genetic resources.

References

Ames, M., Spooner, D.M., 2008. DNA from herbarium specimens settles a controversy about origins of the European potato. American Journal of Botany 95, 252–257.

Apffel-Marglin, F., 1998. The Spirit of Regeneration: Andean Culture Confronting Western Notions of Development. Zed Books, London.

Birch, P.R.J., Bryan, G., Fenton, B., Gilroy, E.M., Hein, I., Jones, J.T., Prashar, H., Taylor, M.A., Torrance, L., Toth, I.K., 2012. Crops that feed the world 8: potato: are the trends of increased global production sustainable? Food Security 4 (4), 477–508.

Blumthaler, M., Ambach, W., Ellinger, R., 1997. Increase in solar UV radiation with altitude. Journal of Photochemistry and Photobiology B: Biology 39, 130–134.

Bonierbale, M., Amoros, W., Landeo, J., 2003. Improved resistance and quality in potatoes for the tropics. Acta Horticulturae 619, 15–22.

Bonierbale, M., Amoros, W., Quiroz, R., Carli, C., Mihovilovich, E., Schafleitner, R., 2008. Breeding potato for wide adaptation: conventional and molecular approaches. In: Global Potato Conference, 9–12 December 2008, New Delhi, India, Conference Summary, pp. 5–6.

Bradeen, J.M., Haynes, K.G., 2011. Introduction to potato. In: Bradeen, J.J., Kole, C. (Eds.), Genetics, Genomics and Breeding of Potato. CRC Press/Science Publishers, Enfield, NH, pp. 1–19.

Bradshaw, J.E., Ramsay, G., 2009. Potato origin and production. In: Singh, J., Kaur, L. (Eds.), Advances in Potato Chemistry and Technology. Academic Press, Burlington, pp. 1–26.

Bretting, P.K., Duvick, D.N., 1997. Dynamic conservation of plant genetic resources. Advances in Agronomy 61, 1–51.

Browman, D.L., Gundersen, N.G., 1996. Altiplano comestible earths: prehistoric and historic geophagy of highland Peru and Bolivia. Geoarcheology 8, 413–425.

Brown, C.R., Culley, D., Bonierbale, M., Amoros, W., 2007. Anthocyanin, carotenoid content, and antioxidant values in native South American potato cultivars. HortScience 42, 1733–1736.

Brush, S.B., 1980. Potato taxonomies in andean agriculture. In: Brokensha, D.W., Warren, D.M., Werner, O. (Eds.), Indigenous Knowledge Systems and Development. University Press of America, Lanham, MD, pp. 37–48.

Brush, S.B., 2000. Genes in the Field: On-farm Conservation of Crop Diversity. International Development Research Centre (IDRC), International Plant Genetic Resources Institute (IPGRI), Lewis Publishers, Boca Raton, FL.

Brush, S.B., 2004. Farmers' Bounty: Locating Crop Diversity in the Contemporary World. Yale University Press, New Haven, CT.

Brush, S.B., Carney, H.J., Huamán, Z., 1980. The Dynamics of Andean Potato Agriculture. Social Science Department Working Paper Series 1980–5. International Potato Center (CIP), Lima.

Brush, S.B., Taylor, J.E., Bellon, M.R., 1992. Technology adoption and biological diversity in Andean potato agriculture. Journal of Development Economics 39, 365–387.

Bukasov, S.M., 1978. Systematics of the potato. Trudy po Prikladnoj Botanike Genetike i Selekcii 62, 3–35.

Burgos, G., Amoros, W., Morote, M., Stangoulis, J., Bonierbale, M., 2007. Iron and zinc concentration of native Andean potato varieties from a human nutrition perspective. Journal of the Science of Food and Agriculture 87 (4), 668–675.

Burgos, G., Auqui, S., Amoros, W., Salas, E., Bonierbale, M., 2009a. Ascorbic acid concentration of native Andean potato varieties as affected by environment, cooking and storage. Journal of Food Composition and Analysis 22, 533–538.

Burgos, G., Salas, E., Amoros, W., Auqui, M., Munoa, L., Kimura, M., Bonbierale, M., 2009b. Total and individual carotenoid profiles in *Solanum phureja* of cultivated potatoes: I. Concentrations and relationships as determined by spectrophotometry and HPLC. Journal of Food Composition and Analysis 22, 503–508.

La Barre, W., 1947. Potato taxonomy among the Aymara indians of Bolivia. Acta Americana 5, 83–103.

van der Berg, R., Groendijk-Wilders, N., 2014. Taxonomy. In: Navarre, R., Pavek, M. (Eds.), The Potato: Botany, Production and Uses. CABI Publishing, Surrey, pp. 12–28.

Cadima, X., 2014. Conserving the Genetic Diversity of Bolivian Wild Potatoes (Ph.D. thesis). Wageningen University, Wageningen.

Camadro, E.L., Erazzú, L.E., Maune, J.F., Bedogni, M.C., 2012. A genetic approach to the species problem in wild potato. Plant Biology 14, 543–554.

Castañeda-Álvarez, N., de Haan, S., Juarez, H., Khoury, C.K., Achicanoy, H.A., Sosa, C.C., Bernau, V., Salas, A., Heider, B., Simon, R., Maxted, N., Spooner, D.M., 2015. Ex situ conservation priorities for the wild relatives of potato (*Solanum* L. section *Petota*). PLoS ONE 10 (6), e0129873. http://dx.doi.org/10.1371/journal.pone.0129873.

Celis, C., Scurrah, M., Cowgill, S., Chumbiauca, S., Green, J., Franco, J., Main, G., Kiezebrink, D., Visser, R.G.F., Atkinson, H.J., 2004. Environmental biosafety and transgenic potato in a centre of diversity for this crop. Nature 432, 222–225.

Centro Internacional de la Papa (CIP), Federación Departamental de Comunidades Campesinas de Huancavelica (FEDECCH), 2006. Catálogo de Variedades de Papa Nativa de Huancavelica – Perú. Metrocolor, Lima, Peru.

Chen, Q., Lynch, D., Platt, W.H., Li, H.Y., Shi, Y., Li, H.J., Beasley, D., Rakosy-Tican, L., Theme, R., 2004. Interspecific crossability and cytogenetic analysis of sexual progenies of Mexican wild diploid 1EBN species *Solanum pinnatisectum* and *S. cardiophyllum*. American Journal of Potato Research 81, 159–169.

Christiansen, J., 1977. The Utilization of Bitter Potatoes to Improve Food Production in the High Altitude of the Tropics (Ph.D. thesis). Cornell University, Ithaca, NY.

Coca Morante, M., 2002. Las Papas en Bolivia: Una Aproximación a la Realidad del Mejoramiento del Cultivo de la Papa en Bolivia. Universidad Mayor de San Simon, Cochabamba, Bolivia.

Coe, S.D., 1994. America's First Cuisines. University of Texas Press, Austin, TX.

Cohen, J.I., Williams, J.T., Plucknett, D.L., Shands, H., 1991. Ex situ conservation of plant genetic resources: global development and environmental concerns. Science 253, 866–872.

Condori, B., Hijmans, R.J., Ledent, J.F., Quiroz, R., 2014. Managing potato biodiversity to cope with frost risk in the high Andes: a modeling perspective. PLoS ONE 9, e81510.

Contreras, A., Castro, I., 2008. Catálogo de Variedades de Papas Nativas de Chile. Universidad Austral de Chile, Validivia, Chile.

Correll, D.S., 1962. The potato and its wild relatives. Contributions from the Texas Research Foundation. Botanical Studies 4, 1–606.

Cosio Cuentas, P., 2006. Variabilidad de Papas Nativas en Seis Comunidades de Calca y Urubamba – Cusco. Asociación Arariwa, Cusco, Peru.

Dereeper, A., Nicolas, S., Cunff, L., Bacilieri, R., Doligez, A., Peros, J.P., Ruiz, M., This, P., 2011. SNiPlay: a web-based tool for detection, management and analysis of SNPs. Application to grapevine diversity projects. BMC Bioinformatics 12, 134.

Devaux, A., Ordinola, M., Hibon, A., Flores, R., 2010. El Sector Papa en la Región Andina: Diagnóstico y Elementos para una Visión Estratégica (Bolivia, Ecuador y Perú). Centro Internacional de la Papa, Lima, Peru.

Dillehay, T.D., 1997. Monte Verde: A Late Pleistocene Settlement in Chile, the Archeological Context and Interpretation, vol. 2. Smithsonian Press, Washington, DC.

Duarte, D., Restrepo, P., Kushalappa, A., Mosquera, T., 2013. Phenotypic characterization of reducing and non-reducing sugars content in *Solanum tuberosum* group Phureja. In: Poster Presented at Solanaceae Conference, Genome vs Phenome, October 1–17, Beijing, China. Available at: http://issuu.com/seguridadalimentarianarinoproyecto/ (accessed in 2015).

Duarte-Delgado, D., Narváez-Cuenca, C.E., Restrepo-Sánchez, L.P., Kushalappa, A., Mosquera-Vásquez, T., 2015. Development and validation of a liquid chromatographic method to quantify sucrose, glucose, and fructose in tubers of *Solanum tuberosum* Group Phureja. Journal of Chromatography B 975, 18–23.

Dueñas, A., Mendivil, R., Lovaton, G., Loaiza, A., 1992. Campesinos y Papas: a Propósito de la Variabilidad y Erosión Genética en Comunidades Campesinas del Cusco. In: Degregori, C.I., Escobal, J., Marticorena, B. (Eds.), Perú: el Problema Aagrario en Debate SEPIA IV, Seminario Permanente de Investigación Agraria (SEPIA). Universidad Nacional de la Amazonía Peruana (UNAP), Iquitos, Peru, pp. 287–309.

Ezekiel, R., Singh, N., Sharma, S., Kaur, A., 2013. Beneficial phytochemicals in potato - a review. Food Research International 50, 487–496.

Feingold, S.E., Massa, G.A., Norero, N.S., Lorenzen, J., 2010. Initiatives on potato functional genetics. The Americas Journal of Plant Science and Biotechnology 4, 79–89.

Food and Agriculture Organization of the United Nations (FAOSTAT), 2013. FAOSTAT database, http://faostat3. fao.org/home/E.

Food and Agriculture Organization of the United Nations (FAO), 2008. International Year of the Potato: The Global Potato Economy. FAO, Rome, Italy.

Food and Agriculture Organization of the United Nations (FAO), 2010. The Second Report on the State of the World's Plant Genetic Resources for Food and Agriculture. FAO, Rome, Italy.

Food and Agriculture Organization of the United Nations (FAO), 2013. FAOSTAT Database. FAO. Available at: http://faostat3.fao.org (accessed in 2015).

Fowler, C., Mooney, P., 1990. Shattering: Food, Politics, and the Loss of Genetic Diversity. University of Arizona Press, Tucson, AZ.

Galván, G., Fraguas, F., Quirici, L., Santos, C., Silvera, E., Siri, M.I., Villanueva, P., Raudiviniche, L., González, M., Torres, D., Castillo, A., Dalla Rizza, M., Vilaró, F., Gepp, V., Ferreira, F., Pianzzola, M.J., 2007. Solanum commersonii: Una especie con gran potencial para el mejoramiento genético de papa por resistencia a *Ralstonia solanacearum*. In: Clausen, A., Condon, F., Berretta, A. (Eds.), Avances de Investigación en Recursos Genéticos en el CONO SUR II, pp. 87–102.

Gavrilenko, T., Antonova, O., Shuvalova, A., Krylova, E., Alpatyeva, N., Spooner, D.M., Novikova, L., 2013. Genetic diversity and origin of cultivated potatoes based on plastid microsatellite polymorphism. Genetic Resources and Crop Evolution 60, 1997–2015.

Ghislain, M., Andrade, D., Rodriguez, F., Hijmans, R.J., Spooner, D.M., 2006. Genetic analysis of the cultivated potato *Solanum tuberosum* L. Phureja Group using RAPDs and nuclear SSRs. Theoretical and Applied Genetics 113, 1515–1527.

Giraldo, D., Juarez, H., Pérez, W., Trebejo, I., Yzarra, W., Forbes, G., 2010. Severity of potato late blight (*Phytophthora infestans*) in agricultural areas of Peru associated with climate change. Revista Peruana Geo-Atmosférica 2, 56–67.

Glendinning, D.R., 1983. Potato introductions and breeding up to the early 20th century. New Phytologist 94, 479–505.

Global Crop Diversity Trust (GCDT), 2006. Global Strategy for the Ex Situ Conservation of Potato. GCDT. Available at: http://www.croptrust.org/documents/web/Potato-Strategy-FINAL-30Jan07.pdf (accessed in 2011).

González, L., 2013. Catálogo de Variedades de Papas Nativa y de Uso Local en el Estado de Mérida, Venezuela. Instituto Nacional de Investigaciones Agrícolas, Maracay, Venezuela.

González, M., Galván, G.A., Siri, M.I., Borges, A., Vilaró, F., 2013. Resistencia a la marchitez bacteriana de la papa en *Solanum commersonii*. Agrociencia (Uruguay) 17, 45–54.

Gorbatenko, L.E., 2006. Potato Species of South America: Ecology, Geography, Introduction, Taxonomy, and Breeding Value, Russian Academy of Agricultural Sciences. State Scientific Centre of the Russian Federation, St. Petersburg, Russia.

Graham, R.D., Welch, R.M., Saunders, D.A., Ortiz-Monasterio, I., Bouis, H.E., Bonierbale, M., de Haan, S., Burgos, G., Thiele, G., Liria, R., Meisner, C.A., Beebe, S.E., Potts, M.J., Kadian, M., Hobbs, P.R., Gupta, R.K., Twomlow, S., 2007. Nutritious subsistence food systems. Advances in Agronomy 92, 1–74.

Graves, C., 2001. The Potato Treasure of the Andes: from Agriculture to Culture. International Potato Center, Lima, Peru.

Gutiérrez, R., Valencia, C., 2010. Las Papas Nativas de Canchis: Un Catálogo de Biodiversidad. Intermediate Technology Development Group, Lima, Peru.

de Haan, S., Juarez, H., 2010. Land use and potato genetic resources in Huancavelica, central Peru. Journal of Land Use Science 5, 179–195.

de Haan, S., Bonierbale, M., Ghislain, M., Núñez, J., Trujillo, G., 2007. Indigenous biosystematics of Andean potatoes: folk taxonomy, descriptors, and nomenclature. Acta Horticulturae 745, 89–134.

de Haan, S., Burgos, G., Arcos, J., Ccanto, R., Scurrah, M., Salas, E., Bonierbale, M., 2010a. Traditional processing of black and white chuño in the Peruvian Andes: regional variants and effect on the mineral content of native potato cultivars. Economic Botany 64, 217–234.

de Haan, S., Núñez, J., Bonierbale, M., Ghislain, M., 2010b. Multilevel agrobiodiversity and conservation of Andean potatoes in central Peru. Mountain Research and Development 30, 222–231.

de Haan, S., Núñez, J., Bonierbale, M., Ghislain, M., van der Maesen, J., 2013. A simple sequence repeat (SSR) marker comparison of a large in- and ex-situ potato landrace cultivar collection from Peru reaffirms the complementary nature of both conservation strategies. Diversity 5, 505–521.

Hamilton, E., 1934. American Treasure and the Price Revolution in Spain, 1501–1650. Harvard University Press, Cambridge, USA.

Hancco, J., Blas, R., Quispe, M., Ugás, R., 2008. Pampacorral: Catálogo de sus Papas Nativas. Universidad Nacional Agraria La Molina, Lima, Peru.

Hanneman, R.E., 1989. The potato germplasm resource. American Potato Journal 66, 655–667.

Hanneman, R.E., 1994. Assignment of endosperm balance numbers to the tuber-bearing Solanums and their close non-tuber-bearing relatives. Euphytica 74, 19–25.

van Harten, A.M., 1998. Mutation Breeding: Theory and Practical Applications. Cambridge University Press, Cambridge.

Harlan, J.R., de Wet, J.M.J., 1971. Toward a rational classification of cultivated plants. Taxon 20, 509–517.

Harris, G.F., Niha, P.P., 1999. Nga Riwai Maori: Maori Potatoes. Working Papers No. 2–99. The Open Polytechnic of New Zealand, Lower Hutt, New Zealand.

Hawkes, J.G., 1956. Taxonomic studies on the tuber-bearing Solanums. 1. *Solanum tuberosum* and the tetraploid species complex. Proceedings of the Linnean Society 166, 97–144.

Hawkes, J.G., 1962. The origin of *Solanum juzepczukii* Buk. and *S. curtilobum* Juz. et Buk. Zeitschrift für Pflanzenzüchtung 47, 1–14.

Hawkes, J.G., 1963. A Revision of the Tuber-Bearing Solanums. II. Scottish Plant Breeding Station Record 1963, pp. 76–181.

Hawkes, J.G., 1973. Potato genetic erosion survey: preliminary report. CIP Bulletin 166 (Lima, Peru).

Hawkes, J.G., 1990. The Potato: Evolution, Biodiversity & Genetic Resources. Belhaven Press, London.

Hawkes, J.G., 1999. The evidence of the extent of N.I. Vavilov's new world Andean centres of cultivated plant origins. Genetic Resources and Crop Evolution 46, 163–168.

Hawkes, J.G., Francisco-Ortega, J., 1992. The potato in Spain during the late 16th century. Economic Botany 46, 86–97.

Hawkes, J.G., Francisco-Ortega, J., 1993. The early history of the potato in Europe. Euphytica 70, 1–7.

Hijmans, R.J., Spooner, D.M., Salas, A.R., Guarino, L., De la Cruz, J., 2002. Atlas of Wild Potatoes. International Plant Genetic Resources Institute, Rome.

Hosaka, K., 2003. T-type chloroplast DNA in *Solanum tuberosum* L. ssp. *tuberosum* was conferred from some populations of *S. tarijense* Hawkes. American Journal of Potato Research 80, 21–32.

Huamán, Z., Spooner, D.M., 2002. Reclassification of landrace populations of cultivated potatoes (*Solanum* sect. *Petota*). American Journal of Botany 89, 947–965.

Huamán, Z., Hawkes, J.G., Rowe, P.R., 1982. A biosystematics study of the origin of the diploid potato *Solanum ajanhuiri*. Euphytica 31, 665–675.

Instituto Nacional de Innovación Agraria (INIA), Red LatinPapa, 2012. Catálogo de Nuevas Variedades de Papa: Sabores y Colores para el Gusto Peruano. Tarea Asociación Gráfica Educativa, Lima, Peru.

International Board for Plant Genetic Resources (IBPGR), 1991. Geneflow: Women and Plant Genetic Resources. International Board for Plant genetic Resources, Rome.

Iriarte, V., Condori, B., Parapo, D., Acuña, D., 2009. Catálogo Etnobotánico de Papas Nativas del Altiplano Norte de La Paz – Bolivia. Fundación PROINPA, Cochabamba, Bolivia.

Jansky, S.H., Dempewolf, H., Camadro, E.L., Simon, R., Zimnoch-Guzowska, E., Bisognin, D.A., Bonierbale, M., 2013. A case for crop wild relative preservation and use in potato. Crop Science 53, 746–754.

Jarvis, A., Lane, A., Hijmans, R.J., 2008. The effect of climate change on crop wild relatives. Agriculture, Ecosystems and Environment 126, 13–23.

Jo, K.R., Kim, C.J., Kim, S.J., Kim, T.K., Bergervoet, M., Jongsma, M.A., Visser, R.G.F., Jacobsen, E., Vossen, J.H., 2014. Development of late blight resistant potatoes by cisgene stacking. BMC Biotechnology 14, 50.

Johns, T., 1985. Chemical Ecology of the Aymara of Western Bolivia: Selection for Glycoalkaloids in the *Solanum×Ajanhuiri* Complex (Ph.D. thesis). University of Michigan, Ann Arbor, MI.

Johns, T., 1986. Detoxification function of geophagy and the domestication of potato. Journal of Chemical Ecology 12, 635–646.

Johns, T., 1989. A chemical-ecological model of roots and tuber domestication in the Andes. In: Harris, D.R., Hillman, G.C. (Eds.), Foraging and Farming. Unwin Hyman, London, pp. 504–519.

Johns, T., 1990. With Bitter Herbs They Shall Eat It: Chemical Ecology and the Origins of Human Diet and Medicine. University of Arizona Press, Tucson, AZ.

Johns, T., Keen, S.L., 1986. Ongoing evolution of the potato on the altiplano of western Bolivia. Economical Botany 40, 409–424.

Johns, T., Huamán, Z., Ochoa, C., Schmiediche, P.E., 1987. Relationships among wild, weed, and cultivated potatoes in the *Solanum ajanhuiri* complex. Systematic Botany 12, 541–552.

Johnston, S.A., den Nijs, T.P.M., Peloquin, S.J., Hanneman, R.E., 1980. The significance of genic balance to endosperm development in interspecific crosses. Theoretical and Applied Genetics 57, 5–9.

King, J.C., Slavin, J.L., 2013. White potatoes, human health, and dietary guidance. Advances in Nutrition 4, 3935–4015.

Kiple, K.F., Ornelas, K.C. (Eds.), 2000. The Cambridge World History of Food. Cambridge University Press, Cambridge, UK.

Khan, M.A., Saravia, D., Munive, S., Lozano, F., Farfan, E., Eyzaguirre, R., Bonierbale, M., 2014. Multiple QTLs linked to agro-morphological and physiological traits related to drought tolerance in potato. Plant Molecular Biology Reporter 1–13.

Kirkman, M.A., 2007. Global markets for processed potato products. In: Vreugdenhil, D. (Ed.), Potato Biology and Biotechnology: Advances and Perspectives. Elsevier, Oxford, pp. 27–44.

Knapp, S., 2008. Species concepts and floras: what are species for? Biological Journal of the Linnean Society 95, 17–25.

Kroschel, J., Sporleder, M., Tonnang, H.E.Z., Juarez, H., Carhuapoma, P., Gonzales, J.C., Simon, R., 2013. Predicting climate change caused changes in global temperature on potato tuber moth *Phthorimaea operculella* (Zeller) distribution and abundance using phenology modeling and GIS mapping. Agricultural and Forest Meteorology 170, 228–241.

Lawrence, D.F., Slack, S.A., Plaisted, R.L., 1994. Russet Bake-King: a uniform russeted sport of Bake-King. American Potato Journal 71, 127–129.

Lawson, H.M., 1983. True potato seeds as arable weeds. Potato Research 26, 237–246.

Lobo-Cabrera, M., 1988. El comercio canario europeo bajo Felipe II. Viceconsejería de Cultura y Deportes del Gobierno de Canarias y Secretaría Regional de Turismo. Cultura e Emigraçao de Governo Regional da Madeira, Funchal, Portugal.

Mamani, M., 1981. El chuño: Preparación, uso, almacenamiento. In: Lechtmand, H., Soldi, A.M. (Eds.), La Tecnología en el Mundo Andino. Universidad Nacional Autónoma de México, México DF, pp. 235–246.

Manzur, M.I., 2012. Cátalogo de Semillas Tradicionales de Chile. Fundación Sociedades Sustentables, Santiago, Chile.

Masuelli, R.W., Camadro, E.L., 1997. Crossability relationships among wild potato species with different ploidies and endosperm balance numbers. Euphytica 94, 227–235.

Masuelli, R.W., Camadro, E.L., Erazzú, L.E., Bedogni, M.C., Marfil, C.F., 2009. Homoploid hybridization in the origin and evolution of wild diploid potato species. Plant Systematics and Evolution 277, 143–151.

Maxted, N., Ford-Lloyd, B.V., Hawkes, J.G., 1997. Plant Genetic Conservation: The In-situ Approach. Chapman & Hall, London.

McKey, D.B., Elias, M., Pujol, B., Duputié, A., 2012. Ecological approaches to crop domestication. In: Gepts, P., Famula, T.R., Bettinger, R.L., Brush, S.B., Damania, A.B., McGuire, P.E., Qualset, C.O. (Eds.), Biodiversity in Agriculture: Domestication, Evolution, and Sustainability. Cambridge University Press, Cambridge, pp. 377–406.

Meinzen-Dick, R.S., Devaux, A., Antezana, I., 2009. Underground assets: potato biodiversity to improve the livelihoods of the poor. International Journal of Agricultural Sustainability 7, 235–248.

Monteros, A.R., 2011. Potato Landraces: Description and Dynamics in Three Areas of Ecuador (Ph.D. thesis). Wageningen University, Wageningen.

Monteros, C., Yumisaca, F., Andrade-Piedra, J., Reinoso, I., 2010. Catálogo: Cultivares de Papas Nativas. Sierra Centro Norte del Ecuador. Instituto Nacional Autónomo de Investigaciones Agropecuarias, Centro Internacional de la Papa, Quito, Ecuador.

Narancio, R., Zorrilla, P., Gonzalez, M., Vilaró, F., Pritsch, C., Dalla Rizza, M., 2013. Insights on gene expression response of a characterized resistant genotype of *Solanum commersonii* against *Ralstonia solanacearum*. European Journal of Plant Pathology 136, 823–835.

Narum, S., Buerkle, C.A., Davey, J.W., Miller, M.R., Hohenlohe, P.A., 2013. Genotyping-by-sequencing in ecological and conservation genomics. Molecular Ecology 22, 2841–2847.

Nesterenko, S., Sink, K.C., 2003. Carotenoid profiles of potato breeding lines and selected cultivars. HortScience 38, 1173–1177.

Nunn, N., Qian, N., 2011. Columbus's contribution to world population and urbanization: a natural experiment examining the introduction of potatoes. Quarterly Journal of Economics 126, 593–650.

Ochoa, C.M., 1975. Potato collecting expeditions in Chile, Bolivia and Peru, and genetic erosion of indigenous cultivars. In: Frankel, O.H., Hawkes, J.G. (Eds.), Crop Genetic Resources for Today and Tomorrow. Cambridge University Press, Cambridge, pp. 167–173.

Ochoa, C.M., 1990. The Potatoes of South America: Bolivia. Cambridge University Press, Cambridge.

Ochoa, C.M., 1999. Las Papas de Sudamérica: Peru. Centro Internacional de la Papa (CIP), Lima, Peru.

Oliemans, W.H., 1988. Het Brood van de Armen: de geschiedenis van de aardappel temidden van ketters, kloosterlingen en kerkvorsten. SDU uitgeverij, The Hague, Holland.

Olivas Weston, R., 2001. La Cocina de los Incas: Costumbres Gastronómicas y Técnicas Culinarias. Universidad San Martín de Porres, Lima, Peru.

Ortiz, R., Ehlenfeldt, M., 1992. The importance of endosperm balance number in potato breeding and the evolution of tuber-bearing *Solanum* species. Euphytica 60, 105–113.

Osman, S.F., Herb, S.F., Fitzpatrick, P.J., Schmiediche, P., 1978. Glycoalkaloid composition of wild and cultivated tuber-bearing *Solanum* species of potential value in potato breeding programs. Journal of Agricultural and Food Chemistry 26, 1246–1248.

Oswald, A., de Haan, S., Sanchez, J., Ccanto, R., 2009. The complexity of simple tillage systems. Journal of Agricultural Science 147, 399–410.

Ovchinnikova, A., Krylova, E., Gavrilenko, T., Smekalova, T., Zhuk, M., Knapp, S., Spooner, D.M., 2011. Taxonomy of cultivated potatoes (*Solanum* section *Petota*: Solanaceae). Botanical Journal of the Linnean Society 165, 107–155.

Paget, M., Amoros, W., Salas, E., Eyzaguirre, R., Alspach, P., Apiolaza, L., Noble, A., Bonierbale, M., 2014. Genetic evaluation of micronutrient traits diploid potato from a base population of Andean landrace cultivars. Crop Science 54, 1–11.

Peña, C., Restrepo-Sánchez, L.P., Kushalappa, A., Rodríguez-Molano, L.E., Mosquera, T., Narváez-Cuenca, C.E., 2015. Nutritional contents of advanced breeding clones of *Solanum tuberosum* group Phureja. LWT—Food Science and Technology 62 (1). http://dx.doi.org/10.1016/j.lwt.2015.01.038.

Pérez Baca, L., 1996. Crianza de la Papa en Paucartambo – Qosqo. Centro de Servicios Agropecuarios, Cusco, Peru.

Piñeros, C., Guateque, A., Peña, C., Cuéllar, D., Kushalappa, A., Narváez, C., Restrepo, P., Mosquera, T., 2013. Genetic diversity in potato breeding for nutritional quality in Colombia. In: Poster Presented at Solanaceae Conference, Genome vs Phenome, October 1–17, Beijing, China. Available at: http://issuu.com/seguridadalimentarianarinoproyecto/ (accessed in 2015).

Piñeros, C., Peña, C., Del Catillo, S., Rodriguez, E., Restrepo, P., Kubow, S., Sabally, K., Kushalappa, A., Mosquera, T., 2014. Exploring biodiversity to introduce nutritional quality criteria in potato breeding programs. In: Oral Presentation Presented at the European Association for Potato Research 19th Triennial Conference, July 6–11, Brussels, Belgium. Available at: http://www.seguridadalimentarianarino.unal.edu.co/sites/default/files/pdf-eventos/EC_2014-07_Introducing.nutritional.criteria.potato.breeding.programs.pdf (accessed in 2015).

Pumisacho, M., Sherwood, S., 2002. El Cultivo de la Papa en Ecuador. Instituto Nacional Autónomo de Investigaciones Agropecuarias, Centro Internacional de la Papa, Quito, Ecuador.

Quiros, C.F., Ortega, R., Van Raamsdonk, L., Herrera-Montoya, M., Cisneros, P., Schmidt, E., Brush, S.B., 1992. Increase of potato genetic resources in their center of diversity: the role of natural outcrossing and selection by Andean farmers. Genetic Resources and Crop Evolution 39, 107–113.

Rabinowitz, D., Linder, C.R., Ortega, R., Begazo, D., Murguia, H., Douches, D.S., Quiros, C.F., 1990. High levels of interspecific hybridization between *Solanum sparsipilum* and *S. stenotomum* in experimental plots in the Andes. American Potato Journal 67, 73–82.

Ramulu, K.S., Dijkhuis, P., Roest, S., 1984. Genetic instability in protoclones of potato (*Solanum tuberosum* L. cv. 'Bintje'): new types of variation after vegetative propagation. Theoretical and Applied Genetics 68, 515–519.

Reader, J., 2008. Propitious Esculent: The Potato in World History. William Heinemann, Portsmouth, England.

Ríos, D., Ghislain, M., Rodriguez, F., Spooner, D.M., 2007. What is the origin of the European potato? Evidence from Canary Island landraces. Crop Science 47, 127–1280.

Rodríguez, F., Ghislain, M., Clausen, A.M., Jansky, S.H., Spooner, D.M., 2010. Hybrid origins of cultivated potatoes. Theoretical and Applied Genetics 121, 1187–1198.

Salaman, R.N., 1937. The potato in its early home and its early introduction into Europe. Journal of the Royal Horticultural Society 62, 61–266.

Salaman, R.N., 1946. The early European potato: its character and place of origin. Journal of the Linnean Society (Botany) 53, 1–27.

Salaman, R.N., 1949. The History and Social Influence of the Potato. Cambridge University Press, Cambridge, England.

Sattaur, O., 1988. A bitter potato that saves lives. New Scientist 1612, 51.

Schmiediche, P.E., Hawkes, J.G., Ochoa, C.M., 1980. The breeding of the cultivated potato species *Solanum × juzepczukii* Buk. and *S. × curtilobum* Juz. et Buk. I. A study of the natural variation of *S. × juzepczukii*, *S. × curtilobum* and their wild progenitor *S. acaule* Bitt. Euphytica 29, 685–704.

Scott, G.J., 2011a. Growth rates for potatoes in Latin America in comparative perspective: 1961–07. American Journal of Potato Research 88, 143–152.

Scott, G.J., 2011b. Tendencias cruzadas: el consumo y utilización de la papa en América Latina entre 1961 y 2007 y sus implicancias para la industria. Revista Latinoamericana de la Papa 16, 1–38.

Scott, G.J., 2011c. Plants, people, and the conservation of biodiversity of potatoes in Peru. Natureza & Conservação 9, 21–38.

Scurrah, M., Celis-Gamboa, C., Chumbiauca, S., Salas, A., Visser, R.G.F., 2008. Hybridization between wild and cultivated potato species in the Peruvian Andes and biosafety implications for the deployment of GM potatoes. Euphytica 164, 881–892.

Seefeldt, H.F., Tønning, E., Wiking, L., Thybo, A.K., 2011. Appropriateness of culinary preparations of potato (*Solanum tuberosum* L.) varieties and relation to sensory and physicochemical properties. Journal of the Science of Food and Agriculture 91, 412–420.

Simmonds, N.W., 1964. Studies on the tetraploid potatoes. II. Factors in the evolution of the *tuberosum* group. Journal of the Linnean Society (Botany) 59, 43–56.

Simmonds, N.W., 1966. Studies on the tetraploid potatoes. III. Progress in the experimental re-creation of the *tuberosum* group. Journal of the Linnean Society (Botany) 59, 279–288.

Singh, J., Kaur, L., 2009. Advances in Potato Chemistry and Technology. Academic Press, Burlington.

Slattery, C.J., Halil Kavakli, I., Okita, T.W., 2010. Engineering starch for increased quantity and quality. Trends in Plant Science 5, 91–298.

Spooner, D.M., 2009. DNA barcoding will frequently fail in complicated groups: an example in wild potatoes. American Journal of Botany 96, 1177–1189.

Spooner, D.M., van den Berg, R.G., 1992. An analysis of recent taxonomic concepts in wild potatoes (*Solanum* sect. *Petota*). Genetic Resources and Crop Evolution 39, 23–37.

Spooner, D.M., van der Berg, R.G., Rodríguez, A., Bamberg, J., Hijmans, R.J., Lara Cabrera, S.I., 2004. Wild potatoes (*Solanum* section *Petota*, Solanaceae) of North and Central America. Systematic Botany Monographs 68.

Spooner, D.M., McLean, K., Ramsay, G., Waugh, R., Bryan, G.J., 2005a. A single domestication for potato based on multilocus amplified fragment length polymorphisms genotyping. Proceedings of the National Academy of Science (PNAS) 102, 14694–14699.

Spooner, D.M., Nuñez, J., Rodríguez, F., Naik, P.S., Ghislain, M., 2005b. Nuclear and chloroplast DNA reassessment of the origin of Indian potato varieties and its implications for the origin of the early European potato. Theoretical Applied Genetics 110, 1020–1026.

Spooner, D.M., Jansky, S.H., Clausen, A., Herrera, M.R., Ghislain, M., 2012. The enigma of *Solanum maglia* in the origin of the Chilean cultivated potato, *Solanum tuberosum* Chilotanum Group. Economic Botany 66, 12–21.

Spooner, D.M., Ghislain, M., Simon, R., Jansky, S.H., Gavrilenko, T., 2014. Systematics, diversity, genetics, and evolution of wild and cultivated potatoes. Botanical Review 80, 283–383.

Tapia, M.E., 2000. Mountain agrobiodiversity in Peru: seed fairs, seed banks, and mountain-to-mountain exchange. Mountain Research and Development 20, 220–225.

Tapia, M.E., Rosas, A., 1993. Seed fairs in the Andes: a strategy for local conservation of plant genetic resources. In: de Boef, W., Amanor, K., Wellard, K., Bebbington, A. (Eds.), Cultivating Knowledge: Genetic Diversity, Farmer Experimentation and Crop Research. Intermediate Technology Publications, London, pp. 111–118.

Terrazas, F., Cadima, X., Garcia, R., Zeballos, J., 2008. Catálogo Etnobotánico de Papas Nativas: Tradición y Cultura de los Ayllus del Norte Potosí y Oruro. Fundación PROINPA, Cochabamba, Bolivia.

Thiele, G., 1999. Informal potato seed systems in the Andes: why are they important and what should we do with them? World Development 27, 83–99.

Towle, M.A., 1961. The Ethnobotany of Pre-Columbian Peru. Viking Fund Publications in Anthropology, Chicago, IL.

Trognitz, B.R., Bonierbale, M., Landeo, J.A., Forbes, G., Bradshaw, J.E., Mackay, G.R., Waugh, R., Huarte, M.A., Colon, L., 2001. Improving potato resistance to disease under the global initiative on late blight. In: Cooper, H.D., Spillane, C., Hodgkin, T. (Eds.), Broadening the Genetic Base of Crop Production. IPGRI/FAO, New York, pp. 385–398.

Ugarte, M.L., Iriarte, V., 2000. Papas Bolivianas: Catálogo de Cien Variedades Nativas. Fundación PROINPA, Cochabamba, Bolivia.

Ugent, D., Dillehay, T., Ramirez, C., 1987. Potato remains from a late pleistocene settlement in Southcentral Chile. Economical Botany 41, 17–27.

United Nations Conference on Environment and Development (UNCED), 1992. Convention on Biological Diversity. United Nations, Geneva, Italy. Available at: https://www.cbd.int/doc/legal/cbd-en.pdf (accessed in 2015).

Väänänen, T., 2007. Glycoalkaloid Content and Starch Structure in *Solanum* Species and Interspecific Somatic Potato Hybrids (Ph.D. thesis). University of Helsinki, Helsinki.

Veilleux, R.E., De Jong, H., 2007. Potato. In: Singh, R.J. (Ed.), Genetic Resources, Chromosome Engineering, and Crop Improvement, vol. 3. CRC Press, Boca Raton, pp. 17–58.

Vinson, J.A., Demkosky, C.A., Navarre, D.A., Smyda, M.A., 2012. High-antioxidant potatoes: acute in vivo antioxidant source and hypotensive agent in humans after supplementation to hypertensive subjects. Journal of Agricultural Food Chemistry 60, 6749–6754.

Walker, T.S., Schmiediche, P.E., Hijmans, R.J., 1999. World trends and patterns in the potato crop: an economic and geographic survey. Potato Research 42 (2), 241–264.

Wang, Q., Zhang, W., 2004. China's potato industry and potential impacts on the global market. American Journal of Potato Research 81 (2), 101–109.

Werge, R.W., 1979. Potato processing in the central highlands of Peru. Ecology of Food and Nutrition 7, 229–234.

Winters, P., Hintze, L.H., Ortiz, O., 2006. Rural development and the diversity of potatoes on farms in Cajamarca, Peru. In: Smale, M. (Ed.), Valuing Crop Biodiversity: On-farm Genetic Resources and Economic Change. CABI Publishing, Wallingford, pp. 146–161.

Woolfe, J.A., 1987. The Potato in the Human Diet. Cambridge University Press, Cambridge.

Zimmerer, K.S., 1991a. Labor shortages and crop diversity in the southern Peruvian sierra. Geographical Review 82, 414–432.

Zimmerer, K.S., 1991b. The regional biogeography of native potato cultivars in highland Peru. Journal of Biogeography 18, 165–178.

Zimmerer, K.S., 1996. Changing Fortunes: Biodiversity and Peasant Livelihood in the Peruvian Andes. University of California Press, Berkeley, CA.

Zimmerer, K.S., 2014. Conserving agrobiodiversity amid global change, migration, and nontraditional livelihood networks: the dynamic use of cultural landscape knowledge. Ecology and Society 19, 1.

CHAPTER 2

Cell Wall Polysaccharides of Potato

Marie-Christine Ralet, Fanny Buffetto, Isabelle Capron, Fabienne Guillon
INRA, UR1268 Biopolymères, Interactions et Assemblages, Nantes, France

1. Introduction

Plant cell walls are the most abundant source of terrestrial biomass. They surround each cell outside the plasmalemma, creating a continuous extracellular matrix, which forms the skeleton of plant tissues. Plant cell walls constitute a highly complex and dynamic entity of extreme importance in plant growth and development. They can be considered fiber composites, in which cellulose microfibrils are embedded in a matrix of complex carbohydrates and lignins. These biopolymer assemblies collectively determine the shape and mechanical properties of plant cells. The plant cell wall is generally composed of three levels of organization:

1. The *middle lamella*, situated between two cells at the most extreme position of each, is mainly made of pectic components. It is shaped during mitosis and ensures the overall cohesion of adjacent cells.
2. Young cells are surrounded by a highly hydrophilic *primary cell wall* that is synthesized during the cell expansion stage. The primary cell wall is made of a loose network of cellulose microfibrils combined with hemicelluloses and pectins.
3. When cell expansion is over and the cell no longer needs to grow, a *secondary cell wall* can exist. Primary and secondary cell walls can be differentiated by their pectin and lignin contents and by the structure of their hemicellulosic compounds. In secondary cell walls, the pectin content decreases dramatically and is mainly replaced by lignin. Cell wall loses its flexibility and becomes thicker to form a consolidated material.

When potatoes are processed for starch production, a low-value, cell wall-rich byproduct known as potato pulp or pressed potato fibers is produced. Meyer et al. (2009) reported that the annual co-production of potato pulp in Europe amounts to at least 1 million tons of wet potato pulp (i.e., 2×10^5 tons of dry matter). Potato pulp mainly consists of cell walls (40–65 wt% dry matter) containing fragments of both flesh and peels, and of residual starch, usually 20–30% dry matter (Meyer et al., 2009). Nowadays, potato pulp is mainly used directly as cattle feed and is not upgraded to any significant degree (Ravn et al., 2015). The huge amounts of pulp available have motivated the reemergence of research devoted to potato cell wall polysaccharide extraction and uses.

Advances in Potato Chemistry and Technology. http://dx.doi.org/10.1016/B978-0-12-800002-1.00002-9

2. Isolation of Potato Cell Walls

Potato cell walls are difficult to purify; the principal problem is starch contamination. There are also problems not specific to potatoes, such as the need to inactivate endogenous cell wall-degrading enzymes and the difficulty of removing adsorbed proteins from cell wall preparations (Jardine et al., 2002).

Raw potato starch and cell walls both are water-insoluble and hence difficult to separate from each other (Jardine et al., 2002). Ring and Selvendran (1978) developed an efficient method in which starch removal was achieved with 90% aqueous dimethylsulfoxide (DMSO) at room temperature. This method involves lengthy incubation in large volumes of DMSO; other comprehensive methods have been specifically implemented to remove starch from fresh potato tubers or from industrial potato pulp for cell wall material purification. A group of three methods was first developed for efficient and convenient cell wall isolation from peeled potato tubers (Jardine et al., 2002). The first method allowed efficient and rapid removal of starch from domestic potatoes but failed to remove starch satisfactorily from industrial starch potatoes. Two alternative methods were therefore implemented. In the first, disruption of the potato cells was achieved using homogenization with Ultra-turrax in a buffer containing 1% (w/v) sodium deoxycholic acid at 4 °C. The residue was extracted with phenol/acetic acid/water (2/1/1 w/v/v) to remove noncovalently bound proteins, lipids, and sodium deoxycholic acid. The residue was then suspended in buffer at pH 4 and the cell wall suspension was heated at 70 °C to gelatinize starch. After cooling, a combination of α-amylase and pullulanase was added to degrade starch. The suspension was incubated at 37 °C before centrifugation, filtration, and washes. In the second method, homogenization with Ultra-turrax was achieved in a mixed-cation buffer containing Triton 100 and extraction was done with a saturated phenol solution. The residue was then frozen into small pellets, cryo-milled, and suspended in mixed-cation buffer. Starch gelatinization and removal were achieved as described previously. Both methods were shown to remove starch efficiently, even from industrial starch potatoes. The first method appeared to be much more efficient for protein removal but is more time-consuming than the second one (Jardine et al., 2002).

Although the methods developed by Jardine et al. (2002) proved to be efficient for starch and/or protein removal, they are not well adapted for large-scale treatment of potato pulp. An enzyme-catalyzed starch removal procedure was optimized for potato pulp (Thomassen and Meyer, 2010). The authors aimed to remove starch efficiently, employing "as few steps, as few enzymatic activities, as low enzyme dosage, as low energy input and as high pulp dry matter as possible." Thus, statistically designed experiments were performed varying the amount of dry matter, incubation time, and incubation temperature, and the thermostable α-amylase (Termamyl) dose. All of the removable starch (based on AOAC method 985.29 and theoretical maxima) (Meyer et al., 2009) could be released from the pulp in one step using 8% w/w of pulp dry matter, 0.2% v/w Termamyl dose/substrate (2.4 KNU/g DW material) at 70 °C for

65–85 min (Thomassen and Meyer, 2010). Byg et al. (2012) similarly treated frozen potato pulp with Termamyl but used different experimental conditions (1% w/w of pulp dry matter, 0.5 KNU Termamyl/ g DW material at 80 °C for 60 min). Finally, Ramaswamy et al. (2013) also used Termamyl (85 °C for 30 min) and further treated the residue with amyloglucosidase (30 °C for 20 h); this destarching procedure was applied twice. After treatment with starch-degrading enzymes, the slurry was vacuum-filtered (Byg et al., 2012) or centrifuged (Thomassen and Meyer, 2010; Ramaswamy et al., 2013). A large proportion of the potato pulp wall polysaccharides (>50%) was solubilized by the Termamyl–amyloglucosidase destarching treatment applied by Ramaswamy et al. (2013) and was lost upon filtration. Cell wall polysaccharide losses in the filtrate after Termamyl treatment were not quantified in Byg et al. (2012). In the method used by Thomassen and Meyer (2010), cell wall polysaccharide losses in the centrifugation supernatant after Termamyl treatment alone were modest compared with losses observed using the Termamyl–protease–amyloglucosidase destarching treatment used as a reference. A more or less large proportion of potato cell wall polysaccharides can be solubilized depending on physical processes (grinding, milling, etc.) applied and on the destarching procedure chosen. This severely affects the cell wall material recovery and its composition; some types of cell wall polysaccharides are more prone to solubilization than others.

3. Cell Wall Polysaccharides

Potato tubers and hence potato pulp are mainly composed of parenchymatous tissues and therefore of primary cell walls. Potato is a typical eudicotyledon encompassing primary cell walls that are mainly made of pectin, hemicellulosic xyloglucan, and cellulose (McCann and Roberts, 1991).

3.1 Cellulose

Cellulose, the most abundant renewable polymer on earth, has gained interest in nanotechnology research for generating sustainable materials. Agricultural waste such as potato pulp constitutes an attractive source of cellulose for industrial use. Cellulose is insoluble and represents about 30–40% of the dry primary cell wall of potato (Dufresne et al., 2000; Oomen et al., 2004).

Cellulose molecules consist of $(1\rightarrow4)$-β-D-glucopyranoses and exhibit an extended stiff ribbon-like conformation. Multiple hydroxyl groups on the glucose units from one chain form hydrogen bonds with oxygen atoms of glucose units from the same or a neighbor chain. Hydrogen linkages enable stabilization of linear parallel orientation of the cellulose molecule (Sugiyama et al., 1991), and interchain connections result in a multichain complex that ensures rigid construction via hydrogen bonds and van der Waals interactions. The whole structure leads to the generation of solid pseudocrystalline fibers that are organized in microfibrils. Using advanced physical methods and modeling, studies showed that the number of chains composing a microfibril is likely to be in the range of 18–25 (Cosgrove, 2014).

As a chemical raw material, cellulose has been used in the form of fibers or derivatives for nearly 150 years for a wide spectrum of products and materials in daily life. More than 60 years ago, it was demonstrated that cellulose fibers were composed of rodlike residues. These particles, known as cellulose nanocrystals (CNCs), are solid crystalline particles that arise from preferential hydrolysis of the amorphous regions of cellulose fibers. This hydrolysis leads to highly crystalline, solid, rodlike particles of nanometer dimensions (Revol et al., 1992; Klemm et al., 2011). Cellulose nanocrystals have reached a tremendous level of attention in the materials community that does not appear to be relenting. Cellulose nanocrystal opportunities in applications correlate to their intrinsic properties: low density, mechanical strength, stiffness, low thermal expansion coefficient, optical transparency and anisotropy, and flexible surface chemistry (Klemm et al., 2011). Consequently, such nanocrystals are being studied for a number of potential applications, including polymer nanocomposites, transparent or chiral films, rheology modifiers and hydrogels, drug delivery vehicles, artificial blood vessels, and wound dressings (Habibi et al., 2010). Cellulose nanocrystals have been deployed as emulsion stabilizers, taking advantage of their self-assembling ability at the oil–water interface, similar to particle-stabilized pickering systems (Kalashnikova et al., 2011, 2012, 2013). The intriguing ability of CNCs to self-organize into a chiral nematic (cholesteric) liquid crystal phase with a helical arrangement has attracted significant interest, resulting in much effort in research, because this arrangement gives dried CNC films a photonic band gap (Lagerwall et al., 2014). The films thus acquire attractive optical properties, creating possibilities for use in applications such as security papers and mirrorless lasing. Also, the layer-by-layer assembly technique was applied to CNC to design nanocellulose/polymer thin films and coatings with advanced functionalities such as semireflective colored films (Martin and Jean, 2014), enzymatic detectors (Cerclier et al., 2011), and conductive films when associated with conductive particles such as carbon nanotubes (Olivier et al., 2012).

Potato cellulose microfibrils have been recovered from potato pulp cell walls after solubilization of hemicelluloses and pectin in a solution of potassium or sodium hydroxide and further bleaching with sodium chlorite (Dufresne and Vignon, 1998; Dufresne et al., 2000; Abe and Yano, 2009). Removal of noncellulosic wall material facilitated the aggregation of adjacent parallel microfibrils (Earl and Vanderhart, 1981; Newman et al., 2013). Recovery of single microfibrils was achieved using mechanical treatment (Dufresne and Vignon, 1998; Dufresne et al., 2000; Abe and Yano, 2009). The microfibrils exhibited lengths of 10–100 μm and diameters on the order of 10 nm and were demonstrated as an effective reinforcing additive in starch–cellulose composites (Dufresne et al., 2000).

Cellulose nanocrystals were recovered from potato peel waste (Chen et al., 2012). Potato-derived CNC exhibited a broad length distribution with many long fibers compared with cotton CNC (Chen et al., 2012). Potato CNC was used to prepare thermoplastic starch and polyvinyl alcohol nanocomposite films and was shown to improve the mechanical properties of both composite films (Chen et al., 2012).

3.2 Hemicelluloses

Hemicelluloses represent around 25% of primary cell wall dry material. They group all polysaccharides in the cell wall, which are neither cellulosic nor pectic. Hemicelluloses are usually extracted using dilute alkaline solutions. Hemicelluloses can have a homopolymeric or heteropolymeric backbone. Homopolymeric backbones always have the same common structure constituted by (1→4)-β-D-glycopyranose units, which can be glucose, xylose, or mannose (Figure 1). These hemicelluloses are called glucans, xylans, or mannans, respectively. Heteropolymer backbones always involve glucopyranose residues, which can be linked to mannose residues (heteromannan) or linked to themselves with hetero-linkages (mixed-linked β-glucans). Hemicelluloses may be substituted by various short side chains, leading to increased structural diversity. Depending on the developmental stages or plant species, different hemicellulosic polymers are present in the cell wall. In the primary cell wall of eudicotyledons, the most commonly encountered hemicelluloses are glucomannans, glucuronoarabinoxylans, and xyloglucans (Harris et al., 2009; Scheller and Ulvskov, 2010).

3.2.1 Galactoglucomannan

In general, mannans are considered the most ancient plant cell wall hemicelluloses and are thought to have been largely replaced by other hemicelluloses in spermaphytes (Scheller and Ulvskov, 2010; Pauly et al., 2013). Glucomannan usually represents 3–5% of the primary wall of eudicotyledons. The hetero-backbone is composed of (1→4)-β-D-Man*p* with interspersed β-D-Glc*p* residues (Figure 1(c)). In some plant species, mannans are in the form of galactoglucomannans, in which Man*p* residues are substituted at *O*-6 by α-D-Gal*p* single units or by α-D-Gal*p*-(1→2)-α-D-Gal*p* disaccharides (Harris et al., 2009). Hetero-mannans have never been isolated and purified from potato. However, from sugar and linkage analysis (Jarvis et al., 1981), their presence is conceivable.

3.2.2 Glucuronoarabinoxylan

Glucuronoarabinoxylan is classically present in primary cell walls of eudicotyledons, where it usually represents approximately 5% of the dry material (Darvill et al., 1980; Scheller and Ulvskov, 2010). Variability in xylan branching depends on the cell tissue and the development stage (Scheller and Ulvskov, 2010). In glucuronoarabinoxylans, α-L-Ara*f* and α-D-Glc*p*-A branching occur at the *O*-2 position of a xylan backbone (Figure 1(b)), which can also be acetylated at the *O*-2 and *O*-3 positions (Gille and Pauly, 2012). In potato parenchyma, the presence of heteroxylans has been hypothesized because small proportions of (1→4)-β-D-Xyl*p* were found in potato cell wall alkali extracts (Ring and Selvendran, 1978).

3.2.3 Xyloglucan

Xyloglucan is the most abundant hemicellulosic polymer in potato primary cell walls, representing about 11% of dry material (Oomen et al., 2003). Xyloglucan has a (1→4)-β-D-Glc*p* backbone

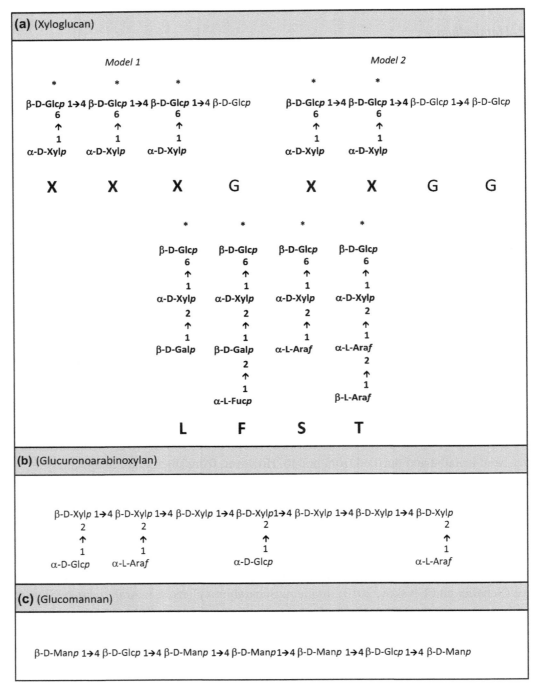

Figure 1

Structure of the most frequently encountered hemicellulose of primary cell wall of eudicotyledons. (a) Xyloglucans. The two models of xyloglucan motifs are represented. For both models, short side chains are variable. The last sugar of the nonreducing end of each side chain is represented by a letter (G, X, L, F, S, and T) (Fry et al., 1993). (b) Glucuronoarabinoxylan. (c) Heterobackbone of glucomannan.

that is branched at *O*-6 by α-D-Xyl*p* residues (Hayashi, 1989). Two types of xyloglucans have been described, based on the degree of substitution of glucan backbone (Vincken et al., 1997; Pauly et al., 2013). Most higher plants produce xyloglucan, in which three of every four Glc*p* residues are substituted with an Xyl*p* residue; this corresponds to the XXXG-type branching pattern. The XXGG-type is less substituted, containing only two Xyl*p* residues per four Glc*p* units (Figure 1(a)). This branching pattern is specific to xyloglucan from *Solanales* (Vincken et al., 1997; Scheller and Ulvskov, 2010; Pauly et al., 2013). Treating xyloglucan from *Solanales* with (1→4)-endo-β-glucanases releases not only XXGG but also low amounts of XGXG, GXXG, GXGG, XXG, and XGG oligosaccharides (Vincken et al., 1996; Jia et al., 2003; Quéméner et al., 2015). These motifs are supposed to be generated for two reasons: the presence of noncontiguous Xyl*p* residues over the backbone and the presence of two possible degrading sites for the endo-glucanase on the glucan main chain. Indeed, the presence of two contiguous unsubstituted Glc-*p* residues would allow the generation of different structures (XXGG↓XXG↓GXXG). The XXXG-type branching pattern, although unusual in *Solanales*, has been detected in *Nicotiana alata* pollen grain and pollen tube walls (Lampugnani et al., 2013).

Xyl*p* residues may have other sugars attached to them; Fry et al. (1993) developed a nomenclature in which single letters represent terminal sugars of lateral chains (Table 1). In *Solanaceae*, several side chains were identified (Table 1), but fucosylated side chains have been detected on side chains only from the XXXG-type xyloglucan found in *Nicotiana alata* (Harris et al., 2009; Lampugnani et al., 2013). Different studies showed that potato xyloglucan contained Glc, Xyl, Gal, and Ara in the molar ratio 6/3/1/1 (Ring and Selvendran, 1981; Vincken et al., 1996; York et al., 1996), in agreement with the hypothesis that the three main side chains—α-D-Xyl*p* (X), β-D-Gal*p*-(1→2)-α-D-Xyl*p* (L), and α-L-Ara*f*-(1→2)-α-D-Xyl*p* (S)—could be present in similar amounts.

Table 1: Nomenclature Defined by Fry et al. (1993) and Description of XXGG-Type Side Chains.

Letter	*Solanales*	Potato	Chain	References
G	+	+	β-D-Glc*p*	Ring and Selvendran (1981)
X	+	+	α-D-Xyl*p*-(1→6)-β-D-Glc*p*	Ring and Selvendran (1981)
L	+	+	β-D-Gal*p*-(1→2)-α-D-Xyl*p*-(1→6)-β-D-Glc*p*	Ring and Selvendran (1981)
S	+	+	α-L-Ara*f*-(1→2)-α-D-Xyl*p*-(1→6)-β-D-Glc*p*	Ring and Selvendran (1981)
T	+	−	β-L-Ara*f*-(1→3)-α-L-Ara*f*-(1→2)-α-D-Xyl*p*-(1→6)-β-D-Glc*p*	York et al. (1996)
U	+	−	β-D-Xyl*p*-(1→2)-α-D-Xyl*p*-(1→6)-β-D-Glc*p*	Quéméner et al. (2015)

(+) Detected; (−) not detected; β-D-Glc*p*: glucose from the xyloglucan backbone.

Xyloglucan can also be acetyl-esterified. Studies on *Solanaceae* showed that acetylation mostly occurs at *O*-6 of the Glc*p* residues (Ring and Selvendran, 1981; Sims et al., 1996; Jia et al., 2005). The presence of acetyl groups is mainly observed on the unsubstituted Glc*p* residue closest to the nonreducing end. Acetyl-esterification was also detected at *O*-5 of Ara*f* residues (Jia et al., 2005).

3.3 Pectin

Pectin is considered a key component for plant primary cell wall architecture in eudicotyledons and is involved in various cell functions and plant processes. Pectin is particularly abundant in potato cell walls.

Pectin is a heterogeneous and complex galacturonic acid (Gal*p*A)-rich polysaccharide composed of three major domains, homogalacturonans (HG), HG analogues, and type I rhamnogalacturonans (RGI), which are covalently linked to each other (Caffall and Mohnen, 2009). The exact molecular arrangement of the different pectic domains relative to each other is still a matter of debate (Vincken et al., 2003; Ralet and Thibault, 2009).

Homogalacturonan is the simplest and most abundant pectic structural domain, usually accounting for approximately 65% of the pectin macromolecule. It consists of a linear backbone of 85–320 (1→4) linked α-D-Gal*p*A residues (Thibault et al., 1993; Hellin et al., 2005; Round et al., 2010, Figure 2(a)). Gal*p*A residues are commonly partly methyl-esterified at *C*-6 (Voragen et al., 1995) and, in some plant species, partly acetyl-esterified at *O*-2 and/or *O*-3 (Ralet et al., 2005, 2008). Both the degree of methyl-esterification (DM) (i.e., the number of methyl-esterified Gal*p*A residues for 100 total Gal*p*A residues) and the degree of acetylation (DA) (i.e., the number of acetyl-esterified Gal*p*A residues for 100 total Ga*p*lA residues) have a profound impact on pectin interaction properties in vitro and *in planta*. Furthermore, not only the degree of the esterification but also the distribution of the esters is important. The degree and distribution of esterification vary according to the plant species, age, and location in the plant cell wall and are implicated in controlling pectin's functional properties (Bonnin et al., 2008; Caffall and Mohnen, 2009).

In ***HG analogues***, HGs can be branched by apiose (Api*f*) or xylose (Xyl*p*) units and/or disaccharide side chains and by longer and more complex side chains (Figure 2(a)). Apiogalacturonans, which are branched with a single β-D-Api*f* unit at *O*-2 and/or *O*-3, have been detected in the walls of aquatic plants (Hart and Kindel, 1970; Ovodov et al., 1971; Ridley et al., 2001). Xylogalacturonans, which are branched at *O*-3 with β-D-Xyl*p* single units or dimers, have been identified in several plant species (Schols et al., 1995; Le Goff et al., 2001; Zandleven et al., 2006). Rhamnogalacturonan II is a low-molecular-weight (5- to 10-kDa) highly complex macromolecule. It is composed of 12 different glycosyl unit types and more than 20 different linkages (O'Neill et al., 2004). Rhamnogalacturonan II contains a short HG backbone substituted with five different side chains including rare sugars such as aceric acid, methyl fucose, methyl xylose, 3-desoxy-D-*manno*-2-octulosonic acid, and 3-desoxy-D-*lyxo*-2-heptulosonic acid.

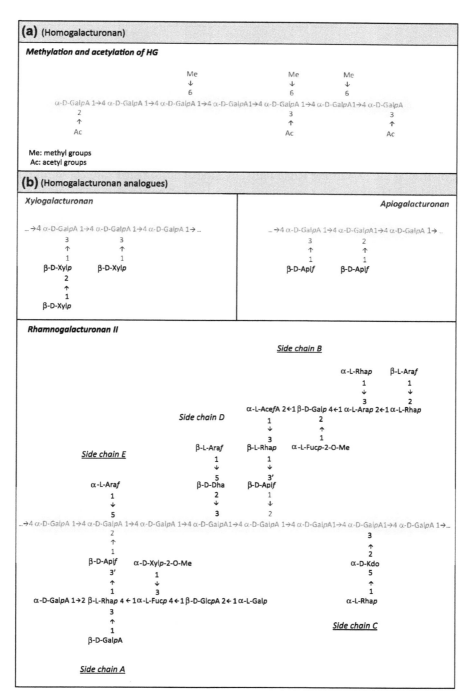

Figure 2

Structural representation of homogalacturonan and homogalacturonan analogues. (a) Position of acetyl- and methyl-esters on homogalacturonan backbone. (b) The three homogalacturonan analogues: xylogalacturonan, apiogalacturonan, and rhamnogalacturonan II. In (b), acetyl- and methyl-esterification is not shown.

Rhamnogalacturonan I usually represents 20–35% of the pectic polysaccharides (Mohnen, 2008). The RGI backbone is composed of GalpA and rhamnosyl (Rhap) units arranged into [→2)-α-L-Rhap-(1→4)-α-D-GalAp-(1→] repeats (Schols et al., 1990; Renard et al., 1995; Albersheim et al., 1996). An average degree of polymerization of 120–240 was reported (Ralet and Thibault, 2009), although higher degrees of polymerization have been hypothesized (McNeil et al., 1984). Partial acetylation often occurs at the *O*-2 and/or *O*-3 positions of the GalpA residues (Komalavilas and Mort, 1989; Ishii, 1997). Rhamnogalacturonan I is enhanced, mainly at the *O*-4 of the Rhap units, by side chain substructures predominantly containing linear and branched α-L-arabinofuranosyl (Araf) and/or β-D-galactopyranosyl (Galp) residues (Figure 3). The degree of substitution of Rhap units and the type, proportion, length, and degree of branching of side chains vary enormously according to plant sources, taxons, organs, and tissues (Ridley et al., 2001; Willats et al., 2001; Caffall and Mohnen, 2009).

Arabinans consist of a (1→5) linked α-L-Araf backbone that can be substituted at *O*-2 and/or *O*-3 by α-L- Araf monomers or short oligomers. Galactans can be subdivided into galactan, type I arabinogalactan (AGI), and type II arabinogalactan (AGII). Galactan and AGI encompass a (1→4)-β-D-Galp backbone. In AGI, this backbone may be branched by Araf and/or Galp monomers or oligomers. Finally, AGII has a main (1→3)-β-D-galactan backbone on which (1→6)-β-D-galactan secondary chains are branched. Araf units can be attached at *O*-3 of the nonreducing termini of the secondary galactan chains. Beside polymeric linear and branched side chains, Rhap units can also be substituted by Galp single units (Schols, 1995).

3.4 Potato Pectin Extraction and Structural Specificities

Potato cell walls are particularly rich in pectin that represents 55–60% DW of the cell wall polysaccharides (Ryden and Selvendran, 1990). The main sugars identified are GalA and Gal, followed by Ara, and Rha and linkage analysis provided evidence that GalA is 4-linked, Gal mainly 4-linked, Rha 2- and 2,4-linked, and Ara mainly 5-linked (Ring and Selvendran, 1978). Other linkages such as 3,5-Ara, 2,3-Gal, 2,4-Gal, 3,4-Gal, and 4,6-Gal were detected in low amounts, evidencing that unbranched arabinan and type I galactan chains are predominant in potato cell wall material. Potato cell wall material was shown to exhibit DM and DA of 45–50% (VanMarle et al., 1997a).

The extractability of potato pectic polysaccharides appears cultivar- and method-dependent (VanMarle et al., 1997a). Globally, smooth extracting agents can extract a large proportion of pectin from potato cell wall material. Jarvis et al. (1981) and VanMarle et al. (1997a) showed that 15–20% of total cell wall material on a starch-free basis (i.e., roughly 30% of total pectin) could be rapidly solubilized from potato cell walls using cold or hot calcium-chelating agents. Ng and Waldron (1997) obtained higher yields (37% of total starch-free cell wall material; i.e., roughly 60% of total pectin) after sequential extraction of DMSO-destarched

(Rhamnogalacturonan I)

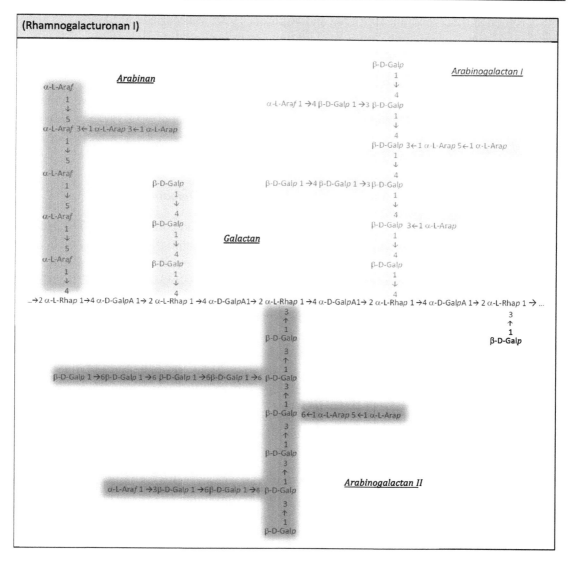

Figure 3

Structural representation of rhamnogalacturonan I. Different side chains may occur on this pectic domain. The length and proportion of arabinan, galactan, arabinogalactan type I, and arabinogalactan type II side chains are plant- and tissue-specific.

potato cell walls with water, NaCl, and a cold calcium-chelating agent. The highest yields were obtained when using industrial potato pulp as raw material. Indeed, sodium acetate buffer (pH 5.2, 85 °C, 30 min followed by pH 5.5, 30 °C, 20 h) could solubilize 46% by weight of cell wall material (i.e., roughly 75% of total pectin) (Ramaswamy et al., 2013). An additional amount of pectic material can be solubilized using cold dilute alkali (Jarvis et al.,

1981; Ryden and Selvendran, 1990; Ng and Waldron, 1997; VanMarle et al., 1997a). Usually, some pectic material cannot be extracted and remains in the cellulose-rich residue. This residual pectic material may be cross-linked and/or entangled with other cell wall polymers (Ryden and Selvendran, 1990; VanMarle et al., 1997a; Øbro et al., 2004).

Whatever the extracting agent, potato pectin's main structural characteristics are its particularly high proportion of RGI and low proportion of HG (Oomen et al., 2003). The presence of RGII and xylogalacturonan has been suspected in some studies but not yet proven (Øbro et al., 2004; Zandleven et al., 2006).

3.5 Specific Recovery and Analysis of Potato RGI Pectic Domain

Direct isolation of RGI from industrial potato waste has received a lot of attention. Indeed, isolated RGI, either intact or enzymatically modified, could potentially be used as high-value functional food ingredient (Meyer et al., 2009) or as a pharmaceutical agent (Øbro et al., 2004) or to improve bone and dental implant biocompatibility (Kokkonen et al., 2012). From a nutritional point of view, increased dietary fiber intake is associated with a decreased risk of several Western diseases. In particular, soluble dietary fiber intake has been associated with hypocholesterolemic and hypoglycemic effects and with the production of short-chain fatty acids that inhibit liver cholesterol synthesis (Lærke et al., 2007). Rhamnogalacturonan-rich soluble fiber extracted from potato pulp was shown to be highly fermentable and highly bifidogenic (Lærke et al., 2007; Thomassen et al., 2011).

To allow direct RGI recovery from cell wall material, one should take advantage of the blockwise arrangement of RGI and HG domains in the pectin molecule (Byg et al., 2012). Indeed, chemical or enzymatic degradation of HG regions only allows solubilization of intact RGI.

Harsh alkaline conditions that promote HG domains degradation by β-elimination reactions have been used to release RGI domains (Khodaei and Karboune, 2013). Several extraction conditions (0.5, 1, or 2 M NaOH or KOH for 24 h at 60 °C) were tested on destarched potato pulp. Alkali extracts were then neutralized and dialyzed. Excellent recovery yields ranging from 22.2% to 55.7% DW of destarched potato pulp were obtained; the harsher the extraction, the higher the recovery yield. Increasing the concentration of alkaline solutions led to the co-extraction of hemicelluloses.

Extraction of RGI has also been achieved using pure HG-degrading enzymes, endo-polygalacturonases, and pectin lyases (Øbro et al., 2004; Thomassen et al., 2011; Byg et al., 2012; Khodaei and Karboune, 2013). Byg et al. (2012) performed large-scale extraction of RGI using a purified endo-polygalacturonase directly on destarched potato pulp. Endo-polygalacturonases are active only on nonesterified or lowly methyl-esterified HGs only (Bonnin et al., 2002). Because the potato cell wall material was not deesterified before endo-polygalacturonase treatment, the enzymatic treatment was not efficient and RGI recovery yields were low:

11% and 9% DW of destarched potato pulp for laboratory scale and large scale, respectively. Øbro et al. (2004) and Khodaei and Karboune (2013) also used a purified endo-polygalacturonase but performed gentle chemical deesterification of the destarched potato pulp before endo-polygalacturonase treatment. Rhamnogalacturonan recovery yields (38–40% DW of destarched potato pulp) were largely increased compared with Byg et al. (2012) findings. Finally, Thomassen et al. (2011) treated destarched potato pulp with a mixture of endo-polygalacturonase and pectin lyase. Pectin lyase degrades methyl-esterified HG domains by β-elimination (Ralet et al., 2012). Within 1 min at pH 6 and 60 °C, 75% dry matter from destarched potato pulp was released. Surprisingly, when destarched potato pulp was incubated at pH 6 and 60 °C with buffer only, 41% dry matter from destarched potato pulp was released within 1 min. Solubilized polysaccharides were recovered by isopropanolic precipitation, and it is possible that some phosphate salts co-precipitated, leading to an artificial increase in recovery yields. The fraction released by buffer only was fractionated by size-exclusion chromatography. The chromatogram showed that most of the fraction was eluted as components of very low molar mass (≤1.3 kDa based on pullulan standards elution time), which supports the hypothesis of the massive presence of residual salts.

Finally, a high-molecular-weight pectic fraction was isolated from potato pulp using an enzymatic preparation containing various pectolytic, hemicellulolytic, and cellulolytic activities (Schols and Voragen, 1994). Notably, this enzymatic preparation contains endo-polygalacturonases, pectin lyases, and pectin methyl-esterases that collectively allow the degradation of HG domains (Bonnin et al., 2014). After enzymatic hydrolysis of potato pulp, enzymes were inactivated and the slurry was centrifuged. Supernatant was then subjected to ultrafiltration and the retentate corresponding to the RGI fraction was recovered. The enzyme preparation used is known to degrade RGI to a certain extent; the authors named the RGI fraction obtained the modified hairy region (MHR). The MHR isolated from potato pulp accounted for 4.1% of the starting dry nondestarched material (i.e., roughly 6% of destarched material).

In potato cell wall material, the GalA/Rha molar ratio varies between 12.3 and 18.2, the Gal/Rha ratio between 8.8 and 26.2, and the Ara/Rha ratio between 1.3 and 6.5 (Table 2). Raw RGI recovered by alkali treatment of potato cell wall material exhibited a GalA/Rha molar ratio much lower than the ratio observed for potato cell wall material (Table 2), which shows that HG domains were efficiently degraded and HG-derived oligomers largely removed upon dialysis. Gal/Rha and Ara/Rha molar ratios were in the same order of magnitude as ratios observed for potato cell wall material (Table 2), evidencing the good preservation of arabinan and galactan side chains upon alkaline extraction. The use of endo-polygalacturonase led to the extraction of RGI fractions that were more or less depleted in HG domains (low GalA/Rha ratio) and roughly unaffected in galactan and arabinan side chains. Thomassen et al. (2011) reported particularly high values for all molar ratios (GalA/Rha, Gal/Rha, and Ara/Rha), which may be the result of an underestimation of Rha content. After purification by

Table 2: GalA, Rha, Gal, and Ara Molar Ratios of Potato Cell Wall Material and Extracted RGI Fractions.

	Extraction Mean	Recovery Yield[a]	GalA/ Rha[b]	Gal/ Rha[b]	Ara/ Rha[b]	References
CWM[c]	–	–	13.9	11.8	5.8	Ryden and Selvendran (1990)
CWM[c]	–	–	18.2	26.2	6.5	Meyer et al. (2009)
CWM[c]	–	–	12.3	8.8	1.3	Khodaei and Kabourne (2013)
Raw RGI	0.5 M NaOH	24.4	4.3	21.7	4.2	Khodaei and Kabourne (2013)
Raw RGI	2 M NaOH	53.6	3.1	19.4	4.4	Khodaei and Kabourne (2013)
Raw RGI	PG[d]	9.0	2.9	11.4	1.5	Byg et al. (2012)
Raw RGI	PG[d]	40.0	3.2	13.2	3.9	Øbro et al. (2004)
Raw RGI	PG[d]	37.9	8.3	15.7	3.2	Khodaei and Kabourne (2013)
Raw RGI	PG + PL[d,e]	75.0	17.5	45.5	6.6	Thomassen et al. (2011)
Purified RGI	PG[d]	–	1.2	14.3	3.5	Øbro et al. (2004)
Raw MHR[f]	Rapidase	6.0[g]	2.2	1.5	1.0	Schols and Voragen (1994)
Purified MHR[f]	Rapidase	–	1.3	1.3	1.0	Schols and Voragen (1994)

[a]Percent DW of destarched cell wall material.
[b]Molar ratio.
[c]CWM, cell wall material.
[d]PG, endo-polygalacturonase.
[e]PL, pectin lyase.
[f]MHR, modified hairy region.
[g]Recalculated from Schols and Voragen (1994) considering that starch represents 30% by weight of the pulp.

size-exclusion chromatography, Øbro et al. (2004) recovered an RGI fraction exhibiting a GalA/Rha molar ratio of 1.2 (3.2 before purification), which is close to the theoretical ratio of 1.0 expected for pure HG-free RGI. Negatively charged oligogalacturonides that do not dialyze out efficiently (Mort et al., 1991) were removed as low-molar-mass material (Øbro et al., 2004). Similarly, potato MHR exhibited a GalA over Rha molar ratio of 2.2. After purification by size-exclusion chromatography (i.e., removal of GalA oligomers that were not eliminated by ultrafiltration), this ratio decreased to 1.3 (Table 2). The low Gal over Rha and Ara over Rha molar ratios found for MHR (Schols and Voragen, 1994) show that arabinan and galactan side chains were highly degraded during the enzymatic extraction; Ara- and Gal-oligosaccharides were further lost during the ultrafiltration process.

The massive presence of linear arabinan and type I galactan side chains in potato RGI is obvious. However, Øbro et al. (2004) and Buffetto et al. (2015) revealed some unexpected structural complexity in potato RGI. Øbro et al. (2004) studied the enzymatic degradation of purified potato RGI with endo-β-1→4 galactanase and endo-α-1→5 arabinanase individually and in combination. The authors concluded that β-(1→4)-galactans may carry few substitutions of short arabinan chains, α-(1→5)-arabinans may be attached to the RGI backbone through short galactan anchor chains, and β-(1→4)-galactans may be present as side chains of the arabinans. Øbro et al. (2004) also evidenced the presence of α-(1→5)-arabinans that resist enzymatic degradation. Buffetto et al. (2015) isolated oligosaccharides made of (1→4)-linked Gal*p* residues interspersed with few internal (1→5)-linked Ara*f* residues. α-(1→5)-Arabinan dimers attached to the RGI backbone through a single Gal anchor chain were also evidenced, in agreement with the hypothesis of Øbro et al. (2004).

4. Effects of Heating on Potato Cell Wall Polysaccharides

Texture is a key quality determinant of cooked potatoes. Several factors may affect cooked potato texture, among which are starch content, cell wall structure and composition, and middle lamella integrity upon cooking (VanMarle et al., 1992). These parameters distinguish firm and mealy cultivars (VanMarle et al., 1992). The cell wall structure and hence types and amounts of pectin solubilized upon cooking differed for mealy (Irene) and nonmealy (Nicola) cultivars. Primary cell walls have a more compact structure and solubilized pectic polymers are more branched and contain fewer HG regions for the Irene cultivar (VanMarle et al., 1997b). More recently, studies were conducted aiming to understand tuber textural properties at a molecular level. A potato microarray was used to identify tuber gene expression profiles that corresponded to differences in various tuber characteristics, among which was texture (Ducreux et al., 2008). Major differences in the expression of genes involved in cell wall biosynthesis were identified between two cultivar groups with contrasted textural properties, Phureja and Tuberosum. In particular, pectin methyl esterase activity was identified as potentially influencing textural properties (Ducreux et al., 2008). The total pectin methyl esterase activity was three- to fivefold lower, the genes encoding two pectin methyl esterase isoforms were expressed at lower levels, and the pectin DM was higher in Phureja (mealy) tubers than in Tuberosum (nonmealy) ones (Ross et al., 2011). Similarly, acetylation level might have a role in tuber textural properties (Orfila et al., 2012).

5. Distribution of Pectin in Potato Tuber and Function in Cell Walls

5.1 Tuber Formation and Histology

The potato tuber develops at the distal end of the stolon. Tuberization is initiated in response to short days and cool temperature and is regulated by hormone and sucrose (Xu et al., 1998a, 1998b; Bush et al., 2001). At the onset of tuber formation, the elongation of stolon stops and cells in pith and cortex enlarge and divide longitudinally, resulting in swelling of the stolon tip. When tubers have a diameter of 0.8 cm, longitudinal division has ceased but randomly oriented division and cell isodiametric expansion occur in the perimedullary region and continue until tubers reach their final diameter. With increasing maturity, the parenchymal cell wall, especially from the cortex, thickens and develops intercellular spaces.

In mature potato tubers, different tissue types can be identified: storage parenchyma together with some vascular tissues in the perimedullary region (flesh) and periderm (skin) (Bush and McCann, 1999). Potato periderm is made up of three layers: phellem, phellogen, and

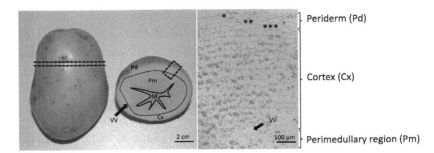

Figure 4
Potato histology. Phellem (*), phellogen (**), and pheloderm (***) are peripheral layers that constitute the periderm (Pd). Cortex (Cx) is a tissue present between the skin and the vascular vessel (VV). Perimedullary region is present in the vascular ring and is delimited by the medulla (M).

phelloderm (Reeve et al., 1969) (Figure 4). Phellem and phelloderm are derived from the periclinal division of the phellogen layer, which operates as lateral meristem during tuber maturation. As phellem cells develop, they become suberized and then die, forming a protective barrier (Sabba and Lulai, 2002). Potato flesh is mainly composed of storage parenchyma that contains thin nonlignified primary cell walls surrounding the starch granules (Ramaswamy et al., 2013). Potato skin contains some secondary cell walls next to primary ones (McDougall et al., 1996; Sabba and Lulai, 2002).

5.2 Distribution of Pectin in Developing and Mature Potato Tuber

The distribution of pectin in developing and mature potato tubers was investigated using monoclonal antibodies (Bush and McCann, 1999; Bush et al., 2001; Oomen et al., 2002; Sabba and Lulai, 2004, 2005; Buffetto et al., 2015) and staining (Sabba and Lulai, 2002). Distribution of pectin epitopes is developmentally regulated during potato tuber formation (Bush et al., 2001). In elongating stolon, the youngest cell population at the stolon tip exhibits walls enriched in highly methylated HG (JIM7 epitope) and α-(1→5)-arabinan (LM6 epitope) whereas lowly methylated HG (JIM5) and β-(1→4)galactan (LM5) become more abundant in walls of elongating cells proximal to the hook. During tuberization, all pectic epitopes become more abundant in parenchymal cell walls. Highly methylated HG is uniformly abundant in all cell walls. Lowly methylated HG is distributed across the width of mature parenchymal cell walls with a higher concentration in the middle lamella and linings of intercellular spaces, and at cell corners. The 2F4 epitope (block of unesterified HG) is mostly restricted to corners and lining intercellular space. β-(1→4)-Galactan, present throughout the walls in recently formed parenchymal walls, is found mainly located in primary walls of mature cells. α-(1→5)-Arabinan is localized across the width of walls except on the expanded middle lamella at cell corners. It is also absent at middle lamellae of the thickest cortical walls. Vascular cell walls contain both highly and partially methylated HG but have relatively low abundances of both galactan and arabinan (Bush and McCann, 1999; Oomen et al., 2002). Buffetto et al. (2015) investigated the occurrence of RGI backbone in potato tuber using INRA-RU1 antibody (Ralet et al., 2010). In the absence of pretreatment with galactanase, there is a gradient in the distribution of the RGI backbone epitope, the most peripheral layer of the cortex reacting more intensely compared with the deep cortex and perimedullar zone. The RGI backbone is mainly visualized in middle lamella and expanded middle lamella at cell corners. Enzymatic removal of galactan unmasks RGI backbone epitope and all cell walls of the cortex and perimedullary zones are labeled equally. The RGI backbone appears to be uniformly distributed across the width of the wall, except in the region closest to plasmalemma.

Pectin epitope distribution in the three cell types composing the periderm was more particularly investigated (Sabba and Lulai, 2002, 2004, 2005). Walls of these three layers labeled differentially for pectin. Highly and lowly methylated HG, galactan (LM5), and arabinan (LM6) epitopes are present in the phelloderm walls whereas the phellem walls appear to be enriched in highly methylated HG and lack low-methylated HG, galactan, and arabinan epitopes. Labeling of the phellogen layer with the different antibodies varies between immature and mature periderm. Cell walls of meristematically active phellogen lack galactan and HG epitopes. Upon maturation, labeling for these epitopes increased dramatically, indicating enrichment in these motifs. In contrast, α-(1→5)-arabinan is present in both meristematically active and inactive phellogen.

Some differences between freshly harvested and short-term, cold-stored tubers have been observed (Bush and McCann, 1999; Bush et al., 2001). In mature stored tubers, low-methylated HG is uniformly distributed throughout cortical and perimedullary walls and there is a gradient of abundance of β-(1→4)-galactan across the cortex. α-(1–5)-Arabinan is almost excluded from the middle lamella of perimedullary walls. These redistributions of pectic epitopes may result from storage-associated modifications involving pectinolytic enzyme activities.

5.3 Pectin Functions in the Wall

Turnover of pectin in relation to plant cell growth has been abundantly described in the literature, and some variation was noticed depending on organs and species. Young and meristematic cells are rich in highly methylated HG compared with elongating cells (Bush et al., 2001; Leboeuf et al., 2005; Sobry et al., 2005). Cell development also has an impact on RGI neutral side chain distribution. Arabinan-rich side chains have been predominantly detected in proliferating cells (Willats et al., 1999), whereas galactan-rich side chains have been observed in the elongating cells and cambium (Willats et al., 1999; Ermel et al., 2000; Bush et al., 2001; McCartney et al., 2003).

As detailed previously, pectic motifs are not uniformly distributed within the cell wall. Highly methylated HG are present throughout the cell wall whereas low- or nonmethylated ones, which are prone to react with calcium, are generally confined to middle lamellae, lining intercellular spaces, and at the cell junction (Willats et al., 2001). Arabinan and galactan are mainly located in the primary cell wall; in most studies, they were reported to be absent at cell junctions. The high concentration of HG at cell junction is thought to contribute to the resistance to stress that tends to separate cells by forming a strong gel via calcium bridges (Jarvis et al., 2003). In potatoes, phellem tissue contains no acidic pectin. This suggests that other mechanisms may be involved in cell adhesion in this tissue. Intermediate forms of soft acidic gels involving H-bonding and hydrophobic interactions have been described in vitro (Tibbits et al., 1998; Gilsenan et al., 2000) and may contribute to cell adhesion *in planta*. The roles of arabinan and galactan side chains in cell wall are more enigmatic. Transgenic potato plants expressing an apoplast-directed endo-$(1\rightarrow5)$-α-arabinanase show reduced amounts of arabinose and severe phenotype: The plants do not produce side shoots, flowers, or stolon and tuber (Skjøt et al., 2002). When the same enzyme is targeted to the Golgi apparatus, plants and tubers contain a reduced level of arabinose but have no detrimental phenotype. Transgenic potato plants expressing a rhamnogalacturonan lyase produce tubers with altered morphology, including radial swelling of the periderm cells and development of intercellular spaces in the cortex (Oomen et al., 2002). The amount of both galactose and arabinose is reduced and the remaining galactan is found in middle lamella at cell corners, in contrast to its normal location in the primary wall of wild type tubers. This suggests that the RGI backbone is involved in anchoring galactan and arabinan at particular regions in the wall and that RGI has an important role in the integrity and function of the wall. It has been suggested that pectin side chains may contribute to the mechanical properties of cell walls (McCartney et al., 2000; McCartney and Knox, 2002). Ulvskov et al. (2005) investigated the significance of RGI structure for wall rheology using potato tubers in which the RGI structure was remodeled in vivo using fungal enzymes. There were fewer tuber RGI molecules from the two transformed lines in linear galactans or branched arabinans. The transformed tuber tissues were more brittle when subjected to uniaxial compression. The authors proposed that pectic matrix may have a role in transmitting stresses to the load-bearing cellulose microfibrils and that small changes to the matrix environment affect the biophysical properties of the wall. It has also been suggested that side chains could act as wall plasticizers and water-binding agents during plant growth and development. ^{13}C-Cross-polarization magic angle spinning experiments were carried out on native RGI isolated from potato and on arabinanase- and galactanase-treated potto RGI (Larsen et al., 2011). Arabinan side chains were shown to hydrate more readily than galactan ones, which suggests that hydration properties may be modulated by the length and fine structure of the side chains (Larsen et al., 2011). Galactan, which accumulates in large amounts in potato tuber during the cold season, may also serve as a source of carbon and energy for plant regrowth during the next growing season.

Clearly, further work is required to understand the complexity of pectin in the context of cell wall architecture and in relation to cell wall function in plant life.

References

Abe, K., Yano, H., 2009. Comparison of the characteristics of cellulose microfibril aggregates of wood, rice straw and potato tuber. Cellulose 16 (6), 1017–1023.

Albersheim, P., Darvill, A.G., O'Neill, M.A., Schols, H.A., Voragen, A.G.J., 1996. An hypothesis: the same six polysaccharides are components of the primary cell wall of all higher plants. In: Visser, J., Voragen, A.G.J. (Eds.), Pectins and Pectinases. Elsevier, Amsterdam, pp. 47–53.

Bonnin, E., Le Goff, A., Korner, R., Vigouroux, J., Roepstorff, P., Thibault, J.F., 2002. Hydrolysis of pectins with different degrees and patterns of methylation by the endopolygalacturonase of *Fusarium moniliforme*. Biochimica et Biophysica Acta 1596, 83–94.

Bonnin, E., Clavurier, K., Daniel, S., Kauppinen, S., Mikkelsen, J.D.M., Thibault, J.F., 2008. Pectin acetylesterases from *Aspergillus* are able to deacetylate homogalacturonan as well as rhamnogalacturonan. Carbohydrate Polymers 74 (3), 411–418.

Bonnin, E., Garnier, C., Ralet, M.C., 2014. Pectin-modifying enzymes and pectin-derived materials: applications and impacts. Applied Microbiology and Biotechnology 98 (2), 519–532.

Buffetto, F., Cornuault, V., Gro Rydahl, M., Ropartz, D., Alvarado, C., Echasserieau, V., Le Gall, S., Bouchet, B., Tranquet, O., Verhertbruggen, Y., Willats, W.G.T., Knox, J.P., Ralet, M.-C., Guillon, F., 2015. The Deconstruction of Pectic Rhamnogalacturonan I Unmasks the Occurrence of a Novel Arabinogalactan Oligosaccharide Epitope, Plant and Cell Physiology, http://dx.doi.org/10.1093/pcp/pcv128 (in press).

Bush, M.S., McCann, M.C., 1999. Pectic epitopes are differentially distributed in the cell walls of potato (*Solanum tuberosum*) tubers. Physiologia Plantarum 107 (2), 201–213.

Bush, M.S., Marry, M., Huxham, I.M., Jarvis, M.C., McCann, M.C., 2001. Developmental regulation of pectic epitopes during potato tuberisation. Planta 213 (6), 869–880.

Byg, I., Diaz, J., Øgendal, L.H., Harholt, J., Jørgensen, B., Rolin, C., Svava, R., Ulvskov, P., 2012. Large-scale extraction of rhamnogalacturonan I from industrial potato waste. Food Chemistry 131 (4), 1207–1216.

Caffall, K.H., Mohnen, D., 2009. The structure, function, and biosynthesis of plant cell wall pectic polysaccharides. Carbohydrate Research 344 (14), 1879–1900.

Cerclier, C., Guyomard-Lack, A., Moreau, C., Cousin, F., Beury, N., Bonnin, E., Jean, B., Cathala, B., 2011. Coloured semi-reflective thin films for biomass-hydrolyzing enzyme detection. Advanced Materials 23 (33), 3791–3795.

Chen, D., Lawton, D., Thompson, M.R., Liu, Q., 2012. Biocomposites reinforced with cellulose nanocrystals derived from potato peel waste. Carbohydrate Polymers 90 (1), 709–716.

Cosgrove, D.J., 2014. Re-constructing our models of cellulose and primary cell wall assembly. Current Opinion in Plant Biology 22, 122–131.

Darvill, J.E., McNeil, M., Darvill, A.G., Albersheim, P., 1980. Structure of plant-cell walls. 11. Glucuronoarabinoxylan, a 2nd hemicellulose in the primary-cell walls of suspension-cultured sycamore cells. Plant Physiology 66 (6), 1135–1139.

Ducreux, L.J.M., Morris, W.L., Prosser, I.M., Morris, J.A., Beale, M.H., Wright, F., Shepherd, T., Bryan, G.J., Hedley, P.E., Taylor, M.A., 2008. Expression profiling of potato germplasm differentiated in quality traits leads to the identification of candidate flavour and texture genes. Journal of Experimental Botany 59 (15), 4219–4231.

Dufresne, A., Vignon, M.R., 1998. Improvement of starch film performances using cellulose microfibrils. Macromolecules 31 (8), 2693–2696.

Dufresne, A., Dupeyre, D., Vignon, M.R., 2000. Cellulose microfibrils from potato tuber cells: processing and characterization of starch-cellulose microfibril composites. Journal of Applied Polymer Science 76 (14), 2080–2092.

Earl, W.L., Vanderhart, D.L., 1981. Observations by high-resolution c-13 nuclear magnetic-resonance of cellulose-i related to morphology and crystal-structure. Macromolecules 14 (3), 570–574.

Ermel, F.F., Follet-Gueye, M.L., Cibert, C., Vian, B., Morvan, C., Catesson, A.M., Goldberg, R., 2000. Differential localization of arabinan and galactan side chains of rhamnogalacturonan 1 in cambial derivatives. Planta 210 (5), 732–740.

Fry, S.C., York, W.S., Albersheim, P., Darvill, A., Hayashi, T., Joseleau, J.P., Kato, Y., Lorences, E.P., Maclachlan, G.A., McNeil, M., Mort, A.J., Reid, J.S.G., Seitz, H.U., Selvendran, R.R., Voragen, A.G.J., White, A.R., 1993. An unambiguous nomenclature for xyloglucan-derived oligosaccharides. Physiologia Plantarum 89, 1–3.

Gille, S., Pauly, M., 2012. O-acetylation of plant cell wall polysaccharides. Frontiers in Plant Science 3.

Gilsenan, P.M., Richardson, R.K., Morris, E.R., 2000. Thermally reversible acid-induced gelation of low-methoxy pectin. Carbohydrate Polymers 41 (4), 339–349.

Habibi, Y., Lucia, L.A., Rojas, O.J., 2010. Cellulose nanocrystals: chemistry, self-assembly, and applications. Chemical Reviews 110, 3479–3500.

Harris, P.J., Singh, J., Kaur, L., 2009. Chapter 3: cell-wall polysaccharides of potatoes. In: Singh, J., Kaur, L. (Eds.), Advances in Potato Chemistry and Technology. Academic Press, San Diego, pp. 63–81.

Hart, D.A., Kindel, P.K., 1970. Isolation and partial characterization of apiogalacturonans from the cell wall of Lemna minor. The Biochemical Journal 116 (4), 569–579.

Hayashi, T., 1989. Xyloglucans in the primary-cell wall. Annual Review of Plant Physiology and Plant Molecular Biology 40, 139–168.

Hellin, P., Ralet, M.C., Bonnin, E., Thibault, J.F., 2005. Homogalacturonans from lime pectins exhibit homogeneous charge density and molar mass distributions. Carbohydrate Polymers 60 (3), 307–317.

Ishii, T., 1997. O-acetylated oligosaccharides from pectins of potato tuber cell walls. Plant Physiology 113 (4), 1265–1272.

Jardine, W.G., Doeswijk-Voragen, C.H.L., MacKinnon, I.M.R., van den Broek, L.A.M., Ha, M.A., Jarvis, M.C., Voragen, A.G.J., 2002. Methods for the preparation of cell walls from potatoes. Journal of the Science of Food and Agriculture 82 (8), 834–839.

Jarvis, M.C., Hall, M.A., Threlfall, D.R., Friend, J., 1981. The polysaccharide structure of potato cell walls: chemical fractionation. Planta 152 (2), 93–100.

Jarvis, M.C., Briggs, S.P.H., Knox, J.P., 2003. Intercellular adhesion and cell separation in plants. Plant Cell and Environment 26, 977–989.

Jia, Z., Qin, Q., Darvill, A.G., York, W.S., 2003. Structure of the xyloglucan produced by suspension-cultured tomato cells. Carbohydrate Research 338, 1197–1208.

Jia, Z.H., Cash, M., Darvill, A.G., York, W.S., 2005. NMR characterization of endogenously O-acetylated oligosaccharides isolated from tomato (*Lycopersicon esculentum*) xyloglucan. Carbohydrate Research 340 (11), 1818–1825.

Kalashnikova, I., Bizot, H., Cathala, B., Capron, I., 2011. New pickering emulsions stabilized by bacterial cellulose nanocrystals. Langmuir 27 (12), 7471–7479.

Kalashnikova, I., Bizot, H., Cathala, B., Capron, I., 2012. Modulation of cellulose nanocrystals amphiphilic properties to stabilize oil/water interface. Biomacromolecules 13 (1), 267–275.

Kalashnikova, I., Bizot, H., Bertoncini, P., Cathala, B., Capron, I., 2013. Cellulosic nanorods of various aspect ratios for oil in water pickering emulsions. Soft Matter 9 (3), 952–959.

Khodaei, N., Karboune, S., 2013. Extraction and structural characterisation of rhamnogalacturonan I-type pectic polysaccharides from potato cell wall. Food Chemistry 139 (1–4), 617–623.

Klemm, D., Kramer, F., Moritz, S., Lindström, T., Ankerfors, M., Gray, D., Dorris, A., 2011. Nanocelluloses: a new family of nature-based materials. Angewandte Chemie-International Edition 50 (24), 5438–5466.

Kokkonen, H., Verhoef, R., Kauppinen, K., Muhonen, V., Jorgensen, B., Damager, I., Schols, H.A., Morra, M., Ulvskov, P., Tuukkanen, J., 2012. Affecting osteoblastic responses with in vivo engineered potato pectin fragments. Journal of Biomedical Materials Research Part A 100 (1), 111–119.

Komalavilas, P., Mort, A.J., 1989. The acetylation at *O*-3 of galacturonic acid in the rhamnose-rich portion of pectins. Carbohydrate Research 189, 261–272.

Lærke, H.N., Meyer, A.S., Kaack, K.V., Larsen, T., 2007. Soluble fiber extracted from potato pulp is highly fermentable but has no effect on risk markers of diabetes and cardiovascular disease in Goto-Kakizaki rats. Nutrition Research 27 (3), 152–160.

Lagerwall, J.P.F., Schutz, C., Salajkova, M., Noh, J., Park, J.H., Scalia, G., Bergstrom, L., 2014. Cellulose nanocrystal-based materials: from liquid crystal self-assembly and glass formation to multifunctional thin films. NPG Asia Materials 6, e80.

Lampugnani, E.R., Moller, I.E., Cassin, A., Jones, D.F., Koh, P.L., Ratnayake, S., Beahan, C.T., Wilson, S.M., Bacic, A., Newbigin, E., 2013. In vitro grown pollen tubes of *Nicotiana alata* actively synthesise a fucosylated xyloglucan. PLoS One 8, e77140.

Larsen, F.H., Byg, I., Damager, I., Diaz, J., Engelsen, S.B., Ulvskov, P., 2011. Residue specific hydration of primary cell wall potato pectin identified by solid-state 13C single-pulse MAS and CP/MAS NMR spectroscopy. Biomacromolecules 12 (5), 1844–1850.

Le Goff, A., Renard, C.M.G.C., Bonnin, E., Thibault, J.F., 2001. Extraction, purification and chemical characterisation of xylogalacturonans from pea hulls. Carbohydrate Polymers 45 (4), 325–334.

Leboeuf, E., Guillon, F., Thoiron, S., Lahaye, M., 2005. Biochemical and immunohistochemical analysis of pectic polysaccharides in the cell walls of Arabidopsis mutant QUASIMODO 1 suspension-cultured cells: implications for cell adhesion. Journal of Experimental Botany 56 (422), 3171–3182.

Martin, C., Jean, B., 2014. Nanocellulose/polymer multilayered thin films: tunable architectures towards tailored physical properties. Nordic Pulp & Paper Research Journal 29 (1), 19–30.

McCann, M., Roberts, K., 1991. Architecture of the primary cell wall. In: Lloyd, C.W. (Ed.), The Cytoskeletal Basis of Plant Growth and Form. Academic Press, London, pp. 109–129.

McCartney, L., Ormerod, A.P., Gidley, M.J., Knox, J.P., 2000. Temporal and spatial regulation of pectic (1→4)-beta-D-galactan in cell walls of developing pea cotyledons: implications for mechanical properties. The Plant Journal 22 (2), 105–113.

McCartney, L., Knox, J.P., 2002. Regulation of pectic polysaccharide domains in relation to cell development and cell properties in the pea testa. Journal of Experimental Botany 53 (369), 707–713.

McCartney, L., Steele-King, C.G., Jordan, E., Knox, J.P., 2003. Cell wall pectic (1→4)-β-d-galactan marks the acceleration of cell elongation in the Arabidopsis seedling root meristem. The Plant Journal 33 (3), 447–454.

McDougall, G.J., Morrison, I.M., Stewart, D., Hillman, J.R., 1996. Plant cell walls as dietary fibre: range, structure, processing and function. Journal of the Science of Food and Agriculture 70, 133–150.

McNeil, M., Darvill, A.G., Fry, S.C., Albersheim, P., 1984. Structure and function of the primary cell walls of plants. Annual Review of Biochemistry 53, 625–663.

Meyer, A.S., Dam, B.R., Laerke, H.N., 2009. Enzymatic solubilization of a pectinaceous dietary fiber fraction from potato pulp: optimization of the fiber extraction process. Biochemical Engineering Journal 43 (1), 106–112.

Mohnen, D., 2008. Pectin structure and biosynthesis. Current Opinion in Plant Biology 11 (3), 266–277.

Mort, A.J., Moerschbacher, B.M., Pierce, M.L., Maness, N.O., 1991. Problems encountered during the extraction, purification, and chromatography of pectic fragments, and some solutions to them. Carbohydrate Research 215, 219–227.

Newman, R.H., Hill, S.J., Harris, P.J., 2013. Wide-angle X-ray scattering and solid-state nuclear magnetic resonance data combined to test models for cellulose microfibrils in mung bean cell walls. Plant Physiology 163 (4), 1558–1567.

Ng, A., Waldron, K.W., 1997. Effect of steaming on cell wall chemistry of potatoes (*Solanum tuberosum* Cv. Bintje) in relation to firmness. Journal of Agricultural and Food Chemistry 45 (9), 3411–3418.

O'Neill, M.A., Ishii, T., Albersheim, P., Darvill, A.G., 2004. Rhamnogalacturonan II: structure and function of a borate cross-linked cell wall pectic polysaccharide. Annual Review of Plant Biology 55, 109–139.

Øbro, J., Harholt, J., Scheller, H.V., Orfila, C., 2004. Rhamnogalacturonan I in Solanum tuberosum tubers contains complex arabinogalactan structures. Phytochemistry 65 (10), 1429–1438.

Olivier, C., Moreau, C., Bertoncini, P., Bizot, H., Chauvet, O., Cathala, B., 2012. Cellulose nanocrystal-assisted dispersion of luminescent single-walled carbon nanotubes for layer-by-layer assembled hybrid thin films. Langmuir 28 (34), 12463–12471.

Oomen, R.F.J., Doeswijk-Voragen, C.H.L., Bush, M.S., Vincken, J.-P., Borkhardt, B., Van Den Broek, L.A.M., Corsar, J., Ulvskov, P., Voragen, A.G.J., McCann, M.C., Visser, R.G.F., 2002. In muro fragmentation of the rhamnogalacturonan I backbone in potato (*Solanum tuberosum* L.) results in a reduction and altered location of the galactan and arabinan side chains and abnormal periderm development. The Plant Journal 30 (4), 403–413.

Oomen, R.F.J., Vincken, J.-P., Bush, M., Skjøt, M., Doeswijk-Voragen, C.L., Ulvskov, P., Voragen, A.G.J., McCann, M.C., Visser, R.G.F., 2003. Towards unravelling the biological significance of the individual components of pectic hairy regions in plants. In: Voragen, F., Henk, S., Richard, V. (Eds.), Advances in Pectin and Pectinase Research. Kluwer Academic Publishers, Dordrecht, pp. 15–34.

Oomen, R.F.J., Tzitzikas, E.N., Bakx, E.J., Straatman-Engelen, I., Bush, M.S., McCann, M.C., Schols, H.A., Visser, R.G., Vincken, J.P., 2004. Modulation of the cellulose content of tuber cell walls by antisense expression of different potato (*Solanum tuberosum* L.) CesA clones. Phytochemistry 65 (5), 535–546.

Orfila, C., Dal Degan, F., Jørgensen, B., Scheller, H.V., Ray, P.M., Ulvskov, P., 2012. Expression of mung bean pectin acetyl esterase in potato tubers: effect on acetylation of cell wall polymers and tuber mechanical properties. Planta 236 (1), 185–196.

Ovodov, Y.S., Ovodova, R.G., Bondarenko, O.D., Krasikova, I.N., 1971. The pectic substances of *zosteraceae* : part IV. Pectinase digestion of zosterine. Carbohydrate Research 18 (2), 311–318.

Pauly, M., Gille, S., Liu, L., Mansoori, N., de Souza, A., Schultink, A., Xiong, G., 2013. Hemicellulose biosynthesis. Planta 238 (4), 627–642.

Quéméner, B., Vigouroux, J., Rathahao, E., Tabet, J.-C., Dimitrijevic, A., Lahaye, M., 2015. Negative electrospray ionization mass spectrometry: a method for sequencing and determining linkage position in oligosaccharides from branched hemicelluloses. Journal of Mass Spectrometry 50 (1), 247–264.

Ralet, M.C., Cabrera, J.C., Bonnin, E., Quéméner, B., Hellin, P., Thibault, J.-F., 2005. Mapping sugar beet pectin acetylation pattern. Phytochemistry 66, 1832–1843.

Ralet, M.C., Crepeau, M.J., Bonnin, E., 2008. Evidence for a blockwise distribution of acetyl groups onto homogalacturonans from a commercial sugar beet (*Beta vulgaris*) pectin. Phytochemistry 69 (9), 1903–1909.

Ralet, M.C., Thibault, J.F., 2009. Hydrodynamic properties of isolated pectin domains: a way to figure out pectin macromolecular structure? In: Schols, H.A., Visser, R.G.F., Voragen, A.G.J. (Eds.), Pectins and Pectinases. Academic Publisher, Wageningen, pp. 35–48.

Ralet, M.C., Tranquet, O., Poulain, D., Moise, A., Guillon, F., 2010. Monoclonal antibodies to rhamnogalacturonan I backbone. Planta 231, 1373–1383.

Ralet, M.C., Williams, M.A.K., Tanhatan-Nasseri, A., Ropartz, D., Quemener, B., Bonnin, E., 2012. Innovative enzymatic approach to resolve homogalacturonans based on their methylesterification pattern. Biomacromolecules 13 (5), 1615–1624.

Ramaswamy, U.R., Kabel, M.A., Schols, H.A., Gruppen, H., 2013. Structural features and water holding capacities of pressed potato fibre polysaccharides. Carbohydrate Polymers 93 (2), 589–596.

Ravn, H.C., Sørensen, B.O., Meyer, A.S., 2015. Time of harvest affects the yield of soluble polysaccharides extracted enzymatically from potato pulp. Food and Bioproducts Processing 93, 77–83.

Reeve, R.M., Hautala, E., Weaver, M.L., 1969. Anatomy and compositional variation within potatoes. American Potato Journal 46 (10), 361–373.

Renard, C.M., Thibault, J.F., Mutter, M., Schols, H.A., Voragen, A.G., 1995. Some preliminary results on the action of rhamnogalacturonase on rhamnogalacturonan oligosaccharides from beet pulp. International Journal of Biological Macromolecules 17 (6), 333–336.

Revol, J.F., Bradford, H., Giasson, J., Marchessault, R.H., Gray, D.G., 1992. Helicoidal self-ordering of cellulose microfibrils in aqueous suspension. International Journal of Biological Macromolecules 14 (3), 170–172.

Ridley, B.L., O'Neill, M.A., Mohnen, D.A., 2001. Pectins: structure, biosynthesis, and oligogalacturonide-related signaling. Phytochemistry 57 (6), 929–967.

Ring, S.G., Selvendran, R.R., 1978. Purification and methylation analysis of cell wall material from *Solanum tuberosum*. Phytochemistry 17 (4), 745–752.

Ring, S.G., Selvendran, R.R., 1981. An arabinogalactoxyloglucan from the cell wall of *Solanum tuberosum*. Phytochemistry 20 (11), 2511–2519.

Ross, H.A., Wright, K.M., McDougall, G.J., Roberts, A.G., Chapman, S.N., Morris, W.L., Hancock, R.D., Stewart, D., Tucker, G.A., James, E.K., Taylor, M.A., 2011. Potato tuber pectin structure is influenced by pectin methyl esterase activity and impacts on cooked potato texture. Journal of Experimental Botany 62 (1), 371–381.

Round, A.N., Rigby, N.M., MacDougall, A.J., Morris, V.J., 2010. A new view of pectin structure revealed by acid hydrolysis and atomic force microscopy. Carbohydrate Research 345 (4), 487–497.

Ryden, P., Selvendran, R.R., 1990. Structural features of cell-wall polysaccharides of potato (*Solanum tuberosum*). Carbohydrate Research 195 (2), 257–272.

Sabba, R.P., Lulai, E.C., 2002. Histological analysis of the maturation of native and wound periderm in potato (*Solanum tuberosum* L.) Tuber. Annals of Botany 90 (1), 1–10.

Sabba, R.P., Lulai, E.C., 2004. Immunocytological comparison native and wound periderm maturation in potato tuber. American Journal of Potato Research 81 (2), 119–124.

Sabba, R.P., Lulai, E.C., 2005. Immunocytological analysis of potato tuber periderm and changes in pectin and extensin epitopes associated with periderm maturation. Journal of the American Society for Horticultural Science 130 (6), 936–942.

Scheller, H.V., Ulvskov, P., 2010. Hemicelluloses. Annual Review of Plant Biology 61, 263–289.

Schols, H.A., Geraeds, C.C.J.M., Searle-van Leeuwen, M.F., Kormelink, F.J.M., Voragen, A.G.J., 1990. Rhamnogalacturonase: a novel enzyme that degrades the hairy regions of pectins. Carbohydrate Research 206 (1), 105–115.

Schols, H.A., Voragen, A.G.J., 1994. Occurrence of pectic hairy regions in various plant cell wall materials and their degradability by rhamnogalacturonase. Carbohydrate Research 256 (1), 83–95.

Schols, H.A., 1995. Structural Characterization of Pectic Hairy Regions Isolated from Apple Cell Walls. Wageningen Universiteit.

Schols, H.A., Bakx, E.J., Schipper, D., Voragen, A.G.J., 1995. A xylogalacturonan subunit present in the modified hairy regions of apple pectin. Carbohydrate Research 279, 265–279.

Sims, I.M., Munro, S.L.A., Currie, G., Craik, D., Bacic, A., 1996. Structural characterisation of xyloglucan secreted by suspension-cultured cells of *Nicotiana plumbaginifolia*. Carbohydrate Research 293 (2), 147–172.

Skjøt, M., Pauly, M., Bush, M.S., Borkhardt, B., McCann, M.C., Ulvskov, P., 2002. Direct interference with rhamnogalacturonan I biosynthesis in golgi vesicles. Plant Physiology 129 (1), 95–102.

Sobry, S., Havelange, A., Van Cutsem, P., 2005. Immunocytochemistry of pectins in shoot apical meristems: consequences for intercellular adhesion. Protoplasma 225 (1–2), 15–22.

Sugiyama, J., Persson, J., Chanzy, H., 1991. Combined infrared and electron diffraction study of the polymorphism of native celluloses. Macromolecules 24 (9), 2461–2466.

Thibault, J.F., Renard, C.M.G.C., Axelos, M.A.V., Roger, P., Crépeau, M.-J., 1993. Studies of the length of homogalacturonic regions in pectins by acid hydrolysis. Carbohydrate Research 238, 271–286.

Thomassen, L.V., Meyer, A.S., 2010. Statistically designed optimisation of enzyme catalysed starch removal from potato pulp. Enzyme and Microbial Technology 46 (3–4), 297–303.

Thomassen, L.V., Vigsns, L.K., Licht, T.R., Mikkelsen, J.D., Meyer, A.S., 2011. Maximal release of highly bifidogenic soluble dietary fibers from industrial potato pulp by minimal enzymatic treatment. Applied Microbiology and Biotechnology 90 (3), 873–884.

Tibbits, C.W., MacDougall, A.J., Ring, S.G., 1998. Calcium binding and swelling behaviour of a high methoxyl pectin gel. Carbohydrate Research 310 (1–2), 101–107.

Ulvskov, P., Wium, H., Bruce, D., Jørgensen, B., Qvist, K.B., Skjøt, M., Hepworth, D., Borkhardt, B., Sørensen, S.O., 2005. Biophysical consequences of remodeling the neutral side chains of rhamnogalacturonan I in tubers of transgenic potatoes. Planta 220, 609–620.

VanMarle, J.T., Clerkx, A.C.M., Boekestein, A., 1992. Cryoscanning electron-microscopy investigation of the texture of cooked potatoes. Food Structure 11 (3), 209–216.

VanMarle, J.T., DeVries, R., Wilkinson, E.C., Yuksel, D., 1997a. Sensory evaluation of the texture of steam-cooked table potatoes. Potato Research 40 (1), 79–90.

VanMarle, J.T., Recourt, K., van Dijk, C., Schols, H.A., Voragen, A.G.J., 1997b. Structural features of cell walls from potato (*Solanum tuberosum* L.) cultivars Irene and Nicola. Journal of Agricultural and Food Chemistry 45 (5), 1686–1693.

Vincken, J.P., Wijsman, A.J., Beldman, G., Niessen, W.M., Voragen, A.G., 1996. Potato xyloglucan is built from XXGG-type subunits. Carbohydrate Research 288, 219–232.

Vincken, J.P., York, W.S., Beldman, G., Voragen, A.G.J., 1997. Two general branching patterns of xyloglucan, XXXG and XXGG. Plant Physiology 114 (1), 9–13.

Vincken, J.P., Schols, H.A., Oomen, R.J., McCann, M.C., Ulvskov, P., Voragen, A.G., Visser, R.G., 2003. If homogalacturonan were a side chain of rhamnogalacturonan I. Implications for cell wall architecture. Plant Physiology 132 (4), 1781–1789.

Voragen, A.G.J., Pilnik, W., Thibault, J.F., Axelos, M.A.V., Renard, C.M.G.C., 1995. Pectin. In: Stephen, A.M. (Ed.), Food Polysaccharides and Their Applications. Marcel Dekker, New York, pp. 287–339.

Willats, W.G., Steele-King, C.G., Marcus, S.E., Knox, J.P., 1999. Side chains of pectic polysaccharides are regulated in relation to cell proliferation and cell differentiation. The Plant Journal 20 (6), 619–628.

Willats, W.G.T., McCartney, L., Mackie, W., Knox, J.P., 2001. Pectin: cell biology and prospects for functional analysis. Plant Molecular Biology 47 (1), 9–27.

Xu, X., van Lammeren, A.A., Vermeer, E., Vreugdenhil, D., 1998a. The role of gibberellin, abscisic acid, and sucrose in the regulation of potato tuber formation in vitro. Plant Physiology 117 (2), 575–584.

Xu, X., Vreugdenhil, D., van Lammeren, A.A.M., 1998b. Cell division and cell enlargement during potato tuber formation. Journal of Experimental Botany 49 (320), 573–582.

York, W.S., Kolli, V.S.K., Orlando, R., Albersheim, P., Darvill, A.G., 1996. The structures of arabinoxyloglucans produced by solanaceous plants. Carbohydrate Research 285, 99–128.

Zandleven, J., Beldman, G., Bosveld, M., Schols, H.A., Voragen, A.G.J., 2006. Enzymatic degradation studies of xylogalacturonans from apple and potato, using xylogalacturonan hydrolase. Carbohydrate Polymers 65 (4), 495–503.

CHAPTER 3

Structure of Potato Starch

Eric Bertoft[1], Andreas Blennow[2]

[1]*Department of Food Science and Nutrition, University of Minnesota, St Paul, MN, USA;* [2]*Department of Plant and Environmental Sciences, University of Copenhagen, Frederiksberg C, Denmark*

1. Introduction

Starch is the major component of potato tubers, amounting to approximately 15–20% of its weight. As an effect, starch is considered a major factor for the functionality of the potato in food applications. For industrial applications, processed starch from potato is considered pure compared with most other starch types. Compared with other commodity starches, potato starch also has some unique properties that are directly attributed to its granular and molecular structures, including very large and smooth granules, a high content of covalently linked phosphate, long amylopectin chains, and high-molecular weight amylose. These characteristics combined make potato starch a tremendous source of functional biopolymer for food and materials science. In particular, potato starch has many applications in the wet end for the manufacture of high-quality paper (Blennow et al., 2003) and for the generation of viscous hydrocolloid systems (Wiesenborn et al., 1994). Moreover, compared with most cereal starches, the well-ordered and dense structure of the native potato starch granule renders it resistant to enzymatic degradation by hydrolytic enzymes such as amyloglucosidases and α-amylases (Sun et al., 2006).

In the potato tuber, starch is found as distinct granules approximately 10–100 μm in diameter (Hoover, 2001). The granules are built up by two polysaccharides consisting exclusively of glucose as the monomer component. The glucopyranosyl residues are connected through α-D-(1,4)-linkages forming chains through α-D-(1,6)-branches at the reducing end side linked to similar other chains. Amylopectin is the major component in starches in general, and in potato it normally constitutes 70–80% by weight (Hoover, 2001; Yusuph et al., 2003) regardless of the size of the granules (Noda et al., 2005). Approximately 4–6% of linkages are of the α-D-(1,6)-type, making it extensively branched. The weight-average molecular size of amylopectin is on the order of 10^7 Da (Aberle et al., 1994; Ratnayake and Jackson, 2007); as a result, the macromolecule consists of a huge number of relatively short chains with an average degree of polymerization (DP) of 21–28 residues (McPherson and Jane, 1999; Zhu and Bertoft, 1996). The minor component of starch is amylose. It is considerably smaller than amylopectin and is essentially a linear polymer consisting of chains of DP on the order of

Advances in Potato Chemistry and Technology. http://dx.doi.org/10.1016/B978-0-12-800002-1.00003-0

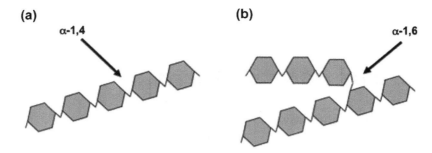

Figure 1
Schematic representation of the structures of (a) linear amylose and (b) branched amylose and amylopectin.

2000–5000 residues (Hoover, 2001). However, a few branches are found in the structure; principally two types of this component exist: linear and branched amylose (Figure 1).

Besides the polysaccharide components, potato starch consists of low amounts of material of a noncarbohydrate nature. Less than 0.5% of the granules are proteins (Yusuph et al., 2003), apparently mostly involved in starch synthesis. Lipids are virtually absent in potato starch, in common with some other tuber and root starches (Hizukuri et al., 1970; Hoover, 2001) but in contrast to many other starches, especially from cereals. Potato starch also contains phosphorus in the form of phosphate covalently linked to the amylopectin component (Hizukuri et al., 1970). It is considered an important factor contributing to potato starch properties. Most other starches also contain covalently bound phosphate, although in only minute amounts. Phosphorus is also a component in cereal starches. However, there it is found in the form of lysophospholipids, which form noncovalent complexes with the amylose component (Morrison, 1995). Trace amounts of different cations, mostly potassium (Blennow et al., 2005a; Noda et al., 2005), were also described as minor components in potato starch granules, apparently coordinated to the phosphate groups.

This chapter considers the molecular structures of the amylose and amylopectin components in potato and how they are organized to form characteristic structures inside the starch granules. The phosphorylation of starch and the synthesis of its components in normal and genetically modified potatoes are also discussed.

2. Polysaccharide Components of Potato Starch

Amylose constitutes the minor, smaller, and mostly linear components of starch. The molecular size distribution of amylose, expressed as weight-average molecular weight (M_w), is $0.2–3.9 \times 10^6$, whereas the number-average degree of polymerization (DP_n) ranges between 840 and 21,800 (Hizukuri and Takagi, 1984). The polydispersity index, DP_w/DP_n (or M_w/M_n), is 1.29–6.9. The slightly branched nature of amylose was known in the 1950s, but it was only comparatively lately that methods to calculate the actual molar fractions of branched and linear amylose were obtained (Takeda et al., 1987). The exo-acting enzyme β-amylase repetitively hydrolyzes maltose residues from the nonreducing end of the glucosyl chains. If the molecule is branched, the enzyme stops at the vicinity of the branches, which it cannot bypass, resulting in a β-limit dextrin containing the reducing end. Thus, the molar amount of branched amylose was detected as tritium-labeled β-limit dextrins (Takeda et al., 1992). In potato, the β-amylolysis limit of the amylose fraction is 68–88% (Hizukuri et al., 1981; Takeda et al., 1984) and the molar ratio of branched–linear amylose is 0.38 (Takeda et al., 1992). The average number of chains per molecule of the whole amylose fraction is only 7.3–12.2 (Hizukuri et al., 1981; Takeda et al., 1984). Because this figure includes both the linear and branched components, the actual chain number of the branched amylose is higher. Although the size of the branched component is unknown (it was detected as the β-limit dextrin), it was shown that the degree of branching increases with size (Takeda et al., 1984).

As mentioned, the molecular weight of amylopectin is much larger than that of amylose. For potato amylopectin, the M_w was reported to be 60.9×10^6, and the radius of gyration (R_g), 224.3 nm (Aberle et al., 1994). Number average values are considerably lower. Gel-permeation chromatography of fluorescent-labeled amylopectin gave a DP_n value of 11,200 (corresponding to $M_n \sim 1.8 \times 10^6$) (Takeda et al., 2003). Moreover, the size-separation technique revealed that the amylopectin component is composed of three molecular species with DP_n values around 16,100, 5500, and 2100, respectively. The molar proportion of the large component predominates (61%). The multiple size components of amylopectin are not unique for potato; similar components are found in maize, rice, and sweet potato (Takeda et al., 2003).

The chains of amylopectin are arranged to facilitate the self-organization of discrete functional crystalline and amorphous sections. Hence, dissecting pools of chain populations and clustering of branch points provide important information on starch architecture at the molecular level. Generally, the unit chains of starch are divided into two groups of long and short chains. Together with the structure of components in acid-treated starch granules, this gave rise to the cluster model (French, 1972; Nikuni, 1978; Robin et al., 1974). In the model, the short chains are found in clusters and the long chains interconnect the clusters (Hizukuri, 1986). The chains in amylopectin were originally divided into A-chains (chains not substituted by other chains), B-chains (substituted with other chains), and the C-chain, which carries the sole reducing-end residue in the macromolecule (and is a special kind of B-chain) (Peat et al., 1952). In the nomenclature of Hizukuri (1986), the B-chains are subdivided. B1-chains are short B-chains that, together with the A-chains, form the major group of short chains. The long chains are subdivided into B2- (spanning two clusters) and B3-chains (spanning three clusters), etc. Details of the cluster model and the exact differences that exist between amylopectins from different sources remain obscure, however.

Numerous investigations describe the unit chain distribution of potato amylopectin. It is obtained by size-exclusion chromatography (Hanashiro et al., 2002), anion-exchange chromatography (McPherson and Jane, 1999), or capillary electrophoresis (Morell et al., 1998) after debranching of the molecule with the debranching enzymes isoamylase and/or pullulanase. In potato, the distribution of the short chains ranges from DP 6 to 35 or 36; the average chain length is 14–17 residues, and in the obtained chromatograms a peak is obtained at DP 13–14. The shortest chains at DP 6–8 typically possess a profile with the highest amounts of DP 6 and lowest of DP 8 (Bertoft, 2004; Hanashiro et al., 2002; McPherson and Jane, 1999; Noda et al., 2005). The DP of the long chains (DP≥36) is 48–57 and the molar ratio of short to long chains is 5.3–6.5 (Bertoft, 2004; Hanashiro et al., 2002; Noda et al., 2005). The long chains consist of at least two subgroups with a DP_w of approximately 58 (B2) and 75 (B3-chains) (Hizukuri, 1986). Their relative molar amounts are 9–14% and 2–3%, respectively (Bertoft, 2004; Hizukuri, 1986). In addition, trace amounts of very long chains with DP>100 were reported (Noda et al., 2005). By debranching fluorescent-labeled amylopectin, it was possible to analyze the size distribution of the C-chain in preparations from starches of different botanical origins (Hanashiro et al., 2002). The chains cover the DP range 10–130; the presence of a peak at DP 42 and a shoulder at DP 25 for potato starch suggests that the C-chains mostly correspond to the lengths of B2- and B1-chains, respectively. However, one-third of the molecules in the labeled C-chain fraction were not debranched, and therefore the average length of the C-chains is probably even shorter (Hanashiro et al., 2002). Thus, the C-chain, from which the synthesis of the macromolecule supposedly begins, is surprisingly short.

In potato amylopectin, the molar ratio of A:B-chains is 1.1–1.5 (Bender et al., 1982; Bertoft, 2004; Zhu and Bertoft, 1996). The size distribution of the B-chains in the limit dextrin of amylopectin, in which the external chains extending from the outermost branches are largely removed, shows the internal unit chain profile of the amylopectin. It was found that the long internal B-chains in a range of different starches are subdivided into the same groups as the long chains of the whole amylopectin (Bertoft et al., 2008). They are also found in practically equal amounts as in the whole amylopectin, which suggests that the long chains of amylopectin are generally composed of only B-chains. However, in an amylose-free potato sample, a small amount (2.6 mol%) of long A-chains were also present (Bertoft, 2004). The short internal B-chains consist also of subgroups. One has internal lengths from 2 to 6 and is quantitatively a minor group. The other possesses DP 7–24 and is the major group. Both groups are found in other starches as well and are therefore not unique to potato. The molar ratio of the short B-chain groups is, however, different in different starches (Bertoft et al., 2008). Because the chains are part of the clusters, it suggests differences in the structure of the clusters in different amylopectins.

The size of the clusters was estimated indirectly from the ratio of short to long chains, which for potato suggests that a single cluster is composed of about six chains (Hanashiro et al., 2002). Depending on the size of the amylopectin, a single amylopectin macromolecule was subsequently estimated to consist of 15–117 clusters (Takeda et al., 2003). Clusters were also isolated as limit dextrins from the macromolecule using endo-acting enzymes. With cyclodextrin transferase from *Klebsiella pneumoniae*, Bender et al. (1982) found clusters in potato of three distinct sizes with DP 40–140, whereas Finch and Sebesta (1992) found only clusters of the larger size (corresponding to M_n 23,000) when using a tetraose-forming amylase from *P. stutzeri*. With α-amylase from *B. amyloliquefaciens* (also referred to as the liquefying α-amylase from *Bacillus subtilis*), the clusters were found to have DP 33–70, corresponding to 5–10 chains (Zhu and Bertoft, 1996).

When discussing the molecular fine structure of amylopectin in more detail, it is useful to refer to three different structural levels. The lowest level, building blocks, is composed of small, tightly branched units and concerns the internal organization of the chains in the second level of clusters. A domain is the third level and is composed of smaller or larger groups of clusters and concerns how they are interconnected and whether clusters of different structures are found in different domains. In a series of experiments (Bertoft, 2007a,b), all three structural levels were isolated from amylose-free potato starch using the α-amylase from *B. amyloliquefaciens* and their structures were analyzed. During initial hydrolysis of amylopectin, the macromolecule depolymerized rapidly and domains containing groups of clusters were isolated when the reaction was interrupted. The domains were then subjected to further hydrolysis with the enzyme until the reaction rate became low (Bertoft, 2007a). At this stage no internal chains long enough to fill all nine subsites of the enzyme active site remained, and the clusters had been released from the domains. In the form of limit dextrins, the size of the clusters for a normal potato starch was DP 127 (Jensen et al., 2013b), whereas in another sample it was between DP 33 and 70 (Zhu and Bertoft, 1996), which suggested that differences exist among potato varieties. For a transgenic low-phosphate starch with suppressed α-glucan, water dikinase (GWD) activity, the average size of the clusters was DP 86 (Jensen et al., 2013b), whereas the clusters of amylose-free potato possessed two major sizes with DP around 54 and 75 (Bertoft, 2007a). Because the external chains roughly contribute 50% of the structure, the actual approximate size of the intact clusters in the amylopectin is almost double. Only small differences in the cluster composition were found in different domains, which suggests that the structure of potato amylopectin is homogeneous (Bertoft, 2007a). The number of long chains found in small domains containing only two clusters suggested that a single long chain is involved in their interconnection, in accordance with previous hypotheses (Hizukuri, 1986). However, as the number of clusters increased, a surplus of long chains tended to be present. This suggested that at higher structural levels alternative models are valid (Bertoft, 2007a). One alternative that allows a variable arrangement of the chains is a backbone model, in which the long chains form a backbone along which building blocks are bound and form clusters (Bertoft, 2013) (Figure 2). In potato a few, short branches are probably connected to the backbone, whereas in other plants, such as cereals, the backbone is more extensively branched (Bertoft et al., 2012).

The composition of branched building blocks in the isolated clusters of the amylose-free potato starch was analyzed by treating the clusters with excess α-amylase (Bertoft, 2007b). Under such conditions, the enzyme is forced to continue the attack at internal chains shorter than about nine residues inside the clusters. As a result, the building blocks, which practically resist further attack (i.e., they are near-limit dextrins), are released. The blocks in potato amylopectin possess short average internal chain lengths (ICL) of approximately 2.5 residues between branches (Bertoft, 2007b). Each cluster consists of five to seven building blocks, depending on its DP. The smallest blocks, which are most (about 60 mol%), constitute a group at DP 5–9 and are singly branched. Blocks at DP 10–14 are doubly branched dextrins, whereas larger blocks are multiply branched and are found only in small number in the clusters (about 20 mol%). In potato, the clusters possess slightly different densities of building blocks, which makes it possible to divide them into two groups with slightly different structural characteristics. The building blocks are interconnected through interblock segments with ICL 7–8. On this basis, it was suggested that a cluster is generally defined as a group of chains interconnected by internal chains shorter than nine residues (Bertoft, 2007b). A hypothetical model of the building block structure of potato amylopectin clusters is shown in Figure 3.

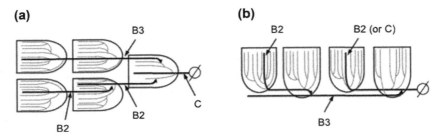

Figure 2

Mode of interconnection of clusters in amylopectin through (a) the traditional cluster structure and (b) the two-directional building block backbone structure. Boxes symbolize clusters with short chains (gray lines). Black lines are B2- and B3-chains involved in cluster interconnection. The C-chain, which otherwise is similar to B-chains, carries the reducing end residue (Ø).

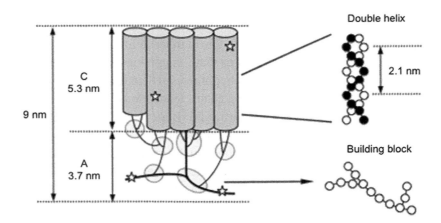

Figure 3

Building block structure of potato amylopectin clusters. Branched building blocks (encircled) are found mainly inside amorphous lamellae (a) of semicrystalline rings in starch granules. Double helices (symbolized as cylinders) extend from the building blocks into the crystalline lamellae (c). Enlargements of a double helix segment, in which the single strands are parallel and left-handed, and a building block are shown to the right. Phosphate groups (symbolized as stars) are characteristic of potato amylopectin.

3. Starch Granules in Potato

Despite the large variety of sizes and shapes of starch granules from different plants, their internal organization is remarkably similar. The starch macromolecules are organized in amorphous and semicrystalline granular rings or shells, commonly called growth rings (Figure 4) (Jenkins et al., 1993). The rings are 100–400 nm thick.

It is generally believed that the physical state of the amylose component is amorphous and therefore is found in the amorphous parts of the granules. However, it is not separated from the amylopectin component. By chemically cross-linking the polymers, it was shown that amylose became

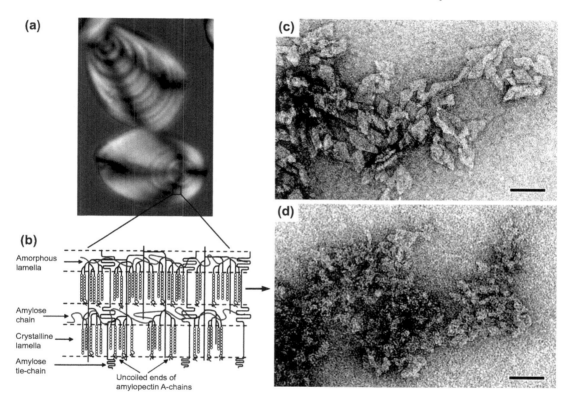

Figure 4
Structures in potato starch granules: (a) Potato starch granules observed under polarized light showing the characteristic Maltese crosses and alternating amorphous and semicrystalline rings. *(Courtesy of J. Wikman, Åbo Akademi, Turku.)* (b) Enlargement showing the structure of a semicrystalline ring with amylose molecules embedded both inside the amorphous lamellae and between the double helices of the amylopectin in the crystalline lamellae (Kozlov et al., 2007). (c) Transmission electron microscopy image of the crystalline residue obtained after hydrolyzing amylopectin-rich native A-type waxy maize starch granules with 2.2 N HCl for 18 days at 36 °C, and (d) B-type potato amylopectin starch granules. The preparations were negatively stained with uranyl acetate. Scale bars = 50 nm. *(Images courtesy of J.-L. Putaux, CERMAV-CNRS, Grenoble.)*

cross-linked to amylopectin, and therefore the two components are found side-by-side inside the granules (Jane et al., 1992). The exact organization of the amylose may differ between granules from different plants and depends on the content of amylose in the granules (Kozlov et al., 2007).

The principal component of the semicrystalline rings is amylopectin, which, through its comparatively short chains, contributes to characteristic stacks of thin, alternating crystalline and amorphous lamellae (Figure 4). The repeat distance of the lamellae is 8.8–9.2 nm, of which the crystalline part is 5.3–5.5 nm (Jenkins et al., 1993; Kozlov et al., 2007). The amorphous lamellae contain most of the branched, inner structure of the clusters of short chains, whereas the external chains of the clusters form left-handed double helices (Figure 3). The pitch distance (one turn of a single strand) is 2.1 nm and consists of six glucosyl residues. In the crystalline lamellae, the double helices are packed in parallel fashion (Imberty et al., 1991). In potato granules, a B-type crystalline allomorph is found (Srichuwong et al., 2005). In this, the double helices are ordered into a hexagonal structure. The internal part of the structure contains a cylindrical room filled with water coordinated to the hydroxyl groups of the carbohydrate residues (Imberty et al., 1991). The B-allomorph is not unique for potato but is typically formed in starch granules containing amylopectin with comparatively long unit chains (Hizukuri, 1985). If starch granules are observed in polarized light under the microscope, a typical Maltese cross is seen (Figure 4). Starch granules have been shown to be positively birefringent (Blennow et al., 2003), which shows that the amylopectin chains are radially oriented inside the granules. Using a microfocus X-ray diffraction beam, it was shown that the components in the core of the potato granules are only slightly oriented, but highly oriented toward the periphery, which seems to contribute to a compact surface structure on the potato granules. In contrast, the crystallites in cereal grain starch granules, like those found in wheat, possess A-type crystals, and are only weakly if at all oriented (Buléon et al., 1997). Observations by atomic force microscopy showed a surface full of round structures with a diameter of 10–50 nm, from which larger protrusions (50–300 nm) extended (Baldwin et al., 1998). The structures, known as blocklets (Gallant et al., 1997), are also found inside the granules (at least in maize granules) (Ridout et al., 2002) and their dimensions suggest that they may be identical to the amylopectin molecules. Images obtained by cryo-electron diffraction and optical tomography from potato granules were interpreted as complex superhelices of the amylopectin component (Oostergetel and Bruggen, 1993). The superhelices form apparently a delicate network around which the granules are built. Later, a superhelical structure of potato amylopectin was suggested based on small-angle X-ray microfocus scattering observations (Waigh et al., 1999). The superhelix was pictured as containing pie-shaped units, possibly clusters. The pitch of the helix is 9 nm, i.e., corresponding to the lamellar repeat distance, and the outer and inner diameters are 19 and 8 nm, respectively (Waigh et al., 1999). Depending on the molecular structure, there are principally two possibilities to form a superhelix from amylopectin. With the traditional cluster model (Figure 2) a superhelix would be a cooperative structure composed of several amylopectin molecules lined parallel to each other. With the backbone model, a superhelix can be formed from a single amylopectin macromolecule (Bertoft, 2004). The connection between the superhelix model and the blocklets is uncertain to date. However, it is possible that a blocklet represents a visual form of the superhelix.

The measurable crystallinity of potato starch granules depends on their water content. In the dry granule, the amylopectin component exists in a glassy nematic state with disorganized double helices. Upon hydration, the amorphous backbone becomes highly plasticized and the double helices are reorganized into a smectic structure equivalent to the 9-nm lamellar repeat distance (Waigh et al., 2000). Differential scanning calorimetry shows that melting (or gelatinization) of the crystals in potato granules appears at a temperature interval of about 59–70 °C, with a peak temperature (T_p) at approximately 63 °C (Yusuph et al., 2003). (The exact interval depends on the sample and experimental conditions.) If the granules are treated at a temperature slightly below the onset temperature of gelatinization (T_o), the double helices reorganize into more perfect structures, resulting in an increase in melting temperature and a narrower interval (Jacobs et al., 1998). This phenomenon, known as annealing, results from reorganization of the double helices through movement of the amorphous internal chains to which they are connected (Tester and Debon, 2000). In particular, longer interblock segments in the clusters promote optimal packing of the double helices (Vamadevan et al., 2013).

By treating the granules in diluted hydrochloric acid or sulfuric acid, the amorphous parts of the granules are hydrolyzed (a process called lintnerization) (Robin et al., 1974). The resulting acid-resistant material (consequently called lintners) consists of remnants of the crystalline lamellae. Analyses of the molecular composition of lintners show that they consist of double helices of the former clusters (Jacobs et al., 1998). The length of the chains in potato lintners is 14–15, which closely corresponds to the thickness of the crystalline lamellae (Genkina et al., 2007; Srichuwong et al., 2005). Only a few branches were found in the lintners from potato starch and other B-crystalline granules compared with A-crystalline granules, which suggests that only little branches are scattered into the crystalline lamellae in potato starch (Jane et al., 1997). When observed using transmission electron microscopy, crystalline remnants from A-crystalline granules possess nanocrystals with amazingly regular parallelepiped blocks with acute angles of 60°–65° (Putaux et al., 2003), in close correspondence with theoretical molecular models (Angellier-Coussy et al., 2009). Similar remnants in the lintners of potato granules are comparatively irregular and fragmented (Figure 4) (Wikman et al., 2014). This is surprising because the overall structural orientation of the components in potato granules is high (Buléon et al., 1997), as noted earlier.

4. Phosphorylated Potato Starch

Most native starch types are slightly phosphorylated with phosphate groups monoesterified to the glucose residues (Blennow et al., 2002). The presence of phosphate esters in starch has been known for more than a century (Fernbach, 1904). The content of phosphate esters in starch is very low. Nevertheless, the effects of these groups in the plant as well as in the purified starch are remarkable. In the plant, phosphate is required for complete degradation of the starch, as described in the next section. Moreover, phosphorylated starches have a tremendous hydration capacity and produce clear and highly viscous pastes (Wiesenborn et al., 1994; Viksø-Nielsen et al., 2001). The most phosphorylated starches, such as those extracted from potato tubers with modified starch, have only one of 200 glucose residues substituted (Schwall et al., 2000; Blennow et al., 2005b). The phosphate esters are found as monoesters linked at the C-6 and C-3 positions of the glucose units (Posternak, 1935; Bay-Smidt et al., 1994; Hizukuri et al., 1970; Tabata and Hizukuri, 1971). The C-3-bound phosphate is a remarkable feature because phosphorylation at this position is rarely seen in nature.

Phosphate monoesters are present almost entirely in the amylopectin fraction (Samec, 1914), preferably in long chains (Takeda and Hizukuri, 1982; Tabata and Hizukuri, 1971; Blennow et al., 1998) with an internal length around DP 20 (Wikman et al., 2011). Isolated phosphory-lated clusters in potato starch were larger than their nonphosphorylated counterparts, and the phosphorylated chains in the clusters were considerably longer than the other chains (Wikman et al., 2011). This is interesting because long-chain amylopectins preferably form the B-type crystalline polymorph found in tubers and in green photosynthetic tissue of the plant, which indicates that there are specific requirements for structuring these types of starches. Supposedly, this is linked to the ability for these starches to be degraded by specific enzymes in the highly hydrated organs, as opposed to dry seeds storing starch with considerably less phosphate.

The presence of phosphate groups in crystalline regions of the starch granule is indicated by effects on starch granule crystallinity, as demonstrated by differential scanning calorimetry (Muhrbeck and Eliasson, 1991), its presence in lintners represents the crystalline lamellae of the starch granule (Blennow et al., 2000), its tight positioning in one of the groves in the double helix as shown by molecular models (Engelsen et al., 2003), its inability to form complexes with copper by electro-paramagnetic resonance (Blennow et al., 2006), and its low mobility as demonstrated by nuclear magnetic resonance imaging (Larsen et al., 2007). Considerable amounts are also indicated to be present in amorphous parts of the starch granule (Blennow et al., 2000; Wikman et al., 2013, 2014). Support for an amorphization effect induced by phosphate was provided by differently phosphorylated lintners: A negative correlation of the melting enthalpy of lintnerized starch and the phosphate content in the native starch was found, as well as in the remaining lintnerized starch (Wikman et al., 2014). Interestingly, phosphate groups possibly affect interactions between branched and linear glucan chains in gelatinized starch. At high (30–40%) starch paste concentrations, phosphate groups repel amylopectin and amylose chains, likely resulting in phase separation between these polysaccharides in gelatinized starch (Jensen et al., 2013a). However, the effects of phosphate monoesters on starch molecular packing have not yet been fully clarified.

The distribution of phosphate in the native starch granule does not seem to be random. The lack of direct evidence for such distributional data stems from the difficulty of analyzing molecular entities and elements in semicrystalline matrices such as those found in the starch granules. As deduced from particle-induced X-ray emission data, phosphate is concentrated at the center of the starch granule (Blennow et al., 2005a). These findings were supported by a chemical gelatinization study (Jane and Shen, 1993) in which lower phosphate concentrations were found at the granule periphery than in the center. In contrast, a confocal laser light scanning microscopy study using a phosphate-specific fluorescent probe found phosphate at the surface, specifically at a rim close to the surface or spread throughout the starch granule depending on the plant genotype (Glaring et al., 2006).

5. Potato Starch Synthesis

Functionally, starch can be considered a polysaccharide synthesized in a manner that permits its efficient degradation. Hence, biosynthesis of the starch granule is a delicate balance between efficient packing of the glucan chains and the possibility of breaking these structures at degradation. To complete this enzymatically catalyzed process in the potato tuber, a multitude of different enzyme activities are required.

These include activation of the glucose residue, elongation of the glucan chain, and transfer of linear backbone chains forming branched structures. This process is generally more complex compared with glycogen biosynthesis involving a multitude of different homologous enzymes supposedly responsible for synthesizing specific structures of the starch granule in different tissues and at different developmental stages. Supposedly, many of these enzymes interact to form enzyme complexes or metabolomes to channel and direct substrates and products (Hennen-Bierwagen et al., 2009) The article by Zeeman et al. (2010) can be consulted for a comprehensive review on this issue and for details as described subsequently.

5.1 Sugar Activation and Regulation of Synthesis

Plants synthesize starch by a pathway starting with the activated sugar adenosine diphosphoglucose (ADP-glucose) from glucose 1-phosphate and adenosine triphosphate in a reaction catalyzed by the enzyme ADP-glucose pyrophosphorylase (AGPase) (Zeeman, et al., 2010). This reaction is the rate-limiting step in starch synthesis. In chloroplasts, AGPase is allosterically activated by 3-phosphoglycerate, the first product of photosynthesis, and inhibited by free phosphate. The rationale for this regulation is the sensing of high photosynthesis (high levels of 3-phosphoglycerate) requiring starch biosynthesis as a means to store assimilated carbon. Likewise, low photosynthesis is revealed by high free phosphate in the cell.

5.2 Chain Elongation

Amylopectin is produced at the surface of the granule by soluble starch synthase (SSS) and starch-branching enzyme (SBE) (Zeeman et al., 2010). Soluble starch synthase catalyzes the elongation of the chain at the nonreducing end in a reaction in which ADP of the ADP-glucose molecule is displaced by the terminal hydroxyl group of the growing glucan chain, creating an elongated linear α-(1,4)-glucan chain. For amylose synthesis only one enzyme is required: the granule-bound starch synthase (GBSS). Just as for SSS, GBSS adds the glucose unit from ADP-glucose to the nonreducing end of a glucose chain. Granule-bound starch synthase is completely bound to the starch granule and elongates the glucan chains processively, i.e., without diffusion between the different substrate chains (Zeeman et al., 2010). Data provide support for the involvement of phosphorylase in chain elongation (Fettke et al., 2012; Kaminski et al., 2012).

5.3 Branching and Maturation Debranching

The formation of α-(1,6)-linkages in starch is catalyzed by the SBE. In this reaction, an α-(1,4)-linkage within the chain is cleaved and an α-(1,6)-linkage is formed between the reducing end of the cleaved glucan chain and a C-6 linked oxygen of an adjacent chain. Starch-branching enzyme activity exists as multiple enzyme isoforms. Family A isoforms use preferentially shorter glucan chains during branch formation than family B. Hence, the two isoforms seem to have distinct but interdependent roles in amylopectin synthesis. The action of debranching activities during biosynthesis seems to be important for correct assembly of the starch granule (Zeeman et al., 2010; Hennen-Bierwagen et al., 2012). Such reactions catalyzed by debranching enzymes trim off branches by hydrolysis that are not correctly positioned in the molecule, generating a branching structure appropriate for crystallization. In the absence of specific debranching enzymes in plant mutants, highly branched phytoglycogen is accumulated. Because of the branched structure of phytoglycogen, it cannot form crystalline arrays and therefore is soluble. The two types of debranching enzymes in plants belong to the isoamylase-type and the pullulanase-type debranching enzymes. Both types efficiently hydrolyze the α-(1,6)-linkage of amylopectin. The balance between branching and debranching most likely forms the fundament for the generation of the well-structured domains, clusters, and building blocks, as explained in detail earlier, but the detailed enzyme actions that take place during synthesis of these are not known.

5.4 Starch Phosphorylation

For tuber starches in general, and for potato starch especially, the B-type crystalline polymorph is the major crystalline unit. Although this crystalline type consists of channels with structured water, these type of starches are generally well ordered and compact and have a smooth starch granule topography. This is apparently a result of the relatively long chains in the amylopectin clusters efficiently filling out the crystallites. Because of this, these types of granules are resistant to amylolytic degradation (Jane et al., 1997). To facilitate and regulate degradation of the starch in the tuber during germination, the starch granules are phosphorylated (Blennow et al., 2002). Not until recently was the enzyme GWD, which catalyzes the phosphorylation, discovered (Lorberth et al., 1998; Ritte et al., 2002). The phosphate groups have some intriguing effects on how the starch granule is metabolized. The phosphate groups provide molecular signals for starch degradation in the plant. It has been suggested (Blennow et al., 2002) that the phosphate groups itself by restructuring the starch granule, which affects granule degradability. Data are now provided indicating that starch phosphate esters stimulate hydrolytic enzyme activity in vitro (Edner et al., 2007), and confirm that phosphate esters themselves can restructure the starch granule stimulating degradation. However, the effect can be more general. (1) Starch granules of *Arabidopsis thaliana* plants with low GWD activity have granule surface glucan chain structures suggested to limit general accessibility to any starch-active enzyme, degradative or biosynthetic (Mahlow et al., 2014). (2) Overexpression of GWD in barley does not reduce starch yield (Shaik et al., 2014). (3) Overexpression of GWD does not stimulate starch degradation in *Arabidopsis* leaves. (4) Starch degradation is not suppressed until GWD is reduced to 30% activity. In support of a more general effect of starch phosphorylation, reduction of GWD inhibits starch biosynthesis, which indicates that starch phosphorylation can stimulate starch biosynthesis as well as starch degradation (Skefflington et al., 2014).

6. Conclusions

Potato starch, and tuberous starch in general, has some unique properties compared with cereal starches. The most important ones include long amylopectin chains forming hydrated and ordered B-type crystallites and the presence of phosphate esters. The clusters of potato amylopectin are comparatively small and are composed of 5–10 short chains. The internal part of the clusters is organized into branched building blocks mainly found in the amorphous lamellae of the semicrystalline granular rings. The external segments of the chains constitute double helices ordered into the characteristic hexagonal structure of the B-type allomorph. In the inner parts of the granules, the order of crystalline orientation is low, but it increases toward the periphery. To date, certain details regarding the structure of potato starch remain unclear. These include the actual mode of interconnection of the cluster units of the amylopectin component and the connection between the proposed superhelical structure and the blocklet structures in the granules. The synthesis of starch is complex and involves a range of enzyme groups. Through genetic engineering, it is possible to control synthesis and develop potato starches with altered structures and functionality. Of special interest is the GWD, which currently provides the only way to biochemically substitute starch directly in the plant.

References

Aberle, T., Burchard, W., Vorwerg, W., Radosta, S., 1994. Conformational contributions of amylose and amylopectin to the structural properties of starches from various sources. Starch/Stärke 46, 329–335.

Angellier-Coussy, H., Putaux, J.-L., Molina-Boisseau, S., Dufresne, A., Bertoft, E., Pérez, S., 2009. The molecular structure of waxy maize starch nanocrystals. Carbohydrate Research 344, 1558–1566.

Baldwin, P.M., Adler, J., Davies, M.C., Melia, C.D., 1998. High resolution imaging of starch granule surfaces by atomic force microscopy. Joural of Cereal Science 27, 255–265.

Bay-Smidt, A.M., Wischmann, B., Olsen, C.-E., Nielsen, T.H., 1994. Starch bound phosphate in potato as studied by a simple method for determination of organic phosphate and ^{31}P-NMR. Stärke 46, 167–172.

Bender, H., Siebert, R., Stadler-Szöke, A., 1982. Can cyclodextrin glycosyltransferase be useful for the investigation of the fine structure of amylopectins? Characterisation of highly branched clusters isolated from digests with potato and maize starches. Carbohydrate Research 110, 245–259.

Bertoft, E., 2004. On the nature of categories of chains in amylopectin and their connection to the super helix model. Carbohydrate Polymer 57, 211–224.

Bertoft, E., 2007a. Composition of clusters and their arrangement in potato amylopectin. Carbohydrate Polymer 68, 433–446.

Bertoft, E., 2007b. Composition of building blocks in clusters from potato amylopectin. Carbohydrate Polymer 70, 123–136.

Bertoft, E., 2013. On the building block and backbone concepts of amylopectin structure. Cereal Chemistry 90, 294–311.

Bertoft, E., Piyachomkwan, K., Chatakanonda, P., Sriroth, K., 2008. Internal unit chain composition in amylopectins. Carbohydrate Polymer 74, 527–543.

Bertoft, E., Koch, K., Åman, P., 2012. Building block organisation of clusters in amylopectin of different structural types. International Journal of Biological Macromolecule 50, 1212–1223.

Blennow, A., Bay-Smidt, A.M., Wischmann, B., Olsen, C.E., Møller, B.L., 1998. The degree of starch phosphorylation is related to the chain length distribution of the neutral and the phosphorylated chains of amylopectin. Carbohydrate Research 307, 45–54.

Blennow, A., Bay-Smidt, A.M., Olsen, C.A., Møller, B.L., 2000. The distribution of covalently bound starch-phosphate in native starch granules. International Journal of Biological Macromolecule 27, 211–218.

Blennow, A., Engelsen, S.B., Nielsen, T.H., Baunsgaard, L., Mikkelsen, R., 2002. Starch phosphorylation—a new front line in starch research. Trends in Plant Science 7, 445–450.

Blennow, A., Bay-Smidt, A.M., Leonhardt, P., Bandsholm, O., Madsen, H.M., 2003. Starch paste stickiness is a relevant native starch selection criterion for wet-end paper manufacturing. Starch 55, 381–389.

Blennow, A., Sjöland, K.A., Andersson, R., Kristiansson, P., 2005a. The distribution of elements in the native starch granule as studied by particle-induced X-ray emission and complementary methods. Analytical Biochemistry 347, 327–329.

Blennow, A., Wischmann, B., Houborg, K., Ahmt, T., Madsen, F., Poulsen, P., Jørgensen, K., Bandsholm, O., 2005b. Structure-function relationships of transgenic starches with engineered phosphate substitution and starch branching. International Journal of Biological Macromolecule 36, 159–168.

Blennow, A., Houborg, K., Andersson, R., Bidzińska, E., Dyrek, K., Łabanowska, M., 2006. Phosphate positioning and availability in the starch granule matrix as studied by EPR. Biomacromolecules 7, 965–974.

Buléon, A., Pontoire, B., Riekel, C., Chanzy, H., Helbert, W., Vuong, R., 1997. Crystalline ultrastructure of starch granules revealed by synchrotron radiation microdiffraction mapping. Macromolecules 30, 3952–3954.

Edner, C., Li, J., Albrecht, T., Mahlow, S., Hejazi, M., Hussain, H., Kaplan, F., Guy, C., Smith, S.M., Steup, M., Ritte, G., 2007. Glucan, water dikinase activity stimulates breakdown of starch granules by plastidial β-amylases. Plant Physiology 145, 17–28.

Engelsen, S.B., Madsen, M.A.O., Blennow, A., Motawia, S., Møller, B.L., Larsen, S., 2003. The phosphorylation site in double helical amylopectin as investigated by a combined approach using chemical synthesis, crystallography and molecular modeling. FEBS Letters 541, 137–144.

Fernbach, A., 1904. Quelques observations sur la composition de l'amidon de pommes de terre. Comptes Rendus de l'Académie des Sciences 138, 428–430.

Fettke, J., Leifels, L., Brust, H., Herbst, K., Steup, M., 2012. Two carbon fluxes to reserve starch in potato (*Solanum tuberosum* L.) tuber cells are closely interconnected but differently modulated by temperature. Journal of Experimental Botany 63, 3011–3029.

Finch, P., Sebesta, D.W., 1992. The amylase of *Pseudomonas stutzeri* as a probe of the structure of amylopectin. Carbohydrate Research 227, c1–c4.

French, D., 1972. Fine structure of starch and its relationship to the organization of starch granules. Journal of Japanese Society of Starch Science 19, 8–25.

Gallant, D.J., Bouchet, B., Baldwin, P.M., 1997. Microscopy of starch: evidence of a new level of granule organization. Carbohydrate Polymer 32, 177–191.

Genkina, N.K., Wikman, J., Bertoft, J., Yuryev, V.P., 2007. Effects of structural imperfection on gelatinization characteristics of amylopectin starches with A- and B-type crystallinity. Biomacromolecules 8, 2329–2335.

Glaring, M.A., Koch, K.B., Blennow, A., 2006. Genotype specific spatial distribution of starch molecules in the starch granule: a combined CLSM and SEM approach. Biomacromolecules 7 (8), 2310–2320.

Hanashiro, I., Tagawa, M., Shibahara, S., Iwata, K., Takeda, Y., 2002. Examination of molar-based distribution of A, B and C chains of amylopectin by fluorescent labeling with 2-aminopyridine. Carbohydrate Research 337, 1211–1215.

Hennen-Bierwagen, T.A., James, M.G., Myers, A.M., 2012. Involvement of debranching enzymes in starch biosynthesis. In: Tetlow, I. (Ed.), Starch, Origins, Structure and Metabolism. Essential Reviews in Experimental Biology, vol. 5. The Society for Experimental Biology, London, pp. 179–215.

Hennen-Bierwagen, T.A., Lin, Q., Grimaud, F., Planchot, V., Keeling, P.L., James, M.G., Myers, A.M., 2009. Proteins from multiple metabolic pathways associate with starch biosynthetic enzymes in high molecular weight complexes: a model for regulation of carbon allocation in maize amyloplasts. Plant Physiology 149, 1541–1559.

Hizukuri, S., Tabata, S., Nikuni, Z., 1970. Studies on starch phosphate. Part 1. Estimation of glucose-6-phosphate residues in starch and the presence of other bound phosphate(s). Stärke 22, 338–343.

Hizukuri, S., Takeda, Y., Yasuda, M., Suzuki, A., 1981. Multi-branched nature of amylose and the action of de-branching enzymes. Carbohydrate Research 94, 205–213.

Hizukuri, S., Takagi, T., 1984. Estimation of the distribution of molecular weight for amylose by the low-angle laser-light-scattering technique combined with high-performance gel chromatography. Carbohydrate Research 134, 1–10.

Hizukuri, S., 1985. Relationship between the distribution of the chain length of amylopectin and the crystalline structure of starch granules. Carbohydrate Research 141, 295–306.

Hizukuri, S., 1986. Polymodal distribution of the chain lengths of amylopectins, and its significance. Carbohydrate Research 147, 342–347.

Hoover, R., 2001. Composition, molecular structure, and physicochemical properties of tuber and root starches: a review. Carbohydrate Polymer 45, 253–267.

Imberty, A., Buléon, A., Tran, V., Pérez, S., 1991. Recent advances in knowledge of starch structure. Starch/Stärke 43, 375–384.

Jacobs, H., Eerlingen, R.C., Rouseu, N., Colonna, P., Delcour, J.A., 1998. Acid hydrolysis of native and annealed wheat, potato and pea starches–DSC melting features and chain length distributions of lintnerised starches. Carbohydrate Research 308, 359–371.

Jane, J.-L., Xu, A., Radosavljevic, M., Seib, P.A., 1992. Location of amylose in normal starch granules. I. Susceptibility of amylose and amylopectin to cross-linking reagents. Cereal Chemistry 69, 405–409.

Jane, J.-L., Shen, J.J., 1993. Internal structure of the potato starch granule revealed by chemical gelatinization. Carbohydrate Research 247, 279–290.

Jane, J.-L., Wong, K.-S., McPherson, A.E., 1997. Branch-structure difference in starches of A- and B-type X-ray patterns revealed by their Naegeli dextrins. Carbohydrate Research 300, 219–227.

Jenkins, P.J., Cameron, R.E., Donald, A.M., 1993. A universal feature in the structure of starch granules from different botanical sources. Starch/Stärke 45, 417–420.

Jensen, S.L., Larsen, F.H., Bandsholm, O., Blennow, A., 2013a. Stabilization of semi-solid-state starch by branching enzyme-assisted chain-transfer catalysis at extreme substrate concentration. Biochemical Engineering Journal 72, 1–10.

Jensen, S.L., Zhu, F., Vamadevan, V., Bertoft, E., Seetharaman, K., Bandsholm, O., Blennow, A., 2013b. Structural and physical properties of granule stabilized starch obtained by branching enzyme treatment. Carbohydrate Polymer 98, 1490–1496.

Kaminski, K.P., Petersen, A.H., Sønderkær, M., Pedersen, L.H., Pedersen, H., Feder, C., Nielsen, K.L., 2012. Transcriptome analysis suggests that starch synthesis may proceed via multiple metabolic routes in high yielding potato cultivars. PLoS One 7, e51248.

Kozlov, S.S., Blennow, A., Krivandin, A.V., Yuryev, V.P., 2007. Structural and thermodynamic properties of starches extracted from GBSS and GWD suppressed potato lines. International Journal of Biological Macromolecule 40, 449–460.

Larsen, F.H., Blennow, A., Engelsen, S.B., 2007. Starch granule hydration—a MAS NMR investigation. Food Biophysics 3, 25–32.

Lorberth, R., Ritte, G., Willmitzer, L., Kossmann, J., 1998. Inhibition of a starch-granule-bound protein leads to modified starch and repression of cold sweetening. Nature Biotechnology 16, 473–477.

Mahlow, S., Hejazi, M., Kuhnert, F., Garz, A., Brust, H., Baumann, O., Fettke, J., 2014. Phosphorylation of transitory starch by α-glucan, water dikinase during starch turnover affects the surface properties and morphology of starch granules. New Phytology 203, 495–507.

McPherson, A.E., Jane, J., 1999. Comparison of waxy potato with other root and tuber starches. Carbohydrate Polymer 40, 57–70.

Morell, M.K., Samuel, M.S., O'Shea, M.G., 1998. Analysis of starch structure using fluorophore-assisted carbohydrate electrophoresis. Electrophoresis 19, 2603–2611.

Morrison, W.R., 1995. Starch lipids and how they relate to starch granule structure and functionality. Cereal Foods World 40, 437–446.

Muhrbeck, P., Eliasson, A.-C., 1991. Influence of the naturally-occurring phosphate-esters on the crystallinity of potato starch. Journal of the Science and Food Agriculture 55, 13–18.

Nikuni, Z., 1978. Studies on starch granules. Stärke 30, 105–111.

Noda, T., Takigawa, S., Matsuura-Endo, C., Kim, S.-J., Hashimoto, N., Yamauchi, H., Hanashiro, I., Takeda, Y., 2005. Physicochemical properties and amylopectin structures of large, small, and extremely small potato starch granules. Carbohydrate Polymer 60, 245–251.

Oostergetel, G.T., Bruggen, E.F.J.V., 1993. The crystalline domains in potato starch granules are arranged in a helical fashion. Carbohydrate Polymer 21, 7–12.

Peat, S., Whelan, W.J., Thomas, G.J., 1952. Evidence of multiple branching in waxy maize starch. Journal of the Chemical Society, Chemical Communications 4546–4548.

Posternak, T., 1935. Sur le phosphore des amidons. Helvetica Chimica Acta 18, 1351–1369.

Putaux, J.-L., Molina-Boisseau, S., Momaur, T., Dufresne, A., 2003. Platelet nanocrystals resulting from the disruption of waxy maize starch granules by acid hydrolysis. Biomacromolecules 4, 1198–1202.

Ratnayake, W.S., Jackson, D.S., 2007. A new insight into the gelatinization process of native starches. Carbohydrate Polymer 67, 511–529.

Ridout, M.J., Gunning, A.P., Parker, M.L., Wilson, R.H., Morris, V.J., 2002. Using AFM to image the internal structure of starch granules. Carbohydrate Polymer 50, 123–132.

Ritte, G., Lloyd, J.R., Eckermann, N., Rottmann, A., Kossmann, J., Steup, M., 2002. The starch-related R1 protein is an alpha-glucan, water dikinase. Proceedings of the National Academy of Sciences of the USA 99, 7166–7171.

Robin, J.P., Mercier, C., Charbonnière, R., Guilbot, A., 1974. Lintnerized starches. Gel filtration and enzymatic studies of insoluble residues from prolonged acid treatment of potato starch. Cereal Chemistry 51, 389–406.

Samec, M., 1914. Studien über Pflanzenkolloide, IV. Die Verschiebungen des Phosphorgehaltes bei der Zustandsänderungen und dem diastatischen Abbau der Stärke. Kolloidchemische Beheife 4, 2–54.

Schwall, G.P., Safford, R., Westcott, R.J., Jeffcoat, R., Tayal, A., Shi, Y.C., Gidley, M.J., 2000. Production of very-high-amylose potato starch by inhibition of SBE A and B. Nature Biotechnology 18, 551–554.

Shaik, S.S., Carciofi, M., Martens, H.J., Hebelstrup, K.H., Blennow, A., 2014. Starch bioengineering affects cereal grain germination and seedling establishment. Journal of Experimental Botany 65, 2257–2270.

Skefflington, A.W., Graf, A., Duxbury, Z., Gruissem, W., Smith, A.M., 2014. Glucan, water dikinase exerts little control over starch degradation in Arabidopsis leaves at night. Plant Physiology 165, 866–879.

Srichuwong, S., Isono, N., Mishima, T., Hisamatsu, M., 2005. Structure of linterized starch is related to X-ray diffraction pattern and susceptibility to acid and enzyme hydrolysis of starch granules. International Journal of Biological Macromolecule 37, 115–121.

Sun, T., Lærke, H.N., Jørgensen, H., Bach Knudsen, K.E., 2006. The effect of heat processing of different starch sources on the in vitro and in vivo digestibility in growing pigs. Animal Feed Science and Technology 131, 66–85.

Tabata, S., Hizukuri, S., 1971. Studies on starch phosphate. Part 2. Isolation of glucose 3-phosphate and maltose phosphate by acid hydrolysis of potato starch. Stärke 23, 267–272.

Takeda, Y., Hizukuri, S., 1982. Location of phosphate groups in potato amylopectin. Carbohydrate Research 102, 321–327.

Takeda, Y., Shirasaka, K., Hizukuri, S., 1984. Examination of the purity and structure of amylose by gel-permeation chromatography. Carbohydrate Research 132, 83–92.

Takeda, Y., Hizukuri, S., Takeda, C., Suzuki, A., 1987. Structures of branched molecules of amyloses of various origins, and molecular fractions of branched and unbranched molecules. Carbohydrate Research 165, 139–145.

Takeda, Y., Maruta, N., Hizukuri, S., 1992. Examination of the structure of amylose by tritium labelling of the reducing terminal. Carbohydrate Research 227, 113–120.

Takeda, Y., Shibahara, S., Hanashiro, I., 2003. Examination of the structure of amylopectin molecules by fluorescent labeling. Carbohydrate Research 338, 471–475.

Tester, R.F., Debon, S.J.J., 2000. Annealing of starch—a review. International Journal of Biological Macromolecule 27, 1–12.

Vamadevan, V., Bertoft, E., Soldatov, D.V., Seetharaman, K., 2013. Impact on molecular organization of amylopectin in starch granules upon annealing. Carbohydrate Polymer 98, 1045–1055.

Viksø-Nielsen, A., Blennow, A., Kristensen, K.H., Jensen, A., Møller, B.L., 2001. Structural, physicochemical, and pasting properties of starches from potato plants with repressed *r1*-gene. Biomacromolecules 3, 836–841.

Waigh, T.A., Donald, A.M., Heidelbach, F., Riekel, C., Gidley, M.J., 1999. Analysis of the native structure of starch granules with small angle X-ray microfocusing scattering. Biopolymers 49, 91–105.

Waigh, T.A., Kato, K.L., Donald, A.M., Gidley, M.J., Clarke, C.J., Riekel, C., 2000. Side-chain liquid-crystalline model for starch. Starch/Stärke 52, 450–460.

Wiesenborn, D.P., Orr, P.H., Casper, H.H., Tacke, B.K., 1994. Potato starch paste behavior as related to some physical/chemical properties. Journal of Food Science 59, 644–648.

Wikman, J., Larsen, F.H., Motawia, M.S., Blennow, A., Bertoft, E., 2011. Phosphate esters in amylopectin clusters of potato tuber starch. International Journal of Biological Macromolecule 48, 639–649.

Wikman, J., Blennow, A., Bertoft, E., 2013. Effect of amylose deposition on potato tuber starch granule architecture and dynamics as studied by lintnerization. Biopolymers 99, 73–83.

Wikman, J., Blennow, A., Buléon, A., Putaux, J.-L., Pérez, S., Seetharaman, K., Bertoft, E., 2014. Influence of amylopectin structure and degree of phosphorylation on the molecular composition of potato starch lintners. Biopolymers 101, 257–271.

Yusuph, M., Tester, R.F., Ansell, R., Snape, C.E., 2003. Composition and properties of starches extracted from tubers of different potato varieties grown under the same environmental conditions. Food Chemistry 82, 283–289.

Zeeman, S.C., Kossmann, J., Smith, A.M., 2010. Starch: its metabolism, evolution, and biotechnological modification in plants. Annual Review of Plant Biology 61, 209–234.

Zhu, Q., Bertoft, E., 1996. Composition and structural analysis of alpha-dextrins from potato amylopectin. Carbohydrate Research 288, 155–174.

CHAPTER 4

Potato Proteins: Functional Food Ingredients

Amanda Waglay, Salwa Karboune

Department of Food Science and Agricultural Chemistry, McGill University, Ste-Anne de Bellevue, Quebec, Canada

1. Introduction

Potatoes (*Solanum tuberosum*) are the fourth most produced crop after rice, wheat, and corn (Kamnerdpetch et al., 2007). Despite their low protein concentration of 1.7%, potatoes are the second highest protein providing crop per hectare grown after wheat. The North American consumer associates potatoes with unhealthy diets because of the poor processing techniques often employed, such as frying. Contrary to this negative connotation, potatoes contain fiber, many high-quality proteins, and several vitamins and minerals (Bártova and Bárta, 2009). These high-quality proteins are composed of a high proportion of lysine (Kamnerdpetch et al., 2007), threonine, tryptophan, and methionine; the latter is an essential amino acid often lacking in cereal and vegetable crops (Kärenlampi and White, 2009). Because of their amino acid composition, potato proteins have been studied and recognized to be nutritionally equivalent to animal protein, lysozyme (Ralet and Guéguen, 2000).

Potato proteins are often classified into three groups, namely patatin, protease inhibitors, and high-molecular-weight proteins. They are associated with several health benefits such as lower allergic response (Fu et al., 2002), antimicrobial effects (Kim et al., 2005), antioxidant potential (Pihlanto et al., 2008), and the regulation of blood pressure and blood serum cholesterol control (Pihlanto et al., 2008; Liyanage et al., 2008) and anticarcinogenic behavior (Blanco-Aparicio et al., 1998). Among these health benefits are several other functional qualities that will help broaden their application in the food industry, such as emulsification and foaming abilities.

Potato proteins have often been overlooked, but their removal from the starch industry byproduct is necessary to help overcome the economic impact coupled to its high polluting capacity (Strolle et al., 1973). Many extraction techniques have been explored to maintain or enhance the functional qualities possessed by the proteins (Cheng et al., 2010; Miedzianka et al., 2011). Further research is necessary to develop an industrial-scale and cost-efficient process to extract potato proteins with minimal functional loss, to broaden their potential applications. For instance, a novel enzymatic extraction approach based on the use of cell wall-degrading enzymes was proven to be potential for the efficient recovery of non-denatured potato proteins (Waglay et al., 2015). This chapter will explore the protein isolates obtained using several extracting agents such as thermal and acidic precipitation, salt precipitation, ethanol precipitation, $(NH_4)_2SO_4$ saturation, and carboxymethyl cellulose (CMC) complexation. To date, many studies have been conducted on examining the recovery yield, functionality, fractionation, and structural impact of the extracting agents on the protein isolates.

2. Potato Proteins

2.1 Chemical and Structural Properties

2.1.1 Patatin

The major protein found in potatoes is patatin, otherwise known as tuberin. It is mostly found in the tuber or stolons of the plant, specifically in the vacuole of the parenchyma tissue (Straetkvern et al., 1999). However, the protein has also been observed in substantial amounts in the small aboveground buds that form only when the tuber and stolons of the plant have been removed. Other studies have also demonstrated that patatin can be found in small quantities in the stem of the plant (Park, 1983).

Patatin represents approximately 40% of the soluble protein found in the tuber and is a group of glycoproteins with molecular weight from 40 to 45 kDa (Kärenlampi and White, 2009). It has been reported that patatin is present as an 88-kDa dimer when analyzed on a native gel. However, in the presence of sodium dodecyl sulfate (SDS), as shown in Figure 1, the protein is broken down into its monomer units (Ralet and Guéguen, 2000). Patatin has been shown to be made up of approximately 366 amino acids. These amino acid residues possess negative and positive charges that are distributed throughout the protein's length (Pots et al., 1998). The isoelectric point of patatin has been established to be at pH 4.9 (van Koningsveld et al., 2001a). Three glycosylation sites have been studied to be specific to patatin. The sites have been examined to occur at the asparagine residues, which have an interesting biological function. These glycosylation sites are named asparagine-linked oligosaccharides and have been linked to signaling for intracellular targeting, protection from proteolytic breakdown, and preservation of protein stability by influencing unfolding patterns (Sonnewald et al., 1989).

Structurally patatin is a tertiary stabilized protein. Pots et al. (1998) determined patatin to contain an estimated 45% β-strand and 33% α-helix. The tertiary structure is stable up to 45 °C; however, after this point, patatin's secondary structure begins to unfold and at temperatures of 55 °C the α-helical part denatures (Pots et al., 1998). Compared with other common protein vegetable sources, patatin is of equal nutritional benefit as egg albumen and has been determined to have better emulsifying properties than soy proteins (Løkra et al., 2008).

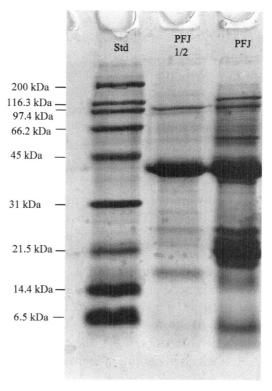

Figure 1
Sodium dodecyl sulfate-PAGE protein profile of imitation PFJ prepared from cultivar Russett.

Patatin is made up of many proteins that are represented by two multigene families (Pots et al., 1999). The class I gene family is represented in large concentrations in the tuber, whereas the class II gene family is represented in smaller concentrations throughout the potato plant (Pots et al., 1999). The literature demonstrates that for the potato variety Bintje, patatin can be separated into four isoforms based on charge. An isoform is defined as "a protein that has the same function as another protein but which is encoded by a different gene and may have a small difference in its sequence" (MedicineNet, 2004). Patatin is represented as four isoforms—A, B, C, and D—which are present in varying amounts: 62%, 26%, 5%, and 7%, respectively, in which all four patatin isoforms are homologous in nature and possess an identical immunological response (Pots et al., 1999). In general, isoforms vary according to differences in charge, molecular mass, or structural properties. The patatin isoforms present in the different potato varieties fluctuate. For example, cv. Désirée was found to have nine isoforms, whereas Bintje was found to have four (Lehesranta et al., 2005; Pots et al., 1999). The different patatin isoform ratios can be observed through experimental techniques that separate based on charge. This is an important parameter that can help with differentiation between patatin families for a range of potato species (Bohac, 1991).

Patatin isoforms are represented by a charge difference, which was demonstrated through anionic exchange (Pots et al., 1999). Pots et al. (1999) found that isoform A had low affinity with the anion-exchange column, and it represented the shortest running distance on a native gel. Conclusions were drawn that isoform A must have the lowest surface charge compared with isoforms B, C, and D. It was also observed that isoform A exhibited the highest isoelectric focusing and shortest capillary electrophoresis. Based on the protein's isoelectric point, at pH 3 isoform A contained the largest amount of positively charged residues, and therefore at pH 8 it contained the largest amount of negatively charged residues compared with the other isoforms, B, C, and D (Pots et al., 1999). Pots et al. (1999), determined that the differences among isoforms were variations in molecular masses of 40.4 and 41.8 kDa. The molecular mass differences of the isoforms were compared with a previous study performed by Mignery et al. (1984) on the potato variety Superior. The isoforms contained 366 amino acids with a variation of 21 amino acids. The varying mutated amino acids between isoforms were represented experimentally with a molar mass difference of 663 Da. However, Mignery et al. (1984) calculated the theoretical difference to be 100 Da. This discrepancy demonstrates that the patatin isoforms may have undergone glycosylation between the protein and carbohydrate present in the potato. Pots et al. (1999) used similar calculations for the isoforms in the Bintje variety. According to calculations, the isoforms for this variety can possess a maximum difference of 198 Da. However, no correlation was made between this calculation and their results attained from matrix-assisted laser desorption-ionization time-of-flight mass spectroscopy (MALDI-TOF MS) (Pots et al., 1999).

Although patatin is present in the tuber as the main storage protein, studies have concluded that patatin and its isoforms must have functions in the tuber other than storage because of its presence during the developing stages of the plant. This has led to the determination that the protein possesses antioxidant activity (Liu et al., 2003; Wang and Xiong, 2005; Kärenlampi and White, 2009). Furthermore, it has been categorized as an esterase enzyme complex (Vreugdenhil et al., 2007). Patatin demonstrates enzymatic activity toward lipid metabolism through lipid acyl hydrolases (LAHs) as well as acyl transferases (Höfgen and Willmitzer, 1990; Liu et al., 2003). It has been hypothesized that these mechanisms allow patatin to participate in plant defenses (Kärenlampi and White, 2009). This LAH has been found to vary according to the potato cultivar, extraction technique, and fatty acid substrate (Table 1). As shown in Table 1, generally, lipid acyl hydrolase activity (LAHA) was higher when the substrate *p*-nitrophenyl laurate was used, as shown by Pots et al. (1999) and Waglay et al. (2014). However, purified patatin examined with its corresponding LAHA substrate *p*-nitrophenyl butyrate provided comparable results. Results showed that patatin is greatly affected by the precipitating agent used, and that ammonium sulfate has the greatest preservation followed by ethanol and ferric chloride (Waglay et al., 2014). Currently employed thermal and acidic precipitation completely denatures patatin, causing it to lose all of its activity (Waglay et al., 2014). This hydrolase activity is because the central core of patatin is composed of a parallel β-sheet and the catalytic serine is located in the nucleophilic elbow loop

Table 1: Comparison of Lipid Acyl Hydrolase (LAH) Activity Corresponding to Patatin Isolates from Different Sources Using Different Extraction Techniques.

Cultivar	Patatin	Extraction Technique	LAH Activity	Substrate	Authors
Beroline	40%	0.1% polyvinyl-polypyrrolidone, DEAE-Sephacel and Concanavalin A–Sepharose chromatography	5[a]	p-Nitrophenyl fatty acid esters (chain length 2–18)	Höfgen and Willmizer (1990)
Désirée	20%		0.5[a]		
Bintje	Purified	DEAE-Sepharose CL-6B, Concanavalin A–Sepharose 4B	4.54[b]	p-Nitrophenol butyrate	Pots et al. (1998)
	Patatin family	Concanavalin A, SourceQ		p-Nitrophenol laurate	Pots et al. (1999)
	62% isoform A		3.72[b]		
	26% isoform B		3.66[b]		
	7% isoform D		3.55[b]		
			3.80[b]		
Elkana	100.00%	Purified patatin: Source 15Q column, Superdex 75 column	1.84[b]	p-Nitrophenyl butyrate	van Koningsveld et al., 2001a
PFJ (Lyckeby Amylex, Czech Republic)	30.70%	Freeze-dried PFJ	0.03[b]	p-Nitrophenyl butyrate	Bártová and Bárta (2009)
	25.60%	Ethanol ppt	0.07[b]		
	20.30%	Ferric chloride ppt	0.03[b]		
	32.30%	Heat-coagulated	0.01[b]		
	n.a.	Purified patatin: Ethanol ppt, DEAE 52-cellulose SERVACEL, Concanavalin A–Sepharose 4B column, Sephadex G-25 gel filtration column	0.5[b]		
Russett	22.90%	PFJ	1.17[c]	p-Nitrophenyl butyrate / 1.76[c] p-Nitrophenol laurate	Waglay et al. (2014)
	37.90%	Heat and acidic	<0.04[c]	<0.04[c]	
	11.10%	Acidic ppt	0.67[c]	0.97[c]	
	37.70%	Ferric chloride	0.15[c]	1.55[c]	
	37.70%	Ethanol	0.15[c]	0.52[c]	
	36.50%	Ammonium sulfate	0.47[c]	1.15[c]	
	36.60%	CMC	<0.04[c]	0.2[c]	

[a]Units activity is expressed in nm/min × mg patatin.
[b]Units activity is expressed in µmol/min/mg protein.
[c]Units activity is expressed in µmol/min/mg patatin.

(Rydel et al., 2003). This core, which consists of a parallel β-sheet with the catalytic residue buried within an elbow loop, is a key attribute belonging to the hydrolase family (Rydel et al., 2003).

2.1.2 Protease Inhibitors

The second group of proteins found in potatoes is the protease inhibitors (tuberinin), which possess molecular weights ranging from 5 to 25 kDa (van Koningsveld et al., 2002). Like patatin, these protease inhibitors represent 30–40% of the total tuber protein (Pouvreau et al., 2001; Jørgensen et al., 2006). The protease inhibitors act by hindering the activity of serine protease, cysteine protease, aspartate protease, and metalloprotease (Kärenlampi and White, 2009). Through these inhibition mechanisms the digestibility and availability of the proteins decrease (Kärenlampi and White, 2009). When patatin and the protease inhibitors are compared, protease inhibitors tend to have more hydrophilic properties; however, both protein fractions tend to coagulate by heat (Kärenlampi and White, 2009).

The protease inhibitors are much more diverse than the patatin family and are able to act on a variety of proteases and other enzymes. A summary of the protease inhibitors is shown in detail in Table 2; general subcategories include protease inhibitor I, protease inhibitor II, potato aspartate protease inhibitor, potato cysteine protease inhibitor, potato Kunitz-type protease inhibitor, other serine protease inhibitor, and potato carboxypeptidase inhibitor (PCI) (Pouvreau et al., 2001). Table 2 shows variations according to molecular weight, isoelectric point, and their proportion within the cultivar Elkana. However, studies demonstrate that these protease inhibitors vary according to chain length, amino acid composition, and inhibitory activities (Pouvreau et al., 2001; Jørgensen et al., 2006). Protease inhibitor I is a pentameric serine protease inhibitor composed of five 7- to 8-kDa isoinhibitor protomers (Pouvreau et al., 2001). Protease inhibitor I has a strong affinity for chymotrypsin, trypsin, and human leukocyte elastase and inhibits their activities (Pouvreau et al., 2001). Protease inhibitor II is a dimeric serine protease inhibitor composed of two 10.2-kDa proteins linked together by disulfide linkages (Pouvreau et al., 2001). They have fluctuating inhibiting effects among the subcategories but generally inhibit trypsin, chymotrypsin, and human leukocyte elastase (Pouvreau et al., 2001). The potato aspartate protease inhibitors also vary in terms of inhibiting activity among their subcategories but they generally inhibit trypsin, chymotrypsin, cathepsin D, and human leukocyte elastase (Pouvreau et al., 2001). The potato cystein protease inhibitors vary among subcategories but most have inhibiting activity toward trypsin, chymotrypsin, papain, and, to a lesser extent, human leukocyte elastase (Pouvreau et al., 2001). Both potato Kunitz-type and other serine protease inhibitors are able to inhibit trypsin and chymotrypsin, whereas only the other serine protease inhibitors are able to inhibit human leukocyte elastase (Pouvreau et al., 2001). Finally, PCI is a thermostable, 4.3-kDa, single-subunit peptide (Pouvreau et al., 2001).

Jørgensen et al. (2006) examined the potato variety cv. Kuras, a variety commonly used in the European potato starch industry. cv. Kuras was found to have a protease inhibitor protein

Table 2: Subcategories of Potato Protease Inhibitors Found in cv. Elkana.[a]

Name	MW	PI	PI Proportion in cv. Elkana PFJ
Potato inhibitor I	7683–7873	5.1–6.3	3.9
	7683–7873	7.2, 7.8	3.1
	7683–7873	5.1, 6.3	1.8
Potato inhibitor II	20,279	6.5	14.5
	20,023	6	4.5
	20,273	6.1	11.3
	20,674	5.8	0.6
	20,676	5.5	2.6
	20,315	5.9	6.9
	20,265	6.9	3.7
Potato aspartate protease inhibitor	19,987	6.2	0.2
	20,039	8.4	2.8
	22,025	8.6	2.8
	19,878	8.7	2.6
	20,141	7.5	2.4
	19,883	8.2	0.8
	22,755	6.7	1.5
	22,769	6.6	1.6
	22,674	5.8	2.6
Potato cysteine protease inhibitor	22,773	7.1	0.9
	20,096	8	2.6
	20,127	8.6	4.1
	20,134	>9.0	3.1
	20,433	8.3	7.2
Potato Kunitz-type protease inhibitor	20,237	>9.0	1.8
	20,194	8	5.3
Other serine protease inhibitor	21,025	8.8	1.3
	21,804	7.5	1.7
Potato carboxypeptidase inhibitor	4274	n.d.	1.8

[a]Adapted from Pouvreau et al. (2001).

fraction that could be separated into five nonhomologous families. Those families were 13A Kunitz peptidase inhibitor, I13 peptidase inhibitor I, I20 peptidase inhibitor II, I25 multicystatin peptidase inhibitor, and I37 carboxypeptidase inhibitor (Jørgensen et al., 2006).

Protease inhibitors have been determined to possess protease inhibition activity. This inhibition activity has been found to increase with long-term storage and sprouting. It is hypothesized that the inhibition activity is required to help in the breakdown of proteins during the developing stages of the tuber. More specifically, the serine protease inhibitors help in protein regulation by removing proteins during chloroplast ontogeny as well as assisting in nitrogen mobilization during germination, and in leaf aging (Weeda et al., 2009). The potato multicystatin has been shown in vitro to degrade patatin, a regulator in protein accumulation in the tuber (Weeda et al., 2009).

2.1.3 Others

The remaining fraction of proteins found in the potato consists of a variety of high-molecular-weight proteins (van Koningsveld et al., 2002). This group of protein has not been extensively studied.

2.2 Nutritional and Health-Promoting Properties

Potatoes are often considered nutritionally beneficial for their high starch content, which supplies the body with energy and provides protein, vitamins, and dietary fiber (Kärenlampi and White, 2009). Overall, carbohydrates represent approximately 75% of the total dry matter. This section will focus primarily on the nutritional benefit of potato proteins.

Potato tubers contain 20 g protein/kg on a fresh weight basis (Kärenlampi and White, 2009). Potato proteins have been proven to be nutritionally superior to most other plant and cereal proteins and relatively close to egg protein (lysozyme) (Ralet and Guéguen, 2000). As a result, potato proteins have high nutritional value, attributed to the amino acids that make up these proteins. Unlike most cereal and plant proteins, potato proteins contain a high proportion of the essential amino acid lysine (Kamnerdpetch et al., 2007; He et al., 2013) and relatively high proportions of sulfur-containing amino acids such as methionine and cysteine (Kärenlampi and White, 2009).

The benefit in using potato proteins compared with other proteins is the lower occurrence of allergenicity (Fu et al., 2002). Other common industry sources, including egg, gluten, soy, fish, and nut proteins, are linked to allergic response in approximately 1–2% of the human population (Løkra et al., 2008). Studies have shown that when 800 infants were given an allergy test, only 5% had an allergic reaction to potatoes, whereas for eggs and cow's milk the allergic reaction was 15% and 9%, respectively (Kärenlampi and White, 2009; He et al., 2013). Most adult allergies to potato can be eliminated through the cooking process (Kärenlampi and White, 2009). Possible allergic response symptoms include eczema, gastrointestinal issues, urticaria and angioedema, wheezing, and anaphylaxis (Kärenlampi and White, 2009).

Protease inhibitors have been shown to exhibit anticarcinogenic behavior and to possess beneficial dietary qualities. Potato protease inhibitors specifically demonstrate anticarcinogenic effects through three mechanisms: interfering with tumor-cell proliferation, forming hydrogen peroxide, and involvement in processes related to solar-ultraviolet (UV) irradiation (Pouvreau et al., 2001). Potato protease inhibitors have also been shown to have a positive effect on satiety through their hunger suppression effect (Schoenbeck et al., 2013).

Kim et al. (2005) found that Kunitz-type protease inhibitors accumulated in the potato leaves and tubers help in mechanical wounding of the crop, UV radiation, and wounds started by insects or phytopathogenic microorganisms. Previous studies on tomatoes found that hytopathogenic microorganisms produce extracellular proteinases, which have a direct influence on the development of diseases (Kim et al., 2005). As a defense mechanism, the plant produces inhibitory polypeptides to deactivate the extracellular proteinases. In the case of potatoes, protease inhibitors of the serine proteinase variety function as the plant's source of defense. The study conducted by Kim et al. (2005) examined the antimicrobial activity of the potato protease inhibitor, potamin-1 (PT-1). This peptide demonstrated antimicrobial activity toward the human pathogenic fungus *Candida albicans*, plant pathogenic fungus *Rhizoctonia solani*, and the pathogenic bacterium *Clavibacter michiganense*. Potamin-1 was tested for its ability to function as a serine protease inhibitor and to inhibit trypsin, chymotrypsin, and papain. All three serine protease enzymes were inhibited dose-dependently by PT-1, hence linking its antimicrobial effect to its serine protease inhibition (Kim et al., 2005).

Patatin has been shown to possess antioxidant activity (Wang and Xiong, 2005). When enzymatic hydrolysis is used, it results in the formation of potato protein hydrolysate, which contains many potato peptides. Wang and Xiong (2005) studied the antioxidant ability of a potato protein hydrolysate fraction in cooked beef patties with the peroxide and thiobarbituric acid reactive substances (TBARS) antioxidant assays. They concluded that potato proteins possess substantial reducing power that can be attributed to peptide cleavage, which results in a product that has a higher availability of hydrogen ions. These hydrogen ions are able to transfer and stabilize the free radicals to a greater extent from the smaller peptide compared with the larger protein. Amino acid composition could be strongly related to the protein's antioxidant activity, because the literature has shown that methionine, histidine, and lysine have been shown to inhibit lipid oxidation. These amino acids are present in potato proteins and may be credited for the improvement in oxidative stability of refrigerated cooked ground beef patties (Wang and Xiong, 2005).

Liu et al. (2003) studied the antioxidative activities of purified patatin. They found that patatin exhibited a dose-dependent response when its antiradical activity was examined with the 1,1-diphenyl-2-picrylhydrazyl antioxidant assay, using butylated hydroxytoluene (BHT) and reduced glutathione as controls. When the particle size was reduced to nanomolar, purified patatin exhibited a response similar to the known antioxidant BHT and had a better antioxidant response than reduced glutathione. The same study determined the ability of patatin to act on low-density lipoprotein (LDL) peroxidation using the TBARS assay, which is particularly important because LDL peroxidation is associated with the development of atherosclerosis in humans. The TBARS assay showed that increasing the concentration of patatin reduced the level of oxidized LDL; therefore, the researchers concluded that patatin possesses a dose-dependent protection response against human LDL oxidation (Liu et al., 2003).

Bioactive peptides are attained from parent protein molecules that have undergone hydrolysis (experimental details of the biogeneration process will be explored in Section 2.4). These peptides usually possess better functional properties than their parent biopolymers and help mediate the mechanisms in which their parent proteins are involved by displaying a regulatory-type response (Liyanage et al., 2008). Many studies have been performed to examine the benefits of soybean-derived peptides, including the reduction of cholesterolemia. Liyanage et al. (2008) compared the characteristics of soybean and potato peptides by examining the lipid metabolism of rats fed a cholesterol-free diet. The proteins that were compared were soybean, casein, and potato. Rats fed solely potato proteins seemed to exhibit the lowest weight gain; however, this may have been because of the hygroscopic ability of potato protein, which would lead to reduced quality of the diet. Their results showed a correlation between the intake of potato and soy peptides and serum cholesterol levels. Potato peptides demonstrated effects similar to soy peptides: a low level of non-HDL-C concentration, high fecal lipid excretion, and lower apoB messenger ribonucleic acid (mRNA) levels compared with the casein-fed group (Liyanage et al., 2008). Hepatic apoB mRNA is a hepatic gene that expresses the LDL receptor. When apoB mRNA is present in low levels, the liver is triggered to increase clearance of LDL cholesterol, which is associated with higher excretion levels in the feces and a reduction in serum cholesterol levels. Serum triglyceride concentration was lowest among rats fed a potato peptide diet, compared with both soy peptide and casein protein diets. The low concentration of serum triglycerides could have been attributed to the low ingestion of the potato peptide diet as well as the increase in total lipid fecal excretion. The study concluded that the potato peptide diet affected the serum cholesterol levels of the potato peptide-fed group based on different mechanisms according to the different lipoproteins and triglycerides (Liyanage et al., 2008).

Studies found that bioactive peptides can participate in a reduction in blood pressure (Liyanage et al., 2008). The bioactive peptides are able to act through the mechanism of angiotensin-converting enzyme I (ACE) inhibition. Mechanisms that raise blood pressure can occur through two processes. The first is catalyzed by converting angiotensin I to angiotensin II, which is a known vasoconstrictor that signals to the kidneys to retain salts and water, therefore increasing the extracellular fluid volume and leading to an increase in blood pressure. The second is by deactivating a vasodilator bradykinin. Angiotensin-converting enzyme inhibition reduces blood pressure by decreasing peripheral vascular resistance and stabilizes renal function (Mäkinen et al., 2008). Mäkinen et al. (2008) studied the effect of potato peptides and ACE inhibition. They found that hydrolysis of potato proteins resulted in increased ACE inhibition, and that autolysis of the proteins resulted in greater ACE inhibition (Mäkinen et al., 2008).

2.3 Functional Properties

The functional properties that proteins possess, including solubility, water-holding ability, gelation, foaming, and emulsification, are important to the food system (McClements et al., 2009). Potato proteins have been reported to possess foaming and emulsification properties (van Koningsveld et al., 2002, 2006). In industry, it is difficult to achieve these desirable characteristics at minimal cost. Potato proteins can be obtained from potato fruit juice (PFJ), which is commonly discarded from the potato starch industry (Ralet and Guéguen, 2001). When extracting the protein, it is important to use appropriate extraction methods to maintain the protein in its raw form, because once the protein is denatured, it will usually lose all functionality.

2.3.1 Solubility

Solubility is affected by many different components present in the solution, including the addition of thermal and acidic treatments, dialysis, metal salts, and organic solvents (van Koningsveld et al., 2001b). Solubility is a parameter that is easily measured in single-protein solutions; however, the measurements become more difficult in protein mixtures because different proteins have varying solubilities. As mentioned previously, potato proteins are made up of many proteins; therefore, it is difficult to assess their solubility. Techniques to circumvent this difficulty include relating protein solubility to the total insoluble protein (van Koningsveld et al., 2001b). In this study, van Koningsveld et al. (2001b) related the ability of the remaining precipitate to resolubilize in a given solution compared with the total protein present in the PFJ. The solubilities were tested in the presence of various acids such as sulfuric acid (H_2SO_4), hydrochloric acid (HCl), and citric and acetic acids. Of the many proteins contained in PFJ, most possess isoelectric points between pH 4.5 and 6.5 (van Koningsveld et al., 2001b).

In the presence of acid, regardless of type, potato proteins seemed to precipitate to a greater extent at pH 3 (van Koningsveld et al., 2001b). In terms of the resolubilization of the precipitates, weak acids (citric and acetic) seemed to have higher resolubility than stronger acids (hydrochloric and sulfuric), because weaker acids were able to resolubilize approximately 4% more total proteins in the PFJ than stronger acids (van Koningsveld et al., 2001b).

When purified patatin and PFJ were compared, the purified patatin seemed to exhibit almost total solubility at pH 4, precipitate at pH 5, and undergo complete resolubilization at pH 6. Conversely, PFJ proteins demonstrated maximum precipitation at a pH < 4, and precipitates collected at pH 5 were only partially resoluble at pH 7. The researchers concluded that purified patatin and purified potato proteins behave differently from PFJ because the proteins present in the PFJ do not behave in relation to their isoelectric point (van Koningsveld et al., 2001b). This trend was also shown by Waglay et al. (2014); proteins present in PFJ had low solubility at pH 3 with increasing resolubility outside that condition (Waglay et al., 2014).

van Koningsveld et al. (2001a) tested the solubility of potato protein in the presence of heat treatments. Results show that when PFJ was heated to above 40 °C, the proteins began to precipitate out of the solution. It was shown that 50% precipitation occurred at 60 °C and complete denaturation occurred at 70 °C. This was confirmed by Waglay et al. (2014), who showed that 60 °C resulted in 50% resolubility whereas at 70 °C there was <10% resolubility

(Waglay et al., 2014). Increasing the ionic strength had a slight influence on the effect of temperature. Ionic strength was varied on a control of $(NH_4)_2SO_4$ precipitated potato protein, which had similar characteristics as undenatured potato protein. When the effect of temperature was tested in the presence of high-ionic strength 200 mM NaCl, the denaturation pattern shifted slightly by 5 °C, whereas complete denaturation took place at 75 °C, as opposed to the 70 °C previously observed. On the contrary, decreasing the ionic strength seemed to have no influence on the solubility curves. van Koningsveld et al. (2001a) also studied the solubility of potato proteins separately and concluded that the protease inhibitors seemed to exhibit a slightly higher precipitation temperature compared with patatin. The protease inhibitors showed insolubility at temperatures of 50–60 °C whereas patatin began to be insoluble at temperatures above 40 °C (van Koningsveld et al., 2001a).

2.3.2 Emulsifying Properties

According to Smith and Culbertson (2000), an emulsion is defined as "mixtures of at least two immiscible liquids, one of which is dispersed in the other in the form of fine droplets." To stabilize an emulsion, it is often beneficial to add an amphiphilic surfactant such as protein. Proteins contain both hydrophobic and hydrophilic components owing to their amino acids; therefore, when added to the emulsion they will orient themselves at the interface of the appropriate molecule based on polarity. This orientation at the interface will help stabilize the emulsion by decreasing the interfacial tension between the two immiscible liquids (Smith and Culbertson, 2000).

Other factors often influence the stability of the emulsion once formed. According to van Koningsveld et al. (2006), common instabilities include creaming and droplet aggregation. Creaming is the result of the low-density droplets layering themselves above the higher-density aqueous phase. Droplet aggregation is when the droplets group together and increase the average size of each droplet; this is described as a colloidal attraction between droplets. Other possible reasons for droplet aggregation include bridging and depletion flocculation. Bridging flocculation involves a molecule that creates a bridge between the interface and the droplets, whereas depletion flocculation occurs in the presence of a high concentration of nonadsorbing polymers (van Koningsveld et al., 2006).

In the study conducted by van Koningsveld et al. (2006), different potato protein samples were observed to establish their emulsion-forming and stabilizing effects under various conditions. Potato fruit juice was collected from the cv. Elkana potato variety. Protein samples were prepared from PFJ with different extraction techniques to yield five unique protein samples: potato protein isolate, $(NH_4)_2SO_4$ precipitate, patatin, ethanol-precipitated patatin, and protease inhibitor pool (van Koningsveld et al., 2006). The emulsions were formed in a neutral environment, and it was observed that the emulsion containing patatin showed no droplet aggregation under microscopic conditions, whereas the other protein samples showed severe droplet aggregation, and therefore the emulsion was unstable.

van Koningsveld et al. (2006) determined the type of aggregation that was occurring with the protein samples. To decipher between the aggregation types, different techniques were used. For depletion aggregation, interactions between the droplets should be weak; therefore, with increasing shear rate the viscosity of the emulsion should decrease. When this test was conducted, the potato protein isolates and protease inhibitors exhibited higher viscosity at a low shear rate, and therefore depletion aggregation was concluded not to cause the instability of the emulsion. Furthermore, the addition of dithiothreitol completely broke apart the aggregates, demonstrating bridging aggregation between the disulfide bonds present in the protease inhibitors and dithiothreitol. The researchers concluded that this could be avoided by simply ensuring that there is no incorporation of air during the preliminary stages of emulsion formation (van Koningsveld et al., 2006).

A key parameter for a protein to maintain its functional characteristics is the pH environment to which the protein is subjected. When a protein is close to its isoelectric point, it possesses low solubility and therefore loses most of its emulsifying ability (Smith and Culbertson, 2000). Ralet and Guéguen (2000) studied the emulsion properties of different potato protein fractions over pH 4–8. The pH range was tested with the addition and absence of NaCl at 1%. They observed that in an acidic environment in the absence of NaCl, the patatin emulsion was slightly more stable compared with the emulsion in the presence of NaCl. Moreover, when the protease inhibitor fraction was examined, it did not have pH dependency (Ralet and Guéguen, 2000).

2.3.3 Foaming Properties

According to Phillips et al. (1990), foam is defined as "a two-phase system in which a distinct gas bubble phase is surrounded by a continuous liquid lamellar phase." Proteins are often incorporated into foams to help stabilize them. The molecular characteristics of the substance attributed to foaming are its solubility, the ability to form interfacial interactions, the ability to unfold at the interface, the ability to react with a gaseous or aqueous phase, the amount of charged and polar subunits that prevent foam formation, and steric effects (Ralet and Guéguen, 2001). When examining the foaming properties of a given protein, it is important to examine both the foam-forming ability and the stability of the foam formed. These parameters are evaluated by measuring the foam volume and the amount of liquid lost over time.

Extraction methods shown to maintain potato protein's functional characteristics in foam stabilization are ultrafiltration, CMC complexation, and anion-exchange chromatography (van Koningsveld et al., 2002).

A study completed by van Koningsveld et al. (2002) examined potato protein's ability to form and stabilize the foam. The foams were formed with two different methods known as the sparging and whipping techniques. The sparging technique is more desirable for highly structured proteins because it allows the rigid protein more time at the interface, which results in the release of bubbles owing to buoyancy forces (van Koningsveld et al., 2002). The whipping technique begins with beating, which causes velocity and pressure fluctuations. As a result, the bubbles interact with each other, causing both the interfacial area and the surface tension to change over time. The foam will form once the whipping speed is high enough to cause the surface tension to become too great, thereby allowing the bubbles to coalesce. The addition of various proteins allows the foam to have an optimum whipping time for formation. This is beneficial because a longer whipping time will have a positive correlation with foam volume; however, the disadvantage is that overwhipping will cause the proteins to denature to the point that they aggregate, which decreases foam volume (van Koningsveld et al., 2002).

Various conditions can occur that lead to decreased stabilization of the foam. A major concern is the presence of hydrophobic molecules, because these fat molecules coalesce, leading to spreading of the bubbles, or their particle size is larger than the foam bubbles, which enhances foam breakdown. Another cause of foam instability is drainage, which is the result of gravity forcing the liquid out of the foam. However, in the presence of proteins drainage is greatly reduced owing to the formation of a stagnant layer. An additional negative impact on foam stabilization is Ostwald ripening. Whereas the ideal foam is composed of many small bubbles, air is more soluble in a liquid phase, and therefore solubility is higher around smaller air bubbles. This leads to the continual growth of larger air bubbles, a process known as Ostwald ripening (van Koningsveld et al., 2002).

van Koningsveld et al. (2002) observed the whipping time at varying whipping speeds for different proteins such as β-lactoglobulin, β-casein, potato protein isolated using 15% ethanol, potato protein isolated using 20% ethanol, $(NH_4)_2SO_4$ potato protein isolates, patatin, patatin extracted using 20% ethanol precipitation, protease inhibitors, and protease inhibitors isolated using 20% ethanol precipitation. The results revealed that at a standard whipping time of 70 s the protease inhibitors showed no optimal whipping speed, whereas $(NH_4)_2SO_4$ potato protein isolates, patatin extracted with 20% ethanol, and protease inhibitors extracted with 20% ethanol all had an optimum whipping speed of 4000 revolutions per minute (rpm) and potato protein isolates extracted using 20% ethanol formed a foam at 3000 rpm. When patatin was examined, it was found that using the standard whipping time (70 s) no foam was formed; therefore, the whipping time was reduced to 30 s and the optimum speed was 3000 rpm. Overall, for most protein samples, as the whipping speeds increased and the whipping times increased, foam formation and foam volume became better, until an optimum level (van Koningsveld et al., 2002).

The foam that was formed from patatin at neutral pH was similar in appearance to the foam formed by lysozyme (egg white protein). Patatin is structurally rigid, and therefore, as predicted, shorter whipping times and speeds were required for the protein to unfold at the interface. Two conclusions can be drawn about patatin with foam formation by the whipping test: Either the rate of unfolding is extremely slow or the rate of refolding is extremely fast, which would explain why longer whipping times did not improve foam formation. When the sparging test was used, the results were similar to those of other proteins. As previously mentioned, sparging is the result of low surface expansion rates. The conclusion of this technique in foam formation was that patatin must exhibit slow unfolding rates, and therefore the foam could not stabilize against the effects of drainage, coalescence, and Ostwald forces (van Koningsveld et al., 2002).

In the study completed by Ralet and Guéguen (2001), raw potato proteins, a patatin fraction, and a 16- to 25-kDa fraction (protease inhibitors) were compared with standard Ovomousse M (commercial spray-dried hen egg white powder). A plastic column and a porous metal disk were used to form the foams. Air was blown through the metal disk into the column and two electrodes located in the columns measured the drainage of the foams once formed. All samples required the same whipping time to reach a volume of 35 ml at a neutral pH and with no NaCl solution added. Under these conditions, the patatin fraction, raw potato protein, and Ovomousse M exhibited the same characteristics over time because the foam structures were not significantly broken. When the three samples were compared, patatin seemed to represent the more stable foam over time whereas the 16- to 25-kDa fraction lost most of its foaming characteristics owing to drainage. When salt solutions were added to the protein samples, the samples seemed to form more stable foams more quickly and they became pH independent. The pH of the solution is important because protein's electrostatic charge is pH dependent. When a protein is in a solution close to its isoelectric point, it should stabilize foams to a greater extent because of the decrease in electrostatic repulsion. Patatin is an exception to this rule because, as observed, patatin is not as soluble at pH 4 (Ralet and Guéguen, 2001).

2.4 Potato Byproduct as a Source of Potato Proteins

Industrial uses of potatoes mostly involve extraction of the starch source, resulting in a waste product referred to as PFJ, at a rate of about 5–12 m^3 for 1 metric ton of potatoes (Ralet and Guéguen, 2000; Miedzianka et al., 2014). One issue that arises is the difficulty of disposing of this product because it possesses a high polluting capacity. As a result, when removing its protein components it is beneficial to discard the effluent as well as to obtain useful proteins that could have applications in the food industry (Ralet and Guéguen, 2001). Many studies have been performed to improve the removal of proteins from the starch industry byproduct because it contains the highest bioavailable oxygen demand compared with other industry byproducts (Strolle et al., 1973).

Potato fruit juice contains many desirable components that can be extracted, such as proteins, amino acids, organic acids, and potassium. Once the proteins are removed from the PFJ, the juice is subjected to ion-exchange chromatography, which separates the other components. However, protein concentration must be below 180 parts per million (ppm) to prevent the proteins from being separated in the column instead of the other minor components. Analysis of raw PFJ shows that the protein level is approximately 1500–4000 ppm

(Strolle et al., 1973) or approximately 25 g protein/kg PFJ (Pastuszewska et al., 2009). The protein extracted from PFJ has been studied to be comparable to that of soy and animal proteins, because similar amino acid content is attained (Pastuszewska et al., 2009). The soluble solids composition of the juice consists of 35% crude protein, 35% total sugar, 20% minerals, 4% organic acids, and 6% other (Strolle et al., 1973). Potato fruit juice contains approximately 1.5% (w/v) soluble proteins and most of the proteins present in PFJ are patatin and protease inhibitor proteins (van Koningsveld et al., 2006). The distribution of molecular weight of the proteins present in PFJ is shown in Figure 1. As shown, the juice is mostly composed of a concentrated band at approximately 43 kDa, which indicates patatin, a large cluster of bands around 15–25 kDa representative of protease inhibitors, and a wide distribution of high-molecular-weight proteins. The bands are further characterized in Table 3, where they are represented in proportions of 22.9%, 53.9%, and 23.7% for patatin, protease inhibitors, and high-molecular-weight proteins, respectively.

The most common method to extract proteins from the PFJ is precipitation with the addition of acidic and heat treatments. However, several improvements have been explored to improve the quality of the potato proteins, because heat and acidic treatments denature the proteins. These denatured proteins no longer possess their functional characteristics such as foaming, emulsification, and water- or oil-holding capacity, and therefore their use in food applications is diminished (van Koningsveld et al., 2006; Miedzianka et al., 2014).

The antinutritional or toxic properties present in the PFJ are solanidine glycoalkaloids and protease inhibitors, specifically trypsin inhibitors. A study conducted by Pastuszewska et al. (2009) found that for both solanidine glycoalkaloids and protease inhibitors, the amount varied greatly and the differences probably resulted from the variation in potatoes. However, both

Table 3: Protein Profiles of Potato Isolates Obtained Using Selected Extraction Techniques.[a]

Extraction Technique	Patatin (%)	Protease Inhibitors (%)			Other
	39–43 kDa	21–25 kDa	15–20 kDa	14 kDa	High-MW (%)
PFJ	22.9	24.6	18.4	10.4	23.7
Combination thermal/acidic[b]	37.8	0	20.2	31.3	10.7
Acidic[b]	9.9	8.8	13.6	15.5	41.4
FeCl$_3$[c]	35.7	0	25.5	38.8	0
Ethanol[d]	36.5	7.7	21.7	25.6	5.2
(NH$_4$)$_2$SO$_4$[e]	36.1	5.2	20.5	25.8	11.2
CMC[f]	34.2	10.8	2.6	21.6	24.1

[a]Adapted from Waglay et al. (2014).
[b]Thermal acidic and acidic precipitation was run at pH 4.8 and 2.5 using H$_2$SO$_4$, respectively.
[c]FeCl$_3$ at a concentration of 5 mM.
[d]Ethanol at a concentration of 20%.
[e](NH$_4$)$_2$SO$_4$ at a saturation of 60%.
[f]CMC-to-potato protein at a ratio of 0.1.

amounts were found in concentrations similar to those found in the tuber itself for solanidine glycoalkaloids, and protease inhibitors were present in similar concentrations as soybean meal (Pastuszewska et al., 2009). The study did not mention the beneficial effect of protease inhibitors, and therefore conclusions varied greatly from other studies.

2.5 Extraction Methods of Proteins

The primary challenge is to extract the proteins from waste materials or industrial byproducts without denaturing or affecting their functional characteristics. This section will focus on advantages and disadvantages of a variety of methods that have been studied for recovery of the proteins.

2.5.1 Combination: Thermal and Acidic Precipitation

Acid and heat coagulation, either combined or separately, has been used for many years to recover proteins (Knorr et al., 1977). Historically, these techniques were used to remove the protein content of the PFJ and therefore decrease the polluting capacity and economic cost. Nowadays, it is often undesirable to employ either technique because acid and heat coagulation lead to protein denaturation. When a protein structure is altered, there is often complete loss of its functional characteristics, which therefore impedes its ability to be incorporated into other food products.

Heat coagulation is the method commonly used in the potato starch industry. Temperatures in excess of 90 °C are often used, which has the major disadvantage of economic cost, and at these temperatures the potato proteins are rendered insoluble, which in turn limits their application in other sectors of the food and pharmaceutical industries (Knorr et al., 1977). Heat coagulation can be performed with various protocols. Strolle et al. (1973) studied steam injection heating by incorporating steam into the PFJ to increase the temperature of the juice to 104.4 °C. The juice was then flash-cooled, allowing the proteins to be collected. This study concluded that steam injection heating is ideal at temperatures above 99 °C with pH adjustments of 5.5. Furthermore, steam injection heating was found to be an efficient and simple method for removing proteins from PFJ. The technique was also found to be economically disadvantageous owing to energy consumption as well as the ion-exchange step required to remove the other components (organic acids, amino acids, etc.) present in the PFJ (Strolle et al., 1973).

As shown in Figure 2, Waglay et al. (2014) demonstrated that combined thermal and acidic methods resulted in high protein recovery yield but extremely low purification. Therefore, this process is not selective to potato proteins, but will coagulate other components present in PFJ with the proteins. Conversely, acidic precipitation resulted in lower yield with higher purification, indicative of the strong affinity of acidic precipitation alone on the potato proteins. On the other hand, as shown in Table 3, the protein profile for combined thermal and acidic methods had a high proportion of patatin, with a weak distribution of protease inhibitors. Indeed, the patatin extracted would have minimal applications because the thermal treatment resulted in complete denaturation, whereas acidic precipitation had a significant negative effect on the patatin extracted and a positive effect on the high-molecular-weight proteins being extracted (Table 3).

Acid coagulation commonly involves the use of HCl, phosphoric acid (H_3PO_4), or H_2SO_4 (Knorr et al., 1977). The same authors (Knorr et al., 1977) examined the use of acid and heat coagulation of potato protein from waste effluent. Results showed that ferric chloride ($FeCl_3$) demonstrated protein recovery similar to HCl at pH 2–4, whereas above this range (pH 5–6) HCl demonstrated better results. When the PFJ proteins are recovered with either acidic or heat coagulation they commonly have a dark appearance owing to the harsh environment, and a strong cooked flavor. Therefore, when proteins are extracted with these techniques they are often used as animal feed (Straetkvern et al., 1999).

2.5.2 Precipitation Methods
2.5.2.1 Salt Precipitation

The most commonly used salt in salt precipitation of potato proteins from PFJ is $FeCl_3$ (Knorr, 1981). These salts function as extraction techniques through ionic strength adjustment (Knorr, 1981). Knorr et al. (1977) found that $FeCl_3$ was able to coagulate potato proteins from PFJ just like acid and heat coagulation. In that study, protein precipitates varied in composition depending on the different extraction techniques, acidic (HCl), acidic/heat treatment, and $FeCl_3$ precipitation, whereas HCl alone seemed to extract more total solids compared with the others (Knorr et al., 1977). The same authors (Knorr et al., 1977) found that when heat was not used during the extraction process, the precipitates tended to have higher vitamin C content and the precipitate extracted with $FeCl_3$ seemed to contain more iron (Knorr et al., 1977). The study then compared the amount of protein denaturation caused by the different extraction techniques by measuring the solubility of nitrogen and relating that to the extent of denaturation, in which the greater the solubility of nitrogen was, the less denatured the protein was (Knorr et al., 1977) and hence the more functional characteristics the protein retained. Of the different techniques, $FeCl_3$ demonstrated the highest nitrogen solubility, followed by HCl, and finally HCl in the presence of heat (Knorr et al., 1977). Waglay et al. (2014) reported minimal concentration dependence on protein recovery yield; however, a strong correlation was seen with purification factor in which at a low concentration (5 mM), a purification factor of 6 resulted (Figure 2). Therefore, $FeCl_3$ at low concentrations has a strong affinity for potato proteins. Indeed, $FeCl_3$ extracted a high proportion of patatin, with a weak distribution of protease inhibitors, and no high-molecular-weight proteins (Table 3).

Ferric chloride precipitation is a simple technique to precipitate proteins out of solution, in which increasing $FeCl_3$ concentrations result in high recovery of the proteins. However, after extraction, this technique introduces a disadvantage in measuring the protein content with

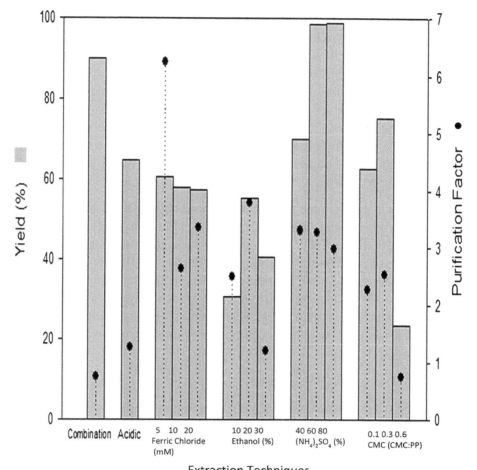

Figure 2

Effect of precipitating agents on recovery yield and purification factor of potato protein isolates: combination thermal/acidic, pH 4.8, 100 °C; acidic, pH 2.5; $FeCl_3$ (5, 10, and 20 mM); ethanol (10%, 20%, and 30%); $(NH_4)_2SO_4$ (40%, 60%, and 80% saturation); and CMC (0.1, 0.3, and 0.6 CMC: potato protein). *Adapted from Waglay et al. (2014).*

simple protein determination methods, because the $FeCl_3$-precipitated proteins require a chelating agent to become soluble. In addition, the high ferric ion content causes interference with the chelating agents and therefore interferes with protein determination using Bradford or BCA methods (Bártova and Bárta, 2009).

2.5.2.2 Ethanol Precipitation

Ethanol precipitation demonstrated results relatively similar to the $FeCl_3$ precipitation method. Bártova and Bárta (2009) compared the two precipitation techniques, ethanol and

$FeCl_3$ precipitation. Results showed that a maximum concentration of ethanol is needed to extract proteins optimally with a high recovery rate, whereas increasing $FeCl_3$ concentrations resulted in high recovery of the proteins. They found that both methods resulted in a precipitate with a lower proportion of patatin. They theorized that the underestimation is due to poor extractability in the SDS buffer, along with the poor thermostability of patatin compared with the more thermostable protease inhibitors that are also present. Compared with the work of Waglay et al. (2014), as observed in Figure 2, ethanol showed a peak concentration at 20% for both protein recovery yield and purification factor. Ethanol at 20% showed a high distribution of patatin, good distribution of protease inhibitors, and a low proportion of high-molecular-weight proteins (Table 3).

Both ethanol and $FeCl_3$ precipitated proteins that possess high nutritional quality, as proven by a high proportion of essential amino acids (Bártova and Bárta, 2009). When the precipitates were compared in their LAHA, the ethanol-precipitated proteins seemed to demonstrate a higher preservation of LAHA (Table 1). However, the low LAHA obtained upon $FeCl_3$ precipitation resulted from the poor resolubility of the precipitated proteins in the buffer (Bártova and Bárta, 2009). The researchers concluded that, compared with $FeCl_3$ precipitation, ethanol precipitation is easier and results in proteins with high nutritional value and retention of LAHA, which indicates the occurrence of minimal protein denaturation (Bártova and Bárta, 2009).

2.5.2.3 Ammonium Sulfate Precipitation

Ammonium sulfate ($[NH_4]_2SO_4$) precipitation is a technique commonly employed to extract proteins based on solubility differences. This technique is normally used to recover undenatured potato proteins. van Koningsveld et al. (2001a) investigated the recovery of potato proteins from acidified PFJ (pH 5.7) with $(NH_4)_2SO_4$ at a saturation of 60%. The $(NH_4)_2SO_4$ recovered protein isolates are often compared with undenatured potato proteins because of the large proportion of protein extracted (75%), possessing a wide variation in molecular weight. Ammonium sulfate precipitation also extracts a large proportion of patatin compared with other techniques (van Koningsveld et al., 2001a). Waglay et al. (2014) reported that patatin extracted exhibited the highest LAHA compared with other precipitating techniques (Waglay et al., 2014).

Figure 2 shows that at saturations of 60% and 80%, a high yield was recovered with good purification factors. Upon further investigation of the protein profile (Table 3), we can see that 60% $(NH_4)_2SO_4$ saturation resulted in a high proportion of patatin, good distribution of protease inhibitors, and a low proportion of high-molecular-weight proteins (Waglay et al., 2014).

van Koningsveld et al. (2001a) further assessed $(NH_4)_2SO_4$ protein isolates and found that solubility was influenced by ionic strength, just like PFJ. When high ionic strength was used, minimum solubility was observed at pH 3, whereas when low ionic strength was used, a broader soluble curve resulted with a minimum around pH 5.0 (van Koningsveld et al., 2001a). When examining the effect of temperature, high ionic strength affected the occurrence of the denaturation temperature, with a shift upward of 5 °C (van Koningsveld et al., 2001a).

2.5.2.4 Carboxymethyl Cellulose Complexation

Another extraction method uses a polysaccharide, CMC, as a precipitating agent. The addition of CMC to a protein solution allows the proteins to coagulate, promoting easier collection through simple centrifugation techniques. The precipitate formation is influenced by the pH of the environment, the ionic strength of the protein and polysaccharide, the net charges of the species, as well as the size, shape, and interaction between molecules (Vikelouda and Kiosseoglou, 2004). The use of a complexing agent such as CMC has been hypothesized as a potential solution to decrease the pollution effect of waste effluents from the potato industry. Vikelouda and Kiosseoglou (2004) studied the effect of CMC extraction on the recovery of the proteins and their functional properties. The precipitation of protein by CMC results from electrostatic interactions occurring between the oppositely charged protein and the carboxyl group of CMC. When present in an acidic environment (pH 2.5), the proteins will interact with the CMC, resulting in an overall decrease in net charge of the molecule. This leads to a change in the isoelectric point of the proteins, which results in a neutral precipitate (Vikelouda and Kiosseoglou, 2004). Adjustment of the pH to 2.5 has been shown to affect several functional characteristics of the proteins. However, the addition of polysaccharide has some benefits because it enhances several functional qualities such as foaming. The protein precipitate recovered upon complexation with CMC was found to possess adequate solubility and foaming properties. Foaming properties could be influenced by the presence of CMC, because the high-surface activity proteins will interact with the polysaccharide, which positively affects foaming ability and stability. In relation to the emulsifying properties, CMC adsorbs at the interface, and the resulting protein–CMC complex protrudes, creating steric hindrance (Vikelouda and Kiosseoglou, 2004).

Gonzalez et al. (1991) reported that CMC helps in the formation of a stable precipitate over a large range of pH, unlike protein solutions alone, which are insoluble at a pH of only about 3. They concluded that for CMC with a lower degree of substitution, pH 2.5 was ideal for protein recovery, and for CMC with a higher degree of substitution, pH 3.5–4.0 was optimal. The CMC-to-protein ratio was also reported to be an important parameter because too much CMC will cause the precipitate to form incorrectly, triggering the supernatant to have a cloudy appearance. Ratios of 0.05:1 and 0.1:1 (CMC to potato protein [w/w]) were found to be optimal amounts for varying degrees of substitution of the CMC (Gonzalez et al., 1991). Indeed, when CMC is added in amounts larger than required for precipitation, the coagulation occurs in two stages (Gonzalez et al., 1991). The first stage involves the formation of the precipitate through protein aggregation; the second involves solubilization of the aggregates.

The second stage occurs when the proteins align and form a stable complex with the CMC, which allows for smaller particle sizes and more active sites on the CMC that can be hydrated (Gonzalez et al., 1991).

Waglay et al. (2014) found that CMC resulted in average recovery yields and purification factors at ratios of 0.1 and 0.3 CMC:potato protein (Figure 2). The protein profile for 0.1 CMC:potato protein resulted in a high proportion of patatin and good distribution of protease inhibitors. However, owing to the pH adjustment (2.5) required for complexation to occur, a high proportion of high-molecular-weight proteins is extracted, which was seen similarly with acidic precipitation alone (Table 3).

Overall, the CMC complexation technique is effective and simple for use in protein recovery from PFJ. However, a disadvantage is the requisite acidic environment, which may lead to loss of the functional characteristics of proteins (Vikelouda and Kiosseoglou, 2004).

2.5.3 Ion Exchange

Løkra et al. (2008) examined the use of chromatographic techniques to separate the water-soluble proteins from PFJ. Contrary to the most popular technique of heat coagulation, which results in the loss of most functional properties of the proteins, adsorption chromatography has many advantages, including the ability to collect specific fractions of proteins, ease in removing other nonnutritional components, and the ability to remove any interfering components (low-molecular-weight components) (Løkra et al., 2008). The expanded bed adsorption technique has been found to be effective at removing impurities from the PFJ, such as fibers, minerals, and pigments (Løkra et al., 2008). Once the protein sample is isolated using expanded bed adsorption, it is important that the proteins undergo drying techniques to be applied in the food industry. The disadvantage to most drying techniques is that they use hot air, which affects the functional characteristics and adds a dark color to the proteins, limiting their application in food products. Therefore, gentle drying techniques are beneficial (Løkra et al., 2008). According to Claussen et al. (2007) the most effective drying method is atmospheric freeze-drying for potato protein concentrates.

Straetkvern et al. (1999) studied the extraction of bioactive patatin from PFJ using slight pH adjustments and mixed-mode, high-density adsorbent as the stationary phase. pH adjustments (to pH 2–3) were needed to elute the patatin from the adsorbent material. Advantages to using mixed-mode adsorbent over other resins include the ability to separate interfering brown products and the nonissue of ionic binding. This mixed-mode adsorbent is made of small glass beads that are coated with agarose substituted with low-molecular-weight affinity ligands. These ligands are composed of an inner core that is hydrophobic and made of aromatic, heteroaromatic, or aliphatic molecules, allowing a variety of hydrophilic or ionic molecules to attach. Among various mixed-mode adsorbents, mixed-mode ES and mixed-mode ExF have shown high binding and specificity for patatin. The difference of these resins is the concentration of ligands, in which ExF variety is substituted at low concentrations whereas ES is substituted at high concentrations (Straetkvern et al., 1999). The mixed-mode ES resin exhibited binding at pH 6.5–8.5, where the elution occurred at 2.3 min. The lower concentration of ligands in mixed-mode ExF was more selective for patatin from the PFJ compared with mixed-mode ES. The benefit of using mixed-mode ExF was the pH environment of 3.5–4.5, which is less acidic than the pH range required for elution with mixed-mode ES, thereby maintaining a higher degree of protein functionality. The resins did not differentiate between the patatin isoforms, which indicates that the elution was not based solely on charge but also on group affinity to the column material (Straetkvern et al., 1999).

A more recent study conducted by Zeng et al. (2013) studied the effect of continuous polymer and pore phase known as Amberlite XAD7HP. This membrane has a high surface area and an aliphatic surface, which allows for absorption of both polar and nonpolar compounds, depending on the system. They concluded that this method resulted in an extract primarily composed of a large proportion of protease inhibitors (Zeng et al., 2013).

Although chromatographic techniques showed promising results, they have two major disadvantages: the difficulty of applying them on an industrial scale and the need for high investment costs. Industrial application would be extremely difficult because the limits to this technique include the high-density resins and their loss of binding sites (Straetkvern et al., 1999), which decrease the efficiency of the column separation over time.

2.6 Separation of Bioactive Proteins and Peptides

Fractionation of potato proteins is often based on differences in size and charge, because the patatin fraction has a higher molecular weight and lower isoelectric point compared with the protease inhibitor fraction. Fractionation to separate the different isoforms present in the patatin fraction was based on the charge because each isoform possesses a different charge (Pots et al., 1998).

Ralet and Guéguen (2000, 2001) isolated the patatin and protease inhibitor fractions from PFJ using anion-exchange chromatography with a diethylaminoethanol (DEAE) covalently linked to Sepharose Fast Flow column (25 mM phosphate buffer, pH 8). The acidic fraction was recovered upon a linear gradient of NaCl at pH 8 whereas the unbound sample was further fractionated using SP-Sepharose Fast Flow (25 mM phosphate buffer, pH 8) and the basic fraction was recovered using a linear gradient of NaCl at pH 8 (Ralet and Guéguen, 2000).

Jørgensen et al. (2006) used a size exclusion column of Superdex 200 HR 10/30 to fractionate the tuber proteins (50 mM sodium phosphate buffer containing 40 mM NaCl at pH 7.0). The recovered protein fractions were further analyzed for their molecular masses using MALDI-TOF MS and their activities were assayed using various activity tests, including protease inhibition, α-mannosidase, LAH, peroxidase, and glyoxalases I and II (Jørgensen et al., 2006).

Racusen and Foote (1980) developed the most widely applied purification technique for the isolation of patatin. They compared the use of DEAE-cellulose with the concanavalin A–Sepharose. Concanavalin A is tetrameric metalloprotein for which sugar molecules that contain α-D-mannopyranosyl and α-D-glucopyranosyl have a strong affinity in the presence of C-3, C-4, and C-5 hydroxyl groups, which are required for concanavalin A (GE Healthcare, 2005). They concluded that the use of the DEAE-cellulose resulted in a more pure sample with no interfering proteins, which was confirmed by their results obtained from silver-stained SDS-polyacrylamide gel electrophoresis (PAGE). However, protein samples collected by concanavalin A–Sepharose resulted in the slight detection of contaminating proteins from SDS-PAGE (Racusen and Foote, 1980). Modification introduced by Bohac (1991) for the purification of potato proteins began with gel filtration using a Bio-Gel P-100 column with a mobile phase of phosphate buffer. The fractions were then subjected to anion-exchange chromatography, using DEAE-Sephacel medium with a mobile phase of linear gradient NaCl followed by a glycoprotein-specific column, concanavalin A–Sepharose column prepared with 25 mM, pH 7.0, and phosphate running buffer with 0.5 M NaCl. Like Racusen and Foote (1980), purification was established using SDS-PAGE and similar conclusions were drawn, in which the concanavalin A–Sepharose fraction still contained many contaminating bands when run on SDS-PAGE; however, affinity chromatography had a direct beneficial effect on specific activity (Bohac, 1991).

Patatin itself is the major storage protein found in tubers; for this reason, it has been studied extensively. When patatin is run on SDS-PAGE, patatin exists as one broad band. Differences between the isoforms are exposed when patatin is studied using isoelectric focusing, which separates proteins based on charge into many bands (Pots et al., 1999). Pots et al. (1999) initially separated the patatin group after initial separation on a concanavalin A–Sepharose column. Pots et al. (1999) and Bártova and Bárta (2009) suspended the protein precipitate in a 25-mM Tris–HCl buffer to pH 8.0. The isoelectric point of patatin is 4.9; therefore, at pH 8 patatin will possess a strong negative charge. This negative charge will interact with the positively charged column depending on the exact charges of the isoforms. Initial separation was performed by anionic chromatography, Pots et al. (1999) and Bártova and Bárta (2009) used columns Source Q and DEAE 52- Cellulose SERVACEL, respectively. Cellulose SERVACEL is a nonporous column consisting of well-separated charge groups that allows for minimal adsorption. As a result, the amount of sites for adsorption can be specified, which is desirable because mild conditions can be used during desorption (SERVACEL, 2007). After anionic chromatography Bártova and Bárta (2009) separated the patatin group on a concanavalin A–Sepharose column, which is opposite in order to that performed by Pots et al. (1999).

2.7 Potential Application of Potato Proteins

2.7.1 Biogeneration of Peptides through Enzymatic Hydrolysis of Potato Proteins

The use of enzymatic hydrolysis has been successful in biogenerating smaller peptides, which have a tendency to possess better functional properties than larger proteins (Moreno and

Cuadrado, 1993; Wang and Xiong, 2005). Taste, viscosity, whip ability, and emulsifying and foaming abilities are all improved when the large proteins undergo a specific set of hydrolysis steps to yield smaller peptides (Moreno and Cuadrado, 1993).

Major advantages in using proteases are that they are selective and specific. Proteases function under mild reaction conditions and are used to partially hydrolyze polypeptide chains. Proteases are classified according to their mechanisms as either endo-proteases or exo-peptidases. Endo-proteases are able to cleave peptide bonds within the protein chain. Conversely exo-peptidases cleave peptide bonds at either the C-terminal or N-terminal ends of the protein chain. To improve enzymatic hydrolysis, they are commonly used in combination to enhance the amount of terminal ends (Kamnerdpetch et al., 2007). It is important to understand the structural properties of the proteins for better control of their hydrolysis into appropriate peptides.

Wang and Xiong (2005) investigated the hydrolysis of potato protein hydrolysates into potato peptides using an endo-protease (Table 4). As expected, the longer the reaction times, the higher degree of hydrolysis was obtained (Kamnerdpetch et al., 2007; Wang and Xiong, 2005). As the degree of hydrolysis increased, the solubility of the resulted hydrolysates was enhanced. Because peptides exhibit more charged groups than proteins, there is enhancement of protein–water interaction as well as increased electrostatic repulsion among the peptide molecules (Wang and Xiong, 2005). Characterization of the peptides present in the substrates was completed by SDS-PAGE. Before hydrolysis, the substrates displayed bands at approximately 45 kDa (patatin) and 18–25 kDa (protease inhibitors). However, after 0.5 h of hydrolysis, the patatin band had vanished and the small-molecular-weight bands appeared, indicating the presence of smaller peptides after hydrolysis (Wang and Xiong, 2005).

Kamnerdpetch et al. (2007) also studied the enzymatic hydrolysis of potato pulp using selected proteases: two endo-proteases (Alcalase 2.41, 415 U/ml; Novo Pro-D, 400 U/ml), one exo-protease (Corolase LAP, 350 LAPU/g), and a combination of endo-protease and exo-peptidase (Flavourzyme, 1000 LAPU/g), as well as their combinations. Upon 26 h of hydrolysis, Flavourzyme (7%, w/w) demonstrated the greatest degree of hydrolysis, 22%, followed by Alcalase (7%, w/w), Novo Pro-D (7%, w/w), and Corolase (7%, w/w), with degrees of hydrolysis of 8%, 3%, and 2%, respectively. Because of the presence of both endo-protease and exo-peptidase in the Flavourzyme, the terminal ends should be enhanced for the exo-peptidase to act upon, yielding more cleavage of either carboxypeptidase (C-terminal) or aminopeptidase (N-terminal) bonds. When comparing the two endo-proteases, Alcalase, which is a serine alkaline protease with a higher specific activity and broader specificity, demonstrated a higher degree of hydrolysis. In all cases, the combination of proteases resulted in a higher degree of hydrolysis compared with the individual enzymes (Kamnerdpetch et al., 2007). The hydrolysate with the highest degree of hydrolysis (44%) was obtained with 2% (w/w) Alcalase and 5% (w/w) Flavourzyme (Kamnerdpetch et al., 2007).

Liyanage et al. (2008) found that potato peptides retrieved from potato pulp using alkaline commercial protease enzymes had a positive impact on lipid metabolism in rats. This enzymatic hydrolysis process resulted in peptides with molecular weights varying from 700 to 1840 Da; the main molecular weight present (90% of the total) was 850 Da. They concluded that enzymatic hydrolysis in this fashion is an economical process that can be easily scaled up into large industrial-scale processes (Liyanage et al., 2008).

A study conducted by Miedzianka et al. (2014) studied the improvement of potato protein isolates by peptide generation, using a commercial product Alcalase (endo-protease from *Bacillus licheniformis*, specific activity of 2.4 AU/g). They used two incubation time conditions of 2 and 4 h (Table 4). It was found that 2 h was sufficient time to improve the solubility of the protein concentrate and functionalities (Miedzianka et al., 2014).

More recently, a study conducted by Pęksa and Miedzianka (2014) investigated the impact of enzymatic hydrolysis on the amino acid composition of peptides obtained using two commercial proteases Alcalase, Flavourzyme, and their combination (Table 4). They found that a combination of both Alcalase and Flavourzyme was ideal, because Alcalase rendered more end terminals to improve the efficiency of Flavourzyme. In turn, more peptide was

Table 4: Literature on Enzymatic Hydrolysis of Potato Proteins.

Starting Material	Enzyme Type	Enzyme Name/ Combination	Incubation Time (h)	Degree of Hydrolysis	Authors
Potato protein concentrate[a]	Endo	Alcalase (1:100 enzyme: substrate)	0.5	0.72	Wang and Xiong (2005)
			1	1.9	
			6	2.3	
Potato pulp[b]	Endo	Alcalase	26	5	Kamnerdpetch et al. (2007)
		Novo Pro-D	26	2.5	
	Endo/Exo	Flavourzyme	26	20	
	Exo	Corolase	26	2	
	Mixtures	(2% Alc + 5% Fla)	26	40	
		(3% Alc + 4% Fla)	26	35	
		(2% Alc + 5% Cor)	26	15	
		(3% Alc + 4% Cor)	26	18	
		(2% NPD + 5% Fla)	26	40	
		(3% NPD + 4% Fla)	26	30	
		(2% NPD + 5% Cor)	26	15	
		(3% NPD + 4% Cor)	26	15	
Potato protein concentrate[c]	Endo	Alcalase	2	3.5	Miedzianka et al. (2014)
		Alcalase	4	7.7	
Potato protein isolate[d]	Endo	Alcalase (5000 ppm)	n.d.	60	Pęksa and Miedzianka (2014)
	Endo/Exo	Flavourzyme (5000 ppm)	n.d.	n.d.	
	Mixture	1 Alc: 1 Fla	n.d.	60	

[a]Potato protein concentrate was obtained from AVEBE B.A. (Veendam, The Netherlands).
[b]Potato pulp was obtained from AVEBE B.A. (Veendam, The Netherlands).
[c]Potato protein concentrate was obtained from a starch factory in Łomża, Poland.
[d]Potato fruit juice was provided by a starch factory in Niechlów, Poland; it then underwent combination thermal treatment (70–80 °C) for 10–20 min.

hydrolyzed into oligopeptides and free amino acids, decreasing the bitterness commonly associated with protein hydrolysates (Pęksa and Miedzianka, 2014). When Alcalase was used alone, there was a lower proportion of amino acids methionine and cysteine (Pęksa and Miedzianka, 2014). The authors concluded that this enzymatic modification to the heat-treated proteins could further take advantage of their use as a value-added ingredient (Pęksa and Miedzianka, 2014).

The hydrolysis of potato proteins to potato peptides is desirable owing to improvement of the functional characteristics. Despite this advantage, the process is limited because of the undesirable flavor profile that results. Research (Ney, 1979) has shown that potato proteins possess high hydrophobicity as a result of the increased presence of hydrophobic amino acids. This hydrophobic character is undesirable because it is linked to a bitter taste that results from the peptides, making incorporation of potato peptides into food products difficult. As a result, the optimal degree of hydrolysis is often limited by the degree of bitterness of these peptides. A study performed by Ney (1979) found that peptides above 6000 Da seemed not to possess the bitterness that smaller peptides had. Addition of gelatin to the potato protein substrate decreased the occurrence of peptides smaller than 6000 Da. The ideal ratio to decrease the development of bitter peptides was determined to be four parts potato protein to one part gelatin (Ney, 1979).

3. Conclusion

Potatoes are a staple food in many countries and provide various nutritional benefits such as carbohydrates, proteins, vitamins, and minerals. Owing to high consumption, potato proteins are of particular interest, because nutritionally they are superior to animal protein lysozyme. The main fractions of proteins present in the tuber are patatin, protease inhibitors, and other high-molecular-weight proteins. As shown, different extraction techniques have been explored to retain functionality and associated health benefits, and in turn improve their potential application.

References

Bártova, V., Bárta, J., 2009. Chemical composition and nutritional value of protein concentrates isolated from potato (*Solanum tuberosum* L.) fruit juice by precipitation with ethanol and ferric chloride. Journal of Agricultural and Food Chemistry 9028–9034.

Blanco-Aparicio, C., Molina, M.A., Fernández-Salas, E., Frazier, M.L., Mas, J.M., Avilés, F.X., et al., 1998. Potato carboxypeptidase inhibitor, a T-knot protein, is an epidermal growth factor antagonist that inhibits tumor cell growth. The Journal of Biological Chemistry 12370–12377.

Bohac, J.R., 1991. A modified method to purify patatin from potato tubers. Journal of Agricultural and Food Chemistry 1411–1415.

Cheng, Y., Xiong, Y., Chen, J., 2010. Antioxidant and emulsifying properties of potato protein hydrolysate in soybean oil-in-water emulsions. Food Chemistry 101–108.

Claussen, I.C., Strømmen, I., Egelandsdal, B., Straetkvern, K.O., 2007. Effects of drying methods on functionality of a native potato protein concentrate. Drying Technology 1101–1108.

Fu, T.J., Abbott, U.R., Hatzos, C., 2002. Digestibility of food allergens and nonallergenic proteins in simulated gastric fluid and simulated intestinal fluid-a comparative study. Journal of Agricultural and Food Chemistry 7154–7160.

GE Healthcare, 2005. GE Healthcare. Retrieved November 22, 2010, from ConA Sepharose 4B: http://www.gelifesciences.com/aptrix/upp00919.nsf/Content/ F766BC24E96220C7C1257628001CFE2A/$file/71707700AF.pdf.

Gonzalez, J.M., Lindamood, J.B., Desai, N., 1991. Recovery of protein from potato plant waste effluents by complexation with carboxymethylcellulose. Food Hydrocolloids 355–363.

He, T., Spelbrink, R.E., Witterman, B.J., Giuseppin, M.L., 2013. Digestion kinetics of potato protein isolates in vitro and in vivio. International Journal of Food Sciences and Nutrition 787–793.

Höfgen, R., Willmitzer, L., 1990. Biochemical and genetic analysis of different patatin isoforms expressed in various organs of potato (*Solanum tuberosum*). Plant Science 221–230.

Jørgensen, M., Bauw, G., Welinder, K.G., 2006. Molecular properties and activities of tuber proteins from starch potato cv. Kuras. Journal of Agricultural and Food Chemistry 9389–9397.

Kamnerdpetch, C., Weiss, M., Kasper, C., Scheper, T., 2007. An improvement of potato pulp protein hydrolyzation process by the combination of protease enzyme systems. Enzyme and Microbial Technology 508–514.

Kärenlampi, S.O., White, P.J., 2009. Potato proteins, lipids, and minerals. In: Singh, J., Kaur, L. (Eds.), Advances in Potato Chemistry and Technology. Elsevier, Burlington, Maine, pp. 99–108.

Kim, J.-Y., Park, S.-C., Kim, M.-H., Lim, H.-T., Park, Y., Hahm, K.-S., 2005. Antimicrobial activity studies on a trypsin-chymotrypsin protease inhibitors obtained from potato. Biochemical and Biophysical Research Communications 921–927.

Knorr, D., 1981. Effects of recovery method on the functionality of protein concentrates from food processing wastes. Journal of Food Process Engineering 215–230.

Knorr, D., Kohler, G.O., Betschart, A.A., 1977. Potato protein concentrates: the influence of various methods of recovery upon yield, compositional and functional characteristics. Journal of Food Processing and Preservation 1, 235–247.

Lehesranta, S.J., Davies, H.V., Shepherd, L.T., Nunan, N., McNicol, J.W., Auriola, S., et al., 2005. Comparison of tuber proteomes of potato (*Solanum* sp.) varieties, landraces, and genetically modified lines. Plant Physiology 138, 1690–1699.

Liu, Y.-W., Han, C.-H., Lee, M.-H., Hsu, F.-L., Hou, W.-C., 2003. Patatin, the tuber storage protein of potato (*Solanum tuberosum* L.), exhibits antioxidant activity in vitro. Journal of Agricultural and Food Chemistry 4389–4393.

Liyanage, R., Han, K.-H., Watanabe, S., Shimada, K.-i., Sekikawa, M., Ohba, K., et al., 2008. Potato and soy peptide diets modulate lipid metabolism in rats. Bioscience, Biotechnology, and Biochemistry 943–950.

Løkra, S., Helland, M.H., Claussen, I.C., Straetkvern, K.O., Egelandsdal, B., 2008. Chemical characterization and functional properties of a potato protein concentrate prepared by large-scale expanded bed adsorption chromatography. Swiss Society of Food Science and Technology 1089–1099.

Mäkinen, S., Kelloniemi, J., Pihlanto, A., Mäkinen, K., Korhonen, H., Hopia, A., et al., 2008. Inhibition of angiotensin converting enzyme I caused by autolysis of potato proteins by enzymatic activities confined to different parts of the potato tuber. Journal of Agricultural and Food Chemistry 9875–9883.

McClements, D.J., Decker, E.A., Park, Y., Weiss, J., 2009. Structural design principles for delivery of bioactive components in nutraceuticals and functional foods. Critical Reviews in Food Science and Nutrition 577–606.

MedicineNet, August 7, 2004. Definition of Isoform. Retrieved November 17, 2010, from MedicineNet.com: http://www.medterms.com/script/main/art.asp?articlekey=38302.

Miedzianka, J., Pęksa, A., Aniolowska, M., 2011. Properties of acetylated potato protein preparations. Food Chemistry. http://dx.doi.org/10.1016/j.foodchem.2011.08.080.

Miedzianka, J., Pęksa, A., Pokora, M., Rytel, E., Tajner-Czopek, A., Kita, A., 2014. Improving the properties of fodder potato protein concentrate by enzymatic hydrolysis. Food Chemistry 512–518.

Mignery, G., Pikaard, C., Hannapel, D., Park, W., 1984. Isolation and sequence analysis of cDNAs for the major potato tuber protein, patatin. Nucleic Acid Research 7987–8000.

Moreno, M.M., Cuadrado, F.V., 1993. Enzymic hydrolysis of vegetable proteins: mechanism and kinetics. Process Biochemistry 481–490.

Ney, K.H., 1979. Taste of potato protein and its derivatives. Journal of the American Oil Chemists Society 295–297.

Park, W.D., 1983. Tuber proteins of potato—a new and surprising molecular system. Plant Molecular Biology Reporter 61–66.

Pastuszewska, B., Tusnio, A., Taciak, M., Mazurczyk, W., 2009. Variability in the composition of potato protein concentrate produced in different starch factories—a preliminary survey. Animal Feed Science and Technology 260–264.

Pęksa, A., Miedzianka, J., 2014. Amino acid composition of enzymatically hydrolysed potato protein preparations. Czech Journal of Food Science 32, 265–272.

Phillips, L.G., German, J.B., O'Neill, T.E., Foregeding, E.A., Harwalkar, V.R., Kilara, A., et al., 1990. Standardized procedure for measuring foaming properties of three proteins, a collaborative study. Journal of Food Science 1441–1453.

Pihlanto, A., Akkanen, S., Korhonen, H.J., 2008. ACE-inhibitory and antioxidant properties of potato (*Solanum tuberosum*). Food Chemistry 104–112.

Pots, A.M., De Jongh, H.H., Gruppen, H., Hamer, R.J., Voragen, A.G., 1998. Heat-induced conformational changes of patatin, the major potato tuber protein. European Journal of Biochemistry 66–72.

Pots, A.M., Gruppen, H., Hessing, M., van Boekel, M.A., Voragen, A.G., 1999. Isolation and characterization of patatin isoforms. Journal of Agricultural and Food Chemistry 4587–4592.

Pouvreau, L., Gruppen, H., Piersma, S.R., van den Broek, L.A., van Koningsveld, G.A., Voragen, A.G., 2001. Relative abundance and inhibitory distribution of protease inhibitors in potato juice from cv. Elkana. Journal of Agricultural and Food Chemistry 2864–2874.

Racusen, D., Foote, M., 1980. A major soluble glycoprotein of potato tubers. Journal of Food Biochemistry 4, 43–52.

Ralet, M.-C., Guéguen, J., 2000. Fractionation of potato proteins: solubility, thermal coagulation and emulsifying properties. Lebensmittle-Wissenschaft Und Technologie 380–387.

Ralet, M.-C., Guéguen, J., 2001. Foaming properties of potato raw proteins and isolated fractions. Lebensmittle-Weissenschaft Und Technologie 266–269.

Rydel, T.J., Williams, J.M., Krieger, E., Moshiri, F., Stallings, W.C., Brown, S.M., et al., 2003. The crystal structure, mutagenesis, and activity studies reveal that patatin is a lipid acyl hydrolase with a Ser-Asp catalytic Dyad. Biochemistry 42, 6696–6708.

Schoenbeck, I., Graf, A., Leuthold, M., Pastor, A., Beutel, S., Scheper, T., 2013. Purification of high value proteins from particle containing potato fruit juice via direct capture membrane adsorption chromatography. Journal of Biotechnology 693–700.

SERVACEL, 2007. Serva Electrophoresis. Retrieved November 22, 2010, from: SERVACEL(R) Cellulose Ion Exchangers: http://www.serva.de/enDE/Catalog/88_Ion_Exchange_Media_SERVACEL_reg_Cellulose_Ion_Exchangers.html.

Smith, D.M., Culbertson, J.D., 2000. Proteins: functional properties. In: Christen, G.L., Smith, J.S. (Eds.), Food Chemistry: Principles and Applications. Science Technology System, West Sacramento, pp. 132–147.

Sonnewald, U., Sturm, A., Chrispeels, M.J., Willmitzer, L., 1989. Targeting and glycosylation of patatin the major potato tuber protein in leaves of transgenic tobacco. Planta 171–180.

Straetkvern, K.O., Schwarz, J.G., Wiesenborn, D.P., Zafirakos, E., Lihme, A., 1999. Expanded bed adsorption for recover of patatin from crude potato juice. Bioseparation 333–345.

Strolle, E.O., Cording, J.J., Aceto, N.C., 1973. Recovering potato proteins coagulated by steam injection heating. Journal of Agricultural and Food Chemistry 974–977.

van Koningsveld, G.A., Gruppen, H., de Jongh, H.H., Wijngaards, G., van Boekel, M.A., Walstra, P., et al., 2001a. Effects of pH and heat treatments on the structure and solubility of potato proteins in different preparations. Journal of Agricultural Food Chemistry 4889–4897.

van Koningsveld, G.A., Gruppen, H., de Jongh, H.H., Wijngaards, G., van Boekel, M.A., Walstra, P., et al., 2001b. The solubility of potato proteins from industrial potato fruit juice as influenced by pH and various additives. Journal of the Science of Food and Agriculture 134–142.

van Koningsveld, G.A., Walstra, P., Gruppen, H., Wijngaards, G., van Boekel, M.A., Voragen, A.G., 2002. Formation and stability of foam made with various potato protein preparations. Journal of Agricultural and Food Chemistry 7651–7659.

van Koningsveld, G.A., Walstra, P., Voragen, A.G., Kuijpers, I.J., van Boekel, M.A., Gruppen, H., 2006. Effects of protein composition and enzymatic activity on formation and properties of potato protein stabilized emulsions. Journal of Agricultural and Food Chemistry 6419–6427.

Vikelouda, M., Kiosseoglou, V., 2004. The use of carboxymthylcellulose to recover potato proteins and control their functional properties. Food Hydrocolloids 21–27.

Vreugdenhil, D., Bradshaw, J., Gebhardt, C., Govers, F., MacKerron, D.K., Taylor, M.A., et al., 2007. Potato Biology and Biotechnology Advances and Perspectives. Elsevier, Oxford.

Waglay, A., Karboune, S., Alli, I., 2014. Potato protein isolates: recovery and characterization of their properties. Food Chemistry 373–382.

Waglay, A., Karboune, S., Khodadadi, M., 2015. Investigation and optimization of a novel enzymatic approach for the isolation of protein from potato pulp. LWT – Food Science and Technology. http://dx.doi.org/10.1016/j.lwt.2015.07.070.

Wang, L.L., Xiong, Y.L., 2005. Inhibition of lipid oxidation in cooked beef patties by hydrolyzed potato protein is reducing and radical scavenging ability. Journal of Agricultural and Food Chemistry 9186–9192.

Weeda, S.M., Kuman, G.M., Knowles, N.R., 2009. Developmentally linked changed in proteases and protease inhibitors suggest a role for potato multicystatin in regulating protein content in potato tubers. Planta 73–84.

Zeng, F.-K., Liu, H., Ma, P.-J., Liu, G., 2013. Recovery of native protein from potato root water by expanded bed adsorption with amberlite XAD7HP. Biotechnology and Bioprocess Engineering 18, 981–988.

CHAPTER 5

Potato Lipids

Mohamed Fawzy Ramadan

Agricultural Biochemistry Department, Faculty of Agriculture, Zagazig University, Zagazig, Egypt

1. Introduction

1.1 Chemical Composition of Potato Tuber

The potato (*Solanum tuberosum*) is a herbaceous annual that produces a tuber, also called a potato. The potato is the fourth most important crop in the world, exceeded only by wheat, rice, and maize, with an annual production approaching 300 million tons (Arvanitoyannis et al., 2008). Potatoes have expanded all over the world and have become the third most widely consumed plant product by humans, after wheat and rice (Kostyn et al., 2013). Commercial potatoes are derived from the inbreeding and selection of wild potatoes domesticated in the Andes mountains of South America (Kubo and Fukuhara, 1996; Fridman, 2006). Wild potatoes (*Papas criolas*) are still consumed by the indigenous population of South America.

The potato tuber, the most important part of the plant, is an excellent source of carbohydrates, proteins, and vitamins. The potato is one of the staples of the human diet and is an important raw material in the starch industry as well. Potato tubers are valued for their high starch content (up to 30.4% of fresh weight (FW)) and digestibility (Jansen et al., 2001). The proteins (up to 2%) are valuable for their amino acids. Potato tubers' biological value may be between 90 and 100, compared with eggs (100), soybeans (84), and beans (73). Potatoes also are a rich source of vitamins and minerals such as vitamin C (0.20 mg/g FW), vitamin B_6 (2.5 µg/g FW), potassium (5.64 mg/g FW), phosphorus (0.30–0.60 mg/g FW), and calcium (0.06–0.18 mg/g FW) (Ridgman, 1989; Buckenhu¨skes, 2005). Reports have mentioned that potatoes may contribute significantly to antioxidant dietary intake and are likely to provide health benefits (Robert et al., 2006). The health-oriented character of the potato is a consequence of its high content of antioxidants (high level of vitamin C) and the presence of antioxidant phenolic compounds such as phenolic acids and flavonoids (Kostyn et al., 2013). Much of the world's production consists of yellow-fleshed potatoes, which have higher total carotenoids than white-fleshed cultivars (Brown, 2005). White- and yellow-fleshed potatoes have a composition similar to carotenoids; however, the yellow color of the latter group is due to higher levels of xanthophylls (Gross, 1991; Brown et al., 1993). Dietary antioxidants include tocopherols, which are believed to have a key role in the body's defense system against reactive oxygen species, which are known to be involved in the pathogenesis of aging and many degenerative diseases such as cardiovascular disease and cancer. Moderate levels of lipophilic tocopherols have been reported in the staple potato (Brown, 2005; Ramadan and Elsanhoty, 2012).

Total lipids (TL) represent approximately 0.1–0.5% of FW of a potato tuber. Lipids in potatoes consist mainly of phospholipids (PL) (47%), glycolipids (GL) (22%), and neutral lipids (NL) (21%). Greater than 94% of tuber lipids are in forms containing esterified fatty acids. Galliard (1973) reported that tuber lipids were found to consist of 47.4% PL, 21.6% galactolipid, 6.4% esterified steryl glucoside (ESG), 1.3% sulfolipid, 2.4% cerebroside (CER), and 15.4% triglyceride. Major lipids and a portion of the triacylglycerol (TAG) are associated with the tuber membranes. Although lipid bodies can occasionally be found within the cytoplasm, it is unlikely that tubers contain appreciable amounts of lipid reserves. Therefore, the fatty acid composition of potato tubers primarily reflects the composition of cell membranes (Spychalla and Desborough, 1990). The composition of the fatty acids of TL isolated from potato tubers is nutritionally advantageous because the essential part of all fatty acids is formed by unsaturated fatty acids with one to three double bonds, mainly linolenic acid (40–50%) (Kolbe et al., 1996; Trevini et al., 1983). However, the small fat content in potatoes in general means that the daily intake of this valuable fat from potatoes is minimal; the amount of linoleic and linolenic acids in 100 g of tubers is 32.13 and 22.75 mg, respectively (Scherz and Senser, 1994; Prescha et al., 2001).

2. Lipids of Potato

Lipids are only a tiny fraction of potato FW, amounting to approximately 0.15 g/150 g FW, less than cooked rice (1.95 g) or pasta (0.5 g) (Priestley, 2006). Lipids, phosphates, and low-molecular-weight proteins are present in the interior of starch granules at relatively low levels. Their presence and interaction within starch granules strongly influence starch behavior toward gelatinization, retrogradation, swelling, viscosity, and leaching of soluble carbohydrates during different technological processing (i.e., cooking, baking, and extrusion) (Kitahara et al., 1997; Lin and Czuchajowska, 1998; Blaszczak et al., 2003).

Lipids were extracted from potato starch (PSt) granules using n-propanol–water (3:1, v/v) employing cold and hot extraction into surface and internal lipid fractions. The PSt lipids (0.53 g/100 g) were characterized as having a high level of surface lipids (0.32 g/100 g) whereas internal lipids accounted for 0.21 g/100 g (Blaszczak et al., 2003). Dhital et al. (2011) separated PSt granules into very small, small, medium, large, and very large fractions, in which the TL amount was higher in smaller PSt granules. Fatty acids of PSt were found in the following order: C16:0 > C18:2 > C18:3 > C18:1. The fatty acid profile from the cold extract of PSt indicated that it contained a high level of saturated fatty acids (42.1%). When the hot solvent mixture was employed to extract lipids from PSt, the fatty acid percentages for C16:O (49.9%), C18:1 (13.6%), and C18:O (7.35%) increased. Vasanthan and Hoover (1992) also observed an increase in the amount of saturated fatty acids from PSt after hot extraction. C20:1 (*n*-9) could be extracted from PSt granules only by hot extraction. Chromatographic profiles of extracted lipids also indicated that fatty acids such as C16:1 (1.05%) and C20:1 (0.27%) could be removed from PSt granules only by employing the hot solvent (Blaszczak et al., 2003).

2.1 Effect of Processing on Potato Lipids

Potato contains low levels of flavor volatiles, and thermal processing influences flavor generation. Precursors of volatiles include sugars and amino acids that are notably involved in the production of pyrazines (which are responsible for an earthy potato-like flavor) formed by Maillard reactions. Alkyl furans are derived by oxidation of unsaturated fatty acids, which are present in the potato. Fatty acid-derived volatiles are considered to contribute to unpleasant notes in potato; for example, hexanal and 2-pentylfuran contribute green flavors, whereas 2,4-decadienal has a fatty character (Dobson et al., 2004).

3. Lipids of Transgenic Potato Cultivars

3.1 Total Lipids and Lipid Classes

The potato is an important target crop for biotechnological applications and a valuable model system for studying signaling processes. Potatoes are susceptible to many pathogens and pests. Because of its tetraploid nature, it is arduous to improve the potato against these diseases through breeding. On the other hand, levels of TL in potato are small (up to 0.5%). Because of this, species of potato that would store more lipids would be important. Thus, improvement of the potato has actively involved genetic engineering. Their manipulation is mainly concerned with carbohydrate metabolism, which provides a carbon skeleton for the synthesis of amino acids and other organic compounds.

Engineering potatoes for resistance to viruses started in 1990 in the United States with the potato cultivar Russet Burbank. The potato was genetically transformed with coat protein genes from both potato virus X and potato virus Y. Resistance against the Colorado potato beetle was introduced in 1993 by expression of the insecticidal *Cry* protein gene from *Bacillus thuringiensis* (Prescha et al., 2001). According to Banerjee et al. (2006), efficient transformation is critical for rapid genetic analyses. Furthermore, there are numerous reports of *Agrobacterium*-mediated transformation with several commercial potato cultivars and species, including protocols for Bintje, Desiree, Russet Burbank, and Superior (Wenzler et al., 1989; Arvanitoyannis et al., 2008). Potato Spunta lines were developed with resistance to tuber moth with a *Cry V* gene. The genetic modification of *Bt* potato Spunta G2 and G3 consists of incorporating *Cry V* gene derived from *B. thuringiensis* strain. Potatoes contain the *Cry V* gene under expression of the 35S CaMV promoter. They also contain a selectable marker gene NPTII kanamycin resistance under expression of the nopaline synthase promoter (El-Sanhoty et al., 2004). El-Sanhoty et al. (2006)

evaluated the safety of genetically modified potato (GMP) Spunta (with *Cry V* gene) lines compared with that of nongenetically modified potato (NGMP) Spunta in rats. No significant differences were found in food intake, daily body weight gain, feed efficiency, and serum biochemical parameters between rat groups.

Prescha et al. (2001) showed that the control potato contains 0.52% TL, in which a 69% increase in TL was observed in tubers of transgen J2. The tubers of the control plants contained a small amount of nonpolar lipids (0.13%), whereas the transgen demonstrated almost a triple increase in nonpolar lipid. In the control potato tuber, polar lipids comprised 0.32% and the increase in this fraction in modified potato was slight (25%). Comparison of the percentage of polar and nonpolar lipids in TL from potato revealed that significantly different ratios of these fractions were observed in J2 potato compared with the wild-type potato. In the control potato, the NL comprised 25.6%, and the polar lipids 61.2% of TL. In J2 potato tubers the fraction of nonpolar lipids was >50% higher, whereas the polar lipid fraction was 20% lower than for the control. These results indicate that the increase in TL content in modified potato was mainly caused by synthesis of greater levels of NL in the tubers.

Genetically modified potato genotypes with overexpressed or underexpressed P14-3-3a (29G) and P14-3-3c (20R) isoforms of 14-3-3 protein were field-trialed (Prescha et al., 2003). The content of protein, starch, sugars, and lipids was determined in the GMP and control tubers. An increase was recorded in TL of potatoes with overexpression of 14-3-3 protein from *Cucurbita pepo* (Arvanitoyannis et al., 2008). Analysis of plants suggested that the function of the isolated 14-3-3 isoform is in the control of carbohydrate and lipid metabolism. Overexpression of the 14-3-3 protein does not affect protein synthesis and vitamin C content, but it does affect lipid amount. Genetically modified potato tubers showed changes in lipid amount and composition. Genetically modified potato tubers contained 69% more TL compared with the wild-type potato. Separation of lipids into polar and nonpolar fractions revealed that the GMP contained almost three times more nonpolar lipids than the control. Fatty acid profile showed that linoleic acid was the main fatty acid of both GMP and NGMP. In the nonpolar fraction of lipids from GMP, the unsaturated fatty acids were found in higher levels (Prescha et al., 2001, 2003).

Ramadan and Elsanhoty (2012) compared fatty acids, sterols, tocopherols distribution, and lipid classes (NL, GL, and PL) as well as unsaponifiable levels in GMP Spunta G2, G3, and NGMP. The NGMP and GMP lines G2 and G3 contained 0.59%, 0.75%, and 0.72% of TL, respectively. Among the lipids present in NGMP (Figure 1), the level of PL was the highest (53%), followed by NL (24%) and GL (23%), respectively. In GMP G2 and G3 lines, levels of NL increased to 40% and 39% of TL, respectively, whereas the levels of total polar lipids (PL and GL) decreased. A significant decrease in PL levels was measured in GMP G2 and G3 (from 53% in NGMP to 40% in GMP), whereas levels of GL decreased from 23% in NGMP to about 20% in GMP.

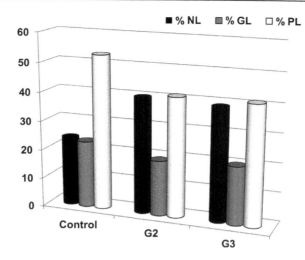

Figure 1
Percentages of nonpolar and polar lipids in GMP and NGMP Spunta.

Table 1: Neutral Lipid Classes (g/100 g of TL) in GMP and NGMP Spunta.

Lipid Class	NGMP	GMP G2	GMP G3
Monoacylglycerols	0.96	1.52	1.32
Diacylglycerols	1.32	2.00	2.18
Free fatty acids	0.91	1.20	1.24
Triacylglycerols	20.2	33.9	33.1
Sterol esters	0.55	1.32	1.05

The proportion of NL classes in GMP and NGMP is shown in Table 1. Total NL in NGMP accounted for 240 g/kg, whereas total NL was measured in higher amounts in GMP G2 and G3 (400 and 390 g/kg, respectively). Classes of NL in GMP and NGMP contained TAG, diacylglycerol (DAG), monoacylglycerol (MAG), free fatty acids (FFA), and esterified sterols (STE), in decreasing order. A significant amount of TAG was found (about 85% of total NL), followed by low levels of DAG (5–5.6% of total NL), whereas FFA, MAG, and STE were recovered in lower amounts (Ramadan and Elsanhoty, 2012).

Classes of GL (Figure 2) present in GMP and NGMP Spunta were sulfoquinovosyldiacylglycerol (SQD), digalactosyldiacylglycerol (DGD), CER, steryl glucoside (SG), monogalactosyldiacylglycerol (MGD), and ESG (Figure 3). Total GL was measured in the highest amounts in NGMP (230 g/kg), followed by GMP G3 (200 g/kg) and GMP G2 (190 g/kg), respectively. Digalactosyldiacylglycerol, MGD, ESG, and SQD were the main components and made up about 90% of the total GL. Cerebrosides and SG were measured in lower

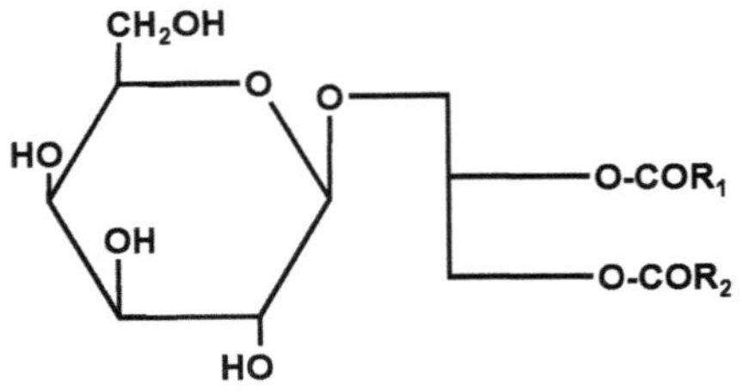

Monogalactosyldiacylglycerol (MGD)

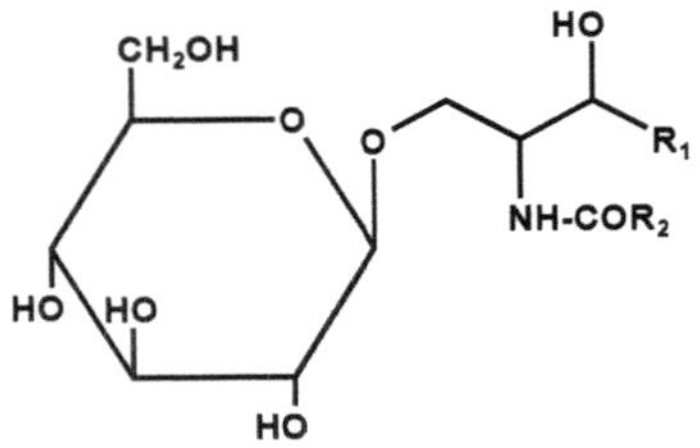

Cerebroside (CER)

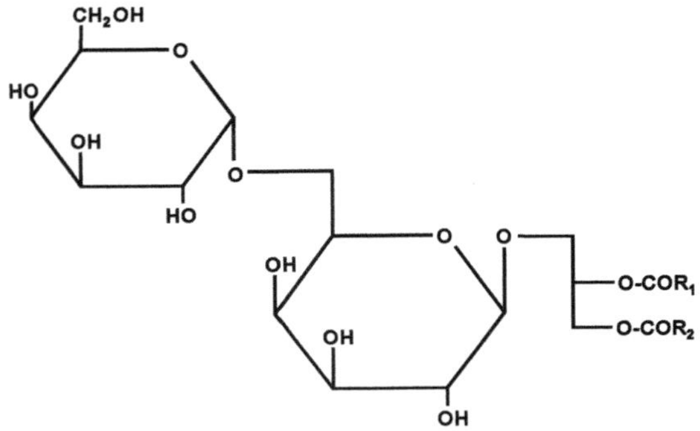

Digalactosyldiacylglycerol (DGD)

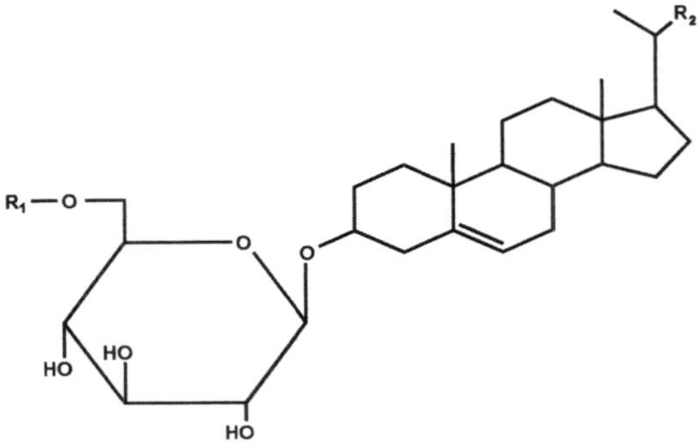

R_1= H, Steryl glucoside (SG)

R_1= acyl, Acylated steryl glucoside
(ASG)

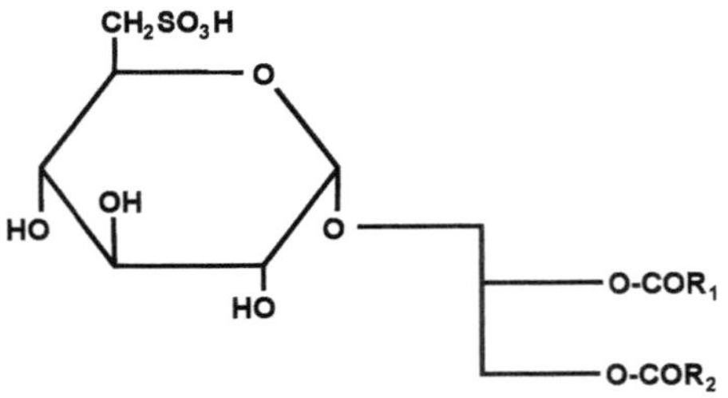

Sulfoquinovosyldiacylglycerol (SQD)

Figure 2
Chemical structures of glycolipids found in potato lipids.

amounts in NGMP and GMP Spunta. Average daily intake of GL in human has been reported to be 140 mg ESG, 65 mg SG, 50 mg CER, 90 mg MGD, and 220 mg DGD (Ramadan and Mörsel, 2003).

Phospholipid classes in GMP and NGMP were separated into four major fractions using HPLC. Phospholipid classes (Figure 4) revealed that predominant PL classes in NGMP and

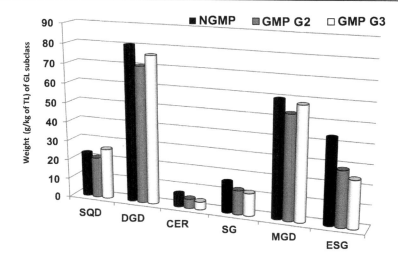

Figure 3
Glycolipid classes (g/kg of TL) in GMP and NGMP Spunta. SQD, sulfoquinovosyldiacylglycerol; DGD, digalactosyldiacylglycerol; CER, cerebrosides; SG, steryl glucoside; MGD, monogalactosyldiacylglycerol; ESG, esterified steryl glucoside.

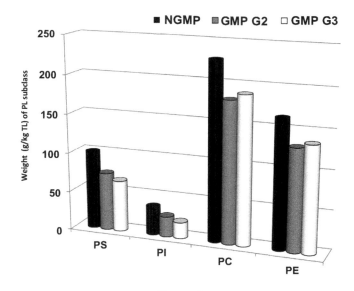

Figure 4
Phospholipid classes (g/kg of TL) in GMP and NGMP Spunta. PS, phosphatidylserine; PI, phosphatidylinositol; PC, phosphatidylcholine; PE, phosphatidylethanolamine.

GMP Spunta were phosphatidylcholine (PC), followed by phosphatidylethanolamine (PE), phosphatidylserine, and phosphatidylinositol (PI), respectively (Ramadan and Elsanhoty, 2012). About 43–46% of total PL was PC, followed by PE (31–33%), whereas phosphatidylserine (16–19% of total PL) and PI (5–7% of total PL) were found in lower levels.

3.2 Fatty Acid Composition

The fatty acid profile of lipids isolated from potatoes is especially nutritionally advantageous because the essential part of all fatty acids is formed by unsaturated fatty acids with one to three double bonds, mainly oleic, linoleic, and linolenic acids. However, the small lipid amounts in potato tubers mean that the daily intake of these valuable lipids is minimal; levels of linoleic and linolenic acids in 100 g of the edible parts of potato tubers are assumed to be 32.1 and 22.7 mg, respectively. Because of this, it would be important to grow species of potato that would store more fat in their tubers (Prescha et al., 2001).

The result of the fatty acid analysis of TL extracted from wild-type (D) and GMP with overexpression of 14-3-3 protein (J2) showed that the percentage of oleic, linoleic, and linolenic acids greatly agrees with the published reports (Kolbe et al., 1996; Trevini et al., 1983; Tevini and Schonecker, 1986). The content of linoleic acid, which is the main fatty acid of potato, came to 49%. Total lipids from the control tubers contained a large amount of palmitic and linolenic acids (10% and 14%, respectively). Fatty acids found in TL of transgenic tubers (J2) were present in proportions similar to those in the control plants. The exception was a 55% increase in oleic acid, but the content of this acid was 4.5% of fatty acids in J2 potato. The main fatty acids of nonpolar fraction were palmitic, linoleic, and linolenic acids. In the case of the control plants, the content of palmitic, linoleic, and linolenic acids was 40%, 21%, and 13% of fatty acids, respectively. In the GMP, significant elevation of the unsaturated fatty acid component of the nonpolar fraction of tuber TL was revealed. The linoleic acid content increased by 48% and linolenic acid increased by 33%. A significant 71% increase in oleic acid was also reported, whereas the palmitic acid content decreased by 43%.

The percent contribution of polar lipids to the TL was more than twice the contribution of the nonpolar lipids. Moreover, in the nonpolar fractions the participation of palmitic acid was observed to be twice as high as in TL. In the case of the transgenic tubers (J2), the increase in nonpolar lipids was accompanied by a significant increase in unsaturated fatty acids in this fraction (Prescha et al., 2001). The fat from both GMP and control potato revealed a nutritionally valuable profile of fatty acids, with a high content of unsaturated acids.

The fatty acid composition in lipids of GMP and NGMP from potato Spunta (Ramadan and Elsanhoty, 2012) is shown in Table 2. The data revealed that oleic and *cis*-α-linoleic acid were the main fatty acid in different potato Spunta lines. In addition, TL contained high levels of palmitic and *cis*-α-linolenic acids. The content of oleic and linoleic acids, which are the main fatty acid of potato, was about 62.5–64.5% of all acids. Fatty acid levels in lipids of transgenic potato Spunta G2 and G3 tubers were relatively different from those in the control plants. Significant changes in the content of main unsaturated fatty acids in lipids were observed between GMP lines and the control cultivar. Polyunsaturated fatty acids were found in higher levels in GMP (31.5–31.9%) than in NGMP (31%). On the other side, saturated fatty acids and monounsaturated fatty acids were found in higher levels in NGMP than in

Table 2: Fatty Acids Composition (%) of GMP and NGMP Spunta.

Fatty Acid	NGMP	GMP G2	GMP G3
C16:0	11.0	10.5	10.7
C18:0	4.00	4.00	3.90
C18:1n-9	45.0	44.5	43.9
C18:2n-6	19.5	20.0	20.6
C18:3n-3	11.5	11.5	11.3
Other acids	9.00	9.50	9.60

Table 3: Levels of Phytosterols (g/100 g of TL) in GMP and NGMP Spunta.

Compound	NGMP	GMP G2	GMP G3
Brassicasterol	0.103	0.129	0.134
Campesterol	0.670	0.679	0.754
Stigmasterol	0.056	0.025	0.024
β-Sitosterol	1.080	1.180	1.260
Δ5-Avenasterol	0.500	0.551	0.574
Δ7-Avenasterol	0.093	0.135	0.153

GMP. The result showed that the percentage of these acids is in agreement with the published results (El-Sanhoty et al., 2004; Dobson et al., 2004). In brief, the profile of the fatty acids in lipids from potatoes revealed the consistent changes in composition of fatty acids in the transgenic lines. The mechanism by which changes in the fatty acid profile in transgenic plants occurred is as yet unknown.

3.3 Sterol Composition

Total unsaponifiables in potato Spunta lines recorded the highest level in GMP G3 (4% of lipids), followed by GMP G2 (3.9% of TL) and NGMP (3.9% of lipids), respectively (Ramadan and Elsanhoty, 2012). The GMP G3 line contained the highest amounts of total sterol (ST) (29 g/kg oil), followed by GMP G2 (26 g/kg oil) and NGMP (25 g/kg oil), respectively. The results pointed to a similarity between the percent proportions of the most abundant ST in the tuber from GMP and NGMP Spunta (Table 3). β-Sitosterol was the main compound and comprised 43.1–43.7% of total ST content in potato Spunta lines (Figure 5). The next major components were campesterol (about 26% of total ST) and Δ5-avenasterol (about 20% of total ST). Other components, e.g., brassicasterol, Δ7-avenasterol, and stigmasterol, were present at lower levels and comprised about 10% of total ST. Lanosterol, sitostanol, and Δ5,24-stigmastadinol were not detected in GMP and NGMP Spunta unsaponifiables (Ramadan and Elsanhoty, 2012). Among different STs, β-sitosterol has been most intensively investigated with respect to its physiological effects in humans. Many beneficial effects have been shown for β-sitosterol.

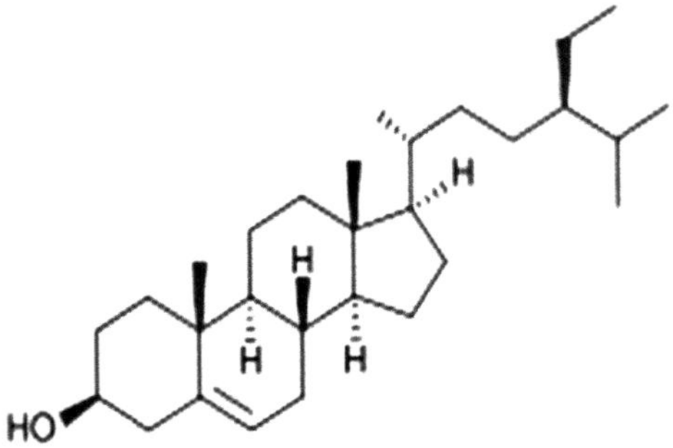

β-Sitosterol

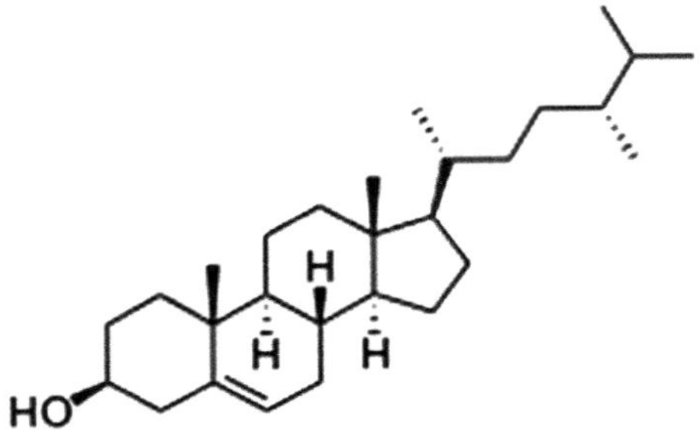

Campesterol

Figure 5
Chemical structures of the main sterols found in potato lipids.

Table 4: Levels of Tocopherols (g/100 g of TL) in GMP and NGMP Spunta.

Compound	NGMP	GMP G2	GMP G3
α-Tocopherol	0.271	0.356	0.400
β-Tocopherol	0.067	0.077	0.081
γ-Tocopherol	0.009	0.013	0.015
δ-Tocopherol	0.004	0.004	0.005

3.4 Tocopherol Composition

The levels and profile of tocopherols in GMP and NGMP from Spunta cultivar are shown in Table 4. Data revealed that GMP G3 contained the highest amounts of total tocopherols (5 g/kg oil), followed by GMP G2 (4.5 g/kg oil) and NGMP (3.5 g/kg oil), respectively (Ramadan and Elsanhoty, 2012). α-Tocopherol was the main compound (77.5–80% of total tocopherols) in GMP and NGMP Spunta, followed by β-tocopherol (16.2–19% of total tocopherols). δ- and γ-tocopherol were detected in lower amounts. The levels of α-tocopherol observed in the Andean potato tubers, ranging from 2.73 to 20.80 mg/kg, were above the quantities reported in the literature for commercial varieties (0.6–3 mg/kg) (Chun et al., 2006; Spychalla and Desborough, 1990). The levels observed for Nicola, which has a brownish skin and pale yellow flesh, as well as Vitelotte, which has a purple skin and partly purple flesh, were 0.8 and 2.3 mg of α-tocopherol kg, respectively (Andre et al., 2007). It is generally assumed that increases in α-tocopherol in the diet contribute to a decreased risk of chronic diseases. Increasing tocopherol content in potatoes is therefore of interest.

References

Andre, C.M., Mouhssin Oufir, M., Guignard, C., Hoffmann, L., Hausman, J., Evers, D., Larondelle, Y., 2007. Antioxidant profiling of native Andean potato tubers (*Solanum tuberosum* L.) reveals cultivars with high levels of β-carotene, α-tocopherol, chlorogenic acid, and petanin. Journal of Agricultural and Food Chemistry 55, 10839–10849.

Arvanitoyannis, I.S., Vaitsi, O., Mavromatis, A., 2008. Potato: a comparative study of the effect of cultivars and cultivation conditions and genetic modification on the physico-chemical properties of potato tubers in conjunction with multivariate analysis towards authenticity. Critical Reviews in Food Science and Nutrition 48, 799–823.

Banerjee, A.K., Prat, S., Hannapel, D.J., 2006. Efficient production of transgenic potato (*S. tuberosum* L. ssp. *andigena*) plants via *Agrobacterium tumefaciens*-mediated transformation. Plant Science 170, 732–738.

Blaszczak, W., Fornal, J., Amarowicz, R., 2003. Lipids of wheat, corn and potato starch. Journal of Food Lipids 10, 301–312.

Brown, C.R., 2005. Antioxidants in potato. Amercain Journal of Potato Research 82, 163–172.

Brown, C.R., Edwards, C.G., Yang, C.-P., Dean, B.B., 1993. Orange flesh trait in potato: inheritance and carotenoid content. Journal of the American Society for Horticultural Science 118, 145–150.

Buckenhüskes, H.J., 2005. Nutritionally relevant aspects of potatoes and potato constituents. In: Haverkort, A.J. (Ed.), Potato in Progress: Science Meets Practice. Culinary and Hospitality Industry Publication Services, Weimar, TX, pp. 17–26.

Chun, J., Lee, J., Ye, L., Exler, J., Eitenmiller, R.R., 2006. Tocopherol and tocotrienol contents of raw and processed fruits and vegetables in the United States diet. Journal of Food Composition and Analyses 19, 196–204.

Dhital, S., Shrestha, A.K., Hasjim, J., Gidley, M.J., 2011. Physicochemical and structural properties of maize and potato starches as a function of granule size. Journal of Agricultural and Food Chemistry 59, 10151–10161.

Dobson, G., Griffiths, D.W., Davies, H.V., Mcnicol, J.W., 2004. Comparison of fatty acid and polar lipid contents of tubers from two potato species, *Solanum tuberosum* and *Solanum phureja*. Journal of Agricultural and Food Chemistry 52, 6306–6314.

El-Sanhoty, R., Abd El-Maged, A.D., Ramadan, M.F., 2006. Safety assessment of genetically modified potato Spunta: degradation of DNA in gastrointestinal track and carry over to rat organs. Journal of Food Biochemistry 30, 556–578.

El-Sanhoty, R., El-Rahman, A.A., Bogl, K.W., 2004. Quality and safety evaluation of genetically modified potatoes spunta with *Cry V* gene: compositional analysis, determination of some toxins, antinutrients compounds and feeding study in rats. Nahrung/Food 48, 13–18.

Fridman, M., 2006. Potato glycoalkaloids and metabolites: roles in the plant and in the diet. Journal of Agricultural and Food Chemistry 54, 8655–8681.

Galliard, T., 1973. Lipids of potato tubers. 1. Lipid and fatty acid composition of tubers from different varieties of potato. Journal of the Science of Food and Agriculture 24, 617–622.

Gross, J., 1991. Pigments in Vegetables: Chlorophylls and Carotenoids. Van Nostrand Reinhold, New York.

Jansen, G., Flamme, W., Schüler, K., Vandrey, M., 2001. Tuber and starch quality of wild and cultivated potato species and cultivars. Potato Research 44, 137–146.

Kitahara, K., Tanaka, T., Suganuma, T., Nagahama, T., 1997. Released of bound lipids in cereal starches upon hydrolysis by glucoamylase. Cereal Chemistry 74, 1–6.

Kolbe, H., Fischer, J., Rogozinska, I., 1996. Einflussfaktoren auf die Inhaltsstoffe der Kartoffel. Teil V: Rohfett und Fettsaeurezusammensetzung. Kartoffelbau 4, 290–296.

Kostyn, K., Szatkowski, M., Kulma, A., Kosieradzka, I., Szopa, J., 2013. Transgenic potato plants with overexpression of dihydroflavonol reductase can serve as efficient nutrition sources. Journal of Agricultural and Food Chemistry 61, 6743–6753.

Kubo, I., Fukuhara, K., 1996. Steroidal glycoalkaloids in Andean potatoes. Advances in Experimental Medicine and Biology 405, 405–417.

Lin, P.-Y., Czuchajowska, Z., 1998. Role of phosphorus in viscosity, gelatinisation, and retrogradation of starch. Cereal Chemistry 75, 705–709.

Prescha, A., Biernat, J., Weber, R., Zuk, M., Szopa, J., 2003. The influence of modified 14-3-3 protein synthesis in potato plants on the nutritional value of the tubers. Food Chemistry 82, 611–617.

Prescha, A., Swiedrych, A., Biernat, J., Szopa, J., 2001. The increase in lipid content in potato tubers modified by 14-3-3 gene overexpression. Journal of Agricultural Food Chemistry 49, 3638–3643.

Priestley, H., 2006. How to think like consumers and win! In: Haase, N.U., Haverkort, A.J. (Eds.), Potato Developments in a Changing Europe. Wageningen Academic Pub, pp. 189–198.

Ramadan, M.F., Elsanhoty, R.M., 2012. Lipid classes, fatty acids and bioactive lipids of genetically modified potato Spunta with *Cry V* gene. Food Chemistry 133, 1169–1176.

Ramadan, M.F., Mörsel, J.-T., 2003. Analysis of glycolipids from black cumin (*Nigella sative* L.), coriander (*Coriandrum sativum* L.) and niger (*Guizotia abyssinica* Cass.) oilseeds. Food Chemistry 80, 197–204.

Ridgman, W.J., 1989. The Potato, by Burton, W.G., xii, 742 pp., third ed. Longman Scientific & Technical, Harlow.

Robert, L., Narcy, A., Rock, E., Demigne, C., Mazur, A., Rémésy, C., 2006. Entire potato consumption improves lipid metabolism and antioxidant status in cholesterol-fed rat. European Journal of Nutrition 45, 267–274.

Scherz, H., Senser, F. (Eds.), 1994. Food Composition and Nutrition Tables. Medpharm Scientific Publishers, Stuttgart, Germany, pp. 170–174.

Spychalla, J.P., Desborough, S.L., 1990. Fatty acids, membrane permeability, and sugars of stored potato tubers. Plant Physiology 94, 1207–1213.

Tevini, M., Schonecker, G., 1986. Occurrence, properties and characterization of potato carotenoids. Potato Research 29, 265.

Trevini, M., Schoenecker, G., Iwanzik, W., Riedmann, M., Stute, R., 1983. Analyse, Vorkommen und Verhalten von Carotenoidestern in Kartoffeln. Kartoffel-Tagung, band 5. Granum Verlag, Detmold, Germany, pp. 42–47.

Vasanthan, T., Hoover, R., 1992. A comparative study of the composition of lipids associated with starch granules from various botanical sources. Food Chemistry 43, 19–27.

Wenzler, H., Mignery, G., May, G., Park, W., 1989. A rapid and efficient transformation method for the production of large numbers of transgenic potato plants. Plant Science 63, 79–85.

CHAPTER 6

Vitamins, Phytonutrients, and Minerals in Potato

Duroy A. Navarre[1], Roshani Shakya[1], Hanjo Hellmann[2]

[1]USDA-ARS, Washington State University, Prosser, WA, USA; [2]School of Biological Sciences, Washington State University, Pullman, WA, USA

1. Introduction

Potatoes are the most grown root crop in the world and the fourth most grown crop overall, after rice, wheat, and maize, and the third most important food crop after wheat and rice. The past several decades saw remarkable changes in global potato production, especially in Asia. Historically, most potato consumption and production occurred in the Americas and Europe, where per capita consumption exceeded several hundred pounds a year in countries including Germany, Poland, and Russia. On the other hand, relatively little potato production occurred in Asia and Africa. The introduction of potatoes into the West from the Andes greatly improved food security because potatoes provide more calories per acre than any other major crop. The cultivation of potatoes is linked to the population increases seen in Europe in the 1700s and 1800s (Nunn and Qian, 2011). The increased food security in premodern Europe may have contributed to the industrial revolution and allowed more resources to be focused on technological development. In recent years potato consumption has fallen markedly in the West, including in countries historically identified with potatoes. However, over the past few decades potato production has soared in the developing world, with China becoming the largest global producer of potatoes. In 2013 China produced 88,987,220 tons, more than Russia, the United States, Poland, and Germany combined, while India produced 45 million tons (FAOSTAT, 2015). As of 2010, the developing world produces over 70% of global potato production. Production decreases about 1% annually in the developed world and increases 1% per year in emerging and developing countries (Bond, 2014). Potato consumption increased 40% in China from 2010 to 2015 (Zienkiewicz, 2015). To boost food security in China, which has 22% of the global population but only 9% of global arable land, the Chinese Academy of Agricultural Sciences in 2015 recommended potatoes be promoted as a staple food, along with rice, wheat, and corn (Hairong, 2015). In the United States, potato acreage decreased from 3.9 million acres in 1922 to 1.0 million in 2015, while the number of potato farms decreased from 51,500 in 1974 to 15,014 in 2007. Despite the dramatic decrease in acreage,

Advances in Potato Chemistry and Technology. http://dx.doi.org/10.1016/B978-0-12-800002-1.00006-6

production increased owing to improved cultivars and superior management (USDA-ERS, 2015).

Potatoes were domesticated between 7000 and 10,000 years ago around Lake Titicaca in the Andes between Peru and Bolivia (Spooner et al., 2005). Potatoes may contain more genetic diversity than any other major crop and this may reflect their ability to grow in remarkably divergent environments (Hawkes, 1990), from arid alpine highlands to tropical rain forests to permafrost soils just below the Arctic Circle, a trait that contributes to the ability of potatoes to provide food security. Potato's genetic diversity is a valuable resource for further improving tuber nutritional content, especially when taking into account that modern cultivars are estimated to contain less than 1% of the available genetic diversity of wild species. About 200 wild potato species exist, in addition to thousands of primitive varieties.

Potato tubers are specialized organs evolved to improve the plant's chances of survival and allow vegetative reproduction. Tubers are not derived from roots, but are modified stems, originating on stolons from axillary buds on the underground part of the stem (Ewing and Struik, 1992; Fernie and Willmitzer, 2001). The fact that tubers are modified stems influences tuber characteristics and chemical composition. For example, the greening of tubers and increase in glycoalkaloids (GAs) triggered by light exposure from which amyloplasts in the tuber parenchyma redifferentiate into chloroplasts (Deng and Gruissem, 1988) may reflect the stem origin of the tuber. Tubers are metabolically active and synthesize a complex mix of metabolites that belies their misperception as simple organs containing only starch. In reality, tubers contain abundant amounts of small molecules and secondary metabolites that have roles in an array of key tuber processes, from regulating tuber organogenesis to mediating responses to the environment. Moreover, many of these compounds have positive effects on human health and are highly desirable in the diet (Abuajah et al., 2015; Katan and De Roos, 2004). Numerous studies document the role of potatoes in human health, including cardiovascular health, yet more remains to be learned (McGill et al., 2013).

2. Potatoes, Nutrition, and the Food Debates

Consumers appear increasingly interested in the relationship between diet and health. One-third of Americans are estimated to take a daily vitamin or dietary supplement and annual sales top US$18 billion. Moreover, the obesity epidemic in the United States and many other countries has resulted in a marked increase in awareness, discussion, and debate about the impact of diet on health. In the United States the response to the obesity epidemic included a controversial attempt to ban large sodas in New York City, overhauled school lunch programs, removal of snack and soda vending machines from schools, and proposals to add "sin taxes" to certain foods. Harvard researchers claimed that potatoes are a leading cause of obesity in the United States and advocated for potatoes to be replaced in the diet (Mozaffarian et al., 2011). According to the Organization for Economic Cooperation and Development,

Mexico and the United States are the first and second most obese countries, while they rank 105th and 44th in per capita potato consumption according to Food and Agriculture Organization (FAO) data (2009). Obesity was not problematic in European countries in the past when potato consumption was much higher than present day. For example, German annual potato consumption in 1900 was 628 lb per person and over 300 lb in the 1950s.

Past years have shown that consumer perception about the nutritional value of potatoes has an impact on sales and strongly suggest that perceived nutritional value is a very important trait for any vegetable, especially potatoes, given recent negative publicity. Generally, crops have been bred and selected primarily for traits such as yield, disease resistance, and appearance. Historically, little effort was directed toward increasing the nutritive value of any crop for reasons that there were more pressing issues, plus the daunting technical difficulty of such an undertaking. With most crops, including potatoes, nutrient profiles are available for only a few varieties. Thus, surprisingly little is known about which varieties have the most or to what extent new varieties can be developed that have even more. Filling in this knowledge gap has been made easier by technological advances such as high-throughput assays, affordable and powerful mass spectrometers, and myriad molecular biological tools.

3. Basic Potato Nutritional Content

Potatoes are approximately 80% water and 20% solids, although this can vary by cultivar, as seen with a high-phytonutrient purple-yellow potato composed of only 9% dry matter (Pillai et al., 2013). Of the 20 g of solids in a 100-g tuber, about 18 g is carbohydrate and 2 g protein. The nutritional quality of potato protein is among the highest in plants. The primary storage proteins in tubers are patatins, which account for 40% of the soluble protein content (Prat et al., 1990). Potatoes contain relatively small amounts of simple sugars, providing mostly complex carbohydrates, which the World Health Organization (WHO) recommends constitute 50–75% of daily calorie intake. There are different types of starch that can be classified as rapidly available, slowly available, or resistant starch. Potato starch varies by cultivar, but typically is about 80% amylopectin and 20% amylose. The greater the percentage of amylose in a potato, the more slowly the starch will be digested. "Resistant starch" is dietarily desirable and functions like soluble fiber to promote glycemic control and gut health. Raw potatoes contain significant amounts of dietarily desirable RS2 resistant starch, which is less digestible than cereal starch, including corn. Upon cooking, potato starch becomes more digestible. Raw potato starch added to foods has been shown to have health-promoting effects, including lowered cholesterol and triglycerides in rats (Raigond et al., 2015). Potatoes cooked and then cooled, such as in potato salads, contain dietarily desirable RS3 retrograde starch that is less digestible and has been shown to have health-promoting effects (Fuentes-Zaragoza et al., 2010). A limited study with only seven cultivars reported that amylose content of tubers did not correlate with either glycemic index values or in vitro starch digestibility and suggested that in vitro digestion procedures should be used to screen potatoes for low glycemic index potential (Ek et al., 2014).

3.1 A Nutrient-Dense Food

Nutrient-dense foods are those that provide an equal or greater amount of nutritional value compared to their calorie content, as opposed to something like a candy bar that provides calories but little nutrition. Potatoes are a nutrient-dense food as seen in Figure 1, which shows the percentage a 100-g (3.5-oz) portion of baked potato with skin provides of the recommended daily values of various nutrients based on a 2000-calorie-per-day diet (data from nutritiondata.self.com and USDA SR-21). This serving would provide 97 calories or 5% of the daily value, but, as seen, provides a greater percentage of various nutrients than it does calories, including 22% of the recommended amounts of vitamin C and 16% of potassium. The nutritional contribution of potatoes to the total British diet was found to be 7% of the total energy, but greater amounts of nutrients, including 15% of vitamin B6, 14% of vitamin C, 13% of fiber, 10% of folate, 9% of magnesium, and only 4% of saturated fatty acids (Gibson and Kurilich, 2012). The authors noted that reducing the amount of added saturated fat and salt improves the nutritional profile of potatoes. Not shown in Figure 1 are phytonutrients, including polyphenols and carotenoids, which unlike vitamins do not have recommended daily allowances but are nonetheless important for health. These phytonutrients are valued by many consumers to the extent that they influence purchasing decisions. In addition to vitamins and minerals, tubers contain a complex

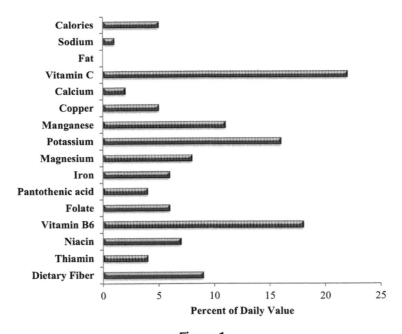

Figure 1
The percentage of the daily value of recommended nutrients supplied by a 100-g serving size of potato.

assortment of small molecules, many of which are phytonutrients. These include polyphenols, flavonols, anthocyanins, phenolic acids, carotenoids, polyamines, GAs, tocopherols, calystegines, and sesquiterpenes.

4. A Survey of Vitamins in Potatoes
4.1 Vitamin C

Potatoes may be best known as a source of vitamin C and potassium. Among 98 vegetables examined, potatoes were found to be the least expensive source of vitamin C when calculating the cost to supply 10% of the recommended dietary allowance (RDA) (Drewnowski and Rehm, 2013). A medium red-skinned potato (173 g) provides about 36% of the RDA according to the U.S. Department of Agriculture (USDA) databases. Scurvy is the best known symptom of vitamin C deficiency, which in severe cases is typified by loss of teeth, liver spots, and bleeding. Tubers can synthesize vitamin C in situ, but also accumulate vitamin C originating in leaves and stems (Tedone et al., 2004; Viola et al., 1998). Vitamin C is a cofactor for many enzymes, functioning as an electron donor, and has a major role in detoxifying reactive oxygen species. Plants are the primary source of vitamin C in the human diet. Leaves and chloroplasts can contain 5 and 25 mM L-ascorbate, respectively (Wheeler et al., 1998). Plants may have multiple vitamin C biosynthetic pathways, with all of the enzymes of the L-galactose pathway having been characterized (Laing et al., 2007; Wolucka and Montagu, 2007). Overexpressing GDP-L-galactose phosphorylase in tubers increased tuber vitamin C up to threefold, whereas sixfold increases were seen in tomato and twofold in strawberries. A study using knockout mice unable to synthesize vitamin C showed that when fed potato chips the mice accumulated vitamin C and had decreased reactive oxygen species, leading to the conclusion that fast-dry-processed chips are a superior source of dietary vitamin C (Kondo et al., 2014). A dietary study with Japanese men showed that vitamin C from mashed potatoes and potato chips was bioavailable and plasma levels increased after eating (Kondo et al., 2012).

Researchers examined tuber vitamin C content in 75 genotypes and found concentrations ranging from 11.5 to 29.8 mg/100 g fresh weight (FW) (Love et al., 2004). This study also reported that some genotypes had more consistent concentrations of vitamin C than others across multiple years or when grown in different locations and proposed that the year may have a bigger effect than the location. A British study measured vitamin C in 33 cultivars grown in three locations around Europe (Dale et al., 2003). If these authors' results in dry weight (DW) are converted to FW assuming potatoes are 80% water, a range of 13–30.8 mg vitamin C per 100 g FW is obtained. Unusually high amounts of vitamin C occurred in somaclones of Russet Burbank, which had levels from 31 to 139 mg/100 g FW among the several hundred screened (Nassar et al., 2014). Why these results differ so dramatically from the extensive potato vitamin C results reported by other labs is unknown. The authors did use a validated procedure and a vitamin C reference material, suggesting that some unknown factor caused levels to markedly increase. If this unknown trigger could be identified, then this could be a way to further increase tuber vitamin C content through management of the crop.

Environment is well known to affect vitamin C levels in potatoes and other plants. Numerous studies show that vitamin C levels decrease rapidly during cold storage of potatoes and can approach a 60% decrease (Keijbets and Ebbenhorst-Seller, 1990). After placing 33 genotypes in cold storage for 15–17 weeks, Dale et al. found substantial decreases in vitamin C compared to prestorage (Dale et al., 2003). Vitamin C decreases ranged from 20 to 60% depending on the genotype. The authors make the important point that breeding efforts to increase vitamin C should focus on poststorage content and that in most cases this is more relevant than fresh-harvest concentrations. This will be truer for countries that place a majority of the potato harvest in cold storage than for developing countries that make limited use of cold storage and for which postharvest loses consequently should be less. Blauer et al. (2013) showed that the magnitude of the loss during cold storage is genotype dependent and suggested that oxidative metabolism has a regulatory role in vitamin C storage stability because storage under reduced O_2 decreased loss. A smaller loss was observed during late tuber maturation under senescing vines, with the authors suggesting this may reflect decreased transport of vitamin C from the senescing foliage to tubers (Blauer et al., 2013). Vitamin C content in 12 genotypes grown in Colorado was measured from harvest through 7 months of storage. After 7 months of cold storage, losses were over 50%. Interestingly, one advanced purple-fleshed breeding line had losses of over 60% after only 2 months of storage, whereas amounts in Yukon Gold after 2 months were not statistically different from those at harvest (Külen et al., 2012).

Wounding can substantially increase vitamin C levels. One study examined changes in vitamin C after storing potatoes for 2 days following slicing or bruising and measured a 400% increase in vitamin C in sliced tubers, but a 347% decrease in bruised tubers (Mondy and Leja, 1986). Vitamin C levels in fresh-cut potatoes stored in air were found to increase, whereas levels decreased in those stored frozen or under a modified atmosphere (Tudela et al., 2002b). These and similar results suggest that wounding of potatoes can be used to substantially increase vitamin C in commercial products; however, before this is likely to be widely adopted as a strategy to increase vitamin C, a method must be found to decrease the browning that can occur in cut tissue.

A Turkish study examined the effects of freeze-storing on peeled, blanched, and then fried potatoes and found a 10% loss of vitamin C after 6 months of storage at −18 °C (Tosun and Yücecan, 2008). However, a 51% loss was caused by the prefreezing operations, which sounds a cautionary note about the importance of how potatoes are handled during processing. Thus, in the absence of cultivars with stable vitamin C levels during cold storage, one solution to minimize postharvest loss of vitamin C for some commercial products would be minimally destructive cooking of tubers shortly after harvest, followed by flash-freezing of the product.

4.2 The Vitamin B Family: Biochemical Function and Content in Potato

Like vitamin C, all members of the vitamin B family are water-soluble compounds, but they are chemically and functionally quite distinct from one another. Historically B vitamins were recognized as factors that maintain growth and health and prevent the development of certain skin diseases in human and animals. Currently, the group comprises thiamin (vitamin B_1), riboflavin (B_2), niacin or niacin amide (B_3), pantothenic acid (B_5), pyridoxine (B_6), biotin (B_7), folic acid (B_9), and cobalamins (B_{12}) (Figure 2) (Hellmann and Mooney, 2010; Roje, 2007). All eight B vitamins are essential parts of the human diet since humans, in contrast to plants, are not able to synthesize these compounds de novo. In addition, three of the B vitamins, thiamin, riboflavin, and pyridoxine, are part of the WHO Model List of Essential Medicines for adults and children (http://www.who. int/medicines/publications/essentialmedicines/en/). This section of the chapter focuses on the five B vitamins for which potato tubers are a good nutritional resource and provides an overview of their general biochemical roles and the beneficial impacts they have for the consumer.

4.2.1 B_1—Thiamin

Thiamin is active as thiamin diphosphate and functions as a cofactor in various metabolic pathways, such as the Krebs cycle and the pentose phosphate pathway. It is associated with

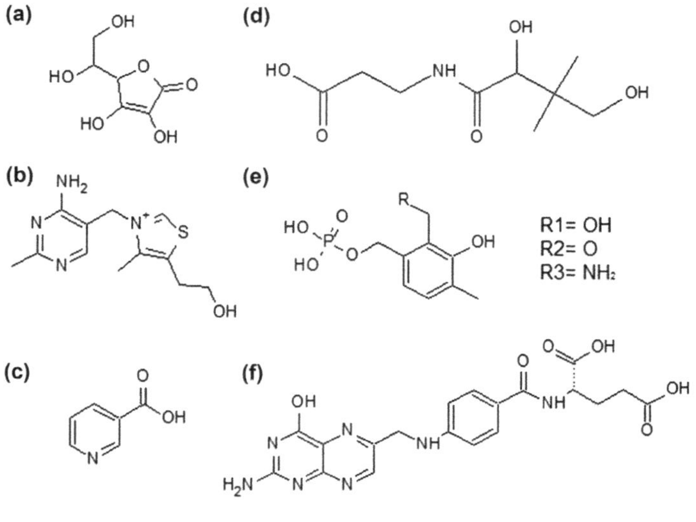

Figure 2
Chemical structures of (a) ascorbate (vitamin C), (b) thiamin (vitamin B_1), (c) niacin (vitamin B_3), (d) pantothenate (vitamin B_5), (e) pyridoxine (vitamin B_6), and (f) folate (vitamin B_9).

carbohydrate and amino acid catabolism, but also the biosynthesis of branched amino acids (Manzetti et al., 2014). In humans, an insufficient supply of thiamin can lead to beriberi, which results in cardiovascular and neurological problems (Abdou and Hazell, 2015). Thiamin deficiency is not uncommon in countries that rely mostly on a white or polished rice-based diet, which is poor in thiamin (Juguan et al., 1999), whereas populations with a balanced diet are unlikely to face this problem. Here, thiamin deficiency mainly occurs in parallel with heavy alcohol consumption, which affects transport, storage, and metabolism of the vitamin (Rees and Gowing, 2013).

Recent work on 33 accessions of primitive cultivars (*Solanum tuberosum* group Andigenum) and three modern potato varieties grown under field conditions in 2009 and 2010 showed a broad variation in thiamin concentrations among the tested clones. One striking aspect was how strongly thiamin amounts varied. For example, in 2009 the lowest levels were 490 ± 91 ng/g FW (PI 225710), while the highest values reached up to 2273 ± 107 ng/g (PI 320377), a more than fourfold difference. However, values in 2010 were in most cases significantly different from those of 2009, revealing a deviation of up to 50%, and changes were connected with either increased or decreased thiamin content. In general, these data emphasize the high breeding potential in potato for increased thiamin content. The approach also demonstrated that under field conditions the vitamin content may vary significantly, a fact that will make selection for specific traits difficult. Nevertheless, the work from Goyer and Sweek emphasizes the necessity of performing field trials and following these over several subsequent years to gain reliable information on the reproducibility of the content of specific metabolites in potato.

The RDA for thiamin provided by the National Institutes of Health for adults is 1.2 mg/day for men and 1.1 mg/day for women. The field trial data from Goyer and Sweek (2011) showed that some potato varieties already contain close to 20% of RDA values in just 100 g of tuber tissue. However, thiamin is known to be sensitive to heat (Kandutsch and Baumann, 1953), and not surprisingly, levels in processed potatoes are significantly lower than 20%. For example, french fries and baked potatoes contain only 0.08 and 0.067 mg thiamin per 100 g of product, respectively, which is less than 10% of the RDA (Table 1). Boiled potatoes still contain around 10% of the RDA (0.106 mg) (Table 1), indicating that the nature of processing is critical for preserving thiamin in the product.

Overall the currently available data accentuate the importance of thiamin as an essential phyto-nutrient, with the method of processing potatoes for food products being a critical parameter for maintaining high thiamin levels. The currently available datasets further demonstrate the potential of potatoes for fortifying thiamin content in tubers through breeding or bioengineering.

4.2.2 B$_3$—Niacin

Niacin refers to the two pyridine derivatives, nicotinamide or nicotinic acid (Figure 2). It is converted in the cell to NAD$^+$ or NADP$^+$ and functions as a cofactor in various redox reactions that are primarily connected with photosynthesis and carbohydrate and fatty acid metabolism

Table 1: Content of Vitamins B_1, B_3, B_5, B_6, and B_9 in Selected Potato Foods.

Vitamin	French Fries Fried in Vegetable Oil	Potato Chips	Baked, Flesh and Skin	Boiled, Cooked in Skin and Flesh
B_1 (thiamin)	0.08 (7.27/6.67)	0.17 (15.45/14.17)	0.067 (5.45/5)	0.106 (10/9.17)
B_3 (niacin)	0.28 (2/1.8)	3.83 (27.4/23.9)	1.41 (10.1/8.8)	1.44 (10.3/9)
B_5 (pantothenic acid)	0.5 (10/10)	0.4 (8/8)	0.38 (7.6/7.6)	0.52 (10.4/10.4)
B_6 (pyridoxine)	0.36 (24–27.69/ 21.18–27.69)	0.78 (52–60/45.88–60)	0.31 (20.7–23.85/ 18.24–23.85)	0.3 (20–23.08/ 17.65–23.08)
B_9 (folate)	0.03 (7.5/7.5)	0.075 (18.75/18.75)	0.026 (6.5/6.5)	0.01 (2.5/2.5)

First values represent mg vitamin/100 g product, while the second values for B_1, B_3, B_6, and B_9 are % RDA for adult women/ men and for B_5 % AI (adequate intake) for adult women/men.
USDA National Nutrient Database for Standard Reference.

(Wahlberg et al., 2000). Niacin itself has been reported to be beneficial for lowering cholesterol levels and may promote vasodilation and hypotension (Gehring, 2004; Rolfe, 2014; Williams and Ramsden, 2005). In addition, niacin deficiencies have been classically connected with pellagra, a disease showing symptoms of diarrhea, dermatitis, and dementia (Fu et al., 2014; Kohn, 1938). Deficiencies are common in countries in which the population mainly depends on a maize-based diet, which requires elaborate processing of the seeds known as "nixtamalization" to make the niacin accessible (Gwirtz and Garcia-Casal, 2014).

Very little research has been done on niacin content in potato, and the available data do not indicate major variations among the few tested genotypes (Wills et al., 1984). The current RDA values for niacin are 14 and 16 mg per day for women and men, respectively. Based on these values, potato actually offers a good supply for the vitamin since 100 g of baked or boiled potatoes contains around 10% of the RDA (Table 1).

Owing to the importance of niacin, and considering the little research that has been done so far on this vitamin in potato, a more detailed profile of the vitamin and how it varies among genotypes would be beneficial to assessing the potential for future breeding efforts that might increase B_3 content in potato products.

4.2.3 B_5—Pantothenic Acid

Pantothenic acid is a precursor of the amino acid β-alanine and a core component of coenzyme A (Figure 2). It is involved in various biosynthetic pathways that are connected with carbohydrate, amino acid, and fatty acid metabolism (Depeint et al., 2006). In addition, the vitamin, like other B vitamins and vitamin C, has a protective function under oxidative stress conditions (Wojtczak and Slyshenkov, 2003). Vitamin B_5 deficiency is rare in humans, most likely because it is produced by intestinal bacteria and is to a degree ubiquitously present in all food products. However, deficiency in mammals and chicken has been reported and is associated with dermatitis, anemia, convulsions, and encephalopathy (Bender, 1999; Depeint et al., 2006; Tahiliani and Beinlich, 1991). Probably because malnutrition with pantothenic acid is rare, no RDA values are established. Currently only adequate intake (AI) values are available, about 5 mg/day for adult men and women (Tahiliani and Beinlich, 1991).

To what extent pantothenic acid content varies in potato germplasm is unknown. The vitamin appears to be stable when heated under moist conditions, but becomes labile under hot and dry conditions and may be sensitive to oxidation and light exposure (Riaz et al., 2009). Consequently, boiled potatoes contain around 10% of the daily AI value per 100 g of product, while in chips only 8% is left (Table 1). Nevertheless, these values show that potato is a good nutritional resource for vitamin B_5 even when different processing methods are applied.

4.2.4 B_6—Pyridoxine

Vitamin B_6 describes a group of six closely related compounds that mainly vary at their 4-position by a hydroxymethyl group, an aldehyde, or an aminomethyl group (Figure 2). The vitamin is probably the most versatile cofactor in living cells since it is required for more than 140 biochemical reactions (Mooney and Hellmann, 2010). It is mainly involved in amino acid biosynthesis and catabolism, but also plays important roles in carbohydrate and fatty acid metabolism, as well as photorespiration. Of note is that independent of its function as a cofactor, vitamin B_6 also serves as a potent antioxidant; and evidence suggests that it is critical for plant pathogen resistance (Denslow et al., 2007; Herrero et al., 2007; Titiz et al., 2006; Zhang et al., 2014).

Like thiamin, vitamin B_6 content varies significantly among potato genotypes (Mooney et al., 2013). Work on mature tubers from 10 different genotypes showed that the total vitamin B_6 content varied from around 18 µg B_6/g DW (PORO07PG63-1) to up to 27 µg B_6/g DW (Clearwater Russet). A similar variation was observed when immature, 60- to 80-day-old tubers of 23 different genotypes were analyzed. Amounts varied between around 16 µg B_6/g DW (PORO07PG63-1) to 22 µg B_6/g DW (Ruby Crescent) (Mooney et al., 2013). Overall vitamin B_6 levels were not significantly different between immature and mature tubers, indicating that both developmental stages represent a good resource for this vitamin. However, it currently remains open to what extent these values are consistent among growing seasons as they are all based on a single harvest year (Mooney et al., 2013).

The current RDA values for vitamin B_6 range between 1.3 mg (young adults) and 1.7 mg (adult males) and can reach 2 mg for lactating women based on recommendations from the Food and Nutrition Board of the Institute of Medicine in the United States. Taking these values into consideration, potato represents an excellent nutritional resource for this vitamin. In addition the vitamin is heat stable and processing steps such as cooking or frying do not affect its content.

This is reflected, for example, by products such as baked potatoes or potato chips, which contain up to 23% and 60% of the RDA per 100 g of product, respectively (Table 1).

One interesting point about vitamin B_6 is its requirement for plastidial α-glucan phosphorylase activity as a cofactor. In addition to α-amylase, the phosphorylase is one of the two key enzymes that participate in the initial breakdown of starch into soluble carbohydrates in tubers (Kossmann and Lloyd, 2000). Under low-temperature (4–8 °C) storage conditions, potato tubers have the tendency to accumulate soluble, reducing carbohydrates over time, such as glucose and fructose, a process known as cold sweetening (Chen et al., 2012; Kossmann and Lloyd, 2000). These sugars are not desired by the processing industry since they can cause brown discoloration and acrylamide formation in fried potato products through the Maillard reaction (Arribas-Lorenzo and Morales, 2009; Lojzova et al., 2009). Of note is that B_6 levels also appear to increase under long-term, low-temperature storage conditions (Mooney et al., 2013). This may have positive impacts on α-glucan phosphorylase activity and starch breakdown. Consequently, preventing a rise in B_6 levels in tubers under such conditions may be a possible approach to slow down starch breakdown and cold sweetening in stored potatoes.

In summary, potato is a very good source of vitamin B_6, and the variation in B_6 content among genotypes demonstrates a strong potential to increase B_6 content in potato tubers. In addition, research on vitamin B_6 has demonstrated that elevated B_6 levels in plants are beneficial for abiotic and biotic stress tolerance (Chen and Xiong, 2005; Denslow et al., 2005; Havaux et al., 2009; Kalbina and Strid, 2006; Moccand et al., 2014; Zhang et al., 2014). Consequently, a further increase in B_6 content may be beneficial not only for the consumer, but also for the overall growth performance and yield of potato plants.

4.2.5 B_9—Folic Acid

Folate is the generic name for 5,6,7,8-tetrahydrofolate (THF) (Figure 2), which is a central compound in 1-carbon metabolism (Hanson and Gregory, 2011). It is required for the synthesis of methionine, pantothenate (vitamin B_5), thymidylate, formyl-Met-tRNA, and purines, as well as for sulfur–iron cluster metabolism (da Silva et al., 2014; Li et al., 2003; Tibbetts and Appling, 2010). Specifically in plants, the vitamin is also required in mitochondria for the activity of glycine decarboxylase, a multimeric enzyme complex that functions in the photorespiratory pathway (Douce and Neuburger, 1999).

Folate deficiency is a worldwide nutritional problem since the vitamin is often not present in high amounts in food, and it is very sensitive to certain food processing steps such as oxidizing conditions or reducing agents (Delchier et al., 2014). Insufficient uptake of folate is connected with a range of disease symptoms, from cardiovascular diseases, anemia, and birth defects such as spina bifida to increased risk for certain cancers (Blancquaert et al., 2010). Plants are probably the best primary source of the vitamin; however, to obviate any deficiencies in the population it is common in industrialized countries to fortify dietary products with folic acid, which can be rapidly metabolized to THF in the cell.

Previous analysis of folate in various types of potato germplasm demonstrated a significant variation in this vitamin, from around 550 ng folate/g DW in A090586-11 to nearly 1400 ng folate in Carola (Goyer and Navarre, 2007). The current RDA value for both men and women is 400 µg of dietary folate equivalents, which includes folic acid. For pregnant women, 400 µg of folic acid per day taken in through fortified foods or supplements is strongly recommended by the Food and Nutrition Board of the Institute of Medicine in the United States to decrease risk of birth defects.

Processed potato products mostly contain relatively low amounts of folate, especially in comparison to unprocessed potato (Goyer and Navarre, 2007, 2009; Goyer and Sweek, 2011), which emphasizes the heat sensitivity of the vitamin. For example, 100 g of boiled or baked potatoes has around 2.5% or 6.5% of the recommended RDA, respectively (Table 1). However, because potato is a product consumed in many countries to a high frequency and amount, it is considered to be a very good daily nutritional source of this vitamin (Alfthan et al., 2003; Brevik et al., 2005; Brussaard et al., 1997). Of note is that the developmental status of the tuber may be an important criterion for potato consumption, with immature tubers having a general tendency to contain higher folate levels in comparison to mature ones (Goyer and Navarre, 2009). Of further interest is that compared to primitive varieties, modern potatoes are significantly enhanced in folate levels, demonstrating the strong potential for breeding efforts in fortifying potato with this important phytonutrient (Goyer and Sweek, 2011).

5. Glycoalkaloids

Potatoes and other Solanaceae, eggplants, and tomatoes contain GAs, which are secondary metabolites that contribute to plant pest and pathogen resistance. From a dietary standpoint, GAs are regarded as antinutritive compounds capable of causing vomiting and other ill effects if ingested in high amounts (Hopkins, 1995; McMillan and Thompson, 1979). Breeders follow voluntary guidelines in the United States that new varieties should contain less than 20 mg/100 g FW of total GAs (Wilson, 1959) but the guidelines established in other countries can vary. GAs can contribute a bitter taste at higher concentrations (Sinden et al., 1976). GAs are found in much higher concentrations in leaves, sprouts, and fruit than in tubers. GA concentrations approaching 18 g/kg FW have been reported in sprouts (Valkonen et al., 1996).

5.1 GAs: Friends or Foes?

The view that GAs are categorically undesirable in potatoes may be simplistic and based on long-standing attitudes arrived at before some of the benefits of GAs were known. For example, in 1984 Sinden wrote: "Glycoalkaloids have no known positive role in human nutrition; the known and suggested effects of even small quantities of these natural toxicants are all negative" (Sinden et al., 1984). Numerous recent studies show potential health-promoting effects of GAs. Obviously, potatoes must contain GA amounts that are below the threshold for having any ill effects in humans, but breeding GAs completely out of potatoes, or breeding them to unnecessarily low levels, could have unintended consequences, including adverse effects on flavor, loss of potential health benefits, and increased pesticide use due to the plant having decreased pest and pathogen resistance. At higher levels GAs can cause potatoes to have a bitter taste, but lower levels are thought to have a positive effect on flavor (Janskey, 2010). New highly replicated and statistically powered studies to better define the threshold at which GAs cause bitterness would be useful for breeders, as older studies reached different conclusions on the bitterness threshold, with some suggesting bitterness is detectable at as low as 11 mg/100 g and others that the threshold is above 25 mg/100 g. A small study using purified GAs reported that chaconine was considerably more bitter than solanine (Zitnak and Filadelfi, 1985), which is a further reminder that the diverse GAs present in potatoes are likely to have quite different properties.

A more nuanced approached to tuber GA content may be worth considering, especially given the advances in identifying key genes in the GA pathway that enable GA concentrations to be greatly reduced. Increased understanding of GA regulation could allow approaches such as high levels of GAs in foliage to provide pest resistance, but lower amounts in tubers. Even so, high leaf concentrations but unnecessarily low levels in tubers could negatively affect flavor, decrease tuber pest resistance, and ameliorate the putative health benefits ascribed to GAs. Critical missing information is needed to allow the scientific community to identify the ideal GA range in potatoes. Historically, potatoes with amounts of GAs much higher than 20 mg/100 g FW were consumed, especially from some of the more primitive germplasm grown in the Andes. The medical evidence for the voluntary 20 mg/100 g FW limit does not seem particularly strong nor is it clear why exactly 20 mg is the threshold. An expert committee reviewed the GA data on behalf of the FAO and WHO and concluded "despite the long history of human consumption of plants containing glycoalkaloids, the available epidemiological and experimental data from human and laboratory animal studies did not permit the determination of a safe level of intake…the development of empirical data to support such a level would require considerable effort" (Kuiper-Goodman and Nawrot, 1992).

Relative to the amount of potato germplasm diversity available, commercial cultivars come from a narrow gene pool and consequently contain primarily solanine and chaconine, yet potato germplasm contains probably over 100 different types of GAs, of which almost nothing is known in terms of their effect on human health or plant disease resistance. This lack of knowledge is potentially preventing a valuable genetic resource from being tapped. Breeding programs can be uncomfortable if a new line is anywhere close to 20 mg and indeed might prefer lines to be below 10 mg, or even 5 mg. In part, this attitude is due to fears that environmental factors in a particular year or location might boost GA amounts above 20 mg/100 g FW. Trends to select breeding lines with ever lower amounts of GAs could potentially make it more difficult to more fully utilize the germplasm diversity available in wild potatoes for crop improvement. A study of the feeding behavior of specialist and generalist insect herbivores between domesticated potato and the wild species *Solanum commersonii* concluded that domestication altered the defensive capacity of *S. tuberosum* and that the altered GA profiles between the two explained the different feeding behavior of the herbivores (Altesor et al., 2014). A comparison of GA profiles between domesticated potatoes susceptible to Colorado potato beetle and six resistant wild potato species observed a correlation between resistance and GAs with a tetrose side chain, such as tomatine and dehydrocommersonine (Tai et al., 2014).

5.2 GA Biosynthesis

Potato GAs are steroidal alkaloids comprising a heterocyclic nitrogen and a C_{27} steroid conjugated to a sugar moiety, most commonly a tri- or tetrasaccharide. The GA biosynthetic pathway is not fully delineated, even for solanine and chaconine, the major potato GAs. Feeding experiments with labeled precursors demonstrated that GAs are derived from the mevalonate pathway via cholesterol (Heftmann, 1983; Johnson et al., 1963; Petersson et al., 2013) and occur throughout the tuber, but are primarily synthesized in the phelloderm (Krits et al., 2007). The GA nitrogen is suggested to be derived from arginine (Kaneko et al., 1976). Much remains to be elucidated about the genes and enzymology involved in conversion of cholesterol into the various GAs. Various glycosylation steps and several glycosyltransferases have been characterized or cloned (McCue et al., 2007; Moehs et al., 1997; Stapleton et al., 1991; Zimowski, 1991).

Identification of these GA biosynthetic genes has enabled transgenic approaches to decreasing potato GA content. Potatoes overexpressing a soybean sterol methyltransferase exhibited decreased amounts of GAs (Arnqvist et al., 2003), while antisense expression of several potato steroidal glycosyltransferases reduced GA levels and also affected metabolites from other pathways (McCue et al., 2007, 2005; Shepherd et al., 2015b). Comparative analysis between tomato and potato allowed identification of 10 genes involved in GA biosynthesis, six of which are present as a cluster on chromosome 7 (Itkin et al., 2013). Silencing of the GA metabolism 4 gene resulted in up to a 74-fold decrease in GAs. Moreover, GAs did not increase in these silenced plants in response to light.

Most efforts to decrease GA content through silencing have focused on biosynthetic genes upstream of cholesterol, potentially resulting in the accumulation of undesirable intermediates according to Sawai et al., who advocate blocking cholesterol synthesis instead. Previously, blocking cholesterol biosynthesis was not possible because the pathway in plants was not fully known, but in 2014 sterol side chain reductase 2 (SSR2) was shown to be a key enzyme for cholesterol biosynthesis (Sawai et al., 2014). Disruption of *SSR2* by transcription activator-like effector nucleases (TALEN) decreased GA content to ~10% of wild type, without decreasing tuber yield or having any obvious negative phenotype.

5.3 Factors That Increase GAs

A wide range of both biotic and abiotic plant stresses have long been known to affect potato GAs, including altitude, soil moisture, development, soil type, climate, mechanical damage, storage temperature, wounding, and disease (Maga, 1980; Sinden et al., 1984). Light-induced greening may be the best known cause of increased GAs and consumers typically know to peel away green portions of skin or to discard tubers that have heavily greened. Greening is considered a serious defect at the retail level, and some markets are less tolerant of greening than others. The extent to which greening and GA synthesis can be unlinked is not completely clear, but one study found genotypes of *Solanum microdontum* that neither greened nor accumulated GAs when exposed to light (Bamberg et al., 2015). Red light induces GA in tubers and increases expression of GA biosynthetic genes, including hydroxymethylglutaryl by coenzyme A reductase, with some evidence suggesting feedback inhibition at the transcriptional level (Cui et al., 2014). Crop fertilization with high nitrogen rates increased GA amounts to 76% compared to plants grown without mineral fertilization (Rytel et al., 2013), whereas another study found no consistent difference between organic or conventional management (Skrabule et al., 2013). Earlier studies also produced conflicting results about the effect of fertilization on GAs (Maga, 1980). Likewise there are conflicting reports about the effect of cold storage on GAs, with some studies finding increased GAs when potatoes were stored at very low temperatures (0–5 °C), while others found no change with cold storage over a range of temperatures and times (Maga, 1994).

5.4 Types of Potato GAs

Potato GAs usually belong to one of two structural types, either solanidanes or spirosolanes (Figure 3). The solanidanes, solanine and chaconine, often comprise upward of 90% of the total GA complement of domesticated potatoes, with chaconine often more abundant than solanine (Griffiths et al., 1997; Sotelo and Serrano, 2000). Solanine and chaconine are the potato GAs most are familiar with, but estimates have been made that the potato family, including wild species, may contain over 90 GAs (Friedman and McDonald, 1997). While characterizing small-molecule diversity in tubers from diverse potato germplasm, we observed using liquid chromatography–mass spectrometry that GAs constituted a major source of diversity. Mass spectrometry is well suited to GA analysis and is much more selective and sensitive than many methods used to analyze GAs. In our study of tubers from four wild potato species and three cultivars, about 100 GAs were tentatively identified (Shakya and Navarre, 2008). This number of GAs was unexpected, especially when considering only seven genotypes were analyzed and that we used only tubers, which have much lower GA concentrations than leaves, sprouts, flowers, or leaves. Consequently, potatoes may have a much greater diversity of GAs than previously appreciated. This GA diversity may offer opportunities for the production of future varieties with a more optimal GA complement. The predominance of solanine and chaconine in modern Western

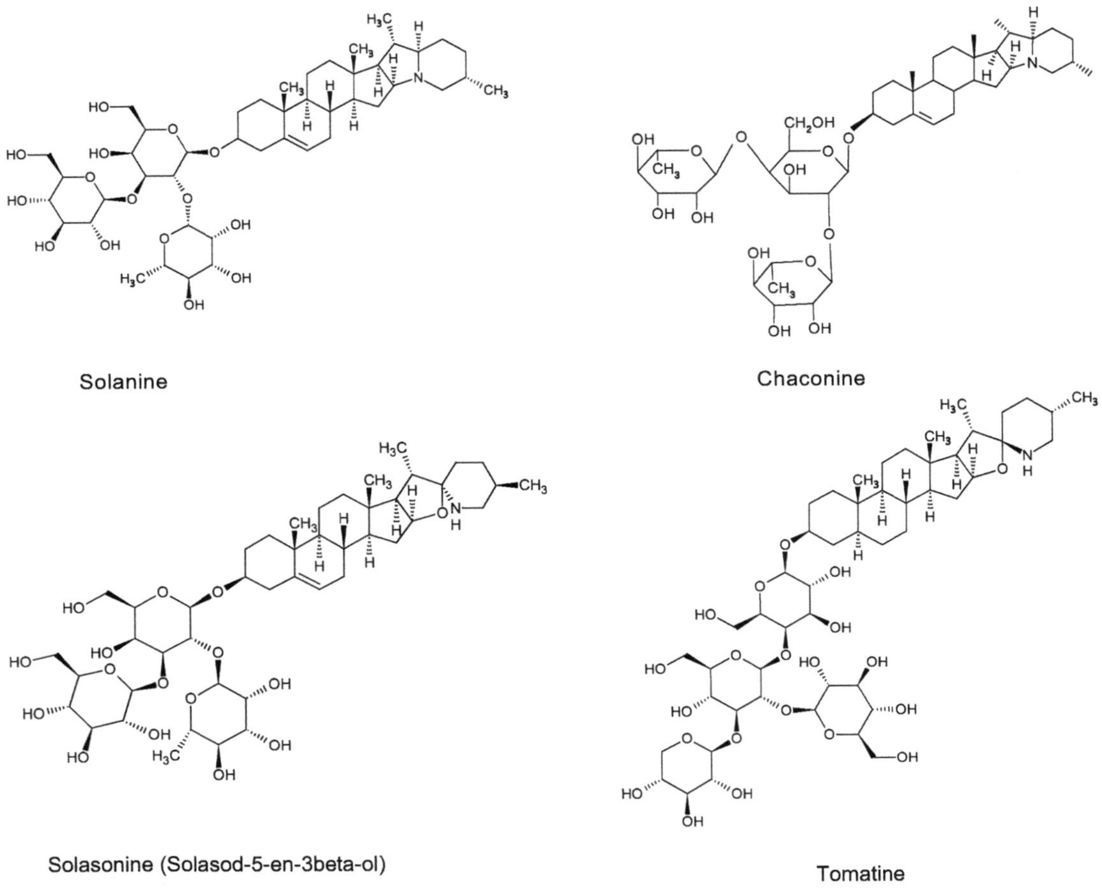

Solanine

Chaconine

Solasonine (Solasod-5-en-3beta-ol)

Tomatine

Figure 3
Four of the GAs present in potatoes.

cultivars may be due to the fact that only a small percentage of available potato germplasm was used in the breeding of these cultivars and reflects a bottleneck in the genetic diversity of commercial cultivars.

5.5 Toxic Effects of GAs

The effects on humans of eating potatoes with high GA concentrations have been well documented (Friedman, 2006). Symptoms can include cramping, diarrhea, vomiting, sweating, rapid pulse, and coma. The physiological effects of GAs are mainly a consequence of their disruption of cell membranes and inhibition of cholinesterase activity. Estimates have varied about the amount of GAs needed to be ingested to have toxic effects, with 1–5 mg/kg of body weight one suggested range, which is roughly equivalent to that of strychnine

(Mensinga et al., 2005; Morris and Lee, 1984). Doses as low as 5–6 mg/kg body weight may be lethal (Morris and Lee, 1984). Chaconine is more toxic than solanine and these two GAs become less toxic with progressive loss of sugars, with the aglycone being the least toxic (Friedman and McDonald, 1997). An important determinant of GA cholinesterase inhibitory activity seems to be the E and F rings of the aglycone (Roddick et al., 2001). In general solanidanes seem to be more toxic than spirosolanes. Friedman has suggested replacing solanidine and chaconine in potatoes with the less toxic tomatine (Friedman, 2002), which also has health-promoting properties. Such a goal could be accomplished by transgenic approaches or by identifying potato genotypes with naturally low solanidine/chaconine and high tomatine.

5.6 Health-Promoting Effects of GAs

In contrast to a few decades ago, when the statement was made that there are no known positive effects of GAs on nutrition, numerous health-promoting effects of GAs have been reported since then. Moreover, there was at least one early study that predated the above comment, in which inhibition of mouse sarcoma tumors by a solamarine was shown (Kupchan et al., 1965). Studies showing health-promoting effects of GAs are now so numerous that they are beyond the scope of this review to cover. Friedman estimates that there are at least six primary GAs ingested in the human diet from consumption of potatoes, tomatoes, and eggplants and suggests that at the concentrations found in potatoes, GAs may help protect against multiple cancers, but epidemiological studies are needed to support this possibility (Friedman, 2015).

More recent studies convincingly show that some GAs have anticancer properties, both in vitro and, importantly, in vivo. Lung cancer is the most frequent cause of cancer-related death, in part because of its propensity to metastasize before the cancer is diagnosed. Using a human lung cancer cell line, α-chaconine was shown to reduce metastasis, and it was suggested that this may allow new chemotherapeutic approaches (Shih et al., 2007). A separate study showed that solamargine, a GA found in some potatoes, increased the susceptibility of two different types of human lung cancer cell lines to several anticancer drugs (Liang et al., 2008).

The spirosolane solasodine may protect against skin cancer (Cham, 1994). GAs including tomatine, solanine, and chaconine were shown to inhibit growth of human colon and liver cancer cells in cell culture assays (Friedman et al., 2005; Lee et al., 2004), with a potency similar to that of the anticancer drug adriamycin. Anticancer effects were also seen in assays using cervical, lymphoma, and stomach cancer cells, and treatments using two or more GAs suggested both synergistic and additive effects (Friedman et al., 2005). Solamargine enhanced the susceptibility of breast cancer cells to anticancer drugs, while various solanidines exhibited cytotoxicity toward multidrug-resistant cancer cell lines (Shiu et al., 2009; Zupko et al., 2014). Micromolar concentrations of solamargine triggered apoptosis in human leukemia cells and squamous cell carcinoma, and other solasodines also had efficacy (Cui et al., 2012; Sun et al., 2011). Chaconine has shown efficacy in cell studies against stomach, colon, liver, and cervical cancer (Friedman, 2015) and prostate cancer (Reddivari et al., 2010).

A key question that cannot be answered using cell culture assays is whether dietary GAs can have similar effects. Solanine, the other major GA in potatoes, showed efficacy against various cancers, including against pancreatic cancer not only in vitro in human pancreatic cancer cells, but also in mice, in which it suppressed proliferation, angiogenesis, and metastasis (Lv et al., 2014). Topical formulations (i.e., creams) showed efficacy against skin cancer in mice and in humans, in which a clinical trial with 86 subjects showed that low doses of solasodines had 100% efficacy (Cham et al., 1991), and the authors suggested further study is warranted to examine whether these formulations can treat other dermatological diseases (Tiossi et al., 2014). Evidence that dietary tomatine is effective against cancer was shown in a feeding study using rainbow trout, in which reduced tumor incidence was found in tomatine-fed trout (Friedman et al., 2007), and in mice tomatine suppressed the growth of prostate cancer cells (Lee et al., 2013). Tomatidine has potential as a chemosensitizing agent, increasing the effectiveness of cancer chemotherapy by inhibiting multidrug resistance in human cancer cells (Lavie et al., 2001).

Beyond potential anticancer efficacy, GAs have been shown to boost the immune response. Mice treated with solasodines underwent total remission of their cancer, and a majority of the mice remained resistant to subsequent injection with terminal doses of cancer cells, suggesting these GAs may prime the immune system for long-term cancer protection (Cham and Chase, 2012). Mice fed GAs were more resistant to infection by *Salmonella* (Gubarev et al., 1998) and tomatine was demonstrated to potentiate the immune response of mice to vaccines (Rajananthanan et al., 1999). GAs are reported to inactivate several types of herpes-viruses (Chataing et al., 1997). Some GAs showed antimalarial activity in vivo (Chen et al., 2010) and were lethal against parasitic flatworms that infest humans (Miranda et al., 2012).

5.7 Future Needs in GA Research

Much remains to be understood about GAs in potato. One need is additional dietary studies to more conclusively establish at what threshold GAs impart a bitter taste. If needed information is missing about solanine and chaconine, then the situation is much worse for the numerous other GAs present in primitive germplasm, as virtually nothing is known about their effect on human health, plant disease resistance, or flavor. These other GAs could potentially be better tolerated by humans at higher concentrations, while at the same time benefiting plant pest and pathogen resistance. The diverse potato GAs beyond solanine and chaconine may represent a valuable genetic resource that is untapped owing to a lack of knowledge. Also needed is a better understanding of those factors that increase GA content in order to allow predictive modeling and reduce the possibility of potatoes with unanticipatedly elevated amounts of GAs reaching the market. Whether some potato genotypes are more prone than others to induced GA biosynthesis, or whether some genotypes have more stable amounts than others, is not well understood. A fear of unanticipated spikes in GAs may be one of the factors driving germplasm development to select genotypes containing below 10 mg/100 g, and such a low limit may exclude new genotypes that present no health hazard and make it even more difficult to incorporate wild germplasm with desirable traits into new domestic lines. Far more medical information about the bioavailability, dietary relevance, and both positive and negative effects on health of individual GAs must be obtained, along with knowledge of the effects of specific GAs on taste and plant stress tolerance, before it will be possible to develop potatoes with an optimal GA complement. If certain amounts and types of GAs are desirable in the diet, then potatoes, tomatoes, and eggplants may have an important role to play in supplying these compounds.

6. Potato Minerals

A wide range of mineral elements occurs in fruits and vegetables, the primary dietary source. The importance of optimal mineral intake to maintain good health is widely recognized (Avioli, 1988). In potatoes, major minerals include potassium, phosphorus, calcium, and magnesium. Minerals can be classified as nutritionally essential major minerals such as calcium (Ca), potassium (K), magnesium (Mg), sodium (Na), phosphorus (P), cobalt (Co), manganese (Mn), nitrogen (N), and chlorine (Cl) and nutritionally essential minor and trace minerals such as iron (Fe), copper (Cu), selenium (Se), Nickel (Ni), lead (Pb), sulfur (S), boron (B), iodine (I), silicon (Si), and bromine (Br).

Potato's most important mineral contribution to the diet may be potassium, of which it has high amounts. USDA databases list potato as providing 18% of the RDA of potassium; 6% of iron, phosphorus, and magnesium; and 2% of calcium and zinc. Avoiding leaching of the minerals during cooking will maximize the amount of minerals ingested from potatoes. Retention of most minerals is high in boiled potatoes cooked with skin (True et al., 1979) and in baked potatoes. There are significant differences in major and trace mineral contents among different genotypes of potato (Randhawa et al., 1984; True et al., 1978). In a study of 74 Andean landraces, the iron content ranged from 29.87 to 157.96 µg/g DW, the zinc content from 12.6 to 28.83 µg/g DW, and the calcium content from 271.09 to 1092.93 µg/g DW (Andre et al., 2007a). Relatively little variation was seen among nine Phureja group genotypes (Peña et al., 2015).

In addition to genotype, many other factors affect the mineral composition of potatoes, for example, location, stage of development, soil type, soil pH, soil organic matter, fertilization, irrigation, and weather (Lombardo et al., 2013). Drought stress caused an increase in most minerals, including potassium, in 21 Andean cultivars (Lefèvre et al., 2012). The same genotypes grown in different locations may have different mineral concentrations due to environmental interactions (Burgos et al., 2007). Location differences could be associated with difference such as soil mineral content, cultural practices, and sampling procedures (Delgado et al., 2001). Potassium, phosphorus, calcium, and magnesium concentrations changed with irrigation and fertilization in physiologically mature tubers (Ilin et al., 2000). The total concentration of iron, calcium, and zinc increased with application of fertilizers, whereas the content of phosphorus and molybdenum was reduced (Bibak et al., 1999; Frossard et al., 2000). Phosphorus, magnesium, and sodium are higher in the tubers of organically grown potatoes and manganese is higher in conventionally grown potatoes (Hajšlová et al., 2005; Warman and Havard, 1998). A survey of potatoes grown in the Mediterranean basin found higher levels of phosphorus in organically grown potatoes, but higher levels of potassium, calcium, and iron in conventionally grown potatoes (Lombardo et al., 2014). A significant positive correlation is found between the levels of nitrates/nitrites and potassium, but not with calcium and magnesium (Cieslik and Sikora, 1998). Mineral concentrations also varied significantly in tuber bud and stem ends, core, and vascular ring tissues. Higher concentrations of minerals are found in the bud end than in the internal layers of a tuber. Iron concentrations are the highest of the trace elements and varied significantly among the different tissues (Ereifej et al., 1998). The wide range of mineral content reported in potatoes may be due not only to genotype and environmental factors, but also to sampling issues.

6.1 Potassium

Increased potassium and reduced sodium in the diet are a priority goal of many nutritionists, who see this as an especially important dietary need to promote cardiovascular health. In this context, the low sodium and high potassium content of potatoes is notable. Potatoes had the lowest Na/K ratio among vegetables examined (Pandino et al., 2011), which is important because *The Dietary Guidelines for Americans* recommends decreased sodium intake, yet many of the foods high in potassium are also high in sodium. Potassium plays a fundamental role in acid–base regulation and fluid balance and is required for optimal functioning of the heart, kidneys, muscles, nerves, and digestive system. Health benefits of sufficient potassium intake include reduced risk of hypokalemia, osteoporosis, high blood pressure, stroke, inflammatory bowel disease, kidney stones, and asthma. A high intake of potassium and low intake of sodium have been hypothesized to reduce the risk of stroke (Larsson et al., 2008; Swain et al., 2008). However, most American women 31–50 years of age consume no more than half of the recommended amount of potassium and men's intake is only moderately higher (Campbell, 2004). A panel of experts evaluated 29 diets in 2015 and ranked the DASH diet the best overall (http://health.usnews.com/best-diet). The DASH (Dietary Approaches to Stop Hypertension) was developed by National Institutes of Health researchers based on the USDA food pyramid. The DASH diet allows potatoes and encourages increased potassium intake along with reduced sodium intake. Foods that increase dietary potassium may be especially valuable in the developed world, in which other mineral deficiencies are not common and cardiovascular disease is a far greater threat to health than insufficient iron.

Potatoes qualify for a health claim approved by the U.S. Food and Drug Administration (FDA), which states: "Diets containing foods that are a good source of potassium and that are low in sodium may reduce the risk of high blood pressure and stroke." Potatoes rank highest for potassium content among 20 most frequently consumed raw vegetables and fruits (source: US Potato Board, DHHS FDA). Potassium varies from 3550 to 8234 µg/g FW (Casanas et al., 2002; Rivero et al., 2003). Potassium content increases during the entire growing season (Lisinska and Leszczynski, 1989). On average one baked potato (156 g) contains 610 mg potassium (USDA, 2005). This is even higher than in banana, a food often recommended by dieticians to people who need to supplement potassium consumption. The Dietary Reference Intake of potassium for adult men and women is 3000–6000 mg per day. The U.S. National Academy of Sciences has increased the recommended intake for potassium to at least 4700 mg per day. Potatoes and beans were found to be the least expensive source of potassium among 98 fresh, frozen, and canned vegetables (Drewnowski and Rehm, 2013).

6.2 Phosphorus

Other than potassium, phosphorus is the main mineral present in the tubers. It has many roles in the human body and is a key player in healthy cells, teeth, and bones. Inadequate phosphorus intake results in abnormally low serum phosphate levels, which result in loss of appetite, anemia, muscle weakness, bone pain, rickets osteomalacia, increased susceptibility to infection, numbness and tingling of the extremities, and difficulty walking. In potatoes phosphorus ranges from ~1300 to 6000 µg/g DW (Lisinska and Leszczynski, 1989; Randhawa et al., 1984; Sanchez-Castillo et al., 1998). Daily requirements are 800–1000 mg.

6.3 Calcium

The potato tuber is considered to be a significant source of calcium, with a very wide range reported. Two studies reported calcium content up to 100 µg/g DW and 459 µg/g FW (Lisinska and Leszczynski, 1989; Randhawa et al., 1984). Among 74 Andean landraces, calcium ranged from 271 to 1093 µg/g DW (Andre et al., 2007a). Wild *Solanum* species vary in the ability to accumulate tuber calcium (Bamberg et al., 1998). High levels of tuber calcium are associated with resistance to pathogens (McGuire and Kelman, 1986) and abiotic stress (Tawfik et al., 1996). Calcium is important for bone and tooth structure, blood clotting, and nerve transmission. Deficiencies are associated with skeletal malformations and blood pressure abnormalities. The RDA for calcium is set at levels to reduce osteoporosis (Bachrach, 2001; Bryant et al., 1999).

6.4 Magnesium and Manganese

Potato magnesium levels range from 142 to 359 µg/g FW (Casanas et al., 2002; Rivero et al., 2003). Magnesium is required for normal functioning of muscles, heart, and the immune system. Magnesium also helps maintain normal blood sugar levels and blood pressure. The RDA for magnesium is 400–600 mg. Potato manganese content varies from 0.73–3.62 µg/g FW (Rivero et al., 2003) to 9–13 µg/g DW (Orphanos, 1980). Manganese has a role in enzyme reactions concerning blood sugar, metabolism, and thyroid hormone function. Recommended daily intake in the United States is 2–10 mg.

6.5 Iron

Iron deficiency affects more than 1.7 billion people worldwide and has been called the most widespread health problem in the world by the WHO. Owing to severe iron deficiency, more than 60,000 women die in pregnancy and childbirth each year, and almost 500 million women of childbearing age suffer from anemia. Dietary iron requirements depend on numerous factors, for example, age, sex, and diet composition. Recommended daily intake is 10–20 mg in the United States. Potato is a modest source of iron. Potato iron should be quite bioavailable because it has very low levels of phytic acid, unlike the cereals. A study of cultivated varieties showed 0.3–2.3 mg of Fe in a 100-g tuber (True et al., 1978). Ranges of iron content from 6 to 158 µg/g of DW have been reported (Andre et al., 2007a; Wills et al., 1984). Some Andean potatoes have iron content comparable to levels found in some cereals (rice, maize, and wheat) (Scurrah et al., 2006). Iron content has broad-sense heritability and ranged from 17 to 62 µg/g DW in a study of potatoes grown in the Pacific Northwest, suggesting that iron levels in potatoes can be further increased by breeding (Brown et al., 2010b; Paget et al., 2014).

6.6 Zinc and Copper

Significant differences in zinc content in potatoes are revealed among the different varieties. The zinc content ranges from 1.8 to 10.2 µg/g FW (Randhawa et al., 1984; Rivero et al., 2003; Andre et al., 2007a). Yellow-fleshed potatoes from various cultivars contain zinc at between 0.5 and 4.6 µg/g FW (Dugo et al., 2004). Zinc is needed for the body's immune system to work properly and is involved in cell division, cell growth, and wound healing. The U.S. RDA is 15–20 mg. Copper is needed for the synthesis of hemoglobin, proper iron metabolism, and maintenance of blood vessels. The U.S. RDA is 1.5–3.0 mg. Copper in potatoes varies from 0.23 to 11.9 mg/kg FW (Rivero et al., 2003; Randhawa et al., 1984). Only a twofold range was found in another study, which showed broad-sense heritability for copper, confirming that levels should be able to be increased by breeding (Brown et al., 2010a). Like zinc, copper is also high in yellow-fleshed potatoes (Dugo et al., 2004).

7. Potato Phenylpropanoids

Phenylpropanoids are secondary metabolites that have complex roles in plants, including promoting biotic and abiotic stress tolerance. Phenylpropanoids are a diverse group of thousands of different compounds, and they are thought to provide numerous diverse health-promoting effects in the human diet, including effects on the gut microbiome, longevity, mental acuity, cardiovascular disease, and eye health (Cardona et al., 2013; Manach et al., 2004; Parr and Bolwell, 2000; Scalbert et al., 2005). Phenolics are the most abundant antioxidants in the diet. Plant phenolics may contain a treasure trove of potential health-promoting compounds. For example, many of the reports in the popular press about the positive health effects of green tea, coffee, or wine are due to phenolic content. The role of phenolics in health is an area of active ongoing medical research with much remaining to be understood. Reflecting the interest in these phytonutrients is that conducting a Google search using phenolics and health as keywords returned over 700,000 links in 2005 and 1.6 million in 2008, whereas a Google Scholar search lists 16,000 publications since 2014.

A *Scientific American* article included the contentious quote "being mainly starch, potatoes do not confer the benefits seen for other vegetables" (Willett and Stampfer, 2003). In addition to not being supported by the vitamin and mineral data described above, this statement of opinion does not jibe with the experience of many plant molecular biologists who faced difficulty with PCR in potatoes because of the high polyphenol content, a well-known issue with tubers (Singh et al., 1998). In fact, potatoes are an important source of dietary phenylpropanoids. One study evaluated the phenolic contribution of 34 fruits and vegetables to the American diet and concluded that potatoes were the third most important source after apples and oranges (Chun et al., 2005). The potatoes used in that study were an unspecified variety bought at a supermarket that almost certainly contained a small amount of phenolics relative to high-phenolic potatoes that are less widely available.

Defining the role of specific phenylpropanoids in human health is a complex undertaking for various reasons, including that there are so many different types, and they are ingested as part of a complex matrix that may have cross-interactions. Moreover, the original ingested compounds are often metabolized in the body, and the resulting metabolites may have health-promoting properties and efficacy different from those of the original compound (Hollman, 2014). In some cases the original compound may be less important than the resulting metabolized products. The influence of the gut microbiome on human health has been one of the most exciting and active areas of health research in recent years. However, much remains to be understood, including the complex and poorly understood relationship between polyphenols and microbiota, each of which influences the other. The specific composition of an individual's gut microbiota may influence the bioefficacy of polyphenols and metabolize them to breakdown products that have the actual health-promoting effects and bioavailability, not the original compounds. This presents an additional layer of complexity, because individuals do not necessarily have the same microbiome, this means the efficacy of dietary phenylpropanoids may vary among individuals, depending in part on their particular microbiome (Bolca et al., 2013; Cardona et al., 2013).

Potatoes may be underappreciated as a source of dietary phenolics. We compared (Figure 4) the total phenolic content and antioxidant content (measured by oxygen radical absorbance capacity (ORAC)) of four potato cultivars to those of 15 common vegetables purchased at a supermarket (Navarre et al., 2011). Russet Burbank and Norkotah Russet are two of the most common white-fleshed potatoes grown in North America, whereas Magic Molly is a purple-fleshed potato developed in Alaska, and Ama Rosa and CO97226 are red-fleshed baby potatoes. Burbank is the most grown potato in North America and contains antioxidants and phenolic amounts that rivaled some of the other vegetables, whereas Norkotah, which has a high amount of phenolics for a white-fleshed line, had amounts comparable or superior to those of several other vegetables including peas and carrots. Notably, the color-fleshed potatoes contained amounts that rivaled or surpassed all the other vegetables. If expressed on a FW basis, the color-fleshed potatoes have higher amounts than the vegetables to which they were compared, including spinach and broccoli. A primitive Phureja group potato was found to have extraordinarily high amounts of phenolics (>40 mg/g DW) and a high ORAC value (>1000) (Pillai et al., 2013), and to our knowledge these are by far the highest amounts ever reported in a potato. These data are not meant to suggest that potatoes are better than other vegetables, but as a counter to the misperception held by some that potatoes provide only starch and little else. Such data also suggest that future breeding efforts can develop potatoes with even higher amounts of phytonutrients than the most popular cultivars currently grown, which already make potatoes the third leading source of phenolics in the American diet.

A study of 74 Andean potato landraces found about an 11-fold variation in total phenolics and a high correlation between phenolics and total antioxidant capacity (Andre et al., 2007a). We screened tubers from thousands of cultivars and wild potato species for phenolics and found over a 15-fold difference in the amount of phenolics in various potato genotypes. Red- and purple-fleshed potatoes tend to have the highest amounts, but are not as widely consumed as white or yellow potatoes. However, other than anthocyanins, most phenolics are colorless and thus are relevant for white-fleshed cultivars, the preferred type of potato in many countries, including the United States.

Many potato nutrients differ in the amounts that accumulate in the skin versus the flesh. The majority of phenolic compounds are found in greater concentrations in the skin, but large quantities are also present in the flesh. Because a sizable majority of the FW of a mature potato is contributed by the flesh, overall the flesh will typically contain more phenolics than the skin on a per-tuber basis. In some cases, such as a yellow-skinned, purple-fleshed potato, the flesh will typically contain a higher concentration of phenolics per unit weight than the skin. Potato skins are well known to be nutritious and consumers who realize this can choose

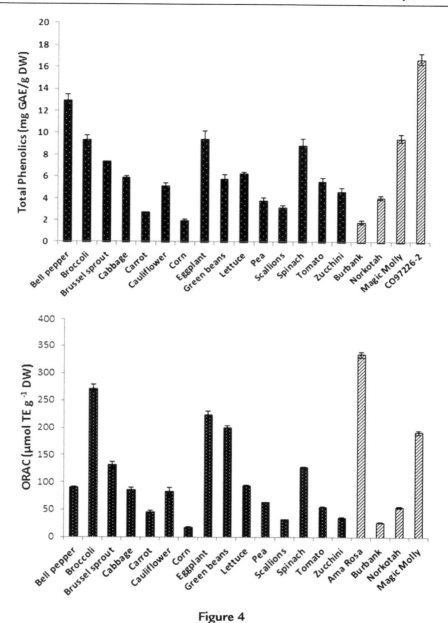

Figure 4

Comparison of the total phenolics (top) and antioxidants (bottom) in four varieties of potatoes with 15 vegetables. Potatoes are represented by striped bars. Total phenolics were measured by the Folin–Ciocalteu method and antioxidants by ORAC. Potatoes were all mature, except for Ama Rosa and CO97226-2, which were baby potatoes harvested 60–80 days after planting.

recipes and products that include skins. Since potato skins are a rich source of phenolics (Nara et al., 2006), the phenolic content in the potato skins generated as waste during french fry processing might be useful as a "value-added" product.

7.1 Chlorogenic Acid

The most abundant phenolics in tubers are the caffeoyl esters, of which chlorogenic acid (CGA) is predominant. CGA (Figure 5) can comprise over 90% of a tuber's total phenolics (Malmberg and Theander, 1985). CGA synthesis in plants was shown to be via hydroxycin-namoyl CoA:quinate hydroxycinnamoyl transferase (HQT), which creates new opportunities to manipulate CGA biosynthesis in potatoes (Niggeweg et al., 2004). To examine ways to increase tuber polyphenol diversity and because HQT expression in tubers did not correlate with CGA concentrations in tubers, we silenced HQT in potatoes to examine whether some of the CGA was supplied by alternative pathways. The results showed that any synthesis by other pathways was minor at best (Payyavula et al., 2015). Interestingly, rerouting phenylpro-panoid metabolism away from CGA did not result in a marked increase in the quantitative diversity of phenylpropanoids present, nor did the plant fully compensate for the greatly reduced amounts of CGA by an equivalent increase in the amount of other phenylpropanoids; instead phenylpropanoid metabolism was downregulated in an organ-specific manner.

Because CGA is the most abundant polyphenol in potatoes, its putative health-promoting effects are of particular interest. CGA supplements are available in health food stores and are typically extracted from artichoke. Dietary CGA is bioavailable in humans and its bioactivity may be

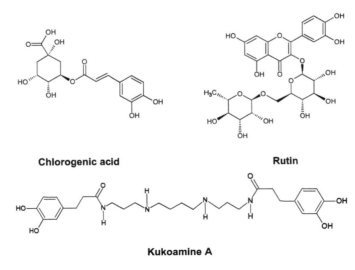

Chlorogenic acid

Rutin

Kukoamine A

Figure 5
Structure of CGA, rutin, and kukoamine A.

modulated by the gut microbiome (Farah et al., 2008; Monteiro et al., 2007; Tomas-Barberan et al., 2014). CGA may promote a healthy gut microbiome; in a batch culture fermentation model of the colon, CGA, but not caffeine, was found to promote growth of *Bifidobacterium* bacterial species that could be beneficial for health (Mills et al., 2015). CGA protected animals against degenerative, age-related diseases when added to their diet and may reduce the risk of some cancers and heart disease (Nogueira and do Lago, 2007). CGA is also thought to be antihypertensive (Onakpoya et al., 2015; Yamaguchi et al., 2007) and may combat metabolic syndrome (Cheng et al., 2014). A small human dietary study with purple potatoes that contained high amounts of CGA significantly decreased the blood pressure of the subjects, but whether this was due to CGA is not known (Vinson et al., 2012). Purple potatoes with high amounts of anthocyanins and CGA decreased inflammation and DNA damage in adult males (Kaspar et al., 2011). CGA is reported to be antiviral and antibacterial. CGA may decrease the risk of type 2 diabetes (Legrand and Scheen, 2007) and has been shown to slow the release of glucose into the bloodstream (Bassoli et al., 2008; Tunnicliffe et al., 2011). If this is the case, then potatoes with high CGA content may give lower glycemic index values. Mice administered CGA had better glucose tolerance, insulin sensitivity, fasting glucose levels, and lipid profiles, apparently through the activation of AMPK (Ong et al., 2013).

7.2 Flavonols, Anthocyanins, and Hydroxycinnamic Acid Amides

Potatoes contain flavonols such as rutin (Figure 5), but have not been thought to be important sources of dietary flavonols because the flavonol amounts in tubers are rather low. One group showed that flavonols increased in fresh-cut tubers up to 14 mg/100 g FW and suggested that because of the large amount of potatoes consumed, they can be a valuable dietary source (Tudela et al., 2002a). We have screened thousands of potato genotypes and found that the amounts of flavonols can vary by more than 30-fold and there is even sizable variation within the same genotype (Navarre et al., 2011). Possibly, of all the compounds we routinely measure, the flavonols vary the most. The reason for such variation may be because the amounts in tubers are very low; sometimes only trace amounts are present so this makes for sizable differences compared to a potato that has higher (but still relatively low) amounts. Interestingly, while purple- and red-fleshed potatoes have high amounts of phenolic acids and anthocyanins that correlate with flesh color, flavonol concentrations do not correlate with flesh color. White potatoes can have just as much as high-phenylpropanoid colored lines. Interestingly, while tubers contained only microgram per gram amounts of flavonols, potato flowers contained 1000-fold higher amounts, showing that potatoes are well capable of synthesizing large amounts of flavonols, despite the low amounts in tubers (Payyavula et al., 2015). Numerous studies suggest that quercetin and related flavonols have multiple health-promoting effects, including reduced risk of heart disease; lowered risk of certain respiratory diseases, such as asthma, bronchitis, and emphysema; and reduced risk of some cancers including prostate and lung cancer (Kawabata et al., 2015).

Potatoes, particularly color-fleshed cultivars, can contain substantial amounts of anthocyanins, compounds that can function as antioxidants and have other health-promoting effects. A gene encoding dihydroflavonol 4-reductase is required for production of pelargonidins in potato, and other candidate genes have been identified (De Jong et al., 2003, 2004). An anthocyanin-enriched fraction from potatoes had anticancer properties (Reddivari et al., 2007). Lewis screened 26 color-fleshed cultivars for anthocyanin content and found up to 7 mg/g FW in the skin and 2 mg/g FW in the flesh (Lewis et al., 1998). Another study evaluated 31 colored genotypes and found a range of 0.5–3 mg/g FW in the skin and up to 1 mg/g FW in the flesh (Jansen and Flamme, 2006). Brown evaluated several genotypes for anthocyanins and found whole tubers that contained up to 4 mg/g FW and that anthocyanin concentration correlated with antioxidant value (Brown et al., 2005).

Tubers contain a diverse panel of hydroxycinnamic acid amides (HCAAs). We have observed over 30 putative HCAAs in a single tuber. Solid proof for the role of specific tuber polyamines remains sparse, but they are implicated in the regulation of starch biosynthesis (Tanemura and Yoshino, 2006), calystegine synthesis (Stenzel et al., 2006), disease resistance (Matsuda et al., 2005), sprouting (Kaur-Sawhney et al., 1982), abiotic stress resistance, and ripening (Fellenberg et al., 2012; Gaquerel et al., 2014). Like CGA, they have a caffeoyl moiety, which might explain why HCAA concentrations increased when CGA biosynthesis was suppressed (Payyavula et al., 2015). A British group detected compounds called kukoamines in potatoes (Parr et al., 2005), which are hydroxycaffeoyl–polyamine conjugates (Figure 5) and had previously been found only in a Chinese medicinal plant, in which they were being studied because they lower blood pressure. It still needs to be established whether enough of these compounds are present and bioavailable enough in potatoes to be bioactive in humans.

7.3 Browning

A concern about developing potatoes with high amounts of polyphenols is whether they would have unacceptable levels of browning or after-cooking darkening as suggested by the older literature. Polyphenol oxidase is the primary enzyme responsible for browning and has multiple substrates; in potato tyrosine and CGA may be the primary two. One more recent study showed that the amounts of neither total phenolics, CGA, nor polyphenol oxidase correlated with the amount of browning observed in fresh-cut potatoes and that these were not rate limiting in the development of browning (Cantos et al., 2002). Additionally, using a QTL approach, another group found no correlation between browning and CGA (Werij et al., 2007). Association genetics using a candidate gene approach identified 21 marker-browning trait associations, the most important of which may have been a polaxa marker that appears to be a specific polyphenol oxidase (PPO) allele and a class III lipase (Urbany et al., 2012, 2011).

Decreasing CGA synthesis in potatoes markedly reduced discoloration in sliced tubers exposed to air for 12 h (Payyavula et al., 2015). Baby potatoes have high amounts of CGA, yet do not seem more prone to browning than mature potatoes. Therefore, other factors including the amounts of vitamin C, organic acids, and iron present all probably have complicated interactions that modulate browning. Potatoes first cured at 16 °C for 10 days underwent markedly less browning when subsequently sliced and the slices were then stored at 2–3 °C for up to 12 days (Wang et al., 2015). PPO activity and phenolic content both increased after wounding, despite the reduced browning, and the authors suggested elevated CGA concentrations may have inhibited browning. Zebra chip is an emerging potato disease that renders potatoes unsuitable for processing, in part owing to vastly accelerated browning of sliced tubers. Tyrosine, polyphenol oxidase, and CGA levels are greatly elevated in infected tubers (Alvarado et al., 2012; Navarre et al., 2009). Polyphenol oxidases in potatoes constitute a large gene family, with some appearing to be much more central to browning than others (Chi et al., 2014). In 2015, the USDA approved a potato genetically engineered by Simplot that uses inhibitory RNA to reduce tuber browning by suppressing polyphenol oxidase 5 expression in tubers (Waltz, 2015). Transgenic potatoes with reduced polyphenol oxidase expression were found to have higher levels of various phenylpropanoids, including CGA (Llorente et al., 2014), while another study with transgenic plants with suppressed PPO showed relatively small differences in the metabolome, much less than that caused by wounding (Shepherd et al., 2015a).

8. Effect of Development on Tuber Phytonutrients

Commercial immature potatoes are called by various names, including new potatoes, baby potatoes, or gourmet potatoes. Baby potatoes are harvested 60–80 days after planting and typically weigh about 28–56 g. These products may be especially important for the potato industry because of their potential appeal to middle-class consumers, not a demographic that currently consumes large amounts of traditional potato products in North America. Potato consumption has been declining in the developed world for complex reasons, including changing lifestyles, greater disposable income leading to altered diets, more food choices, fewer home-cooked meals, and increased demand for fast, easy-to-prepare meals. The perception that potatoes are boring or do not provide much nutrition is also a negative for some consumers. Baby potatoes may have greater appeal to more affluent consumers because they are a more exotic, gourmet item; can be visually striking; cook faster; and, because they are eaten with skin on, require less prep time. Typically they command a price premium and are valued by shoppers because of their perceived superior flavor. Furthermore, baby potatoes contain greater amounts of many phytonutrients than at maturity (Navarre et al., 2013; Payyavula et al., 2013), a trait valued by health-conscious consumers, although still not widely realized. Collectively, these traits can extend the appeal of potatoes into new demographic groups and may provide additional diversification for the industry.

Tuber development has marked effects on various phytonutrients including carotenoids (Morris et al., 2004), vitamin C (Blauer et al., 2013), folate (Goyer and Navarre, 2009), protein (Rosenstock and Zimmermann, 1976), and phenylpropanoids (Hung et al., 1997; Navarre et al., 2013; Payyavula et al., 2013). Baby potatoes also have higher amounts of GAs (Maga, 1980). In a high-phenylpropanoid purple cultivar, phenolics decreased from 14 to 10 mg/g DW during development, carotenoids 30–70%, and the predominant anthocyanin from 6.4 to 4 mg/g. Likewise phytonutrient gene expression tends to decrease with increasing maturity, up to 70% for dihydroflavonol reductase, which encodes the enzyme responsible for the first committed step in anthocyanin biosynthesis (Payyavula et al., 2013). During screening of hundreds of genotypes being evaluated for their potential to be used for baby potato production, we observed anthocyanin content up to 15–19 mg/g DW and ORAC values of over 300 μmol Trolox equivalents/g DW, values that rival spinach and kale. As seen in Figure 4, baby potatoes can contain high amounts of phenolics and antioxidants that equal or surpass other vegetables, including spinach and broccoli. Purple baby potatoes decreased blood pressure in a human dietary study (Vinson et al., 2012).

Future needs for baby potato marketing include helping consumers become more educated about their nutritional merits. This can be accomplished without denigrating mature potatoes, which provide numerous benefits, as documented earlier in this chapter. Also needed is consistently attractive packaging and carefully selected potatoes that are worthy of a premium, gourmet product. Baby potato production requires using cultivars and management that maximize tuber set and the production of a large number of small tubers, as opposed to a smaller amount of large potatoes. Consequently, management is much different from that of a Russet crop and the need remains to develop cultivars and management methods that can increase the yields by a few more tons per acre.

9. Carotenoids

In addition to the hydrophilic compounds described above, potatoes also contain dietarily desirable lipophilic compounds like carotenoids that are synthesized in plastids from isoprenoids (Dellapenna and Pogson, 2006). Carotenoids have numerous health-promoting properties including provitamin A activity and decreased risk of several diseases (Fraser and Bramley, 2004; Gammone et al., 2015). Tuber carotenoid makeup varies by cultivar, but violaxanthin and lutein are usually the most abundant tuber carotenoids. These may be particularly important for eye health and reduced risk of age-related macular degeneration (Abdel-Aal et al., 2013; Chucair et al., 2007; Tan et al., 2008). The yellow/orange flesh color found in some potatoes is due to carotenoids. Orange coloration in potatoes is due to zeaxanthin (Brown et al., 1993), whereas the lutein concentration correlates well with the intensity of yellow coloration. Over a 20-fold range in carotenoid concentrations has been reported in potato germplasm, but results are not in agreement about to what extent the variation is controlled at the transcriptional level (Morris et al., 2004; Payyavula et al., 2012; Zhou et al., 2011). Tuber carotenoid concentrations may be partly controlled by suborganellar compartmentation of the biosynthetic genes (Pasare et al., 2013). Genome-wide QTL analysis of tuber carotenoid content identified one QTL on chromosome 3 that accounted for up to 71% of the variation and that was probably an allele of β-carotene hydroxylase, while a second QTL accounting for up to 20% of the variation was identified on chromosome 9 that did not appear to be a biosynthetic gene (Campbell et al., 2014).

White-fleshed potatoes usually contain fewer carotenoids than the yellow or orange cultivars. A single allele, *Chy2*, may be responsible for changing white flesh to yellow (Wolters et al., 2010). One study found white cultivars had 27–74 µg/100 g FW of carotenoids (Iwanzik et al., 1983). Cultivated diploid potatoes derived from *Solanum stenotomum* and *Solanum phureja* were found to contain up to 2000 µg/100 g FW of zeaxanthin (Brown et al., 1993). A study of 74 Andean landraces found total carotenoid concentrations ranging from 3 to 36 µg/g DW (Andre et al., 2007a). A screen of 24 Andean cultivars identified genotypes with almost 18 µg/g DW each of lutein and zeaxanthin and just over 2 µg/g DW of β-carotene (Andre et al., 2007b). Sixty varieties grown in Ireland over 2 years had total carotenoid amounts that ranged from trace up to 28 and 9 µg/g DW in the skin and flesh, respectively (Valcarcel et al., 2015). Storage can have a significant effect on carotenoid composition and change the type present relative to those present in freshly harvested potatoes (Fernandez-Orozco et al., 2013).

Traditional breeding can increase carotenoid amounts because carotenoids have good broad-sense heritability in potatoes (Haynes et al., 2010). Numerous groups increased potato carotenoids using transgenic strategies. Overexpressing a bacterial phytoene synthase in tubers of the cultivar Desiree increased carotenoids from 5.6 to 35 µg/g DW and changed the ratios of individual carotenoids. β-Carotene concentrations increased from trace amounts to 11 µg/g DW and lutein levels increased 19-fold (Ducreux et al., 2005). A twofold increase in carotenoids was observed in tubers overexpressing *Or* after 6 months of cold storage but no such increase was observed in wild-type or empty-vector-transformed plants (Lopez et al., 2008).

The overexpression of three bacterial genes in Desiree achieved a 20-fold increase in total carotenoids to 114 µg/g DW and a 3600-fold increase in β-carotene to 47 µg/g DW (Diretto et al., 2007). A 250-g serving of these potatoes was estimated to provide 50% of the RDA of vitamin A. Potatoes engineered to have higher zeaxanthin levels were fed to human subjects and the zeaxanthin was found to be readily bioavailable (Bub et al., 2008). Carotenoids from yellow-fleshed potatoes were highly bioaccessible when monitored using an in vitro digestion model, more so than those from corn and red pepper, and these potatoes were suggested to provide 150% of the recommended zeaxanthin concentration based on the mean potato intake in the Andes (Burgos et al., 2013).

10. Effect of Cooking on Phytonutrient Content

Most of the research on potato phytonutrients measures the amounts in raw tubers, whereas it is the amount present after cooking that is important from a dietary standpoint. The literature about the effect of cooking on potatoes and other vegetables reports dramatically different results for reasons that are not clear. Folate is reported to decrease, increase, and stay the same after cooking (Konings et al., 2001), whereas CGA was destroyed in boiled carrots and potatoes (Dao and Friedman, 1992; Miglio et al., 2008), but others report CGA was higher in cooked potatoes and almost double in cooked versus raw artichokes (Ferracane et al., 2008; Navarre et al., 2010). Vitamin C was destroyed in fried carrots but only decreased 14% in fried zucchini in the same experiment (Miglio et al., 2008). A priori, many would have the expectation that cooking would decrease phytonutrients in foods owing to thermal degradation. However, studies clearly show that some foods can be properly cooked without degrading the phytonutrient content. Indeed, the assay for some phytonutrients, such as folate, involves boiling. Similarly, steps can be taken to minimize leaching, by boiling potatoes with their skin on or microwaving instead of boiling. These are important points because they show that many phytonutrients need not be destroyed by cooking and are not inherently labile at normal home-cooking temperatures, i.e., while cooking can considerably degrade nutrient content, it need not. For example, studies have shown that cooking potatoes results in no change or an increase in extractable phytonutrients compared to raw; not only did cooking not decrease the phytonutrient content, it actually made them more extractable (Blessington et al., 2010; Mulinacci et al., 2008; Navarre et al., 2010). Baby potatoes from three different cultivars were baked, boiled, microwaved, steamed, or stir-fried and none showed a decrease in phytonutrients, including vitamin C, rutin, and CGA, and there was a trend to increased extractability of all these compounds after cooking. Others reported about a 50% loss of phenolics in potatoes by all cooking methods, but less with boiling than microwaving or baking (Perla et al., 2012). Similar results were observed in different cultivars, with less loss of CGA and vitamin C occurring with boiling, rather than baking or microwaving (Lachman et al., 2013). Surprisingly, cooking resulted in up to a 15-fold increase in extractable anthocyanins (Lachman et al., 2013), whereas Perla et al. (2012) reported a decrease. Purple Majesty potatoes had decreased total phenolics by all cooking methods, but increased anthocyanins and no decrease in overall antioxidant activity (Lemos et al., 2015). Some anthocyanins may be more heat stable than others (Kim et al., 2012).

Similarly, differing results have been reported for GAs, from no change after cooking to major decreases (Mulinacci et al., 2008; Ponnampalam and Mondy, 1983). An 80–90% reduction in GAs was reported in fried potato chips or french fries (Tajner-Czopek et al., 2012). No difference was seen in GA content when cooked under conditions typical of home cooking, and not until cooking temperatures exceeded 210°C did GA decomposition start to occur (Takagi et al., 1990). Depending on the study, solanine decomposes between 228 and 286°C (Porter, 1972).

Carotenoids are also affected by cooking, with decreases reported for violaxanthin and antheraxanthin, but not lutein and zeaxanthin, after boiling (Burgos et al., 2012). Boiling also increased the amount of carotenoid isomerization in tubers (Burmeister et al., 2011).

11. The Role of Potatoes in Global Food Security

Whereas the new cultivation of potatoes in Europe several centuries ago greatly improved food security, potatoes may be positioned to reprise that role in the coming years as the United Nations predicts the world's population will reach 8.1 billion by 2025. This population increase and concomitant changes in dietary preferences in developing countries, as their middle class develops and consumes increased amounts of meat and dairy, suggest global crop production may need to double by 2050 (Ray et al., 2013). In much of the developed world, potatoes are the most eaten vegetable. Moreover, in the developing world, potato consumption is increasing rapidly and in 2005 the developing world for the first time produced more potatoes than the developed world. The bottom graph in Figure 6 shows the trend in potato production in five continents from 1961 to 2011. Immediately apparent is the decreased per capita production in Europe and soaring production in Asia and Africa, whereas production in North America has been relatively stable. The top shows the amount of potatoes available on a per capita basis, basically a reflection of the amount of potatoes available per person in a country (not the amount actually consumed per capita) and is a reflection of the degree of food security provided by that crop per country. While China now produces more potatoes than Europe, far fewer potatoes are available per person in China versus Europe. Figure 7 uses FAOSTAT data and shows the tons of potatoes produced annually in eight countries between 1963 and 2013, with dramatic increases over this period in China, India, Iran, Nigeria, and South Africa and falling production in Poland and Germany. Such data show that the potato is becoming a global staple crop that can help provide increased food security to food-insecure regions.

In terms of food security it is notable that potatoes yield more calories per acre than any other major crop, a point that becomes increasingly important in light of the planet's increasing population, urban development, increasing consumption of less energy-efficient foods by the emerging middle classes, uncertain effects of climate change, and competition for farmland by biofuel crops.

Food security may be defined as "when all people at all times have access to sufficient, safe, and nutritious food to meet their dietary needs for an active and healthy life." This definition specifies "nutritious food," which potatoes are. As detailed in this chapter, potatoes are a valuable source of dietary vitamins, minerals, and phytonutrients. Moreover, the vitamin and phytonutrient content of potato and other staple foods has more dietary relevance and impact than that from foods eaten in sparse quantities. Potatoes are often cited as producing about 9.2 million calories per acre, but this number does not reflect yields in the Pacific Northwest. Washington State averages 30 tons/acre, and 50 tons/acre (short tons) are not uncommon with Russet potatoes. From the USDA database, a baked Russet potato is listed as having 97 calories per 100 g, which is about 440 calories per pound. Using potato yields in Washington State as a comparison, potatoes produce ~26–44 million calories per acre compared to 7.5, 7.4, 3.0,

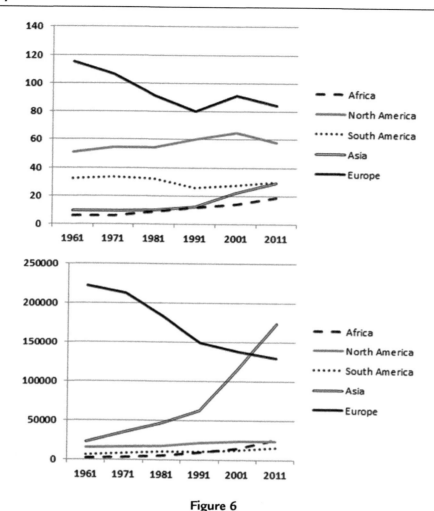

Figure 6
Per capita supply of potatoes in five continents between 1961 and 2011 (top). The *y*-axis represents kilograms of potatoes produced per year on a per capita basis. Production of potatoes in the same five continents (bottom). The *y*-axis scale is 1000 metric tons. *Data are from FAOSTAT.*

and 2.8 million for corn, rice, wheat, and soybeans, respectively (Ensminger and Ensminger, 1993). The cultivar Norkotah Russet would provide about 150,000–249,000 g of potassium per acre, 6700–11,000 g of vitamin C, and 19,000–38,000 g of phenolics based on yields of 30–50 tons/acre. The *Foods & Nutrition Encyclopedia* (Ensminger and Ensminger, 1993) lists potatoes as producing 338 pounds of protein per acre, while corn, rice, wheat, and soybeans produce 409, 304, 216, and 570 pounds of protein per acre, respectively. Based on potatoes containing 2% protein, then in Washington State potatoes provide 1200–2000 pounds of protein per acre. Many locations may not be able to replicate yields reached in the northwest United States with Russet potatoes, but these data show the potential for increased yields with

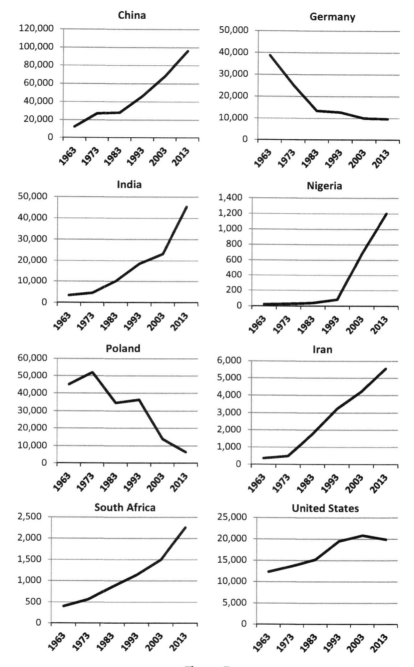

Figure 7
Potato production in eight countries between 1963 and 2013. The *y*-axis scale is 1000 metric tons.
Data from FAOSTAT.

better management. Such data make it clear that potatoes provide much more than just carbohydrates to the diet and emphasize the impact potatoes can have on global nutrition.

The importance of the potato in providing food security is further highlighted by a study that concluded that of 98 vegetable products studied, potatoes and beans provide the most nutrients per dollar (Drewnowski and Rehm, 2013). Curiously, some nutrition and health scientists in the United States have advocated replacing potatoes in the diet with plants like hulled barley, wheat berries, and bulgur because they believe potatoes are a major contributor to obesity. Unfortunately, those suggesting potatoes be replaced in the diet with other foods do not appear to have provided data about whether the substitutes would be affordable by lower-income consumers or how much more land would be required to produce an equivalent amount of energy and nutrition.

Additional factors emphasizing the role of potatoes in providing food security are that they are more water efficient than cereals, yielding $6.2–11.6 \, kg/m^3$ water and producing $150 \, g \, protein/m^3$ of water, and up to 85% of the crop is edible, compared with ~50% in cereals (Birch et al., 2012).

12. Conclusion

The story of potatoes in the human diet continues to evolve, from its origins in the Andes thousands of years ago to its subsequent spread into the West hundreds of years ago, where it became a staple and made possible a new food security that had massive societal ramifications. In recent decades, potato production in Africa and Asia has boomed, while production in the West has declined. Currently, China is by far the world's largest producer of potatoes and has made potatoes a lynchpin of its forward-looking food security strategy. And while a minority of scientists in the health and nutrition fields have claimed that potatoes are a primary contributor to obesity, and that carbohydrates are the villain in the diet, time will tell if this viewpoint is tenable. As detailed in this chapter, potatoes have a complex chemical composition that includes important amounts of minerals, vitamins, and phytonutrients, while contributing only 440 calories per pound. This emphasizes the importance of distinguishing between the nutritional content of the potato itself versus high-calorie additions to the potato. Furthermore, those advocating that potatoes be replaced in the diet with other foods may not have considered the agronomics needed to provide global food security. While already an important nutritional source, advances in crop management, along with traditional breeding and molecular breeding using new technologies like TALEN and CRISPR, have the potential to further increase the phytonutrient content of potato.

References

Abdel-Aal, E.-S., Akhtar, H., Zaheer, K., Ali, R., 2013. Dietary sources of lutein and zeaxanthin carotenoids and their role in eye health. Nutrients 5, 1169.
Abdou, E., Hazell, A.S., 2015. Thiamine deficiency: an update of pathophysiologic mechanisms and future therapeutic considerations. Neurochemical Research 40, 353–361.

Abuajah, C.I., Ogbonna, A.C., Osuji, C.M., 2015. Functional components and medicinal properties of food: a review. Journal of Food Science and Technology 52, 2522–2529.

Alfthan, G., Laurinen, M.S., Valsta, L.M., Pastinen, T., Aro, A., 2003. Folate intake, plasma folate and homocysteine status in a random Finnish population. European Journal of Clinical Nutrition 57, 81–88.

Altesor, P., García, Á., Font, E., Rodríguez-Haralambides, A., Vilaró, F., Oesterheld, M., Soler, R., González, A., 2014. Glycoalkaloids of wild and cultivated solanum: effects on specialist and generalist insect herbivores. Journal of Chemical Ecology 40, 599–608.

Alvarado, V.Y., Odokonyero, D., Duncan, O., Mirkov, T.E., Scholthof, H.B., 2012. Molecular and physiological properties associated with zebra complex disease in potatoes and its relation with *Candidatus Liberibacter* contents in psyllid vectors. PLoS One 7, e37345.

Andre, C.M., Ghislain, M., Bertin, P., Oufir, M., Herrera Mdel, R., Hoffmann, L., Hausman, J.F., Larondelle, Y., Evers, D., 2007a. Andean potato cultivars (*Solanum tuberosum* L.) as a source of antioxidant and mineral micronutrients. Journal of Agricultural and Food Chemistry 55, 366–378.

Andre, C.M., Oufir, M., Guignard, C., Hoffmann, L., Hausman, J.F., Evers, D., Larondelle, Y., 2007b. Antioxidant profiling of native Andean potato tubers (*Solanum tuberosum* L.) reveals cultivars with high levels of beta-carotene, alpha-tocopherol, chlorogenic acid, and petanin. Journal of Agricultural and Food Chemistry 55, 10839–10849.

Arnqvist, L., Dutta, P.C., Jonsson, L., Sitbon, F., 2003. Reduction of cholesterol and glycoalkaloid levels in transgenic potato plants by overexpression of a type 1 sterol methyltransferase cDNA. Plant Physiology 131, 1792–1799.

Arribas-Lorenzo, G., Morales, F.J., 2009. Effect of pyridoxamine on acrylamide formation in a glucose/asparagine model system. Journal of Agricultural and Food Chemistry 57, 901–909.

Avioli, L.V., 1988. Calcium and phosphorus. In: Shils, M.E., Young, E. (Eds.), Modern Nutrition in Health and Disease, seventh ed. Lea & Febiger, Philadelphia, pp. 142–158.

Bachrach, L.K., 2001. Acquisition of optimal bone mass in childhood and adolescence. Trends in Endocrinology and Metabolism 12, 22–28.

Bamberg, J.B., Navarre, D.A., Moehninsi, Suriano, J., 2015. Variation for tuber greening in the diploid wild potato *Solanum microdontum*. American Journal of Potato Research 92 (3), 435–443.

Bamberg, J.B., Palta, J.P., Peterson, L.A., Martin, M., Krueger, A.R., 1998. Fine screening potato (*Solanum*) species germplasm for tuber calcium. American Journal of Potato Research 75, 181–186.

Bassoli, B.K., Cassolla, P., Borba-Murad, G.R., Constantin, J., Salgueiro-Pagadigorria, C.L., Bazotte, R.B., da Silva, R.S., de Souza, H.M., 2008. Chlorogenic acid reduces the plasma glucose peak in the oral glucose tolerance test: effects on hepatic glucose release and glycaemia. Cell Biochemistry and Function 26, 320–328.

Bender, D.A., 1999. Optimum nutrition: thiamin, biotin and pantothenate. Proceedings of the Nutrition Society 58, 427–433.

Bibak, A., Sturup, S., Haahr, V., Gundersen, P., Gundersen, V., 1999. Concentrations of 50 major and trace elements in Danish agricultural crops measured by inductively coupled plasma mass spectrometry. 3. Potato (*Solanum tuberosum* folva). Journal of Agricultural and Food Chemistry 47, 2678–2684.

Birch, P.J., Bryan, G., Fenton, B., Gilroy, E., Hein, I., Jones, J., Prashar, A., Taylor, M., Torrance, L., Toth, I., 2012. Crops that feed the world 8: potato: are the trends of increased global production sustainable? Food Security 4, 477–508.

Blancquaert, D., Storozhenko, S., Loizeau, K., De Steur, H., De Brouwer, V., Viaene, J., Ravanel, S., Rebeille, F., Lambert, W., Van Der Straeten, D., 2010. Folates and folic acid: from fundamental research toward sustainable health. Critical Reviews in Plant Sciences 29, 14–35.

Blauer, J.N., Kumar, M.G.N., Knowles, L.O., Dingra, A., Knowles, N.R., 2013. Changes in ascorbate and associated gene expression during development and storage of potato tubers. Postharvest Biology and Technology 78, 76–91.

Blessington, T., Nzaramba, M.N., Scheuring, D.C., Hale, A.L., Redivari, L., Miller Jr., J.C., 2010. Cooking methods and storage treatments of potato: effects on carotenoids, antioxidant activity, and phenolics. American Journal of Potato Research 87, 479–491.

Bolca, S., Van de Wiele, T., Possemiers, S., 2013. Gut metabotypes govern health effects of dietary polyphenols. Current Opinion in Biotechnology 24, 220–225.

Bond, J., 2014. Potato Utilization and Markets. CABI.

Brevik, A., Vollset, S.E., Tell, G.S., Refsum, H., Ueland, P.M., Loeken, E.B., Drevon, C.A., Andersen, L.F., 2005. Plasma concentration of folate as a biomarker for the intake of fruit and vegetables: the Hordaland Homocysteine Study. American Journal of Clinical Nutrition 81, 434–439.

Brown, C.R., Culley, D., Yang, C.P., Durst, R., Wrolstad, R., 2005. Variation of anthocyanin and carotenoid contents and associated antioxidant values in potato breeding lines. Journal of the American Society for Horticultural Science 130, 174–180.

Brown, C.R., Edwards, C.G., Yang, C.P., Dean, B.B., 1993. Orange flesh trait in potato: inheritance and carotenoid content. Journal of the American Society for Horticultural Science 118, 145–150.

Brown, C.R., Haynes, K.G., Moore, M., Pavek, M.J., Hane, D.C., Love, S.L., Novy, R.G., Miller Jr., J.C., 2010a. Stability and broad-sense heritability of mineral content in potato: copper and sulfur. American Journal of Potato Research 91, 618–624.

Brown, C.R., Haynes, K.G., Moore, M., Pavek, M.J., Hane, D.C., Love, S.L., Novy, R.G., Miller Jr., J.C., 2010b. Stability and broad-sense heritability of mineral content in potato: iron. American Journal of Potato Research 87, 390–396.

Brussaard, J.H., Lowik, M.R., van den Berg, H., Brants, H.A., Goldbohm, R.A., 1997. Folate intake and status among adults in the Netherlands. European Journal of Clinical Nutrition 51 (Suppl. 3), S46–S50.

Bryant, R.J., Cadogan, J., Weaver, C.M., 1999. The new dietary reference intakes for calcium: implications for osteoporosis. Journal of the American College of Nutrition 18, 406S–412S.

Bub, A.M.D., Möseneder, J., Wenzel, G., Rechkemmer, G., Briviba, K., 2008. Zeaxanthin is bioavailable from genetically modified zeaxanthin-rich potatoes. European Journal of Nutrition 47, 99–103.

Burgos, G., Amoros, W., Morote, M., Stangoulis, J., Bonierbale, M., 2007. Iron and zinc concentration of native Andean potato cultivars from a human nutrition perspective. Journal of the Science of Food and Agriculture 87, 668–675.

Burgos, G., Amoros, W., Salas, E., Munoa, L., Sosa, P., Diaz, C., Bonierbale, M., 2012. Carotenoid concentrations of native Andean potatoes as affected by cooking. Food Chemistry 133, 1131–1137.

Burgos, G., Muñoa, L., Sosa, P., Bonierbale, M., zum Felde, T., Díaz, C., 2013. In vitro bioaccessibility of lutein and zeaxanthin of yellow fleshed boiled potatoes. Plant Foods for Human Nutrition 68, 385–390.

Burmeister, A., Bondiek, S., Apel, L., Kuhne, C., Hillebrand, S., Fleischmann, P., 2011. Comparison of carotenoid and anthocyanin profiles of raw and boiled *Solanum tuberosum* and *Solanum phureja* tubers. Journal of Food Composition Analysis 24, 865–872.

Campbell, R., Pont, S.D., Morris, J.A., McKenzie, G., Sharma, S.K., Hedley, P.E., Ramsay, G., Bryan, G.J., Taylor, M.A., 2014. Genome-wide QTL and bulked transcriptomic analysis reveals new candidate genes for the control of tuber carotenoid content in potato (*Solanum tuberosum* L.). Theoretical and Applied Genetics 127, 1917–1933.

Campbell, S., 2004. Dietary reference intakes: water, potassium, sodium, chloride, and sulfate. Clinical Nutrition Insight 30.

Cantos, E., Tudela, J.A., Gil, M.I., Espin, J.C., 2002. Phenolic compounds and related enzymes are not rate-limiting in browning development of fresh-cut potatoes. Journal of Agricultural and Food Chemistry 50, 3015–3023.

Cardona, F., Andrés-Lacueva, C., Tulipani, S., Tinahones, F.J., Queipo-Ortuño, M.I., 2013. Benefits of polyphenols on gut microbiota and implications in human health. The Journal of Nutritional Biochemistry 24, 1415–1422.

Casanas, R., Gonzalez, M., Rodriguez, E., Marrero, A., Diaz, C., 2002. Chemometric studies of chemical compounds in five cultivars of potatoes from Tenerife. Journal of Agricultural and Food Chemistry 50, 2076–2082.

Cham, B.E., 1994. Solasodine glycosides as anti-cancer agents: pre-clinical and clinical studies. South Asia Pacific Journal of Pharmacology 9, 113–118.

Cham, B.E., Chase, T.R., 2012. Solasodine rhamnosyl glycosides cause apoptosis in cancer cells. Do they also prime the immune system resulting in long-term protection against cancer? Planta Medica 78, 349–353.

Cham, B.E., Daunter, B., Evans, R.A., 1991. Topical treatment of malignant and premalignant skin lesions by very low concentrations of a standard mixture (BEC) of solasodine glycosides. Cancer Letters 59, 183–192.

Chataing, B., Concepcion, J.L., de Cristancho, N.B., Usubillaga, A., 1997. Estudio clinico de la efectividad de extractos alcaloides obtenidos de los frutos del *Solanum americanum* Miller sobre el Herpes simplex, Herpes Zoster y Herpes genitalis. Revista de la Facultad de Farmacia 32, 18–25.

Chen, H., Xiong, L., 2005. Pyridoxine is required for post-embryonic root development and tolerance to osmotic and oxidative stresses. Plant Journal 44, 396–408.

Chen, X., Song, B., Liu, J., Yang, J., He, T., Lin, Y., Zhang, H., Xie, C., 2012. Modulation of gene expression in cold-induced sweetening resistant potato species *Solanum berthaultii* exposed to low temperature. Molecular Genetics and Genomics 287, 411–421.

Chen, Y., Li, S., Sun, F., Han, H., Zhang, X., Fan, Y., Tai, G., Zhou, Y., 2010. In vivo antimalarial activities of glycoalkaloids isolated from Solanaceae plants. Pharmaceutical Biology 48, 1018–1024.

Cheng, D.M., Pogrebnyak, N., Kuhn, P., Poulev, A., Waterman, C., Rojas-Silva, P., Johnson, W.D., Raskin, I., 2014. Polyphenol-rich Rutgers Scarlet Lettuce improves glucose metabolism and liver lipid accumulation in diet-induced obese C57BL/6 mice. Nutrition 30, S52–S58.

Chi, M., Bhagwat, B., Lane, W., Tang, G., Su, Y., Sun, R., Oomah, B., Wiersma, P., Xiang, Y., 2014. Reduced polyphenol oxidase gene expression and enzymatic browning in potato (*Solanum tuberosum* L.) with artificial microRNAs. BMC Plant Biology 14, 62.

Chucair, A.J., Rotstein, N.P., Sangiovanni, J.P., During, A., Chew, E.Y., Politi, L.E., 2007. Lutein and zeaxanthin protect photoreceptors from apoptosis induced by oxidative stress: relation with docosahexaenoic acid. Investigative Ophthalmology and Visual Science 48, 5168–5177.

Chun, O.K., Kim, D.O., Smith, N., Schroeder, D., Han, J.T., Lee, C.Y., 2005. Daily consumption of phenolics and total antioxidant capacity from fruit and vegetables in the American diet. Journal of the Science of Food and Agriculture 85, 1715–1724.

Cieslik, E., Sikora, E., 1998. Correlation between the levels of nitrates and nitrites and the contents of potassium, calcium and magnesium in potato tubers. Food Chemistry 63, 525–528.

Cui, C.Z., Wen, X.S., Cui, M., Gao, J., Sun, B., Lou, H.X., 2012. Synthesis of solasodine glycoside derivatives and evaluation of their cytotoxic effects on human cancer cells. Drug Discovery Therapy 6, 9–17.

Cui, T., Bai, J., Zhang, J., Zhang, J., Wang, D., 2014. Transcriptional expression of seven key genes involved in steroidal glycoalkaloid biosynthesis in potato microtubers. New Zealand Journal of Crop and Horticultural Science 42, 118–126.

Dale, M.F.B., Griffiths, D.W., Todd, D.T., 2003. Effects of genotype, environment, and postharvest storage on the total ascorbate content of potato (*Solanum tuberosum*) tubers. Journal of Agricultural and Food Chemistry 51, 244–248.

Dao, L., Friedman, M., 1992. Chlorogenic acid content of fresh and processed potatoes determined by ultraviolet spectrophotometry. Journal of Agricultural and Food Chemistry 40, 2152–2156.

De Jong, W.S., De Jong, D.M., De Jong, H., Kalazich, J., Bodis, M., 2003. An allele of dihydroflavonol 4-reductase associated with the ability to produce red anthocyanin pigments in potato (*Solanum tuberosum* L.). Theoretical and Applied Genetics 107, 1375–1383.

De Jong, W.S., Eannetta, N.T., De Jong, D.M., Bodis, M., 2004. Candidate gene analysis of anthocyanin pigmentation loci in the Solanaceae. Theoretical and Applied Genetics 108, 423–432.

Delchier, N., Ringling, C., Maingonnat, J.F., Rychlik, M., Renard, C.M.G.C., 2014. Mechanisms of folate losses during processing: diffusion vs. heat degradation. Food Chemistry 157, 439–447.

Delgado, E., Pawelzik, E., Poberezny, J., Rogozinska, I., 2001. Effect of location and variety on the content of minerals in German and Polish potato cultivars. In: Horst, W.J., Schenk, M.K., Bürkert, A., Claassen, N., Flessa, H., Frommer, W.B., Goldbach, H., Olfs, H.W., Römheld, V., Sattelmacher, B., Schmidhalter, U., Schubert, S., Wirén, N.V., Wittenmayer, L. (Eds.), Plant Nutrition, vol. 92. Springer, Netherlands, pp. 346–347.

Dellapenna, D., Pogson, B., 2006. Vitamin synthesis in plants: tocopherols and carotenoids. Annual Review of Plant Biology 57, 711–738.

Deng, X.W., Gruissem, W., 1988. Constitutive transcription and regulation of gene expression in non-photosynthetic plastids of higher plants. EMBO Journal 7, 3301–3308.

Denslow, S.A., Rueschhoff, E.E., Daub, M.E., 2007. Regulation of the *Arabidopsis thaliana* vitamin B_6 biosynthesis genes by abiotic stress. Plant Physiology and Biochemistry 45, 152–161.

Denslow, S.A., Walls, A.A., Daub, M.E., 2005. Regulation of biosynthetic genes and antioxidant properties of vitamin B_6 vitamers during plant defense responses. Physiological and Molecular Plant Pathology 66, 244–255.

Depeint, F., Bruce, W.R., Shangari, N., Mehta, R., O'Brien, P.J., 2006. Mitochondrial function and toxicity: role of B vitamins on the one-carbon transfer pathways. Chemico-Biological Interactions 163, 113–132.

Diretto, G., Al-Babili, S., Tavazza, R., Papacchioli, V., Beyer, P., Giuliano, G., 2007. Metabolic engineering of potato carotenoid content through tuber-specific overexpression of a bacterial mini-pathway. PLoS One 2, e350.

Douce, R., Neuburger, M., 1999. Biochemical dissection of photorespiration. Current Opinion in Plant Biology 2, 214–222.

Drewnowski, A., Rehm, C.D., 2013. Vegetable cost metrics show that potatoes and beans provide most nutrients per penny. PLoS One 8, e63277.

Ducreux, L.J.M., Morris, W.L., Hedley, P.E., Shepherd, T., Davies, H.V., Millam, S., Taylor, M.A., 2005. Metabolic engineering of high carotenoid potato tubers containing enhanced levels of beta-carotene and lutein. Journal of Experimental Botany 56, 81–89.

Dugo, G., La Pera, L., Lo Turco, V., Giuffrida, D., Restuccia, S., 2004. Determination of copper, zinc, selenium, lead and cadmium in potatoes (*Solanum tuberosum* L.) using potentiometric stripping methods. Food Additives & Contaminants 21, 649–657.

Ek, K.L., Wang, S., Copeland, L., Brand-Miller, J.C., 2014. Discovery of a low-glycaemic index potato and relationship with starch digestion in vitro. British Journal of Nutrition 111, 699–705.

Ensminger, M.E., Ensminger, A.H., 1993. In: Food & Nutrition Encyclopedia, vol. 1. CRC Press.

Ereifej, K., Shibli, R., Ajlouni, M., Hussein, A., 1998. Mineral contents of whole tubers and selected tissues of ten potato cultivars grown in Jordan. Journal of Food Science and Technology 35, 55–58.

Ewing, E.E., Struik, P.C., 1992. Tuber formation in potato: induction, initiation, and growth. Horticultural Reviews 14, 89–198.

FAOSTAT, 2015. http://faostat.fao.org/site/567/DesktopDefault.aspx?PageID=567#ancor (accessed June, 2015).

Farah, A., Monteiro, M., Donangelo, C.M., Lafay, S., 2008. Chlorogenic acids from green coffee extract are highly bioavailable in humans. Journal of Nutrition 138, 2309–2315.

Fellenberg, C., Ziegler, J., Handrick, V., Vogt, T., 2012. Polyamine homeostasis in wild type and phenolamide deficient *Arabidopsis thaliana* stamens. Frontiers in Plant Science 3, 180.

Fernandez-Orozco, R.B., Gallardo-Guerrero, L., Hornero-Mendez, D., 2013. Carotenoid profiling in tubers of different potato (*Solanum* sp) cultivars: accumulation of carotenoids mediated by xanthophyll esterification. Food Chemistry 141, 2864–2872.

Fernie, A.R., Willmitzer, L., 2001. Molecular and biochemical triggers of potato tuber development. Plant Physiology 127, 1459–1465.

Ferracane, R., Pellegrini, N., Visconti, A., Graziani, G., Chiavaro, E., Miglio, C., Fogliano, V., 2008. Effects of different cooking methods on antioxidant profile, antioxidant capacity, and physical characteristics of artichoke. Journal of Agricultural and Food Chemistry 56, 8601–8608.

Fraser, P.D., Bramley, P.M., 2004. The biosynthesis and nutritional uses of carotenoids. Progress in Lipid Research 43, 228–265.

Friedman, M., 2002. Tomato glycoalkaloids: role in the plant and in the diet. Journal of Agricultural and Food Chemistry 50, 5751–5780.

Friedman, M., 2006. Potato glycoalkaloids and metabolites: roles in the plant and in the diet. Journal of Agricultural and Food Chemistry 54, 8655–8681.

Friedman, M., 2015. Chemistry and anticarcinogenic mechanisms of glycoalkaloids produced by eggplants, potatoes, and tomatoes. Journal of Agricultural and Food Chemistry 63, 3323–3337.

Friedman, M., Lee, K.R., Kim, H.J., Lee, I.S., Kozukue, N., 2005. Anticarcinogenic effects of glycoalkaloids from potatoes against human cervical, liver, lymphoma, and stomach cancer cells. Journal of Agricultural and Food Chemistry 53, 6162–6169.

Friedman, M., McDonald, G.M., 1997. Potato glycoalkaloids: chemistry, analysis, safety, and plant physiology. Critical Reviews in Plant Sciences 16, 55–132.

Friedman, M., McQuistan, T., Hendricks, J.D., Pereira, C., Bailey, G.S., 2007. Protective effect of dietary tomatine against dibenzo[a,l]pyrene (DBP)-induced liver and stomach tumors in rainbow trout. Molecular Nutrition & Food Research 51, 1485–1491.

Frossard, E., Bucher, M., Mächler, F., Mozafar, A., Hurrell, R., 2000. Potential for increasing the content and bioavailability of Fe, Zn and Ca in plants for human nutrition. Journal of the Science of Food and Agriculture 80, 861–879.

Fu, L.S., Doreswamy, V., Prakash, R., 2014. The biochemical pathways of central nervous system neural degeneration in niacin deficiency. Neural Regeneration Research 9, 1509–1513.

Fuentes-Zaragoza, E., Riquelme-Navarrete, M.J., Sánchez-Zapata, E., Pérez-Álvarez, J.A., 2010. Resistant starch as functional ingredient: a review. Food Research International 43, 931–942.

Gammone, M.A., Riccioni, G., D'Orazio, N., 2015. Carotenoids: potential allies of cardiovascular health? Food & Nutrition Research 59.

Gaquerel, E., Gulati, J., Baldwin, I.T., 2014. Revealing insect herbivory-induced phenolamide metabolism: from single genes to metabolic network plasticity analysis. Plant Journal 79, 679–692.

Gehring, W., 2004. Nicotinic acid/niacinamide and the skin. Journal of Cosmetic Dermatology 3, 88–93.

Gibson, S., Kurilich, A.C., 2012. The nutritonal value of potatoes and potato products in the UK diet. Nutrition Bulletin 38, 389–399.

Goyer, A., Navarre, D.A., 2007. Determination of folate concentrations in diverse potato germplasm using a trienzyme extraction and a microbiological assay. Journal of Agricultural and Food Chemistry 55, 3523–3528.

Goyer, A., Navarre, D.A., 2009. Folate is higher in developmentally younger potato tubers. Journal of the Science of Food and Agriculture 89, 579–583.

Goyer, A., Sweek, K., 2011. Genetic diversity of thiamin and folate in primitive cultivated and wild potato (*Solanum*) species. Journal of Agricultural and Food Chemistry 59, 13072–13080.

Griffiths, D.W., Bain, H., Dale, M.F.B., 1997. The effect of low-temperature storage on the glycoalkaloid content of potato (*Solanum tuberosum*) tubers. Journal of the Science of Food and Agriculture 74, 301–307.

Gubarev, M.I., Enioutina, E.Y., Taylor, J.L., Visic, D.M., Daynes, R.A., 1998. Plant derived glycoalkaloids protect mice against lethal infection with *Salmonella typhimurium*. Phytotherapy Research 12, 79–88.

Gwirtz, J.A., Garcia-Casal, M.N., 2014. Processing maize flour and corn meal food products. Technical Considerations for Maize Flour and Corn Meal Fortification in Public Health 1312, 66–75.

Hairong, W., 2015. Rediscovering the value of the potato. In: Beijing Review, vol. 7, pp. 1–2.

Hajšlová, J., Schulzova, V., Slanina, P., Janne, K., Hellenäs, K., Andersson, C., 2005. Quality of organically and conventionally grown potatoes: four-year study of micronutrients, metals, secondary metabolites, enzymic browning and organoleptic properties. Food Additives and Contaminants 22, 514–534.

Hanson, A.D., Gregory, J.F., 2011. Folate biosynthesis, turnover, and transport in plants. Annual Review of Plant Biology 62, 105–125.

Havaux, M., Ksas, B., Szewczyk, A., Rumeau, D., Franck, F., Caffarri, S., Triantaphylides, C., 2009. Vitamin B6 deficient plants display increased sensitivity to high light and photo-oxidative stress. BMC Plant Biology 9, 130.

Hawkes, J.G., 1990. The Potato: Evolution, Biodiversity, and Genetic Resources. Belhaven Press, Oxford.

Haynes, K.G., Clevidence, B.A., Rao, D., Vinyard, B.T., White, J.M., 2010. Genotype × environment interactions for potato tuber carotenoid content. Journal of the American Society for Horticultural Science 135, 250–258.

Heftmann, E., 1983. Biogenesis of steroids in Solanaceae. Phytochemistry 22, 1843–1860.

Hellmann, H., Mooney, S., 2010. Vitamin B_6: a molecule for human health? Molecules 15, 442–459.

Herrero, S., Amnuaykanjanasin, A., Daub, M.E., 2007. Identification of genes differentially expressed in the phytopathogenic fungus *Cercospora nicotianae* between cercosporin toxin-resistant and -susceptible strains. FEMS Microbiology Letters 275, 326–337.

Hollman, P.C.H., 2014. Unravelling of the health effects of polyphenols is a complex puzzle complicated by metabolism. Archives of Biochemistry and Biophysics 559, 100–105.

Hopkins, J., 1995. The glycoalkaloids: naturally of interest (but a hot potato?). Food and Chemical Toxicology 33, 323–328.

Hung, C.-Y., Murray, J.R., Ohmann, S.M., Tong, C.B.S., 1997. Anthocyanin accumulation during potato tuber development. Journal of the American Society for Horticultural Science 122, 20–23.

Ilin, Z., Durovka, M., Markovic, V., 2000. Effect of irrigation and mineral nutrition on the quality of potato. In: II Balkan Symposium on Vegetables and Potatoes, pp. 625–629.

Itkin, M., Heinig, U., Tzfadia, O., Bhide, A.J., Shinde, B., Cardenas, P.D., Bocobza, S.E., Unger, T., Malitsky, S., Finkers, R., Tikunov, Y., Bovy, A., Chikate, Y., Singh, P., Rogachev, I., Beekwilder, J., Giri, A.P., Aharoni, A., 2013. Biosynthesis of antinutritional alkaloids in solanaceous crops is mediated by clustered genes. Science 341, 175–179.

Iwanzik, W., Tevini, M., Stute, R., Hilbert, R., 1983. Carotinoidgehalt und -zusammensetzung verschiedener deutscher Kartoffelsorten und deren Bedeutung fur die Fleischfarbe der Knolle. Potato Research 26, 149–162.

Jansen, G., Flamme, W., 2006. Coloured potatoes (*Solanum tuberosum* L.) - anthocyanin content and tuber quality. Genetic Resources and Crop Evolution 53, 1321–1331.

Janskey, S.H., 2010. Potato flavor. In: Hui, Y.H. (Ed.), Handbook of Fruit and Vegetable Flavors. John Wiley & Sons.

Johnson, D.F., Bennett, R.D., Heftmann, E., 1963. Cholesterol in higher plants. Science 140, 198–199.

Juguan, J.A., Lukito, W., Schultink, W., 1999. Thiamine deficiency is prevalent in a selected group of urban Indonesian elderly people. Journal of Nutrition 129, 366–371.

Kalbina, I., Strid, A., 2006. Supplementary ultraviolet-B irradiation reveals differences in stress responses between *Arabidopsis thaliana* ecotypes. Plant Cell Environment 29, 754–763.

Kandutsch, A.A., Baumann, C.A., 1953. Factors affecting the stability of thiamine in a typical laboratory diet. Journal of Nutrition 49, 209–219.

Kaneko, K., Tanaka, M.W., Mitsuhashi, H., 1976. Origin of nitrogen in the biosynthesis of solanidine by *Veratrum grandiflorum*. Phytochemistry 15, 1391–1393.

Kaspar, K.L., Park, J.S., Brown, C.R., Mathison, B.D., Navarre, D.A., Chew, B.P., 2011. Pigmented potato consumption alters oxidative stress and inflammatory damage in men. Journal of Nutrition 141, 108–111.

Katan, M.B., De Roos, N.M., 2004. Promises and problems of functional foods. Critical Reviews in Food Science and Nutrition 44, 369–377.

Kaur-Sawhney, R., Shih, L.M., Galston, A.W., 1982. Relation of polyamine biosynthesis to the initiation of sprouting in potato tubers. Plant Physiology 69, 411–415.

Kawabata, K., Mukai, R., Ishisaka, A., 2015. Quercetin and related polyphenols: new insights and implications for their bioactivity and bioavailability. Food & Function 6, 1399–1417.

Keijbets, M.J.H., Ebbenhorst-Seller, G., 1990. Loss of vitamin C (L-ascorbic acid) during long-term cold storage of Dutch table potatoes. Potato Research 33, 125–130.

Kim, H.W., Kim, J.B., Cho, S.M., Chung, M.N., Lee, Y.M., Chu, S.M., Che, J.H., Kim, S.N., Kim, S.Y., Cho, Y.S., Kim, J.H., Park, H.J., Lee, D.J., 2012. Anthocyanin changes in the Korean purple-fleshed sweet potato, Shinzami, as affected by steaming and baking. Food Chemistry 130, 966–972.

Kohn, H.I., 1938. The concentration of coenzyme-like substance in blood following the administration of nicotinic acid to normal individuals and pellagrins. Biochemical Journal 32, 2075–2083.

Kondo, Y., Higashi, C., Iwama, M., Ishihara, K., Handa, S., Mugita, H., Maruyama, N., Koga, H., Ishigami, A., 2012. Bioavailability of vitamin C from mashed potatoes and potato chips after oral administration in healthy Japanese men. British Journal of Nutrition 107, 885–892.

Kondo, Y., Sakuma, R., Ichisawa, M., Ishihara, K., Kubo, M., Handa, S., Mugita, H., Maruyama, N., Koga, H., Ishigami, A., 2014. Potato chip intake increases ascorbic acid levels and decreases reactive oxygen species in SMP30/GNL knockout mouse tissues. Journal of Agricultural and Food Chemistry 62, 9286–9295.

Konings, E.J., Roomans, H.H., Dorant, E., Goldbohm, R.A., Saris, W.H., van den Brandt, P.A., 2001. Folate intake of the Dutch population according to newly established liquid chromatography data for foods. American Journal of Clinical Nutrition 73, 765–776.

Kossmann, J., Lloyd, J., 2000. Understanding and influencing starch biochemistry. Critical Reviews in Biochemistry and Molecular Biology 35, 141–196.

Krits, P., Fogelman, E., Ginzberg, I., 2007. Potato steroidal glycoalkaloid levels and the expression of key isoprenoid metabolic genes. Planta 227, 143–150.

Kuiper-Goodman, T., Nawrot, P.S., 1992. JECFA Review (accessed 16.06.15.).

Külen, O., Stushnoff, C., Holm, D.G., 2012. Effect of cold storage on total phenolics content, antioxidant activity and vitamin C level of selected potato clones. Journal of the Science of Food and Agriculture 93, 2437–2444.

Kupchan, S.M., Barboutis, S.J., Knox, J.R., Cam, C.A., 1965. Beta-solamarine: tumor inhibitor isolated from *Solanum dulcamara*. Science 150, 1827–1828.

Lachman, J., Hamouz, K., Musilova, J., Hejtmankova, K., Kotikova, Z., Pazderu, K., Domkarova, J., Pivec, V., Cimr, J., 2013. Effect of peeling and three cooking methods on the content of selected phytochemicals in potato tubers with various colour of flesh. Food Chemistry 138, 1189–1197.

Laing, W.A., Wright, M.A., Cooney, J., Bulley, S.M., 2007. The missing step of the L-galactose pathway of ascorbate biosynthesis in plants, an L-galactose guanyltransferase, increases leaf ascorbate content. Proceedings of the National Academy of Sciences of the United States of America 104 (22), 9534–9539.

Larsson, S.C., Virtanen, M.J., Mars, M., Männistö, S., Pietinen, P., Albanes, D., Virtamo, J., 2008. Magnesium, calcium, potassium, and sodium intakes and risk of stroke in male smokers. Archives of Internal Medicine 168, 459–465.

Lavie, Y., Harel-Orbital, T., Gaffield, W., Liscovitch, M., 2001. Inhibitory effect of steroidal alkaloids on drug transport and multidrug resistance in human cancer cells. Anticancer Research 21, 1189–1194.

Lee, K.R., Kozukue, N., Han, J.S., Park, J.H., Chang, E.Y., Baek, E.J., Chang, J.S., Friedman, M., 2004. Glycoalkaloids and metabolites inhibit the growth of human colon (HT29) and liver (HepG2) cancer cells. Journal of Agricultural and Food Chemistry 52, 2832–2839.

Lee, S.T., Wong, P.F., He, H., Hooper, J.D., Mustafa, M.R., 2013. Alpha-tomatine attenuation of in vivo growth of subcutaneous and orthotopic xenograft tumors of human prostate carcinoma PC-3 cells is accompanied by inactivation of nuclear factor-kappa B signaling. PLoS One 8, e57708.

Lefèvre, I., Ziebel, J., Guignard, C., Hausman, J.F., Gutiérrez Rosales, R.O., Bonierbale, M., Hoffmann, L., Schafleitner, R., Evers, D., 2012. Drought impacts mineral contents in Andean potato cultivars. Journal of Agronomy and Crop Science 198, 196–206.

Legrand, D., Scheen, A.J., 2007. Does coffee protect against type 2 diabetes? Revue Medicale de Liege 62, 554–559.

Lemos, M.A., Aliyu, M.M., Hungerford, G., 2015. Influence of cooking on the levels of bioactive compounds in Purple Majesty potato observed via chemical and spectroscopic means. Food Chemistry 173, 462–467.

Lewis, C.E., Walker, J.R.L., Lancaster, J.E., Sutton, K.H., 1998. Determination of anthocyanins, flavonoids and phenolic acids in potatoes. I. Coloured cultivars of *Solanum tuberosum* L. Journal of the Science of Food and Agriculture 77, 45–57.

Li, G.M., Presnell, S.R., Gu, L., 2003. Folate deficiency, mismatch repair-dependent apoptosis, and human disease. Journal of Nutritional Biochemistry 14, 568–575.

Liang, C.H., Shiu, L.Y., Chang, L.C., Sheu, H.M., Tsai, E.M., Kuo, K.W., 2008. Solamargine enhances HER2 expression and increases the susceptibility of human lung cancer H661 and H69 cells to trastuzumab and epirubicin. Chemical Research in Toxicology 21, 393–399.

Lisinska, G., Leszczynski, W., 1989. Potato Science and Technology. Springer.

Llorente, B., López, M., Carrari, F., Asís, R., Di Paola Naranjo, R., Flawiá, M., Alonso, G., Bravo-Almonacid, F., 2014. Downregulation of polyphenol oxidase in potato tubers redirects phenylpropanoid metabolism enhancing chlorogenate content and late blight resistance. Molecular Breeding 34, 2049–2063.

Lojzova, L., Riddellova, K., Hajslova, J., Zrostlikova, J., Schurek, J., Cajka, T., 2009. Alternative GC-MS approaches in the analysis of substituted pyrazines and other volatile aromatic compounds formed during Maillard reaction in potato chips. Analytica Chimica Acta 641, 101–109.

Lombardo, S., Pandino, G., Mauromicale, G., 2013. The influence of growing environment on the antioxidant and mineral content of "early" crop potato. Journal of Food Composition and Analysis 32, 28–35.

Lombardo, S., Pandino, G., Mauromicale, G., 2014. The mineral profile in organically and conventionally grown "early" crop potato tubers. Scientia Horticulturae 167, 169–173.

Lopez, A.B., Van Eck, J., Conlin, B.J., Paolillo, D.J., O'Neill, J., Li, L., 2008. Effect of the cauliflower or transgene on carotenoid accumulation and chromoplast formation in transgenic potato tubers. Journal of Experimental Botany 59, 213–223.

Love, S.L., Salaiz, T., Shafii, B., Price, W.J., Mosley, A.R., Thornton, R.E., 2004. Stability of expression and concentration of ascorbic acid in North American potato germplasm. HortScience 39, 156–160.

Lv, C., Kong, H., Dong, G., Liu, L., Tong, K., Sun, H., Chen, B., Zhang, C., Zhou, M., 2014. Antitumor efficacy of alpha-solanine against pancreatic cancer in vitro and in vivo. PLoS One 9, e87868.

Maga, J.A., 1980. Potato glycoalkaloids. CRC Critical Reviews in Food Science and Nutrition 12, 371–405.

Maga, J.A., 1994. Glycoalkaloids in solanaceae. Food Reviews International 10, 385–418.

Malmberg, A., Theander, O., 1985. Determination of chlorogenic acid in potato tubers. Journal of Agricultural and Food Chemistry 33, 549–551.

Manach, C., Scalbert, A., Morand, C., Remesy, C., Jimenez, L., 2004. Polyphenols: food sources and bioavailability. American Journal of Clinical Nutrition 79, 727–747.

Manzetti, S., Zhang, J., van der Spoel, D., 2014. Thiamin function, metabolism, uptake, and transport. Biochemistry 53, 821–835.

Matsuda, F., Morino, K., Ano, R., Kuzawa, M., Wakasa, K., Miyagawa, H., 2005. Metabolic flux analysis of the phenylpropanoid pathway in elicitor-treated potato tuber tissue. Plant and Cell Physiology 46, 454–466.

McCue, K.F., Allen, P.V., Shepherd, L.V.T., Blake, A., Maccree, M.M., Rockhold, D.R., Novy, R.G., Stewart, D., Davies, H.V., Belknap, W.R., 2007. Potato glycosterol rhamnosyltransferase, the terminal step in triose side-chain biosynthesis. Phytochemistry 68, 327–334.

McCue, K.F., Shepherd, L.V.T., Allen, P.V., Maccree, M.M., Rockhold, D.R., Corsini, D.L., Davies, H.V., Belknap, W.R., 2005. Metabolic compensation of steroidal glycoalkaloid biosynthesis in transgenic potato tubers: using reverse genetics to confirm the in vivo enzyme function of a steroidal alkaloid galactosyltransferase. Plant Science 168, 267–273.

McGill, C.R., Kurilich, A.C., Davignon, J., 2013. The role of potatoes and potato components in cardiometabolic health: a review. Annals of Medicine 45, 467–473.

McGuire, R.G., Kelman, A., 1986. Calcium in potato tuber cell walls in relation to tissue maceration by *Erwinia carotovora* pv. *atroseptica*. Phytopathology 76, 401–406.

McMillan, M., Thompson, J.C., 1979. An outbreak of suspected solanine poisoning in schoolboys: examinations of criteria of solanine poisoning. Quarterly Journal of Medicine 48, 227–243.

Mensinga, T.T., Sips, A.J., Rompelberg, C.J., van Twillert, K., Meulenbelt, J., van den Top, H.J., van Egmond, H.P., 2005. Potato glycoalkaloids and adverse effects in humans: an ascending dose study. Regulatory Toxicology and Pharmacology 41, 66–72.

Miglio, C., Chiavaro, E., Visconti, A., Fogliano, V., Pellegrini, N., 2008. Effects of different cooking methods on nutritional and physicochemical characteristics of selected vegetables. Journal of Agricultural and Food Chemistry 56, 139–147.

Mills, C.E., Tzounis, X., Oruna-Concha, M.J., Mottram, D.S., Gibson, G.R., Spencer, J.P., 2015. In vitro colonic metabolism of coffee and chlorogenic acid results in selective changes in human faecal microbiota growth. British Journal of Nutrition 113, 1220–1227.

Miranda, M.A., Magalhaes, L.G., Tiossi, R.F., Kuehn, C.C., Oliveira, L.G., Rodrigues, V., McChesney, J.D., Bastos, J.K., 2012. Evaluation of the schistosomicidal activity of the steroidal alkaloids from *Solanum lycocarpum* fruits. Parasitology Research 111, 257–262.

Moccand, C., Boycheva, S., Surriabre, P., Tambasco-Studart, M., Raschke, M., Kaufmann, M., Fitzpatrick, T.B., 2014. The pseudoenzyme PDX1. 2 boosts vitamin B_6 biosynthesis under heat and oxidative stress in Arabidopsis. Journal of Biological Chemistry 289, 8203–8216.

Moehs, C.P., Allen, P.V., Friedman, M., Belknap, W.R., 1997. Cloning and expression of solanidine UDP-glucose glucosyltransferase from potato. Plant Journal 11, 227–236.

Mondy, N.I., Leja, M., 1986. Effect of mechanical injury on the ascorbic acid content of potatoes. Journal of Food Science 51, 355–357.

Monteiro, M., Farah, A., Perrone, D., Trugo, L.C., Donangelo, C., 2007. Chlorogenic acid compounds from coffee are differentially absorbed and metabolized in humans. Journal of Nutrition 137, 2196–2201.

Mooney, S., Chen, L., Kuhn, C., Navarre, R., Knowles, N.R., Hellmann, H., 2013. Genotype-specific changes in vitamin B6 content and the PDX family in potato. BioMed Research International 2013, 389723.

Mooney, S., Hellmann, H., 2010. Vitamin B6: killing two birds with one stone? Phytochemistry 71, 495–501.

Morris, S.C., Lee, T.H., 1984. The toxicity and teratogenicity of Solanaceae glycoalkaloids, particularly those of the potato (*Solanum tuberosum*): a review. Food Technology in Australia 36, 118–124.

Morris, W.L., Ducreux, L., Griffiths, D.W., Stewart, D., Davies, H.V., Taylor, M.A., 2004. Carotenogenesis during tuber development and storage in potato. Journal of Experimental Botany 55, 975–982.

Mozaffarian, D., Hao, T., Rimm, E.B., Willett, W.C., Hu, F.B., 2011. Changes in diet and lifestyle and long-term weight gain in women and men. New England Journal of Medicine 364, 2392–2404.

Mulinacci, N., Ieri, F., Giaccherini, C., Innocenti, M., Andrenelli, L., Canova, G., Saracchi, M., Casiraghi, M.C., 2008. Effect of cooking on the anthocyanins, phenolic acids, glycoalkaloids, and resistant starch content in two pigmented cultivars of *Solanum tuberosum* L. Journal of Agricultural and Food Chemistry 56, 11830–11837.

Nara, K., Miyoshi, T., Honma, T., Koga, H., 2006. Antioxidative activity of bound-form phenolics in potato peel. Bioscience, Biotechnology, and Biochemistry 70, 1489–1491.

Nassar, A.M.K., Kubow, S., Leclerc, Y., Donnelly, D.J., 2014. Somatic mining for phytonutrient improvement of 'Russet Burbank' potato. American Journal of Potato Research 91, 89–100.

Navarre, D.A., Payyavula, R.S., Shakya, R., Knowles, N.R., Pillai, S., 2013. Changes in potato phenylpropanoid metabolism during tuber development. Plant Physiology and Biochemistry 65, 89–101.

Navarre, D.A., Pillai, S., Shakya, R., Holden, M.J., 2011. HPLC profiling of phenolics in diverse potato genotypes. Food Chemistry 127, 34–41.

Navarre, D.A., Shakya, R., Holden, J., Crosslin, J.M., 2009. LC-MS analysis of phenolic compounds in tubers showing Zebra Chip symptoms. American Journal of Potato Research 86, 88–95.

Navarre, D.A., Shakya, R., Holden, J., Kumar, S., 2010. The effect of different cooking methods on phenolics and vitamin C in developmentally young potato tubers. American Journal of Potato Research 87, 350–359.

Niggeweg, R., Michael, A.J., Martin, C., 2004. Engineering plants with increased levels of the antioxidant chlorogenic acid. Nature Biotechnology 22, 746–754.

Nogueira, T., do Lago, C.L., 2007. Determination of caffeine in coffee products by dynamic complexation with 3,4-dimethoxycinnamate and separation by CZE. Electrophoresis 28, 3570–3574.

Nunn, N., Qian, N., 2011. The potato's contribution to population and urbanization: evidence from a historical experiment. Quarterly Journal Economics 126, 593–650.

Onakpoya, I.J., Spencer, E.A., Thompson, M.J., Heneghan, C.J., 2015. The effect of chlorogenic acid on blood pressure: a systematic review and meta-analysis of randomized clinical trials. Journal of Human Hypertension 29, 77–81.

Ong, K.W., Hsu, A., Tan, B.K., 2013. Anti-diabetic and anti-lipidemic effects of chlorogenic acid are mediated by ampk activation. Biochemical Pharmacology 85, 1341–1351.

Orphanos, P., 1980. Dry matter content and mineral composition of potatoes grown in Cyprus. Potato Research 23, 371–375.

Paget, M., Amoros, W., Salas, E., Eyzaguirre, R., Alspach, P., Apiolaza, L., Noble, A., Bonierbale, M., 2014. Genetic evaluation of micronutrient traits in diploid potato from a base population of andean landrace cultivars. Crop Science 54, 1949–1959.

Pandino, G., Lombardo, S., Mauromicale, G., 2011. Mineral profile in globe artichoke as affected by genotype, head part and environment. Journal of the Science of Food and Agriculture 91, 302–308.

Parr, A.J., Bolwell, G.P., 2000. Phenols in the plant and in man. The potential for possible nutritional enhancement of the diet by modifying the phenols content or profile. Journal of the Science of Food and Agriculture 80, 985–1012.

Parr, A.J., Mellon, F.A., Colquhoun, I.J., Davies, H.V., 2005. Dihydrocaffeoyl polyamines (kukoamine and allies) in potato (*Solanum tuberosum*) tubers detected during metabolite profiling. Journal of Agricultural and Food Chemistry 53, 5461–5466.

Pasare, S., Wright, K., Campbell, R., Morris, W., Ducreux, L., Chapman, S., Bramley, P., Fraser, P., Roberts, A., Taylor, M., 2013. The sub-cellular localisation of the potato (*Solanum tuberosum* L.) carotenoid biosynthetic enzymes, CrtRb2 and PSY2. Protoplasma 250. http://dx.doi.org/10.1007/s00709-013-0521-z.

Payyavula, R.S., Navarre, D.A., Kuhl, J., Pantoja, A., 2013. Developmental effects on phenolic, flavonol, anthocyanin, and carotenoid metabolites and gene expression in potatoes. Journal of Agricultural and Food Chemistry 61, 7357–7365.

Payyavula, R.S., Navarre, D.A., Kuhl, J.C., Pantoja, A., Pillai, S.S., 2012. Differential effects of environment on potato phenylpropanoid and carotenoid expression. BMC Plant Biology 12, 39.

Payyavula, R.S., Shakya, R., Sengoda, V.G., Munyaneza, J.E., Swamy, P., Navarre, D.A., 2015. Synthesis and regulation of chlorogenic acid in potato: rerouting phenylpropanoid flux in HQT-silenced lines. Plant Biotechnology Journal 13, 551–564.

Peña, C., Restrepo-Sánchez, L.-P., Kushalappa, A., Rodríguez-Molano, L.-E., Mosquera, T., Narváez-Cuenca, C.-E., 2015. Nutritional contents of advanced breeding clones of Solanum tuberosum group Phureja. LWT - Food Science Technology 62, 76–82.

Perla, V., Holm, D.G., Jayanty, S.S., 2012. Effects of cooking methods on polyphenols, pigments and antioxidant activity in potato tubers. Food Science and Technology 45, 161–171.

Petersson, E.V., Nahar, N., Dahlin, P., Broberg, A., Tröger, R., Dutta, P.C., Jonsson, L., Sitbon, F., 2013. Conversion of exogenous cholesterol into glycoalkaloids in potato shoots, using two methods for sterol solubilisation. PLoS One 8, e82955.

Pillai, S., Navarre, D.A., Bamberg, J.B., 2013. Analysis of polyphenols, anthocyanins and carotenoids in tubers from *Solanum tuberosum* group Phureja, Stenotomum and Andigena. American Journal of Potato Research 90, 440–450.

Ponnampalam, R., Mondy, N.I., 1983. Effect of cooking on the total glycoalkaloid content of potatoes. Journal of Agricultural and Food Chemistry 31, 493–495.

Porter, W.L., 1972. A note on the melting point of alpha-solanine. American Journal of Potato Research 49, 403–406.

Prat, S., Frommer, W.B., Hofgen, R., Keil, M., Kobmann, J., Koster-Topfer, M., Liu, X.-J., Muller, B., Pena-Cortes, H., Rocha-Sosa, M., Sanchez-Serrano, J.J., Sonnewald, U., Willmitzer, L., 1990. Gene expression during tuber development in potato plants. FEBS Letters 268, 334–338.

Raigond, P., Ezekiel, R., Raigond, B., 2015. Resistant starch in food: a review. Journal of the Science of Food and Agriculture 95, 1968–1978.

Rajananthanan, P., Attard, G.S., Sheikh, N.A., Morrow, W.J., 1999. Novel aggregate structure adjuvants modulate lymphocyte proliferation and Th1 and Th2 cytokine profiles in ovalbumin immunized mice. Vaccine 18, 140–152.

Randhawa, K.S., Sandhu, K.S., Kaur, G., Singh, D., 1984. Studies of the evaluation of different genotypes of potato *Solanum tuberosum* for yield and mineral contents. Plant Foods for Human Nutrition (Formerly Qualitas Plantarum) 34, 239–242.

Ray, D.K., Mueller, N.D., West, P.C., Foley, J.A., 2013. Yield trends are insufficient to double global crop production by 2050. PLoS One 8, e66428.

Reddivari, L., Vanamala, J., Chintharlapalli, S., Safe, S.H., Miller Jr., J.C., 2007. Anthocyanin fraction from potato extracts is cytotoxic to prostate cancer cells through activation of caspase-dependent and caspase-independent pathways. Carcinogenesis 28, 2227–2235.

Reddivari, L., Vanamala, J., Safe, S.H., Miller Jr., J.C., 2010. The bioactive compounds alpha-chaconine and gallic acid in potato extracts decrease survival and induce apoptosis in LNCaP and PC3 prostate cancer cells. Nutrition and Cancer 62, 601–610.

Rees, E., Gowing, L.R., 2013. Supplementary thiamine is still important in alcohol dependence. Alcohol and Alcoholism 48, 88–92.

Riaz, M., Asif, M., Ali, R., 2009. Stability of vitamins during extrusion. Critical Reviews in Food Science and Nutrition 49, 361–368.

Rivero, R.C., Hernández, P.S., Rodríguez, E.M.R.g., Martín, J.D., Romero, C.D.a., 2003. Mineral concentrations in cultivars of potatoes. Food Chemistry 83, 247–253.

Roddick, J.G., Weissenberg, M., Leonard, A.L., 2001. Membrane disruption and enzyme inhibition by naturally-occurring and modified chacotriose-containing *Solanum* steroidal glycoalkaloids. Phytochemistry 56, 603–610.

Roje, S., 2007. Vitamin B biosynthesis in plants. Phytochemistry 68, 1904–1921.

Rolfe, H.M., 2014. A review of nicotinamide: treatment of skin diseases and potential side effects. Journal of Cosmetic Dermatology 13, 324–328.

Rosenstock, G., Zimmermann, H.J., 1976. Vergleichende Studien uber den Protein und Nucleinsaurestoffwechsel beim Speicherparenchym von *Solanum tuberosum* L. mit primarer und sekundarer mitotischer Aktivitat. Beitrage zur Biologie der Pflanzen 413–426.

Rytel, E., Lisińska, G., Tajner-Czopek, A., 2013. Toxic compound levels in potatoes are dependent on cultivation methods. Acta Alimentaria 42, 308–317.

Sanchez-Castillo, C.P., Dewey, P.J.S., Aguirre, A., Lara, J.J., Vaca, R., Barra, P.L.d.l., Ortiz, M., Escamilla, I., James, W.P.T., 1998. The mineral content of Mexican fruits and vegetables. Journal of Food Composition Analysis 11, 340–356.

Sawai, S., Ohyama, K., Yasumoto, S., Seki, H., Sakuma, T., Yamamoto, T., Takebayashi, Y., Kojima, M., Sakakibara, H., Aoki, T., Muranaka, T., Saito, K., Umemoto, N., 2014. Sterol side chain reductase 2 is a key enzyme in the biosynthesis of cholesterol, the common precursor of toxic steroidal glycoalkaloids in potato. Plant Cell 26, 3763–3774.

Scalbert, A., Manach, C., Morand, C., Remesy, C., Jimenez, L., 2005. Dietary polyphenols and the prevention of diseases. Critical Reviews in Food Science and Nutrition 45, 287–306.

Scurrah, M., Amoros, W., Burgos, G., Schafleitner, R., Bonierbale, M., 2006. Back to the future: millennium traits in native varieties. In: VI International Solanaceae Conference: Genomics Meets Biodiversity, vol. 745, pp. 369–378.

Shakya, R., Navarre, D.A., 2008. LC-MS analysis of solanidane glycoalkaloid diversity among tubers of four wild potato species and three cultivars (*Solanum tuberosum*). Journal of Agricultural and Food Chemistry 56, 6949–6958.

Shepherd, L., Alexander, C., Hackett, C., McRae, D., Sungurtas, J., Verrall, S., Morris, J., Hedley, P., Rockhold, D., Belknap, W., Davies, H., 2015a. Impacts on the metabolome of down-regulating polyphenol oxidase in potato tubers. Transgenic Research 24, 447–461.

Shepherd, L.V., Hackett, C.A., Alexander, C.J., McNicol, J.W., Sungurtas, J.A., Stewart, D., McCue, K.F., Belknap, W.R., Davies, H.V., 2015b. Modifying glycoalkaloid content in transgenic potato - metabolome impacts. Food Chemistry 187, 437–443.

Shih, Y.W., Chen, P.S., Wu, C.H., Jeng, Y.F., Wang, C.J., 2007. Alpha-chaconine-reduced metastasis involves a PI3K/Akt signaling pathway with downregulation of NF-kappaB in human lung adenocarcinoma A549 cells. Journal of Agricultural and Food Chemistry 55, 11035–11043.

Shiu, L.Y., Liang, C.H., Chang, L.C., Sheu, H.M., Tsai, E.M., Kuo, K.W., 2009. Solamargine induces apoptosis and enhances susceptibility to trastuzumab and epirubicin in breast cancer cells with low or high expression levels of HER2/neu. Bioscience Reports 29, 35–45.

da Silva, R.P., Kelly, K.B., Al Rajabi, A., Jacobs, R.L., 2014. Novel insights on interactions between folate and lipid metabolism. BioFactors 40, 277–283.

Sinden, S.L., Deahl, K.L., Aulenbach, B.B., 1976. Effect of glycoalkaloids and phenolics on potato flavor. Journal of Food Science 41, 520–523.

Sinden, S.L., Sanford, L.L., Webb, R.E., 1984. Genetic and environmental control of potato glycoalkaloids. American Journal of Potato Research 61, 141–156.

Singh, R.P., Singh, M., King, R.R., 1998. Use of citric acid for neutralizing polymerase chain reaction inhibition by chlorogenic acid in potato extracts. Journal of Virological Methods 74, 231–235.

Skrabule, I., Muceniece, R., Kirhnere, I., 2013. Evaluation of vitamins and glycoalkaloids in potato genotypes grown under organic and conventional farming systems. Potato Research 56, 259–276.

Sotelo, A., Serrano, B., 2000. High-performance liquid chromatographic determination of the glycoalkaloids alpha-solanine and alpha-chaconine in 12 commercial varieties of Mexican potato. Journal of Agricultural and Food Chemistry 48, 2472–2475.

Spooner, D.M., McLean, K., Ramsay, G., Waugh, R., Bryan, G.J., 2005. A single domestication for potato based on multilocus amplified fragment length polymorphism genotyping. Proceedings of the National Academy of Sciences of the United States of America 102, 14694–14699.

Stapleton, A., Allen, P.V., Friedman, M., Belknap, W.R., 1991. Purification and characterization of solanidine glycosyltransferase from the potato (*Solanum tuberosum*). Journal of Agricultural and Food Chemistry 39, 1187–1193.

Stenzel, O., Teuber, M., Drager, B., 2006. Putrescine N-methyltransferase in *Solanum tuberosum* L., a calystegine-forming plant. Planta 223, 200–212.

Sun, L., Zhao, Y., Yuan, H., Li, X., Cheng, A., Lou, H., 2011. Solamargine, a steroidal alkaloid glycoside, induces oncosis in human K562 leukemia and squamous cell carcinoma KB cells. Cancer Chemotherapy and Pharmacology 67, 813–821.

Swain, J.F., McCarron, P.B., Hamilton, E.F., Sacks, F.M., Appel, L.J., 2008. Characteristics of the diet patterns tested in the optimal macronutrient intake trial to prevent heart disease (OmniHeart): options for a heart-healthy diet. Journal of the American Dietetic Association 108, 257–265.

Tahiliani, A.G., Beinlich, C.J., 1991. Pantothenic-acid in health and disease. Vitamins and Hormones-Advances in Research and Applications 46, 165–228.

Tai, H.H., Worrall, K., Pelletier, Y., De Koeyer, D., Calhoun, L.A., 2014. Comparative metabolite profiling of Solanum tuberosum against six wild solanum species with Colorado potato beetle resistance. Journal of Agricultural and Food Chemistry 62, 9043–9055.

Tajner-Czopek, A., Rytel, E., Kita, A., Pęksa, A., Hamouz, K., 2012. The influence of thermal process of coloured potatoes on the content of glycoalkaloids in the potato products. Food Chemistry 133, 1117–1122.

Takagi, K., Toyoda, M., Fijiyama, Y., Saito, Y., 1990. Effect of cooking on contents of chaconine and solanine in potatoes. Journal of Food Hygienic Society of Japan 31, 67–73.

Tan, J.S., Wang, J.J., Flood, V., Rochtchina, E., Smith, W., Mitchell, P., 2008. Dietary antioxidants and the long-term incidence of age-related macular degeneration: the Blue Mountains Eye Study. Ophthalmology 115, 334–341.

Tanemura, Y., Yoshino, M., 2006. Regulatory role of polyamine in the acid phosphatase from potato tubers. Plant Physiology Biochemistry 44, 43–48.

Tawfik, A.A., Kleinhenz, M.D., Palta, J.P., 1996. Application of calcium and nitrogen for mitigating heat stress effects on potatoes. American Journal of Potato Research 73, 261–273.

Tedone, L., Hancock, R.D., Alberino, S., Haupt, S., Viola, R., 2004. Long-distance transport of L-ascorbic acid in potato. BMC Plant Biology 4, 16.

Tibbetts, A.S., Appling, D.R., 2010. Compartmentalization of mammalian folate-mediated one-carbon metabolism. Annual Review of Nutrition 30 (30), 57–81.

Tiossi, R.F., Da Costa, J.C., Miranda, M.A., Praca, F.S., McChesney, J.D., Bentley, M.V., Bastos, J.K., 2014. In vitro and in vivo evaluation of the delivery of topical formulations containing glycoalkaloids of Solanum lycocarpum fruits. European Journal of Pharmaceutics and Biopharmaceutics 88, 28–33.

Titiz, O., Tambasco-Studart, M., Warzych, E., Apel, K., Amrhein, N., Laloi, C., Fitzpatrick, T.B., 2006. PDX1 is essential for vitamin B_6 biosynthesis, development and stress tolerance in Arabidopsis. Plant Journal 48, 933–946.

Tomas-Barberan, F., Garcia-Villalba, R., Quartieri, A., Raimondi, S., Amaretti, A., Leonardi, A., Rossi, M., 2014. In vitro transformation of chlorogenic acid by human gut microbiota. Molecular Nutrition & Food Research 58, 1122–1131.

Tosun, B.N., Yücecan, S., 2008. Influence of commercial freezing and storage on vitamin C content of some vegetables. International Journal of Food Science & Technology 43, 316–321.

True, R.H., Hogan, J.M., Augustin, J., Johnson, S.J., Teitzel, C., Toma, R.B., Shaw, R.L., 1978. Mineral composition of freshly harvested potatoes. American Potato Journal 55, 511–519.

True, R.H., Hogan, J.M., Augustin, J., Johnson, S.R., Teitzel, C., Toma, R.B., Orr, P., 1979. Changes in the nutrient composition of potatoes during home preparation. III. Minerals. American Potato Journal 56, 339–350.

Tudela, J.A., Cantos, E., Espin, J.C., Tomas-Barberan, F.A., Gil, M.I., 2002a. Induction of antioxidant flavonol biosynthesis in fresh-cut potatoes. Effect of domestic cooking. Journal of Agricultural and Food Chemistry 50, 5925–5931.

Tudela, J.A., Espin, J.C., Gil, M.I., 2002b. Vitamin C retention in fresh-cut potatoes. Postharvest Biology and Technology 26, 75–84.

Tunnicliffe, J.M., Eller, L.K., Reimer, R.A., Hittel, D.S., Shearer, J., 2011. Chlorogenic acid differentially affects postprandial glucose and glucose-dependent insulinotropic polypeptide response in rats. Applied Physiology, Nutrition, and Metabolism 36, 650–659.

Urbany, C., Colby, T., Stich, B., Schmidt, L., Schmidt, J., Gebhardt, C., 2012. Analysis of natural variation of the potato tuber proteome reveals novel candidate genes for tuber bruising. Journal of Proteome Research 11, 703–716.

Urbany, C., Stich, B., Schmidt, L., Simon, L., Berding, H., Junghans, H., Niehoff, K.H., Braun, A., Tacke, E., Hofferbert, H.R., Lubeck, J., Strahwald, J., Gebhardt, C., 2011. Association genetics in *Solanum tuberosum* provides new insights into potato tuber bruising and enzymatic tissue discoloration. BMC Genomics 12, 7.

USDA-ERS, 2015. Vegetables & Pulses. http://www.ers.usda.gov/topics/crops/vegetables-pulses/potatoes.aspx (accessed June 2015).

USDA, 2005. Dietary Guidelines for Americans World Wide Web. Department of Health & Human Services (USDA/HHS). http://WWW.health.gov/dietaryguidelines/dga2005/document/html/AppendixB.htm#appB.

Valcarcel, J., Reilly, K., Gaffney, M., O'Brien, N., 2015. Total carotenoids and L-ascorbic acid content in 60 varieties of potato (*Solanum tuberosum* L.) grown in Ireland. Potato Research 58, 29–41.

Valkonen, J.P.T., Keskitalo, M., Vasara, T., Pietila, L., 1996. Potato glycoalkaloids: a burden or a blessing? Critical Reviews in Plant Sciences 15 (1), 1–20.

Vinson, J.A., Demkosky, C.A., Navarre, D.A., Smyda, M.A., 2012. High-antioxidant potatoes: acute in vivo antioxidant source and hypotensive agent in humans after supplementation to hypertensive subjects. Journal of Agricultural and Food Chemistry 60, 6749–6754.

Viola, R., Vreugdenhil, D., Davies, H.V., Sommerville, L., 1998. Accumulation of L-ascorbic acid in tuberising stolon tips of potato (*Solanum tuberosum* L.). Journal of Plant Physiology 152, 58–63.

Wahlberg, G., Adamson, U., Svensson, J., 2000. Pyridine nucleotides in glucose metabolism and diabetes: a review. Diabetes/Metabolism Research and Reviews 16, 33–42.

Waltz, E., 2015. USDA approves next-generation GM potato. Nature Biotechnology 33, 12–13.

Wang, Q., Cao, Y., Zhou, L., Jiang, C.-Z., Feng, Y., Wei, S., 2015. Effects of postharvest curing treatment on flesh colour and phenolic metabolism in fresh-cut potato products. Food Chemistry 169, 246–254.

Warman, P.R., Havard, K., 1998. Yield, vitamin and mineral contents of organically and conventionally grown potatoes and sweet corn. Agriculture, Ecosystems & Environment 68, 207–216.

Werij, J.S., Kloosterman, B., Celis-Gamboa, C., de Vos, C.H., America, T., Visser, R.G., Bachem, C.W., 2007. Unravelling enzymatic discoloration in potato through a combined approach of candidate genes, QTL, and expression analysis. Theoretical and Applied Genetics 115, 245–252.

Wheeler, G.L., Jones, M.A., Smirnoff, N., 1998. The biosynthetic pathway of vitamin C in higher plants. Nature 393, 365–369.

Willett, W.C., Stampfer, M.J., 2003. Rebuilding the food pyramid. Scientific American 288, 64–71.

Williams, A., Ramsden, D., 2005. Nicotinamide: a double edged sword. Parkinsonism & Related Disorders 11, 413–420.

Wills, R.B.H., Lim, J.S.K., Greenfield, H., 1984. Variation in nutrient composition of Australian retail potatoes over a 12-month period. Journal of the Science of Food and Agriculture 35, 1012–1017.

Wilson, G.S., 1959. A small outbreak of solanine poisoning. Monthly Bulletin of the Ministry of Health and the Public Health Laboratory Service 18, 207–210.

Wojtczak, L., Slyshenkov, V.S., 2003. Protection by pantothenic acid against apoptosis and cell damage by oxygen free radicals - the role of glutathione (Reprinted from Thiol Metabolism and Redox Regulation of Cellular Functions). BioFactors 17, 61–73.

Wolters, A.M., Uitdewilligen, J.G., Kloosterman, B.A., Hutten, R.C., Visser, R.G., van Eck, H.J., 2010. Identification of alleles of carotenoid pathway genes important for zeaxanthin accumulation in potato tubers. Plant Molecular Biology 73, 659–671.

Wolucka, B.A., van Montagu, M., 2007. The VTC2 cycle and the de novo biosynthesis pathways for vitamin C in plants: an opinion. Phytochemistry 68, 2602–2613.

Yamaguchi, T., Chikama, A., Mori, K., Watanabe, T., Shioya, Y., Katsuragi, Y., Tokimitsu, I., 2007. Hydroxyhydroquinone-free coffee: a double-blind, randomized controlled dose-response study of blood pressure. Nutrition, Metabolism and Cardiovascular Disease 18 (6), 408–414.

Zhang, Y., Jin, X., Ouyang, Z., Li, X., Liu, B., Huang, L., Hong, Y., Zhang, H., Song, F., Li, D., 2014. Vitamin B_6 contributes to disease resistance against *Pseudomonas syringae* pv. tomato DC3000 and *Botrytis cinerea* in *Arabidopsis thaliana*. Journal of Plant Physiology 175C, 21–25.

Zhou, X., McQuinn, R., Fei, Z., Wolters, A.M., van Eck, J., Brown, C., Giovannoni, J.J., Li, L., 2011. Regulatory control of high levels of carotenoid accumulation in potato tubers. Plant, Cell & Environment 34, 1020–1030.

Zienkiewicz, M., 2015. The new staple food in China. Spudsmart Spring 24–26.

Zimowski, J., 1991. Occurrence of a glucosyltransferase specific for solanidine in potato plants. Phytochemistry 30, 1827–1831.

Zitnak, A., Filadelfi, M.A., 1985. Estimation of taste thresholds of three potato glycoalkaloids. Canadian Institute of Food Science and Technology 18, 337–339.

Zupko, I., Molnar, J., Rethy, B., Minorics, R., Frank, E., Wolfling, J., Molnar, J., Ocsovszki, I., Topcu, Z., Bito, T., Puskas, L.G., 2014. Anticancer and multidrug resistance-reversal effects of solanidine analogs synthetized from pregnadienolone acetate. Molecules 19, 2061–2076.

CHAPTER 7

Glycoalkaloids and Calystegine Alkaloids in Potatoes

Mendel Friedman, Carol E. Levin
United States Department of Agriculture, Agricultural Research Service, Western Regional Research Center, Albany, CA, USA

1. Introduction

Potatoes are members of the notorious nightshade (*Solanaceae*) family, well known for their content of natural poisons in the form of alkaloids. Nonetheless, potatoes serve as a major, inexpensive low-fat food source of energy, high-quality protein, fiber, vitamins, and minerals. The alkaloids are produced as biologically active secondary metabolites, which offer some protection against insects and fungi. In fact, the vegetative part of the plant is particularly high in alkaloids, while the edible tuber is relatively low and usually not a concern. However, environmental stressors can stimulate alkaloid levels to increase in the tuber beyond what is considered safe. In addition, levels in different cultivars are variable, and breeding and genetic manipulation programs must be monitored with regard to tuber glycoalkaloid content. Potatoes contain primarily the glycoalkaloids α-solanine and α-chaconine, although smaller levels of the hydrolysis products, the β-, γ-, and aglycone forms, can also be present. Some wild-type potatoes have been found to contain atypical glycoalkaloids as well. Glycoalkaloids are of importance because of their potential toxicity. Although poisonings are rare, a very high intake can be lethal. The *nor*tropane alkaloids, the calystegines, are another category of alkaloids known to be in potatoes. Less is known about these toxic alkaloids, as their presence in potatoes was only just discovered in 1993. In this overview we discuss, with regard to alkaloids, their analytical aspects, prevalence in the plant and the diet, toxicity, and potential health benefits.

2. Glycoalkaloids

Steroidal glycoalkaloids are naturally occurring, secondary plant metabolites that are found in a number of foods in the *Solanum* genus, including potatoes, tomatoes, and eggplants (reviewed in Friedman, 2002, 2015; Friedman and McDonald, 1999a,b, 1997). Although in high doses they are toxic, glycoalkaloids may also have beneficial effects in humans. These include lowering of blood cholesterol (Friedman et al., 2000a,b), protection against infection by

Salmonella Typhimurium (Gubarev et al., 1998), and chemoprevention of cancer (Cham, 1994; Friedman et al., 2005, 2007; Lee et al., 2004). Glycoalkaloids are also known to contribute to flavor, as potatoes with low levels are judged to be bland, while high levels cause bitterness.

In commercial potatoes (*Solanum tuberosum*) there are two major glycoalkaloids, α-chaconine and α-solanine, both triglycosides of the common aglycone solanidine. These two compounds comprise about 95% of the glycoalkaloids present in potato tubers. Their hydrolysis products, the β and γ forms and solanidine, may also be present, but in relatively insignificant concentrations. The structures of these glycoalkaloids and their hydrolysis products are presented in Figure 1. Wild-type potatoes can contain a greater diversity of glycoalkaloids. For example, potato tubers of somatic hybrids, whose progenies were the cultivated potato *S. tuberosum* and the wild-type *Solanum acaule*, contained four glycoalkaloids, including α-tomatine and demissine derived from the fusion parents (Kozukue et al., 1999).

2.1 Abundance in the Tuber

Potatoes accumulate glycoalkaloids in response to stress. For the tuber, these pathways are active both pre- and postharvest. This is relevant information for both farmers and processors alike.

2.1.1 Agronomy

Just as cultivation practices can influence potato quality, so can they influence glycoalkaloid levels. Potatoes synthesize glycoalkaloids in response to stress, so not surprisingly growing conditions can affect both plant and tuber levels of these compounds. Studies of organic farming techniques have shown both an increase (Wszelaki et al., 2005) and a decrease (Rytel et al., 2013) in glycoalkaloid content. Other studies have not shown any association (Hamouz et al., 2005; Skrabule et al., 2013). It seems possible that any correlation between glycoalkaloid levels and organic farming may be coincidental to overall plant stress. The high nitrogen fertilizers used in conventional farming may be one factor in increasing glycoalkaloid content (Najm et al., 2012; Rytel et al., 2013). Stressors known to increase glycoalkaloid content in the field include exposure to light, increased growing temperature (Nitithamyong et al., 1999), and wounding. Response to carbon dioxide levels is equivocal. One study found a 20% increase in total glycoalkaloids when CO_2 was increased from 350 to $1000\,\mu mol/mol$ (Nitithamyong et al., 1999), while other studies found a negative relationship between CO_2 and glycoalkaloids (Donnelly et al., 2001; Högy and Fangmeier, 2009; Vorne et al., 2002). Soil and irrigation management seem to have little effect on glycoalkaloid production (Zhang et al., 1997). However, drought stress increased levels by as much as 50% (Bejarano et al., 2000), as well as dry and warm relative to wet and cool weather (Zarzecka and Gugała, 2007). Potatoes grown in the summer had a higher glycoalkaloid level than spring or winter potatoes (Dimenstein et al., 1997). Light exposure was shown to

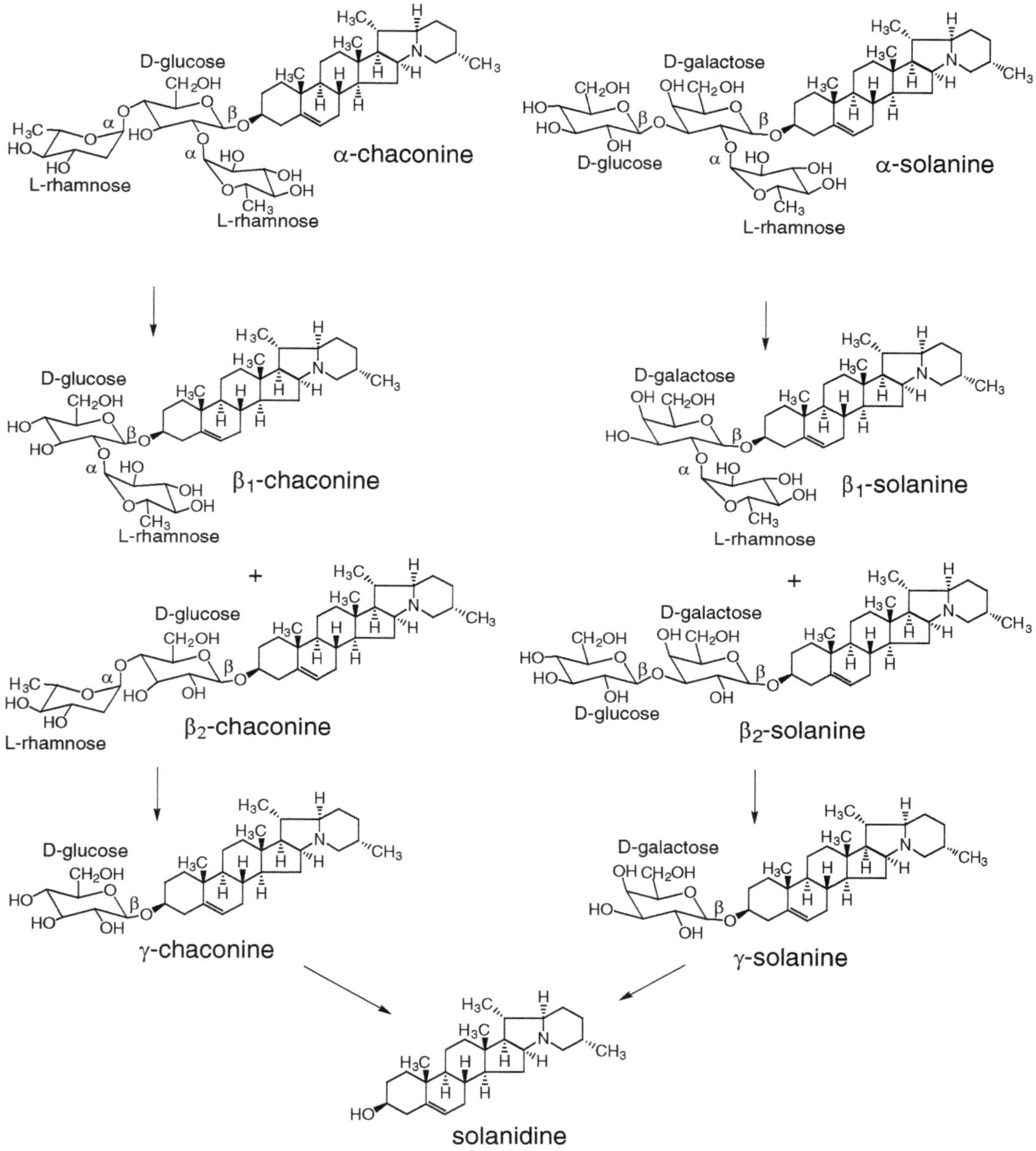

Figure 1

Structures of potato glycoalkaloids α-chaconine and α-solanine and hydrolysis products (metabolites).

increase not just α-solanine and α-chaconine in potatoes, but also tomatidenol-based α- and β-solamarine, commonly found in the herbaceous plant *Solanum dulcamara* (bittersweet nightshade) (Griffiths et al., 2000; Griffiths and Dale, 2001). Glycoalkaloid levels increased in plants grown in growth chambers when temperatures were either higher or lower than 16 °C, with the higher temperatures more stimulating (Nitithamyong et al., 1999). Application of insecticides may increase total glycoalkaloid content of the tuber (Zarzecka et al., 2013). Low-tillage techniques caused a small but significant increase in glycoalkaloids accompanied by lower crop yields (Zarzecka and Gugała, 2007). This increase was insignificant when the potato was peeled, inferring that the additional glycoalkaloids accumulated mostly in the peel (Zarzecka and Gugała, 2007).

Cultivars may respond differently to stress. A study found that cultivars with a high base-level of glycoalkaloids were more susceptible to environmental stimulation of glycoalkaloids (Valcarcel et al., 2014). In another study, of five cultivars exposed to drought, the glycoalkaloids of two were unaffected, while the other three cultivars had a significant two- to fourfold increase (Andre et al., 2009). It is noteworthy that Petersson et al. (2013) found a large response to light stress by the cultivars with the highest base levels of glycoalkaloids. This poses the question as to whether high cultivar levels of glycoalkaloids are static or dynamic, i.e., are levels high simply because the synthesis pathways are continuously more active?

2.1.2 Storage

Potato tubers continue to produce glycoalkaloids in response to postharvest stress. Greening, the production of chlorophyll in the tuber, is associated with glycoalkaloid content, although the two processes are actually coincident and controlled by different pathways (Kozukue et al., 2001). However, greening does give potato producers and consumers a quick evaluation of the amount of light stress the tuber has endured. Light exposure, wounding, and sprouting are the stresses most likely to increase glycoalkaloid production during storage. Interestingly, variations between glycoalkaloids in different potato cultivars induced by growing conditions, in this case drought, were amplified during storage at 10 °C for 4 months (Andre et al., 2009). Within a cultivar, the well-watered potatoes showed little change in glycoalkaloid content during storage, unlike their drought-exposed counterparts.

Glycoalkaloids increased significantly during storage when sprouting inhibitors were not used, regardless of the actual observed amount of sprouting (Haase, 2010). Fluorescent light exposure (as is found at many indoor markets) stimulates glycoalkaloid accumulations more than indirect daylight (Machado et al., 2007). In fact indirect light does not appear to increase glycoalkaloid levels (Kaaber, 1993). Maximum glycoalkaloid production under fluorescent light occurred at light intensity of 8000 to 12,500 lux, while content was reduced at 15,400 lux (Kozukue et al., 1993). Eight days of exposure of tubers during storage to fluorescent light increased glycoalkaloids in 21 varieties by an average of 249%, similar to wounding which increased them by 246% in 48 h (Petersson et al., 2013). Response of tubers to fluorescent light is rapid, with increases in glycoalkaloid levels measured after just 30 min of light exposure (Kozukue et al., 1993). Application of heat (34 °C) during storage for 7 days did not increase the glycoalkaloids (Petersson et al., 2013). However, temperature does appear to affect the response of the tubers to light, as greening caused by illumination was increased during higher temperature (18 and 24, vs 6 °C) storage (Kaaber, 1993).

2.1.3 Processing

Exposure of the tuber to stresses, such as light, mechanical injury, and storage, can enhance glycoalkaloid levels (Kozukue et al., 1993; Machado et al., 2007), so proper handling is key in keeping glycoalkaloid levels low. Removal of the peel and sprouts is the most significant way to reduce glycoalkaloids in potato food products. Glycoalkaloids are not easily leached from potatoes, as they are soluble neither in water (at neutral pH) nor in ether. Thus, losses into boiling water are minimal. Mulinacci et al. (2008) found that boiling or microwaving the intact potato resulted in no significant change in the glycoalkaloid content. However, results were the opposite for Lachman et al. (2013), who found 45% and 51% losses in whole microwaved and baked potatoes, respectively. Lachman et al. (2013) found an average 12% loss from boiling peeled potatoes. High heat appears to be effective at destroying the glycoalkaloids. Frying, such as in the making of french fries and crisps, appears to cause significant losses (~90%) (Rytel et al., 2005; Tajner-Czopek et al., 2014). Two commercial frozen French fry products were found to contain 0.8 and 8.4 mg/kg glycoalkaloids (Friedman and Dao, 1992), although the levels in the original potato flesh before processing, and the processing conditions, are unknown. There is a concern that extracted products from potato could be high in glycoalkaloids. Glycoalkaloids were tracked during the making of potato starch, processed in accordance with industry practices, and found to be absent from the starch and concentrated in the protein (Driedger and Sporns, 1999).

The commercial potato processing industry is considered one of the largest producers of agricultural waste (Elferink et al., 2008), generating an estimated 4.3 million tons of waste a year (Nelson, 2010). We briefly discuss this subject here because much of what is discarded during the processing of the potato is high in glycoalkaloids, namely the peel. Rodriguez de Sotillo (1998) proposed making use of the high antioxidant content of potato peel as an antimicrobial. To prevent possible poisonings, care must be exercised when using the waste in human or animal food (Kling et al., 1986) or when releasing it into the environment as industrial waste. Animal feed is a potential market for this byproduct (Nelson, 2010). Use of waste in lieu of grain for animal feed can potentially lower the environmental impact of meat consumption (Elferink et al., 2008). The quality of beef was not affected by the feeding of potato peels (Nelson, 2010). A study in the Netherlands revealed that use of potato byproducts in pork production was substantial and effective (Elferink et al., 2008). Potato protein feed lowered the microbial count of chicken excreta and cecum (Ohh et al., 2009) and of pig feces (Jin et al., 2008). Peels can also be fermented to produce hydrogen fuels (Djomo et al., 2008).

2.1.4 Distribution

Table 1 contains a survey of glycoalkaloid content in selected potatoes analyzed using a high-performance liquid chromatography (HPLC) NH_2 column technique. None of the whole potatoes exceeded the 200 mg total glycoalkaloids per kilogram of potatoes (see A + B column). However, this was not the case for potato peel. Five of the eight samples exceeded this benchmark. Valcarcel et al. (2014) tested 60 rare, heritage, and commercial potato cultivars with similar results.

Some unusual varieties contain glycoalkaloid compounds other than α-chaconine or α-solanine. Kirui et al. (2009) found that 15 potato cultivars grown in the tropical climate of Nigeria contained high amounts (up to 6.41 mg/100 g fresh weight for the Tigoni clone) of the aglycone solanidine. α-Tomatine, typically found in unripe tomatoes, was found in the wild-type potato *S. acaule* (Kozukue et al., 1999). *S. acaule* also contained demissine (Kozukue et al., 2008). Figure 2 shows the glycoalkaloid patterns of somatic hybrids of *S. tuberosum* and *S. acaule*. The hybrids of *S. tuberosum* crossed with *S. acaule* and another wild-type potato, *Solanum brevidens*, contained α-tomatine and two other glycoalkaloids consisting of tomatidine (the aglycone of α-tomatine) bound to solatriose and chacotriose (the sugars found in α-solanine and α-chaconine, respectively) (Väänänen et al., 2005). Another wild-type potato, *Solanum canasense*, has been found to contain dehydrocommersonine (Kozukue et al., 2008). *Solanum juzepczukii* and *Solanum curtilobum* contained demissine and two newly discovered demissidine glycoalkaloids (Kozukue et al., 2008).

2.1.5 Ratio of α-Chaconine to α-Solanine

The ratios of α-chaconine to α-solanine for the potato samples can vary. Ratios are of interest because, as mentioned earlier, α-chaconine is more toxic than α-solanine. In Table 1 the ratios ranged from 1.2 to 2.6. The ratio in peel, generally in the range of about 2, was higher than in flesh, with values near about 1.5. Previously it was shown that potatoes that had been sliced and incubated increased their content of both glycoalkaloids, but there was a bigger increase for α-chaconine (Fitzpatrick et al., 1977). We can only speculate about possible reasons for the wide variations in these ratios. Since the two glycoalkaloids, which share the common aglycone solanidine but not the same trisaccharide side chain (Figure 1), appear to be synthesized via discrete biosynthetic channels (Choi et al., 1994), it is possible that the rates of biosynthesis of the two glycoalkaloids in the different channels are cultivar-dependent. Another possible rationalization for the varying ratios is that the rate of metabolism of the two glycoalkaloids is also cultivar-dependent. These considerations imply that alteration of the genes encoding enzymes involved in the biosynthesis of α-chaconine and/or α-solanine could have unpredictable results.

Because α-tomatine seems to be nontoxic to humans (Friedman, 2013), the cultivation and commercial availability of α-tomatine-containing potato cultivars have the potential to enhance human health. For example, we reported that the α-tomatine content of the *S. acaule*

Table 1: Glycoalkaloid and Calystegine Content of Potato Flesh, Potato Peel, and Whole Potatoes of Eight Potato Cultivars (in mg/kg).

Potato Cultivar		Glycoalkaloids				Calystegines			
		α-Chaconine (A)	α-Solanine (B)	A + B	A/B	Calystegine A_3	Calystegine B_2	$A_3 + B_2$	B_2/A_3
Atlantic	flesh	4.7	2.9	7.6	1.6	1.1	1.5	2.6	1.4
	peel	8.8	3.6	12.4	2.4	31.2	141	172	4.5
	whole	5	8	8	1.7	3.5	12.9	16.4	3.7
Dark red Norland	flesh	3.4	1.3	4.7	2.6	0	1.3	1.3	–
	peel	128	60.3	188	2.1	6.4	33.3	39.7	5.2
	whole	16.8	7.7	24.5	2.2	0.7	4.7	5.4	6.7
Ranger Russet	flesh	12.8	7	19.8	1.8	1.1	2.3	3.4	2.1
	peel	230	110	340	2.1	87.1	380	467	4.4
	whole	34.3	17.2	51.5	2	9.6	39.7	49.3	4.1
Red Lasoda	flesh	4.4	2.8	7.2	1.6	1.4	4.3	5.7	3.1
	peel	134	96.6	231	1.4	10.5	24.83	35.3	2.4
	whole	16	11.2	27.2	1.4	2.2	6.1	8.3	2.8
Russet Burbank	flesh	25.9	21.4	47.3	1.2	11.1	56.5	67.6	5.1
	peel	182	103	285	1.8	6.6	67.8	74.4	11.8
	whole	36.3	26.8	63.1	1.4	10.8	57.3	68.1	5.3
Russet Norkota	flesh	0.74	0.54	1.3	1.4	0.2	0.8	1	4
	peel	48.1	23	71.1	2.1	33.6	129	163	3.9
	whole	4.8	2.5	7.3	1.9	3	11.9	14.9	4
Shepody	flesh	1.8	1.1	2.9	1.7	2.2	9.1	11.3	4.1
	peel	282	147	429	1.9	44	299	343	6.8
	whole	25.1	13.2	38.3	1.9	5.6	33.1	38.7	5.9
Snowden	flesh	91.5	56.5	148	1.7	0.8	0.8	1.7	1
	peel	372	171	543	2.2	54.2	96.3	150	1.8
	whole	119	67.9	187	1.8	5.8	10.2	16	1.8

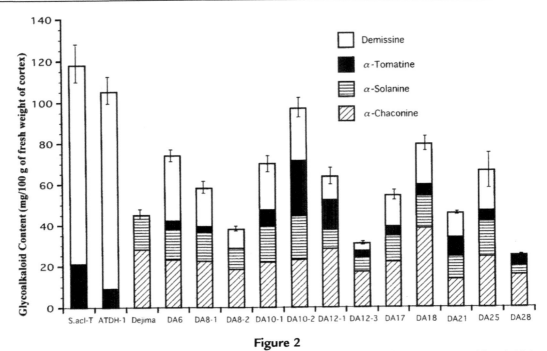

Figure 2
Constituent glycoalkaloids from the tuber of cortex of *S. acaule*-T, the fusion parents (dihaploid *S. acaule*, ATDH-1, and *S. tuberosum* cv. Dejima), and 12 somatic hybrids. Vertical bars indicate standard errors for three separate determinations of the sums of individual glycoalkaloids (Kozukue et al., 1999).

potato Accession No. acl-D-3 is 49.7 µg/100 g fresh weight (Kozukue et al., 2008), a value that is much higher than found in red tomatoes. Such potatoes could serve as a new functional food. We are challenged to accomplish this objective for the benefit of the farm economy and human health. It should be emphasized that safety considerations should guide the use of these compounds as preventative or therapeutic treatments in the diet.

2.2 Toxicity

The toxicity of glycoalkaloids at appropriately high levels appears to be due to such adverse effects as anticholinesterase activity on the central nervous system (Abbott et al., 1960; Heftmann, 1967), induction of liver damage (Caldwell et al., 1991), and disruption of cell membranes adversely affecting the digestive system and general body metabolism (Keukens et al., 1995). At lower doses, the toxicity of glycoalkaloids in humans causes mainly gastrointestinal (GI) disturbances such as vomiting, diarrhea, and abdominal pain. At higher doses, it produces systemic toxicity, including symptoms such as fever, rapid pulse, low blood pressure, rapid respiration, and neurological disorders. Several cases of lethal poisoning have been reported (Korpan et al., 2004). It is believed that as the mucosal cells in the intestine are compromised, absorption through the intestine wall is more rapid. This would account for the differences in symptoms observed for low (GI) and for high (acute systemic) toxicity. This also may explain the observed steep toxicity curve, in which the shift from mild to severe toxicity is over a dose increase of just two- to threefold (Hopkins, 1995). It is worth noting that the apparent nontoxicity of the tomato glycoalkaloid α-tomatine appears to be due to complex formation with cholesterol in the digestive tract followed by fecal elimination (Friedman et al., 2000b, 2007). α-Tomatine is found in some wild-type potatoes, suggesting that breeding programs could increase the levels of this less toxic glycoalkaloid in potatoes.

These concerns of toxicity have led to the establishment of informal guidelines limiting the total glycoalkaloid concentration of new potato cultivars to 200 mg/kg of fresh weight. As discussed elsewhere, these guidelines may be too high (Friedman, 2006; Friedman and McDonald, 1997). In one short-term clinical trial with human volunteers, one test subject experienced GI disturbances after consuming mashed potatoes containing glycoalkaloids at the recommended limit, 200 mg/kg (Mensinga et al., 2005). The Food and Agricultural Organization of the World Health Organization evaluated possible guidelines for solanine intake, but were unable to make a recommendation owing to insufficient data. They stated that "the large body of experience with the consumption of potatoes, frequently on a daily basis, indicated that normal glycoalkaloid levels (20–100 mg/kg) found in properly grown and handled tubers were not of concern" (Anonymous, 2009).

There have been few human studies, most of them anecdotal, so the susceptibility per individual variation and influence of other factors are not well established. The incidence of glycoalkaloid poisoning may be underreported, probably because physicians are more likely to implicate foodborne pathogens (also underreported) or general viral infections as the causative agents of GI illness. We therefore have no real basis to determine the frequency of poisoning caused by glycoalkaloids (Hopkins, 1995).

Susceptibility to illness from glycoalkaloid consumption may be influenced by factors such as diet and general health. Glycoalkaloids are not well absorbed in a healthy GI tract. However, other health issues that may compromise the intestinal wall may allow for them to be absorbed at a faster rate. Hydrolysis products, the β-, γ-, and aglycone forms, are less toxic than the α-form (Rayburn et al., 1994). Therefore, processes that induce hydrolysis of the α-forms, such as exposure to *Aspergillus niger* (Laha and Basu, 1983) and acid pH, may decrease toxicity. Other components and nutrients in the bloodstream and in the diet may affect toxicity as well. Folic acid, glucose 6-phosphate, and nicotine adenine dinucleotide are reported to protect frog embryos against α-chaconine-induced developmental toxicity (Friedman et al., 1997; Rayburn et al., 1995a). Incidentally, folic acid is now widely consumed by pregnant women to protect the fetus from neural tube malformations. It is also worth noting that Renwick (1972) found an epidemiological correlation between the congenital neural malformations anencephaly and spina bifida in fetuses and consumption of blighted potatoes by their mothers. Finally, dietary fiber can generally alter the absorption of nutrients and other compounds thru binding, or by changing the motility of the gut contents.

Other complicating factors regarding the glycoalkaloid content of the diet must be taken into account. α-Chaconine appears to be more biologically active by a factor of about 3–10 than is α-solanine, and certain combinations of the two glycoalkaloids can act synergistically (Rayburn et al., 1995b; Smith et al., 2001). There is some concern that dietary glycoalkaloids, because of their anticholinesterase properties, may adversely influence the actions of anesthetic drugs that are metabolized by acetylcholinesterase and butyrylcholinesterase (Krasowski et al., 1997). Solanidine exhibited estrogenic activity in an in vitro assay (Friedman et al., 2003a), with unknown health effects.

2.3 Healthful Benefits

Glycoalkaloids and hydrolysis products from potatoes, eggplants, and tomatoes were found to be active (antiproliferative activity) against human colon (HT29), liver (HepG2), cervical (HeLa), lymphoma (U937), and stomach (AGS and KATO III) cancer cells (Friedman et al., 2005; Kim et al., 2015; Lee et al., 2004) using a microculture tetrazolium assay. Activity was influenced by the chemical structure of the carbohydrate, the number of the carbohydrate groups making up the carbohydrate side chain attached to the 3-OH positions of the aglycones, and the structure of the aglycone. The relative potency of the anticancer drug Adriamycin against the liver cancer cells was similar to those observed with α-tomatine and α-chaconine. Solanine arrested MCF-7 cells in the S phase, thus inhibiting proliferation, by affecting the cell's microtubular system (Ji et al., 2012). α-Chaconine induced apoptosis in HT29 cells by increasing caspase-3 activity through inhibition of extracellular signal-regulated kinase (Yang et al., 2006)

It is also relevant to note that the glycoalkaloids solamargine and solasonine produced by eggplants and numerous other plant species also possess anticarcinogenic properties (Cham and Chase, 2012; Chang et al., 1998; Cui et al., 2012; Ding et al., 2012, 2013; Kuo et al., 2000; Li et al., 2011; Liang et al., 2007, 2008; Liu et al., 2004; Munari et al., 2014; Shiu et al., 2009; Sun et al., 2010; Zhou et al., 2014).

Glycoalkaloids suppressed the infection of mice with *Plasmodium yoelii* 17XL (a strain of malaria virulent only in mice) (Chen et al., 2010). The carbohydrate moieties appear to be the active sites, particularly 6-OH, with the chacotriose-containing molecules more active than the solatriose-containing molecules.

2.4 Analysis

Because of our long history with glycoalkaloids, a review of their analysis is practically a review of the field of chemical analysis in general, starting with simple gravimetric techniques to the use of the most expensive and sensitive equipment of the day. The toxic nature of glycoalkaloids in the diet suggests the need for accurate methods to measure the content of individual glycoalkaloids and their metabolites in fresh and processed potatoes as well as in body fluids such as plasma and tissues such as liver. Alkaloids are basic compounds, and as such are soluble in acidic aqueous solutions, but highly insoluble in basic aqueous solutions. This characteristic can be exploited during extraction and purification to rid the sample of interference using precipitation, liquid/liquid extraction, and/or solid-phase extraction (Friedman and Levin, 1992). A typical extraction regimen might include extraction into dilute acetic acid followed by treatment with ammonia and either precipitation or liquid/liquid extraction into butanol. Solid-phase extraction can be very selective for alkaloids by washing and eluting with the same solvent, but acidified in one case and basified in the other.

Because wounding of the tuber stimulates synthesis of the glycoalkaloids, sample workup in a timely manner is essential. The glycoalkaloids in potato slices were found to double when left out for 7 h (Maga, 1981).

2.4.1 Total Glycoalkaloid Analysis

For monitoring glycoalkaloid content in agricultural and food products, the following methods provide sufficient information. They cannot, however, give information about the individual glycoalkaloids present.

2.4.1.1 Gravimetric Analysis

Probably the simplest method to determine total glycoalkaloids is to extract them from the substrate and follow it with precipitation. A potato slurried with dilute acetic acid is filtered, and the filtrate treated with ammonia to precipitate the glycoalkaloids. This method does not work well for low glycoalkaloid levels, as the inevitable losses will be a higher percentage of the recovered material. Rodriguez-Saona et al. (1999) warn that anthocyanins in the filtrate will be mostly destroyed by pH greater than 9.2.

2.4.1.2 Colorimetry

Total glycoalkaloid content can be determined colorimetrically. This method requires only simple laboratory equipment and glassware. The glycoalkaloids are hydrolyzed to the aglycone form and titrated with bromphenol blue (Fitzpatrick et al., 1978).

2.4.1.3 Immunoassay

Enzyme-linked immunosorbent assay (ELISA) is a simple method that can analyze a large number of samples in a reasonably short time, as is needed in industrial applications. Interest in monoclonal (Stanker et al., 1994) and polyclonal (Plhak, 1992; Sporns et al., 1996; Ward et al., 1988) antibodies that react with glycoalkaloids and aglycones started some 20+ years ago. Stanker et al. went on to develop an ELISA method (Stanker et al., 1996). Results correlated well with HPLC (Table 2). EnviroLogix, Inc. (Portland, Maine), developed a prototype ELISA kit based on the work of Stanker et al. (Friedman et al., 1998a), but the product never proceeded to market. Solution-phase immunoassays, which combine capillary electrophoresis and

**Table 2: Comparison of Glycoalkaloid Content of the Same
Potatoes and Potato Products Analyzed by HPLC
(Sum of α-Chaconine and α-Solanine) and by ELISA.**

Sample	Assay Method	
	HPLC	ELISA
Whole potatoes:	Fresh (mg/kg)	Fresh (mg/kg)
Russet, organic	5.8	5.1
Russet	22	24
Yukon Gold	40	38
Purple, small	45	37
Red, small	101	128
Gold, small	105	113
White, large	125	132
White, small	203	209
Potato plant parts:	Dehydrated (mg/kg)	Dehydrated (mg/kg)
Flesh, Red Lasoda	45.6	51.6
Peel, Shepody	1432	1251
Sprouts, Shepody	7641	6218
Leaves	9082	8851
Processed potatoes:	Original (mg/kg)	Original (mg/kg)
French fries, A	0	1.2
French fries, B	24.1	22.7
Chips, low fat	15.2	22.7
Skins, A	43.3	35.0
Skins, B	37.2	41.0

laser-induced fluorescence with polyclonal antibodies, claim an even faster analysis than ELISA (Driedger et al., 2000). Immunoassays are simple methods that still have the potential to be used routinely by a broad range of users, but need commercial success to be made available.

2.4.1.4 Biosensors

Biosensors couple products of biological processes with an electronic device, thus performing as a detector. Interactions of glycoalkaloids with cholinesterase coupled with pH-sensitive field-effect transistors are promising biosensors for the rapid determination of glycoalkaloids (Arkhypova et al., 2008; Benilova et al., 2006a,b). The test takes advantage of the anticholinesterase activity of the glycoalkaloids. Espinosa et al. (2014) improved the sensitivity (50 ppb) of the above test by using a genetically modified acetylcholinesterase. These tests could hold great promise, analogous to the ELISA mentioned above.

2.4.1.5 Cell Assay

Solanine was determined, in addition to other naturally occurring toxins, by use of a chemiluminescence assay measuring the impairment of menadione-catalyzed H_2O_2 production in living mammalian cells (Yamashoji et al., 2013).

2.4.2 Analysis of Individual Glycoalkaloids
2.4.2.1 Thin-Layer Chromatography

Before the proliferation of HPLC, separation of glycoalkaloids was commonly done by thin-layer chromatography (TLC) (Fitzpatrick et al., 1978). An extracted sample is spotted onto a TLC plate, allowed to migrate up the plate with solvent, and then stained by reaction with a colorizer such as anisaldehyde or iodine (Friedman et al., 1993). Concurrent spotting of standards allows a quick assessment of glycoalkaloids present as well as approximate concentrations. Necessary equipment and expertise are less demanding than for HPLC. A more modern method, high-performance TLC, has emerged as a more accurate and quantitative analysis (Bodart et al., 2000; Simonovska and Vovk, 2000). However, this method detracts from the simplicity of the TLC method and has not gained widespread acceptance over HPLC.

2.4.2.2 High-Performance Liquid Chromatography

HPLC methods are the most common for laboratory analysis of potato glycoalkaloids and their hydrolysis products (Bodart et al., 2000; Brown et al., 1999; Bushway, 1982; Bushway et al., 1986; Carman et al., 1986; Dao and Friedman, 1994, 1996; Esposito et al., 2002; Fragoyiannis et al., 2001; Friedman et al., 1998a, 1993; Friedman and Dao, 1992; Friedman and Levin, 1992; Friedman and McDonald, 1995; Hellenäs et al., 1992; Houben and Brunt, 1994; Kozukue et al., 1987, 1999, 2001; Kubo and Fukuhara, 1996; Kuronen et al., 1999; Nitithamyong et al., 1999; Panovska et al., 1997; Saito et al., 1990; Simonovska and Vovk, 2000; Sotelo and Serrano, 2000; Väänänen et al., 2000). The glycoalkaloids and their glycosidic hydrolysis products are analyzable in a single isocratic run. The aglycone may also be analyzed along with the glycoalkaloids in a single run using gradient elution. Choices of column packings are numerous, as the molecules have a number of characteristics that can be exploited, including polarity, size, and solubility. A mixed-mode column packing caused by non-end-capped C18-modified silica took advantage of dual interactions of the glycoalkaloids with the packing, with good results (Friedman and Levin, 1992). The primary interaction of the molecules was with the reversed-phase C18 moieties, but the secondary interactions with the exposed silica particles provided an improved separation, allowing a baseline resolution of α-solanine, α-chaconine, and their hydrolysis products (Figure 3(a)). Columns like this with a high amount of silanol interaction can cause tailing, especially for the more hydrophobic molecules, so it is better to use fully end-capped columns to analyze the aglycones (Figure 3(b)). Reversed-phase amino (NH_2) packings have also been used with success (Figure 3(c)) (Kozukue et al., 1999; Meher and Gaur, 2003). Although the NH_2 column technique has longer run times, it is more adaptable to HPLC/mass spectrometry (MS) because buffering salts can be kept to a minimum.

Owing to the lack of a chromophore, low-absorbing solvents must be used to minimize background absorbance in the 200–210λ range for UV detection. Other detectors can be used,

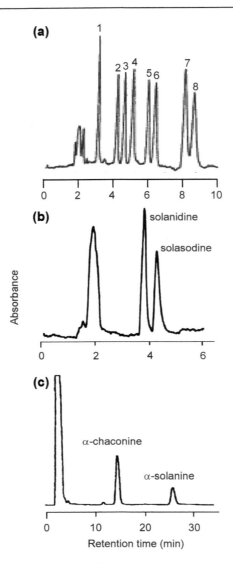

Figure 3

A comparison of HPLC separation methods. (a) HPLC chromatogram of approximately 1 μg each of potato glycoalkaloids and their hydrolysis products: 1, solasonine (internal standard); 2, α-solanine; 3, α-chaconine; 4, $β_2$-solanine; 5, $β_1$-chaconine; 6, $β_2$-chaconine; 7, γ-solanine; 8, γ-chaconine. Conditions: column, resolve C18 (5 μm, 3.9 × 300 mm); mobile phase, 35% acetonitrile/100 mM ammonium phosphate (monobasic) at pH 3; flow rate, 1.0 ml/min; column temperature, ambient; UV detector, 200 nm. (b) HPLC chromatogram of the aglycones solanidine and solasodine. Conditions: column, Supelcosil C18-DB (3 μm, 4.6 × 150 mm); mobile phase, 60% acetonitrile/10 mM ammonium phosphate, pH 2.5; flow rate, 1.0 ml/min; column temperature, ambient; UV detector, 200 nm. (c) HPLC of α-chaconine and α-solanine in the flesh of one variety of potato. Conditions: column, Inertsil NH_2 (5 μm, 4.0 × 250 mm); mobile phase, acetonitrile/20 mM KH_2PO_4 (80/20, v/v); flow rate, 1.0 ml/min; column temperature, 20 °C; UV detector, 208 nm; sample size, 20 μL.

such as the electrochemical (Friedman et al., 1994; Wang et al., 2013), the light scattering, or the refractive index detectors, to overcome this problem, but may not be as robust or selective. Glycoalkaloids can be identified as follows: (1) concurrent analysis with and comparison to standards; (2) collection of peaks and application to TLC along with standards; and (3) HCl hydrolysis of collected HPLC peaks into sugars and the aglycone (usually solanidine), followed by identification by TLC, gas chromatography (GC), HPLC, or GC/MS (Kozukue and Friedman, 2003; Kozukue et al., 1999, 2008).

2.4.2.3 Mass Spectrometry

Glycoalkaloids are not easily volatized, so application of GC/MS is limited. Samples can be hydrolyzed and the aglycones readily analyzed by GC/MS, while the sugars need to be derivatized before analysis (Friedman et al., 1998b). While aglycones need not be derivatized for GC analysis, Laurila et al. (1999) derivatized by trimethylsilylation and pentafluoropropionylation to yield better ion fragments for improved identification. MS can be used directly to identify samples not needing separation, i.e., collected peaks from TLC or HPLC (Friedman et al., 1993).

HPLC/MS can separate and analyze glycoalkaloids with a single injection. Ieri et al. (2011) developed an HPLC/MS method using MSD API–electrospray technology that analyzed potato glycoalkaloids, phenolic acids, and anthocyanins in a single run of 25 min. An ultraperformance liquid chromatography–triple-quadrupole MS method was used to detect glycoalkaloids within 5.5 min in bodily fluids cleaned up by solid-phase extraction (Zhang et al., 2014). Ultra-high-performance chromatography coupled with an electrospray ionization (ESI)–time-of-flight (TOF) mass spectrometer concurrently identified potato organic acids, amino acids, hydroxycinnamic acid derivatives, hydroxycinnamic amides, and flavonols, in addition to glycoalkaloids (Chong et al., 2013). HPLC/MS (LTQ-Orbitrap) was used to analyze the aglycones and glycoalkaloids in a single run (Caprioli et al., 2013). Abell and Sporns (1996) developed a matrix-assisted laser desorption/ionization (MALDI) TOF MS method for potato glycoalkaloids. Ha et al. (2012) further applied this method to microscopic imaging, a technique called MALDI imaging MS, to analyze whole, unextracted, biologic samples. A similar method, which has been applied to tomatoes, uses laser-ablation ESI MS coupled with imaging (Nielen and van Beek, 2014). The authors claim that the sample preparation is simplified, as it does not require very flat surfaces or the addition of matrix as in the MALDI imaging MS method. These methods can measure the distribution of glycoalkaloids in parts of the tuber. They do not require time-consuming and complex sample preparation before analysis. Additionally, problems associated with extraction, such as low extraction efficiency and sample losses, are overcome.

An alternative method, Bianco et al. (2002) used a nonaqueous capillary electrophoresis coupled with ESI–ion trap MS to analyze glycoalkaloids along with their aglycones.

3. Calystegine Alkaloids

Calystegines are polyhydroxylated *nor*tropane alkaloids first discovered in 1988 in the plants *Calystegia sepium, Convolvulus arvensis*, and *Atropa belladonna* (Tepfer et al., 1988). The alkaloids were found in parts of the plant in contact with the soil and appear to provide a carbon and nitrogen source for certain beneficial rhizosphere bacteria (Tepfer et al., 1988). Their structures were elucidated in 1990 (Ducrot and Lallemand, 1990; Goldmann et al., 1990). Subsequently, they have been found in the potato (Dräger et al., 1995; Nash et al., 1993), as well as several other Solanaceae, including eggplant and tomato (Asano et al., 1997). Potato is reported to contain primarily calystegines A_3 and B_2 shown in Figure 4 (Asano et al., 1997; Keiner and Dräger, 2000; Nash et al., 1993), although lesser amounts of other calystegines have been found (Griffiths et al., 2008). At least 14 calystegines are currently known, having A (trihydroxynortropane), B (tetrahydroxynortropane), and C (pentahydroxynortropane) groupings depending on the number of hydroxyl groups (Biastoff and Dräger, 2007). Dihydroxynortropanes have been isolated from a Solanaceae plant, *Duboisia leichhardtii* (Asano et al., 2001b). Calystegine N_1, which has a bridgehead NH_2 group in the place of the bridgehead OH group in calystegine B_2 (Asano et al., 1996), was found in small amounts in potato sprouts, along with B_3, B_4, and an unidentified calystegine-like compound designated X_2 (Griffiths et al., 2008). The hydroxyl groups in calystegines can vary in both position and stereochemistry (Asano et al., 2000). In addition, the alkaloids have a novel aminoketal functionality that generates a tertiary hydroxyl group at the bicyclic ring bridgehead (Asano et al., 2000). *Nor*tropane alkaloids are similar to tropane alkaloids such as atropine found in other Solanaceae plants, but they lack the methyl group substitution on the nitrogen (Asano et al., 2000). The biosynthetic pathway for calystegines appears to be a side branch of the tropane alkaloid pathway starting with the stereospecific reduction of tropinone (Keiner et al., 2000). Stenzel et al. (2006) confirmed that the tropane pathway is expressed and active in potatoes, even though no tropane alkaloids are produced.

3.1 Abundance in the Tuber

Similar to the glycoalkaloids, calystegines accumulate at higher levels in the sprouts and the peel than in the tuber of the potato (Griffiths et al., 2008). Also like glycoalkaloids,

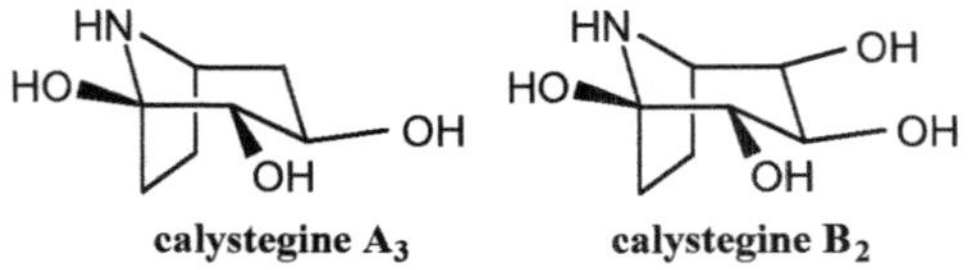

calystegine A_3 calystegine B_2

Figure 4
Structures of the most abundant potato calystegines.

calystegines accumulate in the tuber as it ages and sprouts (Keiner and Dräger, 2000). But unlike glycoalkaloids, calystegines do not seem to increase in response to wounding or light stress (Keiner and Dräger, 2000; Petersson et al., 2013). Griffiths et al. (2008) found that the concentration of calystegines in the potato peel was 13 times greater than in the flesh of five *S. tuberosum* cultivars and four times higher for four *S. phureja* lines (commonly eaten in South America). On average, concentrations in the sprouts of *S. tuberosum* were 100 times higher than in the tuber flesh and 8 times higher than in the peel. The respective values for *S. phureja* were 30 and 7. One sprout sample, of the 22 potato varieties analyzed, had higher levels of calystegine N_1 and calystegine-like X_2 than any of the others. Concentrations of calystegine B_4 in sprouts also varied considerably by cultivar. These results indicate considerable variation in calystegine content within both species as well as within parts of the potato plant. Richter et al. (2007) studied altered carbohydrate metabolism on calystegine synthesis and determined that levels were associated with sucrose availability. Thus there may be a mediated link from variety, to sugar content, to calystegine content.

3.2 Toxicity

No human toxicity data for calystegines have been reported. It is really unknown whether past potato poisonings are a result of glycoalkaloid toxicity alone or a result of the combination of glycoalkaloids and calystegines. Calystegines are known to be biologically active as glycosidase inhibitors (Molyneux et al., 1993). Calystegines are polyhydroxylated alkaloids that are chemically similar to monosaccharides but for the ring oxygen being replaced by a nitrogen (Asano et al., 2000). Because of this similarity, they can act biologically as monosaccharide mimics, substituting for the sugar moiety of the natural substrate of glycosidases (Asano et al., 2000; Molyneux et al., 1993). Binding to the active site inactivates these enzymes that are required for a wide range of biological functions (Asano et al., 2000). B_1 and C_1 calystegines are potent inhibitors of β-glucosidase, while B_2 is a strong inhibitor of α-galactosidase (Asano et al., 1997).

3.3 Healthful Benefits

Calystegines, along with other monosaccharide mimics, have the potential to be used therapeutically against a variety of diseases such as cancer, diabetes, and bacterial and viral infections and to stimulate the immune system (Asano et al., 2001a, 2000; Watson et al., 2001). Also, mimics can be used to induce altered or unusual states in cellular metabolism for the purpose of study. N-linked oligosaccharide glycoproteins in the cell wall are highly specific in part because of the action of glycosidase, and so altering their activities can give insights into cell metabolism (Asano et al., 2000). Also for the benefit of study, a cellular or animal model for genetic lysosomal disease can be produced by mimics creating a deficiency of lysosomal glycosidase (Asano et al., 2000).

There appears to be much interest in the many potential therapies associated with the use of glycosidase inhibitors. Calystegines A_3 and B_2 purified from potatoes have potential to prevent steep increases in blood glucose levels after a carbohydrate-rich meal by inhibiting digestive glycosidases that break down complex sugars (Jockovic et al., 2013). Calystegine B_2 and isomers are potent inhibitors of human lysosomal β-glucocerebrosidase, an enzyme that is associated with the cause of inherited Gaucher disease, which causes skeletal and neurological disorders (Kato et al., 2014). Cancer metastases have been associated with N-linked oligosaccharide glycoprotein changes, and so there is an interest in the use of mimics for cancer therapies or study (Asano et al., 2000). The envelope of many animal viruses is made up of glycoproteins, which could be compromised by glycosidase inhibitors (Asano et al., 2000). There now appears to be interest in synthesizing new and novel drugs based upon the chemistry of calystegines (Kaliappan et al., 2009; Pino-Gonzalez et al., 2012).

3.4 Analysis

Calystegines are hydrophilic molecules and are easily extracted into aqueous/alcoholic solutions and purified by filtration through cationic resins (Friedman et al., 2003b). Use of alcoholic rather than aqueous solutions excludes possible interference from the many saccharides present in potatoes (Dräger, 2002). Nash et al. (1993) initially analyzed calystegines with high-voltage paper electrophoresis using ninhydrin staining to visualize them. As with glycoalkaloids, screening can be quickly achieved using TLC (Dräger, 1995). TLC is a rapid, inexpensive, and simple technique, but lacks the precise quantitative data that other methods can provide (Dräger, 2002). Use of GC requires that the alkaloids be silylated or otherwise made volatile (Dräger, 1995). Dräger (2002) has published a comprehensive review on the analysis of tropane, including *nor*tropane alkaloids. In brief, GC and GC/MS have the advantage of short analysis times and selectivity, but require careful sample preparation, as the sample must be derivatized and is easily destroyed under aggressive treatment. HPLC instrumentation is widely available and can provide good separations. Drawbacks of HPLC are poor differential signal on common detectors and the extra complications presented by the need to elute by gradient. HPLC/MS is a good choice but requires expensive instrumentation and extra skill. Finally, capillary electrophoresis gives good separation and sensitivity, but alternate detection schemes must be provided. Figure 5 illustrates the separation on GC/MS with total ion chromatograms of calystegines A_3 and B_2 extracted from potato powders (Friedman et al., 2003b).

Table 1 compares the calystegine and glycoalkaloid contents of the same potatoes using a GC/MS method (Friedman et al., 2003b). There appears to be no correlation between glycoalkaloid and calystegine content. Since the individual calystegine isomers differ in their biological activities (Asano et al., 1997; Watson et al., 2001), both their ratios and their total amounts present in various potato cultivars may be important in assessing the role of calystegines in the diet.

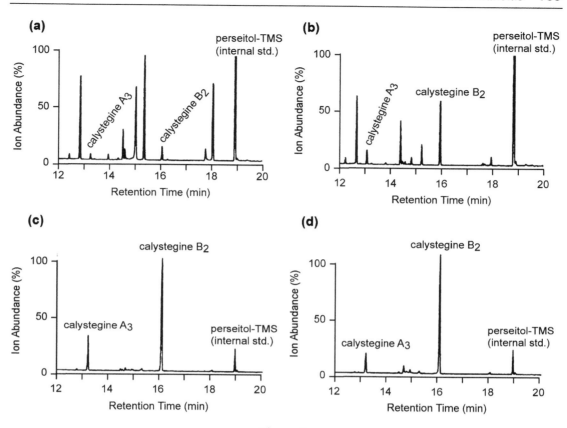

Figure 5

GC/MS total ion chromatograms of the hydrophilic alkaloid fraction extracted from freeze-dried potato flesh and peel. (a) Ranger Russet flesh, (b) Shepody flesh, (c) Ranger Russet peel, and (d) Shepody peel.

4. Conclusions

The referenced methods for the analysis of biologically active calystegine alkaloids and glycoalkaloids in commercial potatoes, in processed potato products, and in new cultivars can lead to improvements in the precision and reliability of analyses for quality control and for safety of final products. Accurate analyses will benefit growers, researchers, processors, and consumers. Analytical studies have also facilitated concurrent studies of toxicities and of beneficial health effects as well as investigations of the biosynthesis of the secondary metabolites. Although potato processors are generally interested in imparting desirable sensory (organoleptic) attributes to potato products, the discovery of health-promoting effects of potato ingredients, balanced against concerns of toxicity, implies that analytical methodology will be paramount in future efforts designed to enhance the levels of these compounds in the human diet. Since the biological activities as well as the roles in host-plant resistance of both glycoalkaloids and calystegines could be interrelated, there is a need to define the levels of both glycoalkaloids and calystegines in various potato cultivars and to study individual and combined effects in animals and humans. Possible therapeutic applications of high-calystegine potato diets also merit study. Future studies should attempt to further simplify and improve the described analytical methodologies.

Acknowledgment

We are most grateful to our colleagues, especially Professor Nobuyuke Kozukue, whose name appears on many of the cited references, for excellent scientific collaboration.

References

Abbott, D.G., Field, K., Johnson, E.I., 1960. Observation on the correlation of anticholinesterase effect with solanine content of potatoes. Analyst 85, 375–377.

Abell, D.C., Sporns, P., 1996. Rapid quantitation of potato glycoalkaloids by matrix-assisted laser desorption/ionization time-of-flight mass spectrometry. Journal of Agricultural and Food Chemistry 44, 2292–2296.

Andre, C.M., Schafleitner, R., Guignard, C., Oufir, M., Aliaga, C.A.A., Nomberto, G., Hoffmann, L., Hausman, J.F., Evers, D., Larondelle, Y., 2009. Modification of the health-promoting value of potato tubers field grown under drought stress: emphasis on dietary antioxidant and glycoalkaloid contents in five native Andean cultivars (*Solanum tuberosum* L.). Journal of Agricultural and Food Chemistry 57, 599–609.

Anonymous, 2009. Appendix: glycoalkaloids in potatoes: public standards concerning safe levels of intake. Journal of Food Composition and Analysis 22, 625.

Arkhypova, V.N., Dzyadevych, S.V., Jaffrezic-Renault, N., Martelet, C., Soldatkin, A.P., 2008. Biosensors for assay of glycoalkaloids in potato tubers. Applied Biochemistry and Microbiology 44, 314–318.

Asano, N., Kato, A., Matsui, K., Watson, A.A., Nash, R.J., Molyneux, R.J., Hackett, L., Topping, J., Winchester, B., 1997. The effects of calystegines isolated from edible fruits and vegetables on mammalian liver glycosidases. Glycobiology 7, 1085–1088.

Asano, N., Kato, A., Watson, A.A., 2001a. Therapeutic applications of sugar-mimicking glycosidase inhibitors. Mini Reviews in Medicinal Chemistry 1, 145–154.

Asano, N., Kato, A., Yokoyama, Y., Miyauchi, M., Yamamoto, M., Kizu, H., Matsui, K., 1996. Calystegin N_1, a novel nortropane alkaloid with a bridgehead amino group from *Hyoscyamus niger*: structure determination and glycosidase inhibitory activities. Carbohydrate Research 284, 169–178.

Asano, N., Nash, R.J., Molyneux, R.J., Fleet, G.W.J., 2000. Sugar-mimic glycosidase inhibitors: natural occurrence, biological activity and prospects for therapeutic application. Tetrahedron-Asymmetry 11, 1645–1680.

Asano, N., Yokoyama, K., Sakurai, M., Ikeda, K., Kizu, H., Kato, A., Arisawa, M., Höke, D., Dräger, B., Watson, A.A., Nash, R.J., 2001b. Dihydroxynortropane alkaloids from calystegine-producing plants. Phytochemistry 57, 721–726.

Bejarano, L., Mignolet, E., Devaux, A., Espinola, N., Carrasco, E., Larondelle, Y., 2000. Glycoalkaloids in potato tubers: the effect of variety and drought stress on the α-solanine and α-chaconine contents of potatoes. Journal of the Science of Food and Agriculture 80, 2096–2100.

Benilova, I.V., Arkhypova, V.M., Dzyadevych, S.V., Jaffrezic-Renault, N., Martelet, C., Soldatkin, A.P., 2006a. Kinetic properties of butyrylcholinesterases immobilised on pH-sensitive field-effect transistor surface and inhibitory action of steroidal glycoalkaloids on these enzymes. Ukrainskii Biokhimicheskii Zhurnal 78, 131–141.

Benilova, I.V., Arkhypova, V.N., Dzyadevych, S.V., Jaffrezic-Renault, N., Martelet, C., Soldatkin, A.P., 2006b. Kinetics of human and horse sera cholinesterases inhibition with solanaceous glycoalkaloids: study by potentiometric biosensor. Pesticide Biochemistry and Physiology 86, 203–210.

Bianco, G., Schmitt-Kopplin, P., de Benedetto, G., Kettrup, A., Cataldi, T.R.I., 2002. Determination of glycoalkaloids and relative aglycones by nonaqueous capillary electrophoresis coupled with electrospray ionization-ion trap mass spectrometry. Electrophoresis 23, 2904–2912.

Biastoff, S., Dräger, B., 2007. Chapter 2 Calystegines, Alkaloids: Chemistry and Biology, pp. 49–102.

Bodart, P., Kabengera, C., Noirfalise, A., Hubert, P., Angenot, L., 2000. Determination of α-solanine and α-chaconine in potatoes by high-performance thin-layer chromatography/densitometry. Journal of AOAC International 83, 1468–1473.

Brown, M.S., McDonald, G.M., Friedman, M., 1999. Sampling leaves of young potato (*Solanum tuberosum*) plants for glycoalkaloid analysis. Journal of Agricultural and Food Chemistry 47, 2331–2334.

Bushway, R.J., 1982. High-performance liquid chromatographic separation of potato glycoalkaloids using a radially compressed amino column. Journal of Chromatography 247, 180–183.

Bushway, R.J., Bureau, J.L., King, J., 1986. Modification of the rapid high-performance liquid chromatographic method for the determination of potato glycoalkaloids. Journal of Agricultural and Food Chemistry 34, 277–279.

Caldwell, K.A., Grosjean, O.K., Henika, P.R., Friedman, M., 1991. Hepatic ornithine decarboxylase induction by potato glycoalkaloids in rats. Food and Chemical Toxicology 29, 531–535.

Caprioli, G., Cahill, M.G., Vittori, S., James, K.J., 2013. Liquid chromatography–hybrid linear ion trap–high-resolution mass spectrometry (LTQ-Orbitrap) method for the determination of glycoalkaloids and their aglycons in potato samples. Food Analytical Methods 7, 1367–1372.

Carman Jr., A.S., Kuan, S.S., Ware, G.M., Francis Jr., O.J., Kirschenheuter, G.P., 1986. Rapid high-performance liquid chromatographic determination of the potato glycoalkaloids α-solanine and α-chaconine. Journal of Agricultural and Food Chemistry 34, 279–282.

Cham, B.E., 1994. Solasodine glycosides as anti-cancer agents: pre-clinical and clinical studies. Asia Pacific Journal of Pharmacology 9, 113–118.

Cham, B.E., Chase, T.R., 2012. Solasodine rhamnosyl glycosides cause apoptosis in cancer cells. Do they also prime the immune system resulting in long-term protection against cancer? Planta Medica 78, 349–353.

Chang, L.-C., Tsai, T.-R., Wang, J.-J., Lin, C.-N., Kuo, K.-W., 1998. The rhamnose moiety of solamargine plays a crucial role in triggering cell death by apoptosis. Biochemical and Biophysical Research Communications 242, 21–25.

Chen, Y., Li, S., Sun, F., Han, H., Zhang, X., Fan, Y., Tai, G., Zhou, Y., 2010. In vivo antimalarial activities of glycoalkaloids isolated from Solanaceae plants. Pharmaceutical Biology 48, 1018–1024.

Choi, D., Bostock, R.M., Avdiushko, S., Hildebrand, D.F., 1994. Lipid-derived signals that discriminate wound- and pathogen-responsive isoprenoid pathways in plants: methyl jasmonate and the fungal elicitor arachidonic acid induce different 3-hydroxy-3-methylglutaryl-coenzyme A reductase genes and antimicrobial isoprenoids in *Solanum tuberosum* L. Proceedings of the National Academy of Sciences of the United States of America 91, 2329–2333.

Chong, E.S., McGhie, T.K., Heyes, J.A., Stowell, K.M., 2013. Metabolite profiling and quantification of phytochemicals in potato extracts using ultra-high-performance liquid chromatography-mass spectrometry. Journal of the Science of Food and Agriculture 93, 3801–3808.

Cui, C.Z., Wen, X.S., Cui, M., Gao, J., Sun, B., Lou, H.X., 2012. Synthesis of solasodine glycoside derivatives and evaluation of their cytotoxic effects on human cancer cells. Drug Discoveries & Therapeutics 6, 9–17.

Dao, L., Friedman, M., 1994. Chlorophyll, chlorogenic acid, glycoalkaloid, and protease inhibitor content of fresh and green potatoes. Journal of Agricultural and Food Chemistry 42, 633–639.

Dao, L., Friedman, M., 1996. Comparison of glycoalkaloid content of fresh and freeze-dried potato leaves determined by HPLC and colorimetry. Journal of Agricultural and Food Chemistry 44, 2287–2291.

Dimenstein, L., Lisker, N., Kedar, N., Levy, D., 1997. Changes in the content of steroidal glycoalkaloids in potato tubers grown in the field and in the greenhouse under different conditions of light, temperature and daylength. Physiological and Molecular Plant Pathology 50, 391–402.

Ding, X., Zhu, F.-S., Li, M., Gao, S.-G., 2012. Induction of apoptosis in human hepatoma SMMC-7721 cells by solamargine from *Solanum nigrum* L. Journal of Ethnopharmacology 139, 599–604.

Ding, X., Zhu, F., Yang, Y., Li, M., 2013. Purification, antitumor activity in vitro of steroidal glycoalkaloids from black nightshade (*Solanum nigrum* L.). Food Chemistry 141, 1181–1186.

Djomo, S.N., Humbert, S., Dagnija, B., 2008. Life cycle assessment of hydrogen produced from potato steam peels. International Journal of Hydrogen Energy 33, 3067–3072.

Donnelly, A., Lawson, T., Craigon, J., Black, C.R., Colls, J.J., Landon, G., 2001. Effects of elevated CO_2 and O_3 on tuber quality in potato (*Solanum tuberosum* L.). Agriculture, Ecosystems and Environment 87, 273–285.

Dräger, B., 1995. Identification and quantification of calystegines, polyhydroxyl nortropane alkaloids. Phytochemical Analysis 6, 31–37.

Dräger, B., 2002. Analysis of tropane and related alkaloids. Journal of Chromatography A 978, 1–35.

Dräger, B., van Almsick, A., Mrachatz, G., 1995. Distribution of calystegines in several Solanaceae. Planta Medica 61, 577–579.

Driedger, D.R., LeBlanc, R.J., LeBlanc, E.L., Sporns, P., 2000. A capillary electrophoresis laser-induced fluorescence method for analysis of potato glycoalkaloids based on a solution-phase immunoassay. 1. Separation and quantification of immunoassay products. Journal of Agricultural and Food Chemistry 48, 1135–1139.

Driedger, D.R., Sporns, P., 1999. Glycoalkaloid concentration in by-products of potato starch extraction as measured by matrix-assisted laser desorption/ionization mass spectrometry. Journal of Food Processing and Preservation 23, 377–390.

Ducrot, P.H., Lallemand, J.Y., 1990. Structure of the calystegines: new alkaloids of the nortropane family. Tetrahedron Letters 31, 3879–3882.

Elferink, E.V., Nonhebel, S., Moll, H.C., 2008. Feeding livestock food residue and the consequences for the environmental impact of meat. Journal of Cleaner Production 16, 1227–1233.

Espinoza, M.A., Istamboulie, G., Chira, A., Noguer, T., Stoytcheva, M., Marty, J.L., 2014. Detection of glycoalkaloids using disposable biosensors based on genetically modified enzymes. Analytical Biochemistry 457, 85–90.

Esposito, F., Fogliano, V., Cardi, T., Carputo, D., Filippone, E., 2002. Glycoalkaloid content and chemical composition of potatoes improved with nonconventional breeding approaches. Journal of Agricultural and Food Chemistry 50, 1553–1561.

Fitzpatrick, T.J., Herb, S.F., Osman, S.F., McDermott, J.A., 1977. Potato glycoalkaloids: increases and variations of ratios in aged slices over prolonged storage. American Potato Journal 54, 539–544.

Fitzpatrick, T.J., Mackenzie, J.D., Gregory, P., 1978. Modifications of the comprehensive method for total glycoalkaloid determination. American Potato Journal 55, 247–248.

Fragoyiannis, D.A., McKinlay, R.G., D'Mello, J.P.F., 2001. Interactions of aphid herbivory and nitrogen availability on the total foliar glycoalkaloid content of potato plants. Journal of Chemical Ecology 27, 1749–1762.

Friedman, M., 2002. Tomato glycoalkaloids: role in the plant and in the diet. Journal of Agricultural and Food Chemistry 50, 5751–5780.

Friedman, M., 2006. Potato glycoalkaloids and metabolites: roles in the plant and in the diet. Journal of Agricultural and Food Chemistry 54, 8655–8681.

Friedman, M., 2013. Anticarcinogenic, cardioprotective, and other health benefits of tomato compounds lycopene, α-tomatine, and tomatidine in pure form and in fresh and processed tomatoes. Journal of Agricultural and Food Chemistry 61, 9534–9550.

Friedman, M., 2015. Chemistry and anticarcinogenic mechanisms of glycoalkaloids produced by eggplants, potatoes, and tomatoes. Journal of Agricultural and Food Chemistry 63, 3323–3337.

Friedman, M., Bautista, F.F., Stanker, L.H., Larkin, K.A., 1998a. Analysis of potato glycoalkaloids by a new ELISA kit. Journal of Agricultural and Food Chemistry 46, 5097–5102.

Friedman, M., Burns, C.F., Butchko, C.A., Blankemeyer, J.T., 1997. Folic acid protects against potato glycoalkaloid α-chaconine-induced disruption of frog embryo cell membranes and developmental toxicity. Journal of Agricultural and Food Chemistry 45, 3991–3994.

Friedman, M., Dao, L., 1992. Distribution of glycoalkaloids in potato plants and commercial potato products. Journal of Agricultural and Food Chemistry 40, 419–423.

Friedman, M., Fitch, T.E., Levin, C.E., Yokoyama, W.H., 2000a. Feeding tomatoes to hamsters reduces their plasma low-density lipoprotein cholesterol and triglycerides. Journal of Food Science 65, 897–900.

Friedman, M., Fitch, T.E., Yokoyama, W.E., 2000b. Lowering of plasma LDL cholesterol in hamsters by the tomato glycoalkaloid tomatine. Food and Chemical Toxicology 38, 549–553.

Friedman, M., Henika, P.R., Mackey, B.E., 2003a. Effect of feeding solanidine, solasodine and tomatidine to non-pregnant and pregnant mice. Food and Chemical Toxicology 41, 61–71.

Friedman, M., Kozukue, N., Harden, L.A., 1998b. Preparation and characterization of acid hydrolysis products of the tomato glycoalkaloid α-tomatine. Journal of Agricultural and Food Chemistry 46, 2096–2101.

Friedman, M., Lee, K.R., Kim, H.J., Lee, I.S., Kozukue, N., 2005. Anticarcinogenic effects of glycoalkaloids from potatoes against human cervical, liver, lymphoma, and stomach cancer cells. Journal of Agricultural and Food Chemistry 53, 6162–6169.

Friedman, M., Levin, C.E., 1992. Reversed-phase high-performance liquid chromatographic separation of potato glycoalkaloids and hydrolysis products on acidic columns. Journal of Agricultural and Food Chemistry 40, 2157–2163.

Friedman, M., Levin, C.E., McDonald, G.M., 1994. α-Tomatine determination in tomatoes by HPLC using pulsed amperometric detection. Journal of Agricultural and Food Chemistry 42, 1959–1964.

Friedman, M., McDonald, G., 1999a. Postharvest changes in glycoalkaloid content of potatoes. In: Jackson, L., Knize, M. (Eds.), Impact of Food Processing on Food Safety. Plenum Press, New York, pp. 121–143.

Friedman, M., McDonald, G., Haddon, W.F., 1993. Kinetics of acid-catalyzed hydrolysis of carbohydrate groups of potato glycoalkaloids α-chaconine and α-solanine. Journal of Agricultural and Food Chemistry 41, 1397–1406.

Friedman, M., McDonald, G.M., 1995. Acid-catalyzed partial hydrolysis of carbohydrate groups of the potato glycoalkaloids α-chaconine in alcoholic solutions. Journal of Agricultural and Food Chemistry 43, 1501–1506.

Friedman, M., McDonald, G.M., 1997. Potato glycoalkaloids: chemistry, analysis, safety, and plant physiology. Critical Reviews in Plant Sciences 16, 55–132.

Friedman, M., McDonald, G.M., 1999b. Steroidal glycoalkaloids. In: Ikan, R. (Ed.), Naturally Occurring Glycosides: Chemistry, Distribution, and Biological Properties. John Wiley and Sons, Ltd, Chichester, pp. 311–343.

Friedman, M., McQuistan, T., Hendricks, J.D., Pereira, C., Bailey, G.S., 2007. Protective effect of dietary tomatine against dibenzo[a,l]pyrene (DBP)-induced liver and stomach tumors in rainbow trout. Molecular Nutrition & Food Research 51, 1485–1491.

Friedman, M., Roitman, J.N., Kozukue, N., 2003b. Glycoalkaloid and calystegine contents of eight potato cultivars. Journal of Agricultural and Food Chemistry 51, 2164–2173.

Goldmann, A., Milat, M.-L., Ducrot, P.-H., Lallemand, J.-Y., Maille, M., Lepingle, A., Charpin, I., Tepfer, D., 1990. Tropane derivatives from *Calystegia sepium*. Phytochemistry 29, 2125–2127.

Griffiths, D.W., Bain, H., Deighton, N., Robertson, G.W., Dale, M.F.B., Finlay, M., 2000. Photo-induced synthesis of tomatidenol-based glycoalkaloids in *Solanum phureja* tubers. Phytochemistry 53, 739–745.

Griffiths, D.W., Dale, M.F.B., 2001. Effect of light exposure on the glycoalkaloid content of *Solanum phureja* tubers. Journal of Agricultural and Food Chemistry 49, 5223–5227.

Griffiths, D.W., Shepherd, T., Stewart, D., 2008. Comparison of the calystegine composition and content of potato sprouts and tubers from *Solanum tuberosum* group *phureja* and *Solanum tuberosum* group tuberosum. Journal of Agricultural and Food Chemistry 56, 5197–5204.

Gubarev, M.I., Enioutina, E.Y., Taylor, J.L., Visic, D.M., Daynes, R.A., 1998. Plant-derived glycoalkaloids protect mice against lethal infection with *Salmonella typhimurium*. Phytotherapy Research 12, 79–88.

Ha, M., Kwak, J.H., Kim, Y., Zee, O.P., 2012. Direct analysis for the distribution of toxic glycoalkaloids in potato tuber tissue using matrix-assisted laser desorption/ionization mass spectrometric imaging. Food Chemistry 133, 1155–1162.

Haase, N.U., 2010. Glycoalkaloid concentration in potato tubers related to storage and consumer offering. Potato Research 53, 297–307.

Hamouz, K., Lachman, J., Dvořák, P., Pivec, V., 2005. The effect of ecological growing on the potatoes yield and quality. Plant, Soil and Environment 51, 397–402.

Heftmann, E., 1967. Biochemistry of steroidal saponins and glycoalkaloids. Lloydia 30, 209–230.

Hellenäs, K.E., Nyman, A., Slanina, P., Lööf, L., Gabrielsson, J., 1992. Determination of potato glycoalkaloids and their aglycone in blood serum by high-performance liquid chromatography. Application to pharmacokinetic studies in humans. Journal of Chromatography 573, 69–78.

Högy, P., Fangmeier, A., 2009. Atmospheric CO_2 enrichment affects potatoes: 2. Tuber quality traits. European Journal of Agronomy 30, 85–94.

Hopkins, J., 1995. The glycoalkaloids: naturally of interest (but a hot potato?). Food and Chemical Toxicology 33, 323–328.

Houben, R.J., Brunt, K., 1994. Determination of glycoalkaloids in potato tubers by reversed-phase high-performance liquid chromatography. Journal of Chromatography A 661, 169–174.

Ieri, F., Innocenti, M., Andrenelli, L., Vecchio, V., Mulinacci, N., 2011. Rapid HPLC/DAD/MS method to determine phenolic acids, glycoalkaloids and anthocyanins in pigmented potatoes (*Solanum tuberosum* L.) and correlations with variety and geographical origin. Food Chemistry 125, 750–759.

Ji, Y.B., Liu, J.Y., Gao, S.Y., 2012. Effects of solanine on microtubules system of human breast cancer MCF-7 cells. Chinese Traditional and Herbal Drugs 43, 111–114.

Jin, Z., Yang, Y.X., Choi, J.Y., Shinde, P.L., Yoon, S.Y., Hahn, T.-W., Lim, H.T., Park, Y., Hahm, K.-S., Joo, J.W., Chae, B.J., 2008. Potato (*Solanum tuberosum* L. cv. Gogu valley) protein as a novel antimicrobial agent in weanling pigs. Journal of Animal Science 86, 1562–1572.

Jockovic, N., Fischer, W., Brandsch, M., Brandt, W., Dräger, B., 2013. Inhibition of human intestinal α-glucosidases by calystegines. Journal of Agricultural and Food Chemistry 61, 5550–5557.

Kaaber, L., 1993. Glycoalkaloids, green discoloration and taste development during storage of some potato varieties (*Solanum tuberosum* L.). Norwegian Journal of Agricultural Sciences 7, 221–229.

Kaliappan, K.P., Das, P., Chavan, S.T., Sabharwal, S.G., 2009. A versatile access to calystegine analogues as potential glycosidases inhibitors. Journal of Organic Chemistry 74, 6266–6274.

Kato, A., Nakagome, I., Nakagawa, S., Koike, Y., Nash, R.J., Adachi, I., Hirono, S., 2014. Docking and SAR studies of calystegines: binding orientation and influence on pharmacological chaperone effects for Gaucher's disease. Bioorganic and Medicinal Chemistry 22, 2435–2441.

Keiner, R., Dräger, B., 2000. Calystegine distribution in potato (*Solanum tuberosum*) tubers and plants. Plant Science 150, 171–179.

Keiner, R., Nakajima, K., Hashimoto, T., Dräger, B., 2000. Accumulation and biosynthesis of calystegines in potato. Journal of Applied Botany 74, 122–125.

Keukens, E.A., de Vrije, T., van den Boom, C., de Waard, P., Plasman, H.H., Thiel, F., Chupin, V., Jongen, W.M., de Kruijff, B., 1995. Molecular basis of glycoalkaloid induced membrane disruption. Biochimica et Biophysica Acta 1240, 216–228.

Kim, S.P., Nam, S.H., Friedman, M., 2015. The tomato glycoalkaloid α-tomatine induces caspase-independent cell death in mouse colon cancer CT-26 cells and transplanted tumors in mice. Journal of Agricultural and Food Chemistry 63, 1142–1150.

Kirui, G.K., Misra, A.K., Olanya, O.M., Friedman, M., El-Bedewy, R., Ewell, P.T., 2009. Glycoalkaloid content of some superior potato (*Solanum tuberosum* L.) clones and commercial cultivars. Archives of Phytopathology and Plant Protection 42, 453–463.

Kling, L.J., Bushway, R.J., Cleale, R.M., Bushway, A.A., 1986. Nutrient characteristics and glycoalkaloid content of potato distiller by-products. Journal of Agricultural and Food Chemistry 34, 54–58.

Korpan, Y.I., Nazarenko, E.A., Skryshevskaya, I.V., Martelet, C., Jaffrezic-Renault, N., El'skaya, A.V., 2004. Potato glycoalkaloids: true safety or false sense of security? Trends in Biotechnology 22, 147–151.

Kozukue, N., Friedman, M., 2003. Tomatine, chlorophyll, β-carotene and lycopene content in tomatoes during growth and maturation. Journal of the Science of Food and Agriculture 83, 195–200.

Kozukue, N., Kozukue, E., Mizuno, S., 1987. Glycoalkaloids in potato plants and tubers. HortScience 22, 294–296.

Kozukue, N., Misoo, S., Yamada, T., Kamijima, O., Friedman, M., 1999. Inheritance of morphological characters and glycoalkaloids in potatoes of somatic hybrids between dihaploid *Solanum acaule* and tetraploid *Solanum tuberosum*. Journal of Agricultural and Food Chemistry 47, 4478–4483.

Kozukue, N., Tsuchida, H., Friedman, M., 2001. Tracer studies on the incorporation of [2-^{14}C]-DL-mevalonate into chlorophylls a and b, α-chaconine, and α-solanine of potato sprouts. Journal of Agricultural and Food Chemistry 49, 92–97.

Kozukue, N., Tsuchida, H., Mizuno, S., 1993. Effect of light intensity, duration, and photoperiod on chlorophyll and glycoalkaloid production by potato tubers. Journal of the Japanese Society for Horticultural Science 62, 669–673.

Kozukue, N., Yoon, K.-S., Byun, G.-I., Misoo, S., Levin, C.E., Friedman, M., 2008. Distribution of glycoalkaloids in potato tubers of 59 accessions of two wild and five cultivated *Solanum* species. Journal of Agricultural and Food Chemistry 56, 11920–11928.

Krasowski, M.D., McGehee, D.S., Moss, J., 1997. Natural inhibitors of cholinesterases: implications for adverse drug reactions. Canadian Journal of Anaesthesia 44, 525–534.

Kubo, I., Fukuhara, K., 1996. Steroidal glycoalkaloids in Andean potatoes. Advances in Experimental Medicine and Biology 405, 405–417.

Kuo, K.-W., Hsu, S.-H., Li, Y.-P., Lin, W.-L., Liu, L.-F., Chang, L.-C., Lin, C.-C., Lin, C.-N., Sheu, H.-M., 2000. Anticancer activity evaluation of the solanum glycoalkaloid solamargine. Triggering apoptosis in human hepatoma cells. Biochemical Pharmacology 60, 1865–1873.

Kuronen, P., Väänänen, T., Pehu, E., 1999. Reversed-phase liquid chromatographic separation and simultaneous profiling of steroidal glycoalkaloids and their aglycones. Journal of Chromatography A 863, 25–35.

Lachman, J., Hamouz, K., Musilová, J., Hejtmánková, K., Kotíková, Z., Pazderů, K., Domkářová, J., Pivec, V., Cimr, J., 2013. Effect of peeling and three cooking methods on the content of selected phytochemicals in potato tubers with various colour of flesh. Food Chemistry 138, 1189–1197.

Laha, M.K., Basu, P.K., 1983. Biological hydrolysis of glycoalkaloids from *Solanum khasianum* by a local strain of *Aspergillus niger*. International Journal of Crude Drug Research 21, 153–155.

Laurila, J., Laakso, I., Vaananen, T., Kuronen, P., Huopalahti, R., Pehu, E., 1999. Determination of solanidine- and tomatidine-type glycoalkaloid aglycons by gas chromatography/mass spectrometry. Journal of Agricultural and Food Chemistry 47, 2738–2742.

Lee, K.-R., Kozukue, N., Han, J.-S., Park, J.-H., Chang, E.-Y., Baek, E.-J., Chang, J.-S., Friedman, M., 2004. Glycoalkaloids and metabolites inhibit the growth of human colon (HT29) and liver (HepG2) cancer cells. Journal of Agricultural and Food Chemistry 52, 2832–2839.

Li, X., Zhao, Y., Wu, W.K.K., Liu, S., Cui, M., Lou, H., 2011. Solamargine induces apoptosis associated with p53 transcription-dependent and transcription-independent pathways in human osteosarcoma U2OS cells. Life Sciences 88, 314–321.

Liang, C.-H., Shiu, L.-Y., Chang, L.-C., Sheu, H.-M., Kuo, K.-W., 2007. Solamargine upregulation of Fas, downregulation of HER2, and enhancement of cytotoxicity using epirubicin in NSCLC cells. Molecular Nutrition & Food Research 51, 999–1005.

Liang, C.-H., Shiu, L.-Y., Chang, L.-C., Sheu, H.-M., Tsai, E.-M., Kuo, K.-W., 2008. Solamargine enhances HER2 expression and increases the susceptibility of human lung cancer H661 and H69 cells to trastuzumab and epirubicin. Chemical Research in Toxicology 21, 393–399.

Liu, L.-F., Liang, C.-H., Shiu, L.-Y., Lin, W.-L., Lin, C.-C., Kuo, K.-W., 2004. Action of solamargine on human lung cancer cells - enhancement of the susceptibility of cancer cells to TNFs. FEBS Letters 577, 67–74.

Machado, R.M.D., Toledo, M.C.F., Garcia, L.C., 2007. Effect of light and temperature on the formation of glycoalkaloids in potato tubers. Food Control 18, 503–508.

Maga, J.A., 1981. Total and individual glycoalkaloid composition of stored potato slices. Journal of Food Processing and Preservation 5, 23–29.

Meher, H.C., Gaur, H.S., 2003. A new UV-LC method for estimation of α-tomatine and α-solanine in tomato. Indian Journal of Nematology 33, 24–28.

Mensinga, T.T., Sips, A.J.A.M., Rompelberg, C.J.M., Van Twillert, K., Meulenbelt, J., Van Den Top, H.J., Van Egmond, H.P., 2005. Potato glycoalkaloids and adverse effects in humans: an ascending dose study. Regulatory Toxicology and Pharmacology 41, 66–72.

Molyneux, R.J., Pan, Y.T., Goldmann, A., Tepfer, D.A., Elbein, A.D., 1993. Calystegins, a novel class of alkaloid glycosidase inhibitors. Archives of Biochemistry and Biophysics 304, 81–88.

Mulinacci, N., Ieri, F., Giaccherini, C., Innocenti, M., Andrenelli, L., Canova, G., Saracchi, M., Casiraghi, M.C., 2008. Effect of cooking on the anthocyanins, phenolic acids, glycoalkaloids, and resistant starch content in two pigmented cultivars of *Solanum tuberosum* L. Journal of Agricultural and Food Chemistry 56, 11830–11837.

Munari, C.C., de Oliveira, P.F., Campos, J.C., Martins Sde, P., Da Costa, J.C., Bastos, J.K., Tavares, D.C., 2014. Antiproliferative activity of *Solanum lycocarpum* alkaloidic extract and their constituents, solamargine and solasonine, in tumor cell lines. Journal of Natural Medicines 68, 236–241.

Najm, A.A., Hadi, M.R.H.S., Fazeli, F., Darzi, M.T., Rahi, A., 2012. Effect of integrated management of nitrogen fertilizer and cattle manure on the leaf chlorophyll, yield, and tuber glycoalkaloids of Agria potato. Communications in Soil Science and Plant Analysis 43, 912–923.

Nash, R.J., Rothschild, M., Porter, E.A., Watson, A.A., Waigh, R.D., Waterman, P.G., 1993. Calystegines in *Solanum* and *Datura* species and the death's-head hawk-moth (*Acherontia atropus*). Phytochemistry 34, 1281–1283.

Nelson, M.L., 2010. Utilization and application of wet potato processing coproducts for finishing cattle. Journal of Animal Science 88, E133–E142.

Nielen, M.W.F., van Beek, T.A., 2014. Macroscopic and microscopic spatially-resolved analysis of food contaminants and constituents using laser-ablation electrospray ionization mass spectrometry imaging. Analytical and Bioanalytical Chemistry 406 (27), 6805–6815. http://dx.doi.org/10.1007/s00216-014-7948-8.

Nitithamyong, A., Vonelbe, J.H., Wheeler, R.M., Tibbitts, T.W., 1999. Glycoalkaloids in potato tubers grown under controlled environments. American Potato Journal 76, 337–343.

Ohh, S.H., Shinde, P.L., Jin, Z., Choi, J.Y., Hahn, T.-W., Lim, H.T., Kim, G.Y., Park, Y., Hahm, K.-S., Chae, B.J., 2009. Potato (*Solanum tuberosum* L. cv. Gogu valley) protein as an antimicrobial agent in the diets of broilers. Poultry Science 88, 1227–1234.

Panovska, Z., Hajslova, J., Kosinkova, P., Cepl, J., 1997. Glycoalkaloid content of potatoes sold in Czechia. Nahrung 41, 146–149.

Petersson, E.V., Arif, U., Schulzova, V., Krtkova, V., Hajslova, J., Meijer, J., Andersson, H.C., Jonsson, L., Sitbon, F., 2013. Glycoalkaloid and calystegine levels in table potato cultivars subjected to wounding, light, and heat treatments. Journal of Agricultural and Food Chemistry 61, 5893–5902.

Pino-Gonzalez, M.S., Oña, N., Romero-Carrasco, A., 2012. Advances in the synthesis of calystegines and related products and their biochemical properties. Mini Reviews in Medicinal Chemistry 12, 1477–1484.

Plhak, L.C., 1992. Enzyme immunoassay for potato glycoalkaloids. Journal of Agricultural and Food Chemistry 40, 2533–2540.

Rayburn, J.R., Bantle, J.A., Friedman, M., 1994. Role of carbohydrate side chains of potato glycoalkaloids in developmental toxicity. Journal of Agricultural and Food Chemistry 42, 1511–1515.

Rayburn, J.R., Bantle, J.A., Qualls Jr., C.W., Friedman, M., 1995a. Protective effects of glucose-6-phosphate and NADP against α-chaconine-induced developmental toxicity in Xenopus embryos. Food and Chemical Toxicology 33, 1021–1025.

Rayburn, J.R., Friedman, M., Bantle, J.A., 1995b. Synergistic interaction of glycoalkaloids α-chaconine and α-solanine on developmental toxicity in Xenopus embryos. Food and Chemical Toxicology 33, 1013–1019.

Renwick, J.H., 1972. Hypothesis: anencephaly and spina bifida are usually preventable by avoidance of a specific but unidentified substance present in certain potato tubers. British Journal of Preventive and Social Medicine 26, 67–88.

Richter, U., Sonnewald, U., Dräger, B., 2007. Calystegines in potatoes with genetically engineered carbohydrate metabolism. Journal of Experimental Botany 58, 1603–1615.

Rodriguez-Saona, L.E., Wrolstad, R.E., Pereira, C., 1999. Glycoalkaloid content and anthocyanin stability to alkaline treatment of red-fleshed potato extracts. Journal of Food Science 64, 445–450.

Rodriguez de Sotillo, D., Hadley, M., Wolf-Hall, C., 1998. Potato peel extract a nonmutagenic antioxidant with potential antimicrobial activity. Journal of Food Science 63, 907–910.

Rytel, E., Golubowska, G., Lisinska, G., Peksa, A., Anilowski, K., 2005. Changes in glycoalkaloid and nitrate contents in potatoes during french fries processing. Journal of the Science of Food and Agriculture 85, 879–882.

Rytel, E., Lisińska, G., Tajner-Czopek, A., 2013. Toxic compound levels in potatoes are dependent on cultivation methods. Acta Alimentaria 42, 308–317.

Saito, K., Horie, M., Hoshino, Y., Nose, N., Nakazawa, H., 1990. High-performance liquid chromatographic determination of glycoalkaloids in potato products. Journal of Chromatography 508, 141–147.

Shiu, L.Y., Liang, C.H., Chang, L.C., Sheu, H.M., Tsai, E.M., Kuo, K.W., 2009. Solamargine induces apoptosis and enhances susceptibility to trastuzumab and epirubicin in breast cancer cells with low or high expression levels of HER2/neu. Bioscience Reports 29, 35–45.

Simonovska, B., Vovk, I., 2000. High-performance thin-layer chromatographic determination of potato glycoalkaloids. Journal of Chromatography A 903, 219–225.

Skrabule, I., Muceniece, R., Kirhnere, I., 2013. Evaluation of vitamins and glycoalkaloids in potato genotypes grown under organic and conventional farming systems. Potato Research 56, 259–276.

Smith, D.B., Roddick, J.G., Jones, J.L., 2001. Synergism between the potato glycoalkaloids α-chaconine and α-solanine in inhibition of snail feeding. Phytochemistry 57, 229–234.

Sotelo, A., Serrano, B., 2000. High-performance liquid chromatographic determination of the glycoalkaloids α-solanine and α-chaconine in 12 commercial varieties of Mexican potato. Journal of Agricultural and Food Chemistry 48, 2472–2475.

Sporns, P., Abell, D.C., Kwok, A.S.K., Plhak, L.C., Thomson, C.A., 1996. Immunoassays for toxic potato glycoalkaloids. ACS Symposium Series 621, 256–272.

Stanker, L.H., Kamps-Holtzapple, C., Beier, R.C., Levin, C.E., Friedman, M., 1996. Detection and quantification of glycoalkaloids: comparison of enzyme-linked immunosorbent assay and high-performance liquid chromatography methods. In: Presented at Immunoassays for Residue Analysis Food Safety, vol. 621. American Chemical Society, Washington, DC, pp. 243–255.

Stanker, L.H., Kamps-Holtzapple, C., Friedman, M., 1994. Development and characterization of monoclonal antibodies that differentiate between potato and tomato glycoalkaloids and aglycons. Journal of Agricultural and Food Chemistry 42, 2360–2366.

Stenzel, O., Teuber, M., Dräger, B., 2006. Putrescine N-methyltransferase in *Solanum tuberosum* L., a calystegine-forming plant. Planta 223, 200–212.

Sun, L., Zhao, Y., Li, X., Yuan, H., Cheng, A., Lou, H., 2010. A lysosomal-mitochondrial death pathway is induced by solamargine in human K562 leukemia cells. Toxicology In Vitro 24, 1504–1511.

Tajner-Czopek, A., Rytel, E., Aniołowska, M., Hamouz, K., 2014. The influence of French fries processing on the glycoalkaloid content in coloured-fleshed potatoes. European Food Research and Technology 238, 895–904.

Tepfer, D., Goldmann, A., Pamboukdjian, N., Maille, M., Lepingle, A., Chevalier, D., Dénarié, J., Rosenberg, C., 1988. A plasmid of *Rhizobium meliloti* 41 encodes catabolism of two compounds from root exudate of *Calystegium sepium*. Journal of Bacteriology 170, 1153–1161.

Väänänen, T., Ikonen, T., Rokka, V.M., Kuronen, P., Serimaa, R., Ollilainen, V., 2005. Influence of incorporated wild *Solanum* genomes on potato properties in terms of starch nanostructure and glycoalkaloid content. Journal of Agricultural and Food Chemistry 53, 5313–5325.

Väänänen, T., Kuronen, P., Pehu, E., 2000. Comparison of commercial solid-phase extraction sorbents for the sample preparation of potato glycoalkaloids. Journal of Chromatography A 869, 301–305.

Valcarcel, J., Reilly, K., Gaffney, M., O'Brien, N., 2014. Effect of genotype and environment on the glycoalkaloid content of rare, heritage, and commercial potato varieties. Journal of Food Science 79, T1039–T1048.

Vorne, V., Ojanperä, K., De Temmerman, L., Bindi, M., Högy, P., Jones, M.B., Lawson, T., Persson, K., 2002. Effects of elevated carbon dioxide and ozone on potato tuber quality in the European multiple-site experiment 'CHIP-project'. European Journal of Agronomy 17, 369–381.

Wang, H.Y., Liu, M.Y., Hu, X.X., Li, M., Xiong, X.Y., 2013. Electrochemical determination of glycoalkaloids using a carbon nanotubes-phenylboronic acid modified glassy carbon electrode. Sensors 13, 16234–16244.

Ward, C.M., Franklin, J.G., Morgan, M.R.A., 1988. Investigations into the visual assessment of ELISA end points: application to determination of potato total glycoalkaloids. Food Additives & Contaminants 5, 621–627.

Watson, A.A., Fleet, G.W., Asano, N., Molyneux, R.J., Nash, R.J., 2001. Polyhydroxylated alkaloids - natural occurrence and therapeutic applications. Phytochemistry 56, 265–295.

Wszelaki, A.L., Delwiche, J.F., Walker, S.D., Liggett, R.E., Scheerens, J.C., Kleinhenz, M.D., 2005. Sensory quality and mineral and glycoalkaloid concentrations in organically and conventionally grown redskin potatoes (*Solanum tuberosum*). Journal of the Science of Food and Agriculture 85, 720–726.

Yamashoji, S., Yoshikawa, N., Kirihara, M., Tsuneyoshi, T., 2013. Screening test for rapid food safety evaluation by menadione-catalysed chemiluminescent assay. Food Chemistry 138, 2146–2151.

Yang, S.A., Paek, S.H., Kozukue, N., Lee, K.R., Kim, J.A., 2006. α-Chaconine, a potato glycoalkaloid, induces apoptosis of HT-29 human colon cancer cells through caspase-3 activation and inhibition of ERK 1/2 phosphorylation. Food and Chemical Toxicology 44, 839–846.

Zarzecka, K., Gugała, M., 2007. Changes in the content of glycoalkaloids in potato tubers according to soil tillage and weed control methods. Plant, Soil and Environment 53, 247–251.

Zarzecka, K., Gugala, M., Mystkowska, I., 2013. Glycoalkaloid contents in potato leaves and tubers as influenced by insecticide application. Plant, Soil and Environment 59, 183–188.

Zhang, L., Porter, G.A., Bushway, R.J., 1997. Ascorbic acid and glycoalkaloid content of atlantic and superior potato tubers as affected by supplemental irrigation and soil amendments. American Potato Journal 74, 285–304.

Zhang, X., Cai, X., Zhang, X., 2014. Determination of α-solanine, α-chaconine and solanidine in plasma and urine by ultra-performance liquid chromatography-triple quadrupole mass spectrometry. Chinese Journal of Chromatography 32, 586–590.

Zhou, Y., Tang, Q., Zhao, S., Zhang, F., Li, L., Wu, W., Wang, Z., Hann, S., 2014. Targeting signal transducer and activator of transcription 3 contributes to the solamargine-inhibited growth and -induced apoptosis of human lung cancer cells. Tumour Biology 35 (8), 8169–8178. http://dx.doi.org/10.1007/s13277-014-2047-1.

CHAPTER 8

Potato Starch and Its Modification

Jaspreet Singh[1], Rosana Colussi[1,2], Owen J. McCarthy[3], Lovedeep Kaur[1]
[1]Riddet Institute and Massey Institute of Food Science and Technology, Massey University, Palmerston North, New Zealand; [2]Departamento de Ciência e Tecnologia Agroindustrial, Universidade Federal de Pelotas, Pelotas, Brazil; [3]Massey Institute of Food Science and Technology, Massey University, Palmerston North, New Zealand

1. Introduction

Among all the carbohydrate polymers, starch is currently enjoying increased attention owing to its usefulness in various food products. Starch has traditionally been used in the food industry to enhance the functional properties of various foods. The physicochemical properties and functional characteristics of starch systems, and their uniqueness in various food products, vary with the starch biological origin. Interest in new value-added products to the industry has resulted in many studies being carried out on the starches. Starch exists naturally in the form of discrete granules within plant cells and is mainly composed of two polymers: amylose and amylopectin. Amylose is a linear polymer composed of glucopyranose units linked through α-D-(1→4) glycosidic linkages, while amylopectin is a highly branched and high-molecular-weight polymer. This composition may affect the physicochemical properties, such as gelatinization, texture, moisture retention, viscosity, and product homogeneity, that are determinants for its applications.

Potatoes in general are an excellent source of starch, which contributes to the textural properties of many foods. Potato starch can be used in food and other industrial applications as a thickener, colloidal stabilizer, gelling agent, bulking agent, and water-holding agent. But limitations, such as low shear and thermal resistance, and high tendency toward retrogradation, restrict its use in some industrial food applications. These limitations are generally overcame by starch modification, which can be achieved through derivatization, such as etherification, esterification, cross-linking, and grafting; decomposition (acid or enzymatic hydrolysis and oxidization); or physical treatment of starch using heat or moisture or pressure, etc. These treatments result in markedly altered gelatinization, pasting, and retrogradation behavior of potato starch. The Food and Drug Administration (FDA) regulates and reinforces the type and amount of each chemical used in starch modification, as well as the maximum percentage of the substitution permitted.

Advances in Potato Chemistry and Technology. http://dx.doi.org/10.1016/B978-0-12-800002-1.00008-X

In this chapter, we present a great deal of information on important physicochemical and functional characteristics of native potato starch in comparison with some cereal starches. In addition, we also discuss various modification techniques and reagents being used to modify potato starch, with an emphasis on the postmodification changes (particularly after derivatization) in its morphological, physicochemical, rheological, and thermal behavior. The various factors that influence potato starch modification have also been discussed in detail.

2. Potato Starch versus Cereal Starches

Potato starch exhibits different granular structure and composition as opposed to cereal starches, which are responsible for the variation in functional behavior of these starches. Cereal starches exhibit the typical A-type X-ray crystalline pattern, whereas potato starch shows the B-form, and legumes show the mixed-state pattern C. The A, B, and C patterns are the polymeric forms of the starch that differ in the packing of amylopectin double helices. The structure of potato starch is discussed in more detail in another chapter.

2.1 Morphology

Starch granule morphology varies with plant genotype and cultural practices. It also depends on the biochemistry of the chloroplast/amyloplast and the physiology of the plant. When viewed under a microscope, the starch granules of potato differ in size and shape from those of cereal starches. Scanning electron micrographs of starch granules from various plant sources are illustrated in Figure 1. Granule size differs considerably among starches and ranges from 1 to 110 μm (Hoover, 2001). For potato starch, the average granule size ranges from 1 to 20 μm for small and 20 to 110 μm for large granules. The average size of individual maize starch granules ranges from 1 to 7 μm for small and 15 to 20 μm for large granules. Rice starch granules generally range from 3 to 5 μm in size. Wheat endosperm at maturity contains two types of starch granules: large A- (diameter 10–35 μm) and small B-type (diameter 1–10 μm).

The extent of variation in the granular structure of starches from cultivar to cultivar is also quite high in potatoes. Granule size distributions of starches of various potato cultivars studied through a particle size analyzer are given in Figure 2 and the distributions in terms of percentages of small, medium, and large granules in Table 2. Interestingly, the small potato starch granules are spherical or oval in shape, but the large ones are generally ellipsoid to cuboid or irregular in shape. This variation in the shape with the size of potato starch granules could be related to granule packing during growth of the storage organs. Limited space availability in tuber cells may lead to an alteration in the shape of a growing granule. The starch granules are angular-shaped for maize, and pentagonal and angular-shaped for rice. A-type granules of wheat starch are disk-like or lenticular in shape and the B-type starch granules are roughly spherical or polygonal in shape.

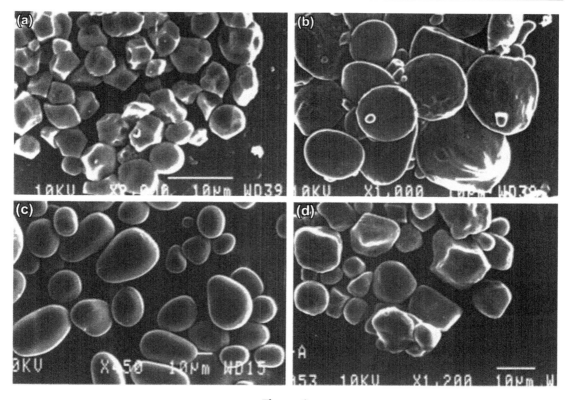

Figure 1
Scanning electron micrographs of starches separated from different sources: (a) rice, (b) wheat, (c) potato, and (d) maize (bar = 10 μm). *Reproduced from Singh et al. (2003), with permission from Elsevier.*

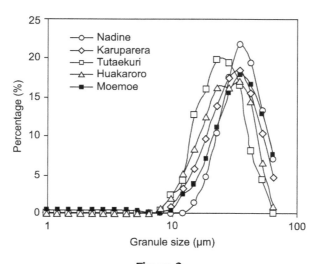

Figure 2
Particle size distribution of starches from some New Zealand potato cultivars. *Reproduced from Singh et al. (2006), with permission from Elsevier.*

Table 1: Physicochemical Properties of Starches from Various Botanical Sources.

Starch Source	Amylose Content (%)	Swelling Power (g/g) (°C)	Solubility (%) (°C)	Organic Phosphorus Contents[a] (% dsb)			Light Transmittance[e] (%, at 650 nm)
				Mono-P[b]	Lipid-P[c]	Inorganic-P	
Normal potato	20.1–31.0	1159 (95)	82 (95)	0.086±0.007	ND[d]	0.0048±0.0003	96
Normal maize	22.4–32.5	22 (95)	22 (95)	0.003±0.001	0.0097±0.0001	0.0013±0.0007	31
Waxy maize	1.4–2.7	–	–	0.0012±0.0006	ND	0.0005±0.0001	46
High-amylose maize	42.6–67.8	6.3 (95)	12.4 (95)	0.005±0.001	0.015±0.003	0.0076±0.0006	–
Normal rice	5–28.4	23–30 (95)	11–18 (95)	0.013	0.048	–	24
Waxy rice	0–2.0	45–50 (95)	2.3–3.2 (95)	0.003	ND	–	–
High-amylose rice	25–33	–	–	–	–	–	–
Normal wheat	18–30	18.3–26.6 (100)	1.55 (100)	0.001	0.058±0.002	Trace	28
Waxy wheat	0.8–0.9	–	–	–	–	–	–

[a]Calculated based on integrated area of P-signals.
[b]Phosphate monoester P-signals located at 4.0–4.5 ppm relative to external 80% *ortho*-phosphoric acid.
[c]Phospholipid P-signals located at -0.4 to 1.2 ppm.
[d]Not detectable.
[e]Calculated using 1% starch paste.
Reproduced from Singh et al. (2003), with permission from Elsevier.

Table 2: Morphological Parameters of Starches from Various New Zealand Potato Cultivars: Proportions of Small, Medium-Size, and Large Granules; Mean Granule Volume; and Granule Specific Surface Area.

Potato Starch Source	Small Granules (%) (1–10 μm)	Medium Granules (%) (11–25 μm)	Large Granules (%) (>25 μm)	Mean Volume (μm³)	Specific Surface Area (m²/g)
Nadine	4.4	16.4	79.2	10,648	0.189
Karuparera	0.9	32.6	66.5	6859	0.198
Tutaekuri	2.7	52.8	48.5	3375	0.248
Huakaroro	2.7	42.1	55.2	4096	0.228
Moemoe	5.1	24.5	70.5	5832	0.188

Reproduced from Singh et al. (2006), with permission from Elsevier.

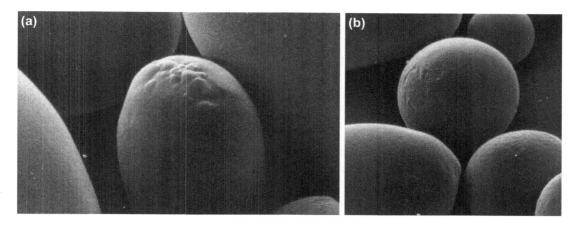

Figure 3
Scanning electron micrographs featuring (a) the presence of some small nodules or protuberances on some potato starch granules and (b) surface fragmentation on some potato starch granules. *Reproduced from Singh et al. (2006), with permission from Elsevier.*

Variations in amylose and amylopectin structure and in their relative amounts in a granule play an important role in controlling starch granule size and shape. Activity of the enzyme granule-bound starch synthase (GBSS) during growth may also affect the starch granule morphology in potatoes (Blennow et al., 2002). The membranes and the physical characteristics of the plastids may also be responsible for providing a particular shape or morphology to starch granules during granule development (Jane et al., 1994; Lindeboom et al., 2004).

When observed under a scanning electron microscope (SEM), the surface of maize, rice, and wheat starch granules appears less smooth than that of potato starch granules. Li et al. (2001) observed the presence of "pinholes" and equatorial grooves or furrows in large maize starch granules. Singh et al. (2006) have shown the presence of small protuberances and fragmentation on the surface of potato starch granules (Figure 3). Physicochemical properties, such as transparency of the starch paste, enzymatic digestibility, amylose content, and swelling power, have been significantly correlated with the average granule size of the starches separated from different potato cultivars (Kaur et al., 2007a,b).

2.2 Composition

Starch paste behavior in aqueous systems depends on the physical and chemical characteristics of the starch granules, such as mean granule size, granule size distribution, amylose/amylopectin ratio, mineral content, etc. The amylose content of the starch granules varies with the botanical source of the starch (Table 1) and is affected by the climatic conditions and soil type during growth. Amylose content of potato starch varies from 23% to 34% for normal potato genotypes (Kim et al., 1995; Wiesenborn et al., 1994; Cai and Wei, 2013). However, waxy potato genotypes, essentially without amylose, have also been reported (Hermansson and Svegmark, 1996). Also, large potato starch granules have higher amylose content than the small granules. The activities of GBSS, involved in the biosynthesis of linear components; and soluble starch synthase (SSS) and starch branching enzymes (SBE), involved in the biosynthesis of branched components within the starch granule, may be responsible for the variation in amylose content among the various starches (Kossmann and Lloyd, 2000; Fulton et al., 2002). Potato mutants lacking GBSS have been reported to synthesize amylose-free starch (Hovenkamp-Hermelink et al., 1987). Also, the amylopectin fraction is synthesized at a faster rate than amylose during the early stages of starch granule growth owing to the high activities of SSS and SBE, which diminish during the later stages, while GBSS retains its activity throughout the growth period in association with the developing starch granule. Thus, the small granules, which are at initial stages of growth, may have higher SSS and SBE activities that result in lower amylose content. The variation in amylose contents among the starches from different and similar plant sources in various studies (Table 1) could be due to the use of different starch isolation procedures and analytical methods used to determine amylose content.

Phosphorus is one of the noncarbohydrate constituents present in starches, and it significantly affects the functional properties of the starches. Phosphorus content varies from 0.003% in waxy maize starch to 0.09% in potato starch (Schoch, 1942; Lim et al., 1994; Table 1). Phosphorus is present as phosphate monoesters and phospholipids in various starches. Phosphate groups, esterified to the amylopectin fraction of potato starch, contribute to its high water-binding capacity, viscosity, transparency, and freeze–thaw stability (Craig et al., 1989; Vamadevan and Bertof, 2015). Haase and Plate (1996) reported that defined phosphorus and amylose contents, paste peak viscosity, and a specific large granule fraction are prominent quality characteristics for the selection of potato lines and cultivars to be utilized in conventional breeding to suit future starch production. Phospholipids present in starch have a tendency to form a complex with amylose and long-branched chains of amylopectin, which limits the starch granule swelling, ultimately resulting in opaque and low-viscosity pastes.

That is why wheat and rice starches, with more phospholipids, produce starch pastes with lower transmittance than potato starches, with fewer phospholipids. Phosphorus content and form in potato starch is influenced by growing conditions, temperature, and postharvest storage of the potato tubers. It has been reported that 61% of the starch phosphate monoesters in potato starch are bound to C-6 of the glucose units, with 38% phosphate monoester on C-3 of the glucose and possibly 1% of monoester on the C-2 position (Jane et al., 1996).

Potato starch contains fewer lipids than the cereal starches. Free fatty acids in rice and maize starches result in amylose–lipid complex formation, thereby contributing to their higher transition temperatures and lower retrogradation.

2.3 Swelling Power and Solubility

When the starch molecules are heated in excess water, their crystalline structure is disrupted and water molecules become linked to the exposed hydroxyl groups of amylose and amylopectin by hydrogen bonding, which causes an increase in granule swelling and solubility. The swelling power and solubility provide an evidence of the magnitude of interaction between starch chains within the amorphous and crystalline domains. The extent of this interaction is influenced by the amylose-to-amylopectin ratio and by the characteristics of amylose and amylopectin in terms of molecular weight/distribution, degree and length of branching, and conformation (Hoover, 2001; Vamadevan and Bertof, 2015). Potato starch exhibits much higher average swelling power than the cereal starches (Table 1). According to deWilligen (1976), maize and wheat granules may swell up to 30 times their original volume and potato starch granules up to 100 times their original volume, without disintegration. The higher swelling power and solubility of potato starch are probably due to the presence of a large number of phosphate groups on the amylopectin molecule. Repulsion between phosphate groups on adjacent chains increases hydration by weakening the extent of bonding within the crystalline domain. Small potato starch granules show greater hydration and swelling power than the large ones, which is mainly due to their higher specific surface area.

The differences in the swelling and solubility behavior of the starches between botanical sources and among the cultivars of any one botanical source are caused by the differences in the amylose and the lipid contents, as well as the granule organization. Amylose plays an important role in restricting initial swelling because swelling proceeds more rapidly after amylose has first been exuded. The increase in starch solubility, with the concomitant increase in suspension clarity, is seen mainly as the result of granule swelling, permitting the exudation of the amylose. The granules become increasingly susceptible to shear disintegration as they swell, and they release soluble material as they disintegrate. The hot starch paste is a mixture of swollen granules and granule fragments, together with colloidally and molecularly dispersed starch granules.

The mixture of the swollen and fragmented granules depends on the botanical source of the starch, water content, temperature, and shearing during heating. The extent of leaching of solubles mainly depends on the lipid content of the starch and the ability of the starch to form amylose–lipid complexes as the amylose involved in complex formation with lipids is prevented from leaching out. The cereal starches contain enough lipids to form lipid-saturated complexes with 7–8% amylose in the starch; hence the maximum amylose leached is about 20% of the total starch (Tester and Morrison, 1990; Ai and Jane, 2015). The higher solubility of the potato starches may be attributed to the lack of starch–lipid inclusion complex owing to the absence of lipids.

2.4 Pasting/Rheological Properties

During gelatinization, the starch granule swells to several times its initial size, ruptures, and simultaneously amylose leaches out and forms a three-dimensional network. Swelling of starch is the property of its amylopectin content, and amylose acts as both a diluent and an inhibitor of swelling (Tester and Morrison, 1990). Starch exhibits unique viscosity behavior with change in temperature, concentration, and shear rate (Nurul et al., 1999). This can be measured by the Brabender Visco-Amylo-Graph rapid visco-analyzer (RVA) pasting curves (Figure 4). The shape of the peak achieved is the reflection of the processes taking place during the pasting cycle. The height of the peak at the given concentration reflects the ability of the granules to swell freely before their physical breakdown. Potato starches are capable of swelling to a higher degree than cereal starches and are also less resistant to breakdown on cooking and hence exhibit viscosity decreases considerably after reaching the maximum

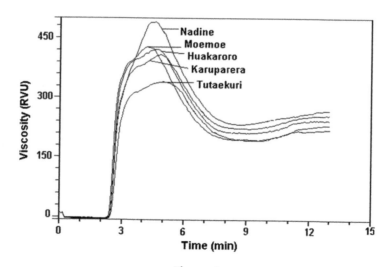

Figure 4
RVA pasting curves of starches isolated from the different New Zealand potato cultivars
(Singh et al., 2006).

Table 3: Gelatinization Parameters (Studied Using DSC) of the Starches from Various Botanical Sources.

Source	Methodology	T_o (°C)	T_p (°C)	T_c (°C)	ΔH_{gel} (J/g)[a]
Potato	S:W[b] 1:2.3	59.72–66.2	62.9–69.6	67.28–75.4	12.55–17.9
Potato	S:W 1:3.3	57.0–68.3	60.6–72.4	66.5–78.0	13.0–15.8
Potato	S:W 1:1.5	57.2	61.4	80.3	17.4
Normal maize	S:W 1:1.5	62.3	67.7	84.3	14.0
Normal maize	S:W 1:3	64.1	69.4	74.9	12.3
Normal maize	S:W 1:9	65.7	71.0	–	12.0
Waxy maize	S:W 1:9	66.6	73.6	–	14.2
Waxy maize	S:W 1:3	64.2	69.2	74.6	15.4
High-amylose maize	S:W 1:9	66.8	73.7	–	13.7
Rice	S:W 1:1.5	62.0	67.4	97.5	11.0
Rice	S:W 1:9	57.7	65.1	–	11.5
Rice	S:W 1:2.3	66.0–67.26	69.74–71.94	74.08–78.04	8.16–10.88
Rice	S:W 1:3	70.3	76.2	80.2	13.2
Waxy rice	DSC	66.1–74.9	70.4–78.8	–	7.7–12.1
Wheat	S:W 1:1.5	51.2	56.0	76.0	9.0
Wheat	S:W 1:2.3	46.0–52.4	52.2–57.6	57.8–66.1	14.8–17.9
Wheat	S:W 1:3	57.1	61.6	66.2	10.7

T_o, onset temperature; T_p, peak temperature; T_c, final temperature; ΔH_{gel}, enthalpy of gelatinization (dsb, based on dry starch weight).
[a]Enthalpy values are expressed in J/g of dry starch.
[b]Starch (S):water (W).
Reproduced from Singh et al. (2003), with permission from Elsevier.

Table 4: Thermal Properties during Retrogradation of Starches from Various Botanical Sources.

Source	T_o (°C)	T_p (°C)	T_c (°C)	ΔH_{ret} (J/g)	R (%)[g]
Potato[a]	59.72–60.70	63.26–64.58	67.28–70.34	6.42–8.61	51.50–62.16
Potato[b]	42.5	55.7	66.9	7.5	43.4
Normal maize[b]	39.0	50.1	59.4	5.8	47.6
Waxy maize[b]	40.2	51.3	60.2	7.3	47.0
High-amylose maize[b]	44.1	ND[f]	115.4	9.9	61.0
Normal rice[b]	40.3	51.0	60.4	5.3	40.5
Normal rice[c]	37.05–38.43	49.80–52.59	62.42–65.92	–	–
Waxy rice[c]	36.72–37.25	50.65–51.26	62.56–62.93	–	–
Waxy rice[b]	43.2	50.6	55.2	0.8	5.0
Normal wheat[b]	38.6	47.6	55.7	3.6	33.7
Normal wheat[d]	29.8–31.7	41.8–42.7	–	7.0–8.5	–
Normal wheat[e]	30.9–32.6	41.2–42.6	–	8.1–9.7	–
Normal wheat[e]	20.4–20.6	33.2–33.7	50.0	10.1–10.6	–
Waxy wheat[e]	19.9–20.5	33.1–33.8	50.4–51.8	11.4–12.6	–

T_o, onset temperature; T_p, peak temperature; T_c, final temperature; ΔH_{ret}, enthalpy of retrogradation (dsb, based on dry starch weight).
[a]Storage at 4 °C for 2 weeks.
[b]Storage at 4 °C for 7 days.
[c]Storage at 4 °C for 4 weeks.
[d]Storage at 5 °C for 2 weeks.
[e]Storage at 5 °C for 4 weeks.
[f]Not detectable.
[g]Retrogradation (%) = $\Delta H_{gel}/\Delta H_{ret}$.
Reproduced from Singh et al. (2003), with permission from Elsevier.

value. The shape of the peak is, however, strongly influenced by the initial concentration of the starch suspension. The increase in viscosity during the cooling period indicates the tendency of various constituents present in the hot paste (swollen granules, fragments of swollen granules, colloidally and molecularly dispersed starch molecules) to associate or retrograde as the temperature of the paste decreases.

The dynamic rheometer allows continuous assessment of dynamic moduli during temperature and frequency sweep testing of starch suspensions. The storage dynamic modulus (G') is a measure of the energy stored in the material and recovered from it per cycle, while the loss modulus (G'') is a measure of the energy dissipated or lost per cycle of sinusoidal deformation (Ferry, 1980). The ratio of the energy lost to the energy stored for each cycle can be defined by a loss factor (tan δ), which is another parameter indicating the physical behavior of a system. The G' of starch progressively increases greatly at a certain temperature (TG') to a maximum (peak G') and then drops with continued heating on a dynamic rheometer. The initial increase in G' is due to granular swelling to fill the entire available volume of the system to form a three-dimensional network of the swollen granules. With further increase in temperature, G' decreases, indicating destruction of the gel structure during prolonged heating. This destruction is due to the melting of the crystalline region remaining in the swollen starch granule, which deforms and loosens the particles (Eliasson, 1986).

The rheological properties of the various starches vary to a large extent with respect to the granular structure (Table 5). Maize starch has a lower peak G' and G'' than potato starch. The presence of higher phosphate monoester content and the absence of lipids and phospholipids in the potato starch may also be responsible for high G' and G''. The phospholipids and the more rigid granules present in maize starch may be responsible for the lower G' of maize starch. The amylose–lipid complex formation during gelatinization of maize starch lowers the G' and G'' (Singh et al., 2002). The protein content of rice starch has been reported to be negatively correlated with peak viscosity and positively correlated with pasting temperature (Lim et al., 1999).

Table 5: Rheological Parameters of Starches from Various Botanical Sources during Heating from 30 to 75 °C, Studied Using a Dynamic Rheometer.

Source	TG' (°C)	Peak G' (Pa)	Peak G'' (Pa)	Breakdown in G' (Pa)	Peak tan δ
Potato[b]	62.7	8519	1580	3606	0.1565
Maize[a]	70.2	6345	1208	2329	0.1905
Rice[a]	72.4	4052	955	2831	0.1972
Wheat[a]	69.6	6935	1370	2730	0.1976

[a]At 20% starch concentration.
[b]At 15% starch concentration.
Reproduced from Singh et al. (2003), with permission from Elsevier.

The extent of breakdown in starch pastes gives a measure of the degree of disintegration of starch granules. The breakdown in G' is the difference between peak G' at TG' and minimum G' at 75 °C. Potato starches generally show higher breakdown in G' than maize, rice, and wheat starches. The differences in the breakdown values of starches may be attributed to the granule rigidity, lipid content, and peak G' values. Amylose content is another factor that affects the rheological and pasting properties of starch. Singh et al. (2007) reported higher G' values for potato starches with higher amylose content during temperature sweep testing. Shewmaker et al. (1994) reported low paste viscosity for starch pastes made from potato genotypes containing low amylose content. Similarly, the starches isolated from waxy potatoes show lower G' and G'' and higher tan δ values (Kaur et al., 2002).

2.5 Gelatinization and Retrogradation: Thermal Properties

The crystalline order in starch granules is often the basic underlying factor influencing its functional properties. Collapse of crystalline order within the starch granules manifests itself as irreversible changes in properties such as granule swelling, pasting, loss of optical birefringence, loss of crystalline order, uncoiling and dissociation of the double helices, and starch solubility (Hoover, 2001). The gelatinization phenomenon starts at the hilum of the granule and swells rapidly to the periphery. Gelatinization occurs initially in the amorphous regions as opposed to the crystalline regions of the granule, because hydrogen bonding is weakened in these areas. The order–disorder transitions that occur on heating an aqueous suspension of starch granules have been extensively investigated using the dynamic scanning calorimeter (DSC). Starch transition temperatures (onset, T_o; peak, T_p; conclusion, T_c) of gelatinization and gelatinization enthalpy (ΔH_{gel}) measured by DSC have been related to degree of crystallinity. The onset temperature reflects the initiation of the gelatinization process, which is followed by a peak and conclusion temperature. After T_c, all amylopectin double helices have dissociated, although swollen granule structures will be retained until more extensive temperature and shear have been applied (Tester and Debon, 2000). A high degree of crystallinity provides structural stability and makes the granule more resistant toward gelatinization, ultimately resulting in higher transition temperatures, and is affected by the chemical composition of the starch (Barichello et al., 1990; Kong et al., 2015).

Starches from different botanical sources differing in composition exhibit different transition temperatures and enthalpies of gelatinization. Singh et al. (2003) reported a lot of research on the gelatinization parameters of starches from various botanical sources (Table 3). Granule shape, percentage of large and small granules, and presence of phosphate esters also affect the gelatinization enthalpy value of various starches (Singh et al., 2006; Yuan et al., 1993). The higher transition temperatures for maize and rice starch could be the result of their more rigid granular structure and the presence of lipids. Research with wheat starch found that the small granules gelatinize at higher temperatures and retrograde at a slower rate than large granules (Ao and Jane, 2007).

The gelatinization and swelling properties are controlled in part by the molecular structure of amylopectin (unit chain length, extent of branching, molecular weight, and polydispersity), starch composition (amylose-to-amylopectin ratio and phosphorus content), and granule architecture (crystalline-to-amorphous ratio) (Tester, 1997; Singh and Singh, 2001; Singh and Singh, 2003; Ai and Jane, 2015). T_p gives a measure of crystallite quality (double-helix length), whereas enthalpy of gelatinization (ΔH_{gel}) gives an overall measure of crystallinity (quality and quantity) and is an indicator of the loss of molecular order within the granule (Tester and Morrison, 1990; Cooke and Gidley, 1992). Gernat et al. (1993) have stated that the amount of double-helical order in native starches is strongly correlated to the amylopectin content and granule crystallinity increases with amylopectin content. This suggests that the ΔH_{gel} should preferably be calculated on an amylopectin basis. However, the ΔH_{gel} for various starches given in Table 3 was not calculated in this manner. Because amylopectin plays a major role in starch granule crystallinity, the presence of amylose lowers the melting point of crystalline regions and the energy for starting gelatinization (Flipse et al., 1996). More energy is needed to initiate melting in the absence of amylose-rich amorphous regions (Krueger et al., 1987). For normal starch, the gelatinization temperature decreases with an increase in amylose content. For high-amylose starch, the B-type crystalline form of amylopectin results in higher gelatinization temperature than normal starch (Richardson et al., 2000a). Amylose double helices also require high temperature and energy to become disordered and therefore lead to a higher gelatinization temperature (Qin et al., 2012).

The potato amylopectin starches exhibit higher endothermic temperatures and enthalpies than the normal potato starches. The amorphous amylose in the normal potato starches decreases the relative amount of crystalline material in the granule, thereby lowering the gelatinization parameters (Svegmark et al., 2002). However, the high-amylose starches with longer average chain length exhibit higher transition temperatures (Jane et al., 1999).

The molecular interactions (hydrogen bonding between starch chains) after cooling of the gelatinized starch paste have been called retrogradation (Hoover, 2001). During retrogradation, amylose forms double-helical associations of 40–70 glucose units (Jane and Robyt, 1984), whereas amylopectin recrystallizes by the association of outermost short branches (Ring et al., 1987). In retrograded starch, the value of enthalpy provides a quantitative measure of the energy transformation that occurs during the melting of recrystallized amylopectin as well as precise measurements of the transition temperatures of the endothermic event (Karim et al., 2000). The endothermic peak of starches after gelatinization and storage at 4 °C appears at lower transition temperatures. Transition temperatures and retrogradation enthalpy (ΔH_{ret}) at the end of the storage period drop down considerably, compared to transition temperatures and enthalpy (ΔH_{gel}) during gelatinization (Table 4). Starch retrogradation enthalpies and transition temperatures are usually 60–80% and 10–26 °C lower, respectively, than those for gelatinization of starch granules (Baker and Rayas-Duarte, 1998). The crystalline forms for retrograded starch are different in nature from those present in the native starch granules and may be weaker than the latter, because recrystallization of amylopectin occurs in a less ordered manner during retrogradation than that in native raw starches.

The extent of decrease in transition temperatures and enthalpy is higher in stored potato starch gels than in the cereal starch gels, which shows its higher tendency toward retrogradation (Singh et al., 2008; Kaur et al., 2007a). The variation in thermal properties of starches after gelatinization and during refrigerated storage may be attributed to the variation in amylose-to-amylopectin ratio and presence/absence of lipids. Amylose content has been reported to be one of the factors that influence starch retrogradation (Kaur et al., 2007a; Fan and Marks, 1998; Shifeng et al., 2009). A greater amount of amylose has traditionally been linked to a greater retrogradation tendency in starches (Whistler and BeMiller, 1996) but amylopectin and intermediate materials also play an important role in starch retrogradation during refrigerated storage (Yamin et al., 1999). The intermediate materials with longer chains than amylopectin may also form longer double helices during reassociation under refrigerated storage conditions. Retrogradation has also been accelerated by amylopectin with longer chain length (Yuan et al., 1993).

3. Potato Starch Modification

Modified starches are native starches that have been altered chemically or physically to improve their functional properties (viscosity, surface activity, enzyme resistance, etc.) for a specific use in the industry (Ortega-Ojeda et al., 2005). Potato starch, like other starches, is modified to overcome limitations of the native starch, such as low shear, acid and thermal resistance, and its high tendency toward retrogradation. The larger size of the native potato starch granules and their high swelling capacity lead to exceptionally large (in volume) swollen granules, which not only result in a high viscosity but also give rise to a less smooth texture. Moreover, the higher fragility of the swollen potato starch granules makes them prone to disperse or solubilize on heating and shearing, resulting in weak-bodied, stringy, and cohesive pastes. The processing of potato starch therefore results in overcooking.

Modification, which involves the alteration of the physical and chemical characteristics of the native potato starch to improve its functional characteristics, can be used to tailor it to specific food applications. The rate and efficacy of any starch modification process depend on the botanical origin of the starch and on the size and structure of its granules. This also includes the surface structure of the granules, which encompasses the outer and inner surface depending on the pores and channels, which cause the development of the so-called specific surface (Juszczak, 2003). Potato starch modification can be achieved in three different ways: physical, conversion, and chemical (derivatization) (Table 6).

Table 6: Some Common Potato Starch Modification Types and Preparation Techniques.

Modification	Types	Preparation
Physical	Heat/moisture/pressure/ sonication treatment	Heat/moisture treatment—heating starch at a temperature above its gelatinization point with insufficient moisture to cause gelatinization
		Annealing—heating a slurry of granular starch at a temperature below its gelatinization point for prolonged periods of time
		High-pressure treatment—treating starch at ultrahigh pressure (above 400 MPa), as an effect of which starch gelatinizes but shows very little swelling and maintains its granular character
		Sonication—this technique applies sounds that are above the human hearing level (20–100 kHz) and can be applied for homogenizing, emulsifying, mixing, extracting, drying, degassing, and cell disruption.
	Pregelatinization	Pregel/instant/cold-water swelling starches prepared using drum drying/spray cooking/extrusion/solvent-based processing
Conversion	Partial acid hydrolysis	Treatment with hydrochloric acid or *ortho*-phosphoric acid or sulfuric acid
	Partial enzymatic hydrolysis	Treatment in an aqueous solution at a temperature below the gelatinization point with one or more food-grade amylolytic enzymes
	Alkali treatment	Treatment with sodium hydroxide or potassium hydroxide
	Oxidation/bleaching	Treatment with peracetic acid and/or hydrogen peroxide, or sodium hypochlorite or sodium chlorite, or sulfur dioxide, or potassium permanganate or ammonium persulfate
	Pyroconversion (dextrinization)	Pyrodextrins—prepared by dry roasting acidified starch
Derivatization	Etherification	Hydroxypropyl starch—esterification with propylene oxide
	Esterification	Starch acetate—esterification with acetic anhydride or vinyl acetate
		Acetylated distarch adipate—esterification with acetic anhydride and adipic anhydride
		Starch sodium octenylsuccinate—esterification by octenylsuccinic anhydride
	Cross-linking	Monostarch phosphate—esterification with *ortho*-phosphoric acid, or sodium or potassium *ortho*-phosphate, or sodium tripolyphosphate
		Distarch phosphate—esterification with sodium trimetaphosphate or phosphorus oxychloride
		Phosphated distarch phosphate—combination of treatments for monostarch phosphate and distarch phosphate
	Dual modification	Acetylated distarch phosphate—esterification by sodium trimetaphosphate or phosphorus oxychloride combined with esterification by acetic anhydride or vinyl acetate
		Hydroxypropyl distarch phosphate—esterification by sodium trimetaphosphate or phosphorus oxychloride combined with etherification by propylene oxide

Reproduced from Singh et al. (2007), with permission from Elsevier.

3.1 Physical Modification

Physically modified potato starch is preferred in processed foods because of its improved functional properties over those of its native counterpart. Moreover, this modification process can be safely used in various food products and other industrial applications. Various physical modification methods of potato starch include annealing (ANN); heat/moisture (HMT), high-pressure (HPT), and osmotic pressure (OPT) treatment; and pregelatinization. Microwave heating, sonication, and irradiation are some other physical modification methods currently employed.

3.1.1 Heat/Moisture/Pressure/Sonication Treatments

ANN represents "physical modification of potato starch slurries in water at temperatures below gelatinization," whereas HMT "refers to the exposure of starch to higher temperatures at very restricted moisture content (18–27%)" (Collado and Corke, 1999). Thus, ANN and HMT are related processes, critically controlled by starch-to-moisture ratio, temperature, and heating time. Both these processes occur at above the glass transition temperature (T_g, a transition of the amorphous regions from a rigid glassy state to a mobile rubbery state, which leads to dissociation of double helices in crystallites) but below the onset of gelatinization (T_o) (Jacobs and Delcour, 1998). In both HMT and ANN, physical reorganization within starch granules is manifested. However, HMT requires higher temperatures to cause this reorganization because of low levels of water in the system, which lead to an elevation of T_g.

HMT leads to an increase in starch gelatinization temperatures and viscosity and narrowing of the gelatinization range ($T_c–T_o$). These properties can be explained on the basis of more glassy amorphous regions within annealed potato starch granules and a more ordered registration of amylopectin double helices restricting the ease of hydration of the starch granules during gelatinization and elevating gelatinization temperatures (Tester and Debon, 2000). Similar effects could be achieved by ANN but because of the lower temperature used, this treatment needs a much longer time; thus, it may have some industrial implications in terms of energy and time. In contrast, the microwave treatment requires a shorter time to modify potato starch characteristics to the same extent as ANN and HMT (Lewandowicz et al., 2000). No major changes occur in granule morphology (size and shape) after these treatments; however, HMT converts the B-type X-ray diffraction pattern of potato starch to the A-type. Examples of HMT of starches include pretreatment of starches for infant foods and processing of potato starch to replace maize starch during shortage, to impart freeze–thaw stability, and to improve its baking quality (Collado and Corke, 1999).

HPT is treating starch at ultrahigh pressure (above 400 MPa), as an effect of which starch gelatinizes but shows very little swelling and maintains its granular character. This results in altered paste and gel properties of the HP-gelatinized starches (Stute et al., 1996). B-type starches such as potato are more pressure resistant and need more pressure to gelatinize completely than the A- and C-type starches (Kudla and Tomasik, 1992). Błaszczak et al. (2005) reported a decrease in transition temperatures after HPT (using 600 MPa for 2 and 3 min) of potato starch. They also observed that the potato starch granule's surface was more resistant to the treatment than its inner part (Figure 5). The inner part of the granule has been reported to be filled with a gel-like network, while the granular form of starch is still retained. Ahmed et al. (2014) studied the effect of HPT on native and modified starches. They observed, through DSC, that the starch gelatinization was pressure dependent. HPT causes reversible hydration of the amorphous phase followed by irreversible distortion of the crystalline region, which destroys the granular structure.

OPT is a new method of physical modification, in which potato starch is suspended in solution saturated with salt such as sodium sulfate and heated (autoclaved) at temperatures above 100 °C for various times. This treatment has been reported to have the same effects on the starch properties as HMT but the starch modified using OPT exhibits better homogeneity (Pukkahuta et al., 2007).

Modification by ultrasound is a technique with growing applications in food processing. This technique applies sounds that are above the human hearing level (20–100 kHz) and can be used for homogenizing, emulsifying, mixing, extracting, drying, degassing, and cell disruption (Chandrapala et al., 2012). A major effect of ultrasound treatment is starch degradation, which is caused by the formation of OH radicals and mechanical damage by the ultrasound (Czechowska-Biskup et al., 2005). Studies have been performed to determine the effects of sonication on starches from varying sources. Sujka and Jamroz (2013) found that the ultrasonic treatment of potato, wheat, corn, and rice starches resulted in depolymerization and changed their functional properties, such as fat and water absorption, least gelation capacity, paste clarity and viscosity, solubility, swelling power, and granule morphology. Luo et al. (2008) studied the properties of waxy and amylo-maize starches treated by ultrasound and also reported that the swelling power, solubility, and granular structure were altered upon ultrasonic treatment. The effect of ultrasound treatment on potato starch granules is shown in Figure 6.

3.1.2 Pregelatinization

Physical modification can be employed to convert native potato starch to cold-water-soluble starch or small-crystallite starch. The process of gelatinization causes changes in both the chemical and the physical nature of granular starch owing to the rearrangement of hydrogen bonding between the water and the starch molecules, resulting in the collapse or disruption of molecular orders within the starch granule (Freitas et al., 2004).

The common methods used in the preparation of this type of starch involve immediate cooking—drying of starch suspensions using drum drying, puffing, continuous cooking–puffing–extruding, and spray drying/injection and nozzle spray drying (Lewandowicz and

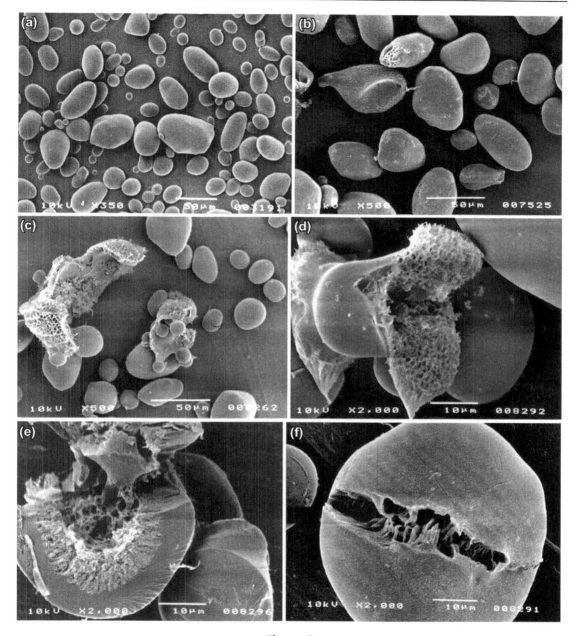

Figure 5

SEM microstructure of potato starch: native (a); treated with high pressure at 600 MPa for 2 min
(b) and 3 min (c); (d)–(f) details of starch structure treated for 3 min. *Reproduced from Błaszczak
(2005), with permission from Elsevier.*

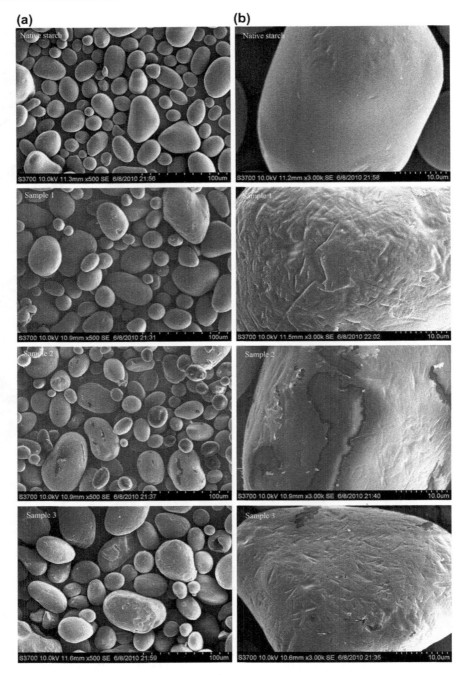

Figure 6
Ultrasound power effect on potato starch granules. Samples 1, 2, and 3 were subjected to 60, 105, and 155 W for 30 min, respectively. *Reproduced from Zhu et al. (2012), with permission from Elsevier.*

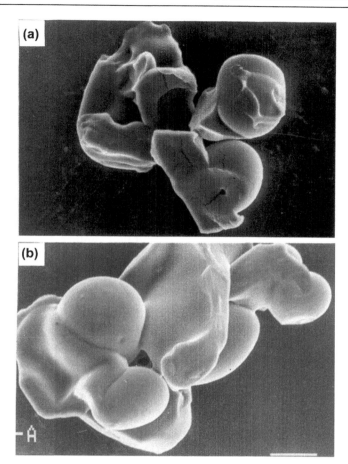

Figure 7
Scanning electron micrographs (a) and (b) showing granular indentation and fragmentation in granular cold-water-soluble starches. *Reproduced from Singh et al. (2003), with permission from Elsevier.*

Soral-Smietana, 2004). Rehydration of drum-cooked starch at room temperature gives a paste of reduced consistency with a dull, grainy appearance and gels of reduced strength (Rajagopalan and Seib, 1992). Slight chemical treatments are also used in the preparation of this type of starch, such as heating of starch in aqueous alcohol solution, alcoholic alkali, or polyhydric alcoholic treatments. The granulated cold-water-soluble starches prepared by these methods exhibit different cold water solubilities, give greater viscosity and smoother texture, and have more processing tolerance than traditional pregelatinized starches. The alcoholic–alkaline treatment causes indentation and distortion of potato starch granular structure (Figure 7). The use of pregelatinized starches includes instant starch products, such as low-fat salad dressings, high-solid fillings, bakery fillings, and dry mixes (Mason, 2009).

3.2 Conversion

3.2.1 Acid/Enzymatic Hydrolysis

The acid-thinned potato starch is normally prepared by controlled hydrolysis of native starch with mineral acid, either at room temperature (for several days) or at elevated temperature (but below gelatinization temperature, for several hours). The derived degradation products (glucose, high-maltose syrups, and maltodextrins) have reduced hot-paste viscosity (HPV), increased solubility and gel strength, and wide applications in the food, paper, and textile industries. Acid modification can be achieved by both dry and wet processes. The dry process involves dry roasting of starch in the presence of limited moisture and certain levels of acid. During the dry roasting process, hydrolysis of the glucosidic bond occurs along with molecular rearrangement that leads to the formation of dextrins. Dextrins are more soluble, have lower viscosity than the wet-hydrolysis products, and find applications in adhesives, gums, and pastes. In wet processes, acids are used at 1–3% concentration, based on starch dry solids. The type of acid used for hydrolysis greatly influences the molecular weight, alkali fluidity number, iodine-binding capacity, and intrinsic viscosity of the degradation product. Molecular weight of starch decreases after modification, with *ortho*-phosphoric acid causing the least and hydrochloric and nitric acids the highest reduction (Singh and Ali, 2000).

Acid-modified starch can be used at higher solids concentration for immediate gelling and provides gum or jelly with shorter texture and flexible properties (Zallie, 1988). In acid modification, the hydroxonium ion attacks the glucosidic oxygen atom and hydrolyzes the glucosidic linkage. Acid depolymerization occurs first in the amorphous regions before hydrolysis of the crystalline regions of the starch granule. Also, acid attacks the granule surface first before entering the granule interior (Wang and Wang, 2001). Because of some of the problems encountered during acid hydrolysis, such as random attack at the branch point (which could lead to an increase in linearity of starch), high glucose yield, and acid removal later on, enzymatic hydrolysis is preferred over acid hydrolysis. Acid modification in the presence of long-chain alcohols helps to reduce the degree of polymerization of amylopectin more effectively and also convert the crystalline regions into more amorphous regions that are prone to acid hydrolysis. Nageli dextrins, lintner starch, etc., are some of the acid-modified starch derivatives.

Enzymatic modification of potato starch on an industrial scale is generally carried using starch-hydrolyzing enzymes such as α-amylase, β-amylase, pullulanase, glucoamylase, and isoamylase (van der Maarel et al., 2002). These enzymes hydrolyze the α-1,4 or α-1,6 glycosidic bonds in amylose and amylopectin by first breaking the glycosidic linkage and subsequently using a water molecule as the acceptor substrate. Maltodextrin properties are classified according to the dextrose equivalent as a measure of degradation and the degree of polymerization. Their solubility varies with dextrose equivalent and with the type of hydrolysis. Enzymatically produced maltodextrins contain fewer high-molecular-weight saccharides and are therefore more water soluble than equivalent products of acid hydrolysis

(Kennedy et al., 1995). Another group of enzymes, such as cyclodextrin glycosyltransferase and amylomaltase, modify starch by using the transferase reaction. These enzymes initially break glycosidic linkages but use another oligosaccharide as the acceptor substrate instead of water and form a new glycosidic linkage. Cyclodextrins, cyclic oligosaccharides composed of six, seven, or eight glucose units linked by an α-1,4 linkage, are produced by enzymatic hydrolysis followed by enzymatic conversion by the action of cyclodextrin glycosyltransferase. These cyclodextrins are called α-, β-, and γ-cyclodextrins, respectively (French et al., 1954). These cyclodextrins have a polar hydrophilic exterior and an apolar hydrophobic interior, which makes them suitable candidates for forming inclusion complexes with hydrophobic molecules of suitable dimension and configuration (Tharanathan, 2005). This property is made use of in food, pharmaceutical, and agrochemical industrial applications in masking unpleasant odors and flavors, removal of undesirable components by their inclusion in the cyclodextrin cavity, etc. A novel thermoreversible gelling product has been prepared by hydrolysis of potato starch using amylomaltase, which may be a good plant-derived substitute for gelatin (van der Maarel et al., 2005; Kurkov and Loftsson, 2013).

3.2.2 Oxidation/Bleaching

Bleached or oxidized starches are produced by reacting starch with a specified amount of reagent under controlled temperature and pH. The starches treated with low levels of reagents, such as hydrogen peroxide, peracetic acid, ammonium persulfate, and sodium hypochlorite, in an aqueous slurry are referred to as bleached starches.

The reagents used to oxidize starch include periodate, chromic acid, permanganate, nitrogen dioxide, and sodium hypochlorite. The pH, temperature, reagent concentration, and starch molecular structure are the main factors controlling oxidation. During oxidation, hydroxyl groups on starch (at C-2, C-3, and C-6 positions) are first oxidized to carbonyl and then to carboxyl groups. These bulky carboxyl groups result in low retrogradation of oxidized starch paste (Wurzburg, 1986a). The number of carboxyl and carbonyl groups on starch determines the extent of modification. Oxidized starches show reduced viscosity, transition temperatures, and enthalpies of gelatinization and retrogradation (Adebowale and Lawal, 2003). Fonseca et al. (2015) modified potato starch with various sodium hypochlorite concentrations and studied its effect on biodegradable film formation. They reported that different hypochlorite concentrations of active chlorine affect the characteristics of potato starches differently. The increase in active chlorine concentration increased the intensity of oxidation of the potato starches. The films produced with oxidized starches had different properties depending on the degree of oxidation.

Oxidation also results in starch depolymerization, which is the cause of low viscosity (Figure 8) and improved clarity and stability exhibited by oxidized starches. Oxidized starches are used in foods as coating and sealing agents in confectionery, as emulsifiers, and as a dough conditioner for bread, whereas bleached starches are used for improved adhesion of batter and breading mixes in fried foods.

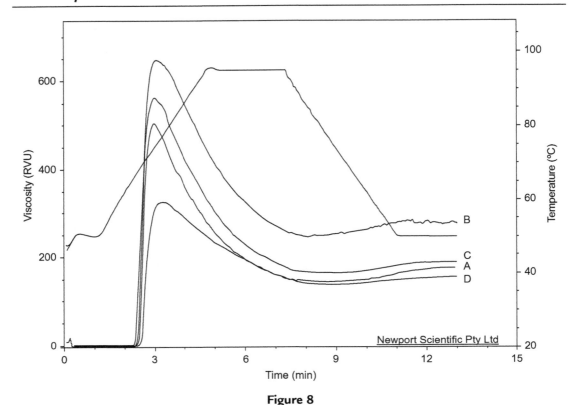

Figure 8

RVA curves of potato starches: native starch (a); starch oxidized with 0.5 g/100 g of active chlorine (b); starch oxidized with 1.0 g/100 g of active chlorine (c); starch oxidized with 1.5 g/100 g of active chlorine (d). *Reproduced from Fonseca et al. (2015), with permission from Elsevier.*

3.3 Derivatization

Derivatization is the most widely employed method for industrial-scale starch modification; therefore we will discuss this modification in more detail.

3.3.1 Common Types of Derivatization

Derivatization involves the introduction of functional groups into starch resulting in markedly altered physicochemical properties. The properties of some modified starches and their applications are presented in Table 7. The chemical and functional properties achieved when modifying starch by chemical substitution depend, inter alia, on starch source, reaction conditions (reactant concentration, reaction time, pH, and the presence of a catalyst), type of substituent, extent of substitution (degree of substitution, DS[1], or molar

---

[1]   DS represents the average number of hydroxyl groups on each anhydroglucose unit, which are derivatized by substituent groups. DS is expressed as average number of moles of substituent per anhydroglucose unit. As each anhydroglucose unit has three hydroxyl groups available for substitution, the maximum possible DS is 3.

Table 7: Some Properties and Applications of Modified Starches.

Types	Properties	Applications
Pregelatinization	Cold water dispersibility	Useful in instant convenience foods
Partial acid or enzymatic hydrolysis	Reduced-molecular-weight polymers, exhibit reduced viscosity, increased retrogradation and setback	Useful in confectionery, batters, and food coatings
Oxidation/bleaching	Low viscosity, high clarity, and low temperature stability	Used in batters and breading for coating various food stuffs, in confectionery as binders and film formers, in dairy as texturizers
Pyroconversion (dextrinization)	Low to high solubility depending on conversion, low viscosity, high reducing sugar content	Used as coating materials for various foods, have good film-forming ability, and used as fat replacers in bakery and dairy products
Etherification	Improved clarity of starch paste, greater viscosity, reduced syneresis and freeze–thaw stability	Used in wide range of food applications such as gravies, dips, sauces, fruit pie fillings, and puddings
Esterification	Lower gelatinization temperature and retrogradation, lower tendency to form gels and higher paste clarity	Used in refrigerated and frozen foods, as emulsion stabilizers and for encapsulation
Cross-linking	Higher stability of granules toward swelling, high temperature, high shear, and acidic conditions	Used as viscosifiers and texturizers in soups, sauces, gravies, bakery and dairy products
Dual modification	Stability against acid, thermal, and mechanical degradation and delayed retrogradation during storage	Used in canned foods, refrigerated and frozen foods, salad dressings, puddings, and gravies

Reproduced from Singh et al. (2007), with permission from Elsevier.

substitution, MS^2), and distribution of the substituent in the starch molecule (Rutenberg and Solarek, 1984; Kavitha and BeMiller, 1998; Richardson et al., 2000b; Luo and Shi, 2012; Colussi et al., 2015).

Some chemical modification reactions of starch are presented in Table 8. The commonly employed methods of modifying potato starch for use in the food industry include etherification, esterification, and cross-linking. Acetylated potato starch with a low DS is commonly obtained by the esterification of native starch with acetic anhydride in the presence of an alkaline catalyst, whereas the food-grade hydroxypropylated starches are generally prepared by etherification of native starch with propylene oxide in the presence of an alkaline catalyst. Food-grade starches are hydroxypropylated to increase paste consistency and clarity and to

---

[2] The substituent moiety on some starch esters and ethers can react further with the modifying reagent during the modification reaction, resulting in the formation of an oligomeric or possibly polymeric substituent. In these cases MS is preferred, which represents the level of substitution as moles of monomeric substituent per mole of anhydroglucose unit. Thus, in contrast to DS, the value of MS can be greater than 3.

Table 8: Some Common Starch Chemical Modification Reactions.

Modification Type	
	Etherification
Hydroxyalkyl starch (with alkylene oxide)	$St-OH + H_2C-\overset{\displaystyle R}{\underset{\displaystyle O}{\diagdown}}C \xrightarrow{NaOH} St-O-CH_2-\overset{\displaystyle R}{\underset{\displaystyle OH}{\mid}}$
	Esterification
Starch acetate (with vinyl acetate)	$St-OH + CH_2=CH-\overset{O}{\overset{\|}{C}}-CH_3 \xrightarrow{NaOH} St-O-\overset{O}{\overset{\|}{C}}-CH_3 + CH-C-OH$
Starch acetate (with acetic anhydride)	$St-OH + CH_3-\overset{O}{\overset{\|}{C}}-O-\overset{O}{\overset{\|}{C}}-CH_3 \xrightarrow{NaOH} St-O-\overset{O}{\overset{\|}{C}}-CH_3 + CH-C-OH$
Starch phosphate (with *ortho*-phosphates)	$St-OH + NaH_2PO_4/Na_2HPO_4 \longrightarrow St-O-\overset{O}{\overset{\|}{P}}-\overset{\bar O Na^+}{\underset{OH}{\mid}}$
Carboxymethyl starch (with monochloroacetic acid)	$St-OH + Cl.CH_2COOH \longrightarrow \overset{ONa}{\underset{St-O-CH_2}{C=o}}$
	Cross-linking
With POCl$_3$	$\overset{O}{\overset{\|}{\underset{Cl\ Cl\ Cl}{P}}} + StOH \xrightarrow{NaOH} St-O-\overset{O}{\overset{\|}{\underset{ONa}{P}}}-O-St + NaCl$
With STMP	$2StOH + Na_3P_3O_9 \xrightarrow{Alkali\ catalyst} St-O-\overset{O}{\overset{\|}{\underset{ONa}{P}}}-O-St$
With EPI	$2StOH + EPI \xrightarrow{Alkali\ catalyst} St-O-CH_2-\overset{OH}{\underset{\mid}{CH}}-CH_2-O-St$

St, starch; POCl$_3$, phosphorus oxychloride; STMP, sodium trimetaphosphate; EPI, epichlorohydrin.
Reproduced from Singh et al. (2007), with permission from Elsevier.

impart freeze–thaw and cold-storage stabilities (Xu and Seib, 1997). The hydrophilic hydroxypropyl groups introduced into the starch chains weaken the granular structure of starch by disrupting inter- and intramolecular hydrogen bonding, leading to an increase in accessibility of the starch granules to water. Chemical modification of native granular starches by etherification also alters the gelatinization and retrogradation behavior of starches. Hydroxypropyl starch finds applications as an instant starch in soups and sauces, cooked meat products, fillings for confectionery, ready-to-eat desserts, canned fruits and jams, and frozen food preparations.

Cross-linking treatment is aimed to add chemical bonds at random locations in a granule, which stabilize and strengthen the granule. Starch pastes from cross-linked potato starches are more viscous, heavily bodied, and less prone to break down with extended cooking times, increased acid content, or severe agitation. Cross-linking minimizes granule rupture, loss of viscosity, and the formation of stringy paste during cooking (Woo and Seib, 1997; Kaur et al., 2006; Prochaska et al., 2009), yielding potato starch that is suitable for canned foods and other food applications (Rutenberg and Solarek, 1984). Nutritional benefits of cross-linked starch as a new source of dietary fiber have also been reported (Wurzburg, 1986c; Woo, 1999).

Cross-linking is generally performed by treatment of granular potato starch with multifunctional reagents capable of forming either ether or ester intermolecular linkages between hydroxyl groups on starch molecules (Rutenberg and Solarek, 1984; Wurzburg, 1986b). Sodium trimetaphosphate (STMP), monosodium phosphate, sodium tripolyphosphate (STPP), epichlorohydrin (EPI), phosphoryl chloride ($POCl_3$), a mixture of adipic acid and acetic anhydride, and vinyl chloride are the main agents used to cross-link food-grade starches (Yeh and Yeh, 1993; Woo and Seib, 1997; Gui-Gie et al., 2006). $POCl_3$ is an efficient cross-linking agent in aqueous slurry at $pH > 11$ in the presence of a neutral salt (Felton and Schopmeyer, 1943). STMP is reported to be an efficient cross-linking agent at high temperature with semidry starch and at warm temperature with hydrated starch in aqueous slurry (Kerr and Cleveland, 1962; Koo et al., 2010; Wongsagonsup et al., 2014). EPI is poorly soluble in water and partly decomposes to glycerol; thus water-soluble cross-linking agents such as $POCl_3$ and STMP are preferred. Moreover, it has also been reported that EPI cross-links are likely to be less uniformly distributed than STMP cross-links (Shiftan et al., 2000). So, the type of cross-linking agent greatly determines the change in functional properties of the treated starches. Cross-linking at low levels, although having a substantial effect on potato starch properties such as granule swelling (Rutenberg and Solarek, 1984), contributes little to water sorption properties (Chilton and Collison, 1974).

Starch phosphates that are conventionally prepared have been reported to give clear pastes of high consistency with good freeze–thaw stability and emulsifying properties, and may be grouped into two classes: monostarch phosphates and distarch phosphates (cross-linked starches). In general, monostarch phosphates (monoesters) can have a higher DS than distarch phosphates (diesters) because even a very few cross-links (in the case of diesters) can drastically alter the paste and gel properties of the starch. Starch phosphates are conventionally prepared through the reaction of potato starch with salts of *ortho-*, *meta-*, pyro-, and tripolyphosphoric acids and phosphorus oxychloride (Paschall, 1964; Nierle, 1969).

Dual modification, a combination of substitution and cross-linking, has been demonstrated to provide stability against acid, thermal, and mechanical degradation of starch and to delay retrogradation during storage. Dual-modified starches are used commonly in salad dressings, canned foods, frozen foods, and puddings. The control of reaction conditions is important during preparation of dual-modified cross-linked/hydroxypropylated starches using different cross-linking reagents with different starch bases such as maize, tapioca, wheat, waxy maize, waxy barley, rice, sago, and potato (Tessler, 1975; Wu and Seib, 1990; Yeh and Yeh, 1993; Yook et al., 1993; Wattanchant et al., 2003; Zhao et al., 2012; Cui et al., 2014; Liu et al., 2014; Shukri et al., 2015). The quantity of cross-linking reagent required to prepare a dual-modified starch (with desirable properties) varies with the source of starch, the type of cross-linking reagent, the efficiency of the cross-linking reaction, the DS required, and the specified range of final modified-starch properties. The various reaction conditions, such as starch base concentration, temperature, pH, and concentration of the catalyst salt, play important roles during preparation of dually modified hydroxypropylated cross-linked starch. The effects of chemical modifications on the thermal, morphological, and pasting/rheological behavior of starches may be quantified using instrumentation such as DSC, SEM, Visco-Amylo-Graph/RVA, and dynamic rheometer (Kaur et al., 2006; Singh et al., 2007; Zhao et al., 2015).

3.3.2 Extent of Derivatization

The rate and efficiency of this chemical modification process depends on the reagent type, botanical origin of the starch, and size and structure of its granules (Huber and BeMiller, 2001; Zhao et al., 2015). This also includes the surface structure of the starch granules, which encompasses the outer and inner surface, depending on the pores and channels, leading to the development of the so-called specific surface (Juszczak, 2003). Channels that open to the granule exterior provide a much larger surface area accessible by chemical reagents and provide easier access by the reagents to the granule interior. However, the reagent may diffuse through the external surface to granule matrix in the absence of channels (BeMiller, 1997). Although starches from various sources exhibit fundamental structural similarities, they differ in the specific details of their microstructure and ultrastructure. These structural differences affect the chemical modification process to a great extent. According to Zhao et al. (2012), the hydroxypropyl substituents are located in the amorphous lamellae (branched zone) and amorphous regions between clusters of side chains in crystalline lamellae, because the amorphous lamellae are flexible and accessible to chemical agents. Internal side chains of the amylopectin molecule, located in the compact interior of the crystalline lamellae, are not (or are less abundantly) substituted with hydroxypropyl groups, since the hydroxypropylation reagent cannot penetrate here.

Table 9: DS[a] and MS[b] of Some Modified Starches from Various Botanical Sources.

Starch Source	DS (Acetylated)	MS (Hydroxypropylated)	DS (Cross-linked)
Normal potato	0.115–0.238[d]	0.098–0.122[h]	0.07–0.26[k]
Normal maize	0.104–0.184[d]	0.091–0.092[i]	0.09–0.25[k]
Normal maize	ND[c]	0.061–0.094[j]	ND
Waxy maize	0.081	0.067–0.127[j]	ND
High-amylose maize	ND	0.078–0.119[j]	ND
Hybrid normal maize	0.030–0.040[e]	ND	ND
Normal wheat	0.035–0.131[f]	0.117–0.123[i]	0.004–0.020[l]
Normal rice	0.087–0.118[g]	>0.03[j]	0.025–0.035[m]
Normal rice	0.018	ND	ND
Waxy rice	0.016	ND	ND

[a]DS = degree of substitution.
[b]MS = molar substitution.
[c]ND = Not detected.
[d]Different levels of acetylation and starches from different potato cultivars.
[e]Two levels of acetylation used.
[f]Different levels of acetylation used.
[g]Starches from different rice cultivars.
[h]Starches from different potato cultivars.
[i]Low MS values from two populations of starch granules.
[j]Two levels of hydroxypropylation.
[k]DS of starches phosphorylated using a mixture of monosodium and disodium phosphate.
[l]MS values; three levels of cross-linking performed.
[m]Different levels of cross-linking performed.
Reproduced from Singh et al. (2007), with permission from Elsevier.

The DS and MS of some chemically modified starches prepared from various sources are presented in Table 9. Potato, maize, and rice starches show significant variation in their DS when acetylated under similar reaction conditions (Singh et al., 2004a,b). Factors such as amylose-to-amylopectin ratio, intragranule packing, and the presence of lipids mainly govern the DS during acetylation of starches from different sources (Phillips et al., 1999; Bello-Pérez et al., 2010; Colussi et al., 2014).

The C=O bond of the acetyl group experiences a different molecular environment depending on whether it is a substituent on amylose or on amylopectin (Phillips et al., 1999). Acetylation occurs in all the amorphous regions and at the outer lamellae of crystalline regions, rather than throughout the crystalline regions of the whole starch granule, owing to poor penetrating ability of acetic anhydride in starch granules (Chen et al., 2004). Studies (Singh et al., 2004a) on acetylated potato starches suggested that the small-size granule population with lower amylose content favors the introduction of acetyl groups and hence results in higher DS. Colussi et al. (2014) studied rice starch with various levels of amylose and observed that low-amylose rice starch was more susceptible to acetylation compared to the medium- and high-amylose rice starches. The authors also found that the higher the DS, the lower the starch crystallinity and, in general, its pasting temperature, breakdown, peak and final viscosities, swelling power, and solubility.

During hydroxypropylation, the hydroxypropyl groups are primarily introduced into the starch chains in the amorphous regions composed mainly of amylose (Steeneken and Smith, 1991). The hydroxypropyl groups are hydrophilic in nature and, when introduced into starch chains, they weaken or disrupt the internal bond structure that holds the granules together and thus influence the physicochemical properties depending on the source of starch, reaction conditions, type of substituent groups employed, and extent of MS (Schmitz et al., 2006). Amylose is modified to a greater extent than amylopectin in hydroxypropylated maize and potato starches, and the modification of amylopectin occurs close to the branch points because of the higher accessibility of the amorphous regions to the modifying reagent (Kavita and BeMiller, 1998). However, Richardson et al. (2001), after investigating the substituent distribution in hydroxypropylated potato amylopectin starch (PAP), suggested that the hydroxypropyl groups are homogeneously distributed on the amylopectin molecule. The modification reaction conditions and starch source may also affect the distribution of hydroxypropyl groups along the starch chain (Steeneken and Woortman, 1994; Kaur et al., 2004). Investigations carried out on hydroxypropylated PAP prepared in granular slurry or solution suggest that more substituents were located in close vicinity to branching points, which constitute the amorphous areas in the semicrystalline granule, than elsewhere. PAP hydroxypropylated in granule slurry had a more heterogeneous substituent distribution compared with starch modified in a polymer "solution" of dissolved starch (Richardson et al., 2001). The reactivity and concentration of reagents influence the DS of cross-linked starches. Also, the type of reagent used and the reaction conditions determine the ratio of mono- and di-type bonds (esters with phosphorus-based agents and glycerols with EPI) during cross-linking (Koch et al., 1982).

Characterization of substitution upon modification is important at both monomer and polymer levels (Richardson et al., 2003). The distribution of substituents at both monomer and polymer levels may be affected by the presence of granule pores and channels, the proportions of amylose and amylopectin and their arrangement, the nature of the granule surface, and granule swelling (BeMiller, 1997; Kavitha and BeMiller, 1998). Techniques such as nuclear magnetic resonance spectroscopy (Xu and Seib, 1997; Heins et al., 1998) or gas chromatography/mass spectrometry (Wilke and Mischnick, 1997; Richardson et al., 2000b) may be helpful for the determination of the distribution of the substituent groups at the monomeric level. The methods used for the analysis of the substituent distribution along the polymer chains and the homogeneity/heterogeneity of substitution are based on partial degradation of the polymer by acid hydrolysis (Arisz et al., 1995; Mischnick and Kuhn, 1996) or by enzymatic degradation (Steeneken and Woortman, 1994; Wilke and Mischnick, 1997; van der Burgt et al., 1998; Kavitha and BeMiller, 1998).

3.3.3 Morphological Properties of Derivatized Potato Starch

Potato starch modification involves physical, chemical, and biochemical phenomena on the surface of contacting phases. Microscopy (light and SEM) has played an important role in increasing the understanding of the granular structure of modified starches. It has been used to detect structural changes caused by chemical modifications and the most substituted regions in starch granules (Kim et al., 1992; Kaur et al., 2004). Most of the structural changes upon hydroxypropylation take place at the relatively less organized central core region of the starch granule, i.e., where the hydroxypropyl groups are most densely deposited (Kim et al., 1992; Ratnayake and Jackson, 2008).

The "pushing-apart effect" exerted by the bulky hydroxypropyl groups, especially in the central region of the granule, might lead to an alteration in granule morphology upon hydroxypropylation. Another possible explanation is that the starch granule itself is not structurally homogeneous from a physical and chemical point of view, since it has different physical natures (amorphous and crystalline regions) as well as different chemical compositions in each region (French, 1984). The treatment of potato starch granules with propylene oxide (10%, dry weight basis) alters granule morphology (Figure 9(a) and (b)). Many of the less affected modified granules developed a depression that resulted later in slight fragmentation, indentation, and the formation of a deep groove in the central core region along the longitudinal axis in highly affected granules. These granules appeared as folded structures with their outer sides drawn inward, giving the appearance of a doughnut (Figure 10(a) and (b)). Moreover, these altered regions were apparently larger in large granules compared with small granules in all the potato starches examined; this may be attributed to differences in the native granule architecture and fragility.

The peripheral regions and also the outer layer of the less affected starch granules remained unaltered, and the changes remained confined to the central core regions. By contrast, highly affected starch granules developed a blister-like appearance, cracks and small protuberances on their surfaces, and a deep groove in the central core region; this suggests that the granule

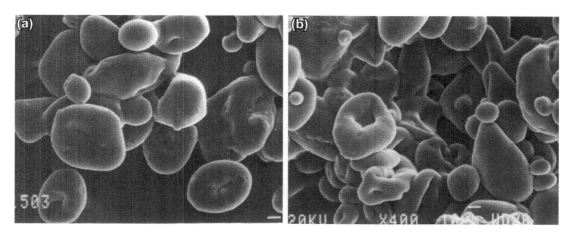

Figure 9
Effects of hydroxypropylation on the granule morphology of potato starches. (a) Potato starch granules after hydroxypropylation (at 10% propylene oxide concentration). (b) Effect of increased concentration of propylene oxide (15%) on the starch granule structure. *Reproduced from Kaur et al. (2004), with permission from Elsevier.*

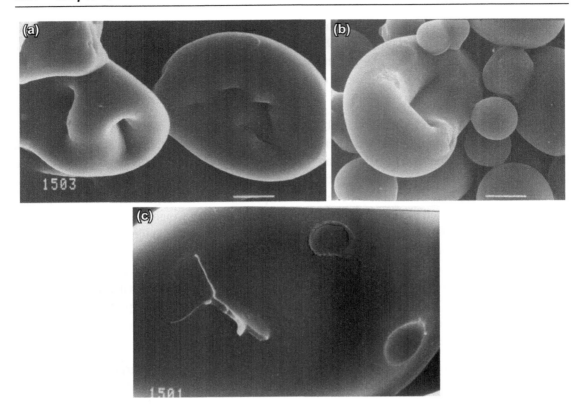

Figure 10

Effects of hydroxypropylation on the granule morphology of potato starches (at 10% propylene oxide concentration). (a and b) Formation of a deep groove in the central core region, folding of starch granules, and formation of a doughnut like appearance. (c) Formation of blister-like appearance and cracks on the starch granules. *Reproduced from Kaur et al. (2004), with permission from Elsevier.*

peripheral regions may be the last to be modified (Figure 10(c)) (Kaur et al., 2004). The granule structure is more substantially altered when the reaction is carried out with a higher concentration of propylene oxide (15% compared with 10%). Highly affected granules appear as if gelatinized, having lost their boundaries and fused together to form a gelatinized mass (Figure 9(c)). This effect was again more pronounced in larger granules (Kaur et al., 2004).

Huber and BeMiller (2001) also reported that the material within the inner regions of potato starch granules was more susceptible to reaction (with propylene oxide) than that in the outer granule layers. Also, as potato starch granules do not possess channels, the reagent diffuses inward through the exterior granule surface. Propylene oxide, being less reactive than other modification reagents, may diffuse into the granule matrix prior to reacting (Huber and BeMiller, 2001). The structural changes in hydroxypropylated starch granules become more evident with increasing MS (Kim et al., 1992; Kaur et al., 2004). Changes in appearance of

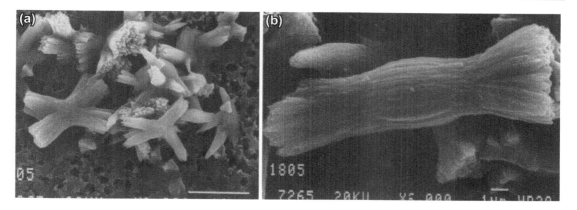

Figure 11

(a) and (b) Rod-shaped microfibrils (showing longitudinal arrangement of crystalline structures) formed during long-term storage of hydroxypropylated gelatinized starch pastes. *Reproduced from Kaur et al. (2004), with permission from Elsevier.*

starch granules after hydroxypropylation have been reported for potato (Kim et al., 1992) and cassava (Jyothi et al., 2007) starches. However, no change in surface or shape characteristics of the granules was reported for canna (Chuenkamol et al., 2007) and pigeon pea (Lawal, 2011) starches. The different observations may be due to different levels of modification, different preparation methods, and the morphologies of the different starches used for the hydroxypropylation process.

Hydroxypropylation leads to alteration in the structure of the gelatinized potato starch gels also, which may be attributed to a weakened granule structure, leading to increased granule disruption during gelatinization. Hydroxypropylated potato starch gels showed extensive aggregation of granule remnants. Kaur et al. (2004) have also reported the appearance of numerous rod-shaped fuzzy clustered microfibrils in potato hydroxypropylated starch gels (after 30 days storage at 4 °C) that could be easily distinguished from the other starchy material (Figure 11). The extensive phase separation occurring during long-term storage at 4 °C may be responsible for the formation of these rod-shaped microfibrils. Slow cooling has also been reported to enhance the formation of nonspherulitic morphologies in starch gels (Nordmark and Ziegler, 2002).

Acetylation treatment has also been found to alter the granule morphology, although to a lesser extent. Singh et al. (2004a) carried out acetylation of maize and potato starches using various concentrations of acetic anhydride and reported that the acetylation treatment caused granule fusion in both maize and potato starches (Figure 12). The granule fusion was also observed to be more pronounced in potato starches with small granules (Singh et al., 2004b). Maize starch granules show a higher resistance toward acetylation treatment than potato starches. The addition of 4% acetic anhydride resulted in the fusion of granules in potato

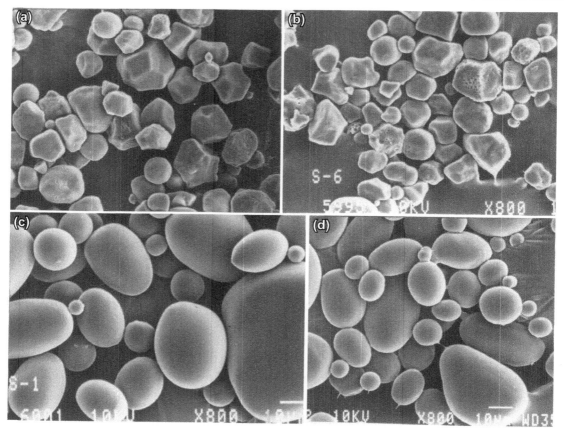

Figure 12

Effects of acetylation on the granule morphology of maize and potato starches. (a) Native maize starch granules. (b) Acetylated maize starch granules. (c) Native potato starch granules. (d) Acetylated potato starch granules. *Reproduced from Singh et al. (2004a), with permission from Elsevier.*

starches, while maize starch granules tended to fuse at concentrations of acetic anhydride of 8% or higher (Singh et al., 2004a). The granule surface of maize and potato starch granules was observed to become slightly rough upon acetylation (Singh et al., 2004b). Gonzalez and Perez (2002) and Sha et al. (2012) verified that the granule surface of rice starches becomes rough and the granules tend to form aggregates upon acetylation; however, other authors (Sodhi and Singh, 2005; Colussi et al., 2015) reported no significant differences between the external morphologies of native and acetylated rice starches.

Some granule fusion and deformation, however, and the rough appearance of the granule surface in the acetylated potato and maize starches might also be the result of surface gelatinization upon addition of NaOH to maintain alkaline conditions during acetic anhydride addition (Singh et al., 2004b). Sha et al. (2012) showed that the granule surface of acetylated starch was less smooth than that of the native starch, but the starch granules still kept a relatively complete particle structure. As the degree of acetylation increased, the intermolecular hydrogen bonds were damaged and starch granule structure was more disrupted.

Phosphorylated potato starches can be prepared using a mixture of mono- and disodium phosphate. The starch granule size is very responsive to the changes in the DS by phosphate groups: the higher the DS, the larger the granule size. The average increase in granule size has been observed to be higher for rice and maize starches than for potato starch for a given DS. The increase may be due to the substitution by phosphate groups inside the modified granules, building certain repulsive forces that increase the sizes of inter- and intramolecular spaces, allowing more water molecules to be absorbed. Phosphorylation does not cause any detectable change in the general granule shape (Sitohi and Ramadan, 2001). After being cross-linked using EPI and $POCl_3$, potato starch granules remain smooth and similar to native starch granules in morphology when viewed under SEM, suggesting that the modification does not cause any detectable morphological change (Kaur et al., 2006). Some chemical modifications of starch have been reported to increase granule surface area and granule porosity (Fortuna et al., 1999).

Studies on waxy maize starches using different reagents ($POCl_3$ and propylene oxide) reported variations in the relative reaction patterns (Huber and BeMiller, 2001). Owing to the higher reactivity of $POCl_3$, the cross-links predominate on the granule surfaces when using highly reactive $POCl_3$ (Gluck-Hirsch and Kokini, 1997; Huber and BeMiller, 2001), while the less reactive propylene oxide generally diffuses into the granule matrix prior to reaction (Huber and BeMiller, 2001; Kaur et al., 2004). Shiftan et al. (2000) reported that EPI cross-linking is not homogeneous and is concentrated in the noncrystalline domain of starch granules. Another factor that may influence the extent of cross-linking is the size distribution of starch granule population (Hung and Morita, 2005). During cross-linking small granules have been reported to be derivatized to a greater extent than the large granules (Bertolini et al., 2003).

3.3.4 Physicochemical Properties of Derivatized Potato Starch

The physicochemical properties of starches such as swelling, solubility, and light transmittance have been reported to be affected considerably by derivatization (Table 10). The change in these properties upon modification depends on the type of chemical modification. Chemical modifications, such as acetylation and hydroxypropylation, increase, while cross-linking has been observed to decrease (depending on the type of cross-linking agent and degree of cross-linking), the swelling power and solubility of starches from various sources. The introduction of (bulky) acetyl groups into starch molecules by acetylation leads to structural reorganization owing to steric hindrance; this results in repulsion between starch molecules, thus facilitating an increase in water percolation within the amorphous regions of granules and a consequent increase in swelling capacity (Lawal, 2004). Studies conducted on acetylated potato starches (Singh et al., 2004a,b) suggest a

Table 10: Swelling Power of Some Modified Starches from Various Botanical Sources.

Starch Source	SP[a] (g/g) (Acetylated)	SP[a] (g/g) (Hydroxypropylated)	SP[a] (g/g) (Cross-linked)
Normal potato	60–71 (Native 58–70)[c,d]	33–39 (Native 28–31)[e]	20–25 (Native 28–31)[h]
Normal potato	ND[b]	ND	23–27 (Native 28–31)[i]
Normal maize	38 (Native 36)	≈20 (Native ≈ 6)[f]	ND
Waxy maize	ND	≈42 (Native ≈ 30)[f]	ND
Normal wheat	ND	9–16 (Native ≈ 6–8)[g]	ND
Normal rice	15–19 (Native 14–18)[e]	ND	≈9 (Native ≈ 18)[e]
Waxy rice	≈31 (Native ≈ 41)	ND	≈14 (Native ≈ 41)[e]

[a]SP = Swelling power (g/g).
[b]Not detected.
[c]The properties of corresponding native (unmodified) starches are given in parentheses.
[d]Starches from different potato cultivars.
[e]Starches from different rice cultivars.
[f]Two levels of hydroxypropylation.
[g]Two wheat starch granule populations.
[h]Starches from different potato cultivars, cross-linking performed using POCl$_3$.
[i]Starches from different potato cultivars; cross-linking performed using EPI.
Reproduced from Singh et al. (2007), with permission from Elsevier.

significant increase in swelling power and solubility upon acetylation in all these starch types. The extent of this increase was observed to be higher for potato starches. The DS introduced after acetylation mainly affects the intensity of change in the swelling power and solubility of starches. The structural disintegration caused by acetylation probably weakens the starch granules, and this enhances amylose leaching from the granule, thus increasing starch solubility (Lawal, 2004; Colussi et al., 2014). Liu et al. (1999b) reported that the waxy starches showed an increased swelling power upon acetylation because of the presence of mainly amylopectin with a more open structure than in nonwaxy starch; this allows rapid water penetration and increased swelling power and solubility. Swelling power and solubility of hydroxypropylated potato starches generally increase with an increase in MS. The decrease in associative forces within the starch granule due to hydroxypropylation may result in an increased penetration of water during heating that leads to the greater swelling (Kaur et al., 2004).

Cross-linking strengthens the bonding between potato starch chains, causing an increase in resistance of the granules to swelling with increasing degree of cross-linking. Higher concentrations of fast-acting cross-linking reagents such as POCl$_3$ result in greater reductions in the swelling potential compared with slower acting agents such as EPI (Gluck-Hirsch and Kokini, 2002). A large concentration of POCl$_3$ cross-links at the surface of the granule causes the formation of a hard outer crust that restricts granule swelling (Huber and BeMiller, 2001). Inagaki and Seib (1992) also reported that the swelling power of cross-linked waxy barley starch declined as the level of cross-linking increased. Cross-linked starches exhibit lower

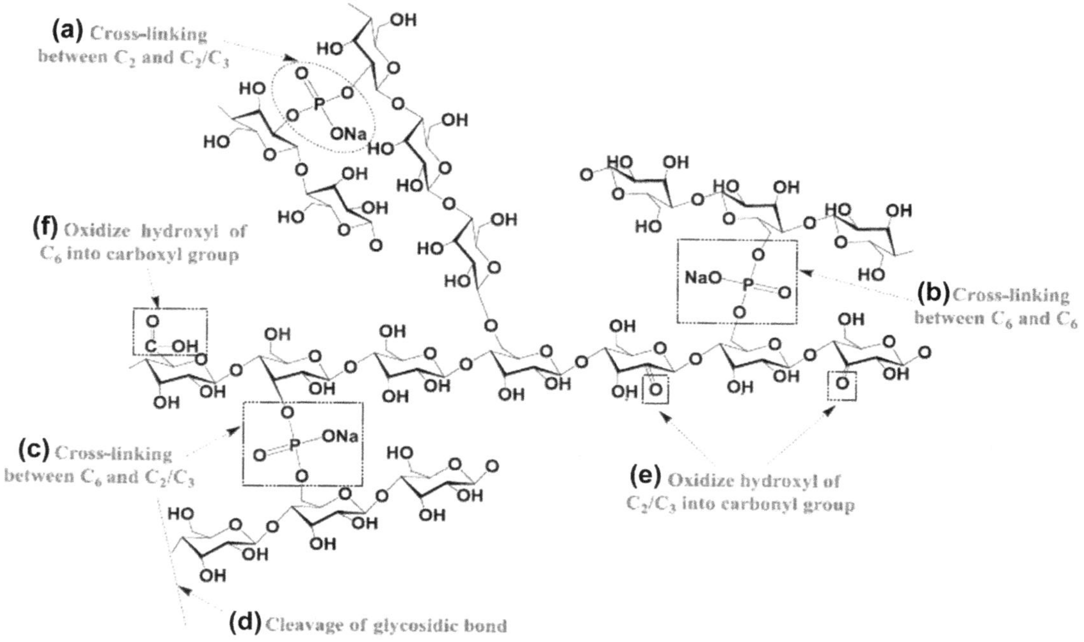

Figure 13
A possible structure of cross-linked oxidized starch. *Reproduced from Liu et al. (2014), with permission from Elsevier.*

solubility than their native equivalents, and solubility decreases further with an increase in the concentration of cross-linking reagent, which may be attributed to an increase in cross-link density (Kaur et al., 2006). Liu et al. (2014) evaluated the functional and physicochemical properties and structure of cross-linked oxidized maize starch and verified that the introduction of cross-links made the oxidation reaction much easier. The introduction of the hydrophilic groups by oxidation enhanced the water-binding capacity and improved the light transmittance and decreased the retrogradation rate. The authors concluded the cross-linked oxidized maize starch showed stronger hydrophilic properties than the other three starches (Figure 13).

Derivatization also alters the light transmittance of potato starch pastes to a considerable extent. Acetylation and hydroxypropylation have been reported to increase the light transmittance of potato starches (Kaur et al., 2004; Singh et al., 2004a,b). The chemical substitution of the —OH groups on the starch molecules by acetyl moieties hampers the formation of an ordered structure following gelatinization and thus retards retrogradation, resulting in a more fluid paste with improved long-term clarity (Lawal, 2004). The high retention of water entering the starch granule results in a greater swelling power and favors the clarity of pastes and gels. Potato starch source, native starch granule size distribution, amylose

Table 11: Solubility (%) in DMSO of Some Modified Starches from Various Botanical Sources.

Starch Source	SB[a] (after 4 h)	SB (after 8 h)	SB (after 16 h)	SB (after 24 h)
Normal potato (hydroxypropylated)[d,b]	≈ 76 (Native ≈ 57)[c]	≈ 83 (Native ≈ 60)	≈ 95 (Native ≈ 65)	≈ 99 (Native ≈ 75)
Normal rice (hydroxypropylated)[e]	≈ 35 (Native ≈ 10)	≈ 80 (Native ≈ 20)	≈ 95 (Native ≈ 35)	≈ 98 (Native ≈ 55)
Normal rice (cross-linked)[f]	≈ 08 (Native ≈ 10)	≈ 10 (Native ≈ 20)	≈ 20 (Native ≈ 35)	≈ 26 (Native ≈ 55)

[a]Solubility in DMSO (%).
[b]Values reported as % transmittance in DMSO.
[c]The properties of corresponding native (unmodified) starches are given in parentheses.
[d]Starches from different potato cultivars used.
[e]Two levels of hydroxypropylation used.
[f]Different levels of cross-linking used.
Reproduced from Singh et al. (2007), with permission from Elsevier.[c]

content, and DS are the important factors affecting light transmittance of acetylated potato starches. Singh et al. (2004b) reported a higher light transmittance of acetylated potato starches compared to maize starches acetylated under similar reaction conditions. Highly cross-linked potato starch pastes generally show lower light transmittance than their counterpart native starches. Incomplete gelatinization and reduced swelling of cross-linked starches are mainly responsible for their reduced paste clarity (Zheng et al., 1999; Morikawa and Nishinari, 2000a; Reddy and Seib, 2000; Kaur et al., 2006; Koo et al., 2010; Wongsagonsup et al., 2014).

The solubility of native, cross-linked, and hydroxypropylated potato starches in dimethyl sulfoxide (DMSO) varies to a significant extent (Table 11). Hydroxypropylation results in a significant increase in the solubility of potato starches in DMSO (Yeh and Yeh, 1993; Kaur et al., 2004). Yeh and Yeh (1993) compared the solubilities (in DMSO) of hydroxypropylated and cross-linked rice starches with those of native rice starch and observed higher and lower solubilities for hydroxypropylated and cross-linked rice starches, respectively. Differences in granule morphology, amylose content, and DS of native and modified potato starches, respectively, have been reported to affect the solubility in DMSO (Kaur et al., 2004). Sahai and Jackson (1996) reported that the solubility of starch in DMSO varies significantly with granule size, presumably reflecting inherent structural heterogeneity within granules.

3.3.5 *Pasting/Rheological Properties of Derivatized Potato Starch*

Derivatization leads to a considerable change in the rheological and pasting properties of potato starch (Table 14). Potato starch paste viscosity can be increased or reduced by applying a suitable chemical modification. Again, modification method, reaction conditions, and starch source are the critical factors that govern the rheological/pasting behavior of potato starch pastes.

Table 12: Transition Temperatures of Some Modified Starches from Various Botanical Sources.

Starch Source	T (°C) (Acetylated)	T (°C) (Hydroxypropylated)	T (°C) (Cross-Linked)
Normal potato	48–67 (Native 57–69)[a,c]	51–62 (Native 57–68)[b]	60–68 (Native 57–68)[h]
Normal potato	55–64 (Native 59–70)[b]	ND	59–68 (Native 57–68)[i]
Normal maize	65–77 (Native 69–77)[a]	59–75 (Native 65–81)[e]	ND
Normal maize	66–75 (Native 70–78)[b]	ND	ND
Waxy maize	64–69 (Native 68–72)	61–79 (Native 63–84)[e]	65–75 (Native 62–76)[j]
Waxy maize	ND	ND	67–72 (Native 68–73)[k]
Hybrid normal maize	71–84 (Native 60–78)	ND	ND
High-amylose maize	ND	66–95 (Native 67–105)[e]	ND
Normal wheat	ND	55–71 (Native 63–84)[f]	63–76 (Native 63–84)[k]
Normal wheat	ND	46–55 (Native 55–67)[g]	61–72 (Native 57–58)[l]
Waxy wheat	ND	ND	62–66 (Native 61–66)[k]
Normal rice	60–72 (Native 66–78)[d]	64–83 (Native 63–84)[e]	70–87 (Native 71–87)[m]
Normal rice	ND	57–92 (Native 63–92)[f]	ND
Waxy rice	60–78 (Native 60–78)	ND	61–78 (Native 60–78)
Waxy rice	ND	ND	55–66 (Native 53–67)[j]

T, transition temperatures of various starches (°C); ND, not detected.
[a]Different levels of acetylation.
[b]Starches from different potato cultivars.
[c]The properties of corresponding native (unmodified) starches are given in parentheses.
[d]Starches from different rice cultivars.
[e]Two levels of hydroxypropylation.
[f]Three levels of hydroxypropylation.
[g]Two wheat starch granule populations.
[h]Starches from different potato cultivars used, cross-linking performed using $POCl_3$.
[i]Starches from different potato cultivars used, cross-linking performed using EPI.
[j]Cross-linking performed using EPI.
[k]Cross-linking performed using $POCl_3$.
[l]Three levels of cross-linking performed using $POCl_3$; onset transition temperatures only.
[m]Different levels of cross-linking performed using $POCl_3$.
Reproduced from Singh et al. (2007), with permission from Elsevier.[c]

Control of reaction conditions such as pH during acetylation may allow reduction of secondary hydrolysis reactions and facilitate the incorporation of acetyl groups. Such modification enhances the water-holding capacity of the starch matrix and the development of more organized structures, leading to a higher resistance to deformation, and thus a higher peak viscosity can be achieved (Betancur-Ancona et al., 1997). Acetylation influences interactions between potato starch chains by steric hindrance, changing starch hydrophilicity and hydrogen bonding and resulting in a lower gelatinization temperature and a greater swelling of granules, the latter resulting in an increased peak viscosity (Liu et al., 1997; Singh et al., 2004a,b; Colussi et al., 2014). The amylose/amylopectin ratio is considered to be the prime determinant of the change in the paste viscosity upon acetylation (Colussi et al., 2014). The substituent groups restrict the tendency of the starch molecules to realign after cooling, thus facilitating lower setback values

Table 13: Enthalpy of Gelatinization of Some Modified Starches from Various Botanical Sources.

Starch Source	ΔH_{gel}J/g (Acetylated)	ΔH_{gel}J/g (Hydroxypropylated)	ΔH_{gel}J/g (Cross-Linked)
Normal potato	10.1–11.4 (Native 12.1)[a,c]	10–11 (Native 11.7–12.9)[b]	12.7–14.7 (Native 11.7–12.9)[h]
Normal potato	10.1–11.8 (Native 12.8–13.2)[b]	ND	12.7–14.4 (Native 11.7–12.9)[i]
Normal maize	08.52 (Native 10.9)[a]	ND	ND
Normal maize	08.9–09.7 (Native 10.6)[b]	07.1–8.4 (Native 11)[e]	ND
Hybrid normal maize	09.8–10.6 (Native 14.3)	ND	ND
High-amylose maize	ND	6.4–8.4 (Native 13.7)[e]	ND
Waxy maize	14.8 (Native 15.6)	08–8.7 (Native 13.6)[e]	15.2 (Native 15.3)[j]
Normal wheat	ND	04.7–05 (Native 6.8)[f]	6.6–7.4 (Native 06.8)[k]
Normal wheat	ND	02.2 (Native 06.2)[g]	–
Waxy wheat	ND	ND	12.8 (Native 13.2)[j]
Normal rice	08.1–11.4 (Native 08.1–11.9)[d]	08–09 (Native 10.4)[e]	11–13 (Native 10.4)[l]
Waxy rice	08.5 (Native 09.8)	ND	10.3 (Native 09.8)

ΔH_{gel}, enthalpy of gelatinization of different starches (J/g); ND, not detected.
[a]Different levels of acetylation.
[b]Starches from different potato cultivars.
[c]The properties of corresponding native (unmodified) starches are given in parentheses.
[d]Starches from different rice cultivars.
[e]Two levels of hydroxypropylation.
[f]Three levels of hydroxypropylation.
[g]Two wheat starch granule populations.
[h]Starches from different potato cultivars; cross-linking performed using $POCl_3$.
[i]Starches from different potato cultivars; cross-linking performed using EPI.
[j]Cross-linking performed using $POCl_3$.
[k]Three levels of cross-linking performed using $POCl_3$.
[l]Different levels of cross-linking performed using $POCl_3$.
Reproduced from Singh et al. (2007), with permission from Elsevier.[c]

(Betancur-Ancona et al., 1997). The HPV and cold paste viscosity (CPV) of the acetylated starches have also been observed to be higher than in the case of the unmodified starch. This may be attributed to physical interaction between more swollen but weaker granules (Gonzalez and Perez, 2002).

Hydroxypropylation can influence interaction between the potato starch chains through different possible mechanisms: (1) by steric hindrance, which prevents the close association of chains and restricts the formation of interchain hydrogen bonds, and (2) by changing the hydrophilicity of the starch molecules and thus altering the bonding with water molecules. The observed effects of hydroxypropylation are consistent with an overall reduction in bonding between starch chains and a consequent increase in the ease of hydration of the starch granule. Gelatinization can thus commence at a lower temperature, and greater swelling of the granule will lead to an increased peak viscosity (Liu

Table 14: Rheological Properties of Some Modified Potato Starches during Heating from 30 to 75 °C Studied Using a Dynamic Rheometer.[a]

Modification Type	TG' (°C)	Peak G' (Pa)	Peak G'' (Pa)	Peak tan δ
Hydroxypropylated[c,d] (using 10% propylene oxide)	–	5951 (3288)	1083 (693)	0.182 (0.211)
Acetylated[b,e] (using 10% acetic anhydride)	54.9 (55.2)	90,220 (85,790)	53,726 (49,529)	0.596 (0.591)
Cross-linked[c,f] (using 0.1% $POCl_3$)	62.2 (63.6)	4450 (3288)	850 (693)	0.191 (0.211)
Cross-linked[c,f] (using 0.2% $POCl_3$)	64.8 (63.6)	2750 (3288)	616 (693)	0.224 (0.211)
Cross-linked[c,f] (using 0.25% EPI)	63.9 (63.6)	3120 (3288)	660 (693)	0.211 (0.211)
Cross-linked[c,f] (using 1% EPI)	64.9 (63.6)	2800 (3288)	630 (693)	0.225 (0.211)

[a]The properties of corresponding native (unmodified) starches are given in parentheses.
[b]At 20% starch concentration.
[c]At 15% starch concentration.
[d]Kaur et al. (2004).
[e]Singh et al. (2004b).
[f]Kaur et al. (2006).

et al., 1999a). Kim et al. (1992) reported that with increase in MS, the pasting temperature of potato hydroxypropylated starches decreased and their peak Brabender viscosity increased. The extent of change in potato starch functional properties upon hydroxypropylation has also been observed to be influenced by the amylose content and starch granule size distribution.

The greater strength of the cross-linked granule limits the breakdown of viscosity under shear, giving a higher HPV and persistence, or resistance to breakdown, of the swollen potato starch granules and, on cooling, results in a higher CPV. Gluck-Hirsch and Kokini (2002) studied the relative effects of various cross-linking agents on the physical properties of starches and reported that $POCl_3$ has the ability to impart a greater viscosity than other agents. $POCl_3$-treated granules have a more rigid external surface than STMP- and EPI-treated granules owing to the concentration of cross-links at the surface of the granule.

The rheological properties of modified potato starches exhibit significant differences from those of native starches when subjected to temperature sweep testing using heating and cooling cycles on a dynamic rheometer (Singh et al., 2004b; Kaur et al., 2004). The parameters such as G' and G'' of acetylated, hydroxypropylated, and cross-linked starches from different sources increase to a maximum and then drop during heating, confirming that these starches follow the same general rheological pattern as native starches. The temperature of maximum G' drops significantly on acetylation or hydroxypropylation, while it increases after cross-linking (Singh et al., 2004b; Kaur et al., 2004, 2006). Changes may be explained on the same basis as for the changes in thermal and pasting properties caused by modification.

Acetylation of maize and potato starches results in increased G' and G'' maxima and a decreased tan δ maximum. Acetylated starches with higher DS exhibit a correspondingly higher increase in G' and G'' maxima during heating. These changes occur for the same reasons that acetylation causes an increase in peak pasting viscosity (see above) (Betancur-Ancona et al., 1997; Singh et al., 2004a).

Acetylated maize and potato starches showed slightly lower G' and G'', compared with their native starch gels, during the cooling cycle of heated starch pastes on the rheometer. This confirms the lower tendency of these modified starches to retrograde (Betancur-Ancona et al., 1997; Singh et al., 2004a). Similarly, increases in the peak G' and G'' of hydroxypropylated potato starches during heating occur, owing to the decrease in associative forces within the starch granules caused by the introduction of hydroxypropyl groups; this introduction results in greater water penetration and swelling and a consequent increase in G'. Hydroxypropylated potato starches with a higher MS and large granules exhibit higher peak G' and G'' during heating than do the native starches (Kaur et al., 2004). Cross-linking of starches leads to a significant alteration in their dynamic rheological behavior. A fairly high degree of cross-linking results in a lower peak G', owing to the lower degree of swelling and consequent lower degree of intergranular interaction. Potato starches treated with a relatively low concentration of $POCl_3$ showed a higher maximum G', and lower tan δ maximum, than their counterpart native starches, whereas those treated with higher concentrations showed an opposite effect (Kaur et al., 2006). Strengthening bonding between starch chains by cross-linking will increase the resistance of the granule toward swelling, leading to lower paste viscosity, which suggests that the concentration of the cross-linking reagent affects the structure within the granule, perhaps by affecting the distribution of the introduced cross-links. Cross-link location has also been reported to have varied effects on different properties of cross-linked potato starches (Muhrbeck and Eliasson, 1991; Muhrbeck and Tellier, 1991; Yoneya et al., 2003). The extent of change in the rheological properties upon cross-linking also varies significantly for starches from different sources depending on the granule size distribution. Potato cultivars with a higher percentage of large irregular starch granules showed a higher susceptibility toward cross-linking and exhibited greater changes in their functional properties upon cross-linking (Kaur et al., 2006). These characteristics could also be explained on the basis of differences in the amylose-to-amylopectin ratio, amylopectin branched-chain length, and degree of crystallinity between larger and smaller granules (Singh and Kaur, 2004; Noda et al., 2005).

Dual modification results in starch pastes with higher peak viscosity and greater stability than those of native starch pastes (Wu and Seib, 1990). However, the effect of dual modification depends upon the preparation procedure. Cross-linking followed by hydroxypropylation has been reported to yield starches that are more stable to shear and heat than the native starch. This may be due to the structural change in the granules after the first modification (cross-linking). Cross-linking reduces the degree of subsequent hydroxypropylation, but prior hydroxypropylation increases the degree of subsequent cross-linking. Moreover, the cross-linked and then hydroxypropylated (XL–HP) starch exhibited lower pasting temperature and viscosity than the hydroxypropylated and then cross-linked(HP–XL) starch samples, suggesting the locations of the cross-links in the two types of starch are different. The cross-links in XL–HP starch were also found to be more resistant to attack by enzymes and chemicals (Reddy and Seib, 2000). The reactivity of the modifying agent during dual modification may vary toward different starch sources. The cross-linked hydroxypropylated waxy wheat starch gave pasting curves showing higher viscosities than those of cross-linked hydroxypropylated waxy maize starch (Reddy and Seib, 2000). Therefore, by appropriate choice of the native starch source (potato, maize, wheat, etc.) and of the type of chemical modification and concentration of the modifying reagent, modified starches with very useful rheological properties can be obtained (Dubois et al., 2001; Singh et al., 2004a,b; Kaur et al., 2004).

3.3.6 Gelatinization and Retrogradation (Thermal Properties) of Derivatized Potato Starch

DSC studies have shown that modification alters thermal transition temperatures and the overall enthalpy (ΔH_{gel}) associated with gelatinization (Tables 12 and 13). Upon hydroxypropylation, the reactive groups introduced into the starch chains are capable of disrupting the inter- and intramolecular hydrogen bonds, leading to an increase in accessibility by water that lowers the temperature of gelatinization. In potato starch, increasing the level of MS results in decreases in ΔH_{gel}, T_o, T_p, and T_c and a widening of the gelatinization temperature range (T_c–T_o) (Perera et al., 1997; Lee and Yoo, 2011). An increase in gelatinization temperature range upon hydroxypropylation of various potato starches has been reported by Kaur et al. (2004), which could be attributed to an increase in homogeneity within both the amorphous and the crystalline regions of the starch granules. A progressive shift in the biphasic gelatinization endotherms to lower temperatures as well as a broadening and shortening of the gelatinization endotherm with increasing MS has also been observed for hydroxypropylated rice starch, indicating increased internal plasticization and destabilization of the amorphous regions of the granules (Seow and Thevamalar, 1993; Woggum et al., 2015).

The decrease in ΔH_{gel} on hydroxypropylation suggests that hydroxypropyl groups disrupt double helices (owing to the rotation of these flexible groups) within the amorphous regions of the potato starch granules. Consequently, the number of double helices that unravel and melt during gelatinization would be lower in hydroxypropylated than in unmodified starches (Perera et al., 1997). Decreases have been recorded in gelatinization temperatures and gelatinization enthalpies of amylose-extender mutant (*ae*), waxy mutant (*wx*), and normal maize starch upon hydroxypropylation (Liu et al., 1999a). The decrease was observed to be greater for high-amylose maize starches (66% amylose) compared with waxy maize starch (3.3% amylose). Substitution on granular starch occurs mainly in the amorphous regions, which promotes swelling in these regions and, thus, disrupts the crystalline phase, which melts at a lower temperature than in the case of the unmodified starch.

Potato, corn, and rice starches with higher degrees of acetylation have been observed to show greater decreases in the transition temperatures and enthalpy of gelatinization (Luo and Shi, 2012; Colussi et al., 2014). Weakening of the starch granules by acetylation leads to early rupture of the amylopectin double helices, which accounts for the lower values of T_o, T_p, and T_c (Adebowale and Lawal, 2003; Wotton and Bamunuarachchi, 1979). Decreases in the thermal parameters are consistent with fewer crystals being present after modification (owing to damage to the crystals during acetylation) and with a cooperative melting process enhanced by added swelling (Singh et al., 2004b). The extent of the lowering of transition temperatures and ΔH_{gel} on acetylation has been observed to be different for starches from different sources. Potato starches show a greater decrease in thermal parameters upon acetylation compared with maize and rice starches (Singh et al., 2004a,b; Sodhi and Singh, 2005). The differences in granule rigidity, presence/absence of lipids, DS, and amylose-to-amylopectin ratio between these starches affect the extent of changes in thermal parameters.

Cross-linking also alters the thermal transition characteristics of starch, the effect depending on the concentration and type of cross-linking reagent, reaction conditions, and botanical source of the starch. An increase in gelatinization temperature has been observed for cross-linked potato starches; these phenomena are related to the reduced mobility of amorphous chains in the starch granule as a result of the formation of intermolecular bridges. The level of phosphate cross-links has a strong influence on the DSC properties of starches. Choi and Kerr (2004) reported that cross-linked starches prepared using a relatively low concentration of $POCl_3$ had gelatinization parameters similar to those of native starches, while cross-linked starches prepared using higher reagent concentrations showed considerably higher T_c and ΔH_{gel} values. Cross-linking at lower levels reduces the proportion of the starch that can be gelatinized, resulting in a lower value of ΔH_{gel} (Yook et al., 1993; Majzoobi et al., 2015). Yoneya et al. (2003) also reported that potato starches treated with 100 ppm of $POCl_3$ displayed significantly lower gelatinization temperatures and lower ΔH_{gel} than other treated samples (with 40–5000 ppm). Decreasing or increasing the degree of phosphate cross-links caused a gradual increase in T_o, T_p, and ΔH_{gel} of cross-linked potato starch samples compared with the values of these properties for the 100-ppm-treated sample.

Chatakanonda et al. (2000) conducted studies on cross-linked rice starches (using as reagent a mixture of STMP and STPP) that showed deep and sharp endotherms, $\approx 30\,°C$ wide. Also, the gelatinization temperature increased significantly (by up to $5\,°C$) with an increase in the degree of cross-linking, while enthalpy showed no significant change ($\approx 15\%$ decrease), suggesting complete melting of crystalline regions despite cross-linking. The introduction of phosphate cross-links into the starch by STMP alone tightened the molecular structure, leading to an increase in the gelatinization temperature (Chatakanonda et al., 2000). The enthalpy of cross-linked normal rice starch was observed to decrease, while waxy rice starch showed an increase in enthalpy after cross-linking (Liu et al., 1999b). These findings suggest that the type and concentration of the reagent and the amylose/amylopectin ratio of starch during cross-linking significantly affect the extent of change in thermal properties. Dual-modified (substituted and cross-linked) normal potato starches are available commercially with varying degrees of hydroxypropylation and cross-linking. The temperatures and enthalpies of gelatinization of the dual-modified (hydroxypropylated/cross-linked or acetylated/cross-linked) starches are generally lower than those of the unmodified starches (Reddy and Seib, 2000; Liu et al., 2014).

The crystallinity of starch granules is disrupted during derivatization (Saroja et al., 2000), and this leads to a greater degree of separation between the outer branches of adjacent amylopectin chain clusters in modified starches compared with those in native starches. Consequently, double-helix formation (during storage) between adjacent amylopectin chains of the modified starches is much slower and less extensive owing to the introduction of functional groups upon chemical modification. Retrogradation of starch is suppressed by cross-linking using an STMP and STPP mixture, as indicated by their lower enthalpy of retrogradation after storage, which may be because of the restricted mobility of cross-linked amylopectin branches due to the presence of phosphate groups (Chung at al., 2004). Thermal behavior during DSC heating of stored native starch gels has also been reported to be more conspicuous than that of hydroxypropylated phosphate cross-linked potato starch (HPS) dispersions, as the retrogradation phenomenon of HPS dispersions was barely observed by DSC measurements even when a high concentration of starch (33%) was used (Morikawa and Nishinari, 2000b). Moreover, it has been reported that the effect of phosphate cross-linking on starch is more pronounced than that of hydroxypropylation with respect to retrogradation. Yook et al. (1993) concluded that the tendency toward retrogradation was greatly reduced by both hydroxypropylation and cross-linking, and they showed a synergistic effect of these treatments in retarding the retrogradation of gelatinized rice starch gels. HP–XL starch exhibited significantly higher onset and peak gelatinization temperatures, enthalpy of gelatinization, and lower retrogradation than XL–HP starch, confirming the previous assumption that the locations of cross-links are different in the two kinds of modified starch (Yook et al., 1993). Hydroxypropylation followed by cross-linking provides starch with better storage stability in food applications.

Liu et al. (2014) studied functional and physicochemical properties and the structure of cross-linked oxidized maize starch. They verified that for the oxidized and cross-linked oxidized starch that the gelatinization temperatures were increased with the oxidation process, but the gelatinization range (T_c–T_o) was decreased. The authors reported that the carboxyl groups introduced into starch stabilize the structure and increase the energy requirement for the starch gelatinization. The retrogradation of cross-linked starch was observed to be slower than the native starch. According to the mechanism of retrogradation, this might be attributed to the introduction of phosphate groups, which were able to combine with water and form hydrogen bonds with hydroxyl groups of amylase, resulting in the hydrogen bonds being prevented from rearranging and reassociating (Chatakanonda et al., 2000).

4. Nutritional and Toxicological Aspects

The demand for starches with special properties, for use in the food processing industry, has led to the introduction of modified starches. Though there is a lot of information available in the literature on chemical modification of starch, limited data are available on the nutritional properties of the resulting starch and their implications on human health. Many modified starches made for food use contain only small amounts of substituent groups and have been used as safe food ingredients. During acetylation and hydroxypropylation of food-grade starches, the level of monosubstitution groups introduced is relatively low. According to the FDA (2015) and the Food Additives and Contaminants Committee (FAC, 1980) the maximum permitted levels of substitution for starch acetates, starch phosphates, and hydroxypropylated starches are 2.5%, 0.4%, and 10%, respectively. Similarly, cross-linked food starches containing one substituent cross-linking group per 1000 or more anhydroglucose units are considered safe (Wurzburg, 1986c). The legislative approval for the use of novel starch derivatives in processed food formulations is still under debate, but several tailor-made starch derivatives with multiple modifications are being prepared and characterized (Tharanathan, 2005). Some of the starch derivatives are being increasingly used as fat replacers or fat substitutes. These derivatives are either partially or totally undigested and therefore contribute zero calories to the food on consumption (Tharanathan, 1995). Many studies have suggested that the physiological effects of chemically modified starches are affected by the type of modification (Ebihara et al., 1998). The chemical modification of starch by acetylation improves the satiating, glycemic, and insulinemic properties of the meal (Raben et al., 1997; Kalpeko et al., 2012). Phosphorylated/cross-linked starches are slowly digestible and are thought to provide nutritional benefits for humans (Sang and Seib, 2006; Sang et al., 2010; Shukri and Shi, 2015; Shukri et al., 2015). Also, the slowly digested modified starches could be used for the treatment of certain medical conditions (e.g., glycogen storage disease and diabetes mellitus) (Wolf et al., 1999).

5. Conclusions

Progress in understanding the high value of chemically modified starches has encouraged the starch industry to produce modified starches using various modification reagents and starch sources. Some factors such as starch composition, concentration and type of reagent, and reaction conditions may affect the reactivity of starch during chemical modifications like acetylation, hydroxypropylation, and cross-linking. The heterogeneity of the granule population within a single starch source may also affect the extent of modification. Although chemical modification is well understood and applied in the food and nonfood industries, physical modification of starch is preferred for environmentally friendly applications, as it is related to the emerging concept of "green technology." The changes observed in the morphological, physicochemical, and pasting/rheological properties and gelatinization and retrogradation (thermal) properties of the starches after modification may provide a crucial basis for understanding the efficiency of the starch modification process at the industrial scale. Various physical modifications and their combinations with chemical modifications are being carried out to create novel starch properties to meet the demand for new food products with improved functional and nutritional properties.

References

Adebowale, K.O., Lawal, O.S., 2003. Functional properties and retrogradation behaviour of native and chemically modified starch of mucuna bean (*Mucuna pruriens*). Journal of the Science of Food and Agriculture 83, 1541–1546.

Ahmed, J., Singh, A., Ramaswamy, H.S., Pandey, P.K., Raghavan, G.S.V., 2014. Effect of high-pressure on calorimetric, rheological and dielectric properties of selected starch dispersions. Carbohydrate Polymers 103, 12–21.

Ai, Y., Jane, J., 2015. Gelatinization and rheological properties of starch. Starch 67, 213–224.

Ao, Z., Jane, J.L., 2007. Characterization and modeling of the A- and B-granule starches of wheat, triticale, and barley. Carbohydrate Polymers 67, 46–55.

Arisz, P.W., Kauw, H.J.J., Boon, J.J., 1995. Substituent distribution along the cellulose backbone on O-methylcelluloses using GC and FAB-MS for monomer and oligimer analysis. Carbohydrate Research 271, 1–14.

Baker, L.A., Rayas-Duarte, P., 1998. Freeze-thaw stability of amaranth starch and the effects of salt and sugars. Cereal Chemistry 75, 301–303.

Barichello, V., Yada, R.Y., Coffin, R.H., Stanley, D.W., 1990. Low temperature sweetening in susceptible and resistant potatoes: starch structure and composition. Journal of Food Science 54, 1054–1059.

Bello-Pérez, L.A., Agama-Acevedo, E., Zamudio-Flores, P.B., Mendez-Montealvo, G., Rodriguez-Ambriz, S.L., 2010. Effect of low and high acetylation degree in the the morphological, physicochemical and structural characteristics of barley starch. LWT-Food Science and Technology 43, 1434–1440.

Bemiller, J.N., 1997. Starch modification: challenges and prospects. Starch 49, 127–131.

Bertolini, A.C., Souza, E., Nelson, J.E., Huber, K.C., 2003. Composition and reactivity of A- and B-type starch granules of normal, partial waxy and waxy wheat. Cereal Chemistry 80, 544–549.

Betancur-Ancona, D., Chel-Guerrero, L., Canizares-Hernandez, E., 1997. Acetylation and characterization of *Canavalia ensiformis* starch. Journal of Agricultural and Food Chemistry 45, 378–382.

Błaszczak, W., Valverde, S., Fornal, J., 2005. Effect of high pressure on the structure of potato starch. Carbohydrate Polymers 59, 377–383.

Blennow, A., Engelsen, S.B., Nielsen, T.H., Baunsgaard, L., Mikkelsen, R., 2002. Starch phosphorylation: a new front line in starch research. Trends in Plant Science 17, 445–450.

van der Burgt, Y.E.M., Bergsma, J., Bleeker, I.P., Mijland, P.J.H.C., van der Kerk-van Hoof, A., Kamerling, J.P., Vliegenthart, J.F.G., 1998. Distribution of methyl substituents over branched and linear regions in methylated starches. Carbohydrate Research 312, 201–208.

Cai, C., Wei, C., 2013. In situ observation of crystallinity disruption patterns during starch gelatinization. Carbohydrate Polymers 92, 469–478.

Chandrapala, J., Oliver, C., Kentish, S., Ashokkumar, M., 2012. Ultrasonics in food processing – food quality assurance and food safety. Trends in Food Science & Technology 26, 88–98.

Chatakanonda, P., Varavinit, S., Chinachoti, P., 2000. Effect of cross-linking on thermal and microscopic transitions of rice starch. LWT-Food Science and Technology 33, 276–284.

Chen, Z., Schols, H.A., Voragen, A.J.G., 2004. Differently sized granules from acetylated potato and sweet potato starches differ in the acetyl substitution pattern of their amylose populations. Carbohydrate Polymers 56, 219–226.

Chilton, W.G., Collison, R., 1974. Hydration and gelation of modified potato starches. Food Technology 9, 87–93.

Choi, S.-G., Kerr, W.L., 2004. Swelling characteristics of native and chemically modified wheat starches as a function of heating temperature and time. Starch 56, 181–189.

Chuenkamol, B., Puttanlek, C., Rungsardthong, V., Uttapap, D., 2007. Characterization of the low-substituted hydroxypropylated cannna starch. Food Hydrocolloids 21, 1123–1132.

Chung, H.-J., Woo, K.-S., Lim, S.-T., 2004. Glass transition and enthalpy relaxation of cross-linked corn starches. Carbohydrate Polymers 55, 9–15.

Collado, L.S., Corke, H., 1999. Heat-moisture treatment effects on sweetpotato starches differing in amylose content. Food Chemistry 65, 339–346.

Colussi, R., El Halal, S.L.M., Pinto, V.Z., Bartz, J., Gutkoski, L.C., Zavareze, E.R., Dias, A.R.G., 2015. Acetylation of rice starch in an aqueous medium for use in food. LWT-Food Science and Technology 62, 1076–1082.

Colussi, R., Pinto, V.Z., Halal, S.L.M., Vanier, N.L., Villanova, F.A., Silva, R.M., Zavareze, E.R., Dias, A.R.G., 2014. Structural, morphological, and physicochemical properties of acetylated high-, medium-, and low-amylose rice starches. Carbohydrate Polymers 103, 405–413.

Cooke, D., Gidley, M.J., 1992. Loss of crystalline and molecular order during starch gelatinization: origin of the enthalpic transition. Carbohydrate Research 227, 103–112.

Craig, S.A.S., Maningat, C.C., Seib, P.A., Hoseney, R.C., 1989. Starch paste clarity. Cereal Chemistry 66, 173–182.

Cui, B., Lu, Y., Tan, C., Wang, G., Li, G., 2014. Effect of cross-linked acetylated starch content on the structure and stability of set yoghurt. Food Hydrocolloids 35, 576–582.

Czechowska-Biskup, R., Rokita, B., Lotfy, S., Ulanski, P., Rosiak, J.M., 2005. Degradation of hitosan and starch by 360-kHz ultrasound. Carbohydrate Polymers 60, 175–184.

Dubois, I., Picton, L., Muller, G., Audibert-Hayet, A., Doublier, J.L., 2001. Structure/rheological properties relations of cross-linked potato starch suspensions. The Journal of Applied Polymer Science 81, 2480–2489.

Ebihara, K., Shiraishi, R., Okuma, K., 1998. Hydroxypropyl-modified potato starch increases fecal bile acid excretion in rats. Journal of Nutrition 128, 848–854.

Eliasson, A.C., 1986. Viscoelastic behaviour during the gelatinization of starch. I. Comparison of wheat, maize, potato and waxy-barley starches. Journal of Texture Studies 17, 253–265.

FAC, 1980. Food Additives and Contaminants Committee Report on Modified Starches Report 31. Ministry of Agriculture, Fishries and Food, London, UK.

Fan, J., Marks, B.P., 1998. Retrogradation kinetics of rice flasks as influenced by cultivor. Cereal Chemistry 75, 53–155.

FDA, 2015. US Food and Drug Administration. https://www.accessdata.fda.gov/scripts/cdrh/cfdocs/cfCFR/CFRSearch.cfm?fr=172.892 (accessed 08.11.15.).

Felton, G.E., Schopmeyer, H.H., 1943. Thick Bodied Starch and Method of Making. U.S. Patent 2,328,537.

Ferry, J.D., 1980. Viscoelastic Properties of Polymers, third ed. John Wiley and Sons, New York.

Flipse, E., Keetels, C.J.A.M., Jacobson, E., Visser, R.G.F., 1996. The dosage effect of the wildtype GBSS allele is linear for GBSS activity, but not for amylose content: absence of amylose has a distinct influence on thew physico-chemical properties of starch. Theoretical and Applied Genetics 92, 121–127.

Fonseca, L.M., Gonçalves, J.R., El Halal, S.L.M., Pinto, V.Z., Dias, A.R.G., Jacques, A.C., Zavareze, E.R., 2015. Oxidation of potato starch with different sodium hypochlorite concentrations and its effect on biodegradable film. LWT-Food Science and Technology 60, 714–720.

Fortuna, I., Juszczak, L., Palasinski, M., 1999. Changes in granule porosity on modification of starch. Food Science Technology Quality 5, 124–130.

Freitas, R.A., Paula, R.C., Feitosa, J.P.A., Rocha, S., Sierakowski, M.-R., 2004. Amylose contents, rheological properties and gelatinization kinetics of yam (*Dioscorea alata*) and cassava (*Manihot utilissima*) starches. Carbohydrate Polymers 55, 3–8.

French, D., 1984. Organization of starch granules. In: Whistler, R.L., et al. (Ed.), Starch Chemistry and Technology. Academic Press, New York, pp. 183–247.

French, D., Levine, M.L., Norberg, E., Nordin, P., Pazur, J.H., Wild, G.M., 1954. Studies on the Schardinger dextrins. VII. Co-substrate specificity in coupling reactions of *Macrerans* amylase. Journal of the American Chemical Society 26, 2387–2391.

Fulton, D.C., Edwards, A., Pilling, E., Robinson, H.L., Fahy, B., Seale, R., Kato, L., Donald, A.M., Geigenberger, P., Martin, C., Smith, A.M., 2002. Role of granule-bound starch synthase in determination of amylopectin structure and starch granule morphology in potato. Journal of Biological Chemistry 29, 10834–10841.

Gernat, C., Radosta, S., Anger, H., Damaschun, G., 1993. Crystalline parts of three different conformations detected in native and enzymatically degraded starches. Starch 45, 309–314.

Gluck-Hirsch, J.B., Kokini, J.L., 2002. Understanding the mechanism of cross-linking agents (POCl₃, STMP, and EPI) through swelling behaviour and pasting properties of cross-linked waxy maize starches. Cereal Chemistry 79, 102–107.

Gluck-Hirsch, J., Kokini, J.L., 1997. Determination of the molecular weight between cross-links of waxy maize starches using the theory of rubber elasticity. Journal of Rheology 41, 129–139.

Gonzalez, Z., Perez, E., 2002. Effect of acetylation on some properties of rice starch. Starch 54, 148–154.

Gui-Jie, M., Peng, W., Xiang-Sheng, M., Xing, Z., Tong, Z., 2006. Crosslinking of corn starch with sodium trimetaphosphate in solid state by microwave irradiation. The Journal of Applied Polymer Science 102, 5854–5860.

Haase, N.U., Plate, J., 1996. Properties of potato starch in relation to varieties and environmental factors. Starch 48, 167–171.

Heins, D., Kulicke, W.-M., Kauper, P., Thielking, H., 1998. Characterization of acetyl starch by means of NMR spectroscopy and SEC/MALLS in composition with hydroxyethyl starch. Starch 50, 431–437.

Hermansson, A.M., Svegmark, K., 1996. Developments in the understanding of starch functionality. Trends in Food Science & Technology 7, 345–353.

Hoover, R., 2001. Composition, molecular structure, and physico-chemical properties of tuber and root starches: a review. Carbohydrate Polymers 45, 253–267.

Hovenkamp-Hermelink, J.H.M., Jacobson, E., Ponstein, A.S., Visser, R.G.F., Vos-Scheperkeuter, G.H., Bijmolt, E.W., DeVries, J.N., Witholt, B., Feenstra, W.J., 1987. Isolation of an amylose-free starch mutant of the potato (*Solanum tuberosum* L.). Theoretical and Applied Genetics 75, 217–221.

Huber, K.C., BeMiller, J.N., 2001. Location of sites of reaction within starch granules. Cereal Chemistry 78, 173–180.

Hung, P.V., Morita, N., 2005. Physicochemical properties of hydroxypropylated and cross-linked starches from A-type and B-type wheat starch granules-review. Carbohydrate Polymers 59, 239–246.

Inagaki, T., Seib, P.A., 1992. Firming of bread crumb with cross-linked waxy barley starch substituted for wheat starch. Cereal Chemistry 69, 321–325.

Jacobs, H., Delcour, J.A., 1998. Hydrothermal modifications of granular starch, with retention of the granular structure: a review. Journal of Agricultural and Food Chemistry 46, 2895–2905.

Jane, J.L., Kasemsuwan, T., Leas, S., Ia, A., Zobel, H., Il, D., 1994. Anthology of starch granule morphology by scanning electron microscopy. Starch 46, 121–129.

Jane, J., Chen, Y.Y., Lee, L.F., McPherson, A.E., Wong, K.S., Radosavljevic, M., Kasemsuwan, T., 1999. Effects of amylopectin branch chain length and amylose content on the gelatinization and pasting properties of starch. Cereal Chemistry 76, 629–637.

Jane, J., Kasemsuwan, T., Chen, J.F., Juliano, B.O., 1996. Phosphorus in rice and other starches. Cereal Foods World 41, 827–832.

Jane, J.L., Robyt, J.F., 1984. Structure studies of amylose V complexes and retrogradaded amylose by action of alpha amylase, a new method for preparing amylodextrins. Carbohydrate Research 132, 105–110.

Juszczak, L., 2003. Surface of triticale starch granules - NC-AFM observations. Electronic Journal of Polish Agricultural Universities. Series Food Science and Technology 6.

Jyothi, A.N., Moorthy, S.N., Rajasekharan, K.N., 2007. Studies on the synthesis and hydroxypropyl derivatives of cassava (*Manihot esculenta Crantz*) starch. Journal of the Science of Food and Agriculture 87, 1964–1972.

Kalpeko, C., Zieba, T., Michalski, A., 2012. Effect of the production method on the properties of RS3/RS4 type resistant starch. Part 2. Effect of a degree of substitution on the selected properties of acetylated retrograded starch. Food Chemistry 135, 2035–2042.

Karim, A.A., Norziah, M.H., Seow, C.C., 2000. Methods for the study of starch retrogradation. Food Chemistry 71, 9–36.

Kaur, L., Singh, J., Singh, N., 2006. Effect of cross-linking on some properties of potato (*Solanum tuberosum* L.) starches. Journal of the Science of Food and Agriculture 86, 1945–1954.

Kaur, L., Singh, J., McCarthy, O.J., Singh, H., 2007a. Physico-chemical, rheological and structural properties of fractionated potato starches. Journal of Food Engineering 82, 383–394.

Kaur, L., Singh, J., Singh, N., Ezekiel, R., 2007b. Textural and pasting properties of potatoes as affected by storage temperature. Journal of the Science of Food and Agriculture 87, 520–526.

Kaur, L., Singh, N., Singh, J., 2004. Factors influencing the properties of hydroxypropylated potato starches. Carbohydrate Polymers 55, 211–223.

Kaur, L., Singh, N., Sodhi, N.S., 2002. Some properties of potatoes and their starches II. Morphological, thermal and rheological properties of starches. Food Chemistry 79, 183–192.

Kavitha, R., BeMiller, J.N., 1998. Characterization of hydroxypropylated potato starch. Carbohydrate Polymers 37, 115–121.

Kennedy, J.F., Knill, C.J., Taylor, D.W., 1995. Maltodextrins. In: Kearsley, M.W., Dziedzic, S.Z. (Eds.), Handbook of Starch Hydrolysis Products and their Derivatives. Blackie Academic & Professional, London, Glasgow, Weinheim, New York, Tokyo, Melbourne, Madras, pp. 65–82.

Kerr, R.W., Cleveland, Jr., F.C., 1962. Thickening Agent and Method of Making the Same. U.S. Patent 3,021,222.

Kim, H.R., Hermansson, A.M., Eriksson, C.E., 1992. Structural characteristics of hydroxypropyl potato starch granules depending on their molar substitution. Starch 44, 111–116.

Kim, S.Y., Wiesenborn, D.P., Orr, P.H., Grant, L.A., 1995. Screening potato starch for novel properties using differential scanning calorimetry. Journal of Food Science 60, 1060–1065.

Koch, V.H., Bommer, H.D., Koppers, J., 1982. Analytical investigations on phosphate cross-linked starches. Starch 34, 16–21.

Kong, X., Zhu, P., Sui, Z., Bao, J., 2015. Physicochemical properties of starches from diverse rice cultivars varying in apparent amylose content and gelatinisation temperature combinations. Food Chemistry 172, 433–440.

Koo, S.H., Lee, K.Y., Lee, H.G., 2010. Effect of cross-linking on the physicochemical and physiological properties of corn starch. Food Hydrocolloids 24, 619–625.

Kossmann, J., Lloyd, J., 2000. Understanding and influencing starch biochemistry. Critical Reviews in Biochemistry and Molecular Biology 35, 141–196.

Krueger, B.R., Knutson, C.A., Inglett, G.E., Walker, C.E., 1987. A differential scanning calorimetry study on the effect of annealing on gelatinization behaviour of corn starch. Journal of Food Science 52, 715–718.

Kudla, E., Tomasik, P., 1992. The modification of starch by high pressure. Part II. Compression of starch with additives. Starch 44, 253–259.

Kurkov, S.V., Loftsson, T., 2013. Cyclodextrinas. International Journal of Pharmaceutics 453, 167–180.

Lawal, O.S., 2004. Succinyl and acetyl starch derivatives of a hybrid maize: physicochemical characteristics and retrogradation properties monitored by differential scanning calorimetry. Carbohydrate Research 339, 2673–2682.

Lawal, O.S., 2011. Hydroxypropylation of pigeon pea (*Cajanus cajan*) starch: preparation, functional characterizations and enzymatic digestibility. LWT-Food Science and Technology 44, 771–778.

Lee, L., Yoo, B., 2011. Effect of hydroxypropylation on physical and rheological properties of sweet potato starch. LWT-Food Science and Technology 44, 765–770.

Lewandowicz, G., Soral-Śmietana, M., 2004. Starch modification by iterated syneresis. Carbohydrate Polymers 56, 403–413.

Lewandowicz, G., Jankowski, T., Fornal, J., 2000. Effect of microwave radiation on physico-chemical properties and structure of cereal starches. Carbohydrate Polymers 42, 193–199.

Li, J.H., Vasanthan, T., Rossnagel, B., Hoover, R., 2001. Starch from hull-less barley: I. Granule morphology, composition and amylopectin structure. Food Chemistry 74, 395–405.

Lim, S.T., Lee, J.H., Shin, D.H., Lim, H.S., 1999. Comparison of protein extraction solutions for rice starch isolation and effects of residual protein content on starch pasting properties. Starch 51, 120–125.

Lim, S.-T., Kasemsuwan, T., Jane, J.-L., 1994. Characterization of phosphorus in starch by ^{31}P-nuclear magnectic resosance spectroscopy. Analytical Techniques and Instrumentation 71, 488–493.

Lindebooma, N., Chang, P.R., Tylera, R.T., 2004. Analytical, biochemical and physicochemical aspects of starch granule size, with emphasis on small granule starches: a review. Starch 56, 89–99.

Liu, H., Ramsden, L., Corke, H., 1997. Physical properties and enzymatic digestibility of acetylated ae, wx, and normal maize starch. Carbohydrate Polymers 34, 283–289.

Liu, H., Ramsden, L., Corke, H., 1999a. Physical properties and enzymatic digestibility of hydroxypropylated ae, wx and normal maize starch. Carbohydrate Polymers 40, 175–182.

Liu, H., Ramsden, L., Corke, H., 1999b. Physical properties of cross-linked and acetylated normal and waxy rice starch. Starch 51, 249–252.

Liu, J., Wang, B., Lin, L., Zhang, J., Liu, W., Xie, J., Ding, Y., 2014. Functional, physicochemical properties and structure of cross-linked oxidized maize starch. Food Hydrocolloids 36, 45–52.

Luo, Z., Fu, X., He, X., Luo, F., Gao, Q., Yu, S., 2008. Effect of ultrasonic treatment on the physicochemical properties of maize starches differing in amylose content. Starch 60, 646–653.

Luo, Z.-G., Shi, Y.-C., 2012. Preparation of acetylated waxy, normal, and high-amylose maize starches with intermediate degrees of substitution in aqueous solution and their properties. Journal of Agricultural and Food Chemistry 60, 9468–9475.

van der Maarel, M.J., van der Veen, B.A., Euverlink, G.J.W., Dijkhuizen, L., 2002. Exploring and exploiting starch-modifying amylomaltases from hyperthermophiles. Biochemical Society Transactions 32, 279–282.

van der Maarel, M.J.E.C., Capron, I., Euverlink, G.-J.W., Bos, H.T., Kaper, T., Binnema, D.J., Steeneken, P.A.M., 2005. A novel thermoreversible gelling product made by enzymatic modification of starch. Starch 57, 465–472.

Majzoobi, M., Seifzadeh, N., Farahnaky, A., Mesbahi, G., 2015. Effects of sonication on physical properties of native and cross-linked wheat starches. Journal of Texture Studies 46, 105–112.

Mason, W.R., 2009. In: BeMiller, J., Whistler, R. (Eds.), Starch: Chemistry and Technology. Academic Press, New York, pp. 745–795.

Mischnick, P., Kuhn, G., 1996. Model studies on methyl amyloses: correlation between reaction conditions and primary structure. Carbohydrate Research 290, 199–207.

Morikawa, K., Nishinari, K., 2000a. Rheological and DSC studies of gelatinization of chemically modified starch heated at various temperatures. Carbohydrate Polymers 43, 241–247.

Morikawa, K., Nishinari, K., 2000b. Effects of concentration dependence of retrogradation behaviour of dispersions for native and chemically modified potato starch. Food Hydrocolloids 14, 395–401.

Muhrbeck, P., Eliasson, A.-C., 1991. Influence of the naturally occurring phosphate esters on the crystallinity of potato starch. Journal of the Science of Food and Agriculture 55, 13–18.

Muhrbeck, P., Tellier, C., 1991. Determination of the phosphorylation of starch from native potato varieties by 31P NMR. Starch 43, 25–27.

Nierle, W., 1969. The influence of the manufacturing conditions on the properties of phosphated corn starch and their applications. Starch 21, 13–15.

Noda, T., Takigawa, S., Matsuura-Endo, C., Kim, S.J., Hashimoto, N., Yamauchi, H., Hanashiro, I., Takeda, Y., 2005. Physicochemical properties and amylopectin structures of large, small, and extremely small potato starch granules. Carbohydrate Polymers 60, 245–251.

Nordmark, T.S., Ziegler, G.R., 2002. Spherulitic crystallization of gelatinized maize starch and its fractions. Carbohydrate Polymers 49, 439–448.

Nurul, I.M., Azemi, B.M.N.M., Manan, D.M.A., 1999. Rheological behaviour of sago (*Metroxylon sagu*) starch paste. Food Chemistry 64, 501–505.

Ortega-Ojeda, F.E., Larsson, H., Eliasson, A.-C., 2005. Gel formation in mixtures of hydrophobically modified potato and high amylopectin potato starch. Carbohydrate Polymers 59, 313–327.

Paschall, E.F., 1964. Phosphorylation with inorganic phosphate salts. Methods in Carbohydrate Chemistry 4, 296–298.

Perera, C., Hoover, R., Martin, A.M., 1997. The effect of hydroxypropylation on the structure and physicochemical properties of native, defatted and heat- moisture treated potato starches. Food Research International 30, 235–247.

Phillips, D.L., Liu, H.L., Pan, D., Corke, H., 1999. General application of Raman spectroscopy for the determination of level of acetylation in modified starches. Cereal Chemistry 76, 439–443.

Prochaska, K., Konowal, E., Sulej-Chojnacka, J., Lewandowicz, G., 2009. Physicochemical properties of cross-linked and acetylated starches and products of their hydrolysis in continuous recycle membrane reactor. Colloids and Surfaces B: Biointerfaces 74, 238–243.

Pukkahuta, C., Shobsngob, S., Varavinit, S., 2007. Effect of osmotic pressure on starch: new method of physical modification of starch. Starch 58, 78–90.

Qin, F., Man, J., Cai, C., Xu, B., Gu, M., Zhu, L., 2012. Physicochemical properties of high-amylose rice starches during kernel development. Carbohydrate Polymers 88, 690–698.

Raben, A., Andersen, K., Karberg, M.A., Holst, J.J., Astrup, A., 1997. Acetylation of or β-cyclodextrin addition to potato starch: beneficial effect on glucose metabolism and appetite sensations. The American Journal of Clinical Nutrition 66, 304–314.

Rajagopalan, S., Seib, P.A., 1992. Properties of granular cold-water soluble starches prepared at atmospheric pressure. Journal of Cereal Science 16, 29–40.

Ratnayake, W.S., Jackson, D.S., 2008. Phase transition of cross-linked and hydroxypropylated corn (*Zea mays* L.) starches. LWT-Food Science and Technology 41, 346–358.

Reddy, I., Seib, P.A., 2000. Modified waxy wheat starch compared to modified waxy corn starch. Journal of Cereal Science 31, 25–39.

Richardson, P.H., Jeffcoat, R., Shi, Y.C., 2000a. High-amylose starches: from biosynthesis to their use as food ingredients. MRS Bulletin 25, 20–24.

Richardson, S., Nilsson, G.S., Bergquist, K., Gorton, L., Mischnick, P., 2000b. Characterisation of the substituent distribution in hydroxypropylated potato amylopectin starch. Carbohydrate Research 328, 365–373.

Richardson, S., Gorton, L., Cohen, A.S., 2001. High-performance anion-exchange chromatography-electrospray mass spectrometry for investigation of the substituent distribution in hydroxypropylated potato amylopectin starch. Journal of Chromatography A 917, 111–121.

Richardson, S., Nilsson, G., Cohen, A., Momcilovic, D., Brinkmalm, G., Gorton, L., 2003. Enzyme-aided investigation of the substituent distribution in cationic potato amylopectin starch. Analytical Chemistry 75, 6499–6508.

Ring, S.G., Collona, P., Panson, K.J., Kalicheversky, M.T., Miles, M.J., Morris, V.J., Oxford, P.D., 1987. The gelation and crystallization of amylopectin. Carbohydrate Research 162, 277–293.

Rutenberg, M.W., Solarek, D., 1984. Starch derivatives: production and uses. In: Whistler, R.L., et al. (Ed.), Starch: Chemistry and Technology. Academic Press, London, pp. 312–388.

Sahai, D., Jackson, D.S., 1996. Structural and chemical properties of native corn starch granules. Starch 48, 249–255.

Sang, Y., Seib, P.A., 2006. Resistant starches from amylose mutants of corn by simultaneous heat-moisture treatment and phosphorylation. Carbohydrate Polymers 63, 167–175.

Sang, Y., Seib, P.A., Herrera, A.I., Prakash, O., Shi, Y.-C., 2010. Effects of alkaline treatment on the structure of phosphorylated wheat starch and its digestibility. Food Chemistry 118, 323–327.

Saroja, N., Shamala, T.R., Tharanathan, R.N., 2000. Biodegration of starch-g-polyacrylonitrile, a packaging material, by *Bacillus cereus*. Process Biochemistry 36, 119–125.

Schmitz, C.S., de Simas, K.N., Santos, K., Joao, J.J., de Mello Castanho Amboni, R.D., Amante, E.R., 2006. Cassava starch functional properties by etherification- hydroxypropylation. International Journal of Food Science & Technology 41, 681–687.

Schoch, T.J., 1942. Non-carbohydrate substance in the cereal starches. Journal of the American Chemical Society 64, 2954.

Seow, C.C., Thevamalar, K., 1993. Internal plasticization of granular rice starch by hydroxypropylation: effects on phase transitions associated with gelatinization. Starch 45, 85–88.

Sha, X.S., Xiang, Z.J., Bin, L., Jing, L., Bin, Z., Jiao, Y.J., Kun, S.R., 2012. Preparation and physical characteristics of resistant starch (type 4) in acetylated indica rice. Food Chemistry 134, 149–154.

Shewmaker, C.K., Boyer, C.D., Wiesenborn, D.P., Thompson, D.B., Boersig, M.R., Oakes, J.V., Stalker, D.M., 1994. Expression of *Escherichia coli* glycogen synthase in the tubers of transgenic potatoes (*Solanum tuberosum*) results in a highly branched starch. Plant Physiology 104, 1159–1166.

Shifeng, Y., Ying, M., Da-Wen, S., 2009. Impact of amylose content on starch retrogradation and texture of cooked milled rice during storage. Journal of Cereal Science 50, 139–144.

Shiftan, D., Ravanelle, F., Alexandre Mateescu, M., Marchessault, R.H., 2000. Change in V/B polymorphratioand T1 relaxation of epichlorohydrin cross-liked high amylose starch excipient. Starch 52, 186–195.

Shukri, R., Shi, Y.-C., 2015. Physiochemical properties of highly cross-linked maize starches and their enzymatic digestibilities by three analytical methods. Journal of Cereal Science 63, 72–80.

Shukri, R., Zhu, L., Seib, P.A., Maningat, C., Shi, Y.-C., 2015. Direct *in-vitro* assay of resistant starch in phosphorylated cross-linked starch. Bioactive Carbohydrates and Dietary Fibre 5, 1–9.

Sinah, V., Ali, S.Z., 2000. Acid degradation of starch. The effect of acid and starch type. Carbohydrate Polymers 41, 191–195.

Singh, J., Singh, N., 2001. Studies on the morphological, thermal and rheological properties of starch separated from some Indian potato cultivars. Food Chemistry 75, 67–77.

Singh, J., Singh, N., 2003. Studies on the morphological and rheological properties of granular cold water soluble corn and potato starches. Food Hydrocolloids 17, 63–72.

Singh, J., Kaur, L., McCarthy, O.J., 2007. Factors influencing the physico-chemical, morphological, thermal and rheological properties of some chemically modified starches for food applications–a review. Food Hydrocolloids 21, 1–22.

Singh, J., Kaur, L., Singh, N., 2004b. Effect of acetylation on some properties of corn and potato starches. Starch 56, 586–601.

Singh, J., McCarthy, O.J., Singh, H., 2006. Physico-chemical and morphological characteristics of New Zealand Taewa (Maori potato) starches. Carbohydrate Polymers 64, 569–581.

Singh, J., McCarthy, O.J., Singh, H., Moughan, P.J., 2008. Low temperature post-harvest storage of New Zealand Taewa (Maori potato): effects on starch physico-chemical and functional characteristics. Food Chemistry 106, 583–596.

Singh, J., Singh, N., Saxena, S.K., 2002. Effect of fatty acids on the rheological properties of corn and potato starch. Journal of Food Engineering 52, 9–16.

Singh, N., Kaur, L., 2004. Morphological, thermal and rheological properties of potato starch fractions varying in granule size. Journal of the Science of Food and Agriculture 84, 1241–1252.

Singh, N., Chawla, D., Singh, J., 2004a. Influence of acetic anhydride on physicochemical, morphological and thermal properties of corn and potato starch. Food Chemistry 86, 601–608.

Singh, N., Singh, J., Kaur, L., Sodhi, N.S., Gill, B.S., 2003. Morphological, thermal and rheological properties of starches from different botanical sources: a review. Food Chemistry 81, 219–231.

Sitohi, M.Z., Ramadan, M.F., 2001. Granular properties of different starch phosphate monoesters. Starch 53, 27–34.

Sodhi, N.S., Singh, N., 2005. Characteristics of acetylated starches prepared using starches separated from different rice cultivars. Journal of Food Engineering 70, 117–127.

Steeneken, P.A.M., Smith, E., 1991. Topochemical effects in the methylation of starch. Carbohydrate Research 209, 239–249.

Steeneken, P.A.M., Woortman, A.J.J., 1994. Substitution pattern in methylated starch as studied by enzymic degradation. Carbohydrate Research 258, 207–221.

Stute, R., Klinger, R.W., Boguslawski, S., Knorr, D., 1996. Effects of high pressures treatment on starches. Starch 48, 399–408.

Sujka, M., Jamroz, J., 2013. Ultrasound-treated starch: SEM and TEM imaging, and functional behaviour. Food Hydrocolloids 31, 413–419.

Svegmark, K., Helmersson, K., Nilsson, G., Nilsson, P.-O., Andersson, R., Svensson, E., 2002. Comparison of potato amylopectin starches and potato starches-influence of year and variety. Carbohydrate Polymers 47, 331–340.

Tessler, M.M., 1975. Hydroxypropylated, Inhibited High Amylose Retort Starches. U.S. Patent 3,904,601.

Tester, R.F., 1997. Starch: the polysaccharide fractions. In: Frazier, P.J., et al. (Ed.), Starch, Structure and Functionality. Royal Society of Chemistry, UK, pp. 163–171.

Tester, R.F., Debon, S.J.J., 2000. Annealing of starch—a review. International Journal of Biological Macromolecules 27, 1–12.

Tester, R.F., Morrison, W.R., 1990. Swelling and gelatinization of cereal starches. Cereal Chemistry 67, 558–563.

Tharanathan, R.N., 1995. Fat substitutes–a new approach. CFTRI Annual Conference Proceedings, p. 63.

Tharanathan, R.N., 2005. Starch-value addition by modification. Critical Reviews in Food Science and Nutrition 45, 371–384.

Vamadevan, V., Bertoft, E., 2015. Structure-function relationships of starch components. Starch 26, 55–68.

Wang, L., Wang, Y.-J., 2001. Structures and physico-chemical properties of acid-thinned corn, potato and rice starches. Starch 53, 570–576.

Wattanchant, S., Muhammad, K., Hashim, D., Rahman, R.A., 2003. Effect of cross-linking reagents and hydroxypropylation levels on dual-modified sago starch properties. Food Chemistry 80, 463–471.

Whistler, R.L., BeMiller, J.N., 1996. Starch. In: Whistler, R.L., BeMiller, J.N. (Eds.), Carbohydrate Chemistry for Food Scientists. Eagan Press, St. Paul, MN, pp. 117–151.

Wiesenborn, D.P., Orr, P.H., Casper, H.H., Tacke, B.K., 1994. Potato starch paste behaviour as related to some physical/chemical properties. Journal of Food Science 59, 644–648.

deWilligen, A.H.A., 1976. The rheology of starch. In: Radley, J.A. (Ed.), Examination and Analysis of Starch and Starch Products. Applied Science Publishers Ltd, London, pp. 61–90.

Wilke, O., Mischnick, P., 1997. Determination of the substitution pattern of cationic starch ethers. Starch 49, 453–458.

Woggum, T., Sirivongpaisal, P., Wittaya, T., 2015. Characteristics and properties of hydroxypropylated rice starch based biodegradable films. Food Hydrocolloids 50, 54–64.

Wolf, W.B., Bauer, L.L., Fahey Jr., G.C., 1999. Effects of chemical modification invitro rate and extent of food starch digestion: an attempt to discover a slowly digested starch. Journal of Food and Agricultural Chemistry 47, 4178–4183.

Wongsagonsup, R., Pujchakarn, T., Jitrakbumrung, S., Chaiwat, W., Fuongfuchat, A., Varavinit, S., Dangtip, S., Suphantharika, M., 2014. Effect of cross-linking on physicochemical properties of tapioca starch and its application in soup product. Carbohydrate Polymers 101, 656–665.

Woo, K.S., Seib, P.A., 1997. Cross-linking of wheat starch and hydroxypropylated wheat starch in alkaline slurry with sodium trimetaphosphate. Carbohydrate Polymers 33, 263–271.

Woo, K.S., 1999. Cross-linked, RS4 Type Resistant Starch: Preparation and Properties (Ph.D. thesis). University Manhattan, KS, Kansas State.

Wotton, M., Bamunuarachchi, A., 1979. Application of DSC to starch gelatinization. Starch 31, 201–204.

Wu, Y., Seib, P.A., 1990. Acetylated and hydroxypropylated distarch phosphates from waxy barley: paste properties and freeze–thaw stability. Cereal Chemistry 67, 202–208.

Wurzburg, O.B., 1986a. Converted starches. In: Wurzburg, O.B. (Ed.), Modified Starches: Properties and Uses. CRC Press, Florida, pp. 17–41.

Wurzburg, O.B., 1986b. Cross-linked starches. In: Wurzburg, O.B. (Ed.), Modified Starches: Properties and Uses. CRC Press, Boca Raton, Florida, pp. 41–53.

Wurzburg, O.B., 1986c. Nutritional aspects and safety of modified food starches. Nutrition Reviews 44, 74–79.

Xu, A., Seib, P.A., 1997. Determination of the level and position of substitution in hydroxypropylated starch by high- resolution 1H-NMR spectroscopy of alpha-limit dextrins. Journal of Cereal Science 25, 17–26.

Yamin, F.F., Lee, M., Pollak, L.M., White, P.J., 1999. Thermal properties of starch in corn variants isolated after chemical mutagenesis of inbred line B73. Cereal Chemistry 76, 175–181.

Yeh, A.I., Yeh, S.L., 1993. Property differences between cross-linked and hydroxypropylated rice starches. Cereal Chemistry 70, 596.

Yoneya, T., Ishibashi, K., Hironaka, K., Yamamoto, K., 2003. Influence of cross-linked potato starch treated with POCl$_3$ on DSC, rheological properties and granule size. Carbohydrate Polymers 53, 447–457.

Yook, C., Pek, U.H., Park, K.H., 1993. Gelatinization and retrogradation characteristics of hydroxypropylated cross-linked rices. Journal of Food Science 58, 405–407.

Yuan, R.C., Thompson, D.B., Boyer, C.D., 1993. Fine structure of amylopectin in relation to gelatinization and retrogradation behaviour of maize starches from three wax-containing genotypes in two inbred lines. Cereal Chemistry 70, 81–89.

Zallie, J., 1988. Benefits of quick setting starches. Manufacturing Confectioner 66, 41–43.

Zhao, J., Chen, Z., Jin, Z., Buwalda, P., Gruppen, H., Schols, H.A., 2015. Effects of granule size of cross-linked and hydroxypropylated sweet potato starches on their physicochemical properties. Journal of Agricultural and Food Chemistry 63, 4646–4654.

Zhao, J., Schols, H.A., Chen, Z., Jin, Z., Buwalda, P., Gruppen, H., 2012. Substituent distribution within cross-linked and hydroxypropylated sweet potato starch and potato starch. Food Chemistry 133, 1333–1340.

Zheng, G.H., Han, H.L., Bhatty, R.S., 1999. Functional properties of cross-linked and hydroxypropylation waxy hull-less barley starches. Cereal Chemistry 76, 182–188.

Zhu, J., Li, L., Chen, L., Li, X., 2012. Study on supramolecular structural changes of ultrasonic treated potato starch granules. Food Hydrocolloids 29, 116–122.

CHAPTER 9

Colored Potatoes

Lachman Jaromír[1], Hamouz Karel[2], Orsák Matyáš[1]

[1]Department of Chemistry, Faculty of Agrobiology, Food and Natural Resources, Czech University of Life Sciences, Prague, Czech Republic; [2]Department of Plant Production, Faculty of Agrobiology, Food and Natural Resources, Czech University of Life Sciences, Prague, Czech Republic

1. Introduction

Purple potato varieties are among hundreds of other varieties found in Parque de la Papa (Potato Park) near Cusco, Peru, a Spanish Colonial city high in the Andes. At Parque de la Papa, roughly 700 varieties native to the region are grown for food, enrichment, research, biodiversity, and preservation. Most of these varieties never see a market, as they are traded among communities or given as gifts. The purple potatoes are native to the Lake Titicaca area within the high plains and mountain slopes of Peru and Bolivia. They are among thousands of varieties that have been cultivated for nearly 8000 years in the Andean regions of Peru, Bolivia, and Ecuador. The diversity of purple potato varieties, their resistance to disease, and their ability to withstand harsh conditions have allowed them to evolve for thousands of years into a twenty-first-century food crop. Purple potatoes are cultivated in potato-growing regions of South America, North America, and Europe. The purple potato, botanical name *Solanum andigenum*, is the name designated to dozens of heirloom and heritage varieties of purple potatoes. Common names of these varieties include Purple Peruvian (fingerling variety), All Blue, Blue Congo, Lion's Paw, Vitilette, Purple Viking, Purple Majesty, and many others. Purple potatoes have deep violet, ink-colored skin and flesh. Depending on the specific variety, their coloring can be opaque or marbled throughout the flesh. Purple Majesty is known as the deepest purple of all the purple varieties, hence the given name. Purple potatoes are grown both for fresh market potatoes and for chipping potatoes. They are available year-round and are typically dry and starchy with a slight earthy and nutty flavor. If left to grow to maturity they become large and oblong, making them suitable for baking and mashing. Unlike white-fleshed potatoes, purple- and red-fleshed potatoes are rich in natural colorants with antioxidant properties—anthocyanins. These pigments are most often found in blue, red, and purple berries and pomegranates and have been shown to be an immune system booster and to aid in the prevention of certain cancers. The purple potato's nutritional value and energy-rich properties have become factors for the potato's explosion in popularity in the late twentieth century and early twenty-first century. Its ability to provide high quantities of vitamins, proteins, and antioxidants has become a valued measure of food security and sovereignty. The rule of thumb with fruits and vegetables is that the deeper and richer the color, the more nutritious the content tends to be. The purple potato is no exception to this rule. Further, these deeply pigmented antioxidants have shown great promise in protecting the integrity and structure of DNA and encouraging the production of cytokines, which are vital to proper immune response. This potato also shows impressive anti-inflammatory properties, helps to protect the health and integrity of the capillaries and strengthen membranes, and may have a role in regulating estrogenic activity, which can help lower the risk of hormone-related disease.

2. Potatoes with Red and Purple Flesh

Purple potato varieties with purple skin and white or yellow flesh have lower concentrations of anthocyanin than all-purple varieties. These potatoes may have purple skin and a variety of flesh tones, from pure white to creamy yellow to gold. There are also closely related varieties that exhibit blue skin and white/yellow flesh. Varieties of purple-skinned, white-fleshed potatoes include AC Domino, Bleue D'Auvergne, Brigus, Caribe, Cowhorn, Lion's Paw, and OAC Royal Gold, among others. Varieties with purple/blue flesh contain higher concentrations of anthocyanins. These potatoes are purple inside and out. The flesh may appear "splotchy" purple and white or pure, deep purple. Some varieties exhibit a white or cream ring near the skin, or even show white and purple "rings" when cut crossways. As with purple/whites, there are blue-skinned, blue-fleshed varieties that are closely related to purple varieties. It is often difficult to draw the line between which should be labeled blue or purple. Varieties of potatoes with purple flesh and skin (sometimes called blue) include Vitelette, Purple Peruvian, All Purple, Purple Majesty, Purple Viking, Davis Purple, Eureka Purple, Fenton Blue, Purple Mountain, Blue Tomcat, and many others. In Europe such varieties as Blaue St. Galler, Rote Emma, Highland Burgundy Red (Red Cardinal), Violette, Vitelotte, Blue Congo, Blaue Ludiano, Salad Blue, Blaue Hindel Bank, Blaue Schweden, Farbe Kartoffel, Salad Red, Shetland Black, British Columbia Blue, Hafija, Blaue Mauritius, and Valfi are becoming popular, especially in Switzerland, Germany, Great Britain, Sweden, and the Czech Republic. Some of their characteristics are given in Table 1.

3. Potato Antioxidants

The potato is currently the fourth most important food crop worldwide after maize, wheat, and rice, with production of more than 323 million tons. The per-capita potato consumption is highest in Europe, where it exceeds 80 kg per year. Potatoes are rich in certain antioxidants, such as polyphenolics (phenolcarboxylic acids), vitamin C, carotenoids, and selenium (Lachman et al., 2006).

Table 1: Characteristics of Selected Colored Potato Cultivars and Yellow-Fleshed Agria Cultivar (Lachman et al., 2012).

Cultivar	Origin of Seed Tubers	Maturity	Skin Color	Flesh Color	Tuber Shape	Yield[a] (t/ha)	Market Yield[a] (%)	Cooking Type[a,b]
Blaue St. Galler	Switzerland	Medium-early	Purple	Purple; partially colored	Oval to long oval	47.1	85.0	B
Rote Emma	Germany	Early to medium-early	Red	Red	Long oval	65.7	63.4	B
Highland Burgundy Red (Red Cardinal)	Germany	Medium-early	Red	Red with white borders	Oval to long oval	43.0	72.4	BC
Valfi	Czech Republic	Medium-early to medium-late	Purple	Purple; partially colored	Oval	50.4	90.4	C
Violette	Czech Republic (Gene Bank)	Medium-late	Purple	Dark purple	Oval, long	8.9	44.3	BC
Agria	Netherlands	Medium-early to medium-late	Yellow	Yellow	Long oval	56.4	95.5	CB
Russet Burbank	Czech Republic (Gene Bank)	Late	Medium russet	White	Long	51.0	95.0	CB

[a]Results from field experiment or tasting tests in 2010.
[b]A, firm; B, fairly firm; C, floury; D, very floury.

3.1 Polyphenols

Potato tubers contain secondary metabolites—polyphenolic compounds—presenting substrates for enzymatic browning of potatoes that occurs during peeling, cutting, or grating of raw potato tubers, which is caused by polyphenol oxidase (Jang and Song, 2004). L-Tyrosine (1–2×10^{-3} mol) and chlorogenic acid (CGA) (2–6×10^{-4} mol) (Dao and Friedman, 1992) are major polyphenolic potato constituents (Matheis, 1987, 1989; Leja, 1989; Niggeweg et al., 2004; Burlingame et al., 2009). The most common phenolic compound in potato tubers is the amino acid tyrosine (770–3900 mg/kg), followed by caffeic acid (280 mg/kg), scopolin (98 mg/kg), CGA (16–71 mg/kg), ferulic acid (28 mg/kg), and cryptochlorogenic acid (11 mg/kg). Caffeic acid may be a product of hydrolysis of CGA (Figure 1) and it exhibits strong antioxidant activity (AOA) as well as its related hydroxycinnamic acid compounds (Chen and Ho, 1997).

Yamamoto et al. (1997) have found caffeic acid levels in the edible parts of potatoes as high as 0.2–3.2 mg/kg; the total polyphenols were 422–834 mg/kg. The skin parts contained double the amount in each case. Some polyphenols are present only at lesser levels, such as neochlorogenic acid (7 mg/kg), p-coumaric acid (4 mg/kg), sinapic acid (3 mg/kg), and 3,4-dicaffeoyl-quinic acid (3 mg/kg). 3,5-Dicaffeoyl-quinic acid, scopoletin, and trans-feruloylputrescine were found only in small levels. Negrel et al. (1996) have found the occurrence of ether-linked ferulic acid amides (feruloyltyramine and/or feruloyloctopamine) in suberin-enriched samples of natural and wound periderms of potato tubers. The major part of the ether bonds involved the ferulic moiety of the amides. In the total plant there were identified glycosides of delphinidin (3-O-rutinoside), quercetin (3-O-glucoside or rutinoside), kaempferol (3-O-diglucoside-7-O-rhamnoside, 3-O-triglucoside-7-O-rhamnoside), and petunidin (3-O-rutinoside).

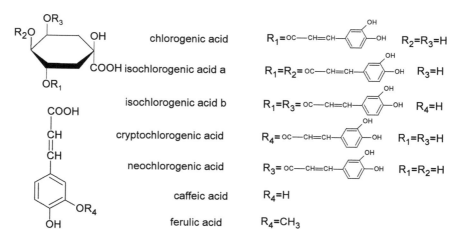

Figure 1

Major phenolic acids in potatoes.

(+)-Catechin is also included among the free phenolics in potatoes (Mendez et al., 2004). In diverse potato genotypes rutin and kaempferol-3-*O*-rutinoside were the most abundant flavonols (Navarre et al., 2011). Polyphenolic compounds, especially flavonoids, are effective antioxidants (Bors and Saran, 1987) owing to their capability to scavenge free radicals of fatty acids and oxygen (Good, 1994).

3.2 Anthocyanin Colorants

It was found that pigmented potato cultivars are a rich source of anthocyanins, in particular acylated derivatives (Eichhorn and Winterhalter, 2005). These phytochemicals have high free radical-scavenging activity, which helps to reduce the risk of chronic diseases and age-related neuronal degeneration. Purple- and red-fleshed potatoes provide a natural source of anthocyanins, which have been associated with health promotion. Anthocyanins in both fresh and processed fruit and vegetables serve two functions—they improve the overall appearance of the food, but also contribute to consumers' health and well-being (Stintzing and Carle, 2004). An important attribute of these pigments is that they are potent antioxidants in the diet (Brown et al., 2003; Brown, 2004). They are widely ingested by humans and their daily intake has been estimated at around 180 mg (Galvano et al., 2004). They are mainly contained in red and purple potato varieties in the skins and flesh of the tubers (Lachman and Hamouz, 2005; Lachman et al., 2005) and protect the human organism against oxidants, free radicals, and low-density lipoprotein cholesterol (Hung et al., 1997). The natural variation of cultivated potato germplasm includes types that are red and purple pigmented owing to the presence of anthocyanins (the structure of aglycones is given in Figure 2) in the skin and/or flesh (Brown et al., 2003). Red potato tubers (skins and flesh) contain pelargonidin glycosides—3-*O*-*p*-coumaroylrutinoside-5-*O*-glucoside (200–2000 mg/kg fresh weight (FW))—and in lesser amounts the glycoside of peonidin—3-*O*-*p*-coumaroylrutinoside-5-*O*-glucoside (20–200 mg/kg FW)

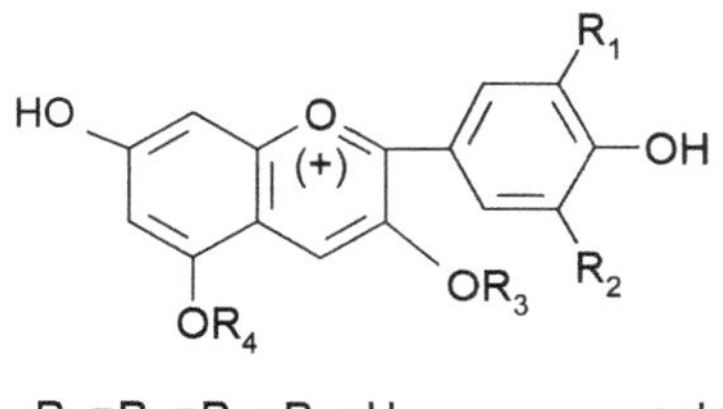

$R_1=R_2=R_3=R_4=H$ — pelargonidin
$R_1=OH$, $R_2=R_3=R_4=H$ — cyanidin
$R_1=R_2=OH$, $R_3=R_4=H$ — delphinidin
$R_1=OCH_3$, $R_2=R_3=R_4=H$ — peonidin
$R_1=OCH_3$, $R_2=OH$, $R_3=R_4=H$ — petunidin
$R_1=R_2=OCH_3$, $R_3=R_4=H$ — malvidin

Figure 2
Main anthocyanin aglycones of purple- and red-fleshed potatoes.

(Lewis et al., 1998). Purple tubers contain similar levels of the glycoside of petunidin—3-*O*-*p*-coumaroylrutinoside-5-*O*-glucoside—and much higher amounts of the malvidin glycoside 3-*O*-*p*-coumaroylrutinoside-5-*O*-glucoside (2000–5000 mg/kg FW). Total anthocyanins range from 69 to 350 mg/kg FW in the red-fleshed and 55 to 171 mg/kg FW in the purple-fleshed clones (Brown et al., 2003). Acylated pigments form more than 98% of the total anthocyanin content (TAC) of potatoes. Individual glycosides differ in acylation pattern by acid type, e.g., caffeic acid is contained in peonidin 3-*O*-[6-*O*-(4-*O*-*E*-caffeoyl-*O*-α-rhamnopyranosyl)-β-glucopyranoside]-5-*O*-β-glucopyranoside (10% anthocyanin content) and petunidin (6%). Naito et al. (1998) found that acylated glycosides of pelargonidin are characteristic of red potato.

The major pigment was identified as pelargonidin 3-*O*-[4″-*O*-(*trans*-*p*-coumaroyl)-α-L-6″-rhamnopyranosyl-β-D-glucopyranoside]-5-*O*-[β-D-glucopyranoside] by chemical and spectral measurements, and pelargonidin 3-*O*-[4″-*O*-(*trans*-feruloyl)-α-L-6″-rhamnopyranosyl-β-D-glucopyranoside]-*O*-[β-D-glucopyranoside] was determined as the minor pigment. In other glycosides *p*-coumaric acid is bound, e.g., in peonarin (25%) and petanin (37%). The same anthocyanins, but in other ratios, are contained in purple potato flesh (4%, 54%, and 32%). The other acylating acid is ferulic acid, e.g., in the purple variety Blue Congo the 3-*O*-[6-(4″-feruloyl-*O*-α-rhamnopyranosyl)-β-*O*-glucopyranoside]-5-*O*-glucopyranosides of petunidin and malvidin are present. The content of anthocyanins is stated as high as 20–400 mg/kg FW of tuber (Rodriguez-Saona et al., 1998). In red varieties pelargonidin 3-*O*-rutinoside-5-*O*-glucoside acylated by *p*-coumaric acid represents about 70% of the TAC. Red-pigmented potatoes contained predominantly acylated pelargonidin glycosides comprising about 80% of the total, while blue-fleshed potatoes contained these compounds and, in addition, acylated petunidin glycosides in a 2 to 1 ratio of the former to the latter (Brown et al., 2005). Structures of the major anthocyanidin glycosides are given in Table 2. Glycosides of peonidin, petunidin, and malvidin are the major anthocyanidin glycosides that contribute to the antioxidant properties of colored potato tubers. In the total plant glycosides of delphinidin (3-*O*-rutinoside), quercetin (3-*O*-glucoside or rutinoside), kaempferol (3-*O*-diglucoside-7-*O*-rhamnoside, 3-*O*-triglucoside-7-*O*-rhamnoside), and petunidin (3-*O*-rutinoside) were identified.

Phenolics are mostly contained in potato tuber peels (De Sotillo et al., 1994a,b). Hung et al. (1997), using the red tuber-producing potato cultivar Norland, observed changes in anthocyanin content and tuber surface color during tuber development—the intensity of redness and anthocyanin content per unit of surface area decreased as tuber weight increased. High-pressure liquid chromatography (HPLC) showed that pelargonidin and peonidin are the major anthocyanidins in the tuber periderm. Previous reports on anthocyanins from various potatoes reveal a dominance of one or more of the *p*-coumaroyl-5-glucoside-3-rhamnoglucosides of pelargonidin, cyanidin, peonidin, delphinidin, petunidin, and malvidin. In some cases the anthocyanins are reported without acylation or the *p*-coumaroyl moiety could be replaced with feruloyl acyl

Table 2: Structure of Anthocyanin Glycosides in Purple and Red Potatoes.

R_1^*	R_2^*	R_3^*	R_4^*	Anthocyanin Glycoside
H	H			Pelargonidin 3-[6-O-(4-O-E-p-coumaroyl-O-α-rhamnopyranosyl)-β-D-glucopyranoside]-5-O-β-D-glucopyranoside
OCH$_3$	H			Peonarin, i.e., peonidin 3-[6-O-(4-O-E-p-coumaroyl-O-α-rhamnopyranosyl)-β-D-glucopyranoside]-5-O-β-D-glucopyranoside
OCH$_3$	H			Peonidin 3-[6-O-(4-O-E-caffeoyl-O-α-rhamnopyranosyl)-β-D-glucopyranoside]-5-O-β-D-glucopyranoside
OCH$_3$	OH			Petanin, i.e., petunidin 3-[6-O-(4-O-E-p-coumaroyl-O-α-rhamnopyranosyl)-β-D-glucopyranoside]-5-O-β-D-glucopyranoside
OCH$_3$	OH			Petunidin 3-[6-O-(4-O-E-caffeoyl-O-α-rhamnopyranosyl)-β-D-glucopyranoside]-5-O-β-D-glucopyranoside

Continued

Table 2: Structure of Anthocyanin Glycosides in Purple and Red Potatoes—cont'd

R_1^*	R_2^*	R_3^*	R_4^*	Anthocyanin Glycoside
OCH_3	OH			Petunidin 3-[6-O-(4-O-E-feruloyl-O-α-rhamnopyranosyl)-β-D-glucopyranoside]-5-O-β-D-glucopyranoside
OCH_3	OCH_3			Malvidin 3-[6-O-(4-O-E-p-coumaroyl-O-α-rhamnopyranosyl)-β-D-glucopyranoside]-5-O-β-D-glucopyranoside
OCH_3	OCH_3			Malvidin 3-[6-O-(4-O-E-feruloyl-O-α-rhamnopyranosyl)-β-D-glucopyranoside]-5-O-β-D-glucopyranoside

(Fossen et al., 2003). Acylated pigments constitute more than 98% of the TACs in both tubers and shoots (Fossen and Andersen, 2000). Colored potatoes may serve as a potential source for natural anthocyanin pigments, since they are a low-cost crop (Jansen and Flamme, 2006), but they are also a powerful source of potato antioxidant micronutrients (André et al., 2007). Thus, purple- and red-fleshed potatoes could be addressed as new sources of natural colorants and antioxidants with added value for the food industry and health of people (Reyes and Cisneros-Zevallos, 2007). Red- and purple-fleshed potatoes have acylated glucosides of pelargonidin, while purple potatoes have, in addition, acylated glucosides of malvidin, petunidin, peonidin, and delphinidin (Brown, 2005; Lachman et al., 2005). Findings of acylated anthocyanins with increased stability have shown that these pigments may impart desirable food characteristics. Among radishes, red potatoes, red cabbage, black carrots, and purple sweet potatoes, red potatoes stand out as a potential source of antioxidant anthocyanins (Reyes et al., 2003). The oxygen radical absorbance capacity (ORAC) and ferric reducing/antioxidant power (FRAP) assays revealed that the antioxidant levels in red- or purple-fleshed potatoes were two to three times higher than those of white- or yellow-fleshed potato. Breeding studies with pigmented potatoes have been performed with the aim of obtaining clones containing high levels of anthocyanins (Hayashi et al., 2003). Because antioxidant and antiviral activities of the potato anthocyanins come from the synergistic effects of each anthocyanin pigment (Lukaszewicz et al., 2004), it seems important nowadays to assess various pigmented potato cultivars regarding their individual anthocyanidin content and the contribution of the anthocyanidin composition to their AOA. Studies have had the main goal of generating potato tubers with increased levels of phenolic compounds and anthocyanins (contained in cell vacuoles, mainly in the epidermis) and thus potatoes with modified and improved antioxidant capacities (Kosieradzka et al., 2004; Lukaszewicz and Szopa, 2005; Stobiecki et al., 2003). Overexpression of DNA encoding dihydroflavonol 4-reductase in the sense orientation could result in an increase in tuber anthocyanins, a fourfold increase in petunidin and pelargonidin (Han et al., 2006). The antioxidant capacities of pigmented fractions from purple potato flakes are related to analyzed potato cultivars and their major anthocyanins (Andersen et al., 2002). By comparison of the color, the anthocyanin content per given surface area, and the phenolic content of the tuber periderm it was found that varieties of red potatoes differed significantly in color (Naito et al., 1998). The increase in periderm phenolics and decrease in anthocyanin content per given surface area could cause darkening of tubers. The ORAC and FRAP assays revealed that the antioxidant levels in red- or purple-fleshed potatoes were two to three times higher than in white- or yellow-fleshed potatoes. Breeding studies with pigmented potatoes and lines of transgenic potato plants (Stobiecki et al., 2003; Lukaszewicz et al., 2004; Kosieradzka et al., 2004; Lukaszewicz and Szopa, 2005) are now being performed with the aim of obtaining clones containing high levels of anthocyanins (Brown et al., 2003; Andersen et al., 2002). Because AOA of potato anthocyanins results from the synergistic effect of each anthocyanin pigment (Hayashi et al., 2003), it is therefore important to assess various pigmented potato cultivars regarding individual anthocyanin content, as well as the contribution of the anthocyanin composition to their AOA. Antioxidant capacities of pigmented fractions from purple potato flakes are related to analyzed potato cultivars and their major anthocyanins (Han et al., 2006). Among other potato phenolic antioxidants such as CGA, gallic acid, catechin, and caffeic acid, especially in purple-fleshed cultivars, malvidin-3-(*p*-coumaryl rutinoside)-5-galactoside mainly contributes to the AOA value (Reddivari et al., 2007).

3.3 Carotenoids

Carotenoids are also efficient antioxidants involved in the antioxidant network (Canfield and Valenzuela, 1993; Järvinen, 1995; Mayne, 1996). In yellow potatoes their concentration is on average 4 mg/kg. Mader (1998) found the total carotenoid content in 35 yellow-fleshed Czech potato varieties to range from 0.16 to 6.36 mg/kg with an average value of 1.94 mg/kg. Van Dokkum et al. (1990) found an average total carotenoid value of 0.75 mg/kg. According to Duke (1992a), the most common among them are β-carotene (1 mg/kg) and its derivative β-carotene-5,6-monoepoxide. But Ong and Tee (1992) found the most common to be lutein (0.13–0.60 mg/kg) and β-carotene (0.03–0.40 mg/kg). Granado et al. (1992) found in an early variety the major components to be lutein (0.12 mg/kg), zeaxanthin (0.04 mg/kg), and β-carotene (0.01 mg/kg) (Figure 3). Globally, potatoes contain up to about 2.7 mg/100 g total carotenoids (Burlingame et al., 2009).

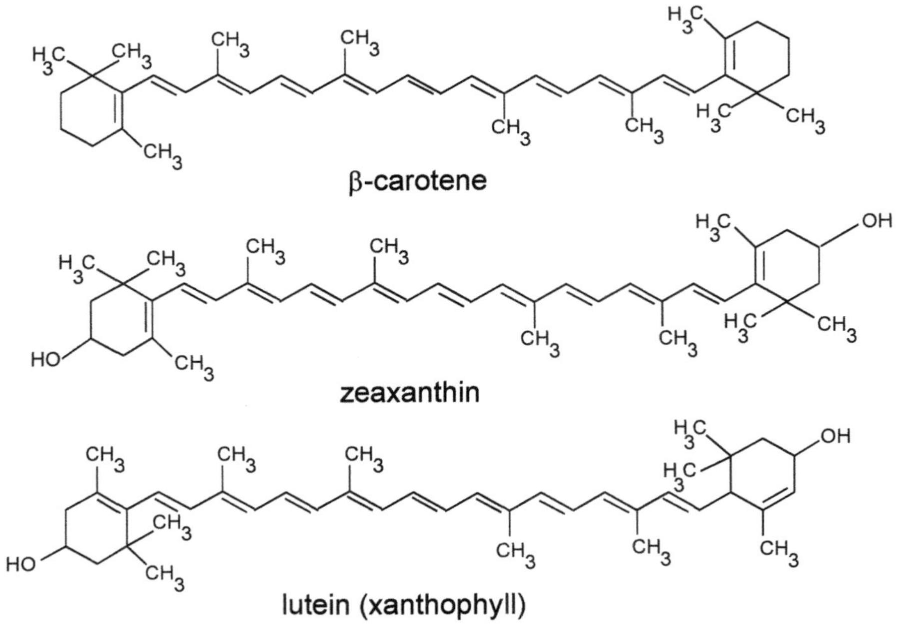

Figure 3
Major potato carotenoids.

Heinonen et al. (1989) found the major carotenoids to be lutein and zeaxanthin (0.13–0.60 mg/kg) and β-carotene (0.032–0.077 mg/kg). They found higher contents in older potato tubers after storage (in March) in comparison with new tubers (in August), which could be explained by changes in water content. Other carotenoids are contained only in minor levels. Among them were found α-carotene, *cis*-antheraxanthin-5,6-monoepoxide, *cis*-neoxanthin, *cis*-violaxanthin, cryptoxanthin, cryptoxanthin-5,6-diepoxide, hypoxanthin, lycopene, and *trans*-zeatine (Bergthaller et al., 1986; Duke, 1992a,b). Müller (1997) found a total carotenoid content in potato tubers of 4.5 mg/kg—this content was constituted by violaxanthin (1.8 mg/kg), antheraxanthin (zeaxanthin-5,6-epoxide, 1.3 mg/kg), lutein (1.0 mg/kg), zeaxanthin (0.16 mg/kg), neoxanthin (0.14 mg/kg), β-carotene (0.05 mg/kg), and β-cryptoxanthin (0.03 mg/kg). As determined by Mader (1998), total carotenoid content is highly dependent on the variety (the highest levels were found in the Agria, Lipta, Albína, Svatava, Zlata, Korela, Tara, Nikola, Lukava, and Karin varieties). Carotenoid content is affected strongly by the year of cultivation, on which semi-early varieties are more dependent in comparison with early varieties. He identified lutein and zeaxanthin (42–66% of peak area) as dominant, and to a lesser level β-carotene (1.1–3%) was present.

3.4 Other Potato Antioxidants

Ascorbic acid (AA) is the major naturally occurring inhibitor of enzymatic browning of potatoes (Almeida and Nogueira, 1995). It reduces the initial oxidation products, the *o*-quinones, back to *o*-diphenols until it is quantitatively oxidized to dehydroascorbic acid. AA also inhibits potato polyphenol oxidase directly by blocking the copper of the active site of the enzyme. AA contained in tubers attracts interest because regarding its content in tubers and the level of consumption, potatoes represent an important source of vitamin C in human nutrition. Potatoes are very rich in AA, containing 170–990 mg/kg (Duke, 1992a). Even in cooked potato tubers on average 130 mg AA/kg remains, and in microwaved potato tubers 151 mg/kg. AA content is affected by extrinsic and intrinsic factors such as variety, year of cultivation, manner of cultivation, environmental conditions, stage of maturity, storage conditions, and many others (Hamouz et al., 1999a,b; Orsák et al., 2001; Lachman et al., 2006). Dipierro and de Leonardis (1997) investigated the changes in the components of the ascorbate–glutathione system during the storage of potato tubers for 40 weeks at both 30 and 90 °C in relation to lipid peroxidation. The AA content of tubers decreased during storage at both 30 and 90 °C. The dehydroascorbate content reached a maximum after about 8 weeks and was significantly higher in tubers stored at 30 °C. Ascorbate free radical reductase, dehydroascorbate reductase, and glutathione reductase, the enzymes involved in the regeneration of AA, were not affected by temperature and remained quite unchanged throughout storage. It can be concluded that the ascorbate system is involved in the scavenging of the free radicals responsible for lipid peroxidation in stored potato tubers at least at low temperatures and in the first period of storage. Potato tubers are also rich in α-tocopherol (0.5–2.8 mg/kg)

(Packer, 1994) and their selenium content is relatively sufficient among vegetables (0.01 mg/kg). Djujic et al. (2000) estimated the average daily dietary intake of selenium as 29.72 µg/day and the contribution of vegetables and potatoes was 6.5%. Another vitamin-like antioxidant contained in potato tubers is α-lipoic acid, which is known as potato growth factor. Inside the cell, α-lipoic acid is readily reduced to dihydrolipoic acid, which neutralizes free radicals (Packer et al., 1995). It directly destroys damaging superoxide radicals, hydroperoxyl radicals, and hydroxyl radicals.

4. Colored Potato AOA

Among vegetables purple- and red-fleshed potatoes stand out as a potential source of antioxidant anthocyanins (Lachman et al., 2000; Reyes et al., 2003). ORAC and FRAP assays revealed that the antioxidant levels in red- or purple-fleshed potatoes were two to three times higher than in white- or yellow-fleshed potato. Breeding studies with pigmented potatoes have been performed with the aim of obtaining clones containing high levels of anthocyanins (Hayashi et al., 2003). Because antioxidant and antiviral activity of the potato anthocyanins comes from the synergistic effect of each anthocyanin pigment (Lukaszewicz et al., 2004), it seems important nowadays to assess various pigmented potato cultivars regarding their individual anthocyanidin content and the contribution of the anthocyanidin composition to their AOA. Several studies have had the main goal of generating potato tubers with increased levels of phenolic compounds and anthocyanins (contained in cell vacuoles, mainly in the epidermis) and thus potatoes with modified and improved antioxidant capacities (Kosieradzka et al., 2004; Lukaszewicz and Szopa, 2005; Stobiecki et al., 2003). Overexpression of DNA encoding dihydroflavonol 4-reductase in the sense orientation could result in an increase in tuber anthocyanins, for example, a fourfold increase in petunidin and pelargonidin (Han et al., 2006). Antioxidant capacities of pigmented fractions from purple potato flakes are related to analyzed potato cultivars and their major anthocyanins (Andersen et al., 2002). By a comparison of AOA values obtained by the individual assays it is evident that the 2,2′-azino-bis(3-ethylbenzothiazoline-6-sulfonic acid) (ABTS) and FRAP assays provided significantly different results in comparison with the 2,2-diphenyl-1-picrylhydrazyl (DPPH) assay. On the basis of plotting the ratios of AOA/TAC expressed as DPPH/TAC, ABTS/TAC, and FRAP/TAC, the highest values were obtained from the cultivars Blue Congo, Shetland Black, Highland Burgundy Red, and British Columbia Blue (Lachman et al., 2009). Plotting the ratios AOA/TAC×AN, where AN is the content of individual anthocyanidins, revealed that the major contributors among individual anthocyanidins to the AOA could be peonidin (Pn), delphinidin (Dp), malvidin (Mv), and cyanidin (Cy) and, to a lesser extent, petunidin (Pt) and pelargonidin (Pg), especially in the cultivars Blue Congo, Highland Burgundy Red, and Shetland Black (Pn > Dp > Mv > Cy > Pt > Pg). Regarding the structure of these anthocyanidins, a high degree of hydroxylation and methoxylation of their aromatic rings suggests their higher contribution to the AOA of potato cultivars. But as Stushnoff et al. (2008) and Im et al. (2008) found, the pigmented genotypes with red and purple skin and flesh contained considerably higher levels of CGA isomers than nonpigmented genotypes, so that the high total phenolic acid content (especially CGA and its isomers and caffeic acid) contributes to higher AOA values in red and purple potatoes. Reddivari et al. (2007) evaluated CGA, gallic acid, catechin, caffeic acid, and malvidin-3-(*p*-coumaryl rutinoside)-5-galactoside as the major polyphenols identified in specialty colored potato selections and estimated that CGA contributed 28%–45% to the AOA.

According to the results obtained by Brown (2005) and Brown et al. (2005), anthocyanins together with phenolic acids and other phenolics and carotenoids contribute to higher AOA of colored potatoes, and later results could support this evidence (Lachman et al., 2009). However, the superoxide anion-scavenging activity of anthocyanin pigments is changed by acylation with phenolic acids (caffeic acid) and the activity of acylated anthocyanins varies depending on the activity of each deacylated one (Moriyama et al., 2003). Reyes et al. (2005) examined the anthocyanin and total phenolic content of various purple- and red-fleshed potato cultivars and reported values ranging from 110 to 1740 mg cyanidin-3-glucoside per kilogram of FW and from 760 to 1810 mg CGA per kilogram of FW. High positive correlations between AOA and the content of these phenolics suggest that these compounds are mostly responsible for the AOA. Previous results (Lachman et al., 2008a,b) indicate that purple-fleshed potatoes have a significantly higher AOA than yellow-fleshed cultivars and that AOA and total polyphenol content are generally linearly correlated. A correlation was found between anthocyanin content and AOA ($r^2 = 0.789$). The greatest positive correlation was found between total polyphenol content and hydroxyl radical-scavenging activity ($r = 0.912$, $p < 0.01$).

5. Factors Influencing Levels of Beneficial Phytochemicals, AOA, and Antinutrients in Colored Potatoes

The main factors affecting phytochemicals and antinutrient contents, their stability, and AOA in potatoes are genotype, agronomic factors, postharvest storage, cooking methods, and processing (Ezekiel et al., 2013).

5.1 Effect of Potato Varieties and Cultivars

The main benefit of colored varieties of potatoes is their anthocyanin content. Owing to the presence of anthocyanins, the AOA of juice from colored potatoes is multiple times higher in comparison to that from yellow-fleshed ones. The majority of colored cultivars contain levels of carotenoids comparable to those of yellow-fleshed ones (Hejtmánková et al., 2013). Moreover, the colored varieties Highland Burgundy Red and Blaue Annelise are relatively high in β-carotene content, the only provitamin present in potatoes. The main carotenoid in all varieties is lutein (54–93%); furthermore, violaxanthin, neoxanthin, zeaxanthin, and β-carotene have been identified in most of the analyzed samples (content of total carotenoid ranged from 0.779 to 13.3 mg/kg dry matter (DM)).

TAC and representation of individual anthocyanidins differ significantly among individual colored potato varieties and cultivars and may be their fingerprint (Lachman et al., 2012). TAC of analyzed cultivars expressed as Cy content varied between 0.7 (Blue Congo) and 74.3 mg/100 g FW (Blaue Ludiano) (Figure 4) and agreed with the variation in anthocyanin contents in potato breeding values expressed as cyanidin-3-glucoside as reported by Brown et al. (2005).

In analyzed tubers of 38 native potato cultivars of various taxonomic groups from South America Brown et al. (2007) determined 0–23 mg Cy equivalents/400 g FW. For Highland Burgundy Red, a high content of Pg is characteristic (98.7%$_{rel}$); therefore, this cultivar differs significantly from other investigated cultivars (Lachman et al., 2009). Similarly British Columbia Blue contains almost exclusively Cy, which is also contained in lesser amounts in the cultivars Blaue Schweden, Farbe Kartoffel, Blaue Mauritius, and Shetland Black. The Violette and Vitelotte cultivars showed a relatively high content of Mv in all three experimental locations (Violette on average 85.2%$_{rel}$ and Vitelotte 81.7%$_{rel}$). In the cultivars Hafija and Blaue Ludiano higher Mv content was also determined. The Shetland Black cultivar differs from others by a higher content of Pn (on average 36.7%$_{rel}$); however, a high content of Pt was also found, similar to the other evaluated cultivars. High Pt content could be found in the cultivars Valfi, Blue Congo, Salad Blue, Blaue St. Galler, Blaue Hindel Bank, Blaue Ludiano, Blaue Schweden, Farbe Kartoffel, and Salad Red.

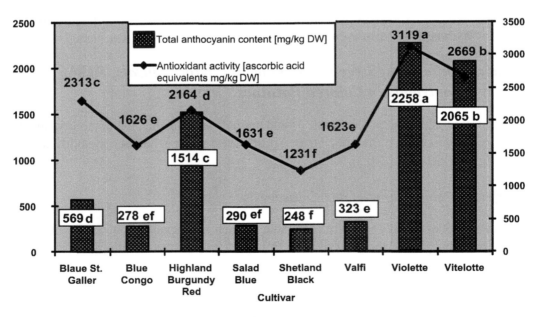

Figure 4
TAC and AOA of selected colored potatoes.

Though the abundance of anthocyanidins in the individual cultivars is significantly different, it can be deduced that on average in red- and purple-fleshed potatoes the most common anthocyanidin is Pt (46.9%$_{rel}$) followed by Mv (22.8%$_{rel}$) and Pg (22.1%$_{rel}$)>Cy (5.38%$_{rel}$)>Pn (2.74%$_{rel}$)>Dp (0.15%$_{rel}$). Individual cultivars significantly differ in their relative proportions of anthocyanidins, and their contents were shown to be specific for a given cultivar (Figure 5). The Violette and Vitelotte cultivars, with low ratios of AOA/TAC, were characterized by an intensive content of Mv (74% and 52% of anthocyanidins, respectively); the Highland Burgundy Red cultivar differed by having Pg as the dominant anthocyanidin (98%). Pt was the dominant anthocyanidin in the cultivars Salad Blue, Blaue St. Galler, Valfi, and Blue Congo (96%, 91%, 90%, and 89%, respectively). For the Shetland Black cultivar with deep blue skin and yellow flesh with small purple vascular rings, high contents of Pt and Pn (52% and 38%) have been found to be typical.

Principal component analysis did not allow the discrimination of clusters when applied to phenolic acids, while the scatter plot obtained with regard to the anthocyanin qualitative–quantitative profiles clustered the cultivars into five distinctive groups (Ieri et al., 2011). The French variety Vitelotte Noire differed significantly, with high anthocyanin content (500 mg/kg) in the pulp from other groups (Bamberger Hornchen; Shetland Black and Skerry Blue; Highland Burgundy Red, Rossa di Cetica, and Desirée; Blauer Schwede, Hermanns Blaue, Odenwalder Blaue, Viola Calabrese, and Quarantina Prugnona). The findings point out that ancient purple–red-flesh potatoes can be considered an interesting source of appreciable amounts of phenolic acids and especially of acylated anthocyanins. Pt derivatives were

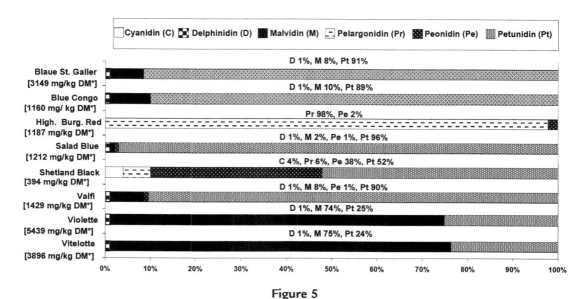

Figure 5

Relative distribution of individual anthocyanidins in percentage and their amounts in mg/kg dry weight (DW) in analyzed cultivars (*amount of individual anthocyanidins in mg/kg DW).

detected in the cultivars Hermanns Blaue, Shetland Black, and Vitelotte, and Pg in Highland Burgundy Red (Eichhorn and Winterhalter, 2005). Mv was the predominant aglycone of the variety Vitelotte, and Shetland Black contained mainly Pn derivatives.

The most important factor responsible for variation in anthocyanin content is genotype. In the study of (Kita et al., 2015), the anthocyanin content of colored potatoes ranged from 6.7 (Valfi cultivar) to 13.0 mg/100 g FW (Vitelotte cultivar) in purple-fleshed tubers and from 3.4 (Rosalinde cultivar) to 6.8 mg/100 g FW (Herbie 26 cultivar) in others. Purple potato cultivars are a richer source of polyphenols than red potato varieties. Among purple tuber varieties Vitelotte had the highest gallic acid equivalent (135.2 mg/100 g FW), followed by Blaue St. Galler, Blue Congo, and Blaue Elise (101.2, 95.8, and 84.1 mg/100 g FW, respectively). In potato cultivars grown in Luxembourg Vitelotte and Luminella had the highest polyphenol contents (5202 and 572 mg/kg DW) in the outer flesh (Deußer et al., 2012). Polyphenol content decreased from the peel via the outer to the inner flesh. Also AOA of purple-fleshed potatoes differed significantly, with the highest value of 10.7 mmol trolox equivalents (TE)/100 g DW (Blaue St. Galler variety) and the lowest antioxidant potential among the purple flesh varieties of 4.4 mmol TE/100 g DW (Valfi). Red-fleshed potato varieties showed similar antioxidant potential between 3.6 and 3.9 mmol TE/100 g DW (Kita et al., 2015).

5.2 Effects of Growing Location, Climatic Conditions, and Cultivation Practice on Major Potato Antioxidants and AOA

Biologically active compounds such as anthocyanins, polyphenols, carotenoids, and AA as well as AOA are also influenced by growing location, climate, and growing and storage conditions of potato tubers. In a study on the relationship of anthocyanins with color in organically and conventionally cultivated potatoes Murniece et al. (2014) obtained results showing that there was no significant influence of different cultivation practices on the amounts of anthocyanins ($p > 0.05$), while between varieties a significant difference was noticed ($p < 0.001$). A strong correlation between the amount of anthocyanins and the color was determined even when the amount of anthocyanins in potatoes was very small. There have been some reports on selenium increasing carbohydrates, organic selenium, total amino acids, AOA, and photosynthesis of potatoes (Cuderman et al., 2008; Germ et al., 2007). In a 2014 study, the main phenolic components in purple potatoes (CGA, caffeic acid, malvidin-5-glu-3-dirhamnose–glucose, caffeic acid–acetylrhamnose ester, and caffeic acid–prenylr-hamnose ester) were significantly increased by selenium (Lei et al., 2014). Total tuber weight was significantly increased by selenium ($r^2 = 0.964$, $p = 0.002$). The reason might be that the higher number of tubers led to smaller tubers. Selenium could increase tuber numbers of purple potatoes, but inhibit tuber expansion.

Anthocyanin and total phenolic contents decrease with tuber growth and maturity (Reyes et al., 2004). Longer days and cooler temperatures favored about 2.5 and 1.4 times higher anthocyanin and phenolic content, respectively. Higher contents of total as well as individual carotenoids and anthocyanins were found in colored potatoes from higher elevation locations with higher average global radiation during the vegetation period (Kotíková et al., 2007; Hejtmánková et al., 2013); however, the differences were statistically nonsignificant. Warmer periods with a higher ratio of sunny days per vegetation period significantly increased carotenoid and anthocyanin content. A statistically significant effect of genotype on CGA in potatoes with different flesh colors with a demonstrable effect of flesh color was found (Hamouz et al., 2013). Higher CGA levels were found in warm locations with frequent periods of drought (Hamouz et al., 2011). In the organically grown potatoes significantly higher levels of CGA were determined compared with conventional treatment. The highest total polyphenol contents were reported in potatoes with various flesh colors in locations with extreme climatic conditions: in those under drought stress in warm land with light sandy soil and also in mountainous areas (Hamouz et al., 2010). Similarly, a pronounced trend of the highest content of AA in the yellow-, purple-, and red-fleshed varieties in the location with the highest temperature averages during the vegetation period was reported (Hamouz et al., 2009).

5.3 Effects of Peeling and Cooking Methods on the Bioactive Compounds of Colored Potatoes

Potatoes are increasingly consumed worldwide, being cooked in diverse ways. The concentration and stability of bioactive compounds, although being present in minor amounts, might represent an important contribution to their daily ingestion. However, their levels are affected by the thermal processes used during home cooking or industrial processing. Independent of the thermal methods used, potato variety and unpeeled versus peeled cooking are also determinant factors. Bioactive compounds include vitamins, carotenoids, anthocyanins, and phenolics with favorable antioxidant properties; on the other hand unfavorable glycoalkaloids are also present. Removal of the peel before or after cooking appears to influence the bioactive compounds in boiled, steam-cooked, fried, baked, or microwaved potatoes (Santos et al., 2015).

Peeling showed the most favorable effect on decreasing unwholesome steroid glycoalkaloids (SGAs), α-solanine and α-chaconine (Lachman et al., 2013). SGAs were decreased by peeling on average to 34.7%, α-solanine to 43.6%, and α-chaconine to 30.5% of levels determined in unpeeled tubers, and similar values were found for red cultivars (decreased to 42.8%, 56.6%, and 36.6%) and purple cultivars (decreased to 39.2%, 46.9%, and 35.7%). These results are in good agreement with those reporting that losses of SGAs were twice as high after peeling than after boiling (Tajner-Czopek et al., 2008; Rytel, 2012). However, in all tested potato cultivars higher losses of α-chaconine than α-solanine were caused by peeling, compared with boiling, whereby conversely a higher decrease was found for α-solanine (Lachman et al., 2013). Remaining levels of SGAs were different depending on the potato cultivar; in the Blaue St. Galler cultivar 57% of the original level remained after peeling, whereas in the Agria cultivar only 5% remained.

Peeling affected AA content only to a small degree, with an average increase to 103.2%, and in red and purple cultivars to 106.2% and 106.1%, respectively. A larger decrease in peeled tubers was determined for CGA content (on average to 65.7%, in red potatoes to 61.3%, and in purple potatoes to 69.6% of levels contained in unpeeled tubers), proving that CGA is concentrated mainly in tuber peel as was reported by Deußer et al. (2012).

The effects of peeling on TAC in individual cultivars were different. For instance, in the purple Violette cultivar a decrease to 58.2% was found, suggesting that major TAC amounts were concentrated in peel and outer tuber flesh, while in red Rote Emma and HB Red cultivars an increase to 286.4% and 127.2% was observed.

AA content was reduced in all cooking treatments. The largest decrease was observed in baked potatoes, while the smallest one was in boiled potatoes. The percentage decrease in boiled potatoes appeared to be more dependent on the cultivar.

It was found that the AA content in fresh unpeeled raw tubers varied in a fairly wide range between 296.7 and 593.0 mg/kg DW, compared to 304.5–552.4 mg/kg DW in fresh peeled tubers. Significantly, the highest content was determined in the yellow-fleshed Agria cultivar, while the traditional white-fleshed Russet Burbank cultivar had the second lowest content of AA. AA content in the cultivars with colored flesh ranged between 296.7 (Violette) and 521.6 mg/kg DW (Highland Burgundy Red) in fresh unpeeled tubers and from 304.5 to 552.4 mg/kg DW in fresh peeled tubers, respectively.

As a result of culinary treatments, AA content decreased on average in all cultivars to 69% (boiling of peeled tubers), 67% (microwave treatment), and 37% (baking) of the original value of the raw unpeeled tubers. The effect of baking can be considered as somewhat less friendly, because baking as the treatment caused the highest AA losses. A somewhat different result, however, was detected for microwave heating in the group of cultivars with red and purple flesh, which, unlike the overall results for the average of all cultivars regardless of the color of flesh, showed no significant difference in the smaller decrease in AA. This is relative to the results for the Highland Burgundy Red and Russet Burbank cultivars, in which a smaller decrease in AA was found after microwave processing in comparison with cooked tubers.

It has been shown that changes in AA content by individual treatments were related to the genotype of each cultivar. The content of AA decreased by boiling and peeling in the traditional Agria and Russet Burbank cultivars to 70% and 65% of its original value, respectively, while in the group of cultivars with colored-flesh AA content decreased to 59% (Highland

Burgundy Red) and to 92% (Blaue St. Galler) of the initial value of fresh unpeeled tubers. After microwave treatment the lowest AA level remained in the cultivar Valfi and the highest in Russet Burbank (55% and 82% of original value, respectively). After baking AA content decreased to 23% (Highland Burgundy Red) or 56% (Agria) of the initial levels of raw unpeeled tubers.

Evaluation of fresh potatoes for AA content revealed that the yellow Agria and red-fleshed High Burgundy Red and Rote Emma cultivars differed significantly from other cultivars, being distinguished by high AA content and also by mutually significantly differing.

Several authors have described a considerable reduction in the quantities of vitamin C during cooking depending on cultivar and cooking method (Murniece et al., 2011). Obtained results differ significantly for each type of cooking method in comparison with uncooked potatoes. Differences in AA concentration were found among cooking methods and storage times (Burgos et al., 2009), and significant noncrossover interactions with genotype were observed for both of these parameters. It was found that AA concentration of boiled tubers of the six evaluated cultivars was higher than in oven-baked and microwaved tubers.

CGA content was reduced by all cooking methods in comparison with fresh unpeeled tubers; however, the largest decrease was caused by microwave treatment. Certain cultivars differed significantly; the highest CGA retention occurred in the red or purple High Burgundy Red and Violette cultivars, while the largest decrease was observed in the yellow-fleshed Agria cultivar (Lachman et al., 2013). CGA ranged from 314.9 to 2401 mg/kg DW in the raw unpeeled tubers and from 102.1 to 1845 mg/kg DW in raw peeled tubers. The lowest CGA was detected in the control yellow-fleshed Agria cultivar, while cultivars with colored flesh contained between 5.0 (Emma Rote with red flesh) and 7.6 times higher (Violette with dark purple flesh) CGA levels. The fundamental difference in CGA between the control yellow-fleshed Agria cultivar and the cultivars with red and purple flesh may be associated with the content of anthocyanins in potatoes with colored flesh. Also, Deußer et al. (2012) found the highest polyphenol content in the colored Vitelotte and Luminella cultivars in the outer flesh (5202 and 572 mg/kg DM, respectively), with CGA and its isomers neo- and cryptochlorogenic acid as major polyphenols, with the highest contents in peels or outer layers.

All types of cooking demonstrably reduced CGA. CGA decreased on average in all cultivars by boiling of peeled potatoes to 52%, microwave treatment to 28%, and baking to 37% of its original value (Figure 6).

The most favorable influence on the content of this valuable antioxidant proved to be from boiling despite the fact that the potatoes were peeled (except the Highland Burgundy Red cultivar, for which it was baking); microwave treatment reduced CGA the most. Some CGA differences from the overall results were found in cultivars with colored flesh in terms of the impact of culinary treatment. Although a consistent trend of the effect of treatment on CGA

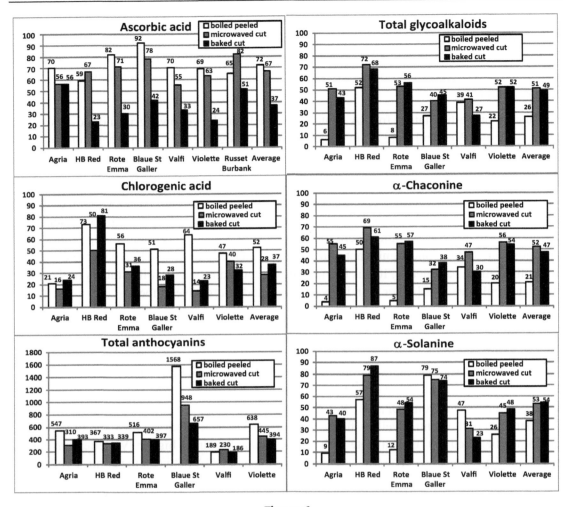

Figure 6

Changes in the content of assessed potato constituents from initial levels in raw unpeeled potatoes (100%) to the final level (%) caused by peeling/cooking, microwaving, and baking.

with the overall results for red-fleshed cultivars was found, the differences between variants were inconclusive owing to the greater internal variability of the group. In the group of cultivars with purple flesh, the difference between the variants of microwave treatment and baking of tubers was inconclusive. The cultivar type had a considerable influence on CGA changes following culinary treatment. Boiling in combination with peeling most demonstrably decreased CGA in the yellow-fleshed Agria cultivar (to 21%), while in the cultivars with colored flesh a decrease to only 47% (Violette) and 73% (Highland Burgundy Red) of original values in the raw tubers occurred. Also, after microwave treatment and baking, the Agria cultivar ranked among the cultivars with the largest CGA decrease (to 16% and 24% of original value). In the cultivars with colored-flesh CGA decreased to 14% (Valfi) and 50% (Highland Burgundy Red) after the microwave treatment and to 23% (Valfi) and 81% (Highland Burgundy Red) after the baking.

Potato tubers are always cooked before being eaten; however, there is very limited information on the effects of cooking on the properties of their constituents (except for AA) in cultivars with different flesh colors. The fresh pulp contains between 30 and 900 mg/kg CGA and minor amounts of other phenolic acids (0–30 mg/kg), but up to 1000–4000 mg/kg CGA can be present in the skin (Ieri et al., 2011). Navarre et al. (2011) evaluated the most abundant phenolic acid, CGA, in the range from 270 to 4730 mg/kg DW; it constitutes about 80% of the total phenolics (Brown, 2005). In a 2012 study, Perla et al. (2012) determined that on average, boiling, microwaving, and baking reduced total phenolics to 54%, 46.5%, and 46%, respectively. There were great differences—red-fleshed advanced selections CO97226-2 R/R and CO97222-1 R/R retained the highest levels of total phenolics (2100–2300 and 1300–2200 mg/kg, respectively) after various cooking treatments. Despite very high total phenolic levels in the raw tubers, Purple Majesty tubers lost approximately 60–70% of the total phenolics after boiling. All other cultivars lost approximately 40–60% of the total phenolics during boiling.

CGA content as well as total polyphenols was also assessed in Bolivian potato cultivars before, during, and after the traditional freezing and sun-drying of potatoes known as chuños (Peñarrieta et al., 2011). The results of this study show that the chuño process resulted in a slight/moderate loss of antioxidants and phenolic compounds; however, CGA remained almost unaltered.

SGAs were the most reduced by boiling and especially by previous peeling of the tubers; the smallest decrease was shown by microwave treatment (Lachman et al., 2013). Significant differences between cultivars have been found, e.g., the lowest glycoalkaloid contents were seen in the cultivar Rote Emma (14.73 ± 0.21 mg/kg DW SGAs, 8.07 ± 0.42 mg/kg DW α-solanine, and 6.67 ± 0.55 mg/kg DW α-chaconine) as well as Blaue St. Galler (36.83 ± 1.1 mg/kg DW SGAs, 19.7 ± 0.82 mg/kg DW α-solanine, and 17 ± 0.78 mg/kg DW α-chaconine) after boiling/peeling. Conversely, the highest retention of SGAs was observed in High Burgundy Red. The higher ratio of α-chaconine/α-solanine was found in potato peel in comparison with the flesh of peeled potatoes (Friedman and Levin, 2009); therefore, when the tubers are peeled before cooking, the ratio α-chaconine/α-solanine decreases.

In a study of the effect of cooking process on SGA content (Lachman et al., 2013), the content of total glycoalkaloids in the raw tubers of experimental cultivars ranged between 137.9 mg/kg DW (Blaue St. Galler cultivar—purple-fleshed) and 474.6 mg/kg DW (Highland Burgundy Red cultivar—red-fleshed). The content of glycoalkaloids of the yellow-fleshed cultivar Agria (282.6 mg/kg DW) slightly exceeded the average SGA content of all cultivars with colored flesh. All three methods of culinary treatment clearly reduced SGA levels.

On average in all cultivars SGAs were most decreased by boiling of peeled potatoes (to 26%), followed by baking (to 49%), and least decreased by microwave treatment (to 51% of the content in the raw tubers), as shown in Figure 6. A higher effect on SGA decrease in peeled and boiled potatoes was shown by peeling, especially in the High Burgundy Red, Agria, and Valfi cultivars. In all thermal treatments SGA changes were significantly influenced by the cultivar—depending on the genotype of each cultivar; the flesh color did not have a significant impact. Boiling decreased SGA level to the greatest extent in the yellow-fleshed Agria (to 6%) and red-fleshed Emma Rote cultivars (to 8% of initial value), whereas in the other red-fleshed cultivar, Highland Burgundy Red, it decreased the least—to 52% of its original value. Much more balanced results were obtained by other treatments of tubers. When the microwave treatment was applied, SGAs decreased in individual cultivars to 40–53% of initial values, except in the Highland Burgundy Red cultivar, in which there was a decline to 72% of original value. Baking decreased SGA content to between 68% (Highland Burgundy Red) and 27% (Valfi) of the levels in raw tubers. On average in all cultivars the ratio of α-solanine to α-chaconine in the raw tubers was determined as 1/2.2. For individual cultivars it ranged from 1/1.7 (Valfi) to 1/4.5 (Blaue St. Galler) and did not correspond with the color of tuber flesh. In all tuber treatments, the ratio of α-solanine to α-chaconine changed, and in most cases there was a greater reduction in the α-chaconine level. Boiling of peeled tubers (with the highest reduction in total glycoalkaloid content) significantly reduced the content of α-chaconine (to 21%) compared to α-solanine (to 38% of the raw unpeeled tubers) (Figure 5), with a resulting ratio of α-solanine to α-chaconine of 1/1.6. The decrease in α-solanine/α-chaconine ratio was almost the same. In baked potatoes, a significant reduction in α-chaconine (to 47%) in comparison with α-solanine (to 54% of initial value) was observed, resulting in a ratio of α-solanine to α-chaconine of 1/2.0.

In the four purple-fleshed Andean potato accessions belonging to *S. andigenum* the effects of boiling on the content of total polyphenols, total anthocyanins, CGA, and AOA were estimated (Burgos et al., 2013). Total phenolic content, CGA, and AOA after boiling increased, whereas TAC was in the accession Chalina the same as in raw tubers, in the accessions Leona and Bolona slightly increased, and in the Guincho accession decreased. In a 2012 study (Tajner-Czopek et al., 2012), the peeling process decreased the SGA content in tubers, independent of the potato cultivar studied, on average by about 20%, boiling of unpeeled potato by about 8%, and boiling of peeled potato by about 39%, compared with the raw material. Glycoalkaloid contents are the highest in the peel and lowest in the inner flesh (Deußer et al., 2012). Higher decreases in SGAs were found in fried products (mean about 83% in crisps, about 92% in french fries after the first stage of frying, and about 94% in ready french fries). The highest decreases in glycoalkaloids were caused by peeling (50%) and blanching (63%), and in dried potato dice production decreases were 33% and 17%, respectively (Rytel, 2012). Significant decreases in glycoalkaloid, particularly α-solanine, and nitrate contents during the process of chip production were observed (Pęksa et al., 2006).

The ratio of α-chaconine to α-solanine content during potato processing was maintained at a similar level during the whole process and was about 2.5/1. The highest amounts of glycoalkaloids were removed during peeling, slicing, washing, and frying. By careful selection of cultivars and boiling and processing methods (peeling, cooking), providing consumers with potatoes generally containing less than 100 mg total SGAs per kilogram FW could be possible.

In processed red- and purple-fleshed potatoes TAC was increased compared with fresh unpeeled or peeled tubers. The highest increase was observed in boiled potatoes, with intervarietal differences. The highest increase was determined in the cultivar with low TAC—Valfi (15.68 times); while in the high-TAC cultivar Violette the increase was relatively low (1.89 times). The content of TAC in the raw unpeeled tubers of five cultivars with colored flesh ranged between 161 (cv. Valfi) and 2420 mg Cy/kg DW (cv. Violette) and apparently corresponded to the intensity and color and the range of pale flesh marbling of individual cultivars.

Cooking treatments resulted in a significant increase in TAC in all cultivars in comparison with raw tubers. On average in all cultivars, the greatest TAC increase was found as a result of boiling of tubers (3.79 times against the value of raw tubers), followed by microwave treatment (3.06 times), and the smallest increase was seen in baked tubers (2.94 times). While the greatest impact on TAC increase by boiling was observed in all cultivars except for Violette, a higher TAC increase by microwaving was found in only three of the five cultivars.

In all cooking treatments the increase in TAC was influenced by cultivar. Boiling caused significantly the highest TAC increase in the cultivar Valfi (15.7 times), compared to the lowest level in the raw tubers, and the smallest increase in the cultivar Violette (1.86 times), which in turn had the highest TAC in raw tubers. During microwaving the TAC increased from 2.30 (cv. Violette) to 9.48 times (cv. Valfi) and during baking from 1.86 to 6.57 times, respectively. As in the case of boiling, the largest TAC increase was determined in the cultivar Valfi and the smallest in the cultivar Violette.

There appears to be a general consensus on the loss of anthocyanins when exposed to heating. However, for the pigmented potatoes, the heating treatment did not cause any changes in the phenolic acid content, while anthocyanins showed only a small decrement (16–29%) in various cultivars of potato with cooking treatment, boiling and microwaving (Mulinacci et al., 2008). Nevertheless, at the same time in three anthocyanin-rich red- and purple-fleshed potato cultivars boiling and microwave heating resulted in an increase or preservation of TAC compared with raw tubers (Brown et al., 2008). Perla et al. (2012) measured the absorption spectra in the visible wavelengths and observed a reduction in the absorption spectra due to cooking and hence suggested that the levels of anthocyanins were affected by cooking. Greater reductions in the absorption spectra were observed for baking and microwaving than for boiling. These changes can be ascribed to the drying temperature, the presence of oxygen during the drying process, and the drying time. Compared with baking and microwaving, boiling probably had less effect on various compounds in the tubers because of the thermal capacity of water. Anthocyanins are very unstable compounds and their stability can be affected by several factors such as temperature, pH, oxygen, and light, and they exist in different forms. It is known that the cooking process can detrimentally interfere with the amount of anthocyanins in a product, as shown by studies using pigmented potatoes (Lachman et al., 2012; Navarre et al., 2011). However, a significant increase in the amount of extracted anthocyanins from micro-waved Purple Majesty tubers, compared with the raw form, has been found, and the microwave process could be promising method for anthocyanin extraction (Lemos et al., 2012). The findings indicate that the structure of petanin, the major anthocyanin present in Purple Majesty potatoes, was unaltered by the microwave process, although the relative ratios between the forms of this anthocyanin were affected.

6. Health Benefits, Nutritional Aspects, and Use of Colored Potatoes

The functional properties of purple- and red-fleshed potatoes and the increased concern regarding their toxicological safety and antioxidant effects indicate that red- and purple-fleshed potatoes have a potential use in the nutraceutical industry. However, there is a need for the selection of useful genotypes and varieties to meet high AOA values.

Potato products made from potato varieties with colored flesh may be an interesting alternative to the traditional products. Consumers may become interested in colored potatoes as an attractive addition to salads or as a raw material for the production of colored crisps and french fries (Tajner-Czopek et al., 2012). In many American and European countries there is now a growing consumer and food producer interest in colored potato varieties in connection with their attractive color and taste as well as with the health-beneficial composition of the tubers. Color-fleshed potato varieties are on average characterized by a three times higher amount of total phenolic content than traditional yellow-fleshed potatoes (Rytel et al., 2014). The colored potato extracts showed higher radical-scavenging activities than the white-fleshed potato extracts and induced significant antiproliferative activities against THP-1 cells (Jang and Yoon, 2012). Similarly, purple potato flakes reduce the serum lipid profile in rats fed a cholesterol-rich diet, whereby the hypocholesterolemic action of dietary purple potato flakes might be related to cecal fermentation and steroid excretion due to phosphorus and polyphenols, including anthocyanins (Han et al., 2013).

The predominating phenolic acids in potato are CGA and its isomers—cryptochlorogenic, neochlorogenic, and isochlorogenic acids—which account for approximately 90% of total phenolic content in tubers. However, the content of phenolic acids in the tubers is not constant and can be changed by technological factors involved in their industrial processing. After peeling, the phenolic acid content was decreased by approximately 80% in the purple-fleshed potatoes, while the yellow varieties showed a decrease of 60% (Rytel et al., 2014). However, the dried potato dice obtained from light-fleshed potatoes had no content of phenolic acids, whereas the dice produced from colored-flesh potatoes contained about 4% of the original phenolic content. There were differences observed between potato varieties with respect to DM, starch, protein content, amino acids concentration, and protein quality, but independent of flesh color (Pęksa et al., 2013). The varieties Blaue Annelise, Highland Burgundy Red, Blue Congo, and Salad Blue were characterized by similar protein quality. Leucine limited the quality of the majority of colored-flesh potato varieties. The best amino acid profiles and protein quality confirmed by chemical scores and EAA index values characterized purple-fleshed Vitelotte and Blaue Annelise; red-fleshed Herbie 26, Highland Burgundy Red, and Rosemarie; and yellow-fleshed Verdi. Potatoes of colored-flesh varieties are characterized by low glycoalkaloid content (5.47 mg/100 g). The production of dehydrated potato dice influences the decrease in glycoalkaloid content in potato products (Rytel et al., 2013). The majority of these compounds are removed during the peeling (70%) and blanching process (29%). The blanching process influences the decrease in glycoalkaloid content more than the predrying process. Studied potatoes of colored-flesh varieties contain less than 300 mg/kg DW glycoalkaloids (Tajner-Czopek et al., 2012). Total glycoalkaloid levels in red-fleshed potato varieties are statistically higher by about 8% compared with the blue-fleshed varieties. The peeling process decreases total glycoalkaloid content on average by about 20%, cooking of unpeeled potato by about 8%, and cooking of peeled potato by about 39% in comparison with raw potatoes. Larger decreases in total glycoalkaloids were found in fried products—83% in crisps, about 92% in french fries after one stage of frying, and about 94% in ready french fries. No significant losses in total antioxidant capacity or total phenolic content were found in flakes using various drying methods (Nayak, 2011; Nayak et al., 2011a,b). However, losses in TAC were observed in potato flakes after drum drying or refractive window drying (41% and 23%, respectively). Most importantly, drum drying, which is mostly used in the potato industry to produce potato flakes, retained most of the phenolic antioxidants and hence, flours from colored potatoes may be used to produce products such as snack foods high in antioxidants and attractive natural colors. Interest in anthocyanin pigments has increased recently, owing to the range of colors, with their possible applications as natural dyes and potential health benefits as dietary antioxidants. The concentration of anthocyanins in purple-fleshed potatoes is similar to the highest anthocyanin production crops, such as blueberries, blackberries, cranberries, or grapes. Hence, purple-fleshed potatoes may be a potential source of natural anthocyanin pigments for industry using pulsed-electric-field-assisted extraction (Puértolas et al., 2013). A significant increase in the amount of extracted anthocyanins from microwaved Purple Majesty potato (rich in anthocyanins, the major one being petanin), compared with the raw form, has been found (Lemos et al., 2012). Thus, the use of microwave techniques (as well as microwave irradiation for extraction) has an effect on the location and form of anthocyanins compared with the raw potato. Time-resolved fluorescence techniques and fluorescence lifetime imaging could bring the potential of these techniques in obtaining pertinent information from extracted (nonpurified) anthocyanins and in situ within the potato tuber cells. Potato chips are among the most popular fried snacks obtained from potato. They are traditionally produced from potatoes featuring white and yellow flesh, to finally result in a product of a golden-yellow color, with characteristic taste and flavor, as well as delicate texture. An interesting alternative can prove to be chips from colored potatoes. In a study of the effects of frying on anthocyanin stability of crisps from red-fleshed (Rosalinde, Herbie 26, Highland Burgundy Red) and purple-fleshed potatoes (Salad Blue, Vitelotte, Valfi, Blue Congo) it was estimated that the process of frying caused degradation of anthocyanin compounds (38–70%), of high levels of total polyphenols (227–845 mg/100 g DW), and of anthocyanins (21–109 mg/100 g DW) in raw tubers (Kita et al., 2013). An HPLC–tandem mass spectrometry analysis showed that Pg and Mv derivatives were more stable during frying than Pt derivatives. Although the frying process affected the anthocyanin and polyphenol levels, the potato crisps obtained exhibited a bright intensive color and good AOA. The process of frying caused almost total degradation of anthocyanin compounds, while polyphenols exhibited good stability, especially in chips obtained from red-fleshed potatoes. The AOA decreased significantly in chips obtained from purple-fleshed potatoes, while better stability and better properties were exhibited by red-fleshed varieties and chips prepared from them (Kita et al., 2015). In the red-fleshed potato varieties, in which the destructive effect of frying on total polyphenols content was small, the TE antioxidant capacity values did not significantly change after frying. Small changes in AOA after frying were observed in Valfi, while in the Blaue St. Galler and Vitelotte varieties a significant reduction (around 75% and 50%, respectively) in radical-scavenging activity was detected. Other compounds formed during frying and associated with browning possess strong AOA. They could be products of the Maillard reaction, caramelization, or chemical oxidation of phenolic compounds. In the case of the Maillard reaction, high AOA is generally associated with the formation of brown melanoidins. No significant correlation of asparagine and reducing sugars with the flesh color of the potato tubers in fried products has been found (Kalita et al., 2013). The levels of asparagine and reducing sugars were positively correlated with acrylamide formation, while total phenolics and CGA correlated negatively. A significant influence of potato variety on experimental flour and snack properties has been reported (Nemś et al., 2015). Flours with the highest antioxidant activities were obtained from Salad Blue and Herbie 26 potatoes; however, the flour prepared from the Blue Congo exhibited a much higher total polyphenol and anthocyanin content. Snacks produced with colored flour had two to three times higher antioxidant activities, higher contents of polyphenols, attractive color, and better expansion.

The potato processing industry generates high amounts of peel as a by-product that is a good source of several beneficial functional ingredients including antioxidant polyphenols (Singh and Rajini, 2004). The red potato varieties Siècle and Purple Majesty had the highest antioxidant potential, total polyphenol content, and chlorogenic and caffeic acid contents compared to other varieties (Al-Weshahy and Rao, 2009). Bound and esterified phenolics contributed as much or even more than the free phenolics to the AOA of the peels; among four potato varieties (Purple, Innovator, Russet, and Yellow), extracts from the Purple variety showed the highest activity (Albishi et al., 2013). The peels of all varieties showed significantly ($p<0.05$) higher phenolic content and antioxidant activities compared to their flesh. Thus, colored potato processing discards may be used in food formulations and their extracts could potentially be employed as an effective source of antioxidants in food systems.

7. Conclusions and Future Trends

Red-, purple-, and blue-fleshed potatoes and their related products are an attractive novelty and an interesting alternative to the traditional white- or cream-colored potato flesh. They are rich in health-promoting antioxidants such as polyphenols (CGA and its isomers, caffeic acid, and flavonoids), anthocyanins, AA, and carotenoids with strong AOA and many functional properties, antiproliferative activities, and hypocholesterolemic action. The antioxidant levels in red- or purple-fleshed potatoes are two to three times higher than in white- or yellow-fleshed potatoes. Levels of antioxidants are mainly affected by variety (for colored potatoes glycosides of Pt, Pn, Pg, Cy, Mv, and Dp may be fingerprints). However, the growing location, climatic conditions, cultivation practice, storage, peeling, and cooking method may also affect their concentrations as well as glycoalkaloid concentrations. Potato products made from colored potatoes are an attractive addition to salads, and production of colored crisps and french fries may be an interesting alternative to the traditional products.

The major interest of scientists studying the pharmacological and nutritional activities of potato anthocyanins is to investigate the activities of the anthocyanins of special colored potato varieties with specific coloration patterns, mainly the red- or purple-skinned ones (Zhao et al., 2009). Different-colored potatoes may contain different contents of various kinds of anthocyanins. More detailed work is needed to determine the pharmacological and nutritional activities of the individual anthocyanins of colored potatoes and phenolic acids and their mutual synergic effects involved in AOA (Wegener et al., 2009). A general opinion is steadily forming that chronic dietary intake of potato antioxidants including anthocyanins has a positive effect on human health, meaning that their anthocyanins have potential in a new trend in the potato and food industries. Currently, potato breeders are attracted to developing new potato varieties with high levels of anthocyanins to increase the nutritional benefits of potatoes and processed potato foods. In addition, colored potatoes may be important alternative sources of natural colorants compared to synthetic ones.

Acknowledgment

Some results given in this chapter were supported by Grant Project NAZV QI101A184 of the Ministry of Agriculture of the Czech Republic.

References

Albishi, T., John, J.A., Al-Khalifa, A.S., Shahidi, F., 2013. Phenolic content and antioxidant activities of selected potato varieties and their processing by-products. Journal of Functional Foods 5, 590–600.

Almeida, M.E.M., Nogueira, J.N., 1995. The control of polyphenoloxidase activity in fruits and vegetables: a study of the interactions between the chemical compounds used and heat treatment. Plant Foods for Human Nutrition 47, 245–256.

Al-Weshahy, A., Rao, A.V., 2009. Isolation and characterization of functional components from peel samples of six potatoes varieties growing in Ontario. Food Research International 42, 1062–1066.

Andersen, A.W., Tong, C.B.S., Krueger, D.E., 2002. Comparison of periderm color and anthocyanins of four red potato varieties. American Journal of Potato Research 79, 249–253.

André, C.M., Ghislain, M., Bertin, P., Oufir, M., Herrera Del Rosario, M., Hoffman, L., Hausman, J.F., Larondelle, Y., Evers, D., 2007. Andean potato cultivars (*Solanum tuberosum* L.) as a source of antioxidant and mineral micronutrients. Journal of Agricultural and Food Chemistry 55, 366–378.

Bergthaller, W., Tegge, G., Hoffmann, W., 1986. Effect of storage temperature on color changes and content of carotenoids of dehydrated diced potatoes. In: Carghill, B.F. (Ed.), Engineering for Potatoes. ASAE, St. Joseph, MI, USA, pp. 456–475.

Bors, W., Saran, M., 1987. Radical scavenging by flavonoid antioxidants. Free Radical Research Communication 2, 289–294.

Brown, C.R., 2004. Nutrient status of potato: assessment of future trends. In: Proceedings of the Washington State Potato Conference, pp. 11–17.

Brown, C.R., 2005. Antioxidants in potato. American Journal of Potato Research 82, 163–172.

Brown, C.R., Wrolstadt, R., Durst, R., Yang, C.P., Clevidence, B., 2003. Breeding studies in potatoes containing high concentrations of anthocyanins. American Journal of Potato Research 80, 241–249.

Brown, C.R., Culley, D., Yang, C.P., Durst, R., Wrolstad, R., 2005. Variation of anthocyanin and carotenoid contents and associated antioxidant values in potato breeding lines. Journal of the American Society for Horticultural Science 130, 174–180.

Brown, C.R., Culley, D., Bonierbale, M., Amoros, W., 2007. Anthocyanin, carotenoid content, and antioxidant values in native South American potato cultivars. Hortscience 42, 1733–1736.

Brown, C.R., Durst, R.W., Wrolstad, R., De Jong, W., 2008. Variability of phytonutrient content of potato in relation to growing location and cooking method. Potato Research 51, 259–270.

Burgos, G., Amoros, W., Muñoa, L., Sosa, P., Cayhualla, E., Sanchez, C., Díaz, C., Bonierbale, M., 2013. Total phenolic, total anthocyanin and phenolic acid concentrations and antioxidant activity of purple-fleshed potatoes as affected by boiling. Journal of Food Composition and Analysis 30, 6–12.

Burgos, G., Auqui, S., Amoros, W., Salas, E., Bonierbale, M., 2009. Ascorbic acid concentration of native Andean potato cultivars as affected by environment, cooking and storage. Journal of Food Composition and Analysis 22, 533–538.

Burlingame, B., Mouillé, B., Charrondière, R., 2009. Nutrients, bioactive non-nutrients and anti-nutrients in potatoes. Journal of Food Composition and Analysis 22, 494–502.

Canfield, L.M., Valenzuela, J.G., 1993. Cooxidations: significance to carotenoid action invivo. In: Canfield, L.M., Krinsky, N.I., Olson, J.A. (Eds.). Canfield, L.M., Krinsky, N.I., Olson, J.A. (Eds.), Carotenoids in Human Health, vol. 691. Annals of the New York Academy of Sciences, New York, pp. 192–199.

Chen, J.H., Ho, C.T., 1997. Antioxidant activities of caffeic acid and its related hydroxycinnamic acid compounds. Journal of Agricultural and Food Chemistry 45, 2374–2378.

Cuderman, P., Kreft, I., Germ, M., Kovacevic, M., Stibilj, V., 2008. Selenium species in selenium-enriched and drought-exposed potatoes. Journal of Agricultural and Food Chemistry 19, 9114–9120.

Dao, L., Friedman, M., 1992. Chlorogenic acid content of fresh and processed potatoes determined by ultraviolet spectrophotometry. Journal of Agricultural and Food Chemistry 40, 2152–2156.

De Sotillo, R.D., Hadley, M., Holm, E.T., 1994a. Phenolics in aqueous potato peel extract: extraction, identification and degradation. Journal of Food Sciences 59, 649–651.

De Sotillo, R.D., Hadley, M., Holm, E.T., 1994b. Potato peel waste: stability and antioxidant activity of a freeze-dried extract. Journal of Food Sciences 59, 1031–1033.

Deußer, H., Guignard, C., Hoffmann, L., Evers, D., 2012. Polyphenol and glycoalkaloid contents in potato cultivars grown in Luxembourg. Food Chemistry 135, 2814–2824.

Dipierro, S., De Leonardis, S., 1997. The ascorbate system and lipid peroxidation in stored potato (*Solanum tuberosum* L.) tubers. Journal of Experimental Botany 48, 779–783.

Djujic, I.S., Jozanov-Stankov, O.N., Milovac, M., Jankovic, V., Djermanovic, V., 2000. Bioavailability and possible benefits of wheat intake naturally enriched with selenium and its products. Biological Trace Element Research 77, 273–285.

Duke, J.A., 1992a. Handbook of Phytochemical Constituents of Grass Herbs and Other Economic Plants. CRC Press, Boca Raton, FL.

Duke, J.A., 1992b. Handbook of Biologically Active Phytochemicals and Their Activities. CRC Press, Boca Raton, FL.

Eichhorn, S., Winterhalter, P., 2005. Anthocyanins from pigmented potato (*Solanum tuberosum* L.) cultivars. Food Research International 38, 943–948.

Ezekiel, R., Singh, N., Sharma, S., Kaur, A., 2013. Beneficial phytochemicals in potato–a review. Food Research International 50, 487–496.

Fossen, T., Andersen, Ø.M., 2000. Anthocyanins from tubers and shoots of the purple potato, Solanum tuberosum. Journal of Horticultural Sciences & Biotechnology 75, 360–363.

Fossen, T., Øvstedal, D.G., Slimestad, R., Andersen Ø, M., 2003. Anthocyanins from a Norwegian potato cultivar. Food Chemistry 81, 433–437.

Friedman, M., Levin, C.E., 2009. Analysis and biological activities of potato glycoalkaloids, calystegine alkaloids, phenolic compounds, and anthocyanins. In: Singh, J., Kaur, L. (Eds.), Advances in Potato Chemistry and Technology. Academic Press, Burlington, pp. 127–162.

Galvano, F., La Fauci, L., Lazzarino, G., Fogliano, V., Ritieni, A., Ciappellano, S., Battistini, N.C., Tavazzi, B., Galvano, G., 2004. Cyanidins: metabolism and biological properties. Journal of Nutritional Biochemistry 15, 2–11.

Germ, M., Kreft, I., Stibilj, V., Urbanc-Berčič, O., 2007. Combined effects of selenium and drought on photosynthesis and mitochondrial respiration in potato. Plant Physiology and Biochemistry 2, 162–167.

Good, D., 1994. The role of antioxidant vitamins. American Journal of Medicine 96, 5–12.

Granado, F., Olmedilla, B., Blanco, I., Rojas-Hidalgo, E., 1992. Carotenoid composition in raw and cooked Spanish vegetables. Journal of Agricultural and Food Chemistry 40, 2135–2140.

Hamouz, K., Lachman, J., Vokál, B., Pivec, V., 1999a. Influence of environmental conditions and type of cultivation on the polyphenol and ascorbic acid content in potato tubers. Rostlinna Vyroba 45, 293–298.

Hamouz, K., Čepl, J., Vokál, B., Lachman, J., 1999b. Influence of locality and way of cultivation on the nitrate and glycoalkaloid content in potato tubers. Rostlinna Vyroba 45, 495–501.

Hamouz, K., Lachman, J., Pazderů, K., Hejtmánková, K., Cimr, J., Musilová, J., Pivec, V., Orsák, M., Svobodová, A., 2013. Effect of cultivar, location and method of cultivation on the content of chlorogenic acid in potatoes with different flesh colour. Plant, Soil and Environment 59, 465–471.

Hamouz, K., Lachman, J., Pazderů, K., Tomášek, J., Hejtmánková, K., Pivec, V., 2011. Differences in anthocyanin content and antioxidant activity of potato tubers with different flesh colour. Plant, Soil and Environment 57, 478–485.

Hamouz, K., Lachman, J., Hejtmánková, K., Pazderů, K., Čížek, M., Dvořák, P., 2010. Effect of natural and growing conditions on the content of phenolics in potatoes with different flesh colour. Plant, Soil and Environment 56, 368–374.

Hamouz, K., Lachman, J., Dvořák, P., Orsák, M., Hejtmánková, K., Čížek, M., 2009. Effect of selected factors on the content of ascorbic acid in potatoes with different tuber flesh colour. Plant, Soil and Environment 55, 281–287.

Han, K.H., Sekikawa, M., Shimada, K., Hashimoto, M., Noda, T., Tanaka, H., Fukushima, M., 2006. Anthocyanin-rich purple potato flake extract has antioxidant capacity and improves antioxidant potential in rats. British Journal of Nutrition 96, 1125–1133.

Han, K.H., Kim, S.J., Shimada, K.I., Hashimoto, N., Yamauchi, H., Fukushima, M., 2013. Purple potato flake reduces serum lipid profile in rats fed a cholesterol-rich diet. Journal of Functional Food 5, 974–980.

Hayashi, K., Mori, M., Knox, Y.M., Suzutan, T., Ogasawara, M., Yoshida, I., Hosokawa, K., Tsukui, A., Azuma, M., 2003. Anti influenza virus activity of a red-fleshed potato anthocyanin. Food Science and Technology Research 9, 242–244.

Heinonen, M.I., Ollilainen, V., Linkola, E.K., Varo, P.T., Koivistoinen, P.E., 1989. Carotenoids in Finnish foods: vegetables, fruits, and berries. Journal of Agricultural and Food Chemistry 37, 655–659.

Hejtmánková, K., Kotíková, Z., Hamouz, K., Pivec, V., Vacek, J., Lachman, J., 2013. Influence of flesh colour, year and growing area on carotenoid and anthocyanin content in potato tubers. Journal of Food Composition and Analysis 32, 20–27.

Hung, C.-Y., Murray, J.R., Ohmann, S.M., Tong, C.B.S., 1997. Anthocyanin accumulation during potato tuber development. Journal of the American Society for Horticultural Science 122, 20–23.

Ieri, F., Innocenti, M., Andrenelli, L., Vecchio, V., Mulinacci, N., 2011. Rapid HPLC/DAD/MS method to determine phenolic acids, glycoalkaloids and anthocyanins in pigmented potatoes (*Solanum tuberosum* L.) and correlations with variety and geographical origin. Food Chemistry 125, 750–759.

Im, H.W., Suh, B.S., Lee, S.U., Kozukue, N., Ohnisi-Kameyama, M., Levin, C.E., Friedman, M., 2008. Analysis of phenolic compounds by high-performance liquid chromatography and liquid chromatography/mass spectrometry in potato plant flowers, leaves, stems, and tubers and in home-processed potatoes. Journal of Agricultural and Food Chemistry 56, 3341–3349.

Jang, J., Song, K.B., 2004. Purification of polyphenoloxidase from the purple-fleshed potato (*Solanum tuberosum* Jasim) and its secondary structure. Journal of Food Science 69, C648–C651.

Jang, H.L., Yoon, K.Y., 2012. Cultivar differences in phenolic contents/biological activities of color-fleshed potatoes and their relationships. Horticulture Environment and Biotechnology 53, 175–181.

Jansen, G., Flamme, W., 2006. Coloured potatoes (*Solanum tuberosum* L.)–anthocyanin content and tuber quality. Genetic Research and Crop Evolution 53, 1321–1331.

Järvinen, R., 1995. Carotenoids, retinoids, tocopherols and tocotrienols in the diet, the Finnish Mobile Clinic Health Examination Survey. International Journal for Vitamin and Nutrition Research 65, 24–30.

Kalita, D., Holm, D.G., Jayanty, S.S., 2013. Role of polyphenols in acrylamide formation in the fried products of potato tubers with coloured flesh. Food Research International 54, 753–759.

Kita, A., Bakowska-Barczak, A., Hamouz, K., Kułakowska, K., Lisińska, G., 2013. The effect of frying on anthocyanin stability and antioxidant activity of crisps from red- and purple-fleshed potatoes (*Solanum tuberosum* L.). Journal of Food Composition and Analysis 32, 169–175.

Kita, A., Bakowska-Barczak, A., Lisińska, G., Hamouz, K., Kułakowska, K., 2015. Antioxidant activity and quality of red and purple flesh potato chips. LWT-Food Science and Technology 62, 525–531.

Kosieradzka, I., Borucki, W., Matysiak-Kata, I., Szopa, J., Sawosz, E., 2004. Transgenic potato tubers as a source of phenolic compounds. Localization of anthocyanins in the peridermis. Journal of Animal and Feed Sciences 13, 87–92.

Kotíková, Z., Hejtmánková, A., Lachman, J., Hamouz, K., Trnková, E., Dvořák, P., 2007. Effect of selected factors on total carotenoid content in potato tubers (*Solanum tuberosum* L.). Plant, Soil and Environment 53, 355–360.

Lachman, J., Hamouz, K., Orsák, M., Pivec, V., 2000. Potato tubers as a significant source of antioxidants in human nutrition. Rostlinná Výroba 46, 231–236.

Lachman, J., Hamouz, K., 2005. Red and purple coloured potatoes as a significant antioxidant source in human nutrition–a review. Plant, Soil and Environment 51, 477–482.

Lachman, J., Hamouz, K., Orsák, M., 2005. Red and purple potatoes–a significant antioxidant source in human nutrition. Chemické Listy 99, 474–482.

Lachman, J., Hamouz, K., Čepl, J., Pivec, V., Šulc, M., Dvořák, P., 2006. The effect of selected factors on polyphenol content and antioxidant activity in potato tubers. Chemické Listy 100, 522–527.

Lachman, J., Hamouz, K., Orsák, M., Pivec, V., Hejtmánková, K., Pazderů, K., Dvořák, P., Čepl, J., 2012. Impact of selected factors–cultivar, storage, cooking and baking on the content of anthocyanins in coloured-flesh potatoes. Food Chemistry 133, 1107–1116.

Lachman, J., Hamouz, K., Musilová, J., Hejtmánková, K., Kotíková, Z., Pazderů, K., Domkářová, J., Pivec, V., Cimr, J., 2013. Effect of peeling and three cooking methods on the content of selected phytochemicals in potato tubers with various colour of flesh. Food Chemistry 138, 1189–1197.

Lachman, J., Hamouz, K., Šulc, M., Orsák, M., Pivec, V., Hejtmánková, A., Dvořák, P., Čepl, J., 2009. Cultivar differences of total anthocyanins and anthocyanidins in red and purple-fleshed potatoes and their relation to antioxidant activity. Food Chemistry 114, 836–843.

Lachman, J., Hamouz, K., Orsák, M., Pivec, V., Dvořák, P., 2008a. The influence of flesh colour and growing locality on polyphenolic content and antioxidant activity in potatoes. Scientia Horticulturae–Amsterdam 117, 109–114.

Lachman, J., Hamouz, K., Šulc, M., Orsák, M., Dvořák, P., 2008b. Differences in phenolic content and antioxidant activity in yellow and purple-fleshed potatoes grown in the Czech Republic. Plant, Soil and Environment 54, 1–6.

Lei, C., Ma, Q., Tang, Q.Y., Ai, X.R., Zhou, Z., Yao, L., Wang, Y., Wang, Q., Dong, J.Z., 2014. Sodium selenite regulates phenolics accumulation and tuber development of purple potatoes. Scientia Horticulturae 165, 142–147.

Leja, M., 1989. Chlorogenic acid as the main phenolic compound of mature and immature potato tubers stored at low and high temperature. Acta Physiologiae Plantarum 11, 201–206.

Lemos, M.A., Aliyu, M.M., Hungerford, G., 2012. Observation of the location and form of anthocyanin in purple potato using time-resolved fluorescence. Innovative Food Science and Emerging Technologies 16, 61–68.

Lewis, C.E., Walker, J.R.L., Lancaster, J.E., Sutton, K.H., 1998. Determination of anthocyanins, flavonoids and phenolic acids in potatoes. I: coloured cultivars of *Solanum tuberosum* L. Journal of Science of Food and Agriculture 77, 45–57.

Lukaszewicz, M., Matysiak-Kata, I., Skala, J., Fecka, I., Cisowski, W., Szopa, J., 2004. Antioxidant capacity manipulation in transgenic potato tuber by changes in phenolic compounds content. Journal of Agricultural and Food Chemistry 52, 1526–1533.

Lukaszewicz, M., Szopa, J., 2005. Pleiotropic effect of flavonoid biosynthesis manipulation in transgenic potato plants. Acta Physiologia Plantarum 27, 221–228.

Mader, P., 1998. Carotenoids in potato tubers of selected varieties of the Czech assortment. In: Book of Abstracts of XXIX. Symposium on New Trends of Production and Evaluation of Foods. Skalský Dvůr, Czech Republic, p. 8.

Matheis, G., 1987. Polyphenol oxidase and enzymatic browning of potatoes (*Solanum tuberosum*). II. Enzymatic browning and potato constituents. Chemie, Mikrobiologie, Technologie der Lebensmittel 11, 33–41.

Matheis, G., 1989. Polyphenol oxidase and enzymatic browning of potatoes (*Solanum tuberosum*). III. Recent progress. Chemie, Mikrobiologie, Technologie der Lebensmittel 12, 86–95.

Mayne, S.T., 1996. Beta-carotene, carotenoids, and disease prevention in humans. FASEB Journal 10, 690–701.

Mendez, C.D.V., Delgado, M.A.R., Rodriguez, E.M.R., Romero, C.D., 2004. Content of free compounds in cultivars of potatoes harvested in Tenerife (Canary Islands). Journal of Agricultural and Food Chemistry 52, 1323–1327.

Moriyama, H., Morita, Y., Ukeda, H., Sawamura, M., Terahara, N., 2003. Superoxide anion-scavenging activity of anthocyanin pigments. Journal of the Japanese Society for Food Science and Technology–Nippon Shokuhin Kagaku Kogaku Kaishi 50, 499–505.

Mulinacci, N., Ieri, F., Giaccherini, C., Innocenti, M., Andrenelli, L., Canova, G., Saracchi, M., Casiraghi, M.C., 2008. Effect of cooking on the anthocyanins, phenolic acids, glycoalkaloids and resistant starch content in two pigmented cultivars of *Solanum tuberosum* L. Journal of Agricultural and Food Chemistry 56, 11830–11837.

Müller, H., 1997. Determination of the carotenoid content in selected vegetables and fruits by HPLC and photodiode array detection. Food Research and Technology 204, 88–94.

Murniece, I., Karklina, D., Galoburda, R., Santare, D., Skrabule, I., Costa, H.S., 2011. Nutritional composition of freshly harvested and stored Latvian potato (*Solanum tuberosum* L.) cultivars depending on traditional cooking methods. Journal of Food Composition and Analysis 24, 699–710.

Murniece, I., Tomsone, L., Skrabule, I., Vaivode, A., 2014. The relationship of anthocyanins with color of organically and conventionally cultivated potatoes. International Journal of Biological, Veterinary, Agricultural and Food Engineering 8, 330–334.

Naito, K., Umemura, Y., Mori, M., Sumida, T., Okada, T., Takamatsu, N., Okawa, Y., Hayashi, K., Saito, N., Honda, T., 1998. Acylated pelargonidin glycosides from red potato. Phytochemistry 47, 109–112.

Navarre, D.A., Pillai, S.S., Shakya, R., Holden, M.J., 2011. HPLC profiling of phenolics in diverse potato genotypes. Food Chemistry 127, 34–41.

Nayak, B., 2011. Effect of Thermal Processing on the Phenolic Antioxidants of Coloured Potatoes (Ph.D. dissertation thesis). Department of Biological Systems Engineering, Washington State University.

Nayak, B., Berrios, J.J., Powers, J.R., Tang, J., Ji, Y., 2011a. Coloured potatoes (*Solanum tuberosum* L.) dried for antioxidant-rich value-added foods. Journal of Food Processing and Preservation 35, 571–580.

Nayak, B., Liu, R.H., Berrios, J.D., Tang, J.M., Derito, C., 2011b. Bioactivity of antioxidants in extruded products prepared from purple potato and dry pea flours. Journal of Agricultural and Food Chemistry 589, 8233–8243.

Negrel, J., Pollet, B., Lapierre, C., 1996. Ether-linked ferulic acid amides in natural and wound periderms of potato tuber. Phytochemistry 43, 1195–1199.

Nemś, A., Pęksa, A., Kucharska, A.Z., Sokół-Łętowska, A., Kita, A., Drożdż, W., Hamouz, K., 2015. Anthocyanin and antioxidant activity of snacks with coloured potato. Food Chemistry 172, 175–182.

Niggeweg, R., Michael, A.J., Martin, C., 2004. Engineering plants with increased levels of the antioxidant chlorogenic acid. Nature Biotechnology 22, 746–754.

Ong, A.S.H., Tee, E.S., 1992. Natural sources of carotenoids from plants and oils. In: Packer, L. (Ed.). Packer, L. (Ed.), Methods in Enzymology, vol. 213. Academic Press Inc., New York, pp. 142–167.

Orsák, M., Lachman, J., Vejdová, M., Pivec, V., Hamouz, K., 2001. Changes of selected secondary metabolites in potatoes and buckwheat caused by UV, γ- and microwave irradiation. Rostlinna Vyroba 47, 493–500.

Packer, L., 1994. Vitamin E is natures' master antioxidant. Science and Medicine 1, 54–63.

Packer, L., Witt, E.H., Tritschler, H.J., 1995. Alpha-Lipoic acid as a biological antioxidant. Free Radical Biology and Medicine 19, 227–250.

Pęksa, A., Gołubowska, G., Aniołowski, K., Lisińska, G., Rytel, E., 2006. Changes of glycoalkaloids and nitrate contents in potatoes during chip processing. Food Chemistry 97, 151–156.

Pęksa, A., Kita, A., Kułakowska, K., Aniołowska, M., Hamouz, K., Nemś, A., 2013. The quality of protein of coloured fleshed potatoes. Food Chemistry 14, 2960–2966.

Peñarrieta, J.M., Salluca, T., Tejeda, L., Alvarado, J.A., Bergenståhl, B., 2011. Changes in phenolic antioxidants during chuño production (traditional Andean freeze and sun-dried potato). Journal of Food Composition and Analysis 24, 580–587.

Perla, V., Holm, D.G., Jayanty, S.S., 2012. Effect of cooking methods on polyphenols, pigments and antioxidant activity in potato tubers. LWT–Food Science and Technology 45, 161–171.

Puértolas, E., Cregenzán, O., Luengo, E., Álvarez, I., Raso, J., 2013. Pulsed-electric-field-assisted extraction of anthocyanins from purple-fleshed potato. Food Chemistry 136, 1330–1336.

Reddivari, L., Hale, A.L., Miller, J.C., 2007. Determination of phenolic content, composition and their contribution to antioxidant activity in specialty potato selections. American Journal of Potato Research 84, 275–282.

Reyes, L.F., Cisneros-Zevallos, L., Brown, C.R., Wrolstad, R., Durst, R., Yang, C.P., Clevidence, B., 2003. Breeding studies in potatoes containing high concentrations of anthocyanins. American Journal of Potato Research 80, 241–250.

Reyes, L.F., Miller, J.C., Cisneros-Zevallos, L., 2004. Environmental conditions influence the content and yield of anthocyanins and total phenolics in purple- and red-flesh potatoes during tuber development. American Journal of Potato Research 81, 187–193.

Reyes, L.F., Miller, J.C., Cisneros-Zevallos, L., 2005. Antioxidant capacity, anthocyanins and total phenolics in purple- and red-fleshed potato (Solanum tuberosum L.) genotypes. American Journal of Potato Research 82, 271–277.

Reyes, L.F., Cisneros-Zevallos, L., 2007. Degradation kinetics and colour of anthocyanins in aqueous extracts of purple- and red-flesh potatoes (*Solanum tuberosum* L.). Food Chemistry 100, 885–894.

Rodriguez-Saona, L.E., Giusti, M.M., Wrolstad, R.E., 1998. Anthocyanin pigment composition of red-fleshed potatoes. Journal of Food Science 63, 458–465.

Rytel, E., 2012. Changes in glycoalkaloid and nitrate content in potatoes during dehydrated dice processing. Food Control 25, 349–354.

Rytel, E., Tajner-Czopek, A., Kita, A., Aniołowska, M., Kucharska, A.Z., Sokol-Letowska, A., Hamouz, K., 2014. Content of polyphenols in coloured and yellow fleshed potatoes during dices processing. Food Chemistry 161, 224–229.

Rytel, E., Tajner-Czopek, A., Aniołowska, M., Hamouz, K., 2013. The influence of dehydrated potatoes processing on the glycoalkaloids content in coloured-fleshed potato. Food Chemistry 141, 2495–2500.

Santos, C.S.P., Cunha, S., Casal, S., 2015. Chapter 14. Bioactive components in potatoes as influenced by thermal processing. In: Preedy, V.R. (Ed.), Processing and Impact on Active Components in Food. Academic Press, Elsevier, Amsterdam, pp. 111–119.

Singh, N., Rajini, P.S., 2004. Free radical scavenging activity of an aqueous extracts of potato peel. Food Chemistry 85, 611–616.

Stintzing, F.C., Carle, R., 2004. Functional properties of anthocyanins and betalains in plants, food, and human nutrition. Trends in Food Science and Technology 15, 19–38.

Stobiecki, M., Matysiak-Kata, I., Frański, R., Skała, J., Szopa, J., 2003. Monitoring changes in anthocyanin and steroid alkaloid glycoside content in lines of transgenic potato plants using liquid chromatography/mass spectrometry. Phytochemistry 62, 959–969.

Stushnoff, C., Holm, D., Thompson, M.D., Jiang, W., Thompson, H.J., Joyce, N.I., Wilson, P., 2008. Antioxidant properties of cultivars and selections from the Colorado potato breeding program. American Journal of Potato Research 85, 267–276.

Tajner-Czopek, A., Jarych-Szyszka, M., Lisińska, G., 2008. Changes in glycoalkaloids content of potatoes destined for consumption. Food Chemistry 106, 706–711.

Tajner-Czopek, A., Rytel, E., Kita, A., Pęksa, A., Hamouz, K., 2012. The influence of thermal process of coloured potatoes on the content of glycoalkaloids in the potato products. Food Chemistry 133, 1117–1122.

Van Dokkum, W., De Vos, R.H., Schrijver, J., 1990. Retinol, total carotenoids, beta-carotene, and tocopherols in total diets of male adolescents in the Netherlands. Journal of Agricultural and Food Chemistry 38, 211–216.

Wegener, C.B., Jansen, G., Jürgens, H.U., Schütze, W., 2009. Special quality traits of coloured potato breeding clones: anthocyanins, soluble phenols and antioxidant capacity. Journal of the Science of Food and Agriculture 89, 206–215.

Yamamoto, I., Takano, K., Sato, H., Kamoi, I., Miamoto, T., 1997. Natural toxic substances polyphenols, limonene, and allyl isothiocyanate, in several edible crops. Journal of Agricultural Science 41, 239–245.

Zhao, L.C., Guo, H.C., Dong, Z.Y., Zhao, Q., 2009. Pharmacological and nutritional activities of potato anthocyanins. African Journal of Pharmacy and Pharmacology 2, 463–468.

CHAPTER 10

Postharvest Storage of Potatoes

Reena Grittle Pinhero[1], Rickey Y. Yada[1,2]

[1]Department of Food Science, University of Guelph, Guelph, Ontario, Canada; [2]Food, Nutrition, and Health Program, Faculty of Land and Food Systems, University of British Columbia, Vancouver, British Columbia, Canada

1. Introduction

Potatoes are the fourth most important vegetable crop in the world and are widely grown and consumed in temperate as well as tropical countries. Potatoes are used as fresh table stock (31%), frozen french fries (30%), and chips (12%), and as dehydrated items (12%) (Miranda and Aguilera, 2006). The amount of land on which potatoes are grown is increasing immensely, especially in developing countries, with a 3.8% increase in developing countries, whereas a 1.8% decrease occurred in industrialized countries from 1996 to 2005 (FAOSTAT, 2014). Postharvest storage of potatoes is imperative since potatoes are a perishable commodity and tubers undergo active metabolism during storage. Continuous supply of potatoes year-round are made available by storage, during which sprout growth is suppressed and disease development is minimized by ambient air-cooling, refrigeration, or use of sprout suppressants. Storage of potatoes for various end uses such as seed purposes, table potatoes, chip, and french fries requires considerable care, attention, and management owing to their high moisture content and metabolic activity. Besides growing conditions, the state of the crop during harvest and any pretreatment to which the tubers are subjected will also affect the quality of potatoes. Therefore, production decisions starting from the selection of healthy seed potatoes to various crop management practices, optimal conditions of harvest, and storage management are crucial to maintain the quality of stored potatoes (Figure 1). This chapter discusses various factors that affect the quality of potatoes during postharvest storage, such as (1) the maturity stage of the crop (early/late) and its intended use (table stock/processing/seed), (2) the preharvest conditions of the crop, (3) harvest and handling conditions, (4) the health of the crop such as the incidence of pests and diseases, (5) biochemical changes, (6) storage preparations and conditions, (7) management of the storage environment, and (8) processing and the nutritional quality of the potatoes.

2. Maturity of Tubers

Potatoes are grown for specific markets such as table stock, salads, processing, and seed purposes. Depending on the end use, quality seed stock should be selected and harvesting of

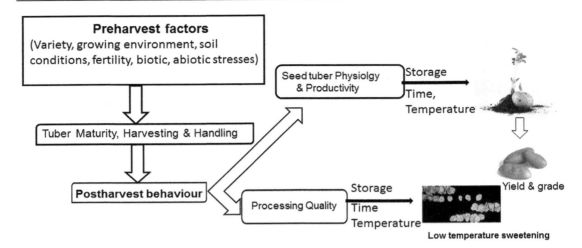

Figure 1

Factors affecting the postharvest behavior and quality of stored tubers. *Adapted and modified from Knowles and Plissey (2008).*

the tuber can be managed. Maturity is a complex physiological and morphological condition affected by various factors such as variety, weather conditions (e.g., temperature, rainfall, light), and biological and environmental stressors. The maturity of the potato can be defined by chronological and physiological age, in which chronological state is defined as the true age in terms of months, whereas physiological age is the physical state of the tuber (for a detailed study, please see Bishop and Clayton, 2009). Besides attaining full size, tuber physiological maturity is indicated by the soluble sugar content, which reaches a minimum and the starch content a maximum, and full development of a thickened periderm layer (skin) below the epidermis (ability of the tuber skin to resist abrasion, i.e., skinning during harvest) (Brecht, 2003). By manipulating the physiological state of the tuber, early emergence, higher early yield, higher senescence and bulking, and higher skin set can be achieved (Bishop and Clayton, 2009). Potatoes used for processing are usually grown under contract, with the processing industry specifying the cultivar, whereas many fresh-market cultivars are grown on the open market. There is a great market for a dual-purpose variety that can be processed or used for the fresh market. From a commercial point of view, early potatoes are harvested before they are fully mature and marketed immediately to obtain a premium price; their skins can be removed easily without peeling. Early potatoes are harvested to meet the demand for the fresh market, and some are used for chip processing. Because of the succulent nature and immature skin of early potatoes, they are easily bruised or damaged and the rate of respiration is about four to five times greater than mature potatoes. Early harvested potatoes are usually stored only briefly. Such early potatoes, which are free from serious bruising and decay, can sometimes be held 4–5 months at 4 °C for table use if they are cured 4 or 5 days or longer at 12–18 °C to heal wounds before storage. If used for chipping, early crop potatoes should be chipped directly from the field because holding these potatoes in cold storage even at a moderate temperature of 10–12 °C for only a few days can cause excessively reduced sugar and undesirable dark chips.

The maturity of the crop has been reported to influence respiration, dormancy, and sweetening behavior during storage (Knowles and Plissey, 2008) (also please see the section on low-temperature sweetening [LTS]). To preserve quality and maximize storability, potatoes should be harvested after reaching maturity, to resist mechanical damage during harvesting and handling operations. The relative maturity of the crop can be estimated by monitoring the sucrose and glucose levels at the end of the growing season to plan and manage the storage temperature. Potatoes with sucrose content greater than 0.15% tuber fresh weight at harvest are prone to premature increases in reducing sugars such as glucose and fructose during storage, which causes darkened processed products.

3. Growing Conditions Affecting Postharvest Storage

The quality of the stored potato is determined primarily by the health and quality of the potato before storage. Good storage management cannot improve the quality of potatoes out of storage if the health of the tubers is compromised during preharvest conditions. The storage behavior of potatoes is primarily dictated by growing conditions such as biotic (pests and diseases) and abiotic factors (soil moisture and temperature) besides production management and practices.

3.1 Abiotic Factors

3.1.1 Moisture

Potatoes grow in most soil types, but deep, well-drained, light-to-medium soil is preferred. Potatoes require an optimum water supply throughout growth, and even moderate water stress affects the various stages of tuber development and will have an effect during storage. The water requirement of potatoes varies based on soil type, maturity, variety, and intended use, whether for table stock or processing. Shortage of water during harvest can result in tubers bruising easily, and clods of earth/clay are conducive to mechanical damage, which may result in disease development and black spot later in storage. Internal bruising is referred to as black spot, which results from a physical impact to the tuber. Moisture stress during tuber growth, particularly during the early stages of development, is associated with the occurrence of sugar-end tubers in some varieties such as Russet Burbank (Iritani and Weller, 1973; Iritani, 1981; Eldredge et al., 1996). The tuber should remain hydrated, especially those in the top 6 inches of soil after vine kill. If the tubers are dehydrated, pressure flattening or internal sprouting during storage can occur, especially those stored for medium-to-long term (personal communication, Dr. Michael D. Lewis). The severity of flattening varies depending on the variety. Excess moisture conditions favor development of several foliage diseases such as late blight and white mold (Gudmestad, 2008), which will affect storage quality. High soil moisture in excess of 90% available soil moisture can lead to storage rot, causing soft rot, pink rot, and leakage as a result of enlargement of lenticels (Gudmestad, 2008).

3.1.2 Nutrient Status

Proper nutrient management is essential for the successful growth and production of quality potatoes. The crop nutritional status and optimum management of water during the potato growing season have large effects on the crop, as well as disease development and storage quality. For example, high nitrogen fertility increases yield but usually decreases storage quality (Booth and Shaw, 1981). High or excessive nitrogen stimulates overly large plant canopies, may delay maturity of the potato crop, lowers the dry matter of the tubers, and results in poorer skin set. Nitrogen (N) fertility can have a role because of its influence on stimulating tuber growth and size. Nitrogen applied late in the season delays maturity, reduces specific gravity, and increases the reducing sugar content. Nitrogen should be applied in amounts that do not exceed the crop's needs, and it should not be applied within 4–6 weeks of vine killing (Knowles and Plissey, 2008). Some studies have shown that N applied at tuber initiation can increase the incidence of hollow heart and brown center (Hiller and Thornton, 1993). Other studies have shown that late applications of N in some years can increase hollow heart incidence (Miller and Rosen, 2005). During a 2-year experiment to study the effect of nitrogen application, it was found that tubers from untreated isopropyl N-(3-chlorophenyl) carbamate (CIPC) (a sprout inhibitor) and treated with low nitrogen rates sprouted earlier and produced more sprout weight than those from the high-N treatments. In CIPC-treated tubers, sprouting was closely related to the N rate, with tubers from the low-N rates sprouting earlier and producing higher sprout weight (Thornton et al., 1994). In a sand culture study, potatoes were grown to maturity in the greenhouse to study the effects of factorial application of four levels, each of potassium (K) (2, 4, 8, and 12 mEq/l) and sulfur (S) (1, 2, 4, and 6 mEq/l), on the yield, quality, and storage behavior of tubers. Besides increasing the yield and other quality parameters such as dry matter and protein contents, K applied with S improved the shelf life of tubers, as determined by the percent weight loss of tubers after the storage of 4 weeks at room temperature. The best interaction with lowest tuber moisture loss was 12 mEq/l K × 6 mEq/l S (Moinuddin and Umar, 2004). There is a direct correlation between dry matter content and internal bruising. Usually potato tubers with high dry matter of more than 22% are more prone to shattering and bruising.

4. Harvesting and Handling Factors Affecting Postharvest Storage

Conditions of harvesting such as the timing of the harvest, tuber health, and soil conditions can significantly affect tuber health throughout storage and the quality of tubers. However, the susceptibility of tubers to external damage during harvesting varies based on cultivars, the stage of maturity of the crop, soil and weather conditions, harvester skills, and the design of harvesting and handling equipment. Depending on the varieties, potatoes intended for long-term storage should not be harvested until the vines have been dead for 10–14 days. Soil moisture at harvest is an important factor with regard to bruising. Optimal harvest conditions are 60–65% available soil moisture and tuber pulp temperatures of 10–18 °C. The soil should be just moist enough (typically between 60% and 75%) to carry the harvested potato and soil to the secondary conveyor on the harvester, where the soil should separate completely from the tubers. Harvesting immature tubers, when the soil conditions are too wet or dry, and during too warm weather conditions, can affect the quality of tubers. When tubers are harvested under very warm conditions, the respiration rate of the tubers increases. Leakage caused by *Pythium* fungi increases when harvesting tubers under warm soil conditions (when the pulp temperature is 21 °C or above) and leads to increased infection. Hence, in warm weather conditions, harvesting should be done during the early morning to avoid the heat. When tubers are harvested under very wet conditions, the chances of getting pink rot also increase. Pink rot is caused by *Phytophthora ethroseptica* fungus and is widespread and often infects tubers under wet soil conditions. Tubers from wet soil may also become oxygen-starved and their lenticels may become enlarged, which increases their susceptibility to internal black heart and decay (Knowles and Plissey, 2008). Too dry soil conditions will result in dehydrated tubers, which in turn will result in pressure flattening or internal sprouting during storage (Figure 2(a) and (b), permission from Dr. Lewis). It is important to consider the soil and crop conditions, weather forecast, and long-term weather records when deciding the harvest date. Internal sprouting often occurs owing to too low a concentration of sprout inhibitor applied to tubers. Within a variety, the degree of damage to outside skin is influenced by dry matter content, maturity, and turgidity of the tubers. Harvesting tubers at an immature stage leads to skinned tubers (Figure 3). A direct correlation exists between the incidence of internal bruising and the content of dry matter; high dry matter leads to high bruising. Dry matter is influenced by growing conditions and variety. Variety, soil type, and temperature also influence tuber shape and skin strength, which in turn greatly influence damage to outside skin. The morphology of the periderm and its susceptibility to damage also may affect the extent of wound and other pathogens. Blight (*Phytophthora infestans*), dry rot (*Fusarium solani var. caeruleum*), and skin spot (*Oospora pustulans*) infections are influenced by skin thickness. Spores are unable to penetrate the intact periderm (Scott and Wilcockson, 1978). Flaccid or flabby, limp tubers are more prone to bruising. Thus, their susceptibility to damage increases with storage time (Booth and Shaw, 1981). Immature tubers will result in skinned tubers and increase the potential to develop dry rot. Tubers sustain mechanical damage during harvesting with poorly adjusted equipment, during transport, and often during grading. Poorly adjusted equipment, such as excessive dropping heights, high belt speeds, and failure to coat the grading riddles with rubber, results in increased bruises and soil on the tubers. Soil on the tuber prevents air circulation and exchange through the pile and at the surface of tuber, and increases rot, black heart, and dehydration. Blue discoloration (internal bruising) or black spot can normally be identified only after peeling; hence, it is not possible to remove the affected tubers from a lot of

(a)   (b)

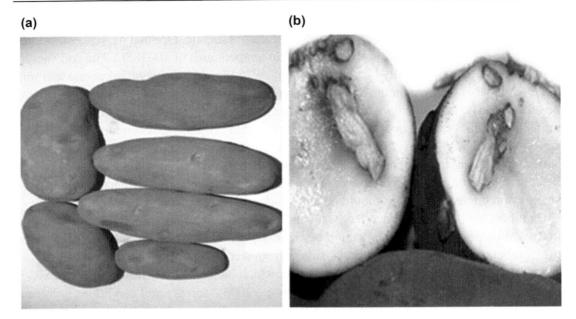

Figure 2
Development of pressure flattening (a) and internal sprouting (b) of stressed/dehydrated tubers
stored for medium- to long-term storage. *Reproduced with permission from Lewis (2007).*

Figure 3
Development of skinned tubers during harvesting as a result of mechanical injury and/or immature
tubers. *Reproduced with permission from Lewis (2007).*

unpeeled potatoes unless external damage has occurred. In the processing industry for
chipping, french fries, and peeled potatoes, this can result in extra costs owing to the elimination of damaged tubers and discolored chips and fries. Dry cleaning (brushing) is required
immediately after lifting of potatoes, but without causing mechanical damage. Avoidance
of direct exposure of tubers to sunlight after harvest may decrease undesirable greening in

potatoes and possible increases in glycoalkaloids, while also decreasing overheating of tubers, which in severe cases results in cell death and blackening. Mechanical damage can also occur during handling operations such as grading, packing, and transporting. Up to three-fourths of total tuber damage occurs during harvest, although significant injuries occur each time tubers are handled. Potatoes are more susceptible to mechanical injury at low temperatures of about 5 °C. Under certain conditions, injury may be reduced by raising the temperature of susceptible tubers before grading. Injury to the surface of the periderm accelerates moisture loss and exposes the interior of the tuber to pathogens, which can result in dry and soft rot during storage. Curing, a prestorage treatment, is important in limiting weight loss and preventing the penetration of microorganisms (discussed elsewhere in this chapter).

5. Pests and Diseases

Diseases are an important source of postharvest loss, particularly combined with mechanical damage resulting from rough handling as well as transportation and poor temperature control. Tubers are initially externally infected with spores in the field. Even a small quantity of diseased tubers in a lot can potentially result in the disease spreading and the whole lot being damaged during storage. Three major bacterial diseases and a greater number of fungal pathogens are responsible for serious postharvest loss. Major bacterial and fungal pathogens that cause postharvest loss in transit and storage and to the consumer are bacterial soft rot (*Pectobacterium atrosepticum* and *Pectobacterium carotovorum* subsp. *carotovorum*), *P. infestans* (late blight), Fusarium rot (*Fusarium* spp.), pink rot (*Phytophthora* spp.), and water rot (*Pythium* spp.). It is recommended that tubers be examined for pests and diseases before storage. Grading of tubers is important before storage to eliminate the diseased or infected tubers. If any substantial amount of one or more tuber diseases is evident during the harvest operation, certain precautions should be taken in storage management when the tubers are put into storage, such as storage management of temperature, RH, and ventilation. Rapid suberization right after tubers are placed in storage, such as high humidity (95%), optimum temperature of 10–12 °C, and good ventilation (25 cubic ft/min/ton) to avoid condensation can minimize *Fusarium*, *Pythium*, and *Phytophthora* infestations. If there is a significant amount of pink rot or leak in tubers going to storage, fans should be operated continuously to increase air movement, and the storage temperature should be dropped below 10 °C, with outside air used as needed to cool the crop as quickly as possible (Booth and Shaw, 1981; Meijers, 1987a; Burton, 1989; Gottschalk and Ezekiel, 2006; Powelson and Rowe, 2008).

6. Biochemical Changes of Tubers During Storage

Many biochemical processes, the incidence of pests and diseases, and factors such as temperature will influence the quality of tubers stored. Even the best storage can only maintain the quality of tubers placed in storage.

6.1 Respiration

As a living organism, potatoes continue to respire in storage. Sugars produced during starch hydrolysis are used for respiration. During starch hydrolysis, loss of dry matter occurs, resulting in weight loss. The respiration rate is affected by temperature and hence will vary depending on storage conditions and the end use of the tubers. For example, most varieties for processing chips and french fries should not be stored below 9 °C owing to LTS (see section) (Burton, 1968, 1978). The respiration rate is minimal at about 5 °C and increases slowly up to about 15 °C, above which respiration begins to increase sharply. Reducing the temperature to 3 °C also results in a sharp increase in respiration because of the high concentration of reducing sugars formed by the breakdown of starch. The activity of the enzyme invertase, which hydrolyzes sucrose into glucose and fructose, is also high at lower storage temperatures. At 0 °C, the rate of respiration is the same as that at 20 °C. At 10 °C, the dry matter loss by respiration represents approximately 1–2% of fresh weight during the first month and about 0.8% per month thereafter, but rises to about 1.5% per month when sprouting is well advanced. However, to keep the respiration rate low, preferably storing potatoes at temperature of around 5 °C is impractical for processing varieties that are susceptible to LTS (Burton, 1963; Booth and Shaw, 1981; Rastovski, 1987a; van Es and Hartmans, 1987a). The heat produced from respiration during storage can increase the storage temperature. To maintain the temperature of the potatoes at a specific level, the heat evolved by the potatoes during respiration must be removed by cooling. It is also important to have a steady supply of fresh air during storage to provide the oxygen needed in respiration and to remove CO_2 released during respiration. Tuber respiration results in CO_2 levels of 0.1% to 4–6% in commercial storage. Accumulation of CO_2 as high as 4% may cause black heart, eventually resulting in rot, and affect the processing quality of stored potatoes by affecting the chip color. Tuber glucose and fructose content may also increase if CO_2 is maintained at 3–4%; however, CO_2-induced sweetening is reversible and variety-dependent. Whereas the CO_2 concentration outdoors is often 360–380 ppm, it often ranges from 1200 to 1500 ppm in a well-maintained potato storage. If concentrations remain above 5000 ppm, it may be an indicator of storage rot and/or insufficient air exchanges/refreshing. Accumulation of CO_2 up to 3% in well-sealed stores is reported to result in unacceptable fry color. Therefore, regular ventilation of stores for 2–4 h/day is recommended (Gottschalk and Ezhekiel, 2006; personal communication, Dr Coffin).

The respiratory quotient, which is the ratio of volume of O_2 absorbed per hour to volume of CO_2 released per hour, should ideally be 1 when the O_2 supply is not limiting. It has been reported that the CO_2-to-O_2 ratio value is approximately 0.8 during the early period of storage and 1.3 when sprout growth starts (Isherwood and Burton, 1975; Burton, 1989; Gottschalk and Ezhekiel, 2006).

6.2 Water Loss

Ninety-eight percent of the moisture that leaves a tuber during storage is lost through its skin by evaporation. Only 2.4% leaves the tuber via the lenticels along with the carbon dioxide produced by respiration (Burton, 1989). Tubers that have lost large amounts of water succumb to mechanical damage such as bruising and blue/black discoloration, greater peeling loss, and a reduction in culinary quality leading to economic loss. The marketability of tubers will be affected if there is water loss in excess of 10% because of the unattractive shriveled appearance. Water loss during potato storage depends on heat production in potatoes, the temperature of the potatoes, relative humidity (RH), the quality of potato skin, the variety, the sprout growth, and the duration of ventilation (Booth and Shaw, 1981; Rastovski, 1987a). During the storage period, the rate of moisture loss from the crop is proportional to the difference in water vapor pressure (WVP) within the cells of the potato skins and the WVP of the air in the voids. The difference is often referred to as the water vapor pressure deficit (WVPD). The lower the RH of the ventilating air, the greater will be the WVPD and there will be greater moisture loss though evaporation. In addition, the colder the ventilating air compared with the tubers, the greater will be the WVPD. Ventilation of the crop with air cooler than the crop, no matter how humid, will always result in moisture loss through evaporation. When the pressure within the cells of the tuber skins and the vapor pressure of the air in the voids surrounding the tubers are the same, no evaporation will take place. For this to occur, the RH of the air in the voids between the tubers has to be 97.8% (Pringle et al., 2009). This assumes that the temperature of the air surrounding the potatoes is the same as that of the tuber. Average water loss from mature undamaged tubers is approximately 0.14–0.17% of the tuber weight per week per mbar vapor pressure deficit, which will increase to 0.5–0.8% per week in damaged tubers (Booth and Shaw, 1981). It has been reported that the removal of the tuber skin resulted in an immediate 300- to 500-fold increase in evaporation (Burton, 1989). The stage of maturity and sprouting also influences the evaporation rate. Freshly harvested immature tubers lose water more rapidly than mature tubers because the immature skin is more permeable to water vapor. The increase in evaporation during sprouting results from the sprout surface being 100–150 times more permeable to water vapor than the periderm of the tuber per unit area and time (van Es and Hartmans, 1987b). Varietal differences in water loss through evaporation are attributed to differences in maturity at harvest, the rate of wound healing, the thickness of periderm, and sprouting characteristics of the variety (Booth and Shaw, 1981; van Es and Hartmans, 1987b; Burton, 1989; Gottschalk and Ezekiel, 2006).

6.3 Sprouting

Natural or innate dormancy is usually defined as "sprouts being unable to develop even in conditions favorable to development" that are under hormonal, environmental, and physiological control (Sonnewald and Sonnewald, 2014). Sprouting, the breaking of dormancy, will affect the quality of the tubers during storage by remobilizing storage compounds, mainly starch and protein, and shrinking tubers owing to loss of water. Sprouting increases both respiration and moisture loss and reduces the quality and value of the crop because of increased levels of toxic glycoalkaloids, accelerated starch breakdown with concomitant accumulation of undesirable reducing sugars, and decreases in vitamin content; it increases physiological aging and affects the appearance of the tuber. Sprouted tubers impede air movement through the potato pile (Burton, 1958). Depending on the cultivar, maturity, soil, and weather conditions, potatoes remain dormant for about 5–9 weeks after harvest (Burton, 1978), with extreme cold wet weather prolonging dormancy, whereas extreme dry warm weather shortens dormancy. For example, Shepody has a shorter dormancy than Russet Burbank. Below 5 °C, sprout growth is slow and increases up to an optimum temperature of 20 °C, above which the growth rate decreases. Higher RH favors sprout growth, which is more pronounced at a higher temperature. Higher CO_2 concentration also favors sprout growth. A storage atmosphere with a CO_2 content of 2.2–9.1% stimulates sprouting in potatoes irrespective of stage of dormancy (Burton, 1958). Burton (1978) observed that the optimum CO_2 content was 2–4% and the stimulatory effect declined at higher concentrations; the result at 7–10% was the same as with normal air storage. The O_2 content required for optimum sprout growth depends on the physiological age of the potatoes. At the beginning of the season, the optimum content of O_2 required for sprouting is 4–5% at a storage temperature of 10–20 °C, whereas it increases to 17–20% around June (Burton, 1968). Sprout inhibitors or other successful management of sprout inhibition such as low temperature or controlled atmosphere (CA) storage is required to manage the quality of tubers successfully during storage. Although chemical treatments are widely used for sprout control, increased awareness of environmental and health concerns has spurred an increase in the use of alternative nonchemical or organic methods for sprout control. Various chemicals include ethylene, nonanol, chlorpropham, maleic hydrazide, carvone, abscissic acid, indol acetic acid, clove oil, mint oils, hydrogen peroxide, and 1,4-dimethylnaphthalene (Buitelaar, 1987; Rastovski, 1987a; van Es and Hartmans, 1987d; Kleinkopf et al., 2003). However, the principal sprout inhibitors used worldwide are isopropyl N-phenylcarbamate (IPC; propham) and CIPC (chloro-IPC; chloropropham). The effectiveness of CIPC depends on the storage conditions, application technology, and variety characteristics (NBuitelaar, 1987; Kleinkopf et al., 2003). Isopropyl N-phenylcarbamate and CIPC cannot be used in seed tubers because they stop cell division and have an irreversible effect on cell division. An Environmental Protection Agency reassessment (2002) within the requirements of the Food Quality Protection Act resulted in a reduction in allowable CIPC residue on fresh potatoes in the US from 50 to 30 ppm. This mandate coincides with tolerance reductions or restrictions for use of CIPC in other parts of the world (Kleinkopf et al., 2003). Hence, there is a need for alternative compounds for sprout suppression. Maleic hydrazide (Buitelaar, 1987), which occurs naturally in volatile compounds such as mono-, di-, and trimethylnapthalenes, benzothiophene, 2,6-diisopropylnapthalene, carvone, and ethylene, has been found to be effective at lower temperatures (7–10 °C) and at higher concentrations compared with CIPC (Lewis et al., 1997; Kalt et al., 1999). Ethylene is an effective sprout inhibitor but its use may result in darkening of fry color (Daniels Lake et al., 2007). Studies on the combined effect of CO_2 and ethylene as a sprout inhibitor have shown that ethylene affected the darkening of french fry color of all cultivars, whereas the CO_2 effect was variety specific (Daniels Lake, 2013).

Nonchemical methods to control sprouting include (1) developing varieties having longer dormancy by breeding, (2) low-temperature storage, (3) CA conditions, and (4) irradiation (Buitelaar, 1987; Kleinkopf et al., 2003). Low-temperature storage is an effective method of controlling sprouting because tubers do not sprout at temperatures below 4 °C. However, low temperature will also result in other disadvantages depending on the end use of the potato. For example, in potatoes destined for processing, a condition called LTS (low-temperature sweetening) will result, depending on the cultivar resistance (Burton, 1978). For table stock and seed potatoes, LTS does not pose a problem. Other advantages of low temperature include the natural control of fungal and bacterial loss and avoidance of chemical use (also see section on LTS). Other nonchemical methods for sprout control include CA storage, in which a particular ratio of reduced O_2 and increased CO_2 concentration is used (Thompson, 1998), and irradiation, which prevents cell division. Experiments using different levels of γ-irradiation (0.04, 0.08, 0.12, and 1 kGy, applied at two different postharvest times, 5 and 30 days after harvest) during storage at 22 °C on the textural behavior, microstructure, reducing sugar, total sugar, and tuber losses of potato cv. Kufri Sindhuri, showed that the lowest dose (0.04 kGy) was sufficient to inhibit sprouting in potatoes exposed on day 5 but not in the tubers exposed on day 30 (Mahto and Das, 2014). The irradiated, nonsprouted potatoes maintained their appearance during storage. Potatoes irradiated early appeared more sensitive to radiation-induced damage, resulting in excessive loss of tubers at 1 kGy, but low doses (up to 0.12 kGy) did not increase the susceptibility of the tubers to rotting. No significant differences between reducing sugar and total sugar content of the control and low-dose irradiated tubers were observed after 120 days. A high dose (1 kGy) induced blackening of the bud tissue and increased rotting percentage and poor textural quality. Increasing low doses (up to 0.12 kGy) progressively reduced the textural deterioration in the tubers during storage. Among the two treatment timings, K. Sindhuri irradiated early after harvest (i.e., on day 5) with 0.08- to 0.12-kGy doses retained higher textural parameters compared with those irradiated after a delay (day 30) (Mahto and Das, 2014). Among three methods (γ-irradiation, volatile oils [caraway, clove, carvone, and eugenol], and iodine vapor) tested for effective inhibition of sprouting of potato cv. Diamond, during storage, γ-irradiation and essential oils maintained quality of potato and inhibited sprouting for 9 weeks whereas iodine vapor was effective only for 6 weeks (Afify et al., 2012).

6.4 Pests and Diseases

Several diseases caused by fungus and bacteria will affect the quality of tubers during storage, resulting in major economic losses. Most storage pathogens originate from either seed tubers or during infections that invade the plant during the growing season. These diseases have been discussed during the section on preharvest conditions (Section 5). Selecting seeds that are disease-free, managing suitable agronomic practices to prevent disease development, and thoroughly cleaning the harvesting equipment and storage facilities will help reduce disease development. Qualitative pathogenic diseases such as common scab, powdery scab, black scurf, and silver scurf affect the appearance of potato and thus affect market value. Quantitative loss from major diseases such as late blight, pink rot, dry rot, and bacterial soft rot are caused by pathogens that are soil and seed borne and affect both field and storage stages, with storage losses tending to be more economically overwhelming for the grower (Gachango et al., 2012). Tuber soft rot is one of the most common causes of storage loss (Figure 4) and can be avoided by storing tubers under dry, cool conditions, avoiding storing with wet tubers, preventing condensation and the development of anaerobic conditions. When soft rot is expected, it is advisable to avoid the curing period because conditions that promote curing are also favorable for soft rot development (Booth and Shaw, 1981; Meijers, 1987a). Seed tubers are affected by skin spot and *Rhizoctonia* that kill the potato eyes (for specific details on diseases and pests, refer to Booth and Shaw, 1981; Rastovski, 1987a; Banks, 2006). Various measures can be taken to control storage diseases, starting from site selection for growing potatoes to preparing the site preparation; using varieties resistant to pests and diseases; carefully harvesting and using improved handling methods to avoid mechanical damage; following good sanitation and cleanliness of implements, containers, and storage facilities; using proper drying and curing of tubers; following good phytosanitary practices such as eliminating infested or infected tubers; and, when necessary, using pesticides in the field

Figure 4
Soft rot of potatoes developed in storage owing to storage of wet tubers, condensation, and anaerobic conditions in storage.

appropriate for the conditions or use as per the manufacturer's recommendations (Booth and Shaw, 1981; Meijers, 1987a). It is also important to avoid conditions favorable for the growth of microorganisms in storage such as excess temperature and humidity.

6.5 Temperature

One key factor controlling storage quality is temperature, through its effects on various biological processes such as respiration, evaporation, sprouting, LTS, freezing effect, curing, mechanical injury, and the incidence of pests and diseases (Burton, 1978; Booth and Shaw, 1981; Rastovski, 1987a; Brook et al., 1995; Gottschalk and Ezekiel, 2006), which have been discussed elsewhere in this chapter.

6.6 Low-Temperature Sweetening (LTS)

The most important aspect of quality for processors and consumers is color. Potato tubers stored at temperatures below 9–10 °C result in high concentrations of reducing sugars such as glucose and fructose, known as LTS (Burton, 1978, van Es and Hartmans, 1987c). These reducing sugars participate in the Maillard browning reaction with free amino acids during frying, resulting in dark brown fries and chips. These darkened chips and fries are unacceptable to consumers and may result in greater amounts of acrylamide production, which has been linked to many cancers (Chuda et al., 2003; Hogervorst et al., 2007). Tubers containing 0.1% reducing sugars are ideal for processing and are unacceptable when over 0.33% (Dale and Mackay, 1994). After harvest, low-temperature storage between 4 and 7 °C can prolong dormancy and reduce shrinkage and diseases; however, temperatures below 9 °C can cause LTS, an often reversible accumulation of reducing sugars. Genotype and environment also affect the storage and processing quality of potatoes. Affleck et al. (2012) studied the stability of sugar levels in eight potato genotypes over four environments (i.e., two locations over 2 years) after 105 and 120 days after planting and 60 and 120 days after storage at 8 °C. The genotype (G) and genotype × environment (GE) biplot analysis (GGE) was used to measure the stability of and association between quality traits and sugar content. Quality and sugar content were measured. The GGE biplots indicated a change in french fry color scores and stability between 105 and 120 days after planting harvest dates. Genotypic differences were noted for french fry color scores and glucose content. Genotypes were identified that were stable for french fry color during the two storage periods indicating low GE interaction. The GGE biplot identified mega-environments that encompassed a group of environments with similar attributes. Rak et al. (2013) evaluated the cold storability (5.5 °C and 8.3 °C for 3, 5, 6, and 9 months' duration) of 47 tetraploid advanced breeding clones selected from progenies produced by crosses between chipping potato germplasm in the Wisconsin potato breeding program along with six well-characterized standard chipping cultivars, including Atlantic, Dakota Pearl, MegaChip, Pike, Snowden, and White Pearl. Their results reinforce that storage environments have an important effect on the color of potato chips fried out of storage and that potato genotypes respond differentially to storage regimes. Genotype×storage environment interaction had a large effect on the variation observed in chip quality evaluations. Certain lines exhibited chip color stability across a range of storage durations and temperatures whereas the chip quality of others was highly variable across storage conditions (Rak et al., 2013). Another study involving varieties with LTS resistance ranging from susceptible (Defender and Russet Burbank) to moderate (Gemstar Russet) to high (Premier Russet) resistance showed that tuber respiration underwent a predictable respiratory acclimation response (RAR) to low temperature and the magnitude of this response was proportional to total sugar accumulation at 4 °C. It was also shown that reducing sugar content reflected the activity of acid invertase and its inhibitors and ratio of fructose to glucose indicated the extent of cold sweetening. This study characterized LTS phenotypes by changes in sugars, invertase, and RARs (Zommick et al., 2014a). It was reported that high soil temperature during bulking and maturation of potatoes alters postharvest carbohydrate metabolism to attenuate genotypic resistance to cold-induced sweetening and accelerates loss of process quality (Zommick et al., 2014b). Research in identifying the pathways and genes responsible for LTS has successfully resulted in the development of cultivars resistant to LTS (Sowokinos, 2001a,b; Pinhero et al., 2007).

Genetic engineering can be used successfully to improve LTS tolerance in potato tubers (Rommens et al., 2006; Pinhero et al., 2011, 2012; Li et al., 2013). It has been reported that silencing the potato tuber–expressed phosphorylase-L gene (PhL) resulted in lowered PhL gene expression and reduced glucose accumulation in cold-stored tubers (Yan et al., 2006; Rommens et al., 2006). Besides lowering glucose accumulation as a result of lowering the expression of starch-associated R1 or PhL genes, french fries derived from the modified tubers of Ranger Russet displayed an improved visual appearance and aroma while accumulating much lower levels of acrylamide (Rommens et al., 2006). Transgenic potatoes developed by overexpression of pyruvate decarboxylase showed LTS tolerance and decreased acrylamide formation (Pinhero et al., 2011, 2012). To combat LTS, a transgenic potato was developed using a synthetic tuber-specific and cold-induced promoter along with acid vacuolar invertase gene from potato expressed in antisense orientation. The transgenic tubers showed a decrease in reduced sugar content during storage at low temperature and acceptable chip color without significant changes in plant morphology and tuberization between the nontransgenic and transgenic lines (Li et al., 2013).

7. Storage Preparations and Conditions

7.1 Storage Methods

The choice of storage method depends on the production system, such as the frequency of harvest, demand, and end use. In countries where more than one crop is feasible, the storage period need not be long (e.g., 3–4 months). Another method is to store the surplus in response to market demand at the time of harvest. In countries where only one crop is feasible, storage becomes imperative to meet the demand for the entire year. Thus, information on the production pattern, marketing system, and tuber demand, whether for the fresh market, seed purposes, or processing or export, will influence the duration of the storage period and the magnitude of the facilities required. Storage methods may be either field storage or storage buildings (Booth and Shaw, 1981; Schouten, 1987; Sparenberg, 1987). Field storage may involve delayed harvest or in-ground storage or variable types of clamps or pits covered with straw and sometimes soil. Storage buildings are either multipurpose or purposely built for potato stores for long-term storage of large quantities of potatoes that are ventilated (Shaw and Booth, 1981; Sparenberg, 1987). In Canada and the US, most potatoes are stored in "bulk," in which large buildings often hold several million kilograms of potatoes individually. In Europe, storage of potatoes in moveable pallet boxes, each holding 700–1000 kg of potatoes, are stored in storage buildings (personal communication, Dr Coffin).

There are nonrefrigerated and refrigerated as well as CA storage methods for storing tubers based on their end use. The optimum temperature required for storage primarily depends on the end use. High-quality tubers can be stored from 2 to 12 months depending on the quality of tubers at harvest, the quality of storage facilities, good storage management, and variety, and whether sprout inhibitors are used (for details on sprout suppression, refer to Section 6.3). The respiration rate of tubers is lowest at 2–3 °C, which will reduce the weight loss. However, storage at 0–2 °C increases the risk of freezing or chilling injury. Usually potatoes that are chilled appear to be normal when removed from low temperature. However, symptoms of chilling become evident in a few days at warmer temperatures (Chourasia and Goswami, 2001). Sprouting increases at storage temperatures above 4–5 °C. Hence, at temperatures above 4 °C, tubers need to be treated for sprout suppression. Tubers for fresh consumption are stored at 7–10 °C to minimize formation of reducing sugars such as glucose and fructose from starch, which results in darkening during cooking. Many chipping cultivars accumulate excessive amounts of reducing sugars if stored below 9–10 °C, a process known as LTS (for a detailed description, see Section 6.6). A detailed description of these methods is described in our previous submission (Pinhero et al., 2009), with only new material discussed in this chapter. Various refrigerated (2–4 and 10–12 °C) and nonrefrigerated storage in heap (17–31 °C; 54–91% RH) and pit (17–27 °C; 67–95% RH) storage of potatoes after CIPC treatment were tested for the suitability of french fries and chip processing of five Indian and two exotic varieties (Mehta et al., 2014). Initial reducing sugar content was low, which increased after 90 days of storage at 2–4 and 10–12 °C and decreased in both types of storage. Sucrose content increased during storage, with higher increases recorded at 2–4 °C. Chip color of potatoes before and after storage at 10–12 °C and in heap and pit storage was highly acceptable in all the varieties except Kennebec, whereas the color was unacceptably dark in potatoes after storage at 2–4 °C. French fries of acceptable color were made from potato varieties except Kennebec at 2–4 °C, with fries of heap and pit-stored potatoes recording highest firmness. Chips made from pit-stored potatoes also recorded highest crispness. All varieties except Kennebec were considered suitable for processing. Heap and pit storage are recommended as economic options for short-term storage of processing potatoes.

The demand for fresh-cut produce has also increased steadily because of its nutritional quality, convenience, and about 2-week shelf-life. Although fresh-cut produce is generally considered to be safe, there have been many food-borne outbreaks in recent years (FDA, 2009), because conventional washing and sanitizing treatments are ineffective at inactivating pathogens on the surface of produce (Sapers et al., 2006). In this regard, the effect of hot water blanching (first blanching at low temperature (60 °C) for 10 or 20 min, and then second blanching at high temperature (~98 °C) for 1, 5, or 10 min) and near-aseptic packaging on the shelf-life of refrigerated potato strips showed no microbial growth within 28 days of refrigerated storage in strips treated for either 10 or 20 min in first blanch followed by 5 or 10 min in second blanch. Fries processed from near-aseptically packaged refrigerated potato strips were significantly lighter in color and higher in textural quality compared with unprocessed fries (neither blanched nor near-aseptically packaged). No significant changes were observed in quality of near-aseptically packaged refrigerated potato strips during 28 days of storage at 7 ± 1 °C. These results indicate that combination of blanching and near-aseptic packaging is the better nonchemical alternative method for potato strips to extend shelf-life (Onera and Walkerb, 2011). Acidulant dip treatment combined with aqueous ozone was tested to extend the shelf-life of fresh-cut potato slices during storage at 4 °C for 28 days (Calder et al., 2011). NatureSeal (NS) and sodium acid sulfate (SAS) were the most effective acidulant treatments in reducing browning regardless of ozone treatment. NatureSeal and SAS also had lower PPO activity compared with other treatments on days 0 and 28, and significantly lower APCs (≤ 2.00 log colony-forming units/g) over refrigerated storage. The SAS treatment was comparable to NS, a commercially available product, and showed promise as an effective antibrowning dip to reduce browning and spoilage in fresh-cut potato products.

CA storage refers to the constant monitoring and adjustment of CO_2 and O_2 levels within gas-tight stores or containers. Controlled atmosphere is most effective when combined with temperature control. There has been great interest in using CA storage on potatoes for fresh, processing, and seed potatoes (Butchbaker et al., 1967; van Es and Hartmans, 1987c; Khanbari and Thompson, 1994). Fellows (1988) recommended a maximum of 10% CO_2 and a minimum of 10% O_2 as the optimum CA storage for potatoes. The amount of O_2 and CO_2 in the atmosphere of the potato store can affect the sprouting of tubers, rotting, physiological disorders, respiration rate, sugar content, and processing quality (Table 1).

Table 1: Sugars (grams per 100 g DW) in Tubers of Three Potato Cultivars Stored for 25 Weeks under Different Controlled Atmospheres at 5 and 10 °C and Reconditioned for 2 Weeks at 20 °C.

Cultivars	Gas Combinations		5 °C			10 °C		
	CO$_2$ (%)	O$_2$ (%)	Sucrose	RS	TS	Sucrose	RS	TS
Record	9.4	3.6	0.757	0.216	0.973	0.910	0.490	1.400
	6.4	3.6	0.761	0.348	1.109	1.385	1.138	2.523
	3.6	3.6	0.622	0.534	1.156	1.600	0.749	2.349
	0.4	3.6	0.789	0.510	1.299	0.652	0.523	1.175
	0.5	21.0	0.323	0.730	1.053	0.998	0.634	1.632
			0.650	**0.488**	**1.138**	**1.109**	**0.707**	**1.816**
Saturna	9.4	3.6	0.897	0.324	1.221	0.685	0.233	0.918
	6.4	3.6	0.327	0.612	0.993	0.643	0.240	0.883
	3.6	3.6	0.440	0.382	0.822	0.725	0.358	1.083
	0.4	3.6	0.291	0.220	0.511	0.803	0.117	0.920
	0.5	21.0	0.216	0.615	0.831	0.789	0.405	1.194
			0.434	**0.473**	**0.907**	**0.729**	**0.271**	**1.000**
Hermes	9.4	3.6	0.371	0.480	0.851	0.256	0.219	0.475
	6.4	3.6	0.215	0.332	0.547	1.364	0.472	1.836
	3.6	3.6	0.494	0.735	1.229	0.882	0.267	1.149
	0.4	3.6	0.287	0.428	0.715	0.585	0.303	0.888
	0.5	21.0	0.695	0.932	1.627	0.617	0.510	1.127
			0.412	**0.682**	**1.094**	**0.741**	**0.354**	**1.095**

RS, reducing sugars; TS, total sugars; Numbers in bold represent mean values.
Reproduced with permission from Thompson (1998).

8. Storage Process

For successful storage, to maintain quality for the end use, factors that need to be carefully monitored and adjusted are preharvest storage preparations, storage and equipment, filling, curing, cooling down and monitoring the pile, maintaining the desired temperature and RH, warming the pile stack for unloading, and unloading the storage. These factors also vary based on the end use of the tubers, such as table, chip, or french fry processing or seed purposes (for a detailed description of storage designs and management, refer to Brook et al., 1995 and Pringle et al., 2009).

8.1 Preharvest Storage Preparations

For optimum storage conditions, before harvest, structural checks should be performed, such as framing to prevent decay and rot, doors with good seals, insulation for intact and dryness, and walls for cleanliness (Meijers, 1987b; Brook et al., 1995; Lewis, 2007). All air equipment system should be checked and repaired. Weight loss variations, sprouting problems, or disease problems often can be traced to an improperly designed or unbalanced air flow system. It is also important that the storage and equipment used for storing should be checked, cleaned, and repaired before storing. Fans, ducts and dampers should be examined. Similarly, humidification system (humidifiers for operation and water flow), ventilation systems and insulation systems, fans, and refrigeration systems should be inspected and corrective measures should be taken as well, and temperature-recording equipment should be fully functional (Meijers, 1987b; Brook et al., 1995; Lewis, 2007).

8.2 Filling the Storage

Harvesting and storage loading are two critical operations undertaken by growers to ensure a good price and avoid disease during storage. Potatoes entering the store should be free from soil, sods, stones, and other foreign materials, for proper circulation of ventilation air. Storage either in bulk or in bins or sacks depends on the need, convenience, and labor cost. Potatoes are stored in jute or polypropylene sacks when the labor cost is low in countries such as India and the sacks are positioned manually. Different varieties or production from different farms should not be mixed in bulk stores. Bulk storage is appropriate when only one variety is to be stored unless temporary storage dividers are erected to allow for the temporary storage of another variety. The storage height of bulk seed potatoes must be limited to 3–3.5 m because greater weight loss will result in bruises at greater heights (Booth and Shaw, 1981; Meijers, 1987b; Gottschalk and Ezekiel, 2006). Many commercial storage facilities in eastern Canada stack up to 5 m high (Dr. Coffin, personal communication). Potatoes can be stacked with bin pilers in bulk up to a height of 4 m with good management, but this can cause bruises or pressure spots as the result of pressure exerted by the upper layer of potatoes on the lower layer. Storing potatoes in bin boxes up to a height of 1–1.5 m will help to avoid bruises or pressure spots. The advantages of using bin box are that (1) it helps to store potatoes of different varieties for different purposes and of different origin, separately; (2) it facilitates drying and cooling; (3) it allows for effective application of sprout suppressants; and (4) unloading is easy (Meijers, 1987b, Gottschalk and Ezekiel, 2006). The drawbacks of a bin system are the relative expense and the difficulty in maintaining uniform temperature and RH in the stacks. Many improvements have been made with correct placement of air exchange ducts and higher-capacity fans to ensure "bin balancing" of air flow. Special care should be taken to store potatoes separately if they are infected with *Phytophthora*. Given good management, and if the affected lots can be blown dried quickly, lots containing 5% tubers affected with *Phytophthora* may be stored relatively well. In the case of soft rot and frozen tubers, the permissible limit is <1% and should be rapidly blown dried to allow for successful storage (Meijers, 1987b).

8.3 Equalization and Drying Phase

Potatoes harvested under wet soil conditions should be dried immediately. Drying involves removal of water present on the outside of the potato or in the soil on the potatoes to eliminate conditions conducive for the multiplication of microorganisms and to prevent the spread of rot and other storage diseases (Sijbring, 1987). Potatoes can be batch dried using natural winds where natural wind is available near the field after harvest, or by using fans. Batch drying using natural winds is possible only in open sites and where the number of windless days is minimal (none), because there are no running costs, and therefore, it is cost-effective. Disadvantages are no control over temperature; if a frost forecast is imminent, tubers must be removed; and condensation occurs on warmer days (Pringle et al., 2009). The ventilation fan should run continuously during this phase while the average potato pile temperature is allowed to settle within 2 °C of the average pulp temperature upon entry into storage. Ventilation should be at the maximum possible rate for the shortest time needed. Excessive ventilation after removal of the surface moisture can dehydrate and soften the stored crop. Careful control of RH is also important. Hence, frequent inspection of potatoes during the drying period is important (Sijbring, 1987; Brook et al., 1995).

8.4 Wound Healing/Curing

Tubers undergo distinct physiological periods during storage, such as a curing period or wound-healing period during which harvest wounds heal, a cooling period when the pulp temperature is lowered to a level that is appropriate for the intended use of the tuber, and a holding period during which respiration is low and the tubers are dormant and then the dormancy ends and tubers are able to sprout (Knowles and Plissey, 2008). Despite all of the precautions taken during harvest and handling, injury to tubers may occur during mechanical harvest, lifting, transportation, and even grading. Weight loss and entry of microorganisms can occur through the injured skin, causing diseases and rot during storage. Lignification, suberization, and periderm formation help the tuber to recover from the damage incurred during mechanical injury (Booth and Shaw, 1981; Meijers, 1987c; Brook et al., 1995). Curing, a prestorage treatment, is important to limit weight loss and prevent the penetration of microorganisms. Under favorable conditions, tuber tissue forms a protective wound periderm over the damaged area. Wound healing and suberization are greatly affected by factors such as temperature, atmospheric humidity, oxygen and carbon dioxide concentration, cultivar, the physiological age of the tuber, and the use of sprout inhibitors. However, the main determinants of the rate of curing are temperature and RH. The usual recommendation of curing for potato tubers is exposure to a temperature of 12–16 °C and an RH of 90–95% for 2 weeks during which the tuber tissue forms a protective layer (wound periderm) over the damaged area (Booth and Shaw, 1981; Meijers, 1987c). Potatoes should be ventilated enough during the curing period to avoid a rise in temperature and humidity and to keep the oxygen concentration at approximately 20%. A thin layer of suberized (corked) cells is first deposited over the damaged area, followed by deposition of cork cambium (phellogen) under the sealing layer, giving rise from the inside to the outside to a dense network of new cells without intercellular spaces, which set rapidly. This wound periderm seems to be even more impermeable than ordinary skin. Suberization is slow at lower temperatures and stops at 2 °C. Conditions required for curing also favor several diseases, especially bacterial rot. To prevent additional respiration loss and conditions conducive for the spread of disease, the temperature of the curing process should not go above 20 °C and should be reduced to the necessary holding temperature as quickly as possible after curing (Burton, 1978; Meijers, 1987c; Gottschalk and Ezekiel, 2006).

8.5 Preconditioning Phase

Preconditioning is used commercially by chip-potato processors to compensate the unpredictable nature of reconditioning of process varieties and to achieve market flexibility (Brook et al., 1995). During this phase, the storage environment is maintained at conditions similar to the wound-healing phase (12–16 °C) with the pulp temperature actively controlled to eliminate pools of reducing sugars in processing potatoes. The duration of this phase depends on the process quality of the potatoes as measured by sugar content and chip color (Brook et al., 1995).

8.6 Cooling Period

The main storage phase of cooling starts after drying and wound healing. Temperature is the principal factor that influences quality and loss of potato during storage through respiration, sprouting, and sweetening and affects the spread of disease. As a general rule, the rate of cooling should be limited to 0.5–3 °C/week. The rate of cooling varies depending on the end use. For processing varieties, it should be 1 °C/week whereas for the fresh market it should be as fast as possible by maintaining a 1–2 °C pile differential from top to bottom. The main ventilation fan should run continuously during the cooling phase to maintain a uniform pile temperature, a temperature differential of −17.5 to −16.6 °C from the bottom to top of the pile during cooling and an RH of 95–99%. Cooling results in weight reduction owing to moisture loss, which can be limited by rapidly cooling the potatoes with humid air (Booth and Shaw, 1981; Rastovski, 1987b; Brook et al., 1995; Gottschalk and Ezekiel, 2006; Lewis, 2007).

8.7 Holding Period

Choice of holding temperature is influenced by the duration of storage period, end use, and variety. Processing potatoes are generally stored between 6 and 10 °C, whereas fresh-market tubers may be stored between 4 and 10 °C and seed tubers are usually stored at 3–4 °C. Once the required temperature has been attained, ventilation should be reduced to a minimum to maintain a uniform pile temperature within one degree of the desired level, to maintain the stack temperature differential from top to bottom of the pile as low as possible, supply O_2, and remove CO_2. Relative humidity should be maintained as high as 90–95% during the holding period to minimize the loss of moisture as a result of evaporation (Booth and Shaw, 1981; Brook et al., 1995).

8.8 Conditioning

Potatoes are most susceptible to mechanical impacts leading to discoloration at low temperatures. Conditioning of tubers is important based on end use. The fry color of chips can be improved by conditioning if the tubers are stored at low temperature and subjected to LTS. Conditioning will also reduce the mechanical damage during unloading because potatoes are more susceptible to mechanical damage at low temperature. Another reason for conditioning is to stimulate sprouting of seed potatoes, which is done by slow warming, thereby allowing the generated heat to remain in the storage to a temperature of up to 12–15 °C for 2–3 weeks. During this period, free sugars will be converted to starch, thereby decreasing the free sugar content. In a series of experiments over 4 years on cultivar Pentland Dell, the Agtron values when stored at 5 °C rose 16 points when the storage temperature was increased to 20 °C (Pringle et al., 2009). However, reconditioning is rarely complete and is often uneven. Senescent sweetening cannot be reversed (Burton, 1978; Booth and Shaw, 1981; Brook et al., 1995; Gottschalk and Ezekiel, 2006). To avoid condensation during the reconditioning phase, ventilation air with a lower than normal RH is recommended. Other precautions to prevent condensation are: (1) stop ventilation when the dew point is reached; (2) seal the storage to prevent the entry of warm air; (3) mix inlet air with storage air to avoid large temperature differences; and (4) use forced air ventilation or refrigeration if sprouts start to elongate (Burton, 1978; Booth and Shaw, 1981; Brook et al., 1995; Pringle, 1996; Gottschalk and Ezekiel, 2006). Before unloading the potatoes, it is important to heat the potatoes stored at low temperature up to 12–15 °C, to minimize the likelihood of handling and processing damage (de Haan, 1987; Gottschalk and Ezekiel, 2006).

9. Management of Storage Environment

Store management should include checking tubers in the tops of boxes or piles for condensation and disease, taking samples from specific boxes or areas to assess development of diseases, and daily recording of temperature. Successful management of temperature, RH, CO_2 level of the storage and air exchange system, and daily monitoring are critical to maintain the quality of stored tubers (Burton, 1978; Booth and Shaw, 1981; Kleinkopf, 1995). The temperature within the storage facility and of the outside ambient air can be measured with a simple minimum and maximum thermometer situated away from external influences, particularly direct sunlight. Air temperatures surrounding the potatoes can be measured by direct reading instruments or remote station indicators. The hottest part of the potato stack is usually 400–500 mm below the top surface, which should be monitored. Temperature from several levels is useful to determine stack temperature gradients and to localize the hot spots. A marked rise in temperature can indicate bacterial soft rot; if identified early, suitable measures can be taken to prevent the spread of rotting by increasing ventilation or removing the tubers from storage (Booth and Shaw, 1981; Figure 4). Relative humidity should be monitored by using a wet and dry bulb thermometer in a sling or battery-operated psychrometer. Portable CO_2 meters are useful tool for growers when monitoring potato storage. Although great advances have been made in automated monitoring of storage, improved success is guaranteed if the grower visits potato storage daily to verify temperature, detect moisture on tubers, and check for fruit flies and/or "early-warning" signals such as the smell of ammonia. Dr. Robert Coffin in Prince Edward Island has developed a checklist referred to as a "storage fitness test." Some key points on the list to check are: (1) balanced air flow through all of the potato pile; (2) temperature gradients in the pile; (3) deteriorating insulation; (4) air leakage around doors; (5) accurate assessment of RH; (6) "free" moisture on potatoes; (7) excessive carbon dioxide concentration; (8) fruit flies; (9) ammonia odor; and (10) accuracy/precision of control panel and monitoring equipment (Dr. Coffin, personal communication).

10. Effect of Postharvest Storage on Processing and Nutritional Quality of Potatoes

The most appropriate definition of quality is the suitability of potatoes for their intended use. It is imperative for both growers and processors to maintain quality based on end use. French fries and potato chips constitute two major processed potato products in the food industry. Production of frozen french fries is the largest sector of the processing industry. Other processed products are dehydrated flakes or granules and canned potatoes in various forms

(Sinha and Hui, 2011). The chemical composition of the tuber, which is influenced by the environment during growth and storage, will greatly affect quality (Mazza, 1983; Salunkhe et al., 1989). The extent of biochemical changes occurring in tubers during storage before processing provides a major influence responsible for finished product discoloration such as LTS (see Section 6.6). Two of the most important physiological processes affecting potato storage and market quality are dormancy/sprouting and wound-healing/skin set. External qualities such as the extent of surface blemishes owing to diseases and pests, sprouting and superficial damage, and internal qualities such as nutritional quality, color of the cooked product (after-cooking blackening and enzymatic and nonenzymatic browning), and eating quality (texture, feel, and crispness) determine the final tuber quality and consumer acceptability. The effect of diseases and pests and sprouting has been discussed elsewhere in this chapter (see Sections 5 and 6.3).

Potatoes are rich sources of carbohydrates, various minerals, vitamins, and phytochemicals. Storage conditions affect nutritional quality. The effect of γ-irradiation and CIPC treatments after storage of potato varieties at 8 °C for 5 months has shown that sprouting was effectively controlled by both treatments. γ-Irradiation increased total free glucose content in two varieties, and immediately after irradiation the rapidly digestible starch and slowly digestible starch increased slightly whereas resistant starch decreased significantly. However, no significant changes were noticed in the in vitro starch digestibility in the control, CIPC, and γ-irradiated varieties up to 3 months (Lu et al., 2012).

Potatoes are the third leading source of dietary antioxidants in the North American diet (Chun et al., 2005). Antioxidant properties are contributed by phenolics, anthocyanins, and carotenoids and are rich, especially in colored potatoes. Storage generally increases total phenolic content in potatoes, but little change or a decrease in phenol content after storage has also been reported in some studies (Ezekiel et al., 2013). The effect of cold storage on 12 Colorado-grown specialty potato clones 2, 4, 6, and 7 months after storage at 4 °C showed that pigmented potato clones had significantly higher total phenolic content and antioxidant activity, whereas the yellow-fleshed potato cultivar "Yukon Gold" had significantly higher vitamin C content. Vitamin C content decreased in all potato clones during cold storage, whereas total phenolic content and antioxidant activity fluctuated during cold storage; after 7 months of cold storage, the levels were slightly higher than at harvest in pigmented clones (Külen et al., 2013). The effect of storage on potatoes at 4 °C or 20 °C for 110 days on phenolic content was studied by Blessington et al. (2010). No significant differences in total phenolic content, chlorogenic acid, caffeic acid, and vanillic acid were observed after storage at 4 °C or 20 °C. There was an increase in rutin, *p*-coumaric acid, and quercetin dehydrate content after storage at 4 °C or 20 °C. When 4 °C stored potatoes were reconditioned for 10 days at 20 °C, there was a significant increase in total phenolic content, chlorogenic acid, caffeic acid, rutin, vanillic acid, *p*-coumaric acid, and quercetin dehydrate levels. All three storage treatments resulted in increased carotenoid content but caused no significant differences in phenolic content and antioxidant activity in most of the eight genotypes studied. In a study in which colored potatoes were stored at 4 °C for 6 months, significant differences were observed in the total anthocyanin content (TAC). Violette and Highland Burgundy red cultivars had 16.8% and 20.3% increases, respectively, in TAC compared with their freshly harvested samples. However, a decrease of 35.9% was observed in Valfi cultivars, whereas Blaue St. Galler and Blue Congo showed no differences in TAC (Lachman et al., 2012). A gas chromatography–mass spectrometry metabolomics study on six commercial potato cultivars (Somogyi kifli, Hopehely, Katica, Lorett, Venusz, and White Lady) at five developmental stages from harvest to sprouting stored at 20–22 °C showed that storage decreased the fructose and sucrose contents and increased proline concentration. Irrespective of length of dormancy, a substantial difference in metabolite composition at each time point in storage was detected in each cultivar except Somogyi kifli (Uri et al., 2014). In a study that evaluated vitamin C content involving 10 advanced clones, it was observed that vitamin C content was higher in younger tubers than in mature ones and its content during storage declined significantly in all clones studied (Cho et al., 2013).

Storage of potatoes can also alter the composition of glycoalkaloid and acrylamide content. Greening of tubers occurs when they are exposed to light intensity as low as 3–11 W/m^2 for a short period of as low as 24 h and is influenced by variety, stage of maturity, and temperature (Salunkhe et al., 1989). At 5 °C, no greening was noticed, whereas it was extensive at 20 °C (Salunkhe et al., 1989). Greening affects the nutritional quality of tubers. Greening in potatoes occurs through the synthesis of chlorophyll in the peridermal layers of tubers exposed to light and is often associated with the formation of glycoalkaloids (Salunkhe et al., 1989). Glycoalkaloids are mainly glycosides of the aglycone, solanidine. Solanidine can cause off-flavors upon cooking at concentrations of 15–20 mg/100 g. Because glycoalkaloids impart a bitter taste and can be toxic above threshold levels, in many countries an official guideline of less than 200 mg/kg fresh weight level has been recommended (Friedman and McDonald, 1997). Symptoms of potato glycoalkaloid poisoning include gastrointestinal disorders, hallucinations, and partial paralysis (Smith et al., 1996) as a result of inhibited acetylcholinesterase activity (Friedman, 2006). It has been reported that the increase in glycoalkaloid content during storage is lower at 10 °C than at 4 °C (Cieslik and Praznik, 1998). Nitithamyong et al. (1999) found that temperature had a larger influence on glycoalkaloid levels compared with light intensity, day length, carbon dioxide concentration, and humidity. It has been shown that an elevated temperature (10 °C) during long-term storage without sprouting inhibitors led to an increase in glycoalkaloid content (up to 518 mg SGA/kg dry matter) in Cilena and Lolita out of three investigated cultivars, Cilena, Lolita, and Marabel, independent of the sprouting level. Cold storage (4 °C) slightly enhanced glycoalkaloid contents in Lolita. Sprout control resulted in a substantial decrease in glycoalkaloid contents in a set of another three cultivars tested, Agria, Karlena, and Tomensa. In addition, the growing location and wet and cool seasons influenced the content. Moderate exposure to light resulted in an increase in glycoalkaloids in autumn, but they decreased in spring. Again, cultivars tested responded differently (Haase, 2010). These factors should be taken into consideration when developing strategies to minimize SGA level in table potatoes. There are various ways to control greening of tubers and the glycoalkaloid content in storage, such as the use of chemical control methods, controlled and modified atmospheric conditions, and ionizing irradiation, which has been discussed under sprout control or storage methods in this chapter. Genetic improvements to reduce glycoalkaloid content in tubers have been reported by breeding *Solanum tuberosum* and *Solanum chacoense* (Sanford et al., 1995) and using antisense technology with the gene for solanidine UDP-glucose glucosyltransferase in Lenape and Desiree lines (Moehs et al., 1997).

There has been increasing concern regarding the production of acrylamide in processed potato products because of its potential carcinogenic properties (Chuda et al., 2003; Hogervorst et al., 2007). Acrylamide has been classified as a probable human carcinogen by the International Agency for Research on Cancer (IARC, 1994). Acrylamide is formed in food by the Maillard browning reaction of reducing sugars with asparagine at temperatures above 120 °C (Zhang and Zhang, 2007); the rate strongly increases from 120 to 170 °C. Consequently, asparagine, glucose, and fructose are considered main precursors, and asparagine, with its amide group, delivers the backbone of the acrylamide molecule. Potatoes contain substantial amounts of glucose, fructose, and asparagine; low-temperature storage has been shown to increase reducing sugar content. A study involving 16 commercial potato varieties from eight countries showed a positive correlation of acrylamide formation between fructose ($r=0.956$), glucose ($r=0.826$), and asparagine ($r=0.842$) (Zhu et al., 2010). Potato chips made from tubers stored at 2 °C contained 10 times more acrylamide than chips from 20 °C and highly correlated with glucose and fructose levels in the tubers (Chuda et al., 2003). All of these studies show a strong correlation between acrylamide formation and reducing sugar concentrations. Thus, substantial improvements can be made in reducing acrylamide formation in chips and french fries by developing varieties resistant to LTS by either breeding or genetic engineering. Significant interactions between varieties (french fry and chipping) and storage time between many of the free amino acids, glucose, fructose, and acrylamide formation were reported in a study using 10 potato varieties composed of french fry types stored at 8.5 °C and chipping types stored at 9.5 °C for 8 months (Halford et al., 2012). There were significant correlations between glucose or total reducing sugar concentration and acrylamide formation in both variety types, although the correlation with fructose was much stronger for chipping than for french fry varieties. Significant correlations were found with acrylamide formation for both total free amino acid and free asparagine concentration in the french fry but not chipping varieties (Halford et al., 2012; Muttucumaru et al., 2014). These studies emphasize the potential of variety selection in preventing unacceptable levels of acrylamide formation in potato products and the variety-dependent effect of long-term storage on acrylamide risk (Halford et al., 2012). Various processing methods and the use of genetic engineering to reduce acrylamide formation in processed foods have been investigated with varying degrees of success. Tuber-specific silencing of the acid invertase gene substantially lowered

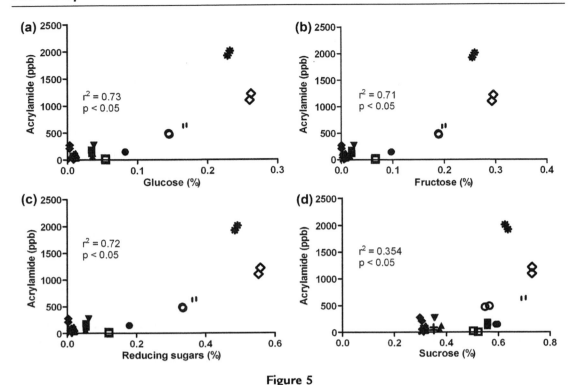

Figure 5

Pearson correlation coefficient for the relationship between acrylamide and glucose (a), fructose (b), total reducing sugars (c), and sucrose (d) measured for control untransformed S and transgenic T tubers before and after storage at 12 °C and 5 °C. S0, untransformed, no storage (■); S12, 14 (untransformed, 14 days after storage at 12 °C, ♦); S12, 28 (untransformed, 28 days after storage at 12 °C); S12, 42 (untransformed, 42 days after storage at 12 °C, X); S5, 14 (untransformed, 14 days after storage at 5 °C, ▼); S5, 28 (untransformed, 28 days after storage at 5 °C, ▲); S5, 42 (untransformed, 42 days after storage at 5 °C, ●); T0 (transgenic, no storage, □); T12, 14 (transgenic, 14 days after storage at 12 °C, ◊); T12, 28, (transgenic, 28 days after storage at 12 °C); T12, 42 (transgenic, 42 days after storage at 12 °C, +); T5, 14, (transgenic, 14 days after storage at 5 °C); T5, 28 (transgenic, 28 days after storage at 5 °C, Δ); T5, 42 (transgenic, 42 days after storage at 5 °C, ○). *Reproduced from Pinhero et al. (2012).*

reducing sugar formation after storage at 5.5 °C for 2 months and reduced acrylamide formation in french fry varieties eightfold (Ye et al., 2010). A study conducted in our laboratory with transgenic potatoes developed by overexpressing the pyruvate decarboxylate gene and cold-inducible promoter showed that chips from transgenic tubers stored at 5 °C produced 69% less acrylamide than corresponding nontransgenic potatoes (Pinhero et al., 2012). A high positive correlation ($p \leq 0.05$) between acrylamide content and glucose, fructose, and reducing sugars and a lower correlation ($p \leq 0.05$) with sucrose were observed (Figure 5). A high negative correlation between chip score and acrylamide was also observed (Figure 6). For a more detailed study on acrylamide reduction, see the review by Foot et al. (2007).

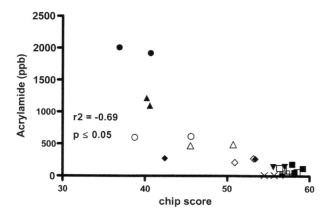

Figure 6

Pearson correlation coefficient for the relationship between acrylamide and chip score measured for control untransformed S and transgenic T tubers before and after storage at 12 °C and 5 °C. S0, untransformed, no storage (■); S12, 14 (untransformed, 14 days after storage at 12 °C, ♦); S12, 28 (untransformed, 28 days after storage at 12 °C); S12, 42 (untransformed, 42 days after storage at 12 °C, X); S5, 14 (untransformed, 14 days after storage at 5 °C, ▼); S5, 28 (untransformed, 28 days after storage at 5 °C, ▲); S5, 42 (untransformed, 42 days after storage at 5 °C, ●); T0 (transgenic, no storage, □); T12, 14 (transgenic, 14 days after storage at 12 °C, ◊); T12, 28, (transgenic, 28 days after storage at 12 °C); T12, 42 (transgenic, 42 days after storage at 12 °C, +); T5, 14 (transgenic, 14 days after storage at 5 °C); T5, 28 (transgenic, 28 days after storage at 5 °C, △); T5, 42 (transgenic, 42 days after storage at 5 °C, ○). *Reproduced from Pinhero et al. (2012).*

For additional readings on potato postharvest storage, the reader is referred to Booth and Shaw (1981), Rastovski (1987a), Banks (2006), Gottschalk and Ezhekiel (2006), Knowles and Plissey (2008), Pinhero et al. (2009), and Pringle et al. (2009).

Acknowledgments

The authors thank Dr Michael D. Lewis, Agri-World Consulting, Fruitland, Idaho, US, for permission to use the figures. Permission to use the table was kindly granted by CAB International, United Kingdom. We also acknowledge financial support from the Ontario Ministry of Agriculture, Food, and Rural Affairs and the Ontario Potato Board.

References

Affleck, I., Sullivan, J.A., Tarn, R., Yada, R., 2012. Stability of eight potato genotypes for sugar content and french fry quality at harvest and after storage. Canadian Journal of Plant Science 92 (1), 87–96.

Afify, A.E.M.R., El-Beltagi, H.S., Aly, A.A., El-Ansary, A.E., 2012. The impact of γ-irradiation, essential oils and iodine on biochemical components and metabolism of potato tubers during storage. Notulae Botanicae Horti Agrobotanici 40 (2), 129–139.

Banks, E., 2006. Potato Field Guide, Insects, Diseases and Defects. Ontario Ministry of Publication 823, p. 170.

Bishop, C.F.H., Clayton, R.C., 2009. Physiology. In: Pringle, R.T., Bishop, C.F.H., Clayton, R.C. (Eds.), Potato Postharvest, pp. 1–29.

Blessington, T., Nzaramba, M.N., Scheuring, D.C., Hale, A.L., Reddivari, L., Miller Jr., J.C., 2010. Cooking methods and storage treatments of potato: effects on carotenoids, antioxidant activity and phenolics. American Journal of Potato Research 87, 479–491.

Booth, R.H., Shaw, R.L., 1981. Principles of Potato Storage. International Potato Centre, Lima, Peru, p. 105.

Brecht, J.K., 2003. Underground storage organs. In: Bartz, J.A., Brecht, J.K. (Eds.), Postharvest Physiology and Pathology of Vegetables, second ed. Marcel Dekker, Inc., pp. 625–647.

Brook, R.C., Fick, R.J., Forbush, T.D., 1995. Potato storage design and management. American Potato Journal 72, 463–479.

Buitelaar, N., 1987. Sprout inhibition in ware potato storage. In: Rastovski, A., van Es, A., et al. (Eds.), Storage of Potatoes. PUDOC, Wageningen, The Netherlands, pp. 331–341.

Burton, W.G., 1958. The effect of the concentrations of carbon dioxide and oxygen in the storage atmosphere upon the sprouting of potatoes at 10 °C. European Potato Journal 1 (2), 47–57.

Burton, W.G., 1963. The basic principles of potato storage as practiced in Great Britain. European Potato Journal 6, 77–92.

Burton, W.G., 1968. The effect of oxygen concentration upon sprout growth on the potato tuber. European Potato Journal 11, 249–265.

Burton, W.G., 1978. The physics and physiology of storage. In: Harris, P.M. (Ed.), The Potato Crop. The Scientific Basis for Improvement. London Chapman and Hall, A Halsted Press Book, John Wiley and Sons, New York, pp. 545–606.

Burton, W.G., 1989. The Potato, third ed. Wiley, New York, USA.

Butchbaker, A.F., Nelson, D.C., Shaw, R., 1967. Controlled-atmosphere storage of potatoes. Transactions of the American Society of Agricultural Engineers 10, 534–538.

Calder, B.L., Skonberg, D.I., Davis–Dentici, K., Hughes, B.H., Bolton, J.C., 2011. The effectiveness of ozone and acidulant treatments in extending the refrigerated shelf life of fresh–cut potatoes. Journal of Food Science 76 (8), S492–S498.

Cho, K.-S., Jeong, H.-J., Cho, J.-H., Park, Y.-E., Hong, S.-Y., Won, H.-S., Kim, H.-J., 2013. Vitamin C content of potato clones from Korean breeding lines and compositional changes during growth and after storage, horticulture. Environment and Biotechnology 54 (1), 70–75.

Chourasia, M.K., Goswami, T.K., 2001. Losses of potatoes in cold storage vis-à-vis types, mechanism and influential factors. Journal of Food Science and Technology 38 (4), 301–313.

Chuda, Y., Ono, H., Yada, H., Takada, A.O., Endo, C.M., Mori, M., 2003. Effects of physiological changes in potato tubers (*Solanum tuberosum* L.) after low temperature storage on the level of acrylamide formed in potato chips. Bioscience, Biotechnology, and Biochemistry 67 (5), 1188–1190.

Cieslik, E., Praznik, W., 1998. Changes of glycoalkaloid content in potato tubers of selected varieties during vegetation and storage. Polish Journal of Food and Nutrition Sciences 7, 417–422.

Chun, O.K., Kim, D.O., Smith, N., Schroeder, D., Han, j.t., Lee, C.Y., 2005. Daily consumption of phenolics and total antioxidant capacity from fruits and vegetables in the American diet. Journal of Science, Food and Agriculture 85, 1715–1724.

Dale, M.F.B., Mackay, G.R., 1994. Inheritance of table and processing quality. In: Bradshaw, J.E., Mackay, G.R. (Eds.), Potato Genetics. CAB International, Wallingford, Oxon, pp. 285–315.

Daniels-Lake, B.J., 2013. The combined effect of CO_2 and ethylene sprout inhibitors on the fry color of stored potatoes (*Solanum tuberosum* L.). Potato Research 56, 115–126. http://dx.doi.org/10.1007/s11540-013-9234-0.

Daniels-Lake, B.J., Prange, R.K., Kalt, W., Walsh, J.R., 2007. Methods to minimize the effect of ethylene sprout inhibitor on potato fry color. Potato Research 49, 303–326.

Eldredge, E.P., Holmes, Z.A., Mosley, A.R., Shock, C.C., Stieber, T.D., 1996. Effects of transitory water stress on potato tuber stem-end reducing sugar and fry color. American Potato Journal 73, 517–530.

EPA, 2002. Report of FQPA Tolerance Reassessment Progress and Interim Risk Management Decision (TRED) for Chlorpropham. http://www.epa.gov/oppsrrd1/REDs/chlorpropham_tred.pdf.

van Es, A., Hartmans, K.J., 1987a. Respiration. In: Rastovski, A., van Es, A., et al. (Eds.), Storage of Potatoes. PUDOC, Wageningen, The Netherlands, pp. 133–140.

van Es, A., Hartmans, K.J., 1987b. Water balance of the potato tuber. In: Rastovski, A., van Es, A., et al. (Eds.), Storage of Potatoes. PUDOC, Wageningen, The Netherlands, pp. 133–140.

van Es, A., Hartmans, K.J., 1987c. Starch and sugars during tuberization, storage and sprouting. In: Rastovski, A., van Es, A., et al. (Eds.), Storage of Potatoes. PUDOC, Wageningen, The Netherlands, pp. 141–147.

van Es, A., Hartmans, K.J., 1987d. Dormancy, sprouting and sprout inhibition. In: Rastovski, A., van Es, A., et al. (Eds.), Storage of Potatoes. PUDOC, Wageningen, The Netherlands, pp. 114–132.

Ezekiel, R., Singh, N., Sharma, S., Kaur, A., 2013. Beneficial phytochemicals in potato—a review. Food Research International 50, 487–496.

FAOSTAT, 2014. http://www.potatopro.com/world/potato-statistics.

FDA, 2009. United States Food and Drug Administration. Analysis and Evaluation of Preventive Control Measures for the Control and Reduction/Elimination of Microbial Hazards on Fresh and Fresh-Cut Produce. Silver Spring, MD. http://www.fda.gov/Food/ScienceResearch/ResearchAreas/SafePracticesforFoodProcesses/ucm091270.htm (accessed 28.12.09.).

Fellows, P.J., 1988. Food Processing Technology. Ellis, Horwood, London.

Foot, R.J., Haase, N.U., Grob, K., Gondé, P., 2007. Acrylamide in fried and roasted potato products. A review on progress in mitigation. Food Additives and Contaminants 24 (Suppl. 1), 37–46.

Friedman, M., 2006. Potato glycoalkaloids and metabolites: roles in the plant and in the diet. Journal of Agricultural Food Chemistry 54, 8655–8681.

Friedman, M., McDonald, G.M., 1997. Potato glycoalkaloids: chemistry, analysis, safety and plant physiology. Critical Reviews in Plant Science 16, 55–132.

Gachango, E., Kirk, W., Schafer, R., Wharton, P., 2012. Evaluation and comparison of biocontrol and conventional fungicides for control of postharvest potato tuber diseases. Biological Control 63, 115–120.

Gottschalk, K., Ezekiel, R., 2006. Storage. In: Gopal, J., Khurana, S.M.P. (Eds.), Handbook of Potato Production, Improvement, and Postharvest Management. Food Products Press, Binghamton, NY, USA, pp. 489–522.

Gudmestad, N.C., 2008. Potato health from sprouting to harvest. In: Johnson, D.A. (Ed.), Potato Health Management, second ed. APS Press, pp. 67–77.

de Haan, P.H., 1987. Unloading of potato stores. In: Rastovski, A., van Es, A., et al. (Eds.), Storage of Potatoes. PUDOC, Wageningen, The Netherlands, pp. 331–341.

Haase, N.U., 2010. Glycoalkaloid concentration in potato tubers related to storage and consumer offering. Potato Research 53 (4), 297–307.

Halford, N.G., Muttucumaru, N., Powers, S.J., Gillatt, P.N., Hartley, L., Elmore, J.S., Mottram, D.S., 2012. Concentrations of free amino acids and sugars in nine potato varieties: effects of storage and relationship with acrylamide formation. Journal of Agricultural Food Chemistry 60, 12044–12055.

Hiller, L.K., Thornton, R.E., 1993. Management of physiological disorders. In: Rowe, R.C. (Ed.), Potato Health Management. APS Press, St. Paul, MN, pp. 87–94.

Hogervorst, J.G., Schouten, L.J., Konings, E.J., Goldbohm, R.A., van den Brandt, P.A., 2007. A prospective study of dietary acrylamide intake and the risk of endometrial, ovarian, and breast cancer. Cancer Epidemiology Biomarkers Prevention 16 (11), 2304–2313.

International Agency for Research on Cancer (IARC), 1994. Monograph on the Evaluation of Carcinogenic Risk to Humans, vol. 61. WHO, p. 243.

Iritani, W.M., 1981. Growth and preharvest stress and processing quality of potatoes. American Potato Journal 58, 71–80.

Iritani, W.M., Weller, L.D., 1973. The development of translucent end potatoes. American Potato Journal 50, 223–233.

Isherwood, F.A., Burton, W.G., 1975. The effect of senescence, handling, sprouting and chemical sprout suppression upon the respiratory quotient of stored potato tubers. Potato Research 18, 98–104.

Kalt, W., Prange, R.K., Daniels-Lake, B.J., Walsh, J., Dean, P., Coffin, R., 1999. Alternative compounds for the maintenance of processing quality of stored potatoes (*Solanum tuberosum*). Journal of Food Processing and Preservation 23, 71–81.

Khanbari, O.S., Thompson, A.K., 1994. The effect of controlled atmosphere storage at 4 °C on crisp color and on sprout growth, rotting and weight loss of potato tubers. Potato Research 37, 291–300.

Külen, O., Stushnoff, C., Holm, D.G., 2013. Effect of cold storage on total phenolics content, antioxidant activity and vitamin C level of selected potato clones. Journal of Science, Food and Agriculture 93 (10), 2437–2444. http://dx.doi.org/10.1002/jsfa.6053.

Kleinkopf, G.E., 1995. Early season storage. American Potato Journal 72, 449–462.

Kleinkopf, G.E., Oberg, N.A., Olsen, N.L., 2003. Sprout inhibition in storage: current status, new chemistries and natural compounds. American Journal of Potato Research 80, 317–327.

Knowles, N.R., Plissey, E.S., 2008. Maintaining tuber health during harvest, storage, and post-storage handling. In: Johnson, D.A. (Ed.), Potato Health Management. APS Press, St. Paul, Minnesota, pp. 79–99.

Lachman, J., Hamouz, K., Orsak, M., Pivec, V., Hejtmankova, K., 2012. Impact of selected factors–cultivar, storage, cooking and baking on the content of anthocyanins in coloured-flesh potatoes. Food Chemistry 133, 1107–1116.

Lewis, M.D., 2007. Practical aspects of potato storage management. In: International Potato Processing and Storage Convention, October 10–12, Calgary, Alberta, Canada.

Lewis, M.D., Kleinkopf, G.E., Shetty, K.K., 1997. Dimethylnaphthalene and diisopropylnaphthalene for potato sprout control in storage: 1. Application methodology and efficacy. American Potato Journal 74, 183–197.

Li, M., Song, B., Zhang, Q., Liu, X., Lin, Y., Ou, Y., Zhang, H., Liu, J., 2013. A synthetic tuber-specific and cold-induced promoter is applicable in controlling potato cold-induced sweetening. Plant Physiology and Biochemistry 67, 41–47.

Lu, Z.-H., Donner, E., Yada, R., Liu, Q., 2012. Impact of γ-irradiation, CIPC treatment, and storage conditions on physicochemical and nutritional properties of potato starches. Food Chemistry 133, 1188–1195.

Mahto, R., Das, M., 2014. Effect of gamma irradiation on the physico-mechanical and chemical properties of potato (*Solanum tuberosum* L.), cv. 'Kufri Sindhuri', in non-refrigerated storage conditions. Postharvest Biology and Technology 92, 37–45. http://dx.doi.org/10.1016/j.postharvbio.2014.01.011.

Mazza, G., 1983. Processing/nutritional quality changes in potato tubers during growth and long term storage. Canadian Institue of Food Science Technology Journal 16 (1), 39–44.

Mehta, A., Singh, B., Ezekiel, R., Minhas, J.S., 2014. Processing quality comparisons in potatoes stored under refrigerated and non-refrigerated conditions. Indian Journal of Plant Physiology 19 (2), 149–155.

Meijers, C.P., 1987a. Diseases and defects liable to affect potatoes during storage. In: Rastovski, A., van Es, A., et al. (Eds.), Storage of Potatoes. PUDOC, Wageningen, The Netherlands, pp. 148–174.

Meijers, C.P., 1987b. Inspection of store and arrival of potatoes. In: Rastovski, A., van Es, A., et al. (Eds.), Storage of Potatoes. PUDOC, Wageningen, The Netherlands, pp. 319–321.

Meijers, C.P., 1987c. Wound healing. In: Rastovski, A., van Es, A. (Eds.), Storage of Potatoes. PUDOC, Wageningen, The Netherlands, pp. 328–330.

Miller, J.S., Rosen, C.J., 2005. Interactive effects of fungicide programs and nitrogen management on potato yield and quality. American Journal of Potato Research 82, 399–409.

Miranda, M.L., Aguilera, J.M., 2006. Structure and texture properties of fried potato products. Food Reviews International 22 (2), 173–201. http://dx.doi.org/10.1080/87559120600574584.

Moehs, C.P., Allen, P.V., Friedman, M., Bellknap, W.R., 1997. Cloning and expression of solanidine UDP-glucose glucosyltransferase from potato. The Plant Journal 11, 227–236.

Moinuddin, Umar, S., 2004. Influence of combined application of potassium and sulfur on yield, quality, and storage behavior of potato. Communication in Soil Science and Plant Analysis 35 (7–8), 1047–1060.

Muttucumaru, N., Powers, S.J., Elmore, J.S., Briddon, A., Mottram, D.S., Halford, N.G., 2014. Evidence for the complex relationship between free amino acid and sugar concentration and acrylamide-forming potential in potato. Annals of Applied Biology 164, 286–300. http://dx.doi.org/10.1111/aab.12101.

Nitithamyong, A., Vonelbe, J.H., Wheeler, R.M., Tibbitts, T.W., 1999. Glycoalkaloids in potato tubers grown under controlled environments. American Journal of Potato Research 76, 337–343.

Onera, M.E., Walkerb, P.N., 2011. Shelf-life of near-aseptically packaged refrigerated potato strips. LWT–Food Science and Technology 44 (7), 1616–1620.

Pinhero, R., Pazhekattu, R., Whitfield, K., Marangoni, A.G., Liu, Q., Yada, R.Y., 2012. Effect of genetic modification and storage on the physico-chemical properties of potato dry matter and acrylamide content of potato chips. Food Research International 49, 7–14.

Pinhero, R., Pazhekattu, R., Marangoni, A.G., Liu, Q., Yada, R.Y., 2011. Alleviation of low temperature sweetening in potato by expressing Arabidopsis pyruvate decarboxylase gene and stress-inducible rd29A: a preliminary study. Physiology and Molecular Biology of Plants 17 (2), 105–114.

Pinhero, R.G., Coffin, R., Yada, R.Y., 2009. Post-harvest storage of potatoes. In: Singh, J., Kaur, L. (Eds.), Advances in Potato Chemistry and Technology, pp. 339–370.

Pinhero, R.G., Copp, L.J., Amaya, C.-L., Marangoni, A.G., Yada, R.Y., 2007. Roles of alcohol dehydrogenase, lactate dehydrogenase and pyruvate decarboxylase in low temperature sweetening in a tolerant and susceptible varieties of potato (*Solanum tuberosum*). Physiologia Plantarum 130 (2), 230–239.

Powelson, M.L., Rowe, H.C., 2008. Managing diseases caused by seedborne and soilborne fungi and fungus-like pathogens. In: Johnson, D.A. (Ed.), Potato Health Management. APS Press, St. Paul, Minnesota, pp. 183–195.

Pringle, B., Bishop, C., Clayton, R., 2009. Potatoes Postharvest. CAB International, Cambridge, MA, USA, p. 448.

Pringle, R.T., 1996. Storage of seed potatoes in pallet boxes. 2. Causes of tuber surface wetting. Potato Research 39, 223–240.

Rak, K., Navarro, F.M., Palta, J.P., 2013. Genotype×storage environment interaction and stability of potato chip color: implications in breeding for cold storage chip quality. Crop Science 53 (5), 1944–1952.

Rastovski, A., 1987a. Storage losses. In: Rastovski, A., van Es, A. (Eds.), Storage of Potatoes. PUDOC, Wageningen, The Netherlands, pp. 177–180.

Rastovski, A., 1987b. Cooling of potatoes and water loss. In: Rastovski, A., van Es, A., et al. (Eds.), Storage of Potatoes, PUDOC, Wageningen, The Netherlands, pp. 183–211.

Rommens, C.M., Ye, J., Richael, C., Swords, K., 2006. Improving potato storage and processing characteristics through all-native DNA transformation. Journal of Agricultural and Food Chemistry 54 (26), 9882–9887.

Salunkhe, D.K., Desai, B.B., Chavan, J.K., 1989. Potatoes. In: Michael, N.A. (Ed.), Quality and Preservation of Vegetables. CRC Press Inc., Boca Raton, Florida, pp. 1–52.

Sanford, L.L., Deahl, K.L., Sinden, S.L., Kobayashi, R.S., 1995. Glycoalkaloid content in tubers of hybrid and backcross populations from a *Solanum tuberosum* x *S. chacoense* cross. American Potato Journal 72, 261–271.

Sapers, G.M., James, R.G., Yousef, A.E., 2006. Microbiology of Fruits and Vegetables. CRC Press Inc., Boca Raton, FL.

Schouten, S.P., 1987. Bulbs and tubers. In: Weichmann, J. (Ed.), Postharvest Physiology of Vegetables. Marcel Dekker, Inc., New York and Basel, pp. 555–581.

Scott, R.K., Wilcockson, S.J., 1978. Application of physiological and agronomic principles to the development of the potato industry. In: Harris, P.M. (Ed.), The Potato Crop. The Scientific Basis for Improvement. London Chapman and Hall, A Halsted Press Book, John Wiley and Sons, New York, pp. 678–704.

Shaw, R.L., Booth, R.H., 1981. Introduction to Potato Storage. International Potato Centre, Lima, Peru, p. 10.

Sijbring, P.H., 1987. Drying of potatoes. In: Rastovski, A., van Es, A., et al. (Eds.), Storage of Potatoes. PUDOC, Wageningen, The Netherlands, pp. 322–327.

Sinha, N.K., Hui, Y.H., 2011. Handbook of Vegetables and Vegetable Processing. Blackwell Pub., Ames, Iowa, p. 772.

Smith, D.B., Roddick, J.G., Jones, J.L., 1996. Potato glycoalkaloids: some unanswered questions. Trends in Food Science and Technology 7, 126–131.

Sonnewald, S., Sonnewald, U., 2014. Regulation of potato sprouting. Planta 239, 27–38. http://dx.doi.org/10.1007/s00425-013-1968-z.

Sowokinos, J.R., 2001a. Pyrophosphorylase in *Solanum tuberosum* L.: allelic and isozyme patterns of UDP-glucose pyrophosphorylase as a marker for cold-sweetening resistance in potatoes. American Journal of Potato Research 78, 57–64.

Sowokinos, J.R., 2001b. Biochemical and molecular control of cold-induced sweetening in potatoes. American Journal of Potato Research 78, 221–236.

Sparenberg, H., 1987. Storage of potatoes at high temperatures. In: Rastovski, A., van Es, A., et al. (Eds.), Storage of Potatoes. PUDOC, Wageningen, The Netherlands, pp. 429–440.

Thompson, A.K., 1998. Controlled Atmosphere Storage of Fruits and Vegetables. CAB International, Biddles Ltd, Guildford and King's Lynn, UK, pp. 201–203.

Thornton, M.K., Lewis, M.D., Barta, J.L., Kleinkopf, G.E., 1994. Effect of nitrogen management on Russet Burbank tuber dormancy and response to CIPC. American Potato Journal 71, 705.

Uri, C., Juhasz, Z., Banfalvi, Z., 2014. A GC–MS-based metabolomics study on the tubers of commercial potato cultivars upon storage. Food Chemistry 159, 287–292.

Yan, H., Chretien, R., Ye, J., Rommens, C.M., 2006. New construct approaches for efficient gene silencing in plants. Plant Physiology 141, 1508–1518.

Ye, J., Shakya, R., Shrestha, P., Rommens, C.M., 2010. Tuber-specific silencing of the acid invertase gene substantially lowers the acrylamide-forming potential of potato. Journal of Agricultural and Food Chemistry 58, 12162–12167.

Zommick, D.H., Knowles, L.O., Knowles, N.R., 2014a. Tuber respiratory profiles during low temperature sweetening (LTS) and reconditioning of LTS-resistant and susceptible potato (*Solanum tuberosum* L.) cultivars. Postharvest Biology and Technology 92, 128–138.

Zommick, D.H., Knowles, L.O., Pavek, M.J., Knowles, N.R., 2014b. In-season heat stress compromises postharvest quality and low-temperature sweetening resistance in potato (*Solanum tuberosum* L.). Planta 239, 1243–1263. http://dx.doi.org/10.1007/s00425-014-2048-8.

Zhang, Y., Zhang, Y., 2007. Formation and reduction of acrylamide in Maillard reaction: a review based on the current state of knowledge. Critical Reviews in Food Science and Nutrition 47, 521–542.

Zhu, F., Cai, Y.-Z., Ke, J., Corke, H., 2010. Composition of phenolic compounds, aminoacids and reducing sugars in commercial potato varieties and their effects on acrylamide formation. Journal of the Science of Food and Agriculture 90, 2254–2262.

CHAPTER 11

Organic Potatoes

Vaiva Bražinskienė[1], Kristina Gaivelytė[2]

[1]Faculty of Business and Technologies, Utena University of Applied Sciences, Utena, Lithuania;
[2]Department of Pharmacognosy, Lithuanian University of Health Science, Kaunas, Lithuania

1. Introduction

Potatoes are able to adapt easily to different growing conditions; they produce high yields and are of high nutritional value, and so they are widespread and are regarded as the most important nongrain agricultural plant in the world. According to data from the United Nations Food and Agriculture Organization regarding the production volume (tons) (Figure 1), the potato ranks fourth after rice, wheat, and corn (FAO).

Currently there are two predominant potato production systems: conventional and organic. Most of the world's consumed potatoes are grown using conventional farming. Although organic farming is becoming more popular, the use of organic farming in potato production is minimal because of low productivity compared with conventionally grown potatoes. It also lacks processing qualities, which leads to very high cost of the raw material and a lower quality of products (Sultana and Siddique, 2014).

The popularity of organic foods may be attributable to the perception that organic foods are healthier, safer, and tastier than conventionally produced foods (Hajšlová et al., 2005; Gilsenan et al., 2010). It is difficult to answer whether this is true, because in trying to answer this question we face many contradicting results and conclusions.

One significant quality study of organically and conventionally grown food was carried out by Dangour et al. (2009). That group of scientists examined more than 150 publications and concluded that there is no evidence that the amount of different accumulated materials in organically and conventionally grown food differs. According to Dangour et al. (2009), differences between organically and conventionally grown crop and livestock production are biologically plausible and likely related to different production conditions of agricultural plants and animals, and soil quality. However, the dispersion of obtained results is thus large enough, as the authors themselves said, to suggest strongly that more comprehensive research in this area be carried out.

Advances in Potato Chemistry and Technology. http://dx.doi.org/10.1016/B978-0-12-800002-1.00011-X

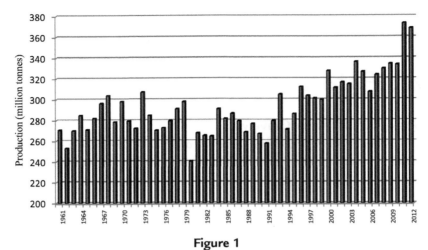

Figure 1
Potato production volume in the world (million ton) in 1961–2012 (FAO).

After a couple of years, the study of Brandt et al. (2011) was published. It analyzed the nutritional value of organically and conventionally grown fruit and vegetables and indicated that the amount of data available on composition differences of fruit and vegetables grown under organic and conventional farming conditions is not only sufficient to prove statistically significant differences but also allows calculation of approximate sizes of such differences. According to those authors, the meta-analysis of secondary metabolites and vitamin content stored in organically and conventionally grown fruits shows that the amount of secondary metabolites in organically grown products is 12% higher than in those conventionally farmed ($p < 0.0001$). This difference varies strongly depending on the subgroup of secondary metabolites, from compounds with a 16% higher concentration ($p < 0.0001$) related to plant defense mechanisms to negligible 2% lower levels of carotenoids.

As examples of large-scale studies show, attempts to generally evaluate the quality of organically and conventionally grown food often end up with unfounded or conflicting conclusions. What result would we get if we tried to examine not all food products together but specifically potatoes? Do organically and conventionally grown potatoes differ?

2. Macroelements (N, P, K, Ca, Mg, S)

2.1 Nitrogen

Nitrogen is a necessary element for all living organisms. Nitrogen is available in many different forms in the soil; the three most abundant forms are nitrate, ammonium, and amino acids (Miller and Cramer, 2005) (Figure 2).

In conventional agriculture, mineral fertilizers are used for potatoes, in which nitrogen often has NH_4^+ and NO_3^- forms, thus, it is easily available to plants compared with nitrogen contained

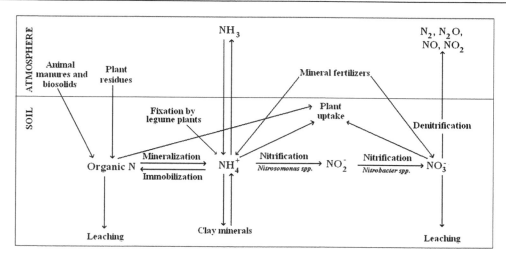

Figure 2
Forms of nitrogen in the soil and their circulation.

in organic fertilizers. Almost all nitrogen is contained in vegetative residues and green manure, and almost half of nitrogen that is found in animal manure is organic and not immediately available to plants. In these forms, nitrogen is released slowly depending on mineralization, immobilization, and nitrification processes ongoing in soil; in these processes, the main role is played by soil microorganisms. Organic nitrogen is less accessible, but potatoes can take it up slowly and there is less risk that the versatile NO_3^- ions will be leached.

Vegetables tend to accumulate reserves of NO_3^- (Maynard et al., 1976), which can have negative consequences to the health of people who use them. Nitrates are especially dangerous for babies, because they cause methemoglobinemia, which often results in death. Much is discussed about nitrate's influence on the development of cancer or even its teratogenic effects (Santamaria, 2006). Because of the negative impact on human health, nitrate content in fruit and vegetables is strictly regulated by law. However, it cannot be stated that nitrates are absolutely harmful, because there is increasing discussion about the benefit of dietary intake of nitrate for the health of people who have cardiovascular disease, and even about the ability to protect against complications of this disease (Bryan and Loscalzo, 2011).

For potato production, it is necessary to ensure that they receive the necessary amount of nitrogen. This chemical element is essential for many biochemical processes, and its deficiency can cause significant physiological changes. Nitrogen is included in the composition of amino acids; it is necessary for the synthesis of nucleotides, which are important as part of the structure of nucleic acids, perform a number of important functions necessary for energy metabolism (included in the adenotriphosphate (ATP) structure), and are involved in signal transduction. Nitrogen is also a part of the chlorophyll molecule structure. The lack of it disturbs photosynthesis, which leads to synthesis changes of organic compounds and, therefore, starch in potatoes.

In the case of conventional farming, together with mineral fertilizers, potatoes obtain a lot of easily available nitrogen. When mineral nitrogen increases in the soil, the total amount of nitrogen and NO_3^- increases as well both in potatoes (Eppendorfer and Eggum, 1994) and in other vegetables (Zhang et al., 2007; Staugaitis et al., 2008). Perhaps this is the main determinant of higher levels of nitrate content in conventionally cultivated potatoes compared with those cultivated organically. More than one group of scientist proved this experimentally (Lairon et al., 1985; Rembiałkowska, 1999; Guziur et al., 2000; Hajšlová et al., 2005; Lombard et al., 2012). The different amounts of nitrates in organically and conventionally cultivated potatoes obtained during some experiments are significant: Lairon et al. (1985) found 46% and Lombardo et al. (2012) found 34% less nitrate in organically grown potatoes than in those cultivated conventionally.

Other opinions may be found: Hamouz et al. (1999a), who analyzed the impact of agricultural systems on nitrate content in potato tubers, concluded that the differences were not statistically significant.

It is likely that the nitrate content in potatoes is determined by mineral nitrogen content in soil and organic fertilizers that have low-to-moderate available nitrogen (especially composts), compared with chemical fertilizers, condition a lower accumulation of nitrate in vegetables (Lairon, 2010). It is observed that the nitrate content in potatoes depends partly on their cultivar (Hajšlová et al., 2005).

2.2 Phosphorus

Phosphorus, like nitrogen, is indispensable to all living organisms. Phosphorus is a part of ATP and other nucleozidphosphate composition. It is an indispensable component of the nucleic acids and is a part of the phospholipid composition that forms the basis of cell membrane structures. Large amounts of phosphorus are stored in plant seeds, where it is important for embryo development, germination, and seedling growth (Marschner, 1995). Because of sugar phosphates used for carbon fixation in the intermediate stages of photosynthesis, the lack of phosphorus leads to a sudden drop in photosynthesis speed (Maathuis, 2009).

More than 90% of phosphorus in the soil is in insoluble form and is not available to plants. The form of phosphorus in which it is found in the soil solution depends on pH; however, it is often $H_2PO_4^-$, and this form of phosphorus can be absorbed by plants (Maathuis, 2009).

Unlike nitrogen, in plants, phosphorus remains in the oxidized form. In plants, one part of phosphorus remains in inorganic form (usually $H_2PO_4^-$) and the other part is composed of a structure of various organic compounds (phospholipids, nucleic acids, sugars, and phosphates) or bound with other phosphates by a high-energy pyrophosphate bond (for example, the formation of ATP) (Marschner, 2012).

The literature that provides an examination of the impact of organic and mineral fertilizers on phosphorus content in potato tubers reflects different opinions. An article published in 1974, which summarizes the work of 12 years and compares the influence of these two different fertilization methods on the amount of phosphorus in potatoes, carrots, and cabbage grown in Germany, found that organically grown vegetables accumulate approximately 13% more phosphorus (Schuphan, 1974). The higher amount of phosphorus was found in organically grown potatoes; the same opinion was shared by Warman and Havard (1998), Colla et al. (2002), Pieper and Barrett (2009), Wszelaki et al. (2005), and Herencia et al. (2011). According to Lombardo et al. (2014), organically grown potatoes can accumulate up to 60% more phosphorus than those grown under conventional farming conditions. However, Phillips et al. (2002) and Hargreaves et al. (2008) argued that the agricultural system has no effect on the amount of phosphorus, and according to Warman (2005), conventionally grown plants accumulate more phosphorus.

It has been observed that fertilization with phosphorus mineral fertilizers can reduce the amount of phosphorus even slightly in potato tubers (Antanaitis and Švedas, 2000; Janušauskaitė and Žėkaitė, 2008), and the increase in mineral phosphorus fertilization does not have a significant effect on the amount of phosphorus in potato tubers (Alvarez-Sánchez et al., 1999). The results of various studies show that organic phosphorus fertilizers may be even more effective than inorganic ones (Nziguheba et al., 1998), because the use of organic fertilizer promotes mineralization of soil phosphorus (Kwabiah et al., 2003), increases its solubility, and thus improves the availability for plants (Nziguheba et al., 1998; Andrews et al., 2002; Herencia et al., 2007).

Phosphorus promotes root growth and so, lacking phosphorus and trying to increase it, plants move phosphorus reserves from older leaves to the roots using phosphorus resources accumulated in vacuoles. In this way, they expand the root surface area and increase the phosphorus absorption surface. However, if there is a sufficient amount of this macroelement in the soil, in contrast, various protective mechanisms start (for example, conversion of excess phosphorus into organic compounds and reduction of its absorption); these impede accumulation of toxic concentrations of phosphates (Schachtman et al., 1998). Starch is also involved in binding phosphates. In potato tubers, even 40% of the total phosphorus can be incorporated into the starch composition (Marschner, 2012). Jacobsen et al. (1998) found that owing to fertilization of phosphorus fertilizers, the content of phosphorylated starch increases in potato tubers, and phosphorus fertilization through the leaves does not cause such a result. This shows that a higher level of phosphorus available to plants in the soil encourages the accumulation of phosphorus in potato tubers.

It is also known that *Mycorrhizal* fungi greatly contribute to improving the absorption of phosphorus in plants (Schachtman et al., 1998).

For all of these reasons, it can be assumed that when potatoes are fertilized with organic fertilizers, the amount of easily available phosphorus increases in the soil and potatoes can accumulate more of this element compared with potatoes grown under conventional farming conditions.

2.3 Potassium

Compared with other agricultural crops, potatoes demand much potassium (Westermann et al., 1994). Potassium activates many enzymes in the form of selective connections. For this reason, it is particularly important for plant metabolism. Potassium-dependent enzymes are pyruvate kinase, phosphofructokinase, and adenosine phosphate–glucose starch synthase involved in the C-metabolism. The dominant role of K^+ in turgor provision and water homeostasis is evident in processes such as pressure-driven solute transport in the xylem and phloem, high levels of vacuolar K^+ accumulation, and large fluxes of K^+ that mediate plant movement. High potassium concentrations are important for ribosome-mediated protein synthesis (Maathuis, 2009). For all of these functions, potassium is especially important for potatoes as an element involved in the regulation of osmosis, sugar flux, and starch synthesis.

In soil, most K^+ is dehydrated and coordinated to oxygen atoms not available to plants (Maathuis, 2009). Plants are able to absorb potassium only in the form of dissolved K^+ ions; for this reason, potatoes are fertilized with various potassium salts (KCl, K_2SO_4, etc.).

The potassium content in potato tubers increases when they are fertilized with both organic (Baniūnienė and Žėkaitė, 2007) and mineral fertilizers (Janušauskaitė and Žėkaitė, 2008); thus, when the amount of potassium in the soil increases, the potassium concentration in potatoes increases as well (Antanaitis and Švedas, 2000). Comparing the amount of potassium accumulated in tubers of organically and conventionally grown potatoes, Schuphan (1974) found 18% and Smith (1993) found 30% higher amount of potassium in organically grown potatoes. Wszelaki et al. (2005) and Rembiałkowska (2007) also believed that organic potatoes accumulate more potassium than those grown under conventional farming conditions.

There are other opinions. Warman and Havard (1998) found a higher amount of potassium in tubers of organically grown potato than in those conventionally cultivated. However, the differences were not statistically significant. Hoefkens et al. (2009) found significantly more potassium in conventionally grown potatoes. Lombardo et al. (2014) compared the amount of potassium in the tubers of organically and conventionally grown potatoes for 2 years; one year, more potassium was found in the tubers of organically grown potatoes and the other year the amount of potassium was higher in the tubers of conventionally grown potatoes.

In the case of conventional agriculture, increased levels of potassium in tubers of conventionally grown potatoes may reflect a higher content of this element in the soil; in other cases, there is no clear link between the soil and composition of mineral elements of the tubers. Although the amount of fertilizers in both systems would be identical, the absorption of elements and accumulation in plants can be determined by the type of fertilizer (organic or inorganic) and solubility in water (Härdter et al., 2005). Another important aspect of analyzing the influence of farming systems on potassium accumulated in potato tubers is that potato potassium increases with increasing soil phosphorus (Antanaitis and Švedas, 2000). It is now known that some potassium ion carriers are activated in rhizodermis by acidic soil pH; some are activated by a decrease in pH in the cytosol (Chen et al., 2008). Phosphate ions are characterized by acidic properties that are likely to reduce the pH and thus activate the transport of potassium into cells of rhizodermis. Based on this assumption, it is more likely that tubers of organically grown potatoes accumulate more potassium (see Section 2.2).

2.4 Magnesium

It is important to ensure the sufficient content of magnesium available for plants in the soil, because this element is necessary for the activities of many enzymes. It helps maintain the secondary structure of ribonucleic acid; thus, gene transcription and translation depend on magnesium content in cells (Misra and Draper, 2000). Magnesium is especially important for photosynthesis because it is the central atom of the chlorophyll molecule. Also, because of its concentration in the chloroplast, it is controlled through the main photosynthetic enzymes. Magnesium is involved in the regulation of ion currents through the chloroplast and vacuole membrane, thus regulating the concentration of ions in the cytoplasm and stomata opening (Shaul, 2002).

Magnesium is relatively weakly absorbed on soil colloids, so it is relatively easily leached and magnesium deficiency is common (Deng et al., 2006), especially in acidic and light granulometric composition soils.

Comparing the content of magnesium accumulated in organically and conventionally grown potatoes, most authors found more of it in organically grown tubers. In organically grown potatoes, Smith (1993) found 50% more magnesium than in potatoes conventionally cultivated. According to the information of Warman and Havard (1998), Wszelaki et al. (2005), Lairon (2010), and Lombardo et al. (2014), organic potatoes accumulated more of this element. According to Wszelaki et al. (2005) and Lombardo et al. (2014), a direct relationship was found between potassium and magnesium content in potato tubers.

It is likely that organically grown potato tubers accumulate more magnesium than those grown conventionally, because magnesium is found in organic fertilizers and mineral nitrogen–phosphorus–potassium (NPK) fertilizers do not supplement magnesium resources in the soil. The magnesium content in the soil directly correlates with the amount of soil microorganisms, which are more active in organic farming cases (Bulluck et al., 2002). Both of these causes increase the concentration of magnesium available to plants in the soil in the case of organic farming. It is known that magnesium is more easily absorbed by plants in an apoplastic way (Shaul, 2002); therefore, it is likely that higher concentrations in the soil determine greater accumulation of this element in potato tubers.

2.5 Sulfur

In soil, sulfur may have inorganic and organic forms. Under aerobic conditions, inorganic sulfur is generally in the form of sulfate (SO_4^{2-}). Sulfates are mostly absorbed by plants.

In addition to the sulfur from the soil, the plants may absorb this element from the atmosphere, in which sulfur can have SO_4^{2-} or H_2S forms (Maathuis, 2009).

Sulfur is commonly found in the reduced form in the composition of amino acids (cysteine and methionine) (Maathuis, 2009). It is included in the composition of glutathione, which is an important part of the cell antioxidant system (Foyer et al., 2007). Moreover, sulfur is a part of sulfolipid composition found in thylakoid membranes. It is believed that they are important for stabilizing the photosystem's components (Ramani et al., 2004).

There is not enough information about the sulfur content in tubers of organically and conventionally grown potatoes. It is known that in nature, sulfur metabolism is supported by microorganisms; owing to their activities, sulfides are oxidized to sulfates. Organic farming is more favorable to microbial activities, and in conventional agriculture sulfur-containing fertilizers are not always used, therefore, it can be assumed that organically cultivated potatoes have more sulfur. This statement is confirmed by Smith's (1993) study, which states that organically grown potatoes contain 160% higher sulfur content than potatoes cultivated conventionally. Wszelaki et al. (2005) stated that organically grown potatoes accumulate more sulfur in both the potato pulp and potato peel.

2.6 Calcium

Potatoes can absorb calcium from the soil only when it is dissolved in Ca^{2+} form; yet the number of free calcium ions in the soil is not high because they tend to adsorb on colloids or form insoluble or nearly insoluble compounds with phosphates, carbonates, and other ions (Maathuis, 2009).

Calcium is important for plants. In cells, calcium has a structural function. It is part of the cell wall structure and it connects negative carboxyl groups of the pectin molecules and thus gives the wall strength. Calcium ions are important as cellular signal transmitters (Maathuis, 2009). Changes in Ca^{2+} concentration in the cytoplasm determine plant responses to abiotic and biotic stress and mechanical damage (Demidchik and Maathuis, 2007).

There is no consensus on the impact of organic and conventional farming on calcium content in potato tubers. In organically grown potatoes, Smith (1993) found 50% more calcium than in those grown under conventional farming conditions. Worthington (2001) stated that organic vegetables have 30% more calcium than those conventionally cultivated. According to Wszelaki et al. (2005), the difference in calcium levels in organically and conventionally cultivated potatoes was not significant. Hoefkens et al. (2009) and Lombardo et al. (2014) argued that tubers of potatoes grown under conventional farming conditions accumulated more calcium.

Bulluck et al. (2002) argued that owing to the application of organic fertilizers, in 2 years the concentration of calcium in the soil doubled but synthetic fertilizers had no effect on calcium concentration in the soil. According to them, compared with mineral fertilizers, organic ones increase calcium ion concentration in the soil. The same opinion was supported by Mäder et al. (2002), but the concentration of calcium in potato tubers may be determined not only by available Ca^{2+} content in the soil but also by the concentration of other ions in it. For example, it is known that with increasing K^+ concentration in the soil, the accumulation of Ca^{2+} in potato tubers is suppressed (Cao and Tibbitts, 1991). Islam and Nahar (2013) confirmed this statement. They carried out an experiment during which potatoes were fertilized with organic fertilizers, NPK fertilizers, and a mixture of organic and mineral NPK fertilizers. It turned out that the most calcium was absorbed by potatoes fertilized with organic fertilizers, whereas potatoes fertilized with mineral fertilizers absorbed the least calcium.

3. *Phenolic Compounds*

According to a survey of nutritionally and pharmacologically valuable compounds found in potato peel, carried out by Schieber and Saldana (2009), more than 20 different phenolic compounds can be found in potato peel (Table 1). It is no secret that phenolic compounds are

Table 1: Phenolic Compounds in Potatoes (Schieber and Saldana, 2009).

Hydroxycinnamic Acids
5-O-Caffeoylquinic acid (chlorogenic acid)
4-O-Caffeoylquinic acid (*crypto*-chlorogenic acid)
3-O-Caffeoylquinic acid (*neo*-chlorogenic acid)
Caffeic acid
p-Coumaric acid
Ferulic acid
Hydroxybenzoic Acids
Gallic acid
Protocatechuic acid
Vanillic acid
Salicylic acid
Non-anthocyanin Flavonoids
Catechin
Epicatechin
Eriodyctiol
Naringenin
Kaempferol glycosides
Quercetin glycosides
Anthocyanins
Petunidin glycosides
Malvidin glycosides
Pelargonidin glycosides
Peonidin glycosides
Dihydrocaffeoyl Polyamines
N^1,N^{12}-*bis*(Dihydrocaffeoyl)spermine (kukoamine A)
N^1,N^8-*bis*(Dihydrocaffeoyl)spermidine
N^1,N^4,N^{12}-*tris*(Dihydrocaffeoyl)spermine
N^1,N^4,N^8-*tris*(Dihydrocaffeoyl)spermidine

seen as powerful antioxidants, with various positive effects on human health. For this reason, they are of particular interest. Their synthesis in plants and the reasons why they increase or reduce the amount of these compounds are widely studied.

There are not many studies that analyze qualitative and quantitative differences in phenolic compounds in tubers of organically and conventionally grown potatoes. Søltoft et al. (2010) examined the influence of agricultural systems, year, and place of growth on the amount of phenolic compounds found in potato tubers, and found that plants fertilized with vegetable fertilizers accumulated more neochlorogenic acid than potatoes grown under conventional farming conditions. Hajšlová et al. (2005) analyzed the influence of agriculture systems on chlorogenic acid content and concluded that organically grown potatoes accumulate more of it. According to the results of Lombard et al. (2012), the total content of phenolic compounds in organically grown potatoes is 18% higher than in those grown under conventional farming conditions. There are contradictory results: Hamouz et al. (2013) stated that, depending on the cultivar, organically grown potatoes can accumulate more, the same amount, or even less chlorogenic acid than potatoes grown under conventional farming conditions. Similar findings were presented by Brazinskiene et al. (2014), but based on their data, the amount of phenolic acids depends not only on the cultivar and agricultural system but also on meteorological conditions. One year more phenolic acids can be found in organically grown potatoes; another year, the opposite happens. What are the reasons for such a result?

The study of Koricheva et al. (1998) on factors that determine secondary metabolites' synthesis of woody plants stated that fertilization of nitrogen-containing fertilizers inhibits the synthesis of secondary metabolites in plants. Brandt and Mølgaard (2001) confirmed this statement, adding that the reduced availability of nitrogen for plants encourages the synthesis of phenolic compounds carrying a protective function, which leads to increased plant resistance to pests and diseases but reduces their growth rate and yield. Musilová et al. (2013) experimentally investigated the influence of the nitrogen content on the amount of phenolic compounds (fertilized with vermicompost) accumulated in potato tubers and confirmed that the increase in the amount of nitrogen in the soil adversely affects the amount of phenolic compounds in potato tubers.

Explaining the influence of nitrogen for the synthesis of secondary metabolites, Brandt and Mølgaard (2001) based their thoughts on the C/N balance theory that appeared in the 1980s (Bryant et al., 1983; Coley et al., 1985), which partly explains the nitrogen effect on the synthesis of phenolic compounds. It says that when the soil is rich with easily accessible nitrogen, at first plants synthesize compounds that require a lot of nitrogen, such as proteins necessary for growth and derivatives that contain nitrogen, such as alkaloids. However, when nitrogen resources are reduced, synthesis processes move to the side of carbon and compounds in which nitrogen is not necessary start to synthesize (starch, cellulose, and secondary metabolites without nitrogen, such as phenolic compounds and terpenoids).

The results of analysis of agricultural systems' influence on potato protein expression carried out by Lehesranta et al. (2007) and Rempelos et al. (2013) confirm the insights of Brandt and Mølgaard (2001). Protein expression of conventionally and organically grown potatoes is different. Organically grown potatoes have a lot of proteins responsible for the response to stress. Conventionally cultivated potatoes have a lot of proteins responsible for a variety of processes related to photosynthesis and synthesis of organic compounds. Lehesranta et al. (2007) and Rempelos et al. (2013) found that fertilization has a decisive influence on this result. Plants that have an easily available source of nitrogen owing to mineral NPK fertilizers tend to devote more resources to the promotion of organic nitrogen compound synthesis; for this reason, their yield is higher. Plants that have smaller resources of nitrogen (the nitrogen that gets into soil with organic fertilizers is not easily available to plants) prepare for responses to abiotic and biotic stressor impact. Therefore, some nitrogen is used to form protein responsible for the synthesis of secondary metabolites. It is likely that for this reason organically grown potatoes synthesize more phenolic compounds, but their yield is lower.

Knowing that phenolic compounds are synthesized by plants in response to various biotic and abiotic stresses (Dixon and Paiva, 1995), it can be assumed that organically grown crops that are not protected using pesticides should accumulate more phenolic compounds than plants not protected by chemical products. Lehesranta et al. (2007) and Rempelos et al. (2013) found that the use of pesticides has no influence on protein expression of potatoes. Steady protein composition during use of pesticides suggests that they do not cause significant changes in the metabolic ways of potatoes. This was confirmed by the experiment carried out by Zhao et al. (2009), although it was done not with potatoes but with Chinese cabbage (*Brassica rapa* L. *chinensis*). During the experiment, plants were grown in a greenhouse under organic and conventional farming conditions and sprayed with insecticide. An examination of accumulated phenolic compounds in Chinese cabbage showed that the use of insecticide in both organic and conventional farming cases did not have a significant effect on them.

In their studies on the response of conventionally and organically grown potatoes to abiotic stress (potato tubers were infected with *Phytophthora infestans*), Daško et al. (2012) observed that in both organically and conventionally grown potato tubers, pathogen contamination increased the content of chlorogenic acid. The interesting thing is that the amount of chlorogenic acid in organic potatoes increased by 1.3 times, and in conventionally cultivated potatoes it increased three times. Thus, growth conditions influenced potato reactions to the pathogen and the whole plant's immunity. The reaction to pathogens of conventionally grown potatoes is particularly strong; the authors named this phytoallergy.

4. Carotenoids

According to Ezekiel et al. (2013), lutein, zeaxanthin, violaxanthin, and neoxanthin are major carotenoids founded in potatoes and β-carotene is founded in trace amounts (Figure 3).

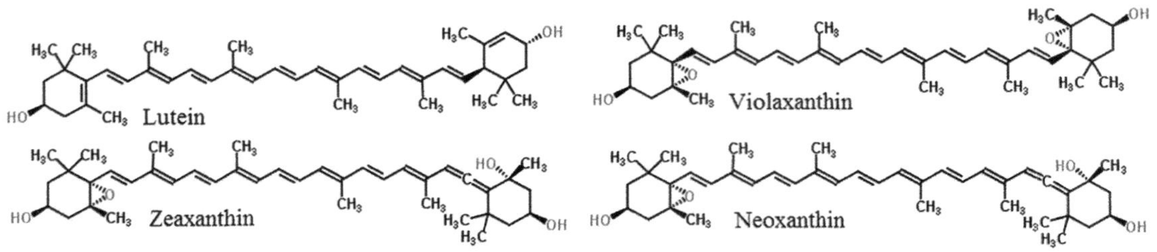

Figure 3
Major carotenoids in potatoes.

Carotenoids have two of the most important functions in plants: photoprotection and light collection (Demmig-Adams et al., 1996). Although carotenoids are inseparable from the process of photosynthesis, they can be accumulated in plants' organs that do not perform photosynthesis (seeds, fruits, blossoms, and roots), where they are important as attractants and antioxidants, and serve as substrates for the production of phytohormones such as abscisic acid and strigolactone as well as other signaling molecules (e.g., blumenin and mycorradicin) (Howitt and Pogson, 2006; Cazzonelli, 2011). Enzymes involved in the synthesis of carotenoids are encoded in the nucleus genes (Bartley and Scolnik, 1995).

Assessing the impact of different agricultural systems on the carotenoid content in vegetables, some authors argued that there are no significant differences between the amount of carotenoids in organically and conventionally grown vegetables (Warman and Havard, 1997; Bourn and Prescott, 2002). Others said that vegetables grown under conventional farming conditions accumulate more carotenoids (Rembiałkowska, 2007). According to Mozafar (1993), nitrogen fertilizers increase the amount of β-carotene in vegetables. This finding is consistent with the C/N theory: When the number of easily available nitrogen increases, the synthesis of organic compounds intensifies. It is likely that the synthesis of carotenoids as pigments necessary for photosynthesis also intensifies.

Few studies have examined the influence of agricultural systems on the amount of carotenoids accumulated by potatoes.

5. Glycoalkaloids

After examining more than 300 *Solanum* species, at least 90 different steroidal alkaloids have been identified; however, in commercial potato cultivars only one form exists: solanidine glycosides, α-solanine and α-chaconine (Friedman et al., 1997). These compounds are toxic to humans. It is known that these glycoalkaloids are able to bind to the active centers of the enzyme acetylcholinesterase and butyrylcholinesterase and inhibit them, in this way disturbing degradation of acetylcholine in synapses and causing cholinergic syndrome, which may lead to coma or even death (Friedman, 2006). With a structure similar to cholesterol, they are able to violate the integrity of cell membranes and can cause damage to digestive system and other organs (Ginzberg et al., 2009). Morris and Lee (1984) estimated that the toxic amount of glycoalkaloids accumulated by potatoes is about 2–5 mg/kg body weight and the fatal amount is about 3–6 mg/kg body weight. In commercially grown potatoes, the amount ranged from 1 to 35 mg/100 g fresh weight, whereas in the wild potatoes the amount of glycoalkaloids may be 100 mg/100 g fresh weight or even many times higher (Sinden et al., 1984).

Although glycoalkaloids accumulated by potatoes and used in high doses are toxic to the human body, their positive effects should be mentioned: antiallergic, antipyretic, and anti-inflammatory effects, potential cholesterol-lowering effect, and antibiotic activities against pathogenic bacteria, viruses, protozoa, and fungi to inhibit the growth of human cancer cells (Friedman, 2006).

Previous studies have shown that differences in levels of glycoalkaloids lead to differences in flavor features (Sinden et al., 1976; Savage et al., 2000), although Skrabule et al. (2013) did not determine a connection between glycoalkaloid content and the taste of boiled tubers. According to Haase (2008), large amounts of glycoalkaloids can result in bitter taste in boiled tubers, but small amounts have a positive effect on the taste of the tubers (Haase, 2010).

All parts of the plant synthesize glycoalkaloids; the highest concentrations are found in the most metabolically active organs: young leaves, fruits, blossoms, and buds (Friedman et al., 1997). There is no evidence that glycoalkaloids are transported from one organ to another. Therefore, it can be assumed that their synthesis and catabolism are regulated at the level of organs or tissues (Arif, 2013). In potato tubers the most glycoalkaloids can be found in the approximately 1-mm-thick outer layer of the tuber; moving toward the center of the tuber, their amount decreases (Kozukue et al., 1987; Friedman et al., 2003). Hence, most glycoalkaloids are removed by peeling potatoes. However, it is important to know how many of these toxic compounds potato tubers can accumulate and whether and to what extent different cropping systems can affect their content.

The amount of steroidal alkaloids in potato tubers is determined genetically and varies greatly depending on potato cultivar, but the conditions under which potatoes are grown and potato tubers are stored have a huge impact on their accumulation (Ginzberg et al., 2009). It is known that light, mechanical damage, elevated temperature, long storage time, and germination have a positive effect on the amount of glycoalkaloids in tubers (Friedman et al., 1997; Petersson et al., 2013). Plants synthesize glycoalkaloids as protection against pests and adverse environmental conditions (Friedman et al., 1997; Friedman, 2006; Ginzberg et al., 2009; Arif, 2013), thus, it is likely that organically grown potatoes should accumulate more of them, because without the use of chemical preservatives they experience more abiotic and biotic stress.

That organically grown potatoes accumulate larger quantities of glycoalkaloids than potatoes grown conventionally has been confirmed by various researchers. According to Hajšlová et al. (2005), the average total quantity of glycoalkaloids is greater in organically grown potato tubers (on average, 80.8 ± 44.5 mg/kg for all cultivars taken together) than in those conventionally grown (58.5 ± 44.1 mg/kg). The significant impact of the year, growth place, and cultivar on the amount of accumulated glycoalkaloids was also determined. No statistically significant change in the total amount of glycoalkaloids between freshly harvested and stored potato tubers was found in study by Hajšlová et al. (2005).

Hoefkens et al. (2009) found significantly larger quantities of glycoalkaloids in tubers of organically grown potatoes. Skrabule et al. (2013) compared the amount of accumulated glycoalkaloids in organically and conventionally grown potatoes for 2 years. In that study, a significant influence of agricultural system in the same geographical area and under similar meteorological conditions on the amount of glycoalkaloids in potatoes was determined in only 1 year. The total amount of glycoalkaloids did not exceed the limits of human health (<20 mg/100 g).

Wszelaki et al. (2005) stated that the concentration of solanidine in flesh of conventionally grown potatoes was half that in potatoes grown organically. Minimum quantities of solanidine were found in tuber peels of conventionally grown potatoes, the average amount of solanidine was found in organically grown and fertilized with compost potatoes, the highest amount of solanidine was found in organically grown and not fertilized potatoes, and the amount of solanidine was almost twice as high as in potatoes cultivated in conventional agricultural areas. This shows that fertilization has a huge impact on the amount of glycoalkaloids in potato tubers. Nowacki et al. (1975), cited in Maga (1994), found a negative relationship between nitrogen and glycoalkaloid quantities. The study noted that the amount of glycoalkaloid increases when the nitrogen fertilization rate decreases. The maximum, average, and minimum rate, respectively, of nitrogen was found in potatoes grown under conventional, organic with compost, and organic without compost farming conditions. On the other hand, Cronk et al. (1974) observed that an excessive amount of nitrogen increases the glycoalkaloid amount compared with normal nitrogen application.

Processing of potato tubers may also affect the amount of glycoalkaloids. In potato tubers, glycoalkaloids are mostly concentrated in the 1.5-mm layer, which has about 100 layers of peel and storage tissue cells beneath the peel. Skin peeling removes about 60–96% of glycoalkaloids (Maga, 1994). Boiled glycoalkaloids go into the flesh or are washed out of the tubers (Mondy and Gosselin, 1988).

Assessing the impact of agricultural systems on the amount of glycoalkaloids in potato tubers, it may be assumed that organically grown potatoes accumulate more glycoalkaloids. This happens for two reasons: smaller amounts of easily available mineral nitrogen in soil in case of organic farming (C/N balance theory) and the influence of abiotic and biotic stress.

6. Ascorbic Acid

Ascorbic acid (vitamin C) is necessary for both plants and animals. In plants, this organic compound performs many different functions and is important as an antioxidant, a cofactor of various enzymes, and a redox buffer (Smirnoff and Wheeler, 2000; Gallie, 2013). Ascorbic acid is particularly important for photosynthesis, especially as an antioxidant that neutralizes hydrogen peroxide that forms in photosystem I during oxygen photo reduction. Also, monodehydroascorbate, which forms when ascorbic acid is influenced by ascorbate peroxidase, acts as a direct acceptor of electron in photosystem I. In addition, ascorbic acid is a cofactor of an enzyme involved in the synthesis of zeaxanthin, which acts as a photoprotectant (Smirnoff, 1996). Ascorbic acid regulates cell division and growth, cell cycle, and flowering time, is involved in signal transduction, regulates the response to abiotic and biotic stress, and is involved in cell wall expansion, synthesis of ethylene, gibberellin, anthocyanins, and hydroxyproline (Love and Pavek, 2008; Gallie, 2013). Ascorbates react rapidly with superoxide, singlet oxygen, ozone, and hydrogen peroxide. They are involved in the neutralization of these reactive oxygen forms that compose during aerobic metabolism and in exposure to certain pollutants and herbicides. Ascorbates are also important for the regeneration of lipophilic antioxidant α-tocopherol (vitamin E) from the α-chromanoxyl radical (Smirnoff, 1996).

In a review, Gallie (2013) discussed ascorbic acid's synthesis in plants; however, it is likely that ascorbic acid's use of L-galactose proposed by Wheeler et al. (1998), compared with other potential synthesis ways in potatoes, is important, perhaps even dominant. This was proven by an experiment carried out by Tedone et al. (2004), in which the amount of ascorbic acid in the leaves of potatoes that received extra L-galacton-1,4-lactone or L-galactose, compared with control, significantly increased. Tedone et al. (2004) found that activation of ascorbic acid's synthesis in leaves determines its larger amount in phloem exudate and tubers. It is known that ascorbic acid is best transferred from the leaves of the plant to the organs in which carbohydrates are stored or actively used (for example, buds, seeds, roots, and tubers) (Smirnoff, 2011).

There is evidence that ascorbic acid is important for growth and large amounts of this compound are found in intensively developing tissues. When growth decelerates, the amount of ascorbic acid decreases and its accumulated reserves are used for protective or other physiological processes (Chen and Gallie, 2006).

There is no consensus on the impact the agricultural system has on the level of ascorbic acid. Hajšlová et al. (2005) noted that compared with conventionally grown potatoes, organically grown ones tend to accumulate more ascorbic acid; however, this tendency was not statistically significant. Hoefkens et al. (2009) argued that organically grown potatoes tend to accumulate more ascorbic acid. Lombardo et al. (2012) objected to this statement. They determined that in conventionally grown potato tubers the content of ascorbic acid was 23% higher than in potatoes cultivated organically. Svec et al. (1976), Warman and Havard (1998), Hamouz et al. (1999b), and Rembiałkowska (1999) did not find a significant difference between the amount of ascorbic acid in potato tubers grown in organic and conventional agricultural systems. Hajšlová et al. (2005) and Skrabule et al. (2013) argued that the amount of ascorbic acid in tubers is significantly determined by the genotype and the influence of cultivation system is not statistically significant. Lombardo et al. (2012) confirmed the influence of variety on the amount of ascorbic acid. A study by Camin et al. (2007) showed that the amount of ascorbic acid varied greatly and the place of cultivation had a greater impact than the agricultural system.

The amount of ascorbic acid in plants is influenced by many factors. Mozafar (1993) reviewed publications related to the influence of nitrogen fertilizers on vitamin content in plants and concluded that nitrogen fertilizers reduce the amount of ascorbic acid in potato tubers. Hamouz et al. (2009) confirmed the negative impact of increasing nitrogen on the amount of ascorbic acid. Nevertheless, Augustin (1975) noted that increases in nitrogen reduce the amount of ascorbic acid in only some varieties of potato tubers. Such an impact of nitrogen fertilizer on the amount of ascorbic acid may be associated with nitrate-driven accumulation of oxalic acid in many vegetables (Liu et al., 2014). It is known that ascorbic acid is a raw material for the synthesis of oxalates and tartrates (Smirnoff, 1996), and oxalic acid is involved in maintaining homeostasis of the cell pH and neutralizes OH^- ions released during the assimilation of nitrates (Raven and Smith, 1976). Whether there is a relationship between nitrate and oxalate amount in potato tubers as bright as in spinach can be answered only by experiments that prove this.

According to some authors, an increase in nitrogen reduces the amount of ascorbic acid in potato tubers, yet its deficit may not have a significant impact on the amount of this organic compound. Kováčik et al. (2014) demonstrated that compared with barley (*Hordeum vulgare* cv. Bojos) receiving a sufficient amount of nitrogen, nitrogen deficiency increases the amount of ascorbic acid in the roots or remains constant, whereas the amount of ascorbic acid in shoots is significantly reduced.

Another factor that has impact on the amount of ascorbic acid is the efficiency of photosynthesis. Although light is not necessary for the synthesis of ascorbic acid, a quantity of light has an indirect effect on synthesis through sugar during photosynthesis (Lee and Kader, 2000). A large amount of nitrogen in the soil can promote the growth of leaves and increase the rate of photosynthesis and thus the amount of sugar necessary for the synthesis of ascorbic acid (Lombardo et al., 2012).

The use of various pesticides may have an impact on the amount of ascorbic acid. Zarzecka and Gugala (2003) demonstrated that the amount of ascorbic acid content in tubers increases with the use of herbicides. Gugała and Zarzecka (2012) found that the amount of ascorbic acid content in potato tubers depended significantly on insecticide treatment, the cultivar, and weather conditions during the study. Using insecticides, the amount of ascorbic acid in potato tubers was higher compared with the control produced without the use of insecticides. Using insecticides, the same tendency was observed in all experimental years; this was confirmed by investigations carried out by other scientists (Fidalgo et al., 2000;

Marwaha, 1988). On the other hand, Antonious et al. (2001) did not find a significant effect of insecticides on the amount of ascorbic acid. That the use of pesticides increases the amount of ascorbic acid could explain two hypotheses: First, pesticides protect against pests and weeds. In this way, a higher leaf area is saved and photosynthesis happens more actively. Second, plants react to pesticides as stressors, and to protect themselves from them, they activate antioxidant systems.

Ascorbic acid is a powerful antioxidant, and its amount in plants is influenced by abiotic and biotic stress. In some studies, a higher amount of ascorbic acid was found in organically grown potatoes, which can be explained by greater exposure to pathogens in organic crops. This encourages the formation of antioxidants characterized by a protective effect. It is believed that the increase in ascorbic acid is a response to various stressors, which have a greater impact in the case of organic farming (Lee and Kader, 2000). The synthesis of nitrogen-free ascorbic acid may also be activated owing to the C/N balance theory.

Ascorbic acid is an organic compound that performs a variety of functions in plants. The influence of agricultural systems on the amount of ascorbic acid is ambiguous, but it is likely that conventionally cultivated potatoes should accumulate more of this compound. It is known that ascorbic acid is synthesized in plants in several different ways; therefore, it is difficult to assess the impact of different factors on the amount of this compound. It is often complex and depends on the intensity of factors.

7. Conclusions

In many cases the amount of macronutrients in potato tubers depends on how much a plant can absorb them from the soil. The availability of most macronutrients to crops is improved by soil microorganisms that are more active in organic farming. Nitrogen is an exception. Conventionally cultivated potatoes accumulate more of this compound owing to the use of synthetic fertilizers because they have a richer source of easily available mineral nitrogen.

Organically grown potatoes should accumulate more phenolic compounds and glycoalkaloids than conventionally cultivated ones. This happens for two reasons: smaller amounts of easily available nitrogen in the soil in the case of organic farming (C/N balance theory) and the influence of abiotic and biotic stress. Because of intensive photosynthesis, potatoes grown under conventional farming conditions should accumulate more carotenoids.

The influence of agricultural systems on the amount of ascorbic acid is ambiguous, but it is likely that conventionally grown potatoes should accumulate more of this compound. It is known that ascorbic acid is synthesized in several different ways; therefore, it is difficult to assess the impact of different factors on the amount of this compound. It is often complex and depends on the intensity of factors.

References

Alvarez-Sánchez, E., Etchers, J.D., Ortiz, J., Nunez, R., Volke, V., Tijerina, L., Martinez, A., 1999. Biomass production and phosphorus accumulation of potato as affected by phosphorus nutrition. Journal of Plant Nutrition 22 (1), 205–217.

Antonious, G.F., Lee, C.M., Snyder, J.C., 2001. Sustainable soil management practices and quality of potato grown on erodible lands. Journal of Environmental Science and Health, Part B 36 (4), 435–444.

Andrews, S.S., Mitchell, J.P., Mancinelli, R., Karlen, D.L., Hartz, T.K., Horwath, W.R., Pettygrove, G.S., Scow, K.M., Munk, D.S., 2002. On-farm assessment of soil quality in California's central valley. Agronomy Journal 94 (1), 12–23.

Antanaitis, Š., Švedas, A., 2000. The relationship between potato yield and concentration of chemical elements in tubers, and soil agrochemical properties. Žemdirbystė, Mokslo Darbai 70, 33–47.

Arif, U., 2013. Effect of wounding and light exposure on sterol, glycoalkaloid, and calystegine levels in potato plants (*Solanum tuberosum* L. group Tuberosum). Acta Universitatis Agriculturae Sueciae 2013 (45), 1652–6880.

Augustin, J., 1975. Variations in the nutritional composition of fresh potatoes. Journal of Food Science 40 (6), 1295–1299.

Baniūnienė, A., Žėkaitė, V., 2007. Organinių trąšų įtaka augalams bulvių sideracinėje grandyje įprastinės ir tausojamosios žemdirbystės sąlygomis. Žemės ūkio mokslai 14 (2), 1–10.

Bartley, G.E., Scolnik, P.A., 1995. Plant carotenoids: pigments for photoprotection, visual attraction, and human health. The Plant Cell 7 (7), 1027.

Bourn, D., Prescott, J., 2002. A comparison of the nutritional value, sensory qualities, and food safety of organically and conventionally produced foods. Critical Reviews in Food Science and Nutrition 42 (1), 1–34.

Brandt, K., Mølgaard, J.P., 2001. Organic agriculture: does it enhance or reduce the nutritional value of plant foods? Journal of the Science of Food and Agriculture 81 (9), 924–931.

Brandt, K., Leifert, C., Sanderson, R., Seal, C.J., 2011. Agroecosystem management and nutritional quality of plant foods: the case of organic fruits and vegetables. Critical Reviews in Plant Sciences 30 (1–2), 177–197.

Brazinskiene, V., Asakaviciute, R., Miezeliene, A., Alencikiene, G., Ivanauskas, L., Jakstas, V., Viskelis, P., Razukas, A., 2014. Effect of farming systems on the yield, quality parameters and sensory properties of conventionally and organically grown potato (*Solanum tuberosum* L.) tubers. Food Chemistry 145, 903–909.

Bryan, N.S., Loscalzo, J., 2011. Nitrite and Nitrate in Human Health and Disease. Humana Press, New York.

Bryant, J.P., Chapin III, F.S., Klein, D.R., 1983. Carbon/nutrient balance of boreal plants in relation to vertebrate herbivory. Oikos 357–368.

Bulluck, L.R., Brosius, M., Evanylo, G.K., Ristaino, J.B., 2002. Organic and synthetic fertility amendments influence soil microbial, physical and chemical properties on organic and conventional farms. Applied Soil Ecology 19 (2), 147–160.

Camin, F., Moschella, A., Miselli, F., Parisi, B., Versini, G., Ranalli, P., Bagnaresi, P., 2007. Evaluation of markers for the traceability of potato tubers grown in an organic versus conventional regime. Journal of the Science of Food and Agriculture 87 (7), 1330–1336.

Cao, W., Tibbitts, T.W., 1991. Potassium concentration effect on growth, gas exchange and mineral accumulation in potatoes. Journal of Plant Nutrition 14 (6), 525–537.

Cazzonelli, C.I., 2011. Goldacre review: carotenoids in nature: insights from plants and beyond. Functional Plant Biology 38 (11), 833–847.

Chen, Z., Gallie, D.R., 2006. Dehydroascorbate reductase affects leaf growth, development, and function. Plant Physiology 142 (2), 775–787.

Chen, Y.F., Wang, Y., Wu, W.H., 2008. Membrane transporters for nitrogen, phosphate and potassium uptake in plants. Journal of Integrative Plant Biology 50 (7), 835–848.

Coley, P.D., Bryant, J.P., Chapin III, F.S., 1985. Resource availability and plant antiherbivore defense. Science (Washington) 230 (4728), 895–899.

Colla, G., Mitchell, J.P., Poudel, D.D., Temple, S.R., 2002. Changes of tomato yield and fruit elemental composition in conventional, low input, and organic systems. Journal of Sustainable Agriculture 20 (2), 53–67.

Cronk, T.C., Kuhn, G.D., McArdle, F.J., 1974. The influence of stage of maturity, level of nitrogen fertilization and storage on the concentration of solanine in tubers of three potato cultivars. Bulletin of Environmental Contamination and Toxicology 11 (2), 163–168.

Dangour, A., Dodhia, M.S., Hayter, A., Aikenhead, M.A., Allen, E., Lock, K., Uauy, R., 2009. Comparison of Composition (Nutrients and Other Substances) of Organically and Conventionally Produced Foodstuffs: A Systematic Review of the Available Literature. Report for Food Standard Agency. London School of Hygiene & Tropical Medicine, London.

Daško, Ľ., Drímal, J., Klimeková, M., Ürgeová, E., 2012. Response of organically and conventionally produced potatoes to a controlled attack of a pathogen. Journal of Food and Nutrition Research (Slovak Republic) 51 (2), 89–95.

Demidchik, V., Maathuis, F.J., 2007. Physiological roles of nonselective cation channels in plants: from salt stress to signalling and development. New Phytologist 175 (3), 387–404.

Demmig-Adams, B., Gilmore, A.M., Adams, W., 1996. Carotenoids 3: in vivo function of carotenoids in higher plants. The FASEB Journal 10 (4), 403–412.

Deng, W., Luo, K., Li, D., Zheng, X., Wei, X., Smith, W., Thammina, C., Lu, L., Li, Y., Pei, Y., 2006. Overexpression of an Arabidopsis magnesium transport gene, AtMGT1, in *Nicotiana benthamiana* confers Al tolerance. Journal of Experimental Botany 57 (15), 4235–4243.

Dixon, R.A., Paiva, N.L., 1995. Stress-induced phenylpropanoid metabolism. The Plant Cell 7 (7), 1085.

Eppendorfer, W.H., Eggum, B.O., 1994. Effects of sulphur, nitrogen, phosphorus, potassium, and water stress on dietary fibre fractions, starch, amino acids and on the biological value of potato protein. Plant Foods for Human Nutrition 45 (4), 299–313.

Ezekiel, R., Singh, N., Sharma, S., Kaur, A., 2013. Beneficial phytochemicals in potato – a review. Food Research International 50 (2), 487–496.

FAO. FAO Statistical Databases. http://faostat3.fao.org/download/Q/QC/E.

Fidalgo, F., Santos, I., Salema, R., 2000. Nutritional value of potato tubers from field grown plants treated with deltamethrin. Potato Research 43 (1), 43–48.

Foyer, C.H., Noctor, G., van Emden, H.F., 2007. An evaluation of the costs of making specific secondary metabolites: does the yield penalty incurred by host plant resistance to insects result from competition for resources? International Journal of Pest Management 53 (3), 175–182.

Friedman, M., McDonald, G.M., Filadelfi-Keszi, M., 1997. Potato glycoalkaloids: chemistry, analysis, safety, and plant physiology. Critical Reviews in Plant Sciences 16 (1), 55–132.

Friedman, M., Roitman, J.N., Kozukue, N., 2003. Glycoalkaloid and calystegine contents of eight potato cultivars. Journal of Agricultural and Food Chemistry 51 (10), 2964–2973.

Friedman, M., 2006. Potato glycoalkaloids and metabolites: roles in the plant and in the diet. Journal of Agricultural and Food Chemistry 54 (23), 8655–8681.

Gallie, D.R., 2013. L-Ascorbic acid: a multifunctional molecule supporting plant growth and development. Scientifica 2013, 1–24.

Gilsenan, C., Burke, R.M., Barry-Ryan, C., 2010. A study of the physicochemical and sensory properties of organic and conventional potatoes (*Solanum tuberosum*) before and after baking. International Journal of Food Science & Technology 45 (3), 475–481.

Ginzberg, I., Tokuhisa, J.G., Veilleux, R.E., 2009. Potato steroidal glycoalkaloids: biosynthesis and genetic manipulation. Potato Research 52 (1), 1–15.

Gugała, M., Zarzecka, K., 2012. Vitamin C content in potato tubers as influenced by insecticide application. Polish Journal of Environmental Studies 21 (4), 1101–1105.

Guziur, J., Schulzová, V., Hajšlová, J., 2000. Vliv lokality a způsobu pěstování na chemické složení hlíz brambor. Bramborářství 8 (1), 6–7.

Haase, N.U., 2008. Healthy aspects of potatoes as part of the human diet. Potato Research 51 (3–4), 239–258.

Haase, N.U., 2010. Glycoalkaloid concentration in potato tubers related to storage and consumer offering. Potato Research 53 (4), 297–307.

Hajšlová, J., Schulzova, V., Slanina, P., Janne, K., Hellenäs, K.E., Andersson, C., 2005. Quality of organically and conventionally grown potatoes: four-year study of micronutrients, metals, secondary metabolites, enzymic browning and organoleptic properties. Food Additives and Contaminants 22 (6), 514–534.

Hamouz, K., Lachman, J., Cepl, J., Vokal, B., 1999a. Influence of locality and way of cultivation on the nitrate and glycoalkaloid content in potato tubers. Rostlinna Vyroba 45, 293–298.

Hamouz, K., Lachman, J., Pivec, V., Vokal, B., 1999b. Influence of environmental conditions and way of cultivation on the polyphenol and ascorbic acid content in potato tubers. Rostlinna Vyroba 45 (7), 293–298.

Hamouz, K., Lachman, J., Dvorak, P., Orsak, M., Hejtmankova, K., Cizek, M., 2009. Effect of selected factors on the content of ascorbic acid in potatoes with different tuber flesh colour. Plant, Soil and Environment 55 (7), 281–287.

Hamouz, K., Lachman, J., Pazderů, K., Hejtmánková, K., Cimr, J., Musilová, J., Pivec, V., Orsák, M., Svobodová, A., 2013. Effect of cultivar, location and method of cultivation on the content of chlorogenic acid in potatoes with different flesh colour. Plant, Soil and Environment 59 (10), 465–471.

Härdter, R., Rex, M., Orlovius, K., 2005. Effects of different Mg fertilizer sources on the magnesium availability in soils. Nutrient Cycling in Agroecosystems 70 (3), 249–259.

Hargreaves, J., Adl, M.S., Warman, P.R., Rupasinghe, H.V., 2008. The effects of organic amendments on mineral element uptake and fruit quality of raspberries. Plant and Soil 308 (1–2), 213–226.

Herencia, J.F., García-Galavís, P.A., Dorado, J.A.R., Maqueda, C., 2011. Comparison of nutritional quality of the crops grown in an organic and conventional fertilized soil. Scientia Horticulturae 129 (4), 882–888.

Herencia, J.F., Ruiz-Porras, J.C., Melero, S., Garcia-Galavis, P.A., Morillo, E., Maqueda, C., 2007. Comparison between organic and mineral fertilization for soil fertility levels, crop macronutrient concentrations, and yield. Agronomy Journal 99 (4), 973–983.

Hoefkens, C., Vandekinderen, I., De Meulenaer, B., Devlieghere, F., Baert, K., Sioen, I., De Henauw, S., Verbeke, W., Van Camp, J., 2009. A literature-based comparison of nutrient and contaminant contents between organic and conventional vegetables and potatoes. British Food Journal 111 (10), 1078–1097.

Howitt, C.A., Pogson, B.J., 2006. Carotenoid accumulation and function in seeds and non-green tissues. Plant, Cell and Environment 29 (3), 435–445.

Islam, M.R., Nahar, B.S., 2013. Effect of organic farming on nutrient uptake and quality of potato. Journal of Environmental Science and Natural Resources 5 (2), 219–224.

Jacobsen, H.B., Madsen, M.H., Christiansen, J., Nielsen, T.H., 1998. The degree of starch phosphorylation as influenced by phosphate deprivation of potato (*Solanum tuberosum* L.) plants. Potato Research 41 (2), 109–116.

Janušauskaitė, D., Žėkaitė, V., 2008. Bulvių cheminės sudėties kitimas vegetacijos metu bei jo ryšys su gumbų derliumi. Žemės ūkio mokslai 15 (4), 53–59.

Koricheva, J., Larsson, S., Haukioja, E., Keinänen, M., 1998. Regulation of woody plant secondary metabolism by resource availability: hypothesis testing by means of meta-analysis. Oikos 83 (2), 212–226.

Kováčik, J., Klejdus, B., Babula, P., Jarošová, M., 2014. Variation of antioxidants and secondary metabolites in nitrogen-deficient barley plants. Journal of Plant Physiology 171 (3), 260–268.

Kozukue, N., Kozukue, E., Mizuno, S., 1987. Glycoalkaloids in potato plants and tubers. HortScience 22 (2), 294–296.

Kwabiah, A.B., Stoskopf, N.C., Palm, C.A., Voroney, R.P., Rao, M.R., Gacheru, E., 2003. Phosphorus availability and maize response to organic and inorganic fertilizer inputs in a short term study in western Kenya. Agriculture, Ecosystems & Environment 95 (1), 49–59.

Lairon, D., 2010. Nutritional quality and safety of organic food. A review. Agronomy for Sustainable Development 30 (1), 33–41.

Lairon, D., Termine, E., Gauthier, S., Lafont, H., 1985. Teneurs en nitrates des productions maraîchères obtenues par des méthodes de l'agriculture biologique. Sciences des Aliments 5, 337–343.

Lee, S.K., Kader, A.A., 2000. Preharvest and postharvest factors influencing vitamin C content of horticultural crops. Postharvest Biology and Technology 20 (3), 207–220.

Lehesranta, S.J., Koistinen, K.M., Massat, N., Davies, H.V., Shepherd, L.V., McNicol, J.W., Cakmak, I., Cooper, J., Lück, L., Kärenlampi, S.O., Leifert, C., 2007. Effects of agricultural production systems and their components on protein profiles of potato tubers. Proteomics 7 (4), 597–604.

Liu, X.X., Zhou, K., Hu, Y., Jin, R., Lu, L.L., Jin, C.W., Lin, X.Y., 2014. Oxalate synthesis in leaves is associated with root uptake of nitrate and its assimilation in spinach (*Spinacia oleracea* L.) plants. Journal of the Science of Food and Agriculture. http://dx.doi.org/10.1002/jsfa.6926.

Lombardo, S., Pandino, G., Mauromicale, G., 2012. Nutritional and sensory characteristics of "early" potato cultivars under organic and conventional cultivation systems. Food Chemistry 133 (4), 1249–1254.

Lombardo, S., Pandino, G., Mauromicale, G., 2014. The mineral profile in organically and conventionally grown "early" crop potato tubers. Scientia Horticulturae 167, 169–173.

Love, S.L., Pavek, J.J., 2008. Positioning the potato as a primary food source of vitamin C. American Journal of Potato Research 85 (4), 277–285.

Maathuis, F.J., 2009. Physiological functions of mineral macronutrients. Current Opinion in Plant Biology 12 (3), 250–258.

Mäder, P., Fliessbach, A., Dubois, D., Gunst, L., Fried, P., Niggli, U., 2002. Soil fertility and biodiversity in organic farming. Science 296 (5573), 1694–1697.

Maga, J.A., 1994. Glycoalkaloids in solanaceae. Food Reviews International 10 (4), 385–418.

Marschner, H., 1995. Mineral Nutrition of Higher Plants. Academic Press, London, pp. 265–270.

Marschner, H., 2012. In: Marschner, P. (Ed.), Marschner's Mineral Nutrition of Higher Plants. Academic Press, pp. 158–165.

Marwaha, R.S., 1988. Nematicides induced changes in the chemical constituents of potato tubers. Plant Foods for Human Nutrition 38 (2), 95–103.

Maynard, D.N., Barker, A.V., Minotti, P.L., Peck, N.H., 1976. Nitrate accumulation in vegetables. Advances in Agronomy 28, 71–118.

Miller, A.J., Cramer, M.D., 2005. Root nitrogen acquisition and assimilation. In: Root Physiology: From Gene to Function. Springer, Netherlands, pp. 1–36.

Misra, V.K., Draper, D.E., 2000. Mg^{2+} binding to tRNA revisited: the nonlinear Poisson-Boltzmann model. Journal of Molecular Biology 299 (3), 813–825.

Mondy, N.I., Gosselin, B., 1988. Effect of peeling on total phenols, total glycoalkaloids, discoloration and flavor of cooked potatoes. Journal of Food Science 53 (3), 756–759.

Morris, S.C., Lee, T.H., 1984. The toxicity and teratogenicity of Solanaceae glycoalkaloids, particularly those of the potato (*Solanum tuberosum*): a review. Food Technology in Australia 36, 118–124.

Mozafar, A., 1993. Nitrogen fertilizers and the amount of vitamins in plants: a review. Journal of Plant Nutrition 16 (12), 2479–2506.

Musilová, J., Lachman, J., Bystrická, J., Poláková, Z., Kováčik, P., Hrabovská, D., 2013. The changes of the polyphenol content and antioxidant activity in potato tubers (*Solanum tuberosum* L.) due to nitrogen fertilization. Potravinarstvo 7 (1), 164–170.

Nowacki, E., Jurzysta, M., Gorski, P., 1975. Effect of availability of nitrogen on alkaloid synthesis in Solanaceae. Bulletin. Serie des sciences biologiques 23, 219.

Nziguheba, G., Palm, C.A., Buresh, R.J., Smithson, P.C., 1998. Soil phosphorus fractions and adsorption as affected by organic and inorganic sources. Plant and Soil 198 (2), 159–168.

Petersson, E.V., Arif, U., Schulzova, V., Krtková, V., Hajslova, J., Meijer, J., Andersson, H.C., Jonsson, L., Sitbon, F., 2013. Glycoalkaloid and calystegine levels in table potato cultivars subjected to wounding, light, and heat treatments. Journal of Agricultural and Food Chemistry 61 (24), 5893–5902.

Phillips, S.B., Mullins, G.L., Donohue, S.J., 2002. Changes in snap bean yield, nutrient composition, and soil chemical characteristics when using broiler litter as fertilizer source. Journal of Plant Nutrition 25 (8), 1607–1620.

Pieper, J.R., Barrett, D.M., 2009. Effects of organic and conventional production systems on quality and nutritional parameters of processing tomatoes. Journal of the Science of Food and Agriculture 89 (2), 177–194.

Ramani, B., Zorn, H., Papenbrock, J., 2004. Quantification and fatty acid profiles of sulfolipids in two halophytes and a glycophyte grown under different salt concentrations. Zeitschrift fur Naturforschung 59c, 835–842.

Raven, J.A., Smith, F.A., 1976. Nitrogen assimilation and transport in vascular land plants in relation to intracellular pH regulation. New Phytologist 76 (3), 415–431.

Rembiałkowska, E., 1999. Comparison of the contents of nitrates, nitrites, lead, cadmium and vitamin C in potatoes from conventional and ecological farms. Polish Journal of Food and Nutrition Sciences 8 (4), 17–26.

Rembiałkowska, E., 2007. Quality of plant products from organic agriculture. Journal of the Science of Food and Agriculture 87 (15), 2757–2762.

Rempelos, L., Cooper, J., Wilcockson, S., Eyre, M., Shotton, P., Volakakis, N., Orr, C.H., Leifert, C., Gatehouse, A.M.R., Tétard-Jones, C., 2013. Quantitative proteomics to study the response of potato to contrasting fertilisation regimes. Molecular Breeding 31 (2), 363–378.

Santamaria, P., 2006. Nitrate in vegetables: toxicity, content, intake and EC regulation. Journal of the Science of Food and Agriculture 86 (1), 10–17.

Savage, G.P., Searle, B.P., Hellenäs, K.E., 2000. Glycoalkaloid content, cooking quality and sensory evaluation of early introductions of potatoes into New Zealand. Potato Research 43 (1), 1–7.

Schachtman, D.P., Reid, R.J., Ayling, S.M., 1998. Phosphorus uptake by plants: from soil to cell. Plant Physiology 116 (2), 447–453.

Schieber, A., Saldana, M.A., 2009. Potato peels: a source of nutritionally and pharmacologically interesting compounds-a review. Food 3 (2), 23–29.

Schuphan, W., 1974. Nutritional value of crops as influenced by organic and inorganic fertilizer treatments. Qualitas Plantarum 23 (4), 333–358.

Shaul, O., 2002. Magnesium transport and function in plants: the tip of the iceberg. Biometals 15 (3), 307–321.

Sinden, S.L., Deahl, K.L., Aulenbach, B.B., 1976. Effect of glycoalkaloids and phenolics on potato flavor. Journal of Food Science 41 (3), 520–523.

Sinden, S.L., Sanford, L.L., Webb, R.E., 1984. Genetic and environmental control of potato glycoalkaloids. American Potato Journal 61 (3), 141–156.

Skrabule, I., Muceniece, R., Kirhnere, I., 2013. Evaluation of vitamins and glycoalkaloids in potato genotypes grown under organic and conventional farming systems. Potato Research 56 (4), 259–276.

Smirnoff, N., Wheeler, G.L., 2000. Ascorbic acid in plants: biosynthesis and function. Critical Reviews in Biochemistry and Molecular Biology 35 (4), 291–314.

Smirnoff, N., 1996. Botanical briefing: the function and metabolism of ascorbic acid in plants. Annals of Botany 78 (6), 661–669.

Smirnoff, N., 2011. Vitamin C: the metabolism and functions of ascorbic acid in plants. Advances in Botanical Research 59, 107–177.

Smith, B.L., 1993. Organic foods vs. supermarket foods: element levels. Journal of Applied Nutrition 45 (1), 35–39.

Søltoft, M., Nielsen, J., Holst Laursen, K., Husted, S., Halekoh, U., Knuthsen, P., 2010. Effects of organic and conventional growth systems on the content of flavonoids in onions and phenolic acids in carrots and potatoes. Journal of Agricultural and Food Chemistry 58 (19), 10323–10329.

Staugaitis, G., Viškelis, P., Rimantas Venskutonis, P., 2008. Optimization of application of nitrogen fertilizers to increase the yield and improve the quality of Chinese cabbage heads. Acta Agriculturae Scandinavica Section B-Soil and Plant Science 58 (2), 176–181.

Sultana, J., Siddique, M.N.E.A., 2014. Organic potato processing status, problem and potentials in the Netherlands: a review. Journal of Science, Technology & Environments Informatics 01 (01), 36–44.

Svec, L.V., Thoroughgood, C.A., Mok, H.C.S., 1976. Chemical evaluation of vegetables grown with conventional or organic soil amendments. Communications in Soil Science & Plant Analysis 7 (2), 213–228.

Tedone, L., Hancock, R.D., Alberino, S., Haupt, S., Viola, R., 2004. Long-distance transport of L-ascorbic acid in potato. BMC Plant Biology 4 (1), 16.

Warman, P.R., 2005. Soil fertility, yield and nutrient contents of vegetable crops after 12 years of compost or fertilizer amendments. Biological Agriculture & Horticulture 23 (1), 85–96.

Warman, P.R., Havard, K.A., 1997. Yield, vitamin and mineral contents of organically and conventionally grown carrots and cabbage. Agriculture, Ecosystems & Environment 61 (2), 155–162.

Warman, P.R., Havard, K.A., 1998. Yield, vitamin and mineral contents of organically and conventionally grown potatoes and sweet corn. Agriculture, Ecosystems & Environment 68 (3), 207–216.

Westermann, D.T., Tindall, T.A., James, D.W., Hurst, R.L., 1994. Nitrogen and potassium fertilization of potatoes: yield and specific gravity. American Potato Journal 71 (7), 417–431.

Wheeler, G.L., Jones, M.A., Smirnoff, N., 1998. The biosynthetic pathway of vitamin C in higher plants. Nature 393 (6683), 365–369.

Worthington, V., 2001. Nutritional quality of organic versus conventional fruits, vegetables, and grains. The Journal of Alternative & Complementary Medicine 7 (2), 161–173.

Wszelaki, A.L., Delwiche, J.F., Walker, S.D., Liggett, R.E., Scheerens, J.C., Kleinhenz, M.D., 2005. Sensory quality and mineral and glycoalkaloid concentrations in organically and conventionally grown redskin potatoes (*Solanum tuberosum*). Journal of the Science of Food and Agriculture 85 (5), 720–726.

Zarzecka, K., Gugala, M., 2003. The effect of herbicide applications on the content of ascorbic acid and glycoalkaloids in potato tubers. Plant, Soil and Environment 49 (5), 237–240.

Zhang, F.C., Kang, S.Z., Li, F.S., Zhang, J.H., 2007. Growth and major nutrient concentrations in *Brassica campestris* supplied with different NH_4^+/NO_3^- ratios. Journal of Integrative Plant Biology 49 (4), 455–462.

Zhao, X., Nechols, J.R., Williams, K.A., Wang, W., Carey, E.E., 2009. Comparison of phenolic acids in organically and conventionally grown pac choi (*Brassica rapa* L. *chinensis*). Journal of the Science of Food and Agriculture 89 (6), 940–946.

CHAPTER 12

Potato Flavor

Marian McKenzie, Virginia Corrigan

The New Zealand Institute for Plant & Food Research Ltd, Palmerston North, New Zealand

1. Introduction

Worldwide, more than a billion people consume potatoes. Global production is estimated to be 370 million metric tonnes a year, which makes them the third most important food crop and the most important nongrain food crop internationally (Cipotato.org, 2015; FAOSTAT, 2014). Furthermore, there is increasing interest in potatoes in countries where other starch sources, such as rice, are traditionally consumed. China and India are now the biggest producers of potatoes, harvesting 86 million and 45 million metric tonnes, respectively, in 2012 (FAOSTAT, 2014).

Flavor makes an important contribution to the success of potato as a food, and the ongoing development of potato varieties with preferred or unusual flavor profiles offers the chance to increase consumption even further. However, flavor discussions among agricultural scientists can often be confusing because of the use of personal and often interchangeable definitions of flavor, odor, and taste. Flavor is generally discussed in the context of human perception and enjoyment of foods and beverages as they are eaten or drunk. However, individual humans vary in their biological ability to perceive specific compounds (Jaeger et al., 2014; McRae et al., 2013), and the psychological factors influencing food selection by consumers are complex (Costell et al., 2010; Lau et al., 1984). Factors such as the setting within which the meal is consumed and the background food culture of the consumer also serve to affect the degree of enjoyment (King et al., 2007; Meiselman et al., 2000; Prescott et al., 2002).

Formally, flavor is defined as the combination of the sense of taste (sweet, sour, salt, bitter, and umami), which is perceived on the tongue, and the sense of smell when volatiles come into contact with receptors located in the nasal passages (reviewed by Auvray and Spence (2008)). These volatiles can be perceived by sniffing the food or beverage, in a process formally described as orthonasal perception of odor or aroma, but also during chewing when the volatiles travel via a retronasal route to receptors in the nose. The simple test is to pinch the nose, which blocks the retronasal route, so that only taste is perceived, and then release the fingers so that the combined experience of taste and odor is perceived. Both taste and odor signals are processed and interpreted in the same region of the brain and it is often difficult for untrained consumers to uncouple the experiences; it is therefore not surprising that taste and odor can have synergistic and antagonistic effects (Prescott, 1999). In other words, it is arguable that the whole (i.e., flavor) is more than the sum of its parts (i.e., taste and odor).

Raw potato tubers have a limited, but distinctive, aroma that is often perceived during preparation of tubers for cooking when the volatiles are released from cut or damaged flesh. In contrast, cooking potato tubers releases a wide range of volatile compounds, and the cooking method—baking, steaming, boiling, roasting, or frying—dictates the type and extent of the flavor-producing reactions that occur and so has a profound effect on perceived flavor. Boiled potatoes have a weak but characteristic aroma, which is produced by chemical transformation of flavor precursor molecules during the cooking process (Ulrich et al., 2000), in particular lipid oxidation and Strecker degradation of amino acids (Petersen et al., 1998). The higher temperatures involved in baking and frying result in a wider range of particular volatiles contributing to baked and fried potato flavors. In baked potatoes, for example, compounds derived from lipids and the Maillard reaction/sugar degradation are dominant on a quantitative basis (Duckham et al., 2002). The flavor of potato products that have undergone high-temperature processing into french fries or crisps is further influenced by the use of a variety of cooking oils and cooking times. Furthermore, production systems and agronomic methods used for tuber production and postharvest storage have been shown to influence the final perception of potato flavor. For example, long-term cold storage of tubers, which is required to maintain a year-round supply, is particularly important, as it is a key contributor to starch breakdown and the accumulation of reducing sugars, both of which can alter tuber flavor.

A comprehensive range of techniques has been used to identify and quantify the key metabolites that contribute to potato flavor. These include gas chromatography–mass spectrometry (GC–MS), GC with specific phosphorus-nitrogen detector (GC–PND), and GC olfactometry (GC–O), and this has resulted in the sensitive detection of flavor profiles from a range of potato cultivars and after various cooking methods and agronomic practices. There has also been some use of taste panels in trying to determine the human responses to flavor compounds in potato, particularly to determine flavor differences between highly variable germplasm such as from cultivars of *Solanum phureja* and *Solanum tuberosum*. However, the use of taste panelists is an inherently slow process, and the rapid screening of large numbers of lines from breeding programs for flavor is not common and to our knowledge has not been attempted for potato thus far. It is clear that potato breeding programs could be assisted by increased understanding of the metabolites contributing to potato flavor and the genetic control of their production, alongside the use of traditional taste and consumer panels, the ultimate aim being to understand the relationship between metabolite production and human perception of flavor fully, leading to improved selections with desirable flavor profiles.

In this chapter we review the major contributors to perceived potato flavor, aroma, taste, and texture; the metabolites that contribute to these; and the methods used to identify them. The influence of various cooking methods, as well as those of storage and production systems, on potato flavor is also discussed. Finally, we consider recent approaches by several groups to determine the key metabolomic or genomic contributors to perceived potato flavor by combining the power of sensory panels and metabolomic and molecular analyses.

There have been several reviews covering these areas, and the reader is directed to these for further information (Comandini et al., 2011; Dresow and Boehm, 2009; Jansky, 2010; Maga, 1994; Taylor et al., 2007).

2. Aroma and Flavor

Although the terms taste and flavor are often used interchangeably, strictly speaking taste perception refers to those sensations that are elicited by the stimulation of the gustatory receptors on the tongue—sweet, sour, salt, bitter, and umami. What we think of as flavor is experienced as a result of the combination of taste, aroma volatiles perceived via the mouth through retronasal ("in the mouth") olfaction, and trigeminal inputs and can be further enhanced by orthonasal olfaction or "sniffing" (Spence, 2013). The perception of flavor occurs as a result of a series of complex interactions between the components of a foodstuff and the person consuming it. For example, the starch and lipid components of french fries may influence flavor perception in the mouth. Amylose/amylopectin ratio influences the retention and perception of aroma compounds during starch gelatinization in starch matrices (Arvisenet et al., 2002), and the presence of lipids has been shown to reduce volatile release from emulsions in in-mouth studies (Doyen et al., 2001).

In this section we look at the influence of volatile compounds on potato flavor. Although over 7000 volatile compounds have been identified in foods, only a relatively small number of these (300–400) contribute to their characteristic odors (Dresow and Boehm, 2009). In addition, aromas are often not due to the presence of a unique characterizing compound, but rather are the result of a specific and balanced blend of a number of aroma compounds (Dresow and Boehm, 2009). In potatoes these compounds come from classes including lipids, aldehydes, alcohols, ketones, acids, esters, hydrocarbons, amines, furans, and sulfur compounds (Table 1). Some flavor compounds are present in the raw tubers, but many are formed from flavor precursor metabolites such as sugars, lipids, and amino acids via a range of enzymatic and nonenzymatic reactions. Unlike fruits, which emit high concentrations of volatile substances, particularly at maturity, potatoes do not emit perceptible volatile compounds during tuber development or at maturity, unless they are undergoing tissue degradation as a result of rotting (Dresow and Boehm, 2009) or cutting (Galliard and Phillips, 1971), for example.

Table 1: Key Compounds Involved in Potato Flavor Production and Their Perceived Sensory Qualities.

Substance	Sensory Quality (Bold = Significantly Correlated with)	References (Bold = Defined as "Key Compounds")
(E)-2-Decenal	Fatty	Mutti and Grosch (1999)
(E)-2-Heptenal	Fatty, marzipan	Petersen et al. (1998) and Thybo et al. (2006)
(E)-2-Hexenal	**Rancidness**	Thybo et al. (2006)
(E)-2-Nonenal	Cucumber, cardboard, fatty, green, library, velvet, sweet, **potato flavor**	Oruna-Concha et al. (2001), Mutti and Grosch (1999), Petersen et al. 1999, and Thybo et al. (2006)
(E)-2-Octenal	Baked potato, bakery, boiled, chips, fatty, **rancidness**	Petersen et al. (1998, 1999), Thybo et al. (2006), and Mutti and Grosch (1999)
(E)-2-Pentenal	Roasty, rubber, unpleasant	Ulrich et al. (2000)
(E)-β-Damascenone	Fruity, sweet	Mutti and Grosch (1999)
(E,E)-2,4-Decadienal	Onions, chips, hot potato, licorice, oily, deep fried-like, fatty, waxy	Petersen et al. (1999), Oruna-Concha et al. (2001), and Mutti and Grosch (1999)
(E,E)-2,4-Heptadienal	**Rancidness, off-flavor**	Thybo et al. (2006)
(E,E)-2,6-Nonadienal	Fatty, cucumber	Ulrich et al. (2000) and Mutti and Grosch (1999)
(E,E)-3,5-Octadienone	Nutty	Ulrich et al. (2000)
(E,Z)-2,4-Decadienal	Fatty, waxy, rancidness	Thybo et al. (2006) and Mutti and Grosch (1999)
(E,Z)-2,6-Nonadienal	Grass, green, vinegar-dressed, cucumber, peas, paint	Petersen et al. (1998, 1999) and Mutti and Grosch (1999)
(Z)-1,5-Octadien-3-one	Geranium-like	Mutti and Grosch (1999)
(Z)-2-Decenal	Fatty	Mutti and Grosch (1999)
(Z)-2-Nonenal	Green, tallowy	Mutti and Grosch (1999)
(Z)-3-Nonenal	Fatty	Mutti and Grosch (1999)
1,2-Ethanediol	Coffee, burnt	Petersen et al. (1999)
1-Hexanol	Hot paper, foot malodor	Petersen et al. (1999)
1-Octen-3-ol	Earthy, mushroom, rancidness	Oruna-Concha et al. (2001), Thybo et al. (2006), and **Duckham et al. (2001, 2002)**
1-Octen-3-one	Mushroom-like	Mutti and Grosch (1999)
1-Pentanol	**Rancidness, intensity, savory**	Thybo et al. (2006) and Morris et al. (2010)
1-Penten-3-ol	**Intensity, savory**	Morris et al. (2010)
1-Penten-3-one	**Sweetness**	Morris et al. (2010)
2,3-Butandione	Buttery	Mutti and Grosch (1999)
2,3-Diethyl-5-methylpyr-azine	Earthy, roasty	Mutti and Grosch (1999)
2,3-Pentanedione		**Duckham et al. (2001, 2002)**
2,4-Decadienal	Earthy, fatty	Oruna-Concha et al. (2001)
2,4-Heptadienal	**Rancidness, off-flavor, sweetness**	Thybo et al. (2006) and Morris et al. (2010)

Table 1: Key Compounds Involved in Potato Flavor Production and Their Perceived Sensory Qualities.—cont'd

Substance	Sensory Quality (Bold = Significantly Correlated with)	References (Bold = Defined as "Key Compounds")
2,4-Mecadienal	Fatty, unpleasant	Ulrich et al. (2000)
2,4-Nonadienal	**Rancidness**	Thybo et al. (2006)
2,6-Dimethylpyrazine	Nutty, warm	Ulrich et al. (2000)
2-Acetyl-1-pyrroline	Roasty, popcorn	Mutti and Grosch (1999)
2-Butylfuran	**Sweetness**	Morris et al. (2010)
2-Decenal	Cardboard-like off-odors and off-flavors	Blanda et al. (2010)
2-Ethyl-3,5-dimethylpyrazine	Nutty, roasted, roasty, coffee-like	Ulrich et al. (2000), Oruna-Concha et al. (2001), and **Duckham et al. (2001)**
2-Ethyl-3,6-dimethylpyrazine	Nutty, roasted, earthy, baked potato-like	Oruna-Concha et al. (2001)
2-Ethylfuran	**Savory, rancidness, off-flavor**	Thybo et al. (2006) and Morris et al. (2010)
2-Heptenal	**Intensity**	Morris et al. (2010)
2-Hexenal	Cardboard-like off-odors and off-flavors, **off-flavor**	Morris et al. (2010) and Blanda et al. (2010)
2-Isobutyl-3-methoxypyrazine		**Duckham et al. (2001, 2002)**
2-Isopropyl-3-methoxypyrazine	Earthy, raw potato, potato-like	Oruna-Concha et al. (2001) and **Duckham et al. (2001, 2002)**
2-Methoxy-3-isobutylpyrazine	Sprouts, peas, green, bell pepper, boiled potato	Petersen et al. (1999)
2-Methoxy-3-isopropylpyrazine	Earthy, green, peas, raw potato, dry	Petersen et al. (1999)
2-Methyl-1-butanol		**Duckham et al. (2002)**
2-Methyl-5-isopropylpyrazine	Nutty, warm, chemical	Ulrich et al. (2000)
2-Methylbutanal	Fruity, **aroma**	Morris et al. (2010) and **Duckham et al. (2001, 2002)**
2-Methylbutanol	Unpleasant, sweat	Ulrich et al. (2000)
2-Methylfuran	**Intensity, savory**	Morris et al. (2010)
2-Methylpropanal	**Aroma**	Morris et al. (2010)
2-Pentenal	Cardboard-like off-odors and off-flavors, **intensity, creaminess**	Morris et al. 2010 and Blanda et al. (2010)
2-Pentylfuran	Unpleasant, green beans, cooked, rubber/cardboard-like off-odors and off-flavors, **sweetness, off-flavor**	Ulrich et al. (2000), Petersen et al. (1998), Thybo et al. (2006), Morris et al. (2010), Blanda et al. (2010), and **Duckham et al. (2001, 2002)**
2-Propanone	**Aroma**	Morris et al. (2010)
2-Propylfuran	**Intensity, savory**	Morris et al. (2010)
3-(Methylthio)propanal		**Duckham et al. (2001, 2002)**
3-Ethyl-2,5-dimethylpyrazine	Nutty, earthy, herbaceous	Ulrich et al. (2000)

Continued

Table 1: Key Compounds Involved in Potato Flavor Production and Their Perceived Sensory Qualities.—cont'd

Substance	Sensory Quality (Bold = Significantly Correlated with)	References (Bold = Defined as "Key Compounds")
3-Ethyl-2-methyl-1,3-hexadiene	**Intensity, savory, creaminess**	Morris et al. (2010)
3-Isobutyl-2-methoxypyrazine	Pepper-like, green	Mutti and Grosch (1999)
3-Methyl-1-butanol		**Duckham et al. (2002)**
3-Methyl-1-pentanol	Burnt, unpleasant, fatty, chips, pungent	Petersen et al. (1999)
3-Methylbutanal	Malty, fruity, **aroma**	Morris et al. (2010), Mutti and Grosch (1999), and **Duckham et al. (2001, 2002)**
3-or-2-Methylbutanoic acid	**Intensity, savory, creaminess**	Morris et al. (2010)
4-Heptenal	**Sweetness**	Morris et al. (2010)
6-Methyl-5-hepten-2-one		**Duckham et al. (2001)**
Acetic acid	Vinegar	Petersen et al. (1998)
Alkyl pyrazines	Earthy, roasted, nutty, buttery	Maga and Holm (1992), Coleman and Ho (1980), Buttery et al. (1971), and Oruna-Concha et al. (2001)
Benzaldehyde	**Aroma**	Morris et al. (2010) and **Duckham et al. (2001, 2002)**
Benzyl alcohol	Flower, citrus, fruit	Petersen et al. (1999)
β-Cyclocitral		**Duckham et al. (2001)**
β-Damascenone		**Duckham et al. (2001)**
Butanedione		**Duckham et al. (2002)**
C4-Heptenal	Boiled potato, stale, earthy	Josephson and Lindsay (1987)
Decanal	Fruity, fatty, floral, burnt plastic, **rancidness**	Petersen et al. (1999), Thybo et al. (2006), and **Duckham et al. (2001, 2002)**
Diacetyl	Buttery, sweet, caramel	Ulrich et al. (2000)
Dimethyl disulfide	Onion-like, cooked cabbage, **aroma**	Oruna-Concha et al. (2001), Morris et al. (2010), and **Duckham et al. (2001, 2002)**
Dimethyl sulfide	**Off-flavor**	Morris et al. (2010)
Dimethyl sulfoxide	**Rancidness**	Thybo et al. (2006)
Dimethyl trisulfide	Cabbage, **aroma**	Morris et al. (2010), Mutti and Grosch (1999), and **Duckham et al. (2001, 2002)**
Dimethyl trisulfide, dimethyl disulfide	Cooked onion	Duckham et al. (2002), Oruna-Concha et al. (2001), and **Duckham et al. (2001)**
E-(or Z)-2-(1-Pentenyl) furan	**Sweetness**	Morris et al. (2010)
E-(or Z)-2-(2-Pentenyl) furan	**Sweetness**	Morris et al. (2010)
Furan	**Aroma, sweetness**	Morris et al. (2010)

Table 1: Key Compounds Involved in Potato Flavor Production and Their Perceived Sensory Qualities.—cont'd

Substance	Sensory Quality (Bold = Significantly Correlated with)	References (Bold = Defined as "Key Compounds")
Furfural	**Aroma**	Morris et al. (2010)
γ-Decalactone	Sweet, peach-like	Mutti and Grosch (1999)
γ-Nonalcatone	Sweet, coconut-like	Mutti and Grosch (1999)
γ-Octalactone	Sweet, coconut-like	Mutti and Grosch (1999)
Geranyl acetone		**Duckham et al. (2001)**
Heptanal	Green	Petersen et al. (1998) and **Duckham et al. (2001, 2002)**
Hexanal	Green, grass, **intensity, savory, creaminess, rancidness**	Ulrich et al. (2000), Petersen et al. (1999), Thybo et al. (2006), Morris et al. (2010), Mutti and Grosch (1999), and **Duckham et al. (2001, 2002)**
Hexanoic acid	Prepared food, soup, cake, raw potato	Petersen et al. (1999)
Linalool	**Potato flavor**	Thybo et al. (2006) and **Duckham et al. (2001)**
Methional	Cooked potato, boiled potato, **potato flavor**	Duckham et al. (2002), Ulrich et al. (2000), Petersen et al. (1998, 1999), Thybo et al. (2006), Oruna-Concha et al. (2001), and Mutti and Grosch (1999)
Methoxypyrazines	Musty, earthy	Duckham et al. (2002)
Methylpropanal	Malty	Mutti and Grosch (1999) and **Duckham et al. (2002)**
Methylpyrazine	Nutty, strong	Ulrich et al. (2000)
Methylsalicylate	**Intensity, savory, creaminess**	Morris et al. (2010)
Nonanal	Rancid, ozone, boiled potato, **rancidness**	Petersen et al. (1999), Thybo et al. (2006), and **Duckham et al. (2001, 2002)**
Octanal	Fatty, citrus-like, **intensity**	Morris et al. (2010), Mutti and Grosch (1999), and **Duckham et al. (2001)**
Oetan-2-one	Mushroom, earthy	Ulrich et al. (2000)
p-Cymene	**Off-flavor**	Thybo et al. (2006)
Pentanal	Green, almond, marzipan, **intensity, savory, creaminess, rancidness**	Petersen et al. (1999), Thybo et al. (2006), Morris et al. (2010), and **Duckham et al. (2001, 2002)**
Phenylacetaldehyde	Green, flower(y), earthy, honey, floral, roses, sweet	Ulrich et al. (2000), Petersen et al. (1998), Oruna-Concha et al. (2001), Mutti and Grosch (1999), and **Duckham et al. (2001, 2002)**
Propanal	**Off-flavor**	Morris et al. (2010)
Pyrrole	Nutty, roasty	Ulrich et al. (2000)
Terpenes (α-copaene, β-damascenone, others)	Fruity, floral	**Duckham et al. (2001)**

Continued

Table 1: Key Compounds Involved in Potato Flavor Production and Their Perceived Sensory Qualities.—cont'd

Substance	Sensory Quality (Bold = Significantly Correlated with)	References (Bold = Defined as "Key Compounds")
tr-4,5-Epoxy-(*E*)-2-decenal	Metallic	Mutti and Grosch (1999)
tr-4,5-Epoxy-(*E*)-2-nonenal	Metallic	Mutti and Grosch (1999)
Undecanal		**Duckham et al. (2001)**
Z-(or *E*)-2-(2-Pentenyl) furan	**Intensity, savory, creaminess**	Morris et al. (2010)

2.1 Raw Potatoes

Approximately 159 volatiles have been identified in raw potatoes, as summarized by Dresow and Boehm (2009). These include those present in the raw potatoes, but more are produced as the tubers are prepared for cooking. Activities such as peeling and slicing disrupt the cell membranes, facilitating the oxidation of fatty acids by the lipoxygenases present in the raw tissue (Galliard and Phillips, 1971). This results in a large increase in the concentrations of hexanal and the isomeric forms of octenal and 2,4-decadienal (Josephson and Lindsay, 1987; Petersen et al., 1998). Significant compounds produced via lipid oxidation of unsaturated fatty acids include 2-octenal, heptanal, pentanol, 2-pentylfuran, 2-methylbutanol, and 3-methylbutanol (Josephson and Lindsay, 1987; Petersen et al., 1998). Other lipoxygenase-initiated reaction products found include 1-penten-3-one, 1-pentanol, 2,4-heptadienal, and 2,6-nonadienal (Gardner, 1996; Ho and Chen, 1994; Josephson and Lindsay, 1987; Petersen et al., 1998; Schieberle and Grosch, 1981; Sok and Kim, 1994). Some studies report finding methoxypyrazines in raw potatoes (Buttery and Ling, 1973; Murray and Whitfield, 1975), in particular 2-methoxy-3-isopropylpyrazine (Murray and Whitfield, 1975). There is some discussion as to whether these are produced by the tuber itself or by soil bacteria (*Pseudomonas taetrolens*) and subsequently absorbed by the tuber (Buttery et al., 1973). Dresow and Boehm (2009) conclude that although raw potatoes include many potent volatiles, no single component could describe the typical aroma of raw potato.

2.2 Effects of Cooking

Volatile compounds identified in cooked potatoes are the result of cooking conditions, including the temperature, time, and moisture content, and the effect of these on enzymatic and nonenzymatic reactions.

The type and concentration of the volatile compounds produced are also highly dependent on the metabolites (i.e., flavor precursors) available in the raw material. These metabolites originate from the basic nutrients in the potatoes and include carbohydrates, particularly the mono- and disaccharides; proteins and free amino acids; the fats; triglycerides or their derivates; and a range of vitamins and minerals (Dresow and Boehm, 2009). The number and concentration of the compounds produced increase with cooking time and temperature, ranging from 182 compounds identified in boiled potato, to almost 400 identified in baked potato (Dresow and Boehm, 2009), and over 500 identified in the volatile fractions of french fries (Maga, 1994). Which of these compounds particularly contribute to potato aroma is yet to be fully resolved, although it is generally agreed that the typical aroma of boiled potato is mainly due to methional and a range of pyrazines (Ulrich et al., 2000).

There are several chemical reactions that occur in potato during cooking and that are responsible for the production and/or degradation of key flavor compounds. Thermal reactions involving protein, sugar, and lipid degradation are considered to be the most important reactions in the formation of volatile compounds in cooked and processed foods (Dresow and Boehm, 2009; Whitfield and Mottram, 1992). In potatoes these reactions include lipid oxidation and thermal degradation reactions such as the Maillard reaction (nonenzymatic browning) and Strecker degradation reactions. The products of these reactions account for from 22% to 69% (lipid degradation products) and from 28% to 77% (sugar degradation and/or the Maillard reaction products) of the total yield of compounds monitored, depending on the cultivar (Duckham et al., 2001, 2002), while concentrations of sulfur-containing compounds, terpenes, and methoxypyrazines are relatively low (Duckham et al., 2001).

The Maillard reaction occurs between carbonyl groups, typically those of reducing sugars and free amino acids, and can occur at quite low temperatures, although the reaction rate increases significantly at normal cooking temperatures (Duckham et al., 2002; Taylor et al., 2007; Whitfield and Mottram, 1992). Methional in particular is formed via this reaction (Duckham et al., 2002; O'Connor et al., 2001; Lindsay, 1996).

The Strecker degradation reactions are generally associated with the Maillard reaction, as they involve the reaction of α-amino acids with α-dicarbonyl compounds derived from concurrent Maillard reaction pathways. An α-amino acid undergoes decarboxylation and is transformed into a structurally related aldehyde (Rizzi, 2008), producing branched-chain carbonyl and alcohol volatiles, which contribute significantly to potato flavor (Duckham et al., 2002). Although the primary reactants are proteins and carbohydrates, compounds such as lipids and polyphenolic compounds can also be involved, producing novel flavor compounds (Rizzi, 2008). The products of the Maillard reaction and Strecker degradations then undergo further reactions, forming a range of compounds that include pyrazines, oxazoles, and thiophenes (Whitfield and Mottram, 1992). The pyrazines formed from amino acids and sugars in the Maillard and Strecker reactions have a typical earthy "potato-like" flavor and the amino acids asparagine and glutamine, which are responsible for pyrazine production, are found in large quantities in potatoes (Martin and Ames, 2001b).

Enzymatic reactions contribute to the formation of precursor compounds, such as amino acids from proteins, which then react nonenzymatically to form flavor compounds. The extent of these reactions is limited by denaturation of the enzymes during the heating process, but can be significant depending on cooking conditions, such as heating profiles and sample sizes.

Degradation of the lipids and fatty acids occurs through thermal degradation reactions and via enzymatic and autoxidation of the unsaturated fatty acids linoleic acid and α-linolenic acid (Dobson et al., 2004). Lipoxygenases contribute to the formation of flavor compounds by catalyzing the oxygenation of polyunsaturated fatty acids to form fatty acid hydroperoxides and may also produce off-flavors (Baysal and Demirdöven, 2007). The end products of the oxidative and enzymatic degradation of lipids include many unsaturated and saturated aldehydes, ketones, and alcohols.

The concentrations of some of these compounds are very low. Although their flavor impact may still be unknown, some may be present at concentrations above the threshold for human perception and are therefore likely to have a significant effect on flavor (Duckham et al., 2002; Petersen et al., 1998).

2.3 Boiled Potatoes

Between 140 and 182 volatile compounds have been identified in boiled potatoes (Ulrich et al., 2000; Maga, 1994; Salinas et al., 1994; Dresow and Boehm, 2009). The amounts and types of volatiles identified vary according to factors such as preparation method (e.g., peeled versus unpeeled), cooking conditions, and analysis methods used, but have been identified as largely due to the formation of lipid oxidation compounds and compounds from Strecker degradation of amino acids into their corresponding aldehydes (Petersen et al., 1998). In particular, boiling favors the production of hexanal and 2-heptenal, and the lipid-derived compounds 2-methylfuran, 2-pentylfuran, 3-hexanone, and 1-octen-3-ol (Oruna-Concha et al., 2002). Unpeeled potatoes generally contain more aromatic compounds, and the proportion of terpenes in particular is higher in unpeeled potatoes than in peeled ones (Nursten and Sheen, 1974).

Ulrich et al. (2000) identified compounds including methional, diacetyl, and at least five different substituted pyrazines as character-impact compounds for boiled potatoes, which was in agreement with results from earlier studies (Salinas et al., 1994; Whitfield and Last, 1991). Other compounds, including hexanal, methyl salicylate, and 2-methylbutanoic acid have also been found to be strongly correlated with potato flavor intensity, creaminess, and savoriness in boiled potatoes (Morris et al., 2010). The sulfur-containing compound methional is often reported as a key component of boiled potato flavor (Lindsay, 1996; Petersen et al., 1998; Ulrich et al., 2000, 1998). Mutti and Grosch (1999) reported methional as being among the most odor active of the compounds they tested and therefore one of the most potent of the boiled potato odorants, although interestingly Duckham et al. (2001) found it in less than half the 11 cultivars they analyzed. Another compound with a high odor activity that was said to contribute to the typical boiled potato flavor was 2,3-diethyl-5-methylpyrazine Mutti and Grosch, 1999). *cis*-4-heptenal has been described as providing a boiled-potato-like characterizing aroma and flavor note in potatoes. When added to freshly boiled mashed potato at low concentrations (0.1–0.4 ppb), the earthy potato-like flavor was enhanced, but when added at more than 7 ppb, an undesirable stale flavor was produced (Josephson and Lindsay, 1987). Other off-flavor compounds identified in boiled potatoes include 2,6-nonadienal (fatty, cucumber), 2,4-decadienal (fatty, unpleasant), and 2-pentyl furan (unpleasant, green beans, cooked) (Ulrich et al., 2000).

2.4 Baked Potatoes

The flavor of the baked potato has been described as very mild but extremely complex (Coleman and Ho, 1980), with reports of more than 150 compounds detected (Duckham et al., 2002), while the literature summarized by Dresow and Boehm (2009) identifies a total of 392 compounds. The conditions for the formation of compounds in unpeeled baked potatoes differ from those in other cooking methods because the flavor volatiles are formed inside an essentially closed system and the metabolites are those from inside the tuber only. There is no interaction with other compounds such as water or frying fats, and compounds normally formed via oxidative reactions as a result of tissue disruption during peeling and chopping do not develop. The major flavor compounds in baked potatoes are formed as a result of the Maillard reaction and/or Strecker degradation, lipid degradation, and the degradation of sulfur amino acids (Duckham et al., 2002).

As for other cooking methods, baked potato flavor is not the result of a single compound, but the number of key compounds characterizing the flavor is relatively small. It is generally agreed in the literature that the pyrazines are likely to be the most important flavor compounds in baked potatoes. These include 2-ethyl-3,6-dimethylpyrazine, 2-isobutyl-3-methoxypyrazine, 2-isopropyl-3-methoxypyrazine, 2-ethyl-3,5-dimethylpyrazine, 2-isobutyl-3-methylpyrazine, 2,3-diethyl-5-methylpyrazine, 3,5-diethyl-2-methylpyrazine, 2-ethyl-6-vinyl-pyrazine, and 2-ethyl-3-methylpyrazine (Buttery et al., 1973; Coleman and Ho, 1980; Coleman et al., 1981; Dresow and Boehm, 2009; Duckham et al., 2001; Pareles and Chang, 1974). In particular, Pareles and Chang (1974) found that a combination of 2-isobutyl-3-methylpyrazine, 2,3-diethyl-5-methylpyrazine, and 3,5-diethyl-2-methylpyrazine resulted in a much more characteristic baked potato aroma than did the presence of any single compound.

The range of descriptors used to describe the aroma of various pyrazines in baked potatoes include characteristic, sweet, earthy, nutty, baked, buttery, pleasant, potato-like (Coleman and Ho, 1980; Coleman et al., 1981), musty, and earthy (Duckham et al., 2002). One reason for the importance of the pyrazines as characteristic baked potato flavor compounds is the relatively high concentration in which they are found in unpeeled baked potatoes compared with concentrations in potatoes cooked using other methods. This is a result of the higher pyrazine-to-aldehyde ratio found in the potato skin compared with that found in the whole baked potato, due to the different temperature and moisture conditions during cooking (Coleman and Ho, 1980).

Other potentially important compounds include methional (Buttery et al., 1973; Dresow and Boehm, 2009; Pareles and Chang, 1974); β-damascenone, dimethyl trisulfide, decanal, and 3-methylbutanal (Duckham et al., 2001); 5-methyl-2-furaldehyde (Pareles and Chang, 1974); and hexanal and 1-octen-3-ol (Oruna-Concha et al., 2002).

2.5 Microwave-Baked Potatoes

Potatoes bake more rapidly in a microwave oven than in a conventional oven (Wilson et al., 2002), but this produces the least flavor of the various cooking methods. This is reflected in the relatively weak volatile isolates extracted from microwave-baked potatoes compared with volatile isolates extracted from potatoes cooked by both conventional baking and boiling (Oruna-Concha et al., 2002). Although trained panelists' ratings and flavor scores were lower than for potatoes cooked by boiling or baking (Brittin and Trevino, 1980; Maga and Twomey, 1977), consumers were less discriminating and showed no differences in preference or acceptability between oven-baked and microwave-baked potatoes (Brittin and Trevino, 1980).

Total moisture loss is similar in conventional- and microwave-baked potatoes but microwaved tubers do not form a crust and therefore lose moisture uniformly during baking compared with conventionally baked potatoes. This may allow for a greater loss of flavor volatiles (Oruna-Concha et al., 2002) through codistillation as water evaporates from the tuber surface during cooking (Jansky, 2010), resulting in a less flavorsome potato. Although evaporative cooling means that the surface of the microwave-baked potatoes remains cooler than in other baking methods, the amounts of Maillard reaction products are still higher than in boiled potatoes, possibly because the moisture loss during cooking creates favorable conditions for this (Van Eijk, 1994).

2.6 French Fries

Most (85%) of the 122 compounds identified in french fries by van Loon et al. (2005) originated from sugar degradation and/or Maillard reactions, with the remainder coming from lipid degradation. Of these, approximately 50 odor-active compounds were identified as being responsible for the 41 odors perceived by sensory profiling. Pyrazines have been identified as the main contributors to the dominant fried potato notes in french fries (Duckham et al., 2002), the key ones being 2-ethyl-3,5-dimethylpyrazine, 3-ethyl-2,5-dimethylpyrazine, 2,3-diethyl-5-methylpyrazine, and 3-isobutyl-2-methyloxypyrazine. Other important french fry flavor compounds include 2,4-decadienal (*E,E*- and *E,Z*-), *trans*-4,5-epoxy-(*E*)-2-decanal, 4-hydroxy-2,5-dimethyl-3(*2H*)-furanone, methylpropanol, 2- and 3-methylbutanal, and methanethiol (Wagner and Grosch, 1998). Frying temperature and time have a significant effect on pyrazine formation (Martin and Ames, 2001b); therefore these factors have a strong influence on french fry flavor. Flavor is further complicated by the type of oil used (Martin and Ames, 2001a), and the effects of repeated oil use, which favor the production of lipid oxidation products such as hexanal (Brewer et al., 1999), and may impart a definite off-flavor.

3. Identifying Flavor Compounds

Ideally, methods used to identify and quantify volatile compounds should equally extract all available compounds, retain them in their original state, and produce no artifact compounds. Methods that have been used to extract volatile compounds from cooked potato include simultaneous distillation and extraction (Nickerson and Likens, 1966; Oruna-Concha et al., 2001; Ulrich et al., 2000), solvent-assisted flavor evaporation (Engel et al., 1999), static headspace extraction (Sides et al., 2000; Wagner and Grosch, 1998), dynamic headspace extraction (DH) (Duckham et al., 2001; Oruna-Concha et al., 2002; Thybo et al., 2006), and solid-phase microextraction (SPME) (Arthur and Pawliszyn, 1990; Morris et al., 2010).

The separation and identification techniques used for potato include standard GC–MS (Duckham et al., 2001; Oruna-Concha et al., 2001; Thybo et al., 2006; Ulrich et al., 2000), GC–PND (Ulrich et al., 2000), and GC–O, whereby volatile compounds are separated by GC and the human nose is used at a "sniffing port" to describe the various aromas as they exit the GC column. A portion can also be diverted for simultaneous identification by MS (GC–MS–O) (Majcher and Jelen, 2009; Ulrich et al., 2000).

Petersen et al. (1998) used GC–O and GC–MS to identify key volatile compound differences between raw and boiled potatoes. Twenty compounds were present in both raw and boiled tubers, but their concentrations varied significantly and generally decreased with cooking.

van Loon et al. (2005) used purge-and-trap GC–MS to identify volatile compounds from french fries released under "mouth conditions" to mimic the release of volatile compounds from the food to the nose epithelia. A trained sensory panel was used to help to determine those compounds responsible for french fry flavor. Compounds identified as potent french fry odorants were quantified using stable isotope dilution assays and dissolved in sunflower oil to check their flavor profiles. Using this method, van Loon et al. (2005) identified approximately 50 odor-active compounds that were responsible for the 41 odors perceived by the sensory panel.

Although many groups have made use of such sensitive compound detection techniques, it is important to note that the detection, nondetection, or even absolute concentration of a compound does not necessarily accurately reflect its contribution to the typical potato aroma and flavor profile, which is the result of a complex mix of aroma volatiles and basic tastants.

Which compounds are the key character-impact compounds for potatoes is still under investigation, although a compound may be considered as a key contributor if it has a high relative perceived intensity when assessed in isolation and a character closely resembling the food's aroma quality (Bult et al., 2002). The higher the perceived intensity and typicality of an odor, the higher the character impact of that component is likely to be in a mixed aroma (Buttery and Ling, 1998). High- and low-character-impact compounds are those that respectively make a positive contribution or are irrelevant or make a negative contribution.

One early methodology used a calculated "aroma value" (Patton and Josephson, 1957; Rothe and Thomas, 1963), the relative aroma impact value (RAV), which was defined as the concentration of a compound divided by its odor threshold value. A compound with a RAV >1.0 should be detectable and one with a RAV <1.0 should not be detectable, although there will be exceptions. Although a useful starting point, the range of threshold values quoted in the later literature has revealed a weakness in this method. As outlined in the introduction, people vary biologically in their ability to detect odors, and this is confounded by the range of approaches used to detect and calculate odor thresholds. Duckham et al. (2001) calculated the RAV by dividing the relative GC peak of several compounds detected in baked potato by their odor threshold value in water. Using this approach, they identified and semiquantified 81 volatile compounds in the baked flesh of 11 potato cultivars using DH collection and GC–MS. Compounds including 2-isobutyl-3-methoxypyrazine, 2-isopropyl-3-methoxypyrazine, β-damascenone, dimethyl trisulfide, decanal, and 3-methylbutanal were confirmed as those contributing most to the aroma of the baked potatoes.

Approaches such as this take into account that key odor compounds such as the pyrazines may have a very low threshold. For example, the odor threshold in water of the methoxypyrazine 2-methoxy-3-isopropylpyrazine is 2 parts in 10^{12} (Duckham et al., 2002). Therefore, although present at concentrations that may even be undetectable in some analyses, changes in methoxypyrazine concentrations could potentially have large effects on the overall flavor profile of potatoes.

A variation on this is the odor activity value, which is determined as a ratio of the measured concentration of a compound in a sample to the sensory threshold of the compound. Compounds that can still be detected after a series of sequential dilutions are deemed to have a high odor activity value and their contribution to the sample aroma is likely to be significant. Application of this method can typically reduce a complex mixture to the ~10 to 20 compounds most important to the overall sample aroma (Grosch, 2001).

Flavor synergies between compounds should also be considered, particularly when identifying compounds using GC–O, which chromatographically decomposes the constituent mixture into a series of compounds that are described by participants in the effluate of the GC column. Using this method, effects of synergistic and antagonistic interactions between components cannot be estimated, but it is well known that the contribution of an odorant to a mixture is not linear because sensory interactions occur, and the intensity of attributes associated with low-character-impact components can be suppressed by the presence of high-impact components (Bult et al., 2002). Alternatively, compounds that may appear to be unimportant to the aroma and flavor profiles because they seem to be odorless, or have "uninteresting" or "irrelevant" aromas, may actually have effects on the perception of other compounds. Indeed, compounds such as 3-methylbutanal, hexanal, phenylacet aldehyde, (*E/E*)-2,4-heptadienal, (*E/E*)-2/4-octadienal, (*E/Z*)-2,6-nonadienal, and β-damascenone, which have relatively low flavor dilution factors, may still contribute to the overall aroma of potatoes through interactions with high-character-impact compounds (Ulrich et al., 2000).

4. Taste

Taste is the sensory impression produced when a chemical reaction occurs between a soluble substance in the mouth and the taste receptor cells on the tongue. The five basic human tastes: sweet, sour, salt, bitter, and umami are all naturally found in potato, though saltiness appears very limited.

4.1 Bitterness

Potato compounds contributing to bitter taste include the glycoalkaloids, phenolic compounds, and organic acids.

The major glycoalkaloids in potato are α-solanine and α-chaconine, which make up about 95% of the total glycoalkaloid content (Slanina, 1990). These are presumably used by the plant as deterrents to herbivory, and although they have apparently been selected against during domestication, all cultivars still retain the capability to produce these compounds in and immediately below the skin, with several wild species eaten historically having a potentially toxic total glycoalkaloid content (Johns and Alonso, 1990). Ingestion of these compounds causes a range of symptoms, including stomach disturbances, and is described (in the mouth) as causing a burning sensation (Sinden et al., 1976). Glycoalkaloid contents vary with cultivar (Ramsay et al., 2004) and also according to growth and storage regime (Sengul et al., 2004). Excessive amounts of glycoalkaloid develop when tubers are exposed to sunlight and therefore they are associated with tissue greening because of chlorophyll production. While consumers often regard green potatoes as bad for them, it is actually the associated glycoalkaloids that are the problem. Total glycoalkaloid contents in potatoes destined for human consumption range from 2.5 to 15 mg/100 g, and amounts over 20 mg/100 g fresh weight (FW) are generally regarded as unpalatable and potentially toxic (Johns and Keen, 1986; Osman, 1983). However, these compounds also potentially contribute to the overall flavor/sensory profile of potato. Glycoalkaloids can be detected by humans at concentrations as low as 14 mg/100 g FW, though individual thresholds may vary. A significant correlation between potato bitterness and glycoalkaloid content has been established when rated by sensory panels (Sinden et al., 1976), but the glycoalkaloids may make an important contribution at lower concentrations, at which they have been reported to improve the flavor of potato (Valkonen et al., 1996). Morris et al. (2010) reported that the content of the glycoalkaloid solanine was positively correlated with aroma intensity, but negatively associated with flavor intensity, flavor creaminess, and flavor savoriness. In bitter leafy vegetables there are demonstrated relationships among preference; genetic ability to taste bitter compounds, such as 6-*n*-propylthiouracil; and amount consumed (Dinehart et al., 2006). Genetic sensitivity to bitterness may also apply to the consumption of potato.

Phenolic compounds are commonly found in the skin and are also believed to contribute to potato bitterness (Mondy and Gosselin, 1988), although the relationship between bitterness and phenolic content in potato does not appear to be as closely correlated when assessed by sensory panels (Sinden et al., 1976).

4.2 Sourness

Organic acids, such as chlorogenic, malic, and citric acids, are products of incomplete sugar oxidation and the deamination of amino acids (Lisinska and Aniolowski, 1990). Total organic acids found in potato tubers range from 0.4% to 1.0%, with citric acid being found in the greatest amount, followed by malic and pyrrolidone carboxylic acids, and their concentrations are dependent on tuber development and maturity. These compounds would be expected to be responsible for producing a sour taste, but they have not been considered major contributors to potato flavor (Vainionpaa et al., 2000). However, samples containing 120 mg/100 g FW of chlorogenic acid have been described as slightly sour by some taste panelists (Sinden et al., 1976).

4.3 Sweetness

In contrast to fruits, sweetness is not a large contributor to flavor in potato tubers and, indeed, is historically regarded as an undesirable trait (Burton, 1965). Glucose and fructose in particular have been selected against in potato breeding programs because of the negative effect of their production during storage at cool temperatures, especially in potato cultivars targeted for further processing into crisps or french fries (Sowokinos, 2001). The high temperatures required during frying or baking result in the nonenzymatic reaction of free amino acids, particularly asparagine, with the reducing sugars in the Maillard reaction. If reducing sugar content is too high, this results in unacceptable rates of browning and also acrylamide production. Therefore, this is not just a product quality issue but also a human health issue (Medeiros Vinci et al., 2012). Ironically, as the consumption of sugars has increased in developed countries, consumers have a greater preference for sweet foods, even in those that are traditionally savory, such as potato. Indeed, sweetness of baked potatoes has been correlated with desirable potato flavor (Jansky, 2008; Jitsuyama et al., 2009); potato chips with added sugar are preferable to consumers (Maier et al., 2007); and some consumers specifically seek out the darker potato products, presumably because of their flavor and odor preferences (Maga, 1973). Unfortunately, these darker potato products also have the highest acrylamide content (McKenzie et al., 2013), and from a human health point of view are the least acceptable.

4.4 Umami

Umami is a Japanese word meaning "delicious" or "tastiness." Umami appears to be independent of the other four "basic" tastes and acts as an enhancer of flavor and mouthfeel in foods, giving the feeling of creaminess and viscosity to savory dishes (Halpern, 2000). The first identified umami factor was glutamic acid, found in Japanese seaweed stock in 1908. Umami is now defined as the taste of monosodium glutamate (MSG) and 5′-ribonucleotides such as the sodium salts of 5′-adenosine monophosphate (AMP), 5′-inosine monophosphate, and 5′-guanosine monophosphate (GMP) (Fuke and Shimizu, 1993). Indeed, AMP and GMP have been found by sensory panels to be closely associated with flavor intensity, flavor creaminess, and flavor savoriness in potatoes (Morris et al., 2010). Glutamate is the most potent amino acid contributing to umami, with aspartate having only 7% of its taste activity. Umami intensity can also be increased by the addition of salts, including sodium, potassium, and magnesium salts (Ugawa and Kurihara, 1994). Interestingly, there are no reports of umami compounds positively enhancing the flavor of sweet, fruity, or bland foods.

Umami is of great importance in determining potato flavor and is generally regarded as crucial to the identification of potato acceptability by the consumer. Early research suggested that the characteristic taste of boiled potato was due almost entirely to the interaction of amino acids and the 5′-ribonucleotides, which are released during the cooking process (Halpern, 2000; Solms, 1971). In raw potatoes the 5′-ribonucleotide content is relatively low, but on cooking, large amounts are released from the enzymatic hydrolysis of RNA, especially during the early stages of cooking when internal temperatures are lower (around 40–60 °C) and the nucleases are more active (Solms and Wyler, 1979). Concentrations of 5′-nucleotides have been shown to increase up to 12–15 and 25–30 nmol/g FW in *S. tuberosum* and *S. phureja* cultivars, respectively (Morris et al., 2007), and a ribonuclease with enhanced expression has been found to be active in Phureja potato lines (Ducreux et al., 2008).

Despite its critical role in potato flavor, little research has been conducted on the sensory effect of umami in cooked potato. An 18-person trained taste panel reported that the supplementation of boiled potato with either glutamic acid alone or a glutamic acid/nucleotide mix resulted in a stronger potato flavor, with the glutamic acid/nucleotide mix providing the strongest flavor overall (Solms, 1971). However, research on a variety of food matrices supplemented with various amounts of MSG and disodium inosate/guanylate indicated a significant increase in palatability of mashed potato in only one of the 15 treatments, compared with palatability of other foods tested (Barylko-Pikielna and Kostyra, 2007).

Morris et al. (2007) compared concentrations of amino acids and ribonucleotides with umami taste in boiled tubers from cultivars of *S. phureja* and *S. tuberosum*. The contents of both glutamate and the 5'-nucleotides were significantly higher in the *S. phureja* cultivars than in the *S. tuberosum* cultivars. Furthermore, the equivalent umami concentration in the cultivars was strongly correlated with flavor attributes and with acceptability as judged by a trained evaluation panel, indicating the importance of umami as a contributor to potato flavor. However, it is important to note that the glutamate, aspartate, GMP, and AMP concentrations have been shown to vary not only according to cultivar but also according to storage temperature and time (Morris et al., 2010); therefore these compounds are not static once a tuber is mature and will change according to postharvest treatment.

5. Texture

Texture is an extremely important sensory attribute in potato, as it has a direct influence on consumer preference and also affects the release of volatile compounds during chewing (Lucas et al., 2002). However, consumers have distinct preferences regarding optimal texture profiles for each food and these will vary from consumer to consumer, making texture a complex trait (further reviewed in Taylor et al. (2007)).

Texture attributes such as hardness, firmness, springiness, adhesiveness, graininess, mealiness, moistness, and chewiness have been used by trained panelists to describe potato (Martens and Thybo, 2000). The trait is controlled by a number of factors, such as dry matter, starch content and distribution, amylase/amylopectin content, cell size, cell wall structure and composition, sugar and protein contents, and nitrogen concentrations (Taylor et al., 2007). These factors all contribute to the mealiness/waxiness spectrum, which is a continuum and is one of the components of texture most readily described by sensory panelists.

In general terms, a mealy potato has a dry, soft, granular feeling in the mouth, while a waxy potato is moist and firm and has a gummy mouthfeel. Mealiness is generally associated with high dry matter (Van Dijk et al., 2002), and waxiness with low dry matter (Bordoloi et al., 2012). Mealy potatoes have also been associated with higher amylase contents, as well as having a greater percentage of larger sized cells and starch granules (Barrios et al., 1963).

Potato cultivars with closely packed cells and high starch content have been reported to have greater fracturability and hardness than those with loosely packed and large cells (Bordoloi et al., 2012; Singh et al., 2005). The structure and quantity of starch grains also play an important role in determining potato texture (Ridley and Hogan, 1976). Gelatinization during cooking causes the starch granules to swell and this puts varying pressure on the cell walls, depending on the amount of starch present. This may even cause "rounding" of the cells and exert pressures of up to 100 kPa inside the cell (Jarvis et al., 1992). Those cultivars with a large amount of gelatinized starch are associated with mealy texture in contrast to waxy cultivars, which tend to have less gelatinized starch and higher water content (Martens and

Thybo, 2000; McComber et al., 1994). Visualization of cooked and raw potato surfaces using scanning electron microscopy and confocal scanning laser microscopy has been used to detect textural qualities such as mealiness, graininess, and moistness and has been shown to reflect instrumental measurements and sensory measurements for determining texture (Bordoloi et al., 2012; Martens and Thybo, 2000).

Interestingly, the physical state of starch plays an important role in determining the release of aroma compounds during cooking, with physical interactions between small volatile molecules and amylose/amylopectin ratio being critical (Arvisenet et al., 2002). The binding of volatile compounds to starch is classified into two types: those in which the starch forms inclusion complexes and traps the volatile molecules by hydrophobic bonding within the amylose helices (Nuessli et al., 1997) and those in which polar interactions occur between the aroma compounds and the hydroxyl groups of starch (Maier, 1972). During the gelatinization of corn starch, aromatic compounds are increasingly retained (Boutboul et al., 2002), likewise in potato starch maximal aromatic release occurs in the first 10 min of cooking, before full gelatinization is complete (Descours et al., 2013).

Cell wall structure is also an important contributor to texture (Jarvis and Duncan, 1992; Martens and Thybo, 2000; McComber et al., 1994; van Marle et al., 1997), and cell walls have been shown to respond differently to cooking depending on cultivar (Bordoloi et al., 2012). In potato, pectin contributes around 55% of the cell wall polysaccharides and strong links have been identified among the degree of methylation of pectin, the pectin methyl esterase (PME) activity, and cooked tuber texture (Ross et al., 2011b). PME plays a major role in cell wall strengthening in fruits and vegetables by removing the methyl groups from pectins, increasing the likelihood of calcium bridges forming between the free acid groups of adjacent chains (Ng and Waldron, 1997). Recently, a gene encoding PME was found to be expressed in the Phureja potato group at 10% of the concentrations observed in the generally firmer Tuberosum group (Ducreux et al., 2008). Furthermore, overexpression of the PME gene in Desiree tubers resulted in reduced levels of pectin methylation and was associated with firmer potato texture (Ross et al., 2011a).

Cell walls and middle lamella have also been shown to be thicker in 'Russet Burbank' (a mealy cultivar) compared with the waxy 'Red Pontiac' (McComber et al., 1994). These cell wall differences could result in more resistance to shearing and therefore a harder, more particulate mouthfeel. This is particularly important in french fry cultivars, in which a crispy exterior crust surrounding a fluffy, mealy interior is preferred. Further investigation into pectin, the main constituent of the potato cell wall, has also found differences between cultivars. 'Irene', a mealy cultivar, was found to have more or longer side chains than 'Nicola', a nonmealy cultivar, which had more highly branched pectin (van Marle et al., 1997). Also, single cells of 'Irene' had significantly more cell wall material than cells of 'Nicola', implying that, like those of 'Russet Burbank', their cell walls are thicker.

6. *Influence of Growth and Storage Environment on Flavor*

The effects of production environment and subsequent storage are also important in terms of flavor development in potatoes. For example, sulfur and potassium fertilization of the crop can influence the tuber content of compounds such as methional and those contributing to umami (Duckham et al., 2002; Morris et al., 2007). In contrast, nitrogen fertilization may influence the production of amides and amines, which are associated with off-flavors (Cieslik, 1997; Jansky, 2008), although other groups have reported increased glutamate content on high rates of nitrogen fertilization (Burton, 1989). Musty off-flavors were detected in several million kilograms of potatoes harvested in 2001 in Canada and were found to be due to the presence of 2,4,6-trichloroanisole. This compound is not known to be produced by potato, but may be a by-product of pesticide treatments in warm growing years (Daniels-Lake et al., 2007). Tubers grown under stressful conditions, exposed to high light during growth or following harvest (Valkonen et al., 1996), or that are bruised at harvest (Dale et al., 1998) can produce the bitter-tasting glycoalkaloids.

Postharvest storage of potato tubers at 10 °C or less is required to maintain a year-round supply. Although the metabolic response of tubers is slowed under these conditions, changes in compounds expected to influence flavor still occur, particularly in the fatty acids (Cherif and Abdelkader, 1970; Mondy et al., 1963), sugars, and amino acids (Finglas and Faulks, 1984). The sugars and amino acids are precursors for most of the flavor compounds formed on cooking, and tubers coming out of storage have been shown to be sweeter, more mealy and have greater overall flavor (Jansky, 2008). The soluble protein and amino acid contents of stored potato have been observed to change particularly during 10 °C storage, with glutamate and asparagine content increasing from 34% of the total amino acid pool to up to 90% following 25 weeks of storage at 10 °C (Brierley et al., 1997). This could be expected to maximize umami perception after cooking. Morris et al. (2010) compared the major umami compounds in Phureja and Tuberosum samples at harvest and after 3 months of storage at 10 °C. Phureja had the highest glutamate and aspartate contents at harvest, but, in contrast to the findings of Brierley et al. (1997), Morris et al. (2010) observed up to 68% loss of these compounds on storage. Smaller declines in glutamate and aspartate content were also observed for Tuberosum samples. At the end of the storage period, the equivalent umami concentrations were similar for both groups. Storage has also been shown to increase the concentrations of the major glycoalkaloids, solanine and chaconine, which increased 13- and 12-fold, respectively, in *S. tuberosum* 'Montrose' after storage at 4 °C (Morris et al., 2010).

Storage temperatures affect glycoalkaloid content, with colder temperatures (4 °C) resulting in a more rapid increase in glycoalkaloid content than warmer temperatures (Griffiths et al., 1998). In contrast, photoinduced chlorophyll accumulation was not affected by storage temperature, although it did decrease slightly with storage time.

Several volatile compounds also increase following storage, including the aldehydes propanal, 5-methylhexanal, and 2-hexenal (Morris et al., 2010), which are all derived from fatty acid metabolism. Previous work by Dobson et al. (2004) concluded that Phureja tubers have higher fatty acid content than those of Tuberosum and that these increase on storage. Off-flavors have been shown to decrease during cold storage of potatoes (Jansky, 2008; Thybo et al., 2006).

Sprout inhibitors are commonly used during potato storage; therefore, their influence on flavor should also be considered. Isopropyl-*N*-chlorophenyl, the major sprout inhibitor used, did not appear to be detected by taste panels. However, this compound is currently being phased out worldwide for use on potatoes, and tubers treated with two alternative sprout inhibitors, 1,8-cineole and salicylaldehyde, were both detectable as different from untreated controls. It is not clear if the detectable difference was due to a perceived flavor difference of the sprout inhibitor(s) itself or to an altered effect on tuber metabolism by the sprout inhibitor(s) (Boylston et al., 2001).

The effects of storage on texture are mainly related to changes in starch, which is degraded during storage, and this, combined with the breakdown on the middle lamella of the cell (Martens and Thybo, 2000), is expected to contribute to the reported decrease in tuber mealiness during storage. Water loss from stored tubers may also contribute to decreased mealiness (Shetty et al., 1992).

The effect of postharvest treatment of potatoes on flavor development also needs to be carefully considered. For example, cold-stored boiled potatoes display an undesirable flavor change described as "cardboard-like" soon after storage begins. This flavor was thought to be the result of lipoxygenase activity, which releases linoleic and linolenic acid during the boiling process, which are subsequently decomposed into aldehydes during the storage process (Petersen et al., 1999). However, in another study the production of typical off-flavor oxidation compounds, such as the aldehydes, did not correlate well with increased lipoxygenase activity under similar conditions (Petersen et al., 2003). Precooked vacuum-packed potatoes produced similar off-flavor compounds, including 2,4-nonadienal and 2,4-decadienal, as well as hexenal, 2-octenal, and 2-nonenal. The amounts of these compounds were also influenced by growing location (Jensen et al., 1999). For the production of prepeeled, vacuum-packed tubers, the quality of the raw material, the potato cultivar, and the physiological maturity of the tubers were found to be critical, and deterioration in the sensory quality of tubers processed this way was due to increased rancidity, off-flavors, and surface hardness (Thybo et al., 2006). The aromatic compounds that characterized differences between potato cultivars treated this way and stored for various lengths of time include methional, linalool, *p*-cymene, nonanal, and decanal (Thybo et al., 2006). In contrast, nonstored cooked potato tubers had high intensities of methional, linalool, and *p*-cymene and low intensities of nonanal and decanal, with high sensory scores for potato flavor and low scores for rancidity.

The flavor characteristics of highly processed potato products, such as dehydrated potato flakes, extruded potato products, and fermented potato slices, have been well reviewed by Comandini et al. (2011).

7. Combining Sensory Panels with Molecular and Metabolomics Approaches

While many previous potato studies have identified a wide range of volatile and nonvolatile components potentially involved in flavor perception, attempts to correlate these with descriptive analysis data from trained taste panels have been limited (Morris et al., 2010; Ulrich et al., 2000). It is not clear why there has not been more research using this methodology to identify the key compounds responsible for both flavor and off-flavors (and also what flavor factors drive consumer acceptance) in potato. However, it may be because of the relative sameness of all potatoes available to sensory scientists in comparison with fruits, for example, and also because of the inherent difficulty of identifying and confirming key compounds in this way. Issues can also include compound odor threshold and/or odor impact and synergistic/antagonistic interactions between important flavor compounds, as discussed earlier. Table 2 summarizes the compound separation and identification approaches that have been used alongside sensory analysis in this area.

Ulrich et al. (2000) asked 15 trained panelists to assess samples of three potato genotypes after boiling and mashing, with an emphasis on aroma. The sensory attributes were limited to eight descriptors: sweet-like, earthy, burnt, fodder, untypical, musty, fruity, and typical. The samples were also analyzed by GC–PND, GC–MS, and GC–O. They reported that the genotypes tested

Table 2: Studies Comparing Metabolomic Content with Sensory Analysis in Potato.

Sample	Extraction Method	Compound Separation/ Identification	Sensory Analysis	References
Three cultivars of *S. tuberosum*, boiled and mashed	SDE[a]	GC-MS, GC-PND, GC-O	Trained panel, focus on odor and retronasal odor	Ulrich et al. (2000)
Two cultivars of *S. phureja* and *S. tuberosum* stored at 4 °C for 6 weeks for umami-related compound analysis or 2 °C (for indeterminate time) before boiling for sensory analysis	Techniques specific for nucleotides and amino acids only	HPAEC, HPLC	Trained panel, focus on four attributes, three of which relate to umami-type descriptors	Morris et al. (2007)
Two cultivars of *S. phureja* and *S. tuberosum*, stored at different temperatures and times, boiled and mashed	SPME	GC-MS	Trained panel, rating six sensory attributes	Morris et al. (2010)
Single *S. tuberosum* cultivar overexpressing α-copaene synthase gene from *S. phureja*, microwaved and boiled	SPME	GC-MS	Untrained panel, focus on aroma followed by five general sensory attributes	Morris et al. (2011)

[a]SDE, simultaneous distillation extraction; GC-MS, gas chromatography–mass spectrometry; GC-PND, gas chromatography with nitrogen-specific detection; GC-O, gas chromatography olfactometry; SPME, solid-phase microextraction; HPLC, high-performance liquid chromatography; HPAEC, high-performance anion-exchange chromatography.

were characterized by a basic pattern of character-impact compounds and suggested that this basic pattern, with quantitative differences, is likely to be seen in other potato varieties. The combination of GC-MS and GC–O identified 19 aroma-impact compounds responsible for typical boiled potato flavor and off-notes, which included compounds such as methional, diacetyl, and at least five substituted pyrazines. This is in good agreement with the findings of others (Buttery et al., 1973; Salinas et al., 1994). It is also worth noting that this group included "earthiness" as a positive flavor attribute, making potatoes one of the few foods, along with beetroot (Mottram, 1998) and ulluco (Busch et al., 2000), in which earthiness is considered part of the characteristic flavor rather than a flavor taint, as it is in drinking water, fish, and wine, for example (Darriet et al., 2000; Robertson et al., 2005; Young et al., 1996).

Morris et al. (2010) used a trained sensory panel for descriptive analysis of cooked samples from four potato cultivars taken at harvest and following 3 months of storage at 10 °C. The sensory responses (aroma, intensity, sweetness, savory, creaminess) were correlated with the concentrations of metabolites measured by SPME GC–MS and presented as a heat map of compounds associated with the sensory attributes. There was a strong positive correlation between a specific set of metabolites, including AMP and GMP, hexanal, pentanal, and 3-ethyl-2-methyl-1,3-hexa-diene, and the attributes flavor intensity, flavor creaminess, and flavor savoriness. Interestingly, these same three flavor attributes are strongly negatively correlated with a different set of metabolites, such as methional, 4-heptanal, benzeneacet aldehyde, chaconine, and solanine. When the sensory attributes aroma and sweetness were considered, the metabolites were correlated in the opposite manner, i.e., those that were positively correlated with flavor intensity, flavor creaminess, and flavor savoriness were negatively correlated with aroma and sweetness, and vice versa. They also used principal components analysis of the dataset to investigate flavor differences between the flavorsome Phureja group and the generally blander Tuberosum group. *S. phureja* cultivars were found to have a more intense flavor and stronger savory flavor than *S. tuberosum* cultivars (Morris et al., 2010). Earlier studies have shown significant differences in the concentrations of the methyl ester of methyl butanoic acid, methyl salicylate, and 10 other volatiles between *S. phureja* and *S. tuberosum* genotypes (Winfield et al., 2005). While this work presents an important attempt to marry the results of sensory analysis directly with chemical compound identification, the sensory descriptions used were a small set of rather broad descriptors: aroma, intensity, sweetness, savory, creaminess, and off-flavors.

In another approach, expression profiling using a 44,000-element potato microarray was used to compare the gene expression profiles of tubers from two *S. phureja* and two *S. tuberosum* cultivars, and clear differences were observed. Increased expression of the gene for α-copaene synthase, the enzyme responsible for the production of α-copaene, which is a key sesquiter-pene, was found in Phureja (Ducreux et al., 2008). Other flavor-related genes with differential gene expression between the two groups included several involved in branched-chain amino acid aminotransferase and a ribonuclease, which indicates increased potential for 5'-ribonucleotide formation in potato tubers of the Phureja group on cooking.

Using a transgenic approach, Morris et al. (2011) produced *S. tuberosum* that overexpressed the α-copaene synthase gene from *S. phureja*. Concentrations of α-copaene were found to be up to 15-fold those of the controls. However, sensory evaluators were not able to detect any aromatic or taste differences in the transgenic lines, suggesting that α-copaene is not in fact a major component of potato flavor.

The use of quantitative trait locus maps in potato populations with a wide variety of flavor profiles could also provide important clues as to which compounds are critical for preferred flavor notes. Crosses made from key cultivars of *S. phureja* and *S. tuberosum*, for example, would be particularly interesting in this regard.

8. Summary

The flavor of cooked potato is of great importance in terms of consumer acceptability and in growing the market share of this important food crop. The perceived flavor of potatoes is a result of the interaction of specific volatile aroma compounds and taste components, modified to varying extents by texture. It is also heavily influenced by cooking method, agronomic production, and storage methods. However, as the major focus of potato breeding programs has traditionally been traits such as yield, appearance, pathogen resistance, and processing quality, this may have resulted in genotypic candidates with potentially promising sensory profiles being discarded before sensory traits were considered. Although consumers are developing increasing interest in the flavor of potato cultivars, and developers of potato cultivars are identifying this as a marketing opportunity, there is still much to be learned about potato flavor and how the many compounds identified in cooked tubers may contribute to aspects of potato flavor. The combination of metabolomic and gene expression analysis with sensory panel use is potentially a powerful approach for identifying the key volatile compounds involved in perceived potato flavor and for assisting with the breeding effort to produce potato cultivars with the flavors that consumers enjoy.

References

Arthur, C.L., Pawliszyn, J., 1990. Solid-phase microextraction with thermal-desorption using fused-silica optical fibers. Analytical Chemistry 62, 2145–2148.

Arvisenet, G., Le Bail, P., Voilley, A., Cayot, N., 2002. Influence of physicochemical interactions between amylose and aroma compounds on the retention of aroma in food-like matrices. Journal of Agricultural and Food Chemistry 50, 7088–7093.

Auvray, M., Spence, C., 2008. The multisensory perception of flavor. Consciousness and Cognition 17, 1016–1031.

Barrios, E.P., Newsom, D., Miller, J., 1963. Some factors influencing the culinary quality of Irish potatoes II. Physical characters. American Journal of Potato Research 40, 200–208.

Barylko-Pikielna, N., Kostyra, E., 2007. Sensory interaction of umami substances with model food matrices and its hedonic effect. Food Quality and Preference 18, 751–758.

Baysal, T., Demirdöven, A., 2007. Lipoxygenase in fruits and vegetables: a review. Enzyme and Microbial Technology 40, 491–496.

Blanda, G., Cerretani, L., Comandini, P., Toschi, T.G., Lercker, G., 2010. Investigation of off-odour and off-flavour development in boiled potatoes. Food Chemistry 118, 283–290.

Bordoloi, A., Kaur, L., Singh, J., 2012. Parenchyma cell microstructure and textural characteristics of raw and cooked potatoes. Food Chemistry 133, 1092–1100.

Boutboul, A., Giampaoli, P., Feigenbaum, A., Ducruet, V., 2002. Influence of the nature and treatment of starch on aroma retention. Carbohydrate Polymers 47, 73–82.

Boylston, T.D., Powers, J.R., Weller, K.M., Yang, J., 2001. Comparison of sensory differences of stored Russet Burbank potatoes treated with CIPC and alternative sprout inhibitors. American Journal of Potato Research 78, 99–107.

Brewer, M.S., Vega, J.D., Perkins, E.G., 1999. Volatile compounds and sensory characteristics of frying fats. Journal of Food Lipids 6, 47–61.

Brierley, E.R., Bonner, P.L.R., Cobb, A.H., 1997. Aspects of amino acid metabolism in stored potato tubers (cv. Pentland Dell). Plant Science 127, 17–24.

Brittin, H.C., Trevino, J.E., 1980. Acceptability of microwave and conventionally baked potatoes. Journal of Food Science 45, 1425–1427.

Bult, J.H.F., Schifferstein, H.N.J., Roozen, J.P., Boronat, E.D., Voragen, A.G.J., Kroeze, J.H.A., 2002. Sensory evaluation of character impact components in an apple model mixture. Chemical Senses 27, 485–494.

Burton, W.G., 1965. The sugar balance in some British potato varieties during storage. I. Preliminary observations. European Potato Journal 8, 80–91.

Burton, W.G., 1989. The Potato.

Busch, J.M., Sangketkit, C., Savage, G.P., Martin, R.J., Halloy, S., Deo, B., 2000. Nutritional analysis and sensory evaluation of ulluco (Ullucus tuberosus Loz) grown in New Zealand. Journal of the Science of Food and Agriculture 80, 2232–2240.

Buttery, R.G., Guadagni, D.G., Ling, L.C., 1973. Volatile components of baked potatoes. Journal of the Science of Food and Agriculture 24, 1125–1131.

Buttery, R.G., Ling, L.C., 1973. Earthy aroma of potatoes. Journal of Agricultural and Food Chemistry 21, 745–746.

Buttery, R.G., Ling, L.C., 1998. Additional studies on flavor components of corn tortilla chips. Journal of Agricultural and Food Chemistry 46, 2764–2769.

Buttery, R.G., Seifert, R.M., Guadagni, D.G., Ling, L.C., 1971. Characterization of volatile pyrazine and pyridine components of potato chips. Journal of Agricultural and Food Chemistry 19, 969–971.

Cherif, A., Abdelkader, A.B., 1970. Quantitative analysis of fatty acids present in different parts of the potato tuber; variation during storage at 10°C. Potato Research 13, 284–295.

Cieslik, E., 1997. Effect of the levels of nitrates and nitrites on the nutritional and sensory quality of potato tubers. Hygiene and Nutrition in Foodservice and Catering 1, 225–230.

Coleman, E.C., Ho, C.T., 1980. Chemistry of baked potato flavor. 1. Pyrazines and thiazoles identified in the volatile flavor of baked potato. Journal of Agricultural and Food Chemistry 28, 66–68.

Coleman, E.C., Ho, C.T., Chang, S.S., 1981. Isolation and identification of volatile compounds from baked potatoes. Journal of Agricultural and Food Chemistry 29, 42–48.

Comandini, P., Cerretani, L., Blanda, G., Bendini, A., Toschi, T.G., 2011. Characterization of potato flavours: an overview of volatile profiles and analytical procedures. Food 5, 1–14.

Costell, E., Tárrega, A., Bayarri, S., 2010. Food acceptance: the role of consumer perception and attitudes. Chemosensory Perception 3, 42–50.

Dale, M.F.B., Griffiths, D.W., Bain, H., 1998. Effect of bruising on the total glycoalkaloid and chlorogenic acid content of potato (*Solanum tuberosum*) tubers of five cultivars. Journal of the Science of Food and Agriculture 77, 499–505.

Daniels-Lake, B.J., Prange, R.K., Gaul, S.O., McRae, K.B., de Antueno, R., 2007. A musty "off" flavor in Nova Scotia potatoes is associated with 2,4,6-trichloroanisole released from pesticide-treated soils and high soil temperature. Journal of the American Society for Horticultural Science 132, 112–119.

Darriet, P., Pons, M., Lamy, S., Dubourdieu, D., 2000. Identification and quantification of geosmin, an earthy odorant contaminating wines. Journal of Agricultural and Food Chemistry 48, 4835–4838.

Descours, E., Hambleton, A., Kurek, M., Debeaufort, F., Voilley, A., Seuvre, A.-M., 2013. Aroma behaviour during steam cooking within a potato starch-based model matrix. Carbohydrate Polymers 95, 560–568.

Dinehart, M.E., Hayes, J.E., Bartoshuk, L.M., Lanier, S.L., Duffy, V.B., 2006. Bitter taste markers explain variability in vegetable sweetness, bitterness, and intake. Physiology & Behavior 87, 304–313.

Dobson, G., Griffiths, D.W., Davies, H.V., McNicol, J.W., 2004. Comparison of fatty acid and polar lipid contents of tubers from two potato species, *Solanum tuberosum* and *Solanum phureja*. Journal of Agricultural and Food Chemistry 52, 6306–6314.

Doyen, K., Carey, M., Linforth, R.S.T., Marin, M., Taylor, A.J., 2001. Volatile release from an emulsion: headspace and in-mouth studies. Journal of Agricultural and Food Chemistry 49, 804–810.

Dresow, J.F., Boehm, H., 2009. The influence of volatile compounds of the flavour of raw, boiled and baked potatoes: impact of agricultural measures on the volatile components. Landbauforschung Volkenrode 59, 309–337.

Duckham, S.C., Dodson, A.T., Bakker, J., Ames, J.M., 2001. Volatile flavour components of baked potato flesh. A comparison of eleven potato cultivars. Nahrung – Food 45, 317–323.

Duckham, S.C., Dodson, A.T., Bakker, J., Ames, J.M., 2002. Effect of cultivar and storage time on the volatile flavor components of baked potato. Journal of Agricultural and Food Chemistry 50, 5640–5648.

Ducreux, L.J.M., Morris, W.L., Prosser, I.M., Morris, J.A., Beale, M.H., Wright, F., Shepherd, T., Bryan, G.J., Hedley, P.E., Taylor, M.A., 2008. Expression profiling of potato germplasm differentiated in quality traits leads to the identification of candidate flavour and texture genes. Journal of Experimental Botany 59, 4219–4231.

Engel, W., Bahr, W., Schieberle, P., 1999. Solvent assisted flavour evaporation - a new and versatile technique for the careful and direct isolation of aroma compounds from complex food matrices. European Food Research and Technology 209, 237–241.

FAOSTAT, 2014. Food and Agriculture Organization of the United Nations. http://faostat3.fao.org/home/E.

Finglas, P.M., Faulks, R.M., 1984. The HPLC analysis of thiamin and riboflavin in potatoes. Food Chemistry 15, 37–44.

Fuke, S., Shimizu, T., 1993. Sensory and preference aspects of umami. Trends in Food Science & Technology 4, 246–251.

Galliard, T., Phillips, D.R., 1971. Lipoxygenase from potato tubers. Partial purification and properties of an enzyme that specifically oxygenates the 9-position of linoleic acid. The Biochemical Journal 124, 431–438.

Gardner, H., 1996. Lipoxygenase as a versatile biocatalyst. Journal of the American Oil Chemists' Society 73, 1347–1357.

Griffiths, D.W., Bain, H., Dale, M.F.B., 1998. Effect of storage temperature on potato (*Solanum tuberosum* L.) tuber glycoalkaloid content and the subsequent accumulation of glycoalkaloids and chlorophyll in response to light exposure. Journal of Agricultural and Food Chemistry 46, 5262–5268.

Grosch, W., 2001. Evaluation of the key odorants of foods by dilution experiments, aroma models and omission. Chemical Senses 26, 533–545.

Halpern, B.P., 2000. Glutamate and the flavor of foods. Journal of Nutrition 130, 910S–914S.

Ho, C.T., Chen, Q.Y., 1994. Lipids in food flavors – an overview. In: Ho, C.T., Hartman, T.G. (Eds.), Lipids in Food Flavors, pp. 2–14.

Jaeger, S.R., de Silva, H.N., Lawless, H.T., 2014. Detection thresholds of 10 odor-active compounds naturally occurring in food using a replicated forced-choice ascending method of limits. Journal of Sensory Studies 29, 43–55.

Jansky, S.H., 2008. Genotypic and environmental contributions to baked potato flavor. American Journal of Potato Research 85, 455–465.

Jansky, S.H., 2010. Potato flavor. American Journal of Potato Research 87, 209–217.

Jarvis, M.C., Duncan, H.J., 1992. The textural analysis of cooked potato. 1. Physical principles of the separate measurement of softness and dryness. Potato Research 35, 83–91.

Jarvis, M.C., Mackenzie, E., Duncan, H.J., 1992. The textural analysis of cooked potato. 2. Swelling pressure of starch during gelatinization. Potato Research 35, 93–102.

Jensen, K., Petersen, M.A., Poll, L., Brockhoff, P.B., 1999. Influence of variety and growing location on the development of off-flavor in precooked vacuum-packed potatoes. Journal of Agricultural and Food Chemistry 47, 1145–1149.

Jitsuyama, Y., Tago, A., Mizukami, C., Iwama, K., Ichikawa, S., 2009. Endogenous components and tissue cell morphological traits of fresh potato tubers affect the flavor of steamed tubers. American Journal of Potato Research 86, 430–441.

Johns, T., Alonso, J.G., 1990. Glycoalkaloid change during the domestication of the potato, *Solanum* section *Petota*. Euphytica 50, 203–210.

Johns, T., Keen, S.L., 1986. Taste evaluation of potato glycoalkaloids by the Aymara: a case study in human chemical ecology. Human Ecology 14, 437–452.

Josephson, D.B., Lindsay, R.C., 1987. C4-heptenal – an influential volatile compound in boiled potato flavor. Journal of Food Science 52, 328–331.

King, S.C., Meiselman, H.L., Hottenstein, A.W., Work, T.M., Cronk, V., 2007. The effects of contextual variables on food acceptability: a confirmatory study. Food Quality and Preference 18, 58–65.

Lau, D., Krondl, M., Coleman, P., 1984. Psycological factors affecting food selection. In: Galler, J. (Ed.), Nutrition and Behavior, first ed. Plenum Press, New York, pp. 397–414.

Lindsay, R.C., 1996. Flavors. In: Fennema, O.R. (Ed.), Food Chemistry. Marcel Dekker, Inc., New York, NY, pp. 723–762.

Lisinska, G., Aniolowski, K., 1990. Organic-acids in potato-tubers. 1. The effect of storage temperatures and time on citric and malic-acid contents of potato-tubers. Food Chemistry 38, 255–261.

Lucas, P.W., Prinz, J.F., Agrawal, K.R., Bruce, I.C., 2002. Food physics and oral physiology. Food Quality and Preference 13, 203–213.

van Loon, W.A.M., Linssen, J.P.H., Legger, A., Posthumus, M.A., Voragen, A.G.J., 2005. Identification and olfactometry of French fries flavour extracted at mouth conditions. Food Chemistry 90, 417–425.

Maga, J.A., 1973. Influence of freshness and color on potato chip sensory preferences. Journal of Food Science 38, 1251–1252.

Maga, J.A., 1994. Potato flavor. Food Reviews International 10, 1–48.

Maga, J.A., Holm, D.G., 1992. Subjective and objective comparison of baked potato aroma as influenced by variety/clone. In: Charalambous, G. (Ed.), Food Science and Human Nutrition. Elsevier Science Publishers, Amsterdam, pp. 537–541.

Maga, J.A., Twomey, J.A., 1977. Sensory comparison of four potato varieties baked conventionally and by microwaves. Journal of Food Science 42, 541–542.

Maier, A., Vickers, Z., Inman, J.J., 2007. Sensory-specific satiety, its crossovers, and subsequent choice of potato chip flavors. Appetite 49, 419–428.

Maier, H.G., 1972. Binding of volatile aroma compounds in foods. Lebensmittel-Wissenschaft und-Technologie 5, 1–6.

Majcher, M., Jelen, H.H., 2009. Comparison of suitability of SPME, SAFE and SDE methods for isolation of flavor compounds from extruded potato snacks. Journal of Food Composition and Analysis 22, 606–612.

Martens, H.J., Thybo, A.K., 2000. An integrated microstructural, sensory and instrumental approach to describe potato texture. Lebensmittel-Wissenschaft und-Technologie – Food Science and Technology 33, 471–482.

Martin, F.L., Ames, J.M., 2001a. Comparison of flavor compounds of potato chips fried in palmolein and silicone fluid. Journal of the American Oil Chemists' Society 78, 863–866.

Martin, F.L., Ames, J.M., 2001b. Formation of Strecker aldehydes and pyrazines in a fried potato model system. Journal of Agricultural and Food Chemistry 49, 3885–3892.

McComber, D.R., Horner, H.T., Chamberlin, M.A., Cox, D.F., 1994. Potato cultivar differences associated with mealiness. Journal of Agricultural and Food Chemistry 42, 2433–2439.

McKenzie, M.J., Chen, R.K.Y., Harris, J.C., Ashworth, M.J., Brummell, D.A., 2013. Post-translational regulation of acid invertase activity by vacuolar invertase inhibitor affects resistance to cold-induced sweetening of potato tubers. Plant Cell and Environment 36, 176–185.

McRae, J.F., Jaeger, S.R., Bava, C.M., Beresford, M.K., Hunter, D., Jia, Y.L., Chheang, S.L., Jin, D., Peng, M., Gamble, J.C., Atkinson, K.R., Axten, L.G., Paisley, A.G., Williams, L., Tooman, L., Pineau, B., Rouse, S.A., Newcomb, R.D., 2013. Identification of regions associated with variation in sensitivity to food-related odors in the human genome. Current Biology 23, 1596–1600.

Medeiros Vinci, R., Mestdagh, F., de Meulenaer, B., 2012. Acrylamide formation in fried potato products – present and future, a critical review on mitigation strategies. Food Chemistry 133, 1138–1154.

Meiselman, H.L., Johnson, J.L., Reeve, W., Crouch, J.E., 2000. Demonstrations of the influence of the eating environment on food acceptance. Appetite 35, 231–237.

Mondy, N.I., Gosselin, B., 1988. Effect of peeling on total phenols, total glycoalkaloids, discoloration and flavor of cooked potatoes. Journal of Food Science 53, 756–759.

Mondy, N.I., Owens, E., Mattick, L.R., 1963. Effect of storage on total lipides and fatty acid composition of potatoes. Journal of Agricultural and Food Chemistry 11, 328.

Morris, W.L., Ducreux, L.J.M., Shepherd, T., Lewinsohn, E., Davidovich-Rikanati, R., Sitrit, Y., Taylor, M.A., 2011. Utilisation of the MVA pathway to produce elevated levels of the sesquiterpene alpha-copaene in potato tubers. Phytochemistry 72, 2288–2293.

Morris, W.L., Ross, H.A., Ducreux, L.J.M., Bradshaw, J.E., Bryan, G.J., Taylor, M.A., 2007. Umami compounds are a determinant of the flavor of potato (*Solanum tuberosum* L.). Journal of Agricultural and Food Chemistry 55, 9627–9633.

Morris, W.L., Shepherd, T., Verrall, S.R., McNicol, J.W., Taylor, M.A., 2010. Relationships between volatile and non-volatile metabolites and attributes of processed potato flavour. Phytochemistry 71, 1765–1773.

Mottram, D.S., 1998. Chemical tainting of foods. International Journal of Food Science & Technology 33, 19–29.

Murray, K.E., Whitfield, F.B., 1975. The occurrence of 3-alkyl-2-methoxypyrazines in raw vegetables. Journal of the Science of Food and Agriculture 26, 973–986.

Mutti, B., Grosch, W., 1999. Potent odorants of boiled potatoes. Nahrung – Food 43, 302–306.

van Marle, J.T., Recourt, K., vanDijk, C., Schols, H.A., Voragen, A.G.J., 1997. Structural features of cell walls from potato (*Solanum tuberosum* L.) cultivars Irene and Nicola. Journal of Agricultural and Food Chemistry 45, 1686–1693.

Ng, A., Waldron, K.W., 1997. Effect of steaming on cell wall chemistry of potatoes (*Solanum tuberosum* cv. Bintje) in relation to firmness. Journal of Agricultural and Food Chemistry 45, 3411–3418.

Nickerson, G.B., Likens, S.T., 1966. Gas chromatographic evidence for the occurrence of hop oil components in beer. Journal of Chromatography 21, 1–5.

Nuessli, J., Sigg, B., CondePetit, B., Escher, F., 1997. Characterization of amylose-flavour complexes by DSC and X-ray diffraction. Food Hydrocolloids 11, 27–34.

Nursten, H.E., Sheen, M.R., 1974. Volatile flavour components of cooked potato. Journal of the Science of Food and Agriculture 25, 643–663.

O'Connor, C.J., Fisk, K.J., Smith, B.G., Melton, L.D., 2001. Fat uptake in French fries as affected by different potato varieties and processing. Journal of Food Science 66, 903–908.

Oruna-Concha, M.J., Bakker, J., Ames, J.M., 2002. Comparison of the volatile components of two cultivars of potato cooked by boiling, conventional baking and microwave baking. Journal of the Science of Food and Agriculture 82, 1080–1087.

Oruna-Concha, M.J., Duckham, S.C., Ames, J.H., 2001. Comparison of volatile compounds isolated from the skin and flesh of four potato cultivars after baking. Journal of Agricultural and Food Chemistry 49, 2414–2421.

Osman, S.F., 1983. Glycoalkaloids in potatoes. Food Chemistry 11, 235–247.

Pareles, S.R., Chang, S.S., 1974. Identification of compounds responsible for baked potato flavor. Journal of Agricultural and Food Chemistry 22, 339–340.

Patton, S., Josephson, D.V., 1957. A method for determining significance of volatile flavor compounds in foods. Journal of Food Science 22, 316–318.

Petersen, M.A., Poll, L., Larsen, L.M., 1998. Comparison of volatiles in raw and boiled potatoes using a mild extraction technique combined with GC odour profiling and GC-MS. Food Chemistry 61, 461–466.

Petersen, M.A., Poll, L., Larsen, L.M., 1999. Identification of compounds contributing to boiled potato off-flavour ('POF'). LWT – Food Science and Technology 32, 32–40.

Petersen, M.A., Poll, L., Larsen, L.M., 2003. Changes in flavor-affecting aroma compounds during potato storage are not associated with lipoxygenase activity. American Journal of Potato Research 80, 397–402.

Prescott, J., 1999. Flavour as a psychological construct: implications for perceiving and measuring the sensory qualities of foods. Food Quality and Preference 10, 349–356.

Prescott, J., Young, O., O'Neill, L., Yau, N.J.N., Stevens, R., 2002. Motives for food choice: a comparison of consumers from Japan, Taiwan, Malaysia and New Zealand. Food Quality and Preference 13, 489–495.

Ramsay, G., Griffiths, D.W., Deighton, N., 2004. Patterns of solanidine glycoalkaloid variation in four gene pools of the cultivated potato. Genetic Resources and Crop Evolution 51, 805–813.

Ridley, S.C., Hogan, J.M., 1976. Effect of storage temperature on tuber composition, extrusion force, and Brabender viscosity. American Potato Journal 53, 343–353.

Rizzi, G.P., 2008. The Strecker degradation of amino acids: newer avenues for flavor formation. Food Reviews International 24, 416–435.

Robertson, R., Jauncey, K., Beveridge, M., Lawton, L., 2005. Depuration rates and the sensory threshold concentration of geosmin responsible for earthy-musty taint in rainbow trout, Onchorhynchus mykiss. Aquaculture 245, 89–99.

Ross, H.A., Morris, W.L., Ducreux, L.J.M., Hancock, R.D., Verrall, S.R., Morris, J.A., Tucker, G.A., Stewart, D., Hedley, P.E., McDougall, G.J., Taylor, M.A., 2011a. Pectin engineering to modify product quality in potato. Plant Biotechnology Journal 9, 848–856.

Ross, H.A., Wright, K.M., McDougall, G.J., Roberts, A.G., Chapman, S.N., Morris, W.L., Hancock, R.D., Stewart, D., Tucker, G.A., James, E.K., Taylor, M.A., 2011b. Potato tuber pectin structure is influenced by pectin methyl esterase activity and impacts on cooked potato texture. Journal of Experimental Botany 62, 371–381.

Rothe, M., Thomas, B., 1963. Aromastoffe des Brotes. Zeitschrift für Lebensmittel-Untersuchung und Forschung 119, 302–310.

Salinas, J.P., Hartman, T.G., Karmas, K., Lech, J., Rosen, R.T., 1994. Lipid-derived aroma compounds in cooked potatoes and reconstituted dehydrated potato granules. In: Ho, C.T., Hartman, T.G. (Eds.), Lipids in Food Flavors, pp. 108–129.

Schieberle, P., Grosch, W., 1981. Model experiments about the formation of volatile carbonyl compounds. Journal of the American Oil Chemists' Society 58, 602–607.

Sengul, M., Keles, F., Keles, M.S., 2004. The effect of storage conditions (temperature, light, time) and variety on the glycoalkaloid content of potato tubers and sprouts. Food Control 15, 281–286.

Shetty, K.K., Dwelle, R.B., Fellman, J.K., 1992. Sensory and cooking quality of individually film wrapped potatoes. American Potato Journal 69, 275–286.

Sides, A., Robards, K., Helliwell, S., 2000. Developments in extraction techniques and their application to analysis of volatiles in foods. TrAC – Trends in Analytical Chemistry 19, 322–329.

Sinden, S.L., Deahl, K.L., Aulenbach, B.B., 1976. Effect of glycoalkaloids and phenolics on potato flavor. Journal of Food Science 41, 520–523.

Singh, N., Kaur, L., Ezekiel, R., Guraya, H.S., 2005. Microstructural, cooking and textural characteristics of potato (*Solanum tuberosum* L.) tubers in relation to physicochemical and functional properties of their flours. Journal of the Science of Food and Agriculture 85, 1275–1284.

Slanina, P., 1990. Solanine (glycoalkaloids) in potatoes – toxicological evaluation. Food and Chemical Toxicology 28, 759–761.

Sok, D.-E., Kim, M.R., 1994. Conversion of alpha-linolenic acid to dihydro(pero)xyoctadecatrienoic acid isomers by soybean and potato lipoxygenases. Journal of Agricultural and Food Chemistry 42, 2703–2708.

Solms, J., 1971. Nonvolatile Compounds and Flavour. Academic Press, London, UK.

Solms, J., Wyler, R., 1979. Taste Components of Potatoes.

Sowokinos, J.R., 2001. Biochemical and molecular control of cold-induced sweetening in potatoes. American Journal of Potato Research 78, 221–236.

Spence, C., 2013. Multisensory flavour perception. Current Biology 23, R365–R369.

Taylor, M., McDougall, G., Stewart, D., 2007. Potato flavour and texture. Potato Biology and Biotechnology: Advances and Perspectives 525–540.

Thybo, A.K., Christiansen, J., Kaack, K., Petersen, M.A., 2006. Effect of cultivars, wound healing and storage on sensory quality and chemical components in pre-peeled potatoes. LWT – Food Science and Technology 39, 166–176.

Ugawa, T., Kurihara, K., 1994. Enhancement of canine taste responses to umami substances by salts. American Journal of Physiology 266, R944–R949.

Ulrich, D., Hoberg, E., Neugebauer, W., Tiemann, H., Darsow, U., 2000. Investigation of the boiled potato flavor by human sensory and instrumental methods. American Journal of Potato Research 77, 111–117.

Ulrich, D., Hoberg, E., Tiemann, H., 1998. The aroma of cooked potatoes. Beitrage zur Zuchtungsforschung – Bundesanstalt fur Zuchtungsforschung an Kulturpflanzen 4, 204–209.

Vainionpaa, J., Kervinen, R., de Prado, M., Laurila, E., Kari, M., Mustonen, L., Ahvenainen, R., 2000. Exploration of storage and process tolerance of different potato cultivars using principal component and canonical correlation analyses. Journal of Food Engineering 44, 47–61.

Valkonen, J.P.T., Keskitalo, M., Vasara, T., Pietila, L., 1996. Potato glycoalkaloids: a burden or a blessing? Critical Reviews in Plant Sciences 15, 1–20.

Van Dijk, C., Fischer, M., Holm, J., Beekhuizen, J.G., Stolle-Smits, T., Boeriu, C., 2002. Texture of cooked potatoes (*Solanum tuberosum*). 1. Relationships between dry matter content, sensory-perceived texture, and near-infrared spectroscopy. Journal of Agricultural and Food Chemistry 50, 5082–5088.

Van Eijk, T., 1994. Flavor and flavorings in microwave foods: an overview. In: Parliament, T.H., Morello, M.J., McGorrin, R.J. (Eds.), Thermally Generated Flavors. American Chemical Society, Washington, DC, pp. 395–404.

Wagner, R.K., Grosch, W., 1998. Key odorants of French fries. Journal of the American Oil Chemists' Society 75, 1385–1392.

Whitfield, F., Last, J., 1991. In: Maarse, H. (Ed.), Vegetables in Volatile Compounds in Foods and Beverages. Marcel Dekker, Inc, New York, NY.

Whitfield, F.B., Mottram, D.S., 1992. Volatiles from interactions of Maillard reactions and lipids. Critical Reviews in Food Science and Nutrition 31, 1–58.

Wilson, W.D., MacKinnon, I.M., Jarvis, M.C., 2002. Transfer of heat and moisture during microwave baking of potatoes. Journal of the Science of Food and Agriculture 82, 1070–1073.

Winfield, M., Lloyd, D., Griffiths, W., Bradshaw, J., Muir, D., Nevison, I., Bryan, G., 2005. Assessing organoleptic attributes of *Solanum tuberosum* and *S. phureja* potatoes. Aspects of Applied Biology 127–135.

Young, W., Horth, H., Crane, R., Ogden, T., Arnott, M., 1996. Taste and odour threshold concentrations of potential potable water contaminants. Water Research 30, 331–340.

CHAPTER 13

Microstructure, Starch Digestion, and Glycemic Index of Potatoes

Lovedeep Kaur, Jaspreet Singh
Riddet Institute and Massey Institute of Food Science and Technology, Massey University, Palmerston North, New Zealand

1. Introduction

Starch, a major storage carbohydrate in plants, consists of two types of molecules: amylose (linear polymer of α-D-glucose units linked by α-1,4 glycosidic linkages) and amylopectin (branched polymer of α-D-glucose units linked by α-1,4 and α-1,6 glycosidic linkages). Starches can be classified according to their digestibility, which is generally characterized by the rate and the duration of the glycemic response (Singh et al., 2010, 2013). Most starches contain a portion that digests rapidly (rapidly digesting starch, RDS), a portion that digests slowly (slowly digesting starch, SDS), and a portion that is not hydrolyzed by the enzymes in the small intestine and passes to the large intestine and therefore is considered as resistant to digestion (resistant starch, RS) (Englyst et al., 1999). One of the most widely used methods to classify the starches was suggested by Englyst et al. (1992) and is based on the kinetics of in vitro starch digestion by simulating stomach and small-intestinal conditions and measuring glucose release at various times. Diets containing higher quantities of RDS raise the glucose levels in the blood more quickly than those containing SDS and RS (Lehmann and Robin, 2007).

The glycemic index (GI) is a measure of how much and how quickly a particular food elevates blood glucose levels (Gagné, 2008). Glucose is assigned a score of 100 and is used as the reference food. Foods with a GI value of over 70 are considered as high GI, foods with a GI of 56–69 as medium GI, and foods that have a GI of 55 and less are considered as low GI (ISO 26642, 2010). The glycemic load (GL) is a related measurement, which is the product of the amount of available carbohydrate in a serving and the GI of that food. The higher the GL, the greater the expected elevation in blood glucose and in the insulinogenic effect of the food (Foster-Powell et al., 2002). A GL of 20 or more is high, a GL of 11–19 is medium, and a GL of 10 or less is low (Gagné, 2008).

The microstructure of a food is highly dependent on the composition, processing, and post-processing storage and plays a vital role in determining the rate of starch digestibility in various foods (Bjorck, 1996; Tester et al., 2004; Singh et al., 2010). The microstructure and properties of cell wall polymers (nature of pectic materials, etc.) of natural foods such as potatoes are two important factors that can influence their processing, starch digestibility, and GI (Waldron et al., 1997; Singh et al., 2010). The physical texture of the food also affects starch digestion and the absorption of its hydrolysis products. Food matrix viscosity has been reported to be one of the major factors affecting enzymatic digestibility of starch and glycemic response (Dartois et al., 2010; Singh et al., 2010). Some food components such as polysaccharide-based gums increase the viscosity of food matrix significantly, even at a very low polymer concentration, and therefore alter the viscosity of digesta. This may decrease the postprandial carbohydrate absorption after ingestion of the starchy food. Additionally, a high viscosity food matrix may influence water availability, which is an important requirement for the enzymatic substrate reaction. Lipids and proteins are sometimes present naturally in starch granules or may be present in the food matrix as a part of its formulation. Both these components interact with starch during processing and influence the rate of glucose formation during starch digestion.

This chapter presents a detailed discussion of the above-mentioned factors that control the starch digestibility in cooked potatoes. This includes a review of the information on potato microstructure, viscosity, composition of food, processing techniques, and their relationship with starch digestion.

2. Starch Digestion and GI of Potatoes

Starch is mainly hydrolyzed by the mammalian amylolytic enzymes into glucose through several steps. Salivary α-amylase acts quite efficiently on starch in the mouth but is rapidly inactivated and degraded in the acidic environment of the stomach and hence plays a very minor role in the process of starch digestion. Starch-degrading enzymes are present in digestive fluids as well as in the brush border of the small intestine (Smith and Morton, 2001). The majority of starch hydrolysis is carried out by the pancreatic amylase, which is released in the small intestine via the pancreatic duct. α-Amylase catalyzes the hydrolysis (endo attack) of α-1,4 glycosidic bonds in amylose and amylopectin of starch (Lehmann and Robin, 2007). Both the linear and the branched (amylose and amylopectin) polymers of starch are hydrolyzed by virtue of the binding of their five glucose residues adjacent to the terminal reducing glucose unit to specific catalytic subsites of the α-amylase, followed by cleavage between the second and third α-1,4-linked glucosyl residue (Gray, 1992). The final hydrolysis products from amylose digestion are mainly maltose, maltotriose, and maltotetraose. However, α-amylase from some microbial sources may produce maltohexose and maltoheptose along with maltotriose (Yook and Robyt, 2002). α-Amylases have no specificity for α-1,6 branch linkage in amylopectin, therefore their capacity to break α-1,4 links adjacent to the branching point is decreased mainly by steric hindrance. The results obtained from analysis of the intestinal contents of humans suggest that hydrolysis products from amylopectin mainly consist of dextrins or branched oligosaccharides. The products obtained from α-amylase starch hydrolysis have been observed to possess α-anomeric configuration of the substrate (Kuriki and Imanka, 1999). The resulting oligosaccharides (maltose, maltotriose, and α-dextrins) are further hydrolyzed efficiently by the action of brush border enzymes of the intestine. The enzymes present in the human body are difficult to extract or expensive to buy, therefore enzymes from other mammals or from microorganisms are usually used in the in vitro systems that attempt to simulate the digestive process that occurs in the gastrointestinal tract of human beings. The mammalian enzymes are very similar to human enzymes, whereas the enzymes from microorganisms may work differently even though they are similarly classified.

Potatoes generally have higher starch digestibility and GI values than other foods (Tables 1 and 2). Patients with diabetes are generally advised to decrease their consumption of potatoes (Figure 1). Holt et al. (1995) calculated the satiety index (SI) score of various foods by dividing the area under the satiety response curve (AUC) for the test food by the group mean satiety AUC for white bread and multiplying by 100. They reported that the highest SI score was produced by boiled potatoes (323±51%, compared to 100% produced by white bread), which was double that produced by the foods known as low GI (baked beans and brown pasta). Most foods (76%) had an SI score greater than or equal to white bread. This shows that although boiled potatoes are known to be a high-GI food, they possess excellent ability to satisfy hunger and may be consumed as a diet food full of vitamins and minerals.

The differences in the starch digestibility and GI values of potatoes or potato products are attributed to different factors, including cultivar and maturity; cooking method (food processing and preparation); cooling after cooking; nature of the starch, particularly the amount of amylopectin; modification of the starch; and composition of the food matrix (Henry et al., 2005; Fernandes et al., 2005; Soh and Brand-Miller, 1999; Singh et al., 2010; Xavier and Sunyer, 2002). Considerable difference occurs in GI values among the potato varieties, from very low (23 for unspecified type) to very high (111 for baked Russets) (Foster-Powell et al., 2002). Henry et al. (2005) studied eight British potato varieties and divided them into high- and medium-GI potato varieties. Ramdath et al. (2014) reported a lower GI for the purple varieties and suggested that the GI of colored potatoes is significantly related to their polyphenol content, possibly mediated through an inhibitory effect of anthocyanins on intestinal α-glucosidase. Some of the studies have also shown that young potatoes have lower GI than mature potatoes due to differences in their starch structure (Soh and Brand-Miller, 1999).

The heat utilized, the amount of water, and the time of cooking all have a significant effect on the GI (Vaaler et al., 1984; Collings et al., 1981; Xavier and Sunyer, 2002). Najjar et al. (2004) and Tahvonen et al. (2006) reported significant differences in the GI of potato, depending on its temperature at consumption. Cooling the potato after cooking produced a significantly lower GI value than consuming it hot, which could be related to amylose retrogradation and an increase in RS content during cooling. Fernandez et al. (2005) also reported that

Table 1: The Average GI of 62 Common Foods Derived from Multiple Studies by Different Laboratories.

High-Carbohydrate Foods		Breakfast Cereals		Fruit and Fruit Products		Vegetables	
White wheat bread[a]	75±2	Cornflakes	81±6	Apple, raw[b]	36±2	Potato, boiled	78±4
Whole wheat/whole meal bread	74±2	Wheat flake biscuits	69±2	Orange, raw[b]	43±3	Potato, instant, mashed	87±3
Specialty grain bread	53±2	Porridge, rolled oats	55±2	Banana, raw[b]	51±3	Potato, french fries	63±5
Unleavened wheat bread	70±5	Instant oat porridge	79±3	Pineapple, raw	59±8	Carrots, boiled	39±4
Wheat roti	62±3	Rice porridge/congee	78±9	Mango, raw[b]	51±5	Sweet potato, boiled	63±6
Chapatti	52±4	Millet porridge	67±5	Watermelon, raw	76±4	Pumpkin, boiled	64±7
Corn tortilla	46±4	Muesli	57±2	Dates, raw	42±4	Plantain/green banana	55±6
White rice, boiled[a]	73±4			Peaches, canned[b]	43±5	Taro, boiled	53±2
Brown rice, boiled	68±4			Strawberry jam/jelly	49±3	Vegetable soup	48±5
Barley	28±2			Apple juice	41±2		
Sweet corn	52±5			Orange juice	50±2		
Spaghetti, white	49±2						
Spaghetti, whole meal	48±5						
Rice noodles[b]	53±7						
Udon noodles	55±7						
Couscous[b]	65±4						

Dairy Products and Alternatives		Legumes		Snack Products		Sugars	
Milk, full fat	39±3	Chickpeas	28±9	Chocolate	40±3	Fructose	15±4
Milk, skim	37±4	Kidney beans	24±4	Popcorn	65±5	Sucrose	65±4
Ice cream	51±3	Lentils	32±5	Potato crisps	56±3	Glucose	103±3
Yogurt, fruit	41±2	Soya beans	16±1	Soft drink/soda	59±3	Honey	61±3
Soy milk	34±4			Rice crackers/crisps	87±2		
Rice milk	86±7						

Data are means±SEM.
[a]Low-GI varieties were also identified.
[b]Average of all available data.
Source: Fiona et al. (2008).

Table 2: Starch Digestibility of Various Starches and Starch-Based Foods.

Source	Starch Digestibility	Reference and Method
Cooked potatoes	80–95[a]	Bordoloi et al. (2012b); in vitro gastric with pepsin, and small intestinal digestion with pancreatin and amyloglucosidase
Sorghum meal	60–85[b] and 40–47[c]	Wong et al. (2009); in vitro starch digestion with pepsin and without pepsin pretreatment using pepsin (porcine stomach mucosa) and α-amylase (bacterial; porcine pancreas; human saliva)
Cooked rice noodle dough	43[d] and 33[d]	Koh et al. (2009); in vitro digestion using α-amylase from *Aspergillus oryzae*
Waxy maize starch	100[e]	Han and BeMiller (2007); in vitro digestion by the method of Englyst et al. (1999) using pepsin, pancreatin, and amyloglucosidase
Chemically modified waxy maize starch	76–87[e]	
Normal maize starch	99[e]	
Chemically modified normal maize starch	71–96[e]	
Potato starch	96[e]	
Chemically modified potato starch	67–76[e]	
Wheat flour	72[f]	Englyst et al. (1999); in vitro digestion using pepsin, pancreatin, and amyloglucosidase
Corn flakes	81[f]	
Cooked rice	70–80[g]	Frei et al. (2003); in vitro digestion by the method of Goni et al. (1997) using pepsin, α-amylase, and amyloglucosidase from *Aspergillus niger*
Extruded amaranth seeds	93[h]	Capriles et al. (2008); in vitro digestion by the method of Goni et al. (1997) using pepsin, α-amylase, and amyloglucosidase
Amylose–lipid complexes	48–71[i]	Crowe et al. (2000); in vitro digestion using α-amylase and amyloglucosidase
Legume starches	80–90[j]	Hoover and Zhou (2003); treatment with porcine pancreatic α-amylase
Extruded beans	290–306[k]	Alonso et al. (2000); in vitro digestion with pancreatic amylase
Autoclaved legumes	87–89[l]	Rehman and Shah (2005); in vitro digestion with pancreatic α-amylase

[a]Expressed as rapidly and slowly digestible starch (%).
[b]Expressed as mg reducing sugar/h (with pepsin pretreatment).
[c]Expressed as mg reducing sugar/h (without pepsin pretreatment).
[d]Expressed as mg of maltose equivalent liberation/g of dough.
[e]Expressed as digestibility (%).
[f]Expressed as total glucose after 120-min incubation with enzymes.
[g]Expressed as digestible starch (%).
[h]Expressed as hydrolysis index.
[i]Expressed as conversion to glucose (%).
[j]Expressed as hydrolysis (%).
[k]Expressed as starch digestibility (mg of maltose/g).
[l]Expressed as starch digestibility (%).

Reproduced and modified from Singh et al. (2010); with permission from Elsevier.

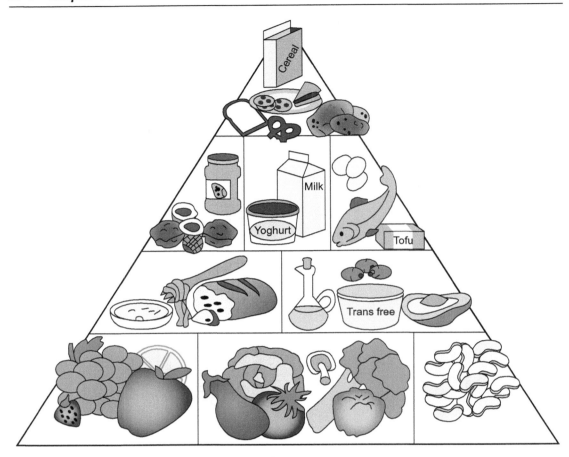

Figure 1
Low-GL pyramid. *Reproduced from Ludwig (2007), with permission from Elsevier.*

precooking and reheating or consuming potatoes cold (e.g., potato salad) may result in a reduced glycemic response. Repeated cycles of heating and cooling have been reported to lead to increased RS content (Kingman and Englyst, 1994).

Starch can also form insoluble complexes with proteins, such as occurs in the browning (Maillard) reaction, making it unavailable for digestion and absorption. Xavier and Sunyer (2002) reported that the more a starch-containing food is heated, moisturized, ground, or pressed, the more it will be amenable to hydrolysis and digestion, except for the portion that forms insoluble complexes. Different methods of cooking, such as boiling, baking, micro-wave cooking, frying, and extrusion, result in different degrees of starch gelatinization and the crystallinity of starch in potato (Table 3).

Tahvonen et al. (2006) reported that although cooling after cooking lowers the GIs of processed potato products, peeling, cubing, slicing, and mashing have only minor effects. Steam boiling or baking in an oven with added water does not significantly affect differences in glycemic response.

Table 3: Incremental AUC and GI Values for 50 g Available Carbohydrate Portions of White Bread and Seven Potatoes Tested by a Cohort of 12 Subjects.

Potatoes Tested	AUC (mmol × min/l)	GI	
		White Bread = 100	Glucose = 100[a]
		←*Mean ± Standard Error of Mean*→	
White bread	174 ± 18[xy]	100[xyz]	71[xyz]
Baked Russet potato	178 ± 25[xy]	107.7 ± 12.3[xyz]	76.5 ± 8.7[xyz]
Instant mashed potato	206 ± 23[x]	123.5 ± 11.3[xy]	87.7 ± 8.0[xy]
Roasted California white potato	165 ± 20[xy]	101.8 ± 11.6[xyz]	72.3 ± 8.2[xyz]
Baked PEI[b] white potato	178 ± 21[xy]	102.5 ± 6.4[xyz]	72.8 ± 4.5[xyz]
Boiled red potato (hot)	208 ± 20[x]	125.9 ± 10.1[x]	89.4 ± 7.2[x]
Boiled red potato (cold)	135 ± 18[y]	79.2 ± 7.4[z]	56.2 ± 5.3[z]
French fried potatoes	155 ± 19[xy]	89.6 ± 7.7[yz]	63.6 ± 5.5[yz]

xyz, means in the same column not sharing the same superscript letter differ significantly ($P < 0.05$).
[a]GI classification (USDA, 2002): low, <55; intermediate, 55–69; high, 70–100.
[b]PEI = Prince Edward Island.
Reproduced from Fernandez et al. (2005), with permission from Elsevier.

Heacock et al. (2002) reported that a small dose of fructose (which is easily obtained in the diet via fruits) consumed 30 or 60 min before consuming mashed potatoes (instant) decreases the glycemic response compared with either immediate or no fructose treatments. Giacco et al. (2001) have evaluated and compared, in type 2 diabetic patients, plasma glucose responses to 50 g available carbohydrate provided by white bread and three other different starchy foods frequently consumed in Italy: pizza, potato dumplings (gnocchi), and bread crisps. The GI of potato dumplings, a typical potato–wheat food that is often used as a substitute for pasta, was found to be significantly lower than the other test foods (GI 74% compared to 114% and 104% for pizza and bread crisps; on the white bread scale, 100). The GI of the potato dumplings was lower than that previously reported for freshly boiled potatoes of the same cultivar, suggesting that mixing of potatoes with wheat starch during gnocchi preparation has an impact on their GI. The authors observed, through scanning electron microscopy, that the structure of gnocchi was compact compared to the porous structure of pizza, white bread, and bread crisps, which might have restricted the action of digestive enzymes, resulting in lower GI (Giacco et al., 2001; Riccardi et al., 2008). Chemically modifying potato starch, through acetylation and the addition of β-cyclodextrin to stabilize the carbohydrate, has also been reported to lower GI (Raben et al., 1997).

3. Potato Microstructure and Starch Digestion

The microstructure of natural foods, such as potatoes, and the nature of their cell wall materials have been reported as important factors that can influence the deformation occurring during mastication or mechanical processing (Waldron et al., 1997). Starch exists in carbohydrate foods in the form of large granules. These granules must be disrupted, through processing, so that the amylose or amylopectin starch macromolecules become available for hydrolysis. Grinding, rolling, pressing, or even thoroughly chewing starchy food has been reported to disrupt the granules (Xavier and Sunyer, 2002).

Depending on the middle lamella characteristics of parenchymatous tissue, the cells can either separate or burst at the point of minimum resistance (Singh et al., 2005). Raw foods generally show cell rupturing, whereas cooked ones show cell separation due to destabilization of pectic materials during thermal processing (Aguilera and Stanley, 1990).

Primary cell walls of growing and fleshy tissues have a conserved general composition of cellulose, hemicelluloses, and pectin (Chanda, 2005). The noncellulosic material acts as a "glue" that holds the microfibrils of cellulose together, which in turn is responsible for the stability of cell walls (Carpita and Gibeaut, 1993). The starch granules in beans are present in the cotyledon cells and are embedded in the protein matrix of the cellular contents (Daussant et al., 1983). This situation restricts the complete swelling of starch during gelatinization owing to steric hindrance and other limiting effects, including restricted water availability. Wursch et al. (1986) pointed out that the thick and mechanically resistant nature of the cotyledon cell walls in legumes prevents complete swelling of starch granules during gelatinization, which may restrict their interaction with digestive enzymes.

Depending on the botanical origin, physicochemical characteristics, and type of processing, starch-based carbohydrates are hydrolyzed at different rates and to different extents in vitro and in vivo (Cummings et al., 1997; Singh et al., 2010). Cooking or thermal treatment during processing of starch leads to an increase in the rate of hydrolysis by gelatinizing the starch and making it more easily available for enzymatic attack during digestion.

3.1 Microstructural Characteristics of Potato and Starch Digestion

3.1.1 Microstructural Characteristics of Potato

Microscopic and rheological techniques provide important information on the structural organization of foods and some researchers have applied these techniques to study the microstructure and rheological properties of potatoes (Singh et al., 2005, 2008a, 2009a,b). In a 2012 study carried out on various potato cultivars, light microscopy was used to reveal apparent differences in the microstructure of tuber parenchyma (Bordoloi et al., 2012a; Figure 2). The cultivar Red Rascal tuber parenchyma cells were elongated and hexagonal, whereas those from Agria appeared roughly spherical. Agria and Nadine tuber parenchyma cell size was generally larger than those of the other two potato cultivars. Moonlight and Red Rascal parenchyma showed a very regular and defined arrangement of cells, in contrast to Agria, which had an irregular cellular arrangement. Raw parenchyma cellular compartments were generally filled with starch granules of a range of shapes and sizes. Each cell also showed few mature starch granules and numerous tiny structures (Figure 2(a), (c) and (g)), resembling the "immature starch granules" as explained by Singh et al. (2005). The number and size of

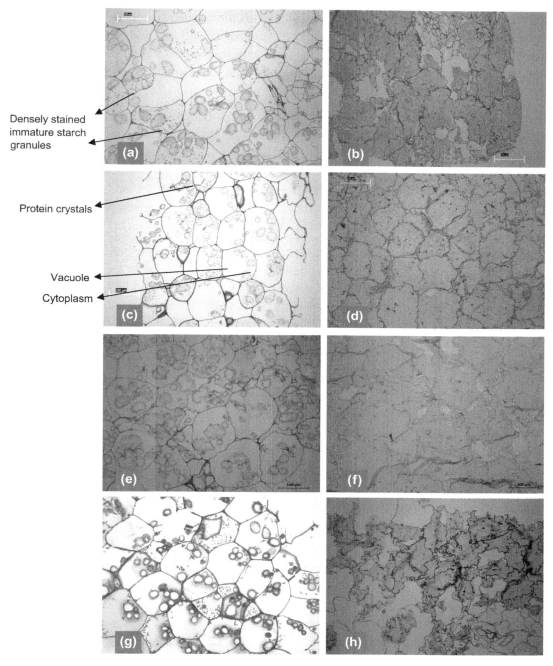

Densely stained immature starch granules

Protein crystals

Vacuole

Cytoplasm

Figure 2
Light micrographs of raw (left) and cooked (right) tuber parenchyma from Nadine (a, b), Moonlight (c, d), Agria (e, f), and Red Rascal (g, h) potato cultivars. *Reproduced from Bordoloi et al. (2012a), with permission from Elsevier.*

mature starch granules per cell vary from cultivar to cultivar. Mealy potatoes have been reported to contain higher starch and amylose contents, as well as a higher percentage of large starch granules (diameter >50 μm), than the waxy cultivars (Barrios et al., 1963). Potatoes with larger cell size have been reported to exhibit larger mean starch granule size and vice versa (Singh et al., 2005).

The swelling and gelatinization of starch granules during cooking exert pressure on the cell walls and thus play an important role in determining potato texture after cooking. Cooking of potatoes also affects noncellulosic matrix, and the extents of deformation and integrity of cell walls vary among cultivars. Potato tuber parenchyma retains the cell wall outline after cooking and the cells are filled with gelatinized starch matrix, as observed for cultivar Moonlight and to some extent for Agria (Figure 2(d) and (f)). However, cooked potato parenchyma in some cultivars also showed disintegrated structures (Figure 2(b) and (h)).

Transmission electron microscopy is a very helpful technique for understanding the changes taking place in the starch and cell wall material during cooking (Figures 3 and 4). The cell wall of raw tuber parenchyma cells was observed to be made up of middle lamella and the primary cell wall (Figure 3(a) and (d)). The middle lamella is mainly composed of pectic substances, whereas the primary cell wall has been reported to be made of cellulose molecules arranged into thin hair-like strands called microfibrils. The microfibrils are arranged in a meshwork pattern along with other components such as hemicellulose, glycans, and pectins, which link them together and help strengthen the cell wall (Raven et al., 2005). Many cytoplasmic organelles were observed in the raw potato parenchyma cells, such as starch granules, mitochondria, Golgi apparatus, amyloplasts, generative cells, and lipid droplets (Figure 3(a)). Plasmodesmata and the pit fields were clearly observed in the cell walls and some densely stained material was also observed along the tonoplast (Figure 3(b) and (c)). Some of the large starch granules showed electron-dense radial "channels" around them, which in some cases crossed the whole granule (Figure 3(a)). Some starch granules also showed broken amyloplast membrane around their surface (figure not shown). Similar starch granules have been reported for other plant sources demonstrating starch degradation (Appenroth et al., 2011).

Upon processing of potatoes, the cell wall material degraded partially, resulting in loosening of the microfibrils. The cell wall of parenchyma cells decreased in thickness after cooking, probably owing to the loss of primary cell wall to a greater extent. Middle lamella and some remains of the primary cell wall were still observed (Figure 3(a)–(c) and (e)). The remains of the primary cell wall along with some electron-dense granular structures were observed floating in the cytoplasmic starchy matrix (Figure 3(d)). Pectic material has been reported to degrade during cooking and be partly solubilized into the cooking medium (Hughes et al., 1975a,b; van Marle et al., 1997). This degradation greatly influences intercellular adhesion and the structure of the remaining cell walls, which are both important texture parameters (van Marle et al., 1992, 1997).

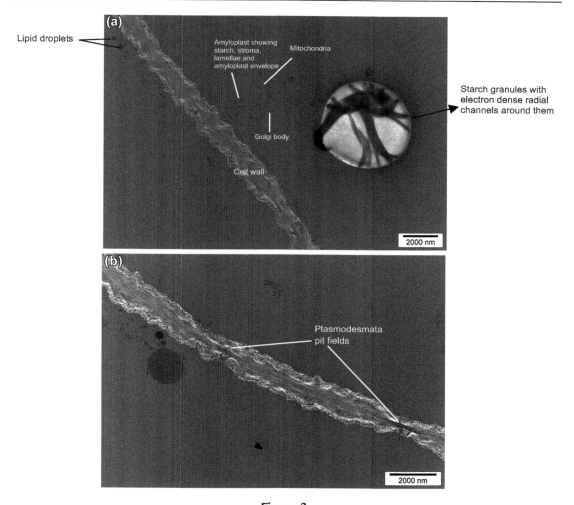

Figure 3

(a–d) Transmission electron micrographs of raw tuber parenchyma cells from Nadine potato cultivar. *Reproduced from Bordoloi et al. (2012a), with permission from Elsevier.*

3.1.2 Starch Digestion In Vitro and Microstructure of Digesta

Starch hydrolysis (%) in cooked potatoes from the above-discussed cultivars was studied by Bordoloi et al. (2012b) during in vitro gastrointestinal digestion and is presented in Figure 5. The first 30 min of hydrolysis represented gastric conditions at pH 1.2, during which some percentage of hydrolysis was observed, possibly due to the acid hydrolysis. When simulated intestinal fluid (SIF; pH 6.8) was added to the reaction mixture, the starch was rapidly digested by the pancreatic amylases. Approximately 70% of the starch in cooked potatoes was digested within the first 10 min of simulated small-intestinal digestion. The starch hydrolysis

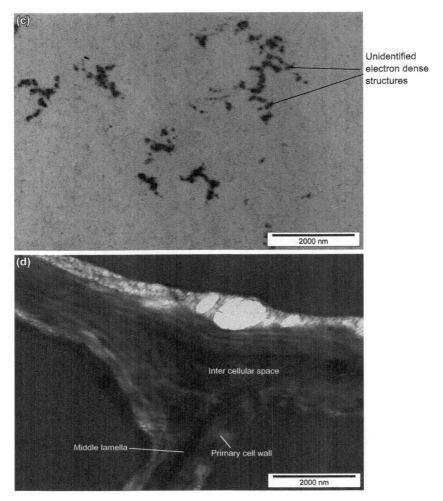

Figure 3   Cont'd.

level ranged between 80 and 95% for different cultivars at the end of small-intestinal digestion. However, this percentage may not be interpreted as the percentage of hydrolysis of cooked potatoes in vivo as the this process is much more complex.

The observed differences among the starch hydrolysis of cooked potato cultivars could be attributed to the interplay of many factors, such as starch characteristics, microstructure of the food, susceptibility of starch to hydrolysis, extent of starch gelatinization, and molecular association between starch components (Snow and O'Dea, 1981; Hoover and Sosulski, 1985; Tester et al., 2004; Singh et al., 2010; Berg et al., 2012). Under simulated small-intestinal conditions, the starch in cooked potatoes was hydrolyzed in a manner similar to that of pure

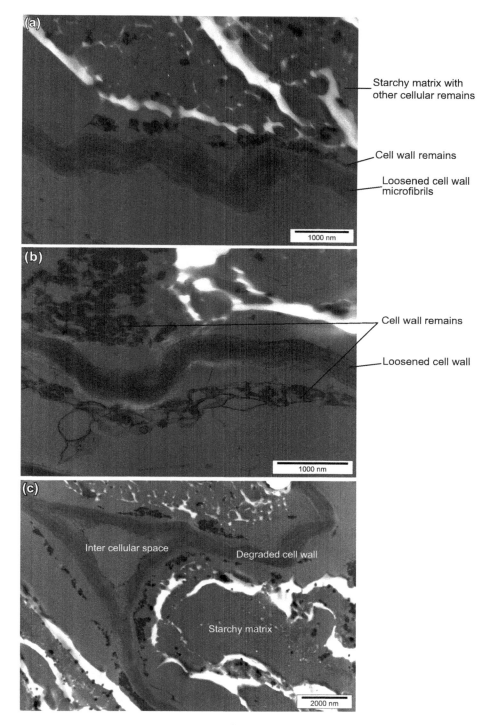

Figure 4

(a–e) Transmission electron micrographs of cooked tuber parenchyma cells from Nadine potato cultivar showing the loss of cell wall integrity after cooking. A decrease in cell wall thickness and loosened cell wall microfibrils are clearly observed. Cytoplasm consists of gelatinized starchy matrix along with cell wall remains and some other granular structures. *Reproduced from Bordoloi et al. (2012a), with permission from Elsevier.*

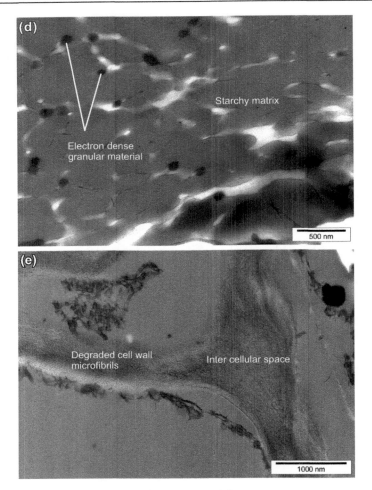

Figure 4 Cont'd.

starch as observed in previous studies (Dartois et al., 2010). However, the starch hydrolysis (%) values in cooked potatoes were slightly lower than those of pure starch throughout the small-intestinal digestion in vitro. The lower levels of hydrolysis could be attributed to the presence of cell wall and other components in the potatoes as reported in the earlier studies (Singh et al., 2010). Other components present in food matrix such as cell wall materials, dietary fiber, polysaccharides, proteins, and viscosity of food matrix have been shown to have an inhibitory effect on starch hydrolysis (Jenkins et al., 1987; Timothy et al., 2000; Rehman and Shah, 2005; Singh et al., 2010).

Still images of the starch hydrolysis in cooked potatoes under simulated small-intestinal conditions are presented in Figure 6(a)–(d). As starch present in potato tuber is quite prone to cooking, most of the granules were dissolved during cooking and only starch

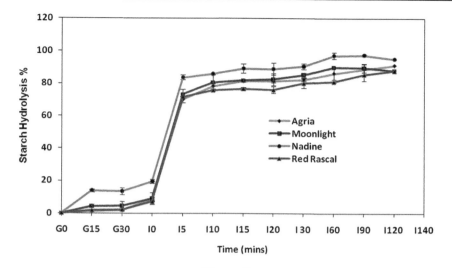

Figure 5

Starch hydrolysis (%) of cooked potatoes from various cultivars during simulated small-intestinal digestion. *Reproduced from Bordoloi et al. (2012b), with permission from Elsevier.*

granule remnants were observed inside the undigested potato tuber cells (Figure 6(a)). However, the cell walls appeared intact and retained their normal morphology after cooking to the optimum levels. Hydrolysis of the starch by SIF-containing pancreatic amylases led to the digestion of starch and its remnants progressively, as evidenced by the homogeneous background of empty cells (Figure 6(b)). After 45 min of simulated small-intestinal digestion, most of the tuber cells showed an empty cavity that is created by the hydrolysis of starch by the digestive enzymes (Figure 6(c) and (d)). The cell walls stayed intact during and after the digestion, which showed that SIF had no effect on the cooked potato tuber cell walls, which are generally made up of cellulose and hemicellulose materials.

Figures 7(a)–(d) and 8(a)–(d) show scanning electron micrographs of freeze-dried samples taken during in vitro small-intestinal digestion of freshly cooked potato. The regular microcellular structure of cooked potato tuber appears to have been maintained to a good extent during and after cooking. A good level of cell integrity is observed in all the samples. The tuber cells appear to have shrunk slightly and show indentation and wrinkles on their surface. The cell wall of the undigested sample shows fewer wrinkles compared to the samples subjected to small-intestinal digestion. Hydrolysis of starch during in vitro digestion and removal of water during freeze-drying might have left an empty space in which the cell wall and some undigested components could have folded in during freeze-drying, causing more wrinkles and indentation. This phenomenon was especially distinctive in cells that underwent the complete 120-min simulated intestinal

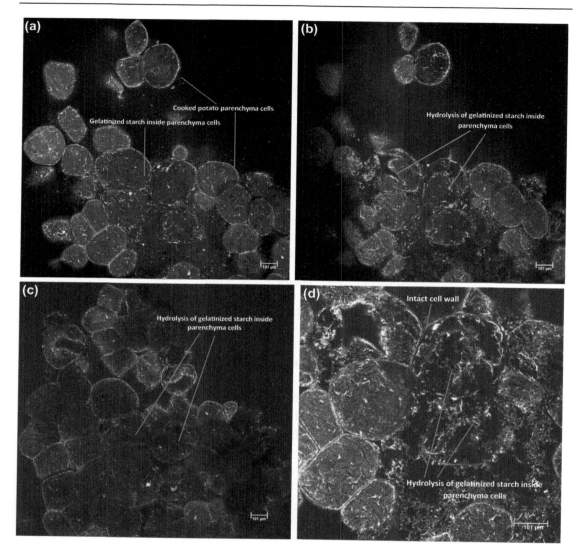

Figure 6
Confocal laser scanning microscopy images of real-time in vitro small-intestinal digestion carried out on cooked Agria potatoes. (a) Image captured before starting in vitro digestion. (b) Image captured during in vitro digestion process. (c, d) Images captured at the end (after 45 min) of simulated small-intestinal digestion, at different magnifications (bar 101 μm). *Reproduced from Bordoloi et al. (2012b), with permission from Elsevier.*

digestive process (Figure 8(d)). However, the possibility of some artifacts produced by freeze-drying cannot be ruled out. The cells stayed mainly intact during the enzymatic action up to 30 min (Figure 7). Figure 7(b) shows the inside of the tuber cell, which appears like a honeycomb showing where starch might have been digested during simulated small-intestinal digestion.

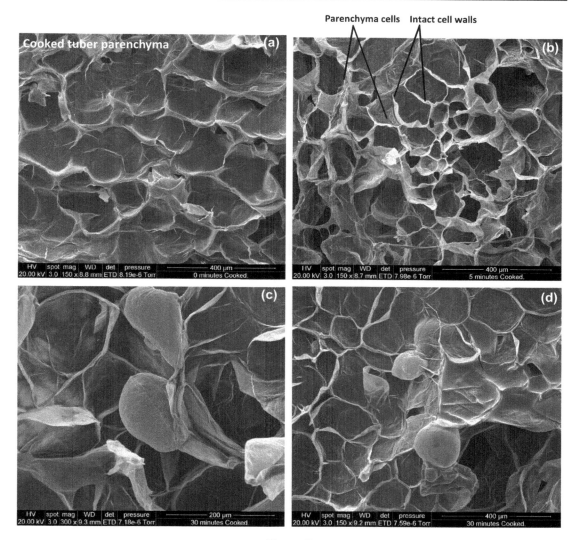

Figure 7

Scanning electron micrographs taken during in vitro small-intestinal digestion of starch in cooked Agria potatoes. (a) Sample taken after 0 min, (b) 5 min, and (c, d) 30 min (at different magnifications). As artifacts produced by freeze-drying are present, these micrographs do not necessarily represent the structure prior to freeze-drying. *Reproduced from Bordoloi et al. (2012b), with permission from Elsevier.*

4. Rheology of Food Matrix and Starch Digestion

Food matrix viscosity has been reported to be one of the major factors affecting enzymatic digestibility of starch and glycemic response (Lightowler and Henry, 2009; Dartois et al., 2010; Singh et al., 2010). The physical texture of the food may affect the digestion of starch

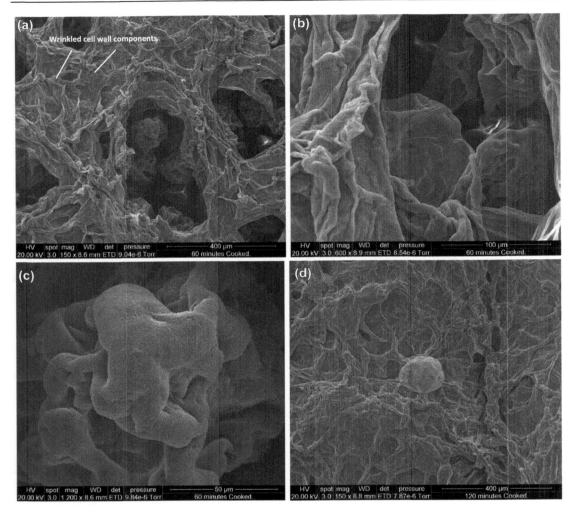

Figure 8

Scanning electron micrographs taken during in vitro small-intestinal digestion of starch in cooked Agria potatoes. (a–c) Samples taken after 60 min (at different magnifications) and (d) 120 min. As artifacts produced by freeze-drying are present, these micrographs do not necessarily represent the structure prior to freeze-drying. *Reproduced from Bordoloi et al. (2012b), with permission from Elsevier.*

and the absorption of hydrolysis products. Plant-derived polysaccharides such as gums are used in food products as thickening, emulsifying, and stabilizing agents and also because of their beneficial effects as soluble dietary fiber (Williams and Phillips, 2003). Soluble dietary fiber like guar gum has the ability to produce high viscosity (at low concentration), thereby significantly affecting the nutrient absorption and postprandial plasma nutrient levels in the gut (Cherbut et al., 1990; Eastwood and Morris, 1992; Ellis et al., 1996; Edwards et al., 1998). The presence of galactomannan-based gums also imposes restrictions on the

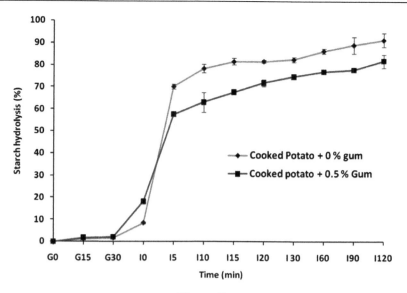

Figure 9
Effect of the addition of guar gum (0.5%) on the starch hydrolysis (%) of cooked Agria potatoes during simulated gastric (for 30 min) followed by small-intestinal (for 2 h) digestion. *Reproduced from Bordoloi et al. (2012b), with permission from Elsevier.*

availability of water molecules and swelling or gelatinization of starch granules and reduces the size of granule remnants in the starch paste, if cooked together (Kaur et al., 2008; Kaur and Singh, 2009).

Bordoloi et al. (2012b) reported that addition of guar gum (0.5%) to the cooked potato matrix led to a significant decrease in both the rate and the extent of final starch hydrolysis (Figure 9). A drop of ~20% in the overall starch hydrolysis (to that of the control) was observed after the first 15 min of hydrolysis under simulated intestinal conditions when guar gum was added. Guar gum also affected the final hydrolysis of the starch significantly ($P < 0.05$), with an ~15% drop in hydrolysis at the end of the simulated digestion period. The rate and extent of starch hydrolysis in the small intestine are dependent upon several intrinsic and extrinsic factors (Englyst et al., 1992). Gums have been reported to produce high viscosity in the gut lumen, which in turn influences the nutrient absorption and postprandial plasma nutrient levels (Edwards, 2003). The slower rate of starch hydrolysis in the presence of guar gum could be attributed to the increase in viscosity of the digesta owing to the enlargement of fully hydrated galactomannan chains of guar gum (Ellis et al., 1995). Dartois et al. (2010) reported that the gum layer around the starch granules could limit access of enzymes to starch, consequently decreasing the enzymatic starch hydrolysis. Hydrocolloids have been reported to form a continuous network by suspending the starch matrix in a coherent gel, which acts as a barrier to the access of enzymes to starch (Koh et al., 2009). Thus, guar gum may act as a physical "barrier" to the interactions of digestive enzymes and starch and/or to the release of hydrolysis products into the aqueous phase of the digesta. This was also evidenced from the hydrolysis levels observed immediately after the addition of simulated intestinal juices in the study of Bordoloi et al. (2012b). The hydrolysis levels for the potato-only sample reach near 70%, whereas the levels observed for the sample containing guar gum were about 55% at 5 min of in vitro intestinal digestion (Figure 9). Gularte and Rosell (2011) reported that a combination of pure potato starch pastes containing guar gum was slowly hydrolyzed, which results in lower glucose liberation under in vitro conditions. Guar gum has also been reported to decrease the abundance of the granule remnants, or ghosts, in the starchy paste by inhibiting the starch components from leaching out of the starch granule (Nagano et al., 2008). Guar gum increased the viscosity of the cooked potato system significantly, which might have affected not only the mass transfer of the molecules (sugars and enzymes), but also the enzyme substrate reactions and hydrolysis kinetics (Bordoloi et al., 2012b; Gularte and Rosell, 2011).

Water availability plays an important role during starch hydrolysis. The hydrophilic nature of guar gum also limits the availability of water for enzyme–substrate reactions, thereby reducing the overall rate of starch hydrolysis. The restrictions imposed on the swelling of the starch granules during gelatinization by the galactomannan-based gums have been observed to result in an alteration of the microstructural and rheological properties of starch gum pastes (Kaur et al., 2008).

The effect of adding guar gum (at various concentrations) on the net apparent glucose absorption has also been reported in in vivo studies on growing pigs (Ellis et al., 1995). Feeding the pigs meals containing guar gum resulted in increased zero-shear viscosity of jejuna digesta along with a significant reduction in the rate of glucose absorption. This postprandial effect of guar gum resulted from the gum's capacity to increase the viscosity of digesta within the gastrointestinal tract owing to the enlargement of fully hydrated galactomannan chains. This whole phenomenon reduces the rate of digestion and absorption of carbohydrates and therefore lowers the postprandial rise in blood glucose. The presence of galactomannan in the starch mixture imposes restrictions on the swelling of starch granules during gelatinization, which results in the size reduction of starch granule remnants in the cooked starch paste, whereas some of the granules may also not gelatinize properly because of less availability of water molecules to starch granules (Kaur et al., 2008). This incomplete gelatinization of starch granules may also increase their resistance to enzymatic hydrolysis. The viscous fiber derived from gums such as guar or tragacanth increases the viscosity even at relatively low polymer concentrations in the food matrix, which may increase the overall viscosity of digesta in the gastrointestinal tract. The consequence is a decrease in the postprandial carbohydrate absorption after ingestion of starchy food.

Brennan et al. (1996) studied the addition of guar gum in white bread and studied its microstructure along with in vitro and in vivo digestibility. They observed that the blood-glucose-lowering action of gum is due to its ability to act as a physical barrier to starch digestion along with increasing the viscosity of digesta. The association between guar galactomannan and starch has also been confirmed through the microstructural analysis of pig digesta.

Lightowler and Henry (2009) investigated mashed potatoes containing 1%, 2%, or 3% levels of high-viscosity hydroxypropylmethylcellulose and observed significant reduction in glycemic responses in all samples compared to mashed potato with no added fiber. The presence of nonstarch polysaccharides from various plant sources may affect the physical properties of the digesta at all sites of the gastrointestinal tract. The high level of viscosity slows down many of the physiological processes associated with the digestion of foods and absorption of nutrients and thus helps in improving the management of glucose intolerance and obesity. The well-documented blood-glucose-reducing effect of dietary supplements of water-soluble nonstarch polysaccharides such as pectin and guar gum also depends on their capacity to increase the viscosity of digesta in the stomach and the small intestine (Johansen et al., 1996). Gastric response to increased meal viscosity has been assessed by techniques such as echoplanar magnetic resonance imaging (Marciani et al., 2000).

4.1 Rheology (Flow Behavior) of Potato Digesta

Detailed measurements of the viscosity of cooked potatoes (with or without the addition of gum) during in vitro small-intestinal digestion have been recorded by Bordoloi et al. (2012b) using a dynamic rheometer (Figures 10 and 11). The viscosity of cooked potatoes dropped considerably as SIF was added, owing to the conversion of highly viscous starch in potatoes

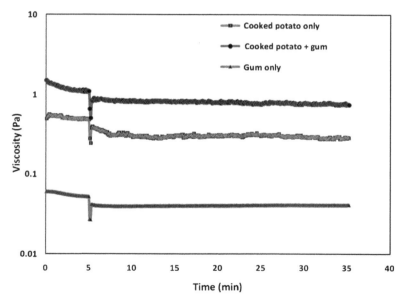

Figure 10

Changes in dynamic rheological properties of cooked Agria potato, cooked potato plus gum, and gum only in the absence of intestinal enzymes at 37 °C and 1 Hz, measured during time-sweep experiments. *Reproduced from Bordoloi et al. (2012b), with permission from Elsevier.*

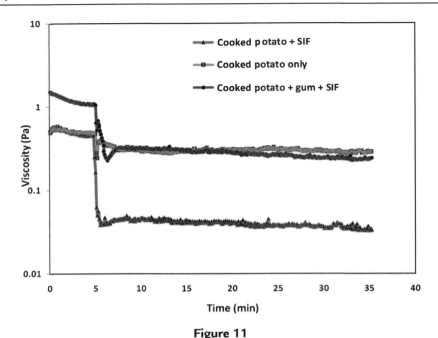

Figure 11

Changes in dynamic rheological properties of cooked Agria potato and cooked potato plus gum in the presence or absence of intestinal enzymes at 37 °C and 1 Hz, measured during time-sweep experiments. *Reproduced from Bordoloi et al. (2012b), with permission from Elsevier.*

to low-viscosity sugars by the digestive enzymes present in SIF (Figure 11). However, the viscosity stayed stable afterward during the whole course of simulated small-intestinal digestion of cooked potatoes.

The addition of guar gum resulted in an increase in viscosity of the cooked potatoes, which could mainly be attributed to the viscosity imparted by the guar galactomannan that stabilized the structure (Figure 10). The viscosity of cooked potato–guar gum mixture was more than 1 Pa throughout the time-sweep experiment. The increase in viscosity can affect gastric function and may inhibit propulsive and mixing effects generated by peristalsis during in vivo digestion (Ellis et al., 1995). The viscosity of cooked potato–guar gum mixture was decreased when SIF (containing enzymes) was added (Figure 11). However, the extent of decrease in viscosity was far less than that observed for cooked potato-only samples. The curves for the control cooked potato sample and cooked potato–guar gum sample (containing SIF with enzymes) overlapped each other (Figure 11). The significant decrease that occurred during the digestion of the cooked potato-only sample was balanced by the presence of gum that helped the matrix to maintain a stable viscosity. Also, in the presence of gums, the interactions between substrates and digestives enzymes are less frequent, which could be another reason for a smaller decrease in viscosity. This may result in a decreased rate of starch digestion and ultimately a slower absorption of the hydrolysis products (e.g., maltose, α-limit dextrins). Solution viscosity has also been reported to influence the enzyme kinetics (Uribe and Sampedro, 2003).

The ionic interactions between starch and gum play an important role in viscosity and visco-elasticity characteristics. The strong electrostatic interactions between cationic starch and anionic gum result in an instantaneous aggregation of granules, whereas nonionic gums form a sheet structure and loosely wrap the granules (Chaisawang and Suphantharika, 2005). These observations are related to the changes in rheological parameters such as storage modulus (G') and loss modulus (G'') of starch–xanthan and the starch–guar systems (Chaisawang and Suphantharika, 2005). However, to our knowledge, no reports are available on the effects of ionic interactions of the gums during in vitro starch hydrolysis. The addition of some cereal-based viscous fiber in meals also influences the digestion and absorption of carbohydrates to a considerable extent. However, the postprandial glucose response may be slightly different as viscous fiber is physically and functionally different from the viscous gums.

In addition, it may be difficult to distinguish among the two effects, whether a decrease in the rate of hydrolysis from the inhibition of enzymes or an increase in the viscosity of gastro-intestinal contents (acting on the mass transfer). An inhibitory action of a very small concentration (0.5%) of galactomannan addition on amylase acting on a range of concentrations of gelatinized starch was observed by Slaughter et al. (2002).

5. Formulated Foods and Starch Digestion

5.1 Influence of Food Matrix Composition on Starch Digestion

The composition of the food matrix also affects the digestion of starch significantly (Singh et al., 2010). Therefore the rate of starch digestion differs with different foods with varying composition. The presence of protein in the food matrix may influence the rate of starch digestion. Digestibility of starches and proteins in various food products is significantly affected by their interaction with one another. The functional properties and starch digestibility have been observed to be influenced by the presence of even small amounts of protein in food products (Ezeogu et al., 2008). Protein fractions such as albumin, globulins, and glutenins help in gluing the protein bodies into a matrix surrounding starch granules that may act as a barrier to starch digestibility (Hamaker and Bugusu, 2003). Cooking or processing may sometimes reduce the starch digestibility, as conformational changes in proteins may occur that could facilitate the formation of disulfide-linked polymers (Oria et al., 1995). The presence of a protein barrier surrounding the starch granule was confirmed by the addition of pronase enzyme to hydrolyze the protein matrix, and a significant enhancement in in vitro starch digestibility was observed afterward owing to the clearance of passage for amylase and amyloglucosidase (Rooney and Pflugfelder, 1986). Wong et al. (2009) reported that a greater abundance of disulfide-bonded proteins and the presence of nonwaxy starch and the granule-bound starch synthase enzyme may affect the digestibility of both starch and protein. Another study by Choi et al. (2008) reported an increase in in vitro starch digestibility when sodium sulfite was added during cooking of waxy sorghum flour. Reducing agents such as sodium sulfite or bisulfite may prevent the formation of enzyme-resistant disulfide-linked plant polymers, which facilitates an easy access of amylolytic enzymes to the starch granule. Effects of the protein matrix on in vitro starch digestibility of processed starch products such as pasta have also been reported (Kim et al., 2008). Jenkins et al. (1987) studied the effects of starch–protein interaction on starch digestibility. Their reports suggested that the occurrence of a starch protein interaction may account for a decreased glycemic response and reduced rate of digestion. Subsequent studies verified that significant amounts of starch may indeed enter the colon (Wolever et al., 1986). It has been observed that removing gluten from wheat flour eliminated starch malabsorption, but this effect was not reversed by subsequently adding back the gluten to the gluten-free flour. This raised the question of whether the natural starch–protein interaction is responsible for the reduced digestibility of starch (Jenkins et al., 1987). The protein network may inhibit the rate of hydrolysis in the lumen of the small intestine (Jenkins et al., 1987). The addition of toppings or fillings based on cottage cheese, baked beans, and tuna has been reported to reduce the GI of potato-, pasta-, and toast-based meals (Henry et al., 2006).

The amylose chain of starch displays a natural twist providing a helical conformation with six anhydroglucose units per turn (Zobel, 1988). Hydroxyl groups of glucose residues are present on the outer surface of the helix, while the internal cavity is a hydrophobic tube (Zhou et al., 2007). The hydrophobic complexing agents can stay or complex within the amylose helix stabilized through van der Waals forces with the adjacent C-hydrogen of amylose (Godet et al., 1993; Zhou et al., 2007). The effects of free fatty acids (lauric, myristic, palmitic, stearic, and oleic acids; lysolecithin and cholesterol) on the hydrolysis of starch, amylose, and amylopectin using α-amylase and amyloglucosidase have been reported in the literature by Crowe et al. (2000). Around 60% amylose was converted to glucose in 1 h and reached up to 90% after 6 h. The addition of lauric, myristic, palmitic, and oleic acids reduced the enzymatic hydrolysis of amylose by 35%. However, neither stearic acid nor cholesterol presented an inhibition. Lauric acid had no effect on the enzymatic breakdown of amylopectin, whereas the breakdown of whole starch was inhibited 12% by lauric acid. These experiments suggest that only the hydrolysis of the amylose fraction (31% of the whole starch) is affected by lauric acid. Amylose presents a helical conformation and can form inclusion complexes with small hydrophobic molecules. Complexes between fatty acids such as lauric acid and amylose can form rapidly under physiological conditions, which contributes to the formation of RS (Seligman et al., 1998). The formation of such complexes with lipids may result in significant changes in the behavior of the starch, including decreased solubility, increased gelatinization temperature, and delayed retrogradation and resistance to the action of digestive enzymes.

Amylose may bind one lauric acid molecule per ~20 glucose units in the glucose chain, but in contrast, very little lauric acid binds under the same conditions to amylopectin and other branched glucans (Crowe et al., 2000).

Enzymatic resistance of the pure amylose and lipid complexes has also been reported in the literature (Holm et al., 1983; Gelders et al., 2005). Using in vitro and in vivo digestibility studies on amylose–lipid complexes, Holm et al. (1983) observed that complexed amylose is hydrolyzed and absorbed in the gastrointestinal tract to the same extent as free amylose but at a somewhat slower rate. After studying the influence of enzymatic action on the digestibility of complexes formed between amylose of various average chain lengths (degree of polymerization) and docosanoic acid/glycerol monostearate, Gelders et al. (2005) suggested that the enzymatic resistance of complexes increases with increasing amylose degree of polymerization, lipid chain length, and complexation temperature. They further reported that enzymatic hydrolysis of these complexes give rise to two or more dextrin subpopulations, which originate from a sequence of lamellar units with interconnecting, amorphous amylose chains. Fats high in saturated, monounsaturated, or polyunsaturated fatty acids have been reported to lower the GI of breads, with no significant difference between the type of fat (Henry et al., 2008). French fried potatoes have been reported to contain higher amounts of RS than boiled and mashed potatoes (Garica-Alonso and Goni, 2000; Nayak et al., 2014), partly because of the formation of amylose–lipid complex in the former.

The more common types of α-amylases are of protein or glycoprotein nature, noncompetitive, nondialyzable, and generally heat labile (Dreher et al., 1984). Many molecules that exist in plant sources are capable of inhibiting the activity of α-amylase. Polyphenols, specifically anthocyanins, have also been reported to inhibit the digestion of starch in cooked potatoes (Ramdath et al., 2014). Some low-molecular-weight plant-derived molecules like luteolin, strawberry extracts, and green tea polyphenols have also been observed to inhibit α-amylase or reduce postprandial hyperglycemia (McDougall et al., 2005; He et al., 2006).

Various commercial starch-blocking products have been manufactured since the early 1940s but many clinical studies have proved that they do not work in vivo, whereas some studies have reported their positive role in in vitro experiments. α-Amylase inhibitors have been found to be unstable in the stomach and active only after preincubation with amylase in the absence of starch (Lajolo and Genovese, 2002).

Dietary fiber has been suggested as the primary factor influencing the slower rate of glucose release in foods through its high viscosity, which slows down gastric emptying absorption of digested products in the small intestine. The rate of starch digestion, however, cannot be explained by the amount of fiber alone (Jenkins et al., 1980).

Phytic acid is the most important phosphate reserve compound in many plants. It can form a complex with proteins and/or metal ions, reducing their biological availability. Yoon et al.

(1983) have showed that the addition of phytic acid after preincubation with saliva decreased the sugar liberation significantly, while little effect was seen on simultaneous addition. Phytic acid may affect starch digestibility through interaction with amylase protein and/or binding with salivary minerals such as calcium, which is known to catalyze amylase activity. The effects of processing on antinutrients (phytic acid, condensed tannins, etc.) acting on starch digestibility are also important to consider (Alonso et al., 2000; Rehman and Shah, 2005).

5.2 Influence of Food Processing on Starch Digestion

A wide range of techniques is being used by the industry for processing various food materials. Processing leads to an alteration in the food structure and also influences the nutritional characteristics of the food, including starch digestibility. When starch molecules are heated in excess water, the crystalline structure is disrupted and the water molecules become linked by hydrogen bonding to the exposed hydroxyl groups of amylose and amylopectin, which causes an increase in granule swelling and solubility. Therefore, the water activity or the availability of water is an important factor that determines the extent of starch digestibility through enzymatic hydrolysis. Processing has been observed to result in an increase in the degree of starch hydrolysis, reaching values higher than 90% (at the end of incubation with pancreatin) (Anguita et al., 2006). Starches in tubers such as potatoes are particularly well protected from the polar environment of luminal fluids, and even cereals such as wheat may not have access to α-amylase in the intestinal lumen unless they have been physically altered through processing.

The principal processes facilitating starch availability for water penetration and consequent α-amylase action are physical processing and cooking by heating to 100 °C for several minutes. Cooking increases the rate of hydrolysis by gelatinizing the starch and making it more easily available for enzymatic attack. Extrusion cooking significantly increases the in vitro digestibility of starches (Alonso et al., 2000; Altan et al., 2009). The increase in digestibility of starch may be explained on the basis that the starch granules lose their structural integrity because of increased shearing action and kneading in the extruder barrel, which ultimately increases their susceptibility to enzymatic attack. Somewhat low values of digestibility of extrusion-cooked starch or starchy foods have also been seen sometimes, which may be attributed to the formation of amylose–lipid complexes, starch–protein interaction, and limited water availability, which prolongs the starch digestibility during enzymatic hydrolysis (Guha et al., 1997). The decrease in the size distribution of the granule results in an increase in the surface area. As a result, processes like grinding lead to a higher percentage of hydrolysis. Anguita et al. (2006) observed that extrusion provoked a decrease in particle size compared to raw samples and affected the digestibility. Traditional and conventional processing methods were compared with extrusion cooking, and their effects on bean starch digestibility were studied by Alonso et al. (2000). Extrusion produced a higher increase in starch digestibility than other processing methods. Similarly, cooking of legumes was carried out using boiling water or autoclaving at 121 °C for up to 90 min and their effects on starch digestibility were compared (Rehman and Shah, 2005). Higher digestibility values were obtained, which could be attributed to the higher degree of starch gelatinization and destruction of antinutrients, when cooked by autoclaving.

Processing of starchy foods may result in an enhancement of digestibility owing to the loss of phytic acid, tannins, and polyphenols, which normally inhibit the activity of α-amylase and thus decrease the starch digestibility. It has been suggested that the removal of tannins and phytic acid creates a large space within the matrix, which increases the susceptibility to enzymatic attack and consequently improves the starch digestibility (Rehman and Shah, 2005). An interesting comparison of processing parameters such as popping, roasting, flaking, and extrusion and their effects on starch digestibility and predicted GI has been reported in the literature (Capriles et al., 2008). The authors reported that starch hydrolysis rate is significantly enhanced by popping, roasting, and flaking compared to extrusion. Similarly, a range of processing types were studied, and their effects on in vitro digestibility of starches were reported by Roopa and Premavalli (2008). The starch digestibility was shown to increase by 35–40% during cooking, autoclaving, and puffing followed by pressure cooking and germination. They also reported that baking, frying, and shallow frying reduced RDS, while roasting and pressure cooking enhanced the RDS to about 23%, followed by cooking, autoclaving, and puffing.

Assimilation of starches may be enhanced by a series of processes used in food processing that promote efficient entry into the luminal polar solution for interaction with the α-amylases (Goni et al., 1997). Effects of microwave heating and other heating methods such as cooking and pressure cooking on the rate of hydrolysis and GI were compared for kudzu starch and maize starch (Geng et al., 2003). The rate of hydrolysis for both the starches increased following heat treatment and to a greater extent following microwave cooking, which may be attributed to the greater penetration of heat during microwave heating. The change in degree of susceptibility of the starches to enzymatic digestion is a function of the extent to which the microwave heating process induced any changes in the crystalline structure of the starches (Anderson and Guraya, 2006). The effects of microwave treatment time on the digestibility of *Canna edulis* starch have been reported by Zhang et al. (2009). Their results indicated that both low (400 W) and high (1000 W) microwave powers are advantageous to the formation of RS. Irradiation has been used to extend the shelf life and safety of food and is permitted now in many countries. Degradation of starch polymers by γ-irradiation may result in reduction in the molecular weight of amylose and amylopectin, decreased viscosity, and increased acidity (Bao et al., 2005; Abu et al., 2006). Chung and Liu (2009) studied the effect of γ-irradiation on the enzymatic digestibility of corn starch and reported that a slower dose rate decreases the RDS and SDS contents and increases the RS content.

During the gelatinization of starch, the crystalline structure of amylopectin disintegrates and the polysaccharide chains take up a random configuration, thus causing the swelling and rupturing of the starch granules (Singh et al., 2007). After gelatinization and cooling of the cooked starch, recrystallization of the starch chains begins. Amylose aggregation and crystallization in the cooked starch pastes have been reported to be complete within the first few hours, while amylopectin aggregation and crystallization occur at later stages during refrigerated storage (Singh et al., 2008b). Linear chains of amylose facilitate the cross-linkages through hydrogen bonds, whereas the branched chains of amylopectin delay its recrystallization. The retrogradation properties of starches are also influenced by the structural arrangement of starch chains within the amorphous and crystalline regions of the ungelatinized granule, because this structural arrangement influences the extent of granule breakdown during gelatinization and also influences the interactions that occur between starch chains during storage (Singh et al., 2004). RS formation in the cooked starches stored at refrigeration temperatures is greatly influenced by the extent of retrogradation. Retrograded amylose from potatoes is highly resistant to enzymatic hydrolysis. It has been observed that storing cooked potatoes at refrigerated temperatures may lead to a reduction in their digestibility and estimated GI. Potatoes were cooked and then cooled at refrigeration temperatures for up to 2 days and their starch digestibility was compared with that of freshly cooked potatoes by Mishra et al. (2008). They reported that the percentage of RDS in the refrigerated potatoes decreased to 45% from the 95% of the freshly cooked potatoes, which could be attributed to the retrogradation of starch and RS formation. Another study reported that with cooling or storing of potatoes for 24 h, the quantity of initial RS (1.18%) in boiled potatoes was increased to 4.63% (Garcia-Alonso and Goni, 2000; Nayak et al., 2014). A similar effect was observed for cooled french fries and retrograded potato flours. The dispersal of polymers of the gelatinized starch during refrigerated storage has observed to undergo retrogradation, which leads to the formation of semicrystalline structures that resist digestion by amylases.

6. Conclusions

Starch is the most common storage carbohydrate in plants and also the largest source of carbohydrates in human food. Starches can be classified according to their digestibility as rapidly digestible, slowly digestible, and resistant to digestion. The microstructure of potatoes, whether natural or created during processing/storage, plays an important role during the digestion of starch in the gastrointestinal tract and affects the GI. The rheological characteristics of foods, either natural or achieved through added ingredients, may influence the digestion of starch by affecting the availability of water. The other constituents of food matrix, such as proteins, lipids, polysaccharides, and added ingredients, play a significant role during processing and thus contribute to the creation of a typical food microstructure that may influence the digestibility of starch and consequently the absorption of digested carbohydrates in the gastrointestinal tract. Although several studies have reported on starch digestibility, there is still a scarcity of literature about the digestibility of starch in potatoes. The digestibility of starch in potatoes can be studied through in vitro digestion models, which have shown a very strong correlation with in vivo models. Sophisticated techniques, such as electron and confocal microscopy, rheology, etc., have been used to gain in-depth knowledge about the characteristics of food microstructure and digesta during and after the starch digestion process. This new knowledge about potato microstructure and its role during starch digestion may help in the manufacturing of new potato-based foods with controlled starch digestibility and low GI.

Acknowledgment

Permission from Elsevier to reproduce the articles (Singh et al., 2013, 2010) is gratefully acknowledged.

References

Abu, J.O., Duodu, K.G., Minnaar, A., 2006. Effect of γ-irradiation on some physicochemical and thermal properties of cowpea (*Vigna unguiculata* L. Walp) starch. Food Chemistry 95, 386–393.

Aguilera, J.M., Stanley, D.W., 1990. Microstructural Principles of Food Processing and Engineering, second ed. Elsevier Applied Science, New York, pp. 175–329.

Alonso, R., Aguirre, A., Marzo, F., 2000. Effect of extrusion and traditional processing methods on antinutrients and in vitro digestibility of protein and starch in faba and kidney beans. Food Chemistry 68, 159–165.

Altan, A., McCarthy, K.L., Maskan, M., 2009. Effect of extrusion cooking on functional properties and in vitro starch digestibility of barley-based extrudates from fruit and vegetable by-products. Journal of Food Science 74, E77–E86.

Anderson, A.K., Guraya, H.S., 2006. Effects of microwave heat-moisture treatment on properties of waxy and non-waxy rice starches. Food Chemistry 97, 318–323.

Anguita, M., Gasa, J., Martín-Orúe, S.M., Pérez, J.F., 2006. Study of the effect of technological processes on starch hydrolysis, non-starch polysaccharides solubilization and physicochemical properties of different ingredients using a two-step in vitro system. Animal Feed Science and Technology 129, 99–115.

Appenroth, K.-J., Keresztes, A., Krzysztofiwicz, E., Gabrys, H., 2011. Light induced degradation of starch granules in turions of *Spirodela polyrhiza* studied by electron microscopy. Plant Cell Physiology 52, 384–391.

Bao, J., Ao, Z., Jane, J.L., 2005. Characterization of physical properties of flour and starch obtained from gamma-irradiated white rice. Starch 57, 480–487.

Barrios, E.P., Newson, D.W., Miller, J.C., 1963. Some factors influencing the culinary quality of Irish potatoes. II. Physical characters. American Potato Journal 40, 200–206.

Berg, T., Singh, J., Hardacre, A., Boland, M.J., 2012. The role of cotyledon cell structure during in vitro digestion of starch in navy beans. Carbohydrate Polymers 87, 1678–1688.

Björck, I., 1996. Starch: nutritional aspects. In: Eliasson, A.C. (Ed.), Carbohydrate in Foods. Lund University, Sweden, pp. 505–553.

Bordoloi, A., Kaur, L., Singh, J., 2012a. Parenchyma cell microstructure and textural characteristics of raw and cooked potatoes. Food Chemistry 133, 1092–1100.

Bordoloi, A., Singh, J., Kaur, L., Singh, H., 2012b. In vitro digestibility of starch in cooked potatoes as affected by guar gum: microstructural and rheological characteristics. Food Chemistry 133, 1206–1213.

Brennan, C.S., Blake, D.E., Ellis, P.R., Schofield, J.D., 1996. Effects of guar galactomannan on wheat bread microstructure and on the in vitro and in vivo digestibility of starch in bread. Journal of Cereal Science 24, 151–160.

Capriles, V.D., Coelho, K.D., Guerra-Matias, A.C., Areas, J.A.G., 2008. Effects of processing methods on amaranth starch digestibility and predicted GI. Journal of Food Science 73, H160–H164.

Carpita, N.C., Gibeaut, D.M., 1993. Structural models of primary-cell walls in flowering plants—consistency of molecular-structure with the physical-properties of the walls during growth. Plant Journal 3, 1–30.

Chaisawang, M., Suphantharika, M., 2005. Effects of guar gum and xanthan gum additions on physical and rheological properties of cationic tapioca starch. Carbohydrate Polymers 61, 288–295.

Chanda, S.V., 2005. Evaluation of effectiveness of the methods for isolation of cell wall polysaccharides during cell elongation in *Phaseolus vulgaris* seedlings. Acta Physiologiae Plantarum 27, 371–378.

Cherbut, C., Albina, E., Champ, M., Doublier, J.L., Lecannu, G., 1990. Action of guar gum on viscosity of digestive contents and on gastrointestinal motor function in pigs. Digestion 4, 205–213.

Choi, S.J., Woo, H.D., Ko, S.H., Moon, T.W., 2008. Confocal scanning laser microscopy to investigate the effect of sodium bisulfite on in vitro digestibility of waxy sorghum flour. Cereal Chemistry 85, 65–69.

Chung, H.-J., Liu, Q., 2009. Effect of gamma irradiation on molecular structure and physicochemical properties of corn starch. Journal of Food Science 74, C353–C361.

Collings, P., Williams, C., MacDonald, I., 1981. Effect of cooking on serum glucose and insulin responses to starch. British Medical Journal 282, 1032–1033.

Crowe, T.C., Seligman, S.A., Copeland, L., 2000. Inhibition of enzymic digestion of amylose by free fatty acids in vitro contributes to resistant starch formation. Journal of Nutrition 130, 2006–2008.

Cummings, J.H., Roberfroid, M.B., Members of the Paris Carbohydrate Group, 1997. A new look at dietary carbohydrates: chemistry, physiology and health. European Journal of Clinical Nutrition 51, 417–423.

Dartois, A., Singh, J., Kaur, L., Singh, H., 2010. The influence of guar gum on the starch digestibility in vitro-rheological and microstructural characteristics. Food Biophysics 5, 149–160.

Daussant, J., Mosse, J., Vaughan, J.G., 1983. Seed Proteins. Academic Press.

Dreher, M.L., Dreher, C.J., Berry, J.W., 1984. Starch digestibility of foods: a nutritional perspective. Critical Reviews in Food Science and Nutrition 20, 47–71.

Eastwood, M.A., Morris, E.R., 1992. Physical properties of dietary fiber that influence physiological function: a model for polymers along the gastrointestinal tract. American Journal of Clinical Nutrition 55, 436–442.

Edwards, C.A., Johnson, I.T., Read, N.W., 1998. Do viscous polysaccharides slow absorption by inhibiting diffusion or convection. European Journal of Clinical Nutrition 42, 307–312.

Edwards, C.A., 2003. Gums: dietary importance. In: Caballero, B., et al. (Ed.), Encyclopedia of Food Sciences and Nutrition. Academic Press, San Diego, pp. 3007–3012.

Ellis, P.R., Rayment, P., Wang, Q.A., 1996. Physico-chemical perspective of plant polysaccharides in relation to glucose absorption, insulin secretion and the entero-insular axis. Proceedings of the Nutrition Society 55, 881–898.

Ellis, P.R., Roberts, F.G., Low, A.G., Morgan, L.M., 1995. The effect of high-molecular-weight guar gum on net apparent glucose absorption and net apparent insulin and gastric inhibitory polypeptide production in the growing pig: relationship to rheological changes in jejunal digesta. British Journal of Nutrition 74, 539–556.

Englyst, H.N., Kingman, S.M., Cummings, J.H., 1992. Classification and measurement of nutritionally important starch fractions. European Journal of Clinical Nutrition 46, S33–S50.

Englyst, K.N., Englyst, H.N., Hudson, G.J., Cole, T.J., Cummings, J.H., 1999. Rapidly available glucose in foods: an in vitro measurement that reflects the glycemic response. American Journal of Clinical Nutrition 69, 448–454.

Ezeogu, L.I., Duodu, K.G., Emmanbux, M.N., Taylor, J.R.N., 2008. Influence of cooking conditions on the protein matrix of sorghum and maize endosperm flours. Cereal Chemistry 85, 397–402.

Fernandes, G., Velangi, A., Wolever, T.M., 2005. Glycemic index of potatoes commonly consumed in North America. Journal of American Dietetic Association 105, 557–562.

Fiona, S., Atkinson, R.D., Kaye Foster-Powell, R.D., Brand-Miller, J.C., 2008. International tables of glycemic index and glycemic load values: 2008. Diabetes Care 31, 2281–2283.

Foster-Powell, K., Holt, S.H., Brand-Miller, J.C., 2002. International table of glycemic index and glycemic load values. American Journal of Clinical Nutrition 76, 5–56.

Frei, M., Siddhuraju, P., Becker, K., 2003. Studies on the in vitro starch digestibility and the glycemic index of six different indigenous rice cultivars from the Philippines. Food Chemistry 83, 395–402.

Gagné, L., 2008. The glycemic index and glycemic load in clinical practice. Explore: The Journal of Science and Healing 4, 66–69.

Garcia-Alonso, A., Goni, I., 2000. Effect of processing on potato starch: in vitro availability and glycemic index. Starch 52, 81–84.

Gelders, G.G., Duyck, J.P., Goesaert, F., Delcour, J.A., 2005. Enzyme and acid resistance of amylose-lipid complexes differing in amylose chain length, lipid and complexation temperature. Carbohydrate Polymers 60, 379–389.

Geng, Z., Zongdao, C., Toledo, R., 2003. Effects of different processing methods on the glycemic index of kudzu starch. Journal of Chinese Cereals and Oils Association 18, 5.

Giacco, R., Brighenti, F., Parillo, M., et al., 2001. Characteristics of some wheat based foods of the Italian diet in relation to their influence on postprandial glucose metabolism in type 2 diabetic patients. British Journal of Nutrition 85, 33–40.

Godet, M.C., Tran, V., Delagw, M.M., 1993. Molecular modelling of the specific interactions in amylose complexation by fatty acids. International Journal of Biological Macromolecules 15, 11–16.

Goni, I., Garcia-Alonso, A., Saura-Calixto, F., 1997. A starch hydrolysis procedure to estimate glycemic index. Nutrition Research 17, 427–437.

Gray, G.M., 1992. Starch digestion and absorption in nonruminants. Journal of Nutrition 122, 172–177.

Guha, M., Ali, S.Z., Bhattacharya, S., 1997. Twin-screw extrusion of rice flour without die: effect of barrel temperature and screw speed on extrusion and extrudate characteristics. Journal of Food Engineering 32, 251–267.

Gularte, M.A., Rosell, C.M., 2011. Physicochemical properties and enzymatic hydrolysis of different starches in the presence of hydrocolloids. Carbohydrate Polymers 85, 237–244.

Hamaker, B.R., Bugusu, B.A., 2003. Overview: sorghum proteins and food quality. In: Paper Presented at: Workshop on the Proteins of Sorghum and Millets: Enhancing Nutritional and Functional Properties for Africa (CD), Pretoria, South Africa.

Han, J.-A., BeMiller, J.N., 2007. Preparation and physical properties of slowly digesting modified food starches. Carbohydrate Polymers 67, 366–374.

He, Q., Lv, Y., Yao, K., 2006. Effects of tea polyphenols on the activities of R-amylase, pepsin, trypsin and lipase. Food Chemistry 101, 1178–1182.

Heacock, P.M., Hertzler, S.R., Wolf, B.W., 2002. Fructose prefeeding reduces the glycemic response to a high-glycemic index, starchy food in humans. Journal of Nutrition 132, 2601–2604.

Henry, C.J.K., Lightowler, H.J., Newens, K.J., Pata, N., 2008. The influence of adding fats of varying saturation on the glycaemic response of white bread. International Journal of Food Sciences and Nutrition 59, 61–69.

Henry, C.J.K., Lightowler, H.J., Kendall, F.L., Storey, M., 2006. The impact of the addition of toppings/fillings on the glycemic response to commonly consumed carbohydrate foods. European Journal of Clinical Nutrition 60, 763–769.

Henry, C.J., Lightowler, H.J., Strik, C.M., Storey, M., 2005. Glycaemic index values for commercially available potatoes in Great Britain. British Journal of Nutrition 94, 917–921.

Holm, J., Bjorck, I., Ostrowska, S., Eliasson, A., Asp, N., Larsson, K., Lundquist, I., 1983. Digestibility of amylose-lipid complexes in vitro and in vivo. Starch 35, 294–297.

Holt, S.H., Miller, J.C., Petocz, P., Farmakalidis, E., 1995. A satiety index of common foods. European Journal of Clinical Nutrition 49, 675–690.

Hoover, R., Sosulski, F., 1985. Studies on the functional characteristics and digestibility of starches from *Phaseolus vulgaris* biotypes. Starch 37, 181–191.

Hoover, R., Zhou, Y., 2003. In vitro and in vivo hydrolysis of legume starches by α-amylase and resistant starch formation in legumes—a review. Carbohydrate Polymers 54, 401–417.

Hughes, J.C., Faulks, R.M., Grant, A., 1975a. Texture of cooked potatoes. relationship between compressive strength, pectic substances and cell size of redskin tubers of different maturity. Potato Research 18, 495–514.

Hughes, J.C., Faulks, R.M., Grant, A., 1975b. Texture of cooked potatoes: relationship between the compressive strength of cooked potato disks and release of pectic substances. Journal of the Science of Food and Agriculture 26, 731–738.

International Standards Organisation ISO 26642, 2010. Food Products—Determination of the Glycaemic Index (GI) and Recommendation for Food Classification. International Standards Organisation.

Jenkins, D.J.A., Thorne, M.J., Wolever, T.M.S., Jenkins, A.L., Rao, A.V., Thompson, L.U., 1987. The effect of starch-protein interaction in wheat on the glycemic response and rate of in vitro digestion. American Journal of Clinical Nutrition 45, 946–951.

Jenkins, D.J.A., Wolever, T.M.S., Taylor, R.H., et al., 1980. Rate of digestion and postprandial glycaemia of foods in normal and diabetic subjects. British Medical Journal 281, 14–17.

Johansen, H.N., Knudsen, K.E.B., Sandström, B., Skjøth, F., 1996. Effects of varying content of soluble dietary fibre from wheat flour and oat milling fractions on gastric emptying in pigs. British Journal of Nutrition 75, 339–351.

Kaur, L., Singh, J., 2009. The role of galactomannan seed gums in diet and health—a review. In: Govil, J.N., Singh, V.K. (Eds.). Govil, J.N., Singh, V.K. (Eds.), Recent Progress in Medicinal Plants. Standardization of Herbal/Ayurvedic Formulations, vol. 24. Stadium Press LLC, Houston, TX, pp. 429–467.

Kaur, L., Singh, J., Singh, H., McCarthy, O.J., 2008. Starch-Cassia gum interactions: a microstructure-rheology study. Food Chemistry 111, 1–10.

Kim, E.H.-J., Petrie, J.R., Motoi, L.M., Morgenstern, M.P., Sutton, K.V., Mishra, S., Simmons, L.D., 2008. Effect of structural and physicochemical characteristics of the protein matrix in pasta on in vitro starch digestibility. Food Biophysics 3, 229–234.

Kingman, S.M., Englyst, H.M., 1994. The influence of food preparation methods on the in vitro digestibility of starch in potatoes. Food Chemistry 49, 181–186.

Koh, L.W., Kasapis, S., Lim, K.M., Foo, C.W., 2009. Structural enhancement leading to retardation of in vitro digestion of rice dough in the presence of alginate. Food Hydrocolloids 23, 1458–1464.

Kuriki, T., Imanka, I., 1999. The concept of the α-amylase family: structural similarity and common catalytic mechanism. Journal of Biosciences and Bioengineering 87, 557–565.

Lajolo, F.M., Genovese, M.I., 2002. Nutritional significance of lectins and enzyme inhibitors from legumes. Journal of Agricultural and Food Chemistry 50, 6592–6598.

Lehmann, U., Robin, F., 2007. Slowly digestible starch—its structure and health implications: a review. Trends in Food Science and Technology 18, 346–355.

Lightowler, H.J., Henry, C.J.K., 2009. Glycemic response of mashed potato containing high-viscosity hydroxypropylmethyl cellulose. Nutrition Research 29, 551–557.

Ludwig, D.S., 2007. Clinical update: the low-glycaemic-index diet. The Lancet 369, 890–892.

Marciani, L., Gowland, P.A., Spiller, R.C., Manoj, P.R., Moore, R.J., Young, P., Al-Sahab, S., Bush, D., Wright, J., Fillery-Travis, A.J., 2000. Gastric response to increased meal viscosity assessed by echo-planar magnetic resonance imaging in humans. Journal of Nutrition 130, 122–127.

McDougall, G.J., Shpiro, F., Dobson, P., Smith, P., Blake, A., Stewart, D., 2005. Different polyphenolic components of soft fruits inhibit alphaamylase and alpha-glucosidase. Journal of Agricultural and Food Chemistry 53, 2760–2766.

Mishra, S., Monro, J., Hedderley, D., 2008. Effect of processing on slowly digestible starch and resistant starch in potato. Starch 60, 500–507.

Nagano, T., Tamaki, E., Funami, T., 2008. Influence of guar gum on granule morphologies and rheological properties of maize starch. Carbohydrate Polymers 72, 95–101.

Najjar, N., Adra, N., Hwalla, N., 2004. Glycemic and insulinemic responses to hot versus cooled potato in males with varied insulin sensitivity. Nutrition Research 24, 993–1004.

Nayak, B., Berrios, J.de J., Tang, J., 2014. Impact of food processing on the glycemic index (GI) of potato products. Food Research International 56, 35–46.

Oria, M.P., Hamaker, B.R., Shull, J.M., 1995. In vitro protein digestibility of developing and mature sorghum grain in relation to a-, b- and g-kafirin disulfide crosslinking. Journal of Cereal Science 22, 85–93.

Raben, A., Andersen, K., Karberg, M.A., Holst, J.J., Astrup, A., 1997. Acetylation of or beta-cyclodextrin addition to potato starch: beneficial effect on glucose metabolism and appetite sensations. American Journal of Clinical Nutrition 66, 304–314.

Ramdath, D.D., Padhi, P., Hawke, A., Sivaramalingam, T., Tsao, R., 2014. The glycemic index of pigmented potatoes is related to their polyphenol content. Food and Function 5, 909–915.

Raven, P.H., Evert, R.F., Eichhorn, S.E., 2005. Biology of Plants, seventh ed. W.H. Freeman and Company, New York.

Rehman, Z.-U., Shah, W.R., 2005. Thermal heat processing effects on antinutrients, protein and starch digestibility of food legumes. Food Chemistry 91, 327–331.

Riccardi, G., Rivellese, A.A., Giacco, R., 2008. Role of glycemic index and glycemic load in the healthy state, in prediabetes, and in diabetes. American Journal of Clinical Nutrition 87 (Suppl. 1), 269S–274S.

Rooney, L.W., Pflugfelder, R.L., 1986. Factors affecting starch digestibility with special emphasis on sorghum and corn. Journal of Animal Science 63, 1607–1623.

Roopa, S., Premavalli, K.S., 2008. Effect of processing on starch fractions in different varieties of finger millet. Food Chemistry 106, 875–882.

Seligman, S.A., Copeland, L., Appels, R., Morell, M.K., 1998. Analysis of lipid binding to starch. In: O'Brien, L., Blakeney, A.B., Ross, A.S., Wrigley, C.W. (Eds.), Cereals. Royal Australian Chemical Institute, North Melbourne, Australia, pp. 87–90.

Singh, J., Dartois, A., Kaur, L., 2010. Starch digestibility in food matrix: a review. Trends in Food Science and Technology 21, 168–180.

Singh, J., Kaur, L., McCarthy, O.J., 2007. Factors influencing the physico-chemical, morphological, thermal and rheological properties of some chemically modified starches for food applications—a review. Food Hydrocolloids 21, 1–22.

Singh, J., Kaur, L., McCarthy, O.J., 2009a. Potato starch and its modification. In: Singh, J., Kaur, L. (Eds.), Advances in Potato Chemistry and Technology. Elsevier Academic Press, USA, pp. 273–318.

Singh, J., Kaur, L., Singh, H., 2013. Food microstructure and starch digestion. In: Henry, J. (Ed.). Henry, J. (Ed.), Advances in Food and Nutrition Research, vol. 70. Academic Press, pp. 137–179.

Singh, J., Kaur, L., McCarthy, O.J., Moughan, P.J., Singh, H., 2008a. Rheological and textural characteristics of raw and par-cooked Taewa (Maori potatoes) of New Zealand. Journal of Texture Studies 39, 210–230.

Singh, J., Kaur, L., McCarthy, O.J., Moughan, P.J., Singh, H., 2009b. Development and characterization of extruded snacks from New Zealand Taewa (Maori potato) flours. Food Research International 42, 666–673.

Singh, J., McCarthy, O.J., Singh, H., Moughan, P.J., 2008b. Low temperature post-harvest storage of New Zealand Taewa (Maori potato): effects on starch physico-chemical and functional characteristics. Food Chemistry 106, 583–596.

Singh, N., Kaur, L., Ezekiel, R., Gurraya, H.S., 2005. Microstructural, cooking and textural characteristics of potato (*Solanum tuberosum* L.) tubers in relation to physico-chemical and functional properties of their flours. Journal of the Science of Food and Agriculture 85, 1275–1284.

Singh, N., Kaur, L., Singh, J., 2004. Relationships between various physicochemical, thermal and rheological properties of starches separated from different potato cultivars. Journal of the Science of Food and Agriculture 84, 714–720.

Slaughter, S.L., Ellis, P.R., Jackson, E.C., Butterworth, P.J., 2002. The effect of guar galactomannan and water availability during hydrothermal processing on the hydrolysis of starch catalysed by pancreatic α-amylase. Biochimica et Biophysica Acta 1571, 55–63.

Smith, M.E., Morton, D.G., 2001. The Digestive System. Churchill Livingstone, Edinburgh.

Snow, P., O'Dea, K., 1981. Factors affecting the rate of hydrolysis of starch in food. American Journal of Clinical Nutrition 34, 2721–2727.

Soh, N.L., Brand-Miller, J., 1999. The glycaemic index of potatoes: the effect of variety, cooking method and maturity. European Journal of Clinical Nutrition 53, 249–254.

Tahvonen, R., Hietanen, R.M., Sihvonen, J., Salmine, E., 2006. Influence of different processing methods on the glycemic index of potato (Nicola). Journal of Food Composition and Analysis 19, 372–378.

Tester, R.F., Karkalas, J., Qi, X., 2004. Starch structure and digestibility enzyme-substrate relationship. World's Poultry Science Journal 60, 186–195.

Timothy, C., Crowe, C., Seligman, S.A., Copeland, L., 2000. Inhibition of enzymic digestion of amylose by free fatty acids in vitro contributes to resistant starch formation. Journal of Nutrition 130, 2006–2008.

US Department of Agriculture, 2002. Agricultural Research Service. USDA National Nutrient Database for Standard Reference. Release 15, 2002. Nutrient Data Laboratory Home Page http://www.nal.usda.gov/fnic/foodcomp (accessed 30.05.04.).

Uribe, S., Sampedro, J.G., 2003. Measuring solution viscosity and its effect on enzyme activity. Biological Procedures Online 5, 108–115.

Vaaler, S., Hanssen, K.E., Aagenaes, O., 1984. The effect of cooking upon the blood glucose response to ingested carrots and potatoes. Diabetes Care 7, 221–223.

van Marle, J.T., Clerkx, A.C.M., Boekestein, A., 1992. Cryo-scanning electron microscopy investigation of the texture of cooked potatoes. Food Structure 11, 209–216.

van Marle, J.T., Recourt, K., van Dijk, C., Schols, H.A., Voragen, A.G.J., 1997. Structural features of cell walls from potato (*Solanum tuberosum* L.) cultivars Irene and Nicola. Journal of Agriculture and Food Chemistry 45, 1686–1693.

Waldron, K.W., Smith, A.C., Parr, A.J., Ng, A., Parker, M.L., 1997. New approaches to understanding and controlling cell separation in relation to fruit and vegetable texture. Trends in Food Science and Technology 8, 213–220.

Williams, P.A., Phillips, G.O., 2003. Gums: properties of individual gums. In: Caballero, B., et al. (Eds.), Encyclopedia of Food Sciences and Nutrition. Academic, San Diego, pp. 2992–3001.

Wolever, T.M.S., Cohen, Z., Thompson, L.U., Thorne, M.J., Jenkins, M.J.A., Prokipchuk, E.J., Jenkins, D.J.A., 1986. Ileal loss of available carbohydrate in man: comparison of a breath hydrogen method with direct measurement using a human ileostomy model. American Journal of Gastroenterology 81, 115–122.

Wong, J.H., Lau, T., Cai, N., Singh, J., Pedersen, J.F., Vensel, W.H., Hurkman, W.J., Wilson, J.D., Lemaux, P.G., Buchanan, B.B., 2009. Digestibility of protein and starch from sorghum (*Sorghum bicolor*) is linked to biochemical and structural features of grain endosperm. Journal of Cereal Science 49, 73–82.

Wursch, P., Delvedovo, S., Koellreutter, B., 1986. Cell structure and starch nature as key determinants of the digestion rate of starch in legume. American Journal of Clinical Nutrition 43, 25–29.

Xavier, F., Sunyer, P., 2002. Glycemic index and disease. American Journal of Clinical Nutrition 76 (Suppl. 1), 290S–298S.

Yook, C., Robyt, J.F., 2002. Reactions of alpha amylases with starch granules in aqueous suspension giving products in solution and in a minimum amount of water giving products inside the granule. Carbohydrate Research 337, 1113–1117.

Yoon, J.H., Thompson, L.U., Jenkins, D.J.A., 1983. The effect of phytic acid on in vitro rate of starch digestibility and blood glucose response. American Journal of Clinical Nutrition 38, 835–842.

Zhang, J., Wang, Z.-W., Shi, X.-M., 2009. Effect of microwave heat/moisture treatment on physicochemical properties of *Canna edulis* Ker starch. Journal of the Science of Food and Agriculture 89, 653–664.

Zhou, Z., Robards, K., Helliwell, S., Blanchard, C., 2007. Effect of the addition of fatty acids on rice starch properties. Food Research International 40, 209–214.

Zobel, H.F., 1988. Molecules to granules: a comprehensive starch review. Starch 40, 44–50.

CHAPTER 14

Thermal Processing of Potatoes

María Dolores Álvarez Torres, Wenceslao Canet Parreño

Department of Characterization, Quality, and Safety, Institute of Food Science, Technology and Nutrition (ICTAN-CSIC), Madrid, Spain

1. Introduction

Nowadays the potato is one of the major food crops in the world. Heavy yields can be grown relatively cheaply in a wide variety of soils and climates, and potatoes are one of the mainstays in the diet of people in many parts of the world.

In 2013 total EU 28 (Europe Union of the 28) potato production reached 63,331,000 metric tonne, in a cultivation area of 1,748,000 ha, with yields of 362.2 (100 kg/ha) (Eurostat, 2013). Thus, potato processing and potato products are popular and acceptable throughout Europe. Figure 1 shows the 2013 production and external trade in processed potatoes for the EU (28 countries). The main item is frozen potatoes, including prefried, at 4,385,214 t; followed by potatoes prepared or preserved as potato chips or crisps, 1,800,000 t; frozen potatoes, uncooked or cooked by steaming or boiling in water, 399,090 t; dried potatoes, whether cut or not, or sliced but not further prepared, and in the form of flour, meal, flakes, granules, and pellets, 398,003 t; and potatoes prepared or preserved in the form of flour, meal, or flakes (excluding frozen, dried, crisps, by vinegar or acetic acid), 119,911 t (Eurostat, 2013). Frozen potato products account for more than 67.36% of all potatoes that are processed, crisps for 25.34%, dried potato products for 5.60%, and miscellaneous products in the form of flour, meal, or flakes for 1.69%. Comparing the processed potato production figures for the 2013 European market (EU 28) with the 2005 figures (EU 25) shown in Alvarez and Canet (2009), dried potatoes have increased 32.9%, frozen potatoes 40%, and crisps 9.2%, and miscellaneous products in the form of flour, meal, or flakes have decreased 43%. Comparing the figures for 2013 of the outdoor European market (EU 28) with the 2005 figures (EU 25) shows a significant decrease in the export and import of processed potatoes to 28% and 2.5%, respectively.

Any food, whether frozen or not, is considered to be of good quality if it meets the following requirements: there must be a total absence of pathogens and compounds toxic to humans (hygiene and health quality); it must be easily digestible, with good nutritional value, meaning high concentrations of vitamins, macronutrients, and minerals, and an appropriate

Advances in Potato Chemistry and Technology. http://dx.doi.org/10.1016/B978-0-12-800002-1.00014-5

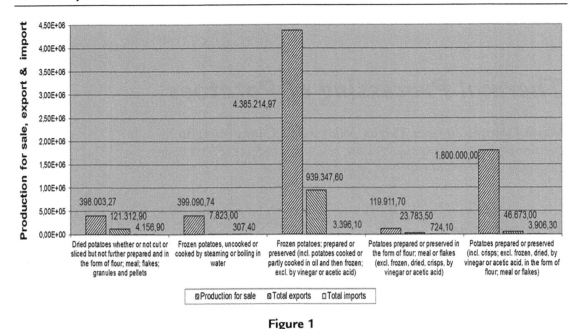

Figure 1

EU (28 countries) production and external trade of processed potatoes (t) January–December 2013. *Prepared with values obtained from Eurostat (2013).*

caloric content (nutritional quality); its sensory attributes, appearance, flavor, aroma, and texture must be constant and, in the case of frozen products, they must be as close as possible to those of fresh produce (sensory quality); and the presentation and mode of preparation must conform to consumer preferences (commercial quality) (Canet and Alvarez, 2011).

Freezing of vegetables immediately postharvest guarantees consumers a minimum loss of nutritional value and higher vitamin C content than could be attained by any other form of preservation and distribution. Furthermore, if properly handled before freezing and correctly processed and distributed, there is no possibility of growth of microbial contaminants, which also guarantees hygiene and health quality.

The quality of potato tubers depends upon genetic, climatic, biotic, chemical, and edaphic factors; varietal characteristics; precipitation, temperature, and sunshine conditions; competition with other plants; the use of chemicals; and the physical, chemical, and biological properties of the soil, which influence the capacity of the crop to take up the necessary water and nutrients to ensure success.

The production of high-quality potato products depends on the initial quality of the tuber; on the conditions and time lapse between harvesting and processing; on preparatory operations such as washing, peeling, trimming, and cutting; in the case of frozen potato products, on prefreezing treatments like blanching or prefrying, freezing per se, storage temperature, time and tolerance of frozen storage; and on final preparation procedures such as boiling, cooking, or microwaving. It is this set of factors, known as P-P (product–processing) factors, and the achievement of optimum interaction among them, that defines quality. This quality can be considerably diminished by adverse storage times and temperatures, that is, T-T-T (time–temperature–tolerance) factors. The large number of influencing factors makes it difficult first, to optimize the final quality, and second, to assess and quantify the loss of quality during storage.

In frozen vegetables, health quality, nutritional quality, and aspects of sensory quality like color and texture can be objectively assessed and controlled; also, in frozen potato products the effects of the thermal treatments included in the process have to be assessed because of their influence on texture, color, and nutritional value. However, in the case of overall assessment of sensory quality, only the consumer can perceive and process the overall blend of sensations that denote quality and cause consumers to prefer, accept, or reject a product.

In view of the daily growing importance of frozen potato products, after dealing with raw composition as it relates to processing and quality, technological product factors and the suitability of varieties for processing, and quality assessment of raw and processed material, we review how the final quality of the potato is affected by preparatory operations such as washing, peeling, and cutting. Here, we place particular emphasis on findings on the way in which the thermal treatments entailed in the process—blanching and prefrying, stepwise blanching, freezing and thawing cycles, frozen storage conditions, thawing, and cooking—affect the structure and the texture (which is considered the main attribute of potato quality) and on ways to optimize it.

2. Product versus Quality

The raw material used in the preparation of frozen potato products is an important conditioning factor affecting both the quality and the nutritional value of the final product. A product's suitability for processing is determined by its composition, by agrotechnical practices and conditions, by the cultivar (cultivated variety) involved, and by technological factors such as ripeness and the time elapsing between harvesting and processing.

2.1 Composition

The product factors briefly mentioned in the previous section constitute an important field of research given the considerable influence they have on the final quality of potato products, a subject comprehensively dealt with in other chapters of this book and treated here only as it relates to the quality of frozen potato products. There is an extensive literature on proximate potato composition, which depends on and varies with the variety, growing area, cultural practices, storage history, and methods of analysis used.

The main components in terms of their importance for the final quality of frozen potato products are starch, sugars, nonstarch polysaccharides, and phenol and related substances. Starch comprises between 65% and 80% of the dry weight of the potato tuber and is the most important nutritional component. Starch is present as microscopic ellipsoid granules inside the parenchyma tissue cells, measuring about 100 by 60 μm on average on the longitudinal axis. The largest granules are found in the vascular areas, and small grain sizes are associated with dry seasons, small tubers, immaturity, potassium deficiency, and prolonged postharvest storage. Granule size and size distribution contribute to the character and quality of potato products. Starch content is accepted as a varietal characteristic. When the temperature of the potato tissue rises to 50 °C, the starch granules absorb the cell water and swell; in the range 64–71 °C the starch begins to gelatinize and the cells tend to separate and become rounded. Excessive cell separation results in sloughing, a tendency that is important not only in the cooking of the potato but also in commercial peeling methods using heat and lye scald, in which excessive rupture of cells and escaping gelled starch produce a gummy texture in the processed potato products. The starch cell together with the cell wall constituents, especially pectin substances, plays a very important role in determining the final texture of frozen potato products, as discussed later on.

The main factors influencing the sugar content of potatoes during postharvest storage are variety and temperature: varieties with low specific gravity tend to accumulate more sugar than varieties with high specific gravity; freshly mature tubers may contain only traces of sugars, while tubers harvested before maturity may have 1.5% sugar; and small tubers usually contain higher sugar percentages than large ones. At storage temperatures below 10 °C the total and reducing sugars increase, the rate and extent of the increase being greater the lower the temperature down to the freezing point. The sugar contents of tubers stored at low temperature and reconditioned at higher temperatures gradually decrease over a period of 3–4 weeks, while the starch percentage increases during conditioning. Poor texture after cooking is associated with low starch and high sugar content. Potatoes containing more than 2% reducing sugars (dry weight) are unacceptable for processing; it is general practice to keep potatoes that are poor sugar formers in storage and process them while their sugar level is low and they have not sprouted or to condition them (2–3 weeks) as cold storage tubers at room temperature or by storing them at 10 °C.

Potatoes contain nonstarch polysaccharides, which make up the cell wall and the intercellular cementing substances, among which we may distinguish crude fiber, cellulose, pectin substances, hemicelluloses, and other polysaccharides. Dry matter (DM) after removal of all the soluble, starch, and nitrogenous constituents is considered the crude fiber (1% of dry weight), which includes cell wall components, lignin, and suberin. Cellulose is part of the supporting membrane of the cell wall, a mixture of high-molecular-weight polymers of glucose residues, together with hemicelluloses and other polysaccharides (10–20% of nonstarch potato polysaccharides).

Pectin substances, polymers of galacturonic acid, with the carboxyl groups more or less methylated, are usually divided into protopectin, soluble pectin, and pectic acid. Protopectin is an insoluble, highly polymerized form of pectin associated with the cell wall structure. Water-soluble pectin is the product of partial depolymerization of protopectin. Freshly harvested potatoes are relatively high in protopectin, and protopectin content decreases and the soluble pectin content increases during storage. The pectic acid fraction in the form of calcium and magnesium salts is believed to be the cementing substance in the middle lamella of potato tissues. In the development of potato product texture, swelling of the gelatinized starch is the major factor causing cell rounding and separation, a tendency that is opposed by the molecular size and calcium content of the pectic material of the middle lamella and cell wall. Heating at 70 °C produces changes in the pectic substances of the middle lamella, and addition of calcium results in a firmer cooked potato. These two mechanisms, the swelling of the gelatinized starch and the precipitation of pectic substances on cell walls, both induced by thermal treatments, are the key to achieving the optimum desirable final texture for frozen potato products.

Phenolic compounds are associated with the color of raw potatoes and certain types of discoloring in processed potato products. When potato tubers are injured by bruising, cutting, or peeling, the phenols are rapidly converted to colored melanins owing to oxidation of phenolic compounds by enzyme phenolase; this is determined by the concentration of tyrosine, which is considered the most important factor in controlling the extent of pigmentation caused by enzymatic browning. The levels of enzymatic browning and tyrosine and phenolase activity decrease when the plants are treated with higher concentrations of potassium.

2.2 Varietal Suitability for Processing

Potatoes used for processing must possess certain quality characteristics, and as a result potato varieties have been developed specifically for the purpose of processing. The sixth edition of the *World Catalogue of Potato Varieties*, edited by Pieterse and Judd (2014), was published in 2014. The *Catalogue* contains descriptions of more than 4500 potato varieties that are cultivated in over 100 countries worldwide; it also contains descriptions of approximately 1900 wild potatoes. As already noted, the biochemical composition, especially the DM content, carbohydrates (reducing and nonreducing sugars), textural characteristics, and color-related alterations (discoloration of raw flesh, browning, and postcooking blackening) are the key quality parameters for processing (Kadam et al., 1991). DM content is one of the most important factors in processing qualities over a range of uses, which substantially affect the texture in practice and the potato's suitability for processing. The specific potato cultivar requirements for processing chips and french fries are moderately high DM content with low reducing sugar and large, long, oval tubers. For crisps, high DM and low reducing sugars are essential and moderate-sized oval tubers are preferred. Moderately low DM and small tubers are the essential requirements for canned potatoes. And finally, medium-firm, slightly mealy potatoes that do not disintegrate are the best suited to multipurpose and domestic use.

2.3 Agrotechnical Practices and Conditions

There are a number of cultural and environmental conditions that prevail during the growing season and strongly affect the processing quality of potatoes: climate, rainfall, irrigation and soil moisture, soil type, time of planting and harvesting, kind and amount of fertilizers used, and control of insects and diseases.

Climatic factors that influence the potato yield are temperature, light intensity and photoperiod, rainfall, and length of growing season. Potato is classified as a "cool-season crop." Temperature influences the rate of absorption of plant nutrients and translocation within the plant and the rate of respiration, and it also facilitates early growth. Temperature determines to a great extent the length of the growing season, the maturity of the tubers at harvest, and also their specific gravity. At low temperatures, the rate of respiration is less than the rate of photosynthesis, resulting in more accumulation of carbohydrates in the tubers and an increase in the tuber's specific gravity; low and moderate temperatures during the growing season result in higher yields and high-specific-gravity tubers. Although other factors than temperature are involved, the influence of light on growth and yield is dependent upon light intensity and quality and day length. During the stage of tuber formation, potato plants devote a large portion of the carbohydrates produced by photosynthesis to growth of the tubers and less to vegetative growth. Differences in soil moisture as a result of rainfall, irrigation, and type of soil affect the DM content of potatoes and their quality for processing. The most critical period as regards water requirements is the onset of tuberization. It is desirable to have uniform moisture supply in the soil at all times during the growing season. The soil moisture should be about 60% of field capacity for optimum yield.

The soil type in which potatoes are grown may affect the specific gravity of tubers because of its water-holding capacity, drainage, aeration, structure, temperature, and fertility. Potato plants respond well in soils that have 50% or more pore or air space; sandy loam and loamy soils rich in organic matter are the most suitable for potato cultivation. The optimum soil pH for potatoes is 5.0–5.5 so as to limit potato scab disease, which is favored by alkalinity.

Potatoes require large quantities of mineral nutrients for maximum growth. The recommended ratios of nitrogen, phosphorus, and potassium are 1/2/1, 1/2/2, 2/3/3, and 1/1/1 (Sawant et al., 1991). High-yield potato cultivars require large amounts of potassium for growth and development, and potassium sulfate (K_2SO_4) is recommended because it causes less reduction of specific gravity than potassium chloride (KCl) and potash (K_2O). Micronutrients do also affect potato quality and yield; applications of B, Mn, and Zn increase the starch grain size, while applications of iodine or ammonium molybdate increase the DM and starch content.

2.4 Technological Factors

The industry needs continuous supplies of raw material, which means that it is essential to properly organize the varieties used, planting times, growing zones, mechanical harvesting, transportation, and handling to avoid rapid loss of quality after harvesting.

Because processing causes complete tissue death, potatoes that are to be processed need to be harvested at the exact moment of ripening. Sucrose or chemical maturity monitoring (Gould, 1999a) can be used to optimize harvest timing and make correct decisions on storage temperature protocols. Stark et al. (2003) made the following suggestions for quality maintenance during production and storage.

Potatoes intended for chip production should have a reducing sugar level below 0.35 mg/g (or 0.035%) of fresh tuber weight at harvest time. Potatoes intended for processing as french fries should have less than 1.2 mg/g (or 0.12%) of tuber fresh weight. In both cases sucrose content should be below 1.5 mg/g (or 0.15) of fresh tuber weight.

If sucrose is below the indicated levels, harvesting can proceed and the tubers can be stored in a normal fashion at the appropriate final holding temperature for the variety and the intended market. Potatoes intended for chip processing are typically stored at 10–12.7 °C, for french fry processing 8.3–10 °C, for fresh use 5.5–7.2 °C, and for seed 2.7–4.4 °C.

If tubers come out of the field with sucrose above the indicated levels, they will show color problems after cooking. Storage managers may consider one of the strategies based on manipulation of early storage temperatures shown by Alvarez and Canet (2009).

Sugar accumulation in storage can generally be attributed to cold temperatures, inadequate supply of air to the pile, or senescent sweetening. These conditions can be detected by sugar monitoring, usually before any obvious decline in quality.

Problems with low-temperature sweetening can usually be solved with a 2- to 4-week period of reconditioning at 12.7–15.5 °C, followed by a slow return to the desired holding temperature.

Inadequate air movement in storage also causes sucrose levels to slowly increase. Early detection of the rise in sucrose levels allows this problem to be resolved by increasing the frequency or length of ventilation to the pile.

After 5–8 months of storage, a slow increase in sucrose levels over the next several months may be an indication of senescent sweetening. This is a permanent condition and only gets worse with time. If a sample of potatoes removed from the storage does not respond to reconditioning, then the potatoes should be marketed as quickly as possible. A more detailed discussion of potato tuber quality during storage can be found in Chapter 16 of the book *Potato Production Systems* (Stark and Love, 2003) and in the chapter of this book on postharvest storage of potatoes.

2.5 Quality Assessments of Raw and Processed Potatoes

Only raw material of high nutritional, safety, and sensory qualities should be selected for processing. At the factory, the processor has the choice of accepting or rejecting incoming loads of raw material. The decision is made after measuring samples from each consignment against a raw material specification, which should describe precisely what is required and the extent to which the quality may deviate from the standard before it is rejected (Arthey, 1993).

The tuber samples should be selected at random and be representative of the entire batch in question. The samples should be evaluated for specific gravity and graded for size and absence of external and internal defects. A subsample should be fried, and if the fry color is not satisfactory, a reducing sugar evaluation should be made and the strategies for storage and end use followed as recommended above. The size is important to determine the usage of a given lot of potatoes; medium-large round tubers are preferred by the chip industry, as their shape facilitates peeling with minimal loss; the long, oblong type of potato is preferred by the french fry industry; and size uniformity is preferred for processing in all cases to obviate the need for size grading, rejection of tubers, or cutting large specimens before processing.

Gould (1999b) devised and presented a potato receiving inspection report, a methodology for evaluating quality and grading of potatoes for making chips and the specifications of potatoes for the chip market. He also cites the U.S. standards for grades of potatoes for processing, canned white potatoes, frozen french fried potatoes, and frozen hash brown potatoes. For frozen vegetables in general (including potato products) Canet and Alvarez (2011) cite quality indices or specifications laid down by the industry's own quality control laboratory or produced by public or private institutions.

A wide variety of methods and instruments have been employed to control the quality of raw and processed potatoes. The key characteristics of importance for potatoes for processing are DM, reducing sugar content, color, size, and defects. PotatoPro, the online information source (Potatopro.com) for the global potato industry, offers a directory of suppliers, manufacturers, brands, and products in the global potato industry; also, Martin Lishman are manufacturers and distributors of equipment for evaluating all these parameters of potato quality (http://www.martinlishman.com/agricultural/home/potato-quality-equipmt). Apparatuses for direct measurement of sugar are based on the action of glucose oxidase on the glucose in potato juice; this is oxidized to gluconic acid, and the *ortho*-tolidine indicator changes from yellow to various shades of green depending upon the concentration of glucose in the juice (Kadam et al., 1991). This is a suitable method given the high correlation between the glucose determined and the resulting color of the potato product. The industry standard for instrumental color measurement of par-fried or finish fried french fries is the Agtron-E30, a direct reading reflectance-abridged spectrophotometer designed to provide the ratio of reflectance of a product in two spectral modes, near infrared and green. For chips the Agtron M-series, with a large-area (26 square inch) reflectance spectrophotometer, is designed to deal with the measuring of color and process-related changes associated with foods of irregular geometry with little or no sample preparation. The benefit of instrumental measurements compared to the use of color cards is their objectivity; the Agtron equipment is therefore the preferred method for product specifications (Agtron, Inc., Reno, NV, USA). To control potato damage and bruising, a wirelessly monitored device simulating a potato (Smart Spud) reduces storage diseases and allows the user to see, in a matter of seconds, exactly how harvesters, windrowers, grading, washing, and packing facilities are causing potato damage. The Smart Spud from Sensor Wireless (Sensor Wireless, Inc., Charlottetown, PEI, Canada) cuts down on potato bruising and damage by at least 10% and improves the quality control routine and record keeping by pinpointing where damage is occurring through impact, vibration, and temperature analysis.

Texture is a key component of the quality and palatability of potato products. Texture is generally quantified by measuring the resistance of a product to an applied force. A number of rheological parameters can be used to evaluate a range of tuber characteristics such as firmness, hardness, softness, adhesiveness, fracturability, etc. There is a considerable amount of literature on methods for objectively measuring potato and potato product texture and related chemical and sensory parameters. Owing to space constraints, we can refer here only to those habitually used by the authors. To evaluate the effects of various thermal treatments included in the freezing process, like blanching, freezing, thawing, etc., Canet (1980) used compression, shear, and stress–relaxation tests performed with an Instron food testing instrument Model 1140 (Instron, Norwood, MA, USA) on cylindrical specimens of potatoes (cv. Jaerla) (Canet et al., 1982a,b). Alvarez (1996) characterized the rheological behavior and established the softness kinetics of potato tissues (cv. Monalisa) by means of compression, shear testing (Alvarez et al., 1997, 1999; Alvarez and Canet, 1997, 1998a,b), and creep testing (Alvarez et al., 1998) on cylindrical specimens and tensile tests using an Instron food testing instrument Model 4501 (Alvarez and Canet, 1998b; Alvarez et al., 1999). Also, stress–relaxation tests and texture profile analyses (TPAs) have been performed using the TA.HDi texture analyzer (Stable Micro Systems Ltd, Godalming, UK). An International Technical Specification describes a method for the determination of rheological properties by uniaxial compression tests at a constant displacement rate (ISO/TS 17996/IDF/RM 205:2006(E)). It was devised for hard and semihard cheeses, but the protocol is perfectly adaptable to potato tissues. Canet et al. (2005c) studied the optimization of low-temperature blanching for maintenance of potato firmness by means of such uniaxial compression tests on cylindrical potato (cv. Kennebec) samples.

For mashed potatoes, TPA, CP, and back-extrusion tests were also performed with a TA.HDi texture analyzer. During tests, mashed potatoes were kept at 55 °C by means of a temperature-controlled Peltier cabinet (XT/PC). For performance of TPAs, an aluminum cylindrical probe (P/35) with 35-mm diameter is driven into a 60-mm-diameter stainless steel vessel containing 50 ± 1 g of sample (Alvarez et al., 2009; Fernández et al., 2009a,b;

Jiménez et al., 2013). Samples are compressed to 33% of initial height at a compression speed of 3 mm/s, with a rest period of 5 s between cycles. Canet et al. (2011) showed that under these conditions the action of the jaw is properly imitated. For the CP tests, a spreadability rig is used, consisting of a 45° conical Perspex probe (P/45C) that penetrates a conical sample holder containing 7 ± 0.1 g of mashed potatoes (Alvarez et al., 2005, 2009, 2011a, 2012a,b; Canet et al., 2005b; Fernández et al., 2006). For performance of back-extrusion tests, a rig (Model A/BE, Stable Micro Systems) is used consisting of a flat 45-mm-diameter Perspex disk plunger that moves within a 50-mm-inner-diameter Perspex cylinder sample holder containing 50 ± 1 g of mashed potatoes. The product is extruded to a distance of 20 mm at 2-mm/s compression rate (Alvarez et al., 2011a,b, 2012a,b, 2013a).

For more detailed discussion of potato product quality assessment and assurance, the book *Total Quality Assurance for the Food Industries* (Gould, 2001) is essential reading, and for frozen french fried potatoes we especially recommend Hui's chapter of the *Handbook of Vegetable Preservation and Processing* (Hui, 2004).

3. Processing versus Quality

Processing consists of a series of stages from reception of the products at the plant to their final dispatch. The first main stage comprises a number of preliminary operations to prepare the product for subsequent frying, drying, canning, or freezing. These operations are blow-dry cleaning, removal of stones and washing, inspection and selection, classification, peeling, and, depending on the variety and the final use, chopping and slicing. Especially in the case of frozen potatoes, the second stage consists of blanching (heating for a short time to inactivate the enzymatic systems responsible for off-odors and flavors and changes in color during frozen storage), plus other prefreezing treatments (use of coadjuvants to improve blanching action before frying, or prefrying), and finally cooling and draining to prevent yield and energy loss during freezing. The third and fourth stages are freezing and frozen storage, respectively. Potato products are usually frozen individually, that is, individually quick-frozen. They are then stored frozen, packaged in small containers weighing anything from 400 g to several kilograms for direct dispatch or in bulk in polyethylene-lined pallet boxes, which can contain several hundred kilos of product, thus helping to optimize the utilization of storage space (Canet and Alvarez, 2000).

3.1 Main Preparatory Procedures

The purpose of these procedures is to take raw potatoes as received by processors and from them make a product that, once processed and frozen, is "ready-to-eat" with minimal final preparation by the consumer. Following an initial selection process to meet the quality standards required, the tubers are cleaned by vibration or air blast to remove unwanted materials (such as leaves, husks, etc.) and brushed to remove earth, stones, and excess dirt.

3.1.1 Washing and Peeling

Washing cleans the potatoes of dirt and impurities (soil and waste matter), of pesticide residues, and of up to 90% of the microbial flora. There are various types of washers that can be used (rotary and high-pressure sprays). Light chlorination by adding gaseous chlorine or sodium hypochlorite enhances the action of water, preventing the formation of sludges of bacterial origin in the equipment and the development of unpleasant odors. Free chlorine contents in the region of 5–10 ppm do not adversely affect product flavor or corrode equipment.

Peeling, one of the most delicate pretreatments, is achieved industrially by abrasion, high-pressure steam, treatment with sodium hydroxide solution, or mechanically. Abrasion is effected by rough, moving surfaces, which remove the outer surface of the product, but it has the drawback of considerable loss of raw material. Steam peeling consists in heating the product to a temperature of up to 80 °C and subjecting it to pressures of 392–686 kPa for between 30 s and 3 min. In sodium hydroxide peeling, the product is preheated and then immersed in a 10–20% solution at a temperature of 60–90 °C for between 1 and 5 min depending on the type of product. Using sodium hydroxide with infrared heating can cut sodium hydroxide solution consumption by 80%, reducing water consumption by up to 95% (Canet and Alvarez, 2011).

3.1.2 Cutting

After washing and peeling, the product may be subjected to any of a number of procedures (e.g., sorting, paring, trimming, cutting, and pulping) depending on variety and type of the final product. There is a wide range of equipment for high-yield performance of these operations. A selection of potato-processing machines offering various kinds of cuts can be found at http://www.urschel.com/Potatoes_15f12ffb6ca0ab8f43b1b54.html (2015; Urschel Laboratories, Inc.).

Figure 2 depicts the operation of a DiversaCut 2110® dicer by Urschel Laboratories, Inc. This equipment uniformly dices, strip cuts, and slices at high production capacities and is specially adapted for potatoes. Flat and crinkle strips in a wide variety of widths (1.6–25.4 mm) can be made (by removing either the crosscut knife spindle or the circular knife spindle); also, minimum flat strips such as 3.2 × 3.2 mm × length of product, crinkle strips such as 7.1 × 7.1 mm × length of product, or maximum flat strips, for instance, 25.4 × 25.4 mm × length of product. A slicing knife, circular knife spindle, and crosscut knife spindle are used for dicing. The dice size is changed by using the required cutting spindles and adjusting the slice thickness. Other options are circular knife cuts (3.2–76.2 mm), crosscut knife cuts (3.2–38.1 mm), and crosscut knife crinkle cuts (7.1–15.9 mm).

After the preparatory procedures and before blanching, or immediately after blanching and after cooling and draining, depending on the product, thorough inspections are essential to

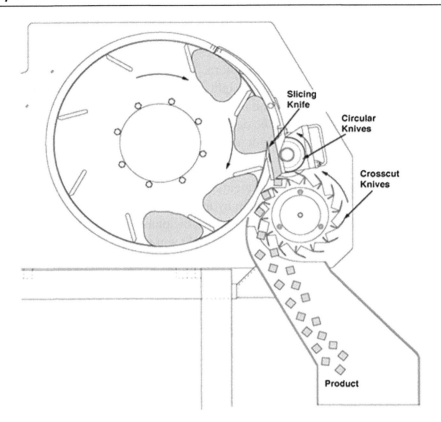

Figure 2
Diagram showing the operation schema of the DiversaCut 2110® dicer by Urschel Laboratories, Inc. This equipment uniformly dices, strip cuts, and slices a wide variety of vegetables. *From Anonymous. How to cut fruits and vegetable products. Urschel Laboratories, Inc., Valparaiso, Indiana, USA, with permission.*

eliminate unwanted material from the line. If done manually, such inspections either reduce the line output or require a lot of manpower. There are now computerized inspection systems using visible spectrum, infrared, or X-ray detectors or TV images, which help to raise output. Nowadays, inspection is faster, more precise, and more economical, and only a very small part of the inspection process is manual (Canet and Alvarez, 2011).

3.2 Thermal Treatments Effects

As mentioned above, there are various stages in the production of frozen potatoes during which there may be significant loss of product quality: initial processing and preparation prior to freezing, the freezing step itself, and the frozen storage that follows freezing. For example, it is well known that blanching leads to a loss of nutrients and other product quality characteristics such as texture, flavor, and color. Recrystallization and surface drying are accelerated by temperature fluctuations during frozen storage of vegetable products (Alvarez and Canet, 1998a, 2000b). In addition, the effects of freezing depend on whether the tissue has been blanched. Rigorous blanching accentuates the damage caused by freezing, producing structural changes that are detectable even after cooking (Canet, 1980; Alvarez, 1996). The freezing process itself causes damage to cell structures, but more appropriate methods can be used to optimize quality. Alvarez et al. (1997) showed that when potato tissue was cooled at a low temperature (3 °C) for a long time (30 min) prior to freezing, the mechanical strength of the tissue increased at the various freezing rates tested. Lasztity et al. (1992) reported that when blanched and then frozen, potato tissue was no longer capable of maintaining differences in concentration and pressure in its cells owing to the damage to the cell walls.

All those processes that can compromise final quality during freezing should be considered separately, but it has to be realized that there are significant interactions between them. For example, low-temperature long-time (LTLT) blanching of potatoes (cv. Kennebec), both without further processing and prior to cooking or freezing + cooking, significantly increased firmness retention as measured from compression parameters (Canet et al., 2005c). Figure 3 shows the rheological behavior of potato samples subjected to various blanching treatments without further cooking, cooked samples, and frozen and cooked blanched samples. Engineering stress value of potato tissue blanched at 60 °C increased by only about 6% of its original value after 60 min (Figure 3(a)). When blanched tissue was cooked, or frozen and cooked, firmness relative to that of the blanched tissue decreased. However, there was a significantly more notable increase in firmness as a function of blanching time at 60 °C.

When tissue blanched at 60 °C was cooked, engineering stress (Figure 3(b)) increased after 60 min by about 270% of its original value (corresponding to the unblanched cooked control). When tissue blanched at 60 °C was frozen and cooked (Figure 3(c)), engineering stress increased after 60 min by about 85% of its original value (corresponding to the unblanched, frozen, cooked control). A strong "gel" was formed in the potatoes blanched at 60 °C; this was especially apparent after further processing. This result suggests that it is important to consider all the different stages involved in the frozen potato production process so as to determine the true effect of LTLT on their texture.

3.2.1 Blanching and/or Drying Methods

Blanching is a thermal treatment commonly applied in a variety of vegetable preservation treatments and is particularly important in freezing because it has a very strong impact on quality (Canet and Alvarez, 2011). The product is heated, typically by brief immersion in water at 85–100 °C or by steaming at 100 °C. The primary objective is to inactivate enzymes responsible for alterations in sensory quality attributes (off-flavors and off-odors) and in nutritional value (loss of vitamins) during storage. The chief factor affecting processing time is blanching temperature. Oxidases, peroxidases, catalases, and lipoxygenases are destroyed

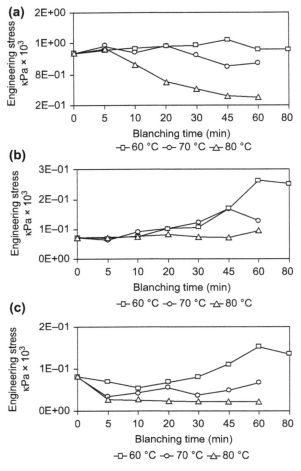

Figure 3
Effect of low-temperature blanching on engineering stress in potato samples. (a) Engineering stress in blanched potato samples. (b) Engineering stress in blanched-and-cooked potato samples. (c) Engineering stress in blanched-frozen-and-cooked potato samples.

by the heat of blanching, and blanching effectiveness is usually monitored by measuring peroxidase activity, in view of its high heat resistance (Canet, 1989).

Some authors have investigated the possibility of improving the textural quality (in this case firmness) of whole new potatoes, applying low-temperature blanching (LTB) at 60, 65, 70, 75, 80, 90, and 100 °C for times up to 1 h (Abu-Ghannam and Crowley, 2006). The activity of pectin methyl esterase (PME) was determined for whole new potatoes with an optimum activity of 2.92 μmol/min/g at 65 °C for 15 min. Processing was by immersion in a thermostatically controlled water bath at 90 or 100 °C for times up to 25 min and with and without blanching at 65 or 75 °C for 15 min. Firmness was significantly higher ($P < 0.05$) in processed potatoes blanched at 65 °C than in those cooked at 95 or 100 °C without blanching.

LTB of sweet potatoes before steam cooking has also a shown significant increase in tissue firmness and cell wall strengthening (He et al., 2014). PME activity decreased by 87.8% after 30 min of blanching in water at 60 °C, while polygalacturonase and β-amylase activity decreased 69.4% and 7.44%, respectively, under the same condition. Both PME and β-amylase play important roles in tissue firmness.

LTLT blanching has been used as a thermal pretreatment prior to various processing operations such as freezing, dehydration, canning, sterilizing, and drying. Potatoes are among the most popular dried vegetables used for snacks. A comprehensive review of the progress in the production of dried potato snacks was presented by Huang and Zhang (2012), and recommendations were made for future research.

The literature gave different optimum blanching conditions for different potato-processing operations. French fry texture was found to improve when LTB was carried out before frying at 200 °C for 4 min, indicating stronger intercellular bonds in response to changes produced in the pectin substances by PME (Aguilera, 1997).

Pedreschi and Moyano (2005) studied the effect of predrying on the texture and oil uptake of potato chips. After blanching, the potato slices were air dried at 60 °C until a final moisture content of 60% (wet basis [wb]) was attained. Blanched potato slices were used as the control. Predrying dramatically decreased the oil absorption and significantly increased the crispness of the blanched potato slices after frying.

In current frozen french fry manufacturing practice, achieving the target texture in potato strips from the blancher relies on "to-bite" assessment by line operators on a trial-and-error basis. Liu and Scanlon (2007) presented a quantitative description of the texture changes in terms of mechanical properties of the strips as affected by the blanching conditions (temperature and time). Strips from the central planes of each sorted tuber were excised and blanched in a steam-heated kettle at temperatures ranging from 62.8 to 90.6 °C for 2–20 min. The texture of the blanched strips was evaluated using an indentation method. The authors found that at lower temperatures (<74 °C), blanching time had little effect on the texture of blanched strips. However, at higher temperatures (≥74 °C) the texture, as expected, softened with increasing temperature and time.

In response to food quality and safety requirements, the SAFES (systematic approach to food engineering systems) methodology has been developed to precisely describe and quantify changes taking place throughout the operations involved in food processing (Barrera et al., 2007). The application of SAFES methodology to french fry manufacture has proven to be a useful tool for describing some textural attributes produced by quantitative changes in water and oil content and in the volume and state of aggregation of the starch. The study also shows the need for additional information in certain areas of knowledge, such as the inactivation rate of the enzymes that catalyze oxidation and the extraction rate of reducing carbohydrates during the blanching step.

Owing to increasing demand from health-conscious consumers, more emphasis has been placed on investigating alternative techniques to replace conventional deep-fat frying in order to produce health-friendly snack products, including potato chips. Low-pressure superheated steam drying (LPSSD) has been proven to have potential for producing fat-free potato chips if performed in combination with appropriate predrying treatments (Pimpaporn et al., 2007). LPSSD at 90 °C with combined blanching and freezing pretreatments has been proposed as the most favorable combination for drying potato chips in terms of drying behavior and dried product quality.

The announcement by Swedish researchers in April 2002 that they had detected acrylamide in many different foods has generated a large number of scientific publications from various parts of the world. Masson et al. (2007) have reviewed the state of the art as regards the occurrence of acrylamide in foods subjected to heat treatments in industrial operations such as frying, baking, toasting, and extrusion, considering the application of potato pretreatments such as washing, blanching, and immersion in acid solutions. Special emphasis is placed on the differences between atmospheric frying and vacuum frying and on the health risks associated with acrylamide intake through common foods. Color formation in predried potato slices during frying and acrylamide formation in the final potato chips has also been addressed (Pedreschi et al., 2007a). When fried at 180 °C, predried potato chips possessed 44% lower acrylamide content than that in the control sample. Pedreschi et al. (2007b) studied the kinetics of color development in blanched and blanched–NaCl-impregnated potato slices during frying by using the first-order dynamic method and evaluated the effect of NaCl on reducing acrylamide formation in potato chips.

A potato genetically engineered to reduce the amounts of acrylamide has been approved for commercial planting in the United States. The potato's DNA has been altered so that less acrylamide is produced when the potato is fried; the new potato also resists bruising, avoiding losses during harvesting, shipping, or storage. The biotech tubers were developed by J.R. Simplot Co., a privately held company based in Boise, Idaho. This potato is one of a new wave of genetically modified crops that aim to provide benefits to consumers, not just to farmers. Some consumers are questioning the safety of genetically engineered crops and demanding that the foods made from them be labeled; the question now is whether the potatoes will be adopted by food companies and restaurant chains. Genetically modified potatoes failed once before. In the late 1990s, Monsanto began selling potatoes genetically engineered to resist the Colorado potato beetle, but the market collapsed after big potato users, fearing consumer resistance, told farmers not to grow them. This time around could be different because the Simplot product called the Innate potato does not contain genes from other species like bacteria, as do many biotech crops, and promises potential health benefits to consumers.

The U.S. National Potato Council, which represents potato farmers, welcomed the approval, albeit with reservations; growers wanted new technology but have expressed concern that exports could be disrupted for countries that have not approved the genetically engineered potato varieties (Polak, 2014).

3.2.2 Stepwise Blanching and PME Activity

One of the procedures used to palliate the negative effects of blanching on texture is stepwise blanching. Previous research involving stepwise blanching was conducted to study the effects of temperature and time variables on the PME activity and texture modifications in potatoes (Canet et al., 1982b; Alvarez et al., 1999; Alvarez and Canet, 1999b). The literature gives different optimum blanching conditions for different potato-processing operations and for different potato varieties. Brown and Morales (1970) recommended 80 °C, 15 min, for the first step and 95 °C, 1 min, for the second for blanching of potatoes prior to frying. Canet et al. (1982b) compared blanching of cylindrical potato specimens at boiling point (97 °C, 2 min) with both one-step blanching (80 °C, 6 min) and stepwise blanching at 50, 60, and 70 °C, followed by cooling and a second step at boiling point (97 °C, 2 min); they concluded that stepwise blanching consisting of LTLT pretreatment (70 °C, 10–15 min) followed by cooling and high-temperature short-time (HTST) blanching (97 °C) reduced damage to the tissue structure. This stepwise blanching has produced substantial improvements in final product textures of potato cultivars Jaerla (Canet et al., 1982b), Monalisa (Alvarez, 1996; Alvarez and Canet, 1999b; Alvarez et al., 1999), and Kennebec (Canet et al., 2005c), including after freezing and final preparation (Canet et al., 1982a, 2005c; Canet and Espinosa, 1984; Alvarez and Canet, 1999b; Alvarez et al., 1999).

Several theories have been put forward in the literature (Andersson et al., 1994) to explain this firming effect in potato: retrogradation of starch, leaching of amylose, and stabilization of the middle lamellae and cell walls by activation of the PME enzyme. What the enzyme does is demethylate the carboxymethyl groups of pectic polysaccharide chains. The reduction in the degree of methylation may in turn trigger various processes related to texture and firmness, such as cross-linking by Ca^{2+} ions, increased hydration at the demethylated sites, reduced susceptibility to heat-induced β-degradation of pectins, and enhanced shielding and repulsion forces by the electric charges within the biopolymer matrix of the cell walls. As such, PME has a long record of confusing effects on observed firmness.

The increase in firmness with respect to that of unblanched potato samples has been found to diminish in the following order: blanched at 60 °C for 60 min and cooked > blanched at 60 °C for 60 min, frozen, and cooked > blanched at 60 °C for 60 min (Canet et al., 2005c). The blanching temperature, which resulted in increased firmness, was also in the region of 60 °C for sweet potatoes (Truong et al., 1998). Compared to the unblanched cooked sweet potatoes, the steam-cooked sweet potato samples blanched at 60 and 70 °C for 45–150 min were respectively about 4.2–13.6 and 3.8–6.2 times firmer. The effect of deesterification of pectin by endogenous PME on firmness becomes evident only after heating at high temperatures, as in sterilization and cooking, and there are various results that confirm previous findings. Deesterified pectin is less susceptible to subsequent β-eliminative degradation and is therefore more heat-stable. Postprocessing results in greater pectin insolubility, which is generally thought to increase cell–cell adhesion (Stolle-Smits et al., 1998).

It has been stated that the differences in optimum conditions for stepwise blanching proposed by various researchers are due to the levels of PME enzyme activity occurring in the different potato varieties used, and that this is affected by the maturity of the potato and the season. Canet et al. (2005c) kept potato tubers (cv. Kennebec) in refrigerated storage, and firmness and PME activity were periodically sampled over 80 days. The PME activity of potato tissues blanched at 60, 70, and 80 °C for varying periods of time was measured (Figure 4(a)). PME activity increased by about 65% in the fresh tubers after 20 min at 60 °C, but after 60 min it declined to about 40% of its original value. At 60 °C, PME in potato tissue appears to have a relatively short inactivation time, or possibly its optimum activation temperature is <60 °C. PME activity declined to less than 10% of its original value after 5 min at 70 °C, and no activity at all was detected when the time at this temperature was prolonged. Similarly, no PME activity was detected in samples blanched at 80 °C. In potato tubers (cvs Nicola and Irene), Van Dijk et al. (2002) found that the PME activity tended to remain constant during preheating at 60 °C for 60 min.

The effect of storage time on the PME activity of fresh tissue is shown in Figure 4(b). PME activity was about 95% greater than its original activity (at 0 days) after 35 days in storage, but after 50 days it had declined to about 42% of the original activity. Moreover, after 65 and 80 days in storage, no PME activity was detected in fresh potato tissues. As Tijskens et al. (1997) pointed out, the relationship between measured PME activity and observed firmness is inherently complex. Substantial compositional changes have been observed in the pectic polymers of fresh potatoes during storage (Van Dijk et al., 2002). It seems reasonable to

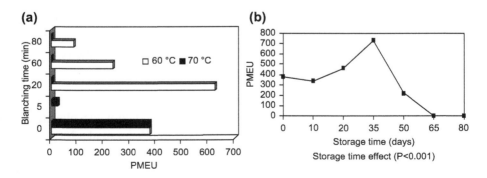

Figure 4
PME activity. (a) Effect of LTB on PME activity. (b) Effect of storage time on PME activity.

assume that the increase in PME activity after 35 days would cause changes in the pectic polymers, leading to a firmer texture.

The expression of a gene encoding an isoform of PME (PEST1) was associated with cooked tuber textural properties, and a transgenic approach was undertaken to investigate further the impact of the PEST1 gene (Ross et al., 2011). Antisense and overexpressing potato lines were generated. In overexpressing lines, tuber PME activity was enhanced by up to 2.3-fold, whereas in antisense lines, PME activity was decreased by up to 62%. PME isoform analysis indicated that the PEST1 gene encoded one isoform of PME. Analysis of cell walls from tubers of the overexpressing lines indicated that the changes in PME activity resulted in a decrease in pectin methylation.

3.2.3 Freezing, Freezing–Thawing Cycles, and Thawing–Cooking

The freezing process as such consists of lowering the product temperature to −18 °C at the thermal center, resulting in crystallization of most of the water and some solutes. Ice crystallization occurs only after a degree of supercooling—i.e., reduction of the temperature between −5 and −9 °C in a matter of seconds. In the freezing stage, most of the water in the product undergoes a phase change to ice; this change is not complete until the final temperature at the thermal center is at least as low as the storage temperature.

The duration of the freezing process depends on the freezing rate (°C/h). This is defined by the International Institute of Refrigeration (2006) as the difference between initial temperature and final temperature divided by freezing time, freezing time being defined as the time elapsing from the start of the prefreezing stage until the final temperature has been attained. This will be affected by product size (particularly thickness) and shape, as well as by the parameters of the heat-transfer process and the temperature of the cooling medium.

The beneficial effect of rapid freezing rates and stepwise blanching on final structure and texture, even after cooking, has been reflected in the results of texture analysis by various methods (histological, sensory, imitative, and objective) in studies of potatoes (Canet, 1980; Canet et al., 1982a; Alvarez and Canet, 1995; Alvarez, 1996). The aspect of freezing that really produces irreversible negative effects on textural quality is crystallization (Canet, 1989); therefore, the freezing rate is particularly important in the phase of maximum ice crystal formation. Freezing rates can now be modulated in the various processing stages using programmed freezing. This is a possibility that the food industry ought to consider seriously. Although in theory the temperature fluctuations to which vegetables are subjected rarely go as far as total thawing and subsequent refreezing, the effect of such fluctuations at various rates can usefully be examined to assess the mechanical damage inflicted on tissue structure and the cumulative effect.

Alvarez et al. (1997) studied the effects of the freezing rate and programmed freezing on the rheological parameters and tissue structure of potato (cv. Monalisa). In particular they analyzed the effect on potato tissues of three different freezing rates (0.5, 1.25, and 2 °C/min) down to −18 °C, thawing up to +20 °C at the same rates, and one, two, three, or four successive freeze–thaw cycles (FTCs). The effect of the freezing rate on the zone of maximum crystallization was also examined, along with various combinations of programmed freezing, and the effect of prior cooling was assessed. The alterations in rheological behavior of slow-thawed potato tissues were minimized by precooling (3 °C for 30 min), slow cooling (0.5 °C/min) before and after the maximum ice crystallization phase, and quick freezing (2 °C/min). This is thought to occur because precooling cuts down on the time lapse between freezing at the surface and freezing at the product's thermal center, so that freezing-induced expansion takes place before too rigid a crust forms on the surface. Scanning electron microscopy (SEM) examination of the tissues showed differing degrees of mechanical damage to tissue structure and a linear increase in the tissue's mechanical strength caused by precooling.

The structural rigidity of samples subjected to four FTCs was less than half that of samples subjected to only one cycle (Alvarez and Canet, 1995). The authors concluded the following: (1) precooling and a high freezing rate during the phase of maximum ice crystal formation have a positive effect on potato texture and tissue structure, (2) slow thawing has a positive effect, and (3) it is essential not to subject potatoes to more than one FTC.

Alvarez and Canet (1999a) froze mashed potatoes made from dehydrated potato flakes to determine the changes that occurred in the cell structure. Breakdown of the cell wall by freezing reduced the oscillatory rheological properties, and it was found that the frozen–thawed mashed potatoes consisted of dilute dispersions of swollen and disrupted intracellular starch granules. The viscoelastic behavior of starch pastes is likely to be related to the volume fraction, number, and size distribution of the swollen particles they contain, and such factors increase with increasing flake concentration.

Freezing of mashed potatoes involves initial processing, freezing, frozen storage, and thawing. Freezing of foods can have a detrimental effect on their sensory and water-holding properties as a result of physical disruption of cells or cell components (Jul, 1984) or of changes in the structure of certain macromolecules (Suzuki et al., 1994). Technological solutions adopted to minimize the effects of freezing include the use of rapid freezing methods. The effect of freezing conditions on the quality of freeze-chilled reconstituted mashed potato has been examined by O'Leary et al. (2000) and Redmond et al. (2002). Freeze chilling led to a significantly darker, firmer product (higher probe penetration force) and increased centrifugal drip loss (DL) compared with chilled product.

The structure and quality of commercial mash made from flakes are also detrimentally affected by freezing (Alvarez et al., 2005; Canet et al., 2005b), and therefore the authors do not recommend either freezing or frozen storage of this type of mashed potato. However, the effects of freezing temperature (−80, −40, or −24 °C) and thawing mode (microwave or overnight at 4 °C) on quality parameters of mashed potatoes made from tubers (cv. Kennebec) were also examined in the cited study. TPA hardness and oscillatory parameters were lower in the samples thawed at 4 °C than in those thawed by microwaving at all the freezing temperatures used, which may be ascribed to gelatinization of the starch released from damaged cells. It was concluded that mash made from Kennebec potatoes should be frozen quickly and thawed by microwave under the conditions described to obtain a product more like freshly made mashed potato.

Another strategy to minimize damage from freezing and thawing is to incorporate compounds that interact with water and offer protection against the deleterious effects of thawing, in particular, for example, cryoprotectants (Downey, 2003). Hydrocolloids and proteins, the two kinds of biopolymers used by food technologists to control structure, texture, and stability (Dickinson, 1998), both possess cryoprotectant properties. In addition, hydrocolloids are known specifically for their water-holding characteristics and are used in starch-based products to influence the gelatinization and rheological properties of starches.

It is generally accepted that each hydrocolloid affects the pasting and rheological properties of starch-based systems, like mashed potato, in a different way (Chaisawang and Suphantharika, 2006). There are many possible factors involved in this, the most important being the molecular structure of the hydrocolloids and/or the ionic charges of both starches and hydrocolloids (Shi and BeMiller, 2002; Chaisawang and Suphantharika, 2005). Starch is also widely used in the food industry as a thickening, stabilizing, and gelling agent (Morikawa and Nishinari, 2000; Mandala et al., 2004), but uses of native starch are limited since pastes present problems including retrogradation, syneresis, and slow resistance to shear treatment (Korus et al., 2004). To improve the physical and chemical properties of these pastes, starches have been chemically modified, for example by acid hydrolysis, oxidation, etherification, and cross-linking.

Downey (2002) studied the effects of addition of hydrocolloids (xanthan gum (XG), guar gum, pectin, carrageenan) and dairy proteins (sodium caseinate (SC), whey protein (WP) concentrate) on centrifugal DL and maximum resistance to penetration force in frozen and thawed, cooked puréed vegetables (potatoes (cv. Rooster), carrots, and turnips). However, mixtures of cryoprotectants may offer specific advantages with regard to alterations in the physical properties of food ingredients. These can arise from either synergistic interactions or phase-separation phenomena (Doublier, 1997).

It is also the case that consumers at home normally use microwave appliances to thaw and heat frozen foods. In view of the importance of temperature history for product functional and sensory properties, it is clearly essential to obtain information on the effects of such thermal treatment (Downey, 2003). The temperature-dependent dielectric and thermal properties can be used in microwave heat-transfer models for improving heating performance of mashed potato in domestic microwave ovens (Chen et al., 2013).

Cryoprotectants such as hydrocolloids (amidated low-methoxyl (ALM) and high-methoxyl pectins), κ- and ii-carrageenan (κ-C and ii-C), XG, and dairy proteins (WP, SC) have been added to mashed potatoes to investigate ways of improving the effects of freezing and thawing using a domestic microwave oven (Alvarez et al., 2008). The authors found that each hydrocolloid and protein, depending on concentration, affected the mechanical properties (instrumental textural profile analysis (ITPA), cone penetration (CP) test), the total color difference (ΔE^*) with respect to fresh control (FC), and the sensory attributes of fresh (F) and frozen–thawed (F/T) mashed potatoes in a different way. In the F/T samples, adding 5 and 8 g/kg ALM; 3, 5, and 8 g/kg κ-C; 1.5, 3, 5, and 8 g/kg ii-C; and 1.5, 5, and 15 g/kg WP significantly increased ITPA consistency. Also, adding 2.5 and 5 g/kg XG significantly enhanced ITPA consistency of the F/T product. In both F and F/T samples, κ-C produced the highest ITPA consistency and also a high CP average force; this indicated a stronger synergistic effect in κ-C/denatured milk protein systems, although the excessive thickening and stickiness it conferred was judged undesirable by the panelists.

Fernández et al. (2008) characterized the rheological behavior of the above-mentioned mashed potatoes with added biopolymers using steady shear measurements. F and F/T mashed potatoes present shear thinning with yield stress (Canet et al., 2005a), and dynamic shear data reveal weak gel-like behavior in potato purées (PPs) (Alvarez et al., 2004, 2010b). F and F/T samples with added κ-C and ii-C were more structured. Freezing and thawing did not significantly affect the steady data of the product with added carrageenans and xanthan, highlighting the potential of these gums to stabilize the texture of frozen and microwave-thawed mashed potatoes compared with F counterparts. Figures 5 and 6 show radar plots constructed to represent perceived sensory texture attributes in both F and F/T mashed potatoes enriched with the various cryoprotectants. The product yielded by xanthan was softer than controls without added cryoprotectants in both F and F/T samples, but samples containing 0.5 and 1.5 g/kg added xanthan were judged more acceptable than the FC because of the creamy mouthfeel it produced (Alvarez et al., 2010b). Following this, the authors suggested research into blends of various concentrations to achieve optimum effects, as shown later on.

The effects of added functional ingredients on the rheological behavior and texture of F and F/T mashed potatoes have also been studied (Alvarez et al., 2010a,b, 2011a,b,d, 2012a,b, 2013a,b). The main purpose of this applied research was to develop, rheologically characterize, and optimize the texture of F and F/T purées made from potato, so as to provide both individual consumers and caterers with purées that can be rapidly made ready (microwaving) without tedious conventional preparation and incorporate various functional properties demanded by consumers.

Extra virgin olive oil (EVOO) has interesting nutritional characteristics, which are linked to its biophenol content, with very important antioxidant properties. The effects of adding EVOO on the textural, rheological, physical, and sensory characteristics of F and F/T mashed

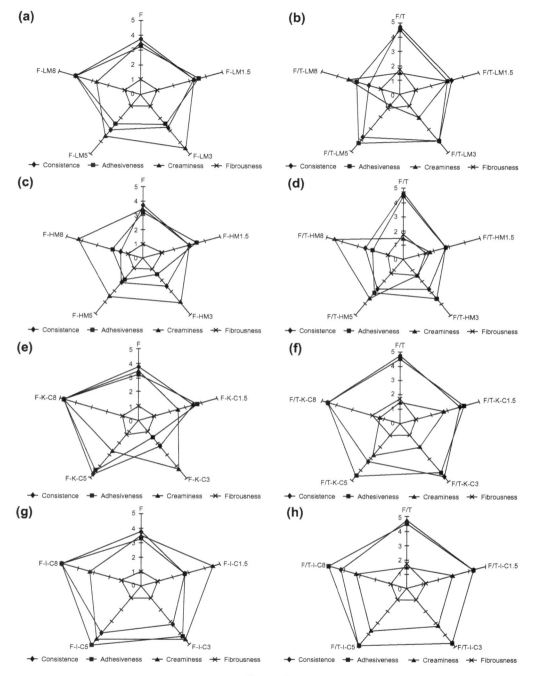

Figure 5

Effects of biopolymer concentration and freezing and thawing processes on the texture sensory attributes of mashed potatoes with added pectins and carrageenans. Identification of system notation is in the text.

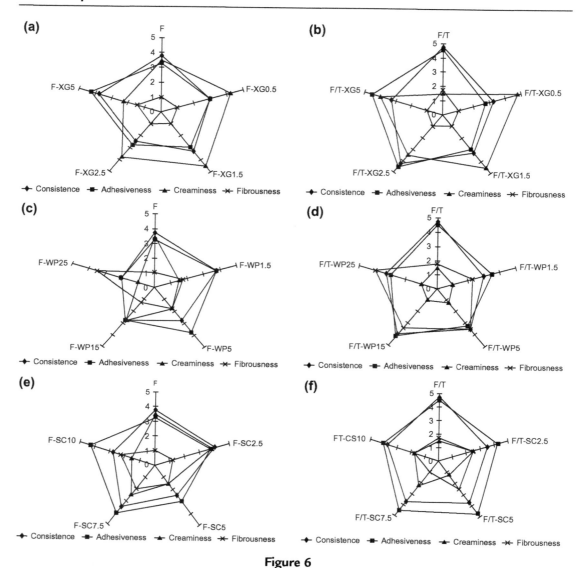

Figure 6
Effects of biopolymer concentration and freezing and thawing processes on the texture sensory
attributes of mashed potatoes with added XG and dairy proteins. Identification of system notation
is in the text.

potatoes formulated without and with added cryoprotectants were investigated (Alvarez et al.,
2011a,d). Increased EVOO concentration resulted in less structured systems and enhancement
of color due to an increase in overall light scattering and pigment content. Addition of EVOO
in increasing concentrations enhanced the sensory quality of mashed potatoes in terms of
reduced granularity, denseness, cohesiveness, adhesiveness, and fibrousness and increased

homogeneity, ease of swallowing, and palate coating. Creaminess was the most crucial factor for overall acceptability (OA) of the products and could be explained by the presence of EVOO aggregates observed by microstructure analysis. Samples with 50 g/kg added EVOO were judged the best of all.

Inulin (INL) and its hydrolyzed product (oligofructose) are both recognized as functional food ingredients and are particularly attractive to people pursuing a healthy diet (Tseng and Xiong, 2009). In particular, long-chain INL can act as a fat mimetic thanks to its capacity to form microcrystals, which interact with one another to form small aggregates that occlude a great amount of water; this creates a fine and creamy texture that gives a mouthfeel similar to that of fat (Villegas et al., 2007). F and F/T mashed potatoes formulated without and with added κ-C and XG, and with added INL (1.5, 3, 4.5, and 6%), were perceived as significantly softer (smoother) and creamier than the samples without INL (Alvarez et al., 2011b).

Following this, the effects of the addition of INL and EVOO blends at various INL/EVOO ratios (0/0, 0/60, 15/45, 30/30, 45/15, 60/0, 30/45, and 45/30) on the rheological, physical, sensory, and structural properties of F and F/T mashed potatoes formulated without and with added cryoprotectants were analyzed and compared (Alvarez et al., 2010a, 2013a). Addition of INL and EVOO (either alone or blended) reduced apparent viscosity and pseudoplasticity, producing softer systems, indicating that both ingredients behave as soft fillers entrapped and suspended in a rigid amylose/amylopectin matrix, respectively. Development of gelling properties of INL-mashed potato systems was observed when the products were frozen and thawed, which was related to the formation of smaller INL particles.

The potential use of commercial fibers (pea fiber, INL, and their blends) as fiber-enriching agents in F/T mashed potatoes was also reported (Alvarez et al., 2012b). The association of pea fiber at low concentration (<15 g/kg mashed potatoes) and INL at high concentration (>45 g/kg) is strongly encouraged to fortify the diet without promoting negative effects on textural and rheological properties of F/T mashed potatoes or color and OA of the resulting products.

The effects of the addition of soy protein isolate (SPI) (0, 15, 30, 45, and 60 g/kg) on viscoelastic properties, large deformation measurements, and microstructure of F and F/T mashed potatoes were investigated (Alvarez et al., 2012a). Rheological data showed weak gel behavior for both F and F/T mashed potatoes without and with added SPI, together with a significant decrease in system viscoelasticity (G' and G'') with increasing SPI volume fraction, primarily attributed to the lack of interaction between the amylose/amylopectin matrix and the dispersed SPI particles or aggregates as revealed by SEM.

A PP serving of 200 g with added SPI and/or INL concentrations of 1.5–6% provides from 3 to 12 g of SPI and/or INL, respectively (Alvarez et al., 2012c, 2014). The presence of SPI strands is a dominant factor in texture perception of PP with regard to mouthfeel, geometric attributes (visual graininess and fibrousness), and the after-feel attribute (mouth coating), while INL crystallites are the most important feature influencing creaminess. Figure 7 shows the morphological differences between six selected F/T samples rated by both trained and untrained panels. It is clear that at the 0/6 ratio (Figure 7(b)), INL crystallites were dispersed throughout the amylose/amylopectin matrix, forming a strong gel with characteristics of a one-phase system. With addition of SPI alone at 6%, proteins formed a network composed predominantly of filamentous SPI aggregates (Figure 7(d)). Adding small amounts of SPI

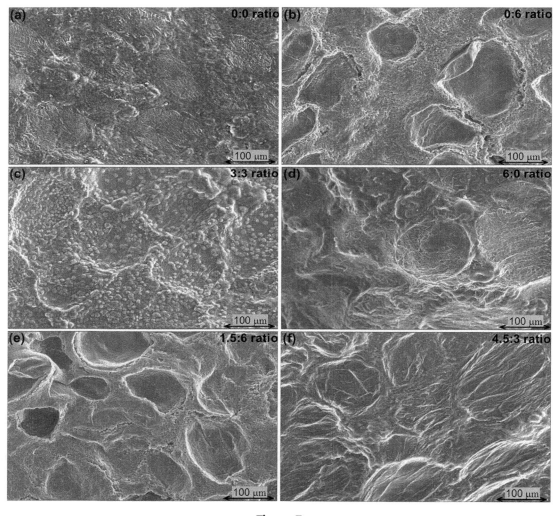

Figure 7

Micrographs of F/T PP with added SPI and INL blends at six selected SPI/INL ratios. (a) 0/0 ratio. (b) 0/6 ratio. (c) 3/3 ratio. (d) 6/0 ratio. (e) 1.5/6 ratio. (f) 4.5/3 ratio. Original magnification was 200 (bar = 100 mm).

(1.5%) together with INL at >3% increases the intensity of perceived creaminess, which was the most important descriptor in practical terms. In view of the foregoing, PP with added SPI and INL is a promising foodstuff for incorporation into a diet and, in addition, even improves product structure.

Anion-exchange high-performance liquid chromatography (HPLC) with pulsed amperometric detection (HPAEC-PAD) analyses have also shown that the FTC did not influence the degree of polymerization of the INL (Alvarez et al., 2013b). Neither was INL chemical structure affected by SPI enrichment. Consequently, F/T mashed potatoes with added blend ratios of SPI/INL retain the INL prebiotic effect. Nevertheless, SPI significantly increased the differences in the rheological behavior between F samples and their F/T counterparts, therefore reducing the freeze–thaw stability conferred by XG and κ-C. An increase in the daily intake of soluble fiber and/or soy protein can be achieved, depending on one's requirements, by adding 0/60, 15/45, 45/15, 45/30, 60/15, and 60/60 blend ratios to the products destined to be frozen and thawed, while leading to a less structured system with reduced increase in viscoelastic function values after the FTC.

In addition, a study compared trained, semitrained, and untrained sensory profiles for seven frozen PP samples by using quantitative descriptive analysis (QDA), flash profiling (FP), and projective mapping (PM) (Jiménez et al., 2013). Data analysis methods for QDA, FP, and PM were principal component analysis, generalized procrustes analysis, and multiple factor analysis (MFA). For the three profiling techniques, hierarchical clustering analysis was applied to group PPs as per their sensory profiles. Assessors' consensus was analyzed through MFA and correlation coefficients. For each technique, MFA also allowed assessing correlations between sensory and instrumental variables. Results showed that both FP and PM methodologies were able to identify differences in sensory characteristics of PPs, providing very similar sensory maps. Nevertheless, higher correlations between sensory attributes evaluated by semitrained and trained assessors were found, indicating that preferably FP should be used as a rapid choice as opposed to QDA in hot-served PPs.

The results of these studies expanded the basic knowledge and technological know-how regarding the effects on final texture of adding various functional ingredients and at the same time identified developments and specifications that the industry can apply to produce fresh and frozen PP with a given functional characteristic and an optimum standard final texture.

On the other hand, frozen purées made from mixtures of vegetables (containing potato) have a good potential market in Europe. In the manufacture of these frozen vegetable purées, potato is added before freezing. These products may therefore be considered starchy foods and as such may present quality problems such as syneresis and organoleptic and textural changes, especially if they are subjected to FTCs. These problems have been ascribed in hydroxypropyl potato starch pastes to phase separation caused by retrogradation of the starch (Eliasson and Kim, 1992; Kim and Eliasson, 1993; Kim et al., 1993).

Alvarez and Canet (2001c) analyzed the influence of freeze–thawing cycles (up to six) on the viscoelastic properties of three commercial frozen vegetable purées made from mixtures of potato with other vegetables (broccoli–potato, carrot–potato, and celery–potato). The purpose of the study was to ascertain the changes occurring in dynamic properties of frozen purées during heating (20–90 °C) and cooling to 20 °C. Repeated cycles led to concentration of dissolved starch components and changes in the cell and tissue structure of the vegetable components, although the effect of cycling on rheological behavior depends on the measurement temperature. When the number of FTCs was increased, phase angle (δ) peaks were reached in shorter times (~1000 s at 80 °C), possibly indicating slower relaxation processes in these systems with successive freezing and thawing. Similar behavior has been reported for amorphous amylopectin in this dynamic property (Kalichevsky et al., 1992). Upon cooling, irrespective of the number of FTCs, structural rearrangements can involve local aggregation phenomena that lead to an increased degree of phase separation and/or a more coarsely aggregated system, which is accompanied by an increase in the phase angle as the gel becomes more nonhomogeneous (Hermansson, 1997). Clearly, the effect of the successive FTCs on rheological behavior of the purées is dependent on either the measurement temperature or the type of product, so that the effect conferred by the cycles at room temperature is altered if the product is subjected to heating and/or cooling.

Physicochemical properties of chestnut, potato, and yam starches after repeated cycles of freeze–thaw treatments (FTTs) were comparatively studied by Zhang et al. (2014). Freeze–thawed starches formed a honeycomb-like network structure and the spongy structures were broken after the fifth cycle of FTT. The crystallinity of freeze–thawed starch was destroyed at the first cycle and then increased with the increase of cycles of FTT, which might be associated with the influence of stress and subzero temperature. Thermodynamic parameter values (T_0, T_p, T_c, and ΔH) of the freeze–thawed starches were increased with the increase in FTCs. Potato starch had the highest resistant starch (RS) content after the fifth cycle of FTT, which might be a good resource for RS production.

The effects of repeated FTCs (up to seven) on microstructure, thermal, and textural properties of potato starch gel were also investigated by SEM, differential scanning calorimetry, and texture analyzer (Wang et al., 2013). The chemical composition and molecular structure of potato starch were also examined. The potato starch gel formed a honeycomb structure after seven FTCs. The seven FTCs decreased the transition temperature and enthalpy of potato starch in comparison with the native starch, and the retrogradation percentage was 23.6%. The seven FTCs increased the hardness of potato starch gel and decreased springiness and cohesiveness.

For native starches, the freeze–thaw stability is one of the key determinants for their use as the clean-label ingredients in frozen food products. Freeze–thaw stability of native starches would be predicted from their structural features (Srichuwong et al., 2012). Greater syneresis was reported with higher amylose content for corn and potato starches (Singh et al., 2004).

A study on genetically modified potatoes showed that waxy starch having shorter amylopectin branch chains was more resistant to syneresis (Jobling et al., 2002). Freeze–thaw stability of potato starch gel, measured as percentage syneresis, was determined after the first, third, and fifth FTC, achieving the syneresis of 60%, 72%, and 76%, respectively (Srichuwong et al., 2012). Accordingly, this instability would limit the use of potato starch in frozen food applications.

A key factor to consider when using starch in food or industrial applications is that the starch must be properly cooked at an adequate concentration, because the rheological properties of starch pastes are highly dependent on the concentration and cooking conditions (Kim et al., 1993). Alvarez and Canet (2001b) also examined the combined effect of the number of successive FTCs and various final thawing conditions on rheological properties in frozen vegetable purées. The number of FTCs ranged from 0 (that is, thawing only) to 4 and they were applied to three types of commercial product containing potato (broccoli–potato, carrot–potato, and celery–potato). After the FTCs, the purées were thawed at room temperature by microwaving at three different settings or in a saucepan. Rheological properties were affected more by the thawing conditions than by the number of cycles applied. Saucepan thawing raised the values of these properties, apparently as a result of considerable water loss during heating. After one or two cycles (depending on thawing conditions), broccoli–potato presented much greater elasticity and apparent freeze–thaw stability than the others, and celery–potato purée presented the highest fluidity. The final thawing played a significant role in determining the dynamic properties of the products.

3.2.4 Storage and Temperature Oscillation

Frozen vegetable tissue is unstable during storage and the quality deteriorates to a variable extent depending on the type of product and the storage temperature (Canet, 1989). Loss of quality is caused solely by physical and chemical alterations taking place within the product itself, as a result of recrystallization and sublimation of water. The number, size, shape, and orientation of the ice crystals change during storage, resulting in successive melting on the surface of smaller crystals and recrystallization on larger ones. Long periods of frozen storage are not necessarily harmful if a constant low temperature is maintained. For example, Canet (1980) reported that a number of mechanical properties of blanched and frozen potatoes were unaffected by storage at $-24\,°C$. The length of time in frozen storage had no effect on DL, firmness/adhesiveness, vitamin C content, total viable count, or the sensory score in freeze-chilled and frozen mashed potato from three potato cultivars (Redmond et al., 2003).

Alvarez and Canet (1998a, 2000a,b) examined the effects of various ranges of temperature fluctuation (-24 to $-18\,°C$, -18 to $-12\,°C$, -12 to $-6\,°C$, -24 to $-12\,°C$, and -18 to $-6\,°C$) on the compression, shear, and tension parameters of packaged and unpackaged frozen potato tissue. The initial temperature, duration, and number (0, 2, 4, 8, 16, 24, and 32) of fluctuations were varied. Figure 8 shows the thermal history of the product and fluctuations in the air

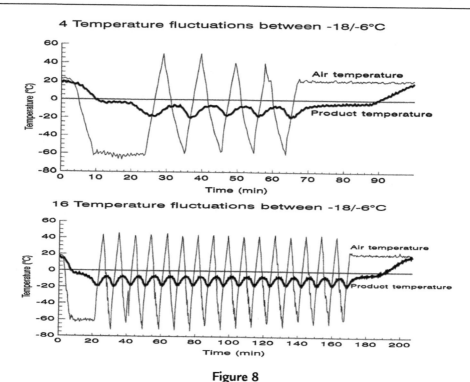

Figure 8

Thermal history of the product and of air temperature in unpackaged samples subjected to four and 16 fluctuations in the range −18 to −6 °C.

temperature in two series performed between −18 and −6 °C on unpackaged potatoes. The highest parameter values occurred in samples subjected to fluctuations between −24 and −18 °C, and the lowest values in the range −18 to −6 °C. The mechanical strength of the frozen tissue decreased as the number of fluctuations increased and in most cases was lower in the packaged samples. At −12 to −6 °C and −18 to −6 °C, the tissues attained temperatures very close to the zone of maximum ice crystal formation. Fluctuations of up to −6 °C accelerated the melting of small ice crystals, thus increasing the amount of available water; this refroze immediately, causing an increase in the size but a decrease in the number of ice crystals.

On the other hand, plots of log (rheological parameters and moisture content) versus number of temperature fluctuations in the ranges −24 to −18 °C and −18 to −6 °C showed two distinct regions; the first was a rectilinear plot with a steep negative slope up to four fluctuations. The second was also a rectilinear plot with a shallow negative slope beyond four fluctuations (Alvarez and Canet, 2000a). Figure 9 shows the softening curves obtained for unpackaged and prepackaged samples subjected to temperature fluctuations in both ranges by plotting log shear force. These two-stage softening rate curves were consistent with the biphasic model

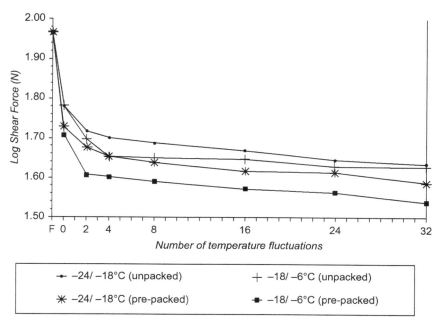

Figure 9

Shear force (log scale) versus number of temperature fluctuations applied in the ranges −24 to −18 °C and −18 to −6 °C for unpackaged and prepackaged frozen potato tissue.

and qualitatively similar to those for thermal softening of the vegetables. The authors found that two substrates, S_a and S_b, may be involved in lending firmness to potato tissue in freezing and frozen storage conditions. Also, principal component analysis clearly separated samples subjected to −18 °C/−6 °C from those subjected to −24 °C/−18 °C (Alvarez and Canet, 2000b). Frozen samples undergoing up to four fluctuations formed a separate cluster from those undergoing a higher number. It is a fact of practical interest that after four fluctuations, an increase in the number of fluctuations during simulated frozen storage caused little further deterioration of potato firmness.

Cryoprotectant mixtures were added to F/T mashed potatoes in the form of ALM pectin and XG, κ-C and XG, and SC and XG, and the effect of long-term frozen storage (0, 3, 6, 9, and 12 months) was examined (Fernández et al., 2009a). Notations used to refer to each of the samples were F/T, mashed potatoes without added biopolymers; ALM1.5/XG1.5, κ-C1.5/XG1.5, and SC1.5/XG1.5, F/T mashed potatoes with 1.5 g/kg added ALM pectin, κ-C, or SC, respectively, and 1.5 g/kg added XG. ALM1.5/XG1.5, κ-C1.5/XG1.5, and SC1.5/XG1.5 mixtures were effective in stabilizing mashed potatoes against freezing, long-term frozen storage, and thawing treatments, as amylose retrogradation and syneresis were prevented.

Possible intermolecular interactions (between different hydrocolloids and protein) and ice recrystallization and sublimation by physical modifications could increase the values of the rheological and textural properties after 1 year in frozen storage, although the hardening detected by the panelist at final storage time was not considered adverse. It was possible to develop a hierarchy for the preference of cryoprotectant mixtures in mashed potatoes: κ-C1.5/XG1.5 > SC1.5/XG1.5 > ALM1.5/XG1.5. These results have important implications for the production of F/T mashed potatoes with improved sensory quality and freeze–thaw stability.

4. Softening Kinetics

Optimal thermal process design relies on relevant and accurate kinetic data for quality evolution (Van Loey et al., 1995). The various thermal processes involved in the production of frozen potatoes affect overall textural quality in different ways. Tissue softening occurs at different rates and is governed by different physicochemical mechanisms. Tissue softening induced by thermal treatments in different media depends on the temperature reached at the thermal center of the product in the heating medium and on the heating rate attained. These two parameters determine the shape of the tissue softening curves and hence the associated kinetic parameters.

It was found that applying the theories of chemical kinetics to the rate of thermal softening of vegetable tissue could provide useful insights into softening mechanisms and could point the way to developing technologies that produce firmer-textured processed products, even though the progress of the reaction is measured by a physical test (firmness) rather than a chemical test (Huang and Bourne, 1983; Bourne, 1987, 1989).

In most studies quantifying loss of firmness, the thermal softening of vegetable tissues has been described by one or two first-order kinetic rate processes (Kozempel, 1988; Rahardjo and Sastry, 1993). Different authors have used different mechanical tests and generally only one rheological parameter as an indicator of product texture. Huang and Bourne (1983, 1987) and Kozempel (1988) used a back-extrusion test cell, taking maximum force readings as a texture measurement. Harada et al. (1985a,b) and Harada and Paulus (1986) used maximum shear force to characterize the behavior of potatoes and three other low-starch tubers during cooking. Verlinden et al. (1995) used a uniaxial compression test in which the rupture force was taken as a measure of texture.

Changes in texture that occur during processing are the result of changes in the chemistry of cell wall and middle lamella hydrophilic polymer material that affect the physical properties (Canet, 1980; Bourne, 1989). However, in potato tissue, in which starch is the major component of the DM, it can be assumed that the phenomena associated with gelatinization are involved in the texture changes that occur during cooking (Pravisani et al., 1985; Verlinden et al., 1995).

Alvarez et al. (2001) determined the kinetic parameters for characterization of the softening of potato tissue (cv. Monalisa) by heating in water. They used compression, shear, tension, and stress–relaxation rheological parameters to represent tissue firmness and to determine what structural components and changes in such components could be contributing to potato tissue firmness during water heating. Kinetics of thermal softening of potato tissue have also been studied in relation to various heating and cooking methods (Alvarez and Canet, 2001a, 2002). For example, the rate of thermal softening of potato tissue by water treatment at 50, 90, and 100 °C was consistent with one pseudo-first-order kinetic mechanism, while at 70 and 80 °C the rate of softening was consistent with two simultaneous pseudo-first-order kinetic mechanisms.

Kinetic theory was successfully used to detect an increase in firmness caused by heating at 60 °C, mainly between 20 and 40 min, presumably by PME activation (Alvarez et al., 2001). Figure 10 shows the softening curves obtained for a rheological parameter derived from each of the mechanical tests performed by plotting $\log F_c$ (maximum compression force), $\log G_s$

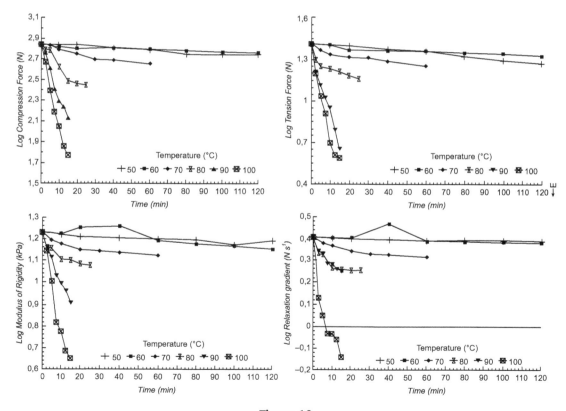

Figure 10
Softening curves for rheological parameters in water heating.

(modulus of rigidity), $\log F_t$ (maximum tension force), and $\log S_r$ (relaxation gradient) versus process time in the range of temperatures studied. A comparison of the plots indicated that two substrates, S_a and S_b, may be involved in lending firmness to potato tissue at 70 and 80 °C. At these temperatures, mechanism 1 is more probably due to gelatinization and light cooking, whereas mechanism 2 is more likely to represent the changes in the pectic substances in the cell wall and interlamellar region. At 90 and 100 °C the gelatinization process was fast and therefore the simple mechanism that was fitted presumably reflects the degree of solubilization of the pectic substances. At 50 and 60 °C there was practically no gelatinization, so that the simple mechanism that was fitted presumably represented incipient solubilization of pectic material.

Figure 11 shows typical Arrhenius plots for rheological parameters from each mechanical test. The rate constants of the first mechanisms at 60 °C ($3 \times 10^{-3}\,\mathrm{K}^{-1}$) were the least linear; this effect is mainly apparent in plots of maximum compression and tension forces (in the plots, rate constants at 60 °C are circled). This could be another consequence and evidence of

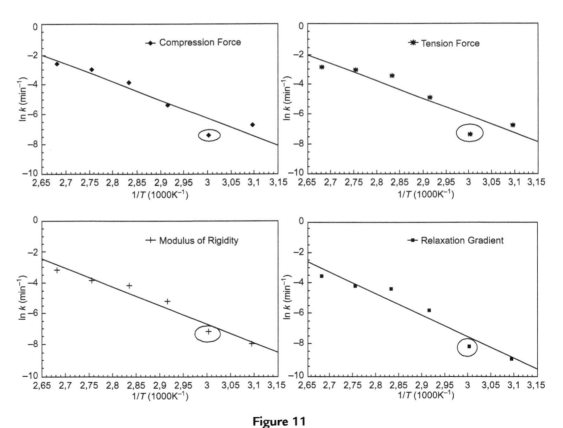

Figure 11

Arrhenius plots of ln apparent rate constants versus reciprocal absolute temperature for potato tissue treated in water.

the fact that in the experimental potato variety the optimum temperature for activation of the PME enzyme is close to 60 °C. These results were consistent with the findings of other studies. Stationary points for the first step of stepwise blanching with maximum PME activity exhibited critical temperature and time values at 64.39 °C and 28.02 min after freezing and steaming of potato tissue (Alvarez et al., 1999). Also, Moledina et al. (1981) reported 83% retained PME activity in water blanching at 60 °C for 30 min and none at 75 °C for 10 min.

Another study was carried out to compare the effect of three different heating methods (Alvarez and Canet, 2001a) on the rate of softening of potato tissue (cv. Monalisa). On the basis of chemical kinetics, the kinetics of thermal softening of potato tissue heated by steaming, steaming + hot air, and microwave exposure were evaluated using the rheological properties from four objective methods as firmness indicators. In the three heat treatments, the rate of thermal softening of tissue could be described by two simultaneous first-order kinetic mechanisms. A comparison of kinetic parameters showed that steaming produced a greater degree of softening than the other two heating methods used. The firmness ratios for shear force showed that approximately 16% of the firmness of fresh potato is retained after steaming treatments compared to 46% and 36%, respectively, for steaming + hot air and microwave. With these heating methods, gelatinization contributes less than cell wall structure to potato tissue softening as determined either by kinetic parameters or by microscopic observations.

Color, as a quality attribute of cooked potatoes, is affected by the extent and nature of the heat during thermal processing. Analysis of kinetic data allows processors to minimize undesirable changes and optimize color retention. The kinetics of color change during cooking of potatoes have been evaluated by Nourian and Ramaswamy (2003a). Potatoes were cut into cylinders and cooked in a temperature-controlled water bath at 80–100 °C for selected times. Color changes associated with cooked potatoes were evaluated using a tristimulus colorimeter in the L, a, and b mode. L and b values decreased, while ΔE and a values increased with time at each cooking temperature. A modified first-order model was used to characterize color change kinetics based on changes occurring between the initial and a maximum or minimum value. Temperature sensitivity of rate constants was adequately described by the Arrhenius and z-value models. In addition, test samples were subjected to a compression test and three textural properties (hardness, stiffness, and firmness) were derived from the resulting force–deformation curves (Nourian and Ramaswamy, 2003b). Texture parameters of cooked potatoes declined as cooking time progressed, and the rate of texture change associated with each temperature was consistent with two pseudo-first-order kinetic mechanisms, one more rapid than the other. Temperature sensitivity of rate constants was also adequately described by Arrhenius and z-value models.

The kinetics of textural changes during convection drying of potato slices found by using various mathematical models was also studied (Troncoso and Pedreschi, 2007). Four models were proposed to validate texture experimental data: (1) two irreversible serial chemical reactions; (2) two irreversible serial chemical reactions with $k_2 = 0$, which means that only the tissue softening stage is considered; (3) fractional conversion; and (4) first-order kinetics. These models linked the dimensionless textural parameter, maximum force $\left(F_{MAX}^{*}\right)$ with drying time. Experimental textural changes were followed during drying of potato slices at various air temperatures (50, 60, 70, and 80 °C). Regardless of the drying temperature, the trend of F_{MAX}^{*} with drying time was almost the same, showing a progressive and significant decrease in its value as the exposure time increased. However, F_{MAX}^{*} of potato slices during drying decreases as the moisture content and temperature decreases, reflecting a clear softening of the potato tissue. A satisfactory agreement between experimental and predicted values was observed for two irreversible serial chemical reactions, with $k_2 \neq 0$ and $k_2 = 0$, and fractional conversion models, with root mean square error values in the ranges 4.0–9.1%, 3.0–7.4%, and 2.0–7.3%, respectively.

Cooking kinetics of potato tubers determined by vibration techniques have been reported by Blahovec et al. (2007). Nondestructive determinations of characteristic tuber resonant frequencies (modes M-1 and M-2) displayed in the amplitude–frequency plots before, during, and after steam cooking of the whole tubers were used to establish cooking kinetic coefficients for four different potato varieties. The results suggested that this method is suitable only for smaller tubers; the sloughing and cracking of larger tubers induced by longer cooking times confound the measurements.

In turn, potato cooking was analyzed on the basis of kinetic theory of potato softening by Blahovec and Hejlova (2010). The application of the theory makes it possible to connect cooking time, i.e., time of cooking prior to the start of disintegration of the potato, with some terms of the kinetic theory, mainly kinetic coefficient and critical concentration of the intracellular bonds. Simple models were developed for description of the disintegration part in terms of the cooking part parameters. Two years of data from three different varieties, in one case in six different cultivation modes, show that deep discrepancies exist between parameters for the nonsalad varieties, which correlated with one another, and the parameters obtained for the salad variety Nicola.

Another study on various textural, rheological, and gelatinization properties showed a significant difference among the varieties of cassava (Sajeev et al., 2010). A dual-mechanism first-order kinetic model was applied to explain the reduction in firmness of the cooked tubers, and based on the value of rate constants, the cooking quality of the tubers can be compared objectively and grouping done as Group I—Sree Rekha, Koliakodan, H740, Adukkumuttan, and Venjaramoodan; Group II—H152, Narayanakappa, Kaliamanja, and Sree Prabha. Degree of cooking can be predicted by using a linear model developed between degree of cooking and relative firmness. Principal component analysis of the texture profile parameters, dynamic mechanical spectra, differential scanning calorimetric properties, pasting properties, and physicochemical properties could be used to group the varieties, and the groups obtained by this method were almost similar to those obtained by the dual-mechanism first-order kinetic model.

Adhikari et al. (2010) investigated the change in textural attributes (hardness, chewiness, and rate of softening) of three potato cultivars (Russet Burbank, Desiree, and Sebago). Uniform cylindrical samples were exposed to various thermal processing regimes including heating at 85 or 95 °C in water and steaming (100.2 °C), also using various salt concentrations (1.5%, 3%, and 6% (w/w) NaCl). The change in textural attributes upon steaming was only marginally different compared to that at 95 °C in water. Low concentrations of salt (1–3%) were found to accelerate the softening of the texture in these cultivars, especially at lower temperatures, which can be attributed to the increased osmotic pressure and the presence of the sodium ion and their combined effect in swelling and rupture of starch granules. The textural attributes were modeled using a two-parameter reaction kinetics model. There was reasonable agreement by the model findings on the textural attributes prepared from all the thermal processing regimes and in the presence and absence of salt (within an average absolute error of 1.9–7%).

5. Quality Optimization

Because of the different variables involved in the freezing process and the above-mentioned interactions, therefore, a statistical technique that takes these into account should be used to determine optimum freezing conditions. One such technique particularly suited to this application is response surface methodology (RSM), which combines the methods of planned and efficient experimental designs with least squares modeling to identify optimum conditions for the process response. RSM is a popular and effective method for solving multivariate problems and optimizing several responses in many types of experimentation (Montgomery, 1991; Myers and Montgomery, 1995), because it can simultaneously consider several factors at many different levels and corresponding interactions among these factors on the basis of a small number of observations. Nwabueze (2010) identified key process variables, building the model and searching the solution through multivariate regression analysis and interpretation of resulting polynomial equations and response surface contour plots as basic steps in adapting the central composite design to achieve process optimization.

5.1 Frozen Potatoes

For optimization of stepwise blanching of F/T potato tissues (cv. Monalisa), Alvarez and Canet (1999b) used RSM to compare the effects of temperature and time of the first step of blanching (LTLT) on compression, shear, tension, and stress–relaxation parameters of F/T potato tissues. Contour plots of apparent modulus of elasticity in compression (E_c), maximum tension force (F_t), and first-cycle relaxed force (F_{rl}) were overlaid (Figure 12). These mechanical properties were chosen to narrow down the optimum zone by means of the significance of the second-order models fitted for these parameters. E_c and F_{rl} set the limits on the optimum zone because of their importance for the study, and F_t could be optimized within the band demarcated by the other two parameters. The shaded area defines the optimum ranges of blanching temperature (60–65 °C) and time (25–35 min).

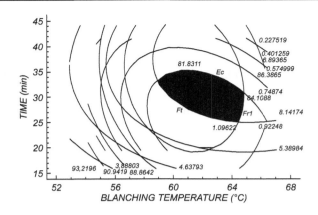

Figure 12

Overlay of contour plots for apparent modulus of elasticity in compression (E_c), maximum tension force (F_t), and first-cycle relaxed force (F_{r1}). The shaded area represents optimum conditions.

RSM was also used to determine optimum freezing conditions for PME activity, rheological parameters, and textural properties in the same potato tissues (Alvarez et al., 1999). The effects of variation in levels of temperature (57.93–72.07 °C) and time (15.86–44.14 min) in the first blanching step on PME activity were studied using a central composite rotatable design. Then, a Box–Behnken factorial design was used to investigate the effects on rheological parameters and textural properties of simultaneous variation of temperature (60–70 °C) and time (20–40 min) in the first blanching step and of steaming temperature (112–122 °C) and time (1–3 min). PME activity was expressed as specific activity (units of PME activity per milligram of protein) and was determined after cooling following the first blanching step (PME_1) and after freezing of previously thawed samples (PME_2). PME_2 was less than PME_1. The highest levels of activity before and after freezing were detected in the samples treated at 65 °C for 30 min. Comparison of activities in the control samples (97 °C, 2 min) before and after freezing showed that the freezing process caused the loss of almost 55% PME activity, the rest of the loss being caused by the second blanching step. The retained enzymatic activity with respect to the value of activity determined in the fresh potato tissue was 15.33 units of activity per milligram of protein. Activity after treatment at 65 °C, 30 min, was 30% higher than in fresh tissue and was very close to the latter after freezing.

Stationary points showing maximum PME activity had critical temperature and time values of 64.22 °C and 29.37 min before freezing and 64.39 °C and 28.02 min after freezing and steaming of the tissues (Figure 13), being very close to the values obtained for some mechanical and textural properties, confirming earlier studies (Alvarez and Canet, 1999b). Results showed a high correlation between increases in PME activity and potato tissue firmness under optimum experimental freezing conditions, demonstrating that the enzyme is one of the main contributors to that firmness, which determines the textural quality of frozen potato tissues (cv. Monalisa).

(a)

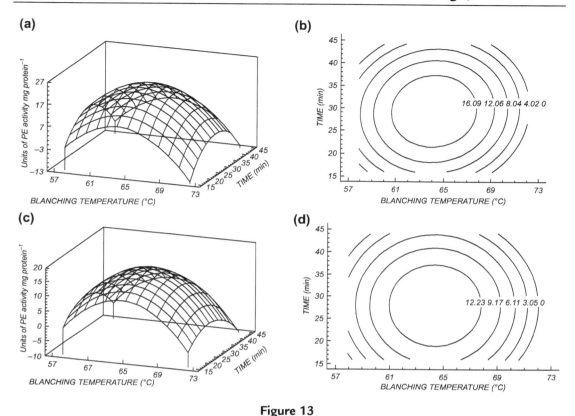

Figure 13

PME (PE) activity response surfaces and contour plots as functions of blanching temperature and time. (a) PE activity after first blanching step (PE$_1$), (b) PE activity after blanching step (PE$_1$), and (c, d) PE activity after freezing (PE$_2$).

Following this, the optimization of LTLT blanching prior to freezing + cooking was also studied using RSM, to find out how the complete process affected the texture of another variety of potatoes (cv. Kennebec) (Canet et al., 2005c). A central composite rotatable design was used to study the effects of variation in levels of LTLT temperature (52.93–67.07 °C) and time (31.72–88.28 min) on compression parameters and PME activity. The highest compression values were recorded at the replicated center points (60 °C, 60 min) and in blanched samples at the axial point (60 °C, 88.28 min), whereas the highest PME activity was found in samples blanched at 55 °C for 80 min; however, the levels of activity at the replicated center points (60 °C, 60 min) and in the samples blanched at 55 °C for 40 min were also very high. Stationary points showing maximum mechanical resistance had critical temperatures and times in the ranges 58–60 °C and 66–75 min. From the shape of the plots it seems reasonable to assume the existence of a relationship between the increase in PME activity and firmer texture (Figure 14).

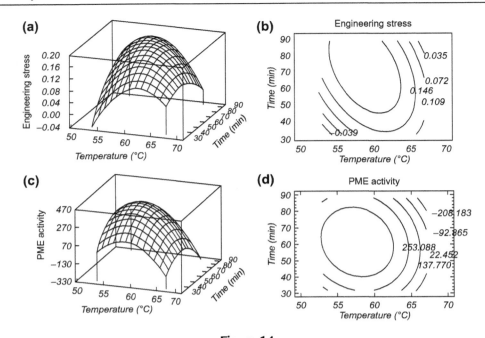

Figure 14

Engineering stress and PME activity response surfaces and contour plots as functions of blanching temperature and time. (a) Engineering stress response surface. (b) Engineering stress contour plot. (c) PME activity response surface. (d) PME activity contour plot.

Severini et al. (2004a) used RSM to investigate variables affecting the firmness of blanched potato slices, such as type of blanching, time of treatment, NaCl or $CaCl_2$ concentration, and lactic acid concentration. The results showed that the mathematical models provided a good estimate of the effects of individual and interactive factors on the retention of firmness in processed potato. The best results were obtained with microwave blanching of slices dipped in a solution containing calcium chloride. RSM was also used to investigate the way in which variables such as treatment time, sodium or calcium chloride concentration, and lactic acid concentration affect microwave blanching of potato slices dipped in these solutions (Severini et al., 2004b). Two three-factor five-level second-order central composite designs were developed to analyze the target variables.

In 2014, Shahraki et al. (2014) used RSM to optimize microwave and frying conditions of french fries with respect to quality attributes. Oil content of pretreated samples was significantly reduced. For dried samples, lower initial moisture content before frying reduced oil absorption. A linear model described the effect of factors on variation of color parameters. The study suggested that french fries with acceptable quality attributes can be obtained by microwave prefrying at 400–500 W for 3–4 min and frying at 180 °C for 6–7 min. Using genetic algorithms optimization did not improve the obtained models significantly.

The digital imaging method allows measurements and analyses of the color of food surfaces that are adequate for food engineering research.

The food processing industries aim at achieving balance between the conducive and the harmful effects of thermal processing. Hence, for designing a thermal process, time–temperature combinations are chosen to impart required lethal effect to ensure microbiological safety and to simultaneously reduce the extent of thermal damage to color, texture, and nutritional attributes. Optimization on the basis of color and texture becomes even more important because acceptance or rejection of most foods relies on the mouthfeel and external appearance (Singh Rattan and Ramaswamy, 2014). Hence, the color and textural attributes were studied during retorting of potatoes and the optimal conditions for preparing canned potatoes were identified. It was concluded that processing at 125 °C and 20 rpm under free axial mode are the optimal conditions for maximum retention of color and textural quality of potatoes (cv. Goldrush).

5.2 Frozen Mashed Potatoes

Downey (2003) studied the effects of cryoprotectant mixtures on physical properties of frozen and thawed puréed cooked potatoes (cv. Rooster). Ingredients were incorporated as mixtures of: (1) guar, pectin, and WP concentrate and (2) xanthan, carrageenan, and SC. Experimental design techniques were used to design the experiments and evaluate the results. Both ingredient mixtures significantly reduced maximum resistance to penetration and centrifugal DL. Modeling of resistance to penetration was quite successful, and the experimental approach has potential for tailoring these physical properties to predetermined levels to meet specific consumer expectations in a range of food products.

Fernández et al. (2006) studied the effect of LTLT blanching prior to cooking on color, textural, firmness, and oscillatory parameters; sensory attributes; and OA of either F or F/T mashed potatoes using RSM to establish the optimum temperature and time for blanching in both mashed potatoes. A central composite rotatable design was used to determine the effects of variation in levels of blanching temperature (57.93–72.07 °C) and time (15.86–44.14 min) on quality parameters. LTLT prior to cooking produced a lighter-colored F mashed potato (higher L^*/b^* ratio) than cooking alone. In both F and F/T mashed potatoes, blanching at 65 °C for 30 min had the effect of thickening the mashed potatoes, which could be the result of PME enzyme activation rendering the cell binding less degradable. Scores for OA were higher in the F/T blanched mashed potatoes than in their F counterparts, highlighting the potential of blanching to improve the quality of mashed potatoes that are subjected to freezing and thawing.

Blends of ALM pectin and XG were added to F and F/T mashed potatoes (Canet et al., 2011). The product was tested by ITPA, color, expressible water (E_w), and sensory analyses. A central composite rotatable experimental design was used to study the effects of variation of

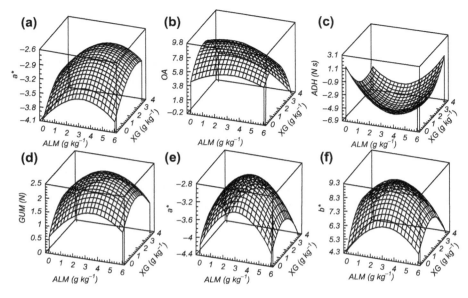

Figure 15

Mechanical, color parameter, and OA response surfaces as functions of ALM pectin and XG concentrations in F and F/T mashed potatoes. (a) Greenness ($a*$) for F product. (b) OA for F product. (c) Adhesiveness (ADH; N s) for F/T product. (d) Gumminess (GUM; N) for F/T product. (e) Greenness ($a*$) for F/T product. (f) Yellowness ($b*$) for F/T product.

ALM pectin (1.5–4.5 g/kg) and XG (0.5–2.5 g/kg) concentrations. The effects were highly dependent on the levels of ALM and XG added, although XG influenced the mechanical behavior, color, and water-holding capacity of the mashed potatoes more than ALM pectin. As Figure 15 shows, in both F and F/T mashed potatoes, $a*$ (Figure 15(a) and (e), respectively) decreased with ALM and XG concentrations up to certain levels, beyond which the independent variable increased, thus defining an obvious optimum region. In contrast, the OA of the F product (Figure 15(b)) and the adhesiveness (Figure 15(c)), gumminess (Figure 15(d)), and $b*$ parameter (Figure 15(f)) of the F/T product increased with ALM and XG concentrations up to certain levels, beyond which the independent variable decreased, consequently defining clear optimum regions. On that basis, values (stationary points) were calculated to give optimum combinations for these responses. Instrumental and sensory tests showed that, when 3 g/kg of ALM pectin and 1.5 g/kg of XG were included in the formulation, the mechanical and functional properties of the F/T product were very acceptable, with almost no detectable changes in color attributes, and very good OA was associated with a pronounced creamy sensation conferred by the XG.

Following this, F and F/T mashed potatoes made with mixtures of SC and XG were assessed for rheological properties as evaluated with two kinds of deformation tests, large deformation (ITPA and CP) and small deformation (oscillatory measurements); color; DL; pH; and two different

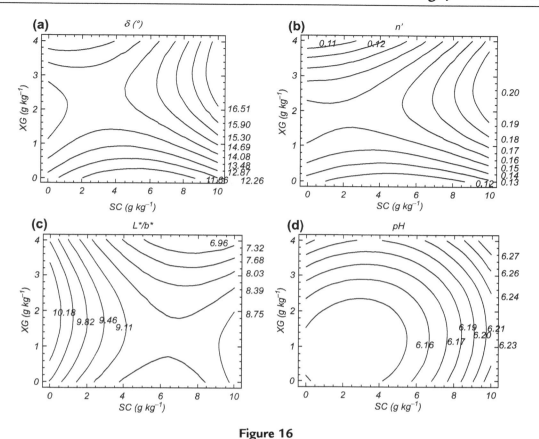

Figure 16
Quality parameter contour plots as functions of SC and XG concentrations in F/T mashed potatoes. (a) Phase angle (δ). (b) Slope of storage modulus (n'). (c) L*/b* ratio from color measurement. (d) pH.

sensory analyses (Fernández et al., 2010). The influence of SC and XG concentrations on rheological properties and quality characteristics was studied by means of RSM using a central composite rotatable experimental design with SC concentrations ranging between 2.5 and 7.5 g/kg and XG concentrations ranging between 0.5 and 2.5 g/kg, as independent variables. Models fitted for the response variables were used to generate contour plots to identify the main effects of SC and XG concentrations (Figure 16). F mashed potatoes with added biopolymer mixtures behaved like a weaker gel structure; this fact is attributed to repelling forces between negatively charged phosphate groups on potato starch granules and negative charges on XG molecules. Processing significantly reduced the effect of SC and XG mixtures on the gel strength and the viscoelastic behavior of the mashed potatoes due to high local starch concentrations and retrogradation of both amylose and amylopectin. DL values revealed that addition of XG at concentrations of 1.5 g/kg or higher, either individually or mixed with SC, was effective in stabilizing mashed potatoes against syneresis.

Next, blends of κ-C and XG were added to mashed potatoes (Fernández et al., 2009b). Product was tested by ITPA and CP tests, oscillatory and steady rheology, color, DL, total soluble solids, and sensory analyses. A central composite rotatable design was used to study the effects of variation in the levels of κ-C (1.5–4.5 g/kg) and XG (0.5–2.5 g/kg). Addition of κ-C had a major impact on gel strength, viscoelastic behavior, sensory attributes, and OA, whereas addition of XG influenced textural and steady properties and color. Mainly, elastic modulus (G') was strongly dependent on the κ-C concentration in mashed potatoes containing amylose. Compared to mashed potatoes with 1.5 g/kg added XG, additional incorporation of 5.12 g/kg κ-C increased G' approximately twofold, possibly owing to an exclusion effect of the swollen starch granules and synergistic effect of κ-C and denatured protein. κ-C provided the appropriate texture, and again XG imparted creaminess and mouthfeel to the product.

The next aim was to ascertain the functionality of κ-C and XG blends on the quality parameters of F/T mashed potatoes (Alvarez et al., 2009). The main purpose was to investigate how the addition of κ-C (anionic polysaccharide) and XG (anionic gum) blends affects the texture, rheology, and sensory characteristics of F/T mashed potatoes with native anionic potato starch (containing anionic phosphate groups). RSM was used to interpret the experimental data. Once the regression equation associated with the response surface was established, a visual presentation of the responses to the variable inputs (κ-C and XG concentrations) was prepared for the two-variable case (Figure 17). The OA of F/T mashed potatoes was mostly improved by the addition of cryoprotectant blends, which is ascribed to improvement of F/T mashed potato texture by retarding starch retrogradation, increasing water-holding capacity, and enhancing the principal characteristics determining consumer acceptance. Creaminess was the most crucial factor for OA of the products. On the other hand, shear stress–shear rate data indicated a positive effect imparting freeze–thawing resistance and stability to mashed potatoes when mainly XG was added, either alone or combined with κ-C (Alvarez et al., 2011c).

The results of these studies had important implications for the production of F/T mashed potatoes with improved sensory quality and freeze–thaw stability over time. The addition of both polysaccharides at a low concentration (each cryoprotectant at 1.5 g/kg) was recommended and included in the formulation of mashed potatoes in subsequent research as the product was intended to be frozen.

5.3 Other Heat-Treated Potatoes

Nath and Chattopadhyay (2008) examined HTST air puffing to produce a ready-to-eat potato–soy snack. Potatoes were peeled, boiled, sliced, dried, and shredded and soybeans were dried and ground. Samples were prepared from potato flour blended with soy flour (5–25%) by adding the required amount of chilled water (5 °C) and common salt. The effects

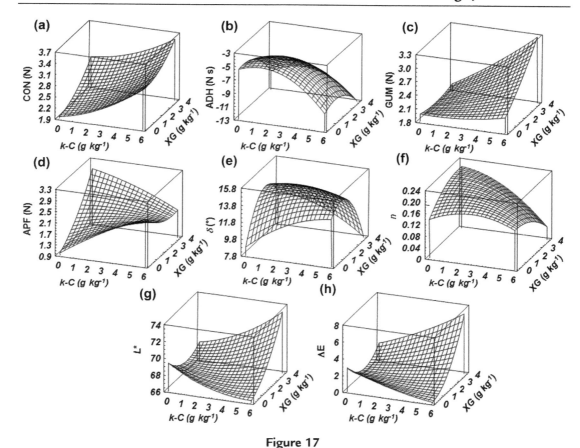

Figure 17
Instrumental measurement response surfaces as functions of κ-C and XG concentrations in F/T mashed potatoes. (a–c) Consistency, adhesiveness, and gumminess from TPA test. (d) Average penetration force from CP test. (e) Phase angle from oscillatory rheological test. (f) Flow behavior index from power law. (g, h) Lightness and total color difference from color measurements.

of drying process parameters and soy flour concentration on quality attributes and microstructural changes were studied. RSM was used to identify optimal conditions. The optimum conditions were as follows: 10.3% soy flour, puffing temperature 230 °C, and puffing time 25.5 s in terms of maximum expansion ratio (3.7) and minimum hardness (2754 g) with a maximum OA index (7.3). Highly porous structures were observed owing to the formation of large air vacuoles and highly expanded starch granules in the potato–soy snack when air puffed at 230 °C for 20 s Nath and Chattopadhyay (2007) adopted oven toasting after HTST air puffing to improve crispness and other quality attributes. The optimum product quality in terms of crispness (38.7), moisture content (3.35%, db), ascorbic acid loss (20.9%, db), *L* value (52), Δ*E* (8.60), and OA (7.8) were obtained at a temperature of 104 °C and drying time of 28 min using RSM.

6. Novel or Emerging Methods and Outlook for the Future

Trends in consumer preferences have stimulated the development of novel concept-driven technologies that provide the required processing through non- or mildly thermal means (Welti-Chanes et al., 2005). Accordingly, much of the recent scientific research for the food industry has focused on nonthermal processing techniques, of which high-pressure processing (HPP) is one of the few that are considered to have real potential in commercial settings (Sun, 2005). HPP is a technology that potentially addresses many, if not all, of the most recent challenges faced by the food industry. It can facilitate the preparation of food products that have the quality of fresh foods but the convenience and profitability associated with pro-longed shelf life (McClements et al., 2001). Many aspects associated with the use of high pressure as a processing method in the food industry were reviewed by Norton and Sun (2008).

As early as 1994, Eshtiaghi et al. dried high-pressure-pretreated potato cubes in a fluidized bed and compared them to untreated, pressure-treated, or water-blanched dried samples. Drying rates varied with pretreatment. Freezing produced the highest drying rates. Pressure-treated and water-blanched samples retained highly acceptable colors. Freezing or hot-water blanching or high-pressure pretreatment, followed by freezing, gave good rehydration. High-pressure treatment resulted in incomplete rehydration, but when combined with freezing, the water uptake was between 2.1 and 4.8 ml/g. Retention of cell wall structures of frozen samples during drying was presumed to be responsible for more efficient mass transfer. Pressure-treated samples had the texture nearest to that of the raw material, while no major differences in color were observed.

Luscher et al. (2005) published studies of pressure-supported freezing and subzero cooling. They calculated the true compressive stress and strain in treated potatoes from force–deformation curves, and their results showed that freezing to the ice III metastable zone enhanced the texture of the potato. They also showed that although pressure-shift freezing (PSF: a sample is frozen by means of pressure release, causing instantaneous crystallization of ice, homogeneously distributed throughout the sample) preserved the skeletal cell structure of the tissue, it promoted cell permeability. Volume changes during FTCs at a pressure of 200 MPa were found to have a detrimental effect on membrane permeability. During subzero cooling, membrane permeability was also adversely influenced by phase transition within the membrane itself, which gave rise to cell lysis.

Demonstration of the existence of liquid and solid metastable phases within the phase dia-gram of water and water-containing food products has opened up a new range of possibilities in PSF and pressure-induced thawing (PIT: a frozen sample can be forced to a phase transi-tion from ice to liquid water by applying pressure along the melting curve of ice I) processes for potatoes (Urrutia-Benet et al., 2007). A reduction in processing time was obtained in a metastable region (for pressures above 200 MPa and temperatures below $-20\,^{\circ}$C) thanks to the incrementing of temperature gradients (in food material, before and after pressure release, in the case of PSF, and between sample and heating medium for PIT). The enzymatic activity of polyphenol oxidase (PPO) was chosen to evaluate the effectiveness of PSF and PIT processes for food preservation, since it is dependent on the cell disruption caused by ice nucleation during freezing and the crystal growth that occurs during storing and thawing. The results of laboratory and pilot-scale experiments showed that the activity of PPO was no greater after freezing and thawing processes when pressure was applied and was even slightly reduced in the metastable region. Additionally, pilot-scale evaluations of key quality-related parameters (color, DL, texture, and microstructure) showed better responses for PSF and PIT than for atmospheric freezing and thawing. However, using relative fail stress and strain values, the authors showed that supercooling in the metastable zone (i.e., around 12 °C below the triple point) produced no additional benefits as regards texture or DL in the samples compared to traditional PSF—i.e., 200 MPa and 20 °C, followed by PIT treatment at 200 MPa. In fact, with this traditional pressure–temperature combination, which supports the formation of ice I, overall sample texture was better than in samples that underwent PSF–PIT treatments with pressure–temperature combinations of 240 MPa/−28 °C, 280 MPa/−20 °C, and 280 MPa/−28 °C. Other than better color retention and slight enzyme inactivation, the only obvious benefit presented by supercooling in the metastable zone was that the swelling of cells in the potato tissue was reduced thanks to the development of smaller ice crystals.

As suggested by Schluter et al. (2004), however, some considerations must be taken into account before freezing to other ice modifications, or cooling to the ice III metastable zone, such as volume changes, overall freezing time, and phase-transition time. As regards process efficiency, owing to the protracted precooling time, cooling to the ice III zone does not benefit PSF; with PIT, on the other hand, efficiency is increased.

On the other hand, Park et al. (2013) showed that in situ electrical conductivity measurement is a useful tool to document the extent of textural changes during HPP. These authors investigated the pressure-induced textural changes in selected vegetables (carrot, potato, and red radish). Electrical conductivity of the vegetables was recorded in situ up to 10 min under elevated pressures (200, 400, 600 MPa at 25 °C) using a custom-fabricated experimental setup. The tissue disintegration index (Z) was determined for raw, processed, and F/T samples. The hardness and stiffness of the samples were evaluated using the instrumental texture analyzer. This information was related to the crunchiness index (CI). Pressure treatment increased electrical conductivity values of all the samples as a function of pressure and holding time. Beyond a certain threshold of pressure holding time, the electrical conductivity did not change further. Empirical model fitting demonstrated the inverse relationship for potato between Z and the CI.

The ohmic tempering method was successful in shortening the tempering time of frozen PP significantly compared to conventional tempering (Seyhun et al., 2013). Increasing the salt concentration or frequency decreased the tempering time for frozen PP samples.

The temperature differences within the samples were observed to be smaller with 0.50% salt concentration and 10 kHz. Ohmic heating can be successfully used for tempering frozen foods, reducing the conventional tempering times by a margin of 90%, but the parameters affecting the electrical conductivity should be carefully controlled.

The effect of pulsed electric field (PEF) pretreatments prior to deep-fat frying on quality aspects of potato fries has been investigated by Ignat et al. (2015). Potato cubes were submitted to 18.9 kJ/kg PEF treatments by applying 9000 pulses at 0.75 kV/cm electric field or 810 pulses at 2.50 kV/cm electric field. Regardless of the process conditions applied, PEF treatments increased moisture and softened potato cubes with no modification on their fresh-like appearance. Potato cubes submitted to PEF also showed a lower oil uptake upon frying than blanched and water-dipped controls. These effects were attributed to the structural modification of potato tissue upon PEF-induced electroporation. This could also favor leaching of reducing sugar from the potato cube surface, accounting for a lower browning tendency during frying.

Although fried vegetable products contain a lot of oil and are not as healthy as dried products, they are popular because of their better flavor and texture. Vacuum frying can produce quality products, although the cost is somewhat higher than that of traditional frying. However, it appears that vacuum frying will be applied widely to produce high-quality fried vegetable/ fruit products (including dried potato products) in the future because of its significant advantages (Huang and Zhang, 2012).

The utilization of microwave heating instead of traditional conduction or radiant heating has been proven to provide better heat and mass transfer rates. For microwave-assisted drying, further research and development to resolve the nonuniformity during microwave heating is necessary for large-scale application. Capital and operating costs of microwave units need to be improved as well. Trends in frozen potato products include the development of improved microwaveable products, more flexible manufacturing techniques, improved defect removal, and reduction in fat content of fast-serve products.

References

Abu-Ghannam, N., Crowley, H., 2006. The effect of low temperature blanching on the texture of whole processed new potatoes. Journal of Food Engineering 74, 335–344.

Adhikari, B.P., Purcell, T., Ristrom, A., Chaudhary, D.S., 2010. Effect of cultivars and thermal processing with salt solutions on the textural attributes (hardness, chewiness and rate of softening) of potatoes (*Solanum Tuberosum* L.). International Journal of Food Properties 13, 1161–1177.

Aguilera, J.M., 1997. Fritura de alimentos. In: Aguilera, J.M. (Ed.), Temas en tecnología de alimentos. Instituto Politécnico Nacional, México, DF, pp. 187–215.

Alvarez, M.D., Canet, W., 1995. Efecto de las velocidades de congelación, descongelación y congelaciones y descongelaciones sucesivas en distintos parámetros reológicos definitorios de la textura de patata (cv. Monalisa). In: Gallegos, C., Guerrero, A., Muñoz, J. (Eds.), Reología y textura de productos alimentarios de interés industrial. Grupo Especializado de Reología. Real Sociedad Española de Química, Sevilla, pp. 3–14.

Alvarez, M.D., 1996. Caracterización reológica de tejidos de patata tratados térmicamente. Cinéticas de ablandamiento Ph.D. dissertation. Universidad Politécnica de Madrid.

Alvarez, M.D., Canet, W., Tortosa, M.E., 1997. Effect of freezing rate and programmed freezing on rheological parameters and tissue structure of potato (cv. Monalisa). Zeitschrift für Lebensmittel-Untersuchung und-Forschung A 204, 356–364.

Alvarez, M.D., Canet, W., 1997. Effect of pre-cooling and freezing rate on mechanical strength of potato tissues (cv *Monalisa*) at freezing temperatures. Zeitschrift für Lebensmittel-Untersuchung und-Forschung A 205, 282–289.

Alvarez, M.D., Canet, W., 1998a. Effect of temperature fluctuations during frozen storage on the quality of potato tissues (cv. Monalisa). Zeitschrift für Lebensmittel-Untersuchung und-Forschung A 206, 52–57.

Alvarez, M.D., Canet, W., 1998b. Rheological characterization of fresh and cooked potato tissues (cv. *Monalisa*). Zeitschrift für Lebensmittel-Untersuchung und-Forschung A 207, 55–65.

Alvarez, M.D., Canet, W., Cuesta, F., Lamúa, M., 1998. Viscoelastic characterization of solid foods from creep compliance data: application to potato tissues. Zeitschrift für Lebensmittel-Untersuchung und-Forschung A 207, 356–362.

Alvarez, M.D., Morillo, M.J., Canet, W., 1999. Optimisation of freezing process with pressure steaming of potato tissues (cv *Monalisa*). Journal of the Science of Food and Agriculture 79, 1237–1248.

Alvarez, M.D., Canet, W., 1999a. Rheological properties of mashed potatoes made from dehydrated flakes: effect of ingredients and freezing. European Food Research and Technology 209, 335–342.

Alvarez, M.D., Canet, W., 1999b. Optimization of stepwise blanching of frozen-thawed potato tissues (cv. Monalisa). European Food Research and Technology 210, 102–108.

Alvarez, M.D., Canet, W., 2000a. Kinetics of softening of potato tissue by temperature fluctuations in frozen storage. European Food Research and Technology 210, 273–279.

Alvarez, M.D., Canet, W., 2000b. Principal component analysis to study the effect of temperature fluctuations during storage of frozen potato. European Food Research and Technology 211, 415–421.

Alvarez, M.D., Canet, W., Tortosa, M.E., 2001. Kinetics of thermal softening of potato tissue (cv. Monalisa) by water heating. European Food Research and Technology 212, 588–596.

Alvarez, M.D., Canet, W., 2001a. Kinetics of thermal softening of potato tissue heated by different methods. European Food Research and Technology 212, 454–464.

Alvarez, M.D., Canet, W., 2001b. Rheological properties of frozen vegetable purees. Effect of freeze-thaw cycles and thawing conditions. European Food Research and Technology 213, 30–37.

Alvarez, M.D., Canet, W., 2001c. Influence of cooking and freeze-thaw cycles on viscoelastic properties of vegetable purees. LWT-Food Science and Technology 34, 549–555.

Alvarez, M.D., Canet, W., 2002. A comparison of various rheological properties for modelling the kinetics of thermal softening of potato tissue (cv. Monalisa) by water cooking and pressure steaming. International Journal of Food Science and Technology 37, 41–55.

Alvarez, M.D., Fernández, C., Canet, W., 2004. Rheological behaviour of fresh and frozen potato puree in steady and dynamic shear at different temperatures. European Food Research and Technology 218, 544–553.

Alvarez, M.D., Fernández, C., Canet, W., 2005. Effect of freezing/thawing conditions and long-term frozen storage on the quality of mashed potatoes. Journal of the Science of Food and Agriculture 85, 2327–2340.

Alvarez, M.D., Canet, W., Fernández, C., 2008. Effect of addition of cryoprotectants on the mechanical properties, colour and sensory attributes of fresh and frozen/thawed mashed. European Food Research and Technology 226, 1525–1544.

Alvarez, M.D., Fernández, C., Canet, W., 2009. Enhancement of freezing stability in mashed potatoes by the incorporation of kappa-carrageenan and xanthan gum blends. Journal of the Science of Food and Agriculture 89, 2115–2127.

Alvarez, M.D., Canet, W., 2009. Thermal processing and quality optimization. In: Singh, J., Kaur, L. (Eds.), Advances in Potato Chemistry and Technology. Academic Press, Burlington, pp. 163–219 (Chapter 7).

Alvarez, M.D., Fernández, C., Olivares, M.D., Canet, W., 2010a. Rheological behaviour and functionality of inulin-extra virgin olive oil-based mashed potatoes. International Journal of Food Science and Technology 45, 2108–2118.

Alvarez, M.D., Fernández, C., Canet, W., 2010b. Oscillatory rheological properties of fresh and frozen/thawed mashed potatoes as modified by different cryoprotectants. Food and Bioprocess Technology 3, 55–70.

Alvarez, M.D., Fernández, C., Jiménez, M.J., Canet, W., 2011a. Texture of extra virgin olive oil-enriched mashed potatoes: sensory, instrumental and structural relationships. Journal of Texture Studies 42, 413–429.

Alvarez, M.D., Fernández, C., Solas, M.T., Canet, W., 2011b. Viscoelasticity and microstructure of inulin-enriched mashed potatoes: influence of freezing and cryoprotectants. Journal of Food Engineering 102, 66–76.

Alvarez, M.D., Fernández, C., Canet, W., 2011c. Effect of cryoprotectant mixtures on rheological properties of fresh and frozen/thawed mashed potatoes. Journal of Food Process Engineering 34, 224–250.

Alvarez, M.D., Fernández, C., Canet, W., 2011d. Flow behaviour and sensory properties of extra virgin olive oil-enriched mashed potatoes: influence of cryoprotectants and freezing. In: Benkeblia, N. (Ed.), Potato VI. Food 5 (Special Issue 1), 27–39. Global Science Books (GSB), UK.

Alvarez, M.D., Fernández, C., Olivares, M.D., Canet, W., 2012a. A rheological characterisation of mashed potatoes enriched with soy protein isolate. Food Chemistry 133, 1274–1282.

Alvarez, M.D., Fernández, C., Olivares, M.D., Canet, W., 2012b. Comparative characterization of dietary fibre enriched frozen/thawed mashed potatoes. International Journal of Food Properties 15, 1022–1041.

Alvarez, M.D., Jiménez, M.J., Olivares, M.D., Barrios, L., Canet, W., 2012c. Texture perception determined by soy protein isolate and inulin addition in potato puree: links with mechanical and microstructural features. Journal of Texture Studies 43, 361–374.

Alvarez, M.D., Fernández, C., Olivares, M.D., Jiménez, M.J., Canet, W., 2013a. Sensory and texture properties of mashed potatoes incorporated with inulin and olive oil blends. International Journal of Food Properties 16, 1–21.

Alvarez, M.D., Olivares, M.D., Blanch, M., Canet, W., 2013b. Mashed potatoes enriched with soy protein isolate and inulin: chemical, rheological and structural basis. Food Science and Technology International 19, 447–460.

Alvarez, M.D., Canet, W., Jiménez, M.J., 2014. Rheology, texture and functionality of soy protein isolate-based potato puree: instrumental, sensory and structural relationships. In: Casamides, J.M., Gonzalez, H. (Eds.), Soy Protein: Production Methods, Functional Properties and Food Sources. Nova Science Publishers, Inc, NY, USA, pp. 1–46.

Andersson, A., Gekas, V., Lind, I., Oliveira, F., Östel, R., 1994. Effect of preheating on potato texture. Critical Reviews in Food Science and Nutrition 34, 229–251.

Arthey, D., 1993. Freezing of vegetables and fruits. In: Mallet, C.P. (Ed.), Frozen Food Technology. Blackie Academic and Professional, London, pp. 237–269.

Barrera, C., Chenoll, C., Andres, A., Fito, P., 2007. Application of SAFES (systematic approach to food engineering systems) methodology to French fries manufacture. Journal of Food Engineering 83, 201–210.

Blahovec, J., Kuroki, S., Sakurai, N., 2007. Cooking kinetics of potato tubers determined by vibration techniques. Food Research International 40, 576–584.

Blahovec, J., Hejlova, A., 2010. Simple kinetic models of potato sloughing. International Journal of Food Properties 13, 51–64.

Bourne, M.C., 1987. Effect of blanch temperature on kinetics of thermal softening of carrots and green beans. Journal of Food Science 52, 667–668 690.

Bourne, M.C., 1989. Applications of chemical kinetic theory to the rate of thermal softening of vegetable tissue. In: Jen, J.J. (Ed.), Quality Factors of Fruits and Vegetables. Chemistry and Technology. American Chemical Society, Washington, pp. 98–110 (Chapter 9).

Brown, A.M., Morales, J.A.W., 1970. Determination of blanching conditions for frozen parfried potatoes. American Potato Journal 47, 321–325.

Canet, W., 1980. Estudio de la influencia de los tratamientos térmicos de escaldado, congelación y descongelación en la textura y estructura de patata Ph.D. dissertation. Universidad Politécnica de Madrid.

Canet, W., Espinosa, J., Ruiz-Altisent, M., 1982a. Effects of blanching and rate of freezing on the texture of potatoes measured by mechanical tests. Refrigeration Science and Technology 4, 277–283.

Canet, W., Espinosa, J., Ruiz-Altisent, M., 1982b. Effects of the stepwise blanching on the texture of frozen potatoes measured by mechanical tests. Refrigeration Science Technology 4, 284–289.

Canet, W., Espinosa, J., 1984. The effect of blanching and freezing rate on the texture of potatoes, carrots and peas, measured by mechanical tests. In: Zeuthen, P., Cheftel, J.C., Eriksson, C., Jul, M., Leniger, H., Linko, P., Varela, G. (Eds.), Thermal Processing and Quality of Foods. Elsevier Applied Science, London, pp. 678–683.

Canet, W., 1989. Quality and stability of frozen vegetables. In: Thorne, S. (Ed.), Developments in Food Preservation. Elsevier Applied Science Ltd, London & New York, pp. 1–50 (Chapter 5).

Canet, W., Alvarez, M.D., 2000. Congelación de alimentos vegetales. In: Madrid Vicente, A. (Ed.), Aplicaciones del frío a los alimentos. AMV y Mundi Prensa, Madrid, pp. 201–258.

Canet, W., Alvarez, M.D., Fernández, C., Luna, P., 2005a. Comparisons of methods for measuring yield stresses in potato puree: effect of temperature and freezing. Journal of Food Engineering 68, 143–153.

Canet, W., Alvarez, M.D., Fernández, C., Tortosa, M.E., 2005b. The effect of sample temperature on instrumental and sensory properties of mashed potato products. International Journal of Food Science and Technology 40, 481–493.

Canet, W., Alvarez, M.D., Fernández, C., 2005c. Optimization of low-temperature blanching for retention of potato firmness: effect of previous storage on compression properties. European Food Research and Technology 221, 423–433.

Canet, W., Alvarez, M.D., 2011. Quality and safety of frozen vegetables. In: Sun, D.-W. (Ed.), Handbook of Frozen Food Processing and Packaging, second ed. CRC Taylor & Francis Group, Boca Raton, London, New York, pp. 387–434 (Chapter 17).

Canet, W., Alvarez, M.D., Fernández, C., 2011. Functionality of amidated low methoxyl pectin and xanthan gum blends on the texture and color of mashed potatoes. Journal of Food Process Engineering 34, 746–776.

Chaisawang, M., Suphantharika, M., 2005. Effect of guar gum and xanthan gum additions on physical and rheological properties of cationic tapioca starch. Carbohydrate Polymers 61, 288–295.

Chaisawang, M., Suphantharika, M., 2006. Pasting and rheological properties of native and anionic tapioca starches as modified by guar gum and xanthan gum. Food Hydrocolloids 20, 641–649.

Chen, J., Pitchai, K., Birla, S., Gonzalez, R., Jones, D., Subbiah, J., 2013. Temperature-dependent dielectric and thermal properties of whey protein gel and mashed potato. Transactions of the ASAE 56, 1457–1467.

Dickinson, E., 1998. Stability and rheological implications of electrostatic milk protein–polysaccharide interactions. Trends in Food Science and Technology 9, 347–354.

Doublier, J.L., 1997. Rheological properties of mixed biopolymer systems. In: UPM (Ed.), Fundamentos de Reología. Los materiales viscoelásticos-3. Universidad Internacional Menéndez Pelayo, Valencia, Spain.

Downey, G., 2002. Quality changes in frozen and thawed, cooked puréed vegetables containing hydrocolloids, gums and dairy powders. International Journal of Food Science and Technology 37, 869–877.

Downey, G., 2003. Effects of cryoprotectant mixtures on physical properties of frozen and thawed puréed cooked potatoes: some introductory studies. International Journal of Food Science and Technology 38, 857–868.

Eliasson, A.C., Kim, H.R., 1992. Changes in rheological properties of hydroxypropyl potato starch pastes during freeze-thaw treatments I. A rheological approach for evaluation of freeze-thaw stability. Journal of Texture Studies 23, 279–295.

Eurostat, 2013. http://ec.europa.eu/eurostat (Copyright © Eurostat).

Fernández, C., Alvarez, M.D., Canet, W., 2006. The effect of low-temperature blanching on the quality of fresh and frozen/thawed mashed potatoes. International Journal of Food Science and Technology 41, 577–595.

Fernández, C., Alvarez, M.D., Canet, W., 2008. Steady shear and yield stress data of fresh and frozen/thawed mashed potatoes: effect of biopolymers addition. Food Hydrocolloids 22, 1381–1395.

Fernández, C., Canet, W., Alvarez, M.D., 2009a. The effect of long-term frozen storage on the quality of frozen and thawed mashed potatoes with added cryoprotectant mixtures. International Journal of Food Science and Technology 44, 1373–1387.

Fernández, C., Canet, W., Alvarez, M.D., 2009b. Quality of mashed potatoes: effect of adding blends of kappa-carrageenan and xanthan gum. European Food Research and Technology 229, 205–222.

Fernández, C., Canet, W., Alvarez, M.D., 2010. Rheological properties and sensory attributes of mashed potatoes made with mixtures of sodium caseinate and xanthan gum. In: Hollingworth, C.S. (Ed.), Food Hydrocolloids: Characteristics, Properties and Structures. Nova Science Publishers, Inc, New York, pp. 81–117.

Gould, W.A., 1999a. Potato production. In: Gould, W.A. (Ed.), Potato Production, Processing and Technology. CTI Publications, Inc., Timonium, Maryland, USA, pp. 11–27.

Gould, W.A., 1999b. Receiving and grading for quality. In: Gould, W.A. (Ed.), Potato Production, Processing and Technology. CTI Publications, Inc., Timonium, Maryland, USA, pp. 42–49.

Gould, W.A., 2001. Total Quality Assurance for the Food Industries. CTI Publications Inc., Timonium, MD.

Harada, T., Tirtohusodo, H., Paulus, K., 1985a. Influence of temperature and time on cooking kinetics of potatoes. Journal of Food Science 50, 459–462 472.

Harada, T., Tirtohusodo, H., Paulus, K., 1985b. Influence of the composition of potatoes on their kinetics. Journal of Food Science 50, 463–468.

Harada, T., Paulus, K., 1986. Analytical methods to characterize the cooking behaviour of potatoes. LWT-Food Science and Technology 19, 39–43.

He, J., Cheng, L., Gu, Z., Hong, Y., Li, Z., 2014. Effects of low-temperature blanching on tissue firmness and cell wall strengthening during sweet potato flour processing. International Journal of Food Science and Technology 49, 1360–1366.

Hermansson, A.M., 1997. Biopolymer microstructure and rheology. In: Windhab, E. (Ed.), Proceedings of the 1st International Symposium on Food Rheology and Structure. Vincentz, Hanover, Zurich, pp. 18–26.

Huang, Y.C., Bourne, M.C., 1983. Kinetics of thermal softening of vegetables. Journal of Texture Studies 14, 1–9.

Huang, L.L., Zhang, M., 2012. Trends in development of dried vegetable products as snacks. Drying Technology 30, 448–461.

Hui, Y.H., 2004. Frozen French fried potatoes and quality assurance. In: Hui, Y.H., Ghazala, S., Graham, D.M., Murrel, K.D., Nip, W.K. (Eds.), Handbook of Vegetable Preservation and Processing. Marcel Dekker Inc., New York, pp. 309–324 (Chapter 19).

Ignat, A., Manzocco, L., Brunton, N.P., Nicoli, M.C., Lyng, J.G., 2015. The effect of pulsed electric field pre-treatments prior to deep-fat frying on quality aspects of potato fries. Innovative Food Science and Emerging Technologies 29, 65–69.

International Institute of Refrigeration, 2006. In: International Institute of Refrigeration (Ed.), Recommendations for the Processing and Handling of Frozen Foods, fourth ed. , p. 10. Paris.

ISO. TS 17996 IDF.RM 205, 2006(E). Cheese-determination of Rheological Properties by Uniaxial Compression at Constant Displacement Rate. ISO, Geneva, Switzerland.

Jiménez, M.J., Canet, W., Alvarez, M.D., 2013. Sensory description of potato puree enriched with individual functional ingredients and their blends. Journal of Texture Studies 44, 301–316.

Jobling, A., Westcott, R.J., Tayal, A., Jeffcoat, R., Schwall, G.P., 2002. Production of a freeze–thaw-stable potato starch by antisense inhibition of three starch synthase genes. Nature Biotechnology 20, 295–299.

Jul, L.M., 1984. The Quality of Frozen Foods. Academic Press, London, pp. 1–80.

Kadam, S.S., Wankier, B.N., Adsule, R.N., 1991. Processing. In: Salunkhe, D.K., Kadam, S.S., Jadhav, S.J. (Eds.), Potato: Production, Processing, and Products. CRC Press, Boca Raton, Ann Arbor, Boston, USA, pp. 112–147.

Kalichevsky, M.T., Jaroszkiewicz, E.M., Ablett, S., Blanshard, J.M.V., Lillford, P.J., 1992. The glass transition of amylopectin measured by DSC, DMTA and NMR. Carbohydrate Polymers 18, 77–88.

Kim, H.R., Eliasson, A.C., 1993. Changes in rheological properties of hydroxypropyl potato starch pastes during freeze-thaw treatments II. Effect of molar substitution and cross-linking. Journal of Texture Studies 24, 199–213.

Kim, M.H.R., Muhrbeck, P., Eliasson, A.C., 1993. Changes in rheological properties of hydroxypropyl potato starch pastes during freeze-thaw treatments III. Effect of cooking conditions and concentration of the starch paste. Journal of the Science of Food and Agriculture 61, 109–116.

Korus, J., Juszczak, L., Witczak, M., Achremowicz, B., 2004. Influence of selected hydrocolloids on triticale starch rheological properties. International Journal of Food Science and Technology 39, 641–652.

Kozempel, M.F., 1988. Modeling the kinetics of cooking and precooking potatoes. Journal of Food Science 53, 753–755.

Lasztity, R., Sebök, A., Major, J., 1992. Textural properties of fruits and vegetables and their changes during freezing and storage at low temperature. Periodica Polytechnica Chemical Engineering 36, 225–238.

Liu, E.Z., Scanlon, M.G., 2007. Modeling the effect of blanching conditions on the texture of potato strips. Journal of Food Engineering 81, 292–297.

Luscher, C., Schluter, O., Knorr, D., 2005. High pressure-low temperature processing of foods: impact on cell membranes, texture, color and visual appearance of potato tissue. Innovative Food Science and Emerging Technologies 6, 59–71.

Mandala, I., Michon, C., Launay, B., 2004. Phase and rheological behaviors of xanthan/amylose and xanthan/starch mixed systems. Carbohydrate Polymers 58, 285–292.

Masson, L., Munoz, J.R., Romero, N., Camilo, C., Encina, C., Hernandez, L., Castro, J., Robert, P., 2007. Acrylamide in fried potatoes: an updated review. Grasas y aceites 58, 185–193.

McClements, J.M.J., Patterson, M.F., Linton, M., 2001. The effect of growth stage and growth temperature on high hydrostatic pressure inactivation of some psychrotrophic bacteria in milk. Journal of Food Protection 64, 514–522.

Moledina, K.H., Haydar, M., Ooraikul, B., Hadziyev, D., 1981. Pectin changes in the precooking step of dehydrated mashed potato production. Journal of the Science of Food and Agriculture 32, 1091–1102.

Montgomery, D.C., 1991. Diseño y Análisis de Experimentos. Grupo Editorial Iberoamérica, Mexico, pp. 467–509.

Morikawa, K., Nishinari, K., 2000. Effects of concentration dependence of retrogradation behaviour of dispersions for native and chemically modified potato starch. Food Hydrocolloids 14, 395–401.

Myers, R.H., Montgomery, D.C., 1995. Response Surface Methodology: Process and Product Optimization Using Designed Experiments. Wiley, New York.

Nath, A., Chattopadhyay, P.K., 2007. Optimization of oven toasting for improving crispness and other quality attributes of ready to eat potato–soy snack using response surface methodology. Journal of Food Engineering 80, 1282–1292.

Nath, A., Chattopadhyay, P.K., 2008. Effect of process parameters and soy flour concentration on quality attributes and microstructural changes in ready-to-eat potato–soy snack using high-temperature short time air puffing. LWT-Food Science and Technology 41, 707–715.

Norton, T., Sun, D.-W., 2008. Recent advances in the use of high pressure as an effective processing technique in the food industry. Food and Bioprocess Technology 1, 2–34.

Nourian, F., Ramaswamy, H.S., 2003a. Kinetics of quality change during cooking and frying of potatoes: Part II. Colour. Journal of Food Process Engineering 26, 395–411.

Nourian, F., Ramaswamy, H.S., 2003b. Kinetics of quality change during cooking and frying of potatoes: Part I. Texture. Journal of Food Process Engineering 26, 377–394.

Nwabueze, T.U., 2010. Basic steps in adapting response surface methodology as mathematical modelling for bioprocess optimisation in the food systems. International Journal of Food Science and Technology 45, 1768–1776.

O'Leary, E., Gormley, T.R., Butler, F., Shilton, N., 2000. The effect of freeze-chilling on the quality of ready-meal components. LWT-Food Science and Technology 33, 217–224.

Park, S.H., Balasubramaniam, V.M., Sastry, S.K., 2013. Estimating pressure induced changes in vegetable tissue using in situ electrical conductivity measurement and instrumental analysis. Journal of Food Engineering 114, 47–56.

Pedreschi, F., Moyano, P., 2005. Effect of pre-drying on texture and oil uptake of potato chips. LWT-Food Science and Technology 38, 599–604.

Pedreschi, F., Leon, J., Mery, D., Moyano, P., Pedreschi, R., Kaack, K., Granby, K., 2007a. Color development and acrylamide content of pre-dried potato chips. Journal of Food Engineering 79, 786–793.

Pedreschi, F., Bustos, O., Mery, D., Moyano, P., Kaack, K., Granby, K., 2007b. Color kinetics and acrylamide formation in NaCl soaked potato chips. Journal of Food Engineering 79, 989–997.

Pimpaporn, P., Devahastin, S., Chiewchan, N., 2007. Effects of combined pretreatments on drying kinetics and quality of potato chips undergoing low-pressure superheated steam drying. Journal of Food Engineering 81, 318–329.

Polak, A., November 17, 2014. Concern after US approves GM potato. Deccan Herald. http://www.deccanherald.com/content/442200/concern-us-approves-gm-potato.html.

Pravisani, C.I., Califano, A.N., Calvelo, A., 1985. Kinetics of starch gelatinization in potato. Journal of Food Science 50, 657–660.

Pieterse, L., Judd, J. (Eds.), 2014. World Catalogue of Potato Varieties, sixth ed. Agrimedia Gmbh, Germany, p. 430.

Rahardjo, B., Sastry, K., 1993. Kinetics of softening of potato tissue during thermal treatment. Transactions of IChemE 71, 235–241.

Redmond, G.A., Butler, F., Gormley, T.R., 2002. The effect of freezing conditions on the quality of freeze-chilled reconstituted mashed potato. LWT-Food Science and Technology 35, 201–204.

Redmond, G.A., Gormley, T.R., Butler, F., 2003. The effect of short and long-term freeze-chilling on the quality of mashed potato. Innovative Food Science and Emerging Technologies 4, 85–97.

Ross, H.A., Morris, W.L., Ducreux, L.J.M., Hancock, R.D., Verrall, S.R., Morris, J.A., Tucker, G.A., Stewart, D., Hedley, P.E., McDougall, G.J., Taylor, M.A., 2011. Pectin engineering to modify product quality in potato. Plant Biotechnology Journal 9, 848–856.

Sajeev, M.S., Sreekumar, J., Unnikrishnan, M., Moorthy, S.N., Shanavas, S., 2010. Kinetics of thermal softening of cassava tubers and rheological modeling of the starch. Journal of Food Science and Technology 47, 507–518.

Sawant, D.M., Dhumal, S.S., Kadam, S.S., 1991. Production. In: Salunkhe, D.K., Kadam, S.S., Jadhav, S.J. (Eds.), Potato: Production, Processing, and Products. CRC Press, Boca Raton, Ann Arbor, Boston, USA, pp. 37–68.

Schluter, O., Urrutia-Benet, G.U., Heinz, V., Knorr, D., 2004. Metastable states of water and ice during pressure-supported freezing of potato tissue. Biotechnology Progress 20, 799–810.

Severini, C., Baiano, A., Del Nobile, M.A., Mocci, G., De Pilli, T., 2004a. Effects of blanching on firmness of sliced potatoes. Italian Journal of Food Science 16, 31–44.

Severini, C., Baiano, A., De Pilli, T., Romaniello, R., Derossi, A., 2004b. Microwave blanching of sliced potatoes dipped in saline solutions to prevent enzymatic browning. Journal of Food Biochemistry 28, 75–89.

Seyhun, N., Ramaswamy, H.S., Zhu, S.M., Sumnu, G., Sahin, S., 2013. Ohmic tempering of frozen potato puree. Food and Bioprocess Technology 6, 3200–3205.

Shahraki, M.H., Ziaiifar, A.M., Kashaninejadi, S.M., Ghorbani, M., 2014. Optimization of pre-fry microwave drying of French fries using response surface methodology and genetic algorithms. Journal of Food Processing and Preservation 38, 535–550.

Shi, X.H., BeMiller, J.M., 2002. Effects of food gums on viscosities of starch suspensions during pasting. Carbohydrate Polymers 50, 7–18.

Singh, N., Kaur, L., Singh, J., 2004. Relationships between physicochemical, thermal and rheological properties of starches separated from different potato cultivars. Journal of the Science of Food and Agriculture 84, 714–720.

Singh Rattan, N., Ramaswamy, H.S., 2014. Quality optimization of canned potatoes during rotary autoclaving. Journal of Food Quality 37, 168–176.

Srichuwong, S., Isonoa, N., Jiangb, H., Mishimaa, T., Hisamatsua, M., 2012. Freeze–thaw stability of starches from different botanical sources: correlation with structural features. Carbohydrate Polymers 87, 1275–1279.

Stark, J.C., Olsen, N., Kleinkopf, G.E., Love, S.L., 2003. Tuber Quality. www.ag.uidaho.edu/potato/production/files/TUBERQUALITY-web.pdf.

Stark, J.C., Love, S.L., 2003. Tuber quality. In: Stark, J.C., Love, S.L. (Eds.), Potato Production Systems. Educational Communications, University of Idaho, Moscow, Idaho, USA, pp. 329–342 (Chapter 16).

Stolle-Smits, T., Donkers, J., Van Dijk, C., Derksen, J., Sassen, M.M.A., 1998. An electron microscopy study on the texture of fresh, blanched and sterilized green bean pods (*Phaseolus vulgaris* L.). LWT-Food Science and Technology 31, 237–244.

Sun, D.W., 2005. In: Sun, D.W. (Ed.), Emerging Technologies for Food Processing. Elsevier, London.

Suzuki, A., Shibanuma, K., Takeda, Y., Abe, J., Hizukuri, S., 1994. Structures and pasting properties of potato starches from Jaga Kids Purple '90 and Red '90. Journal of Applied Glycoscience 41, 425–432.

Tijskens, L.M.M., Waldron, K.W., Ng, A., Ingham, L., Van Dijk, C., 1997. The kinetics of pectin methyl esterase in potatoes and carrots during blanching. Journal of Food Engineering 34, 371–385.

Troncoso, E., Pedreschi, F., 2007. Modeling of textural changes during drying of potato slices. Journal of Food Engineering 82, 577–584.

Truong, V.D., Walter, W.M., Bett, K.L., 1998. Textural properties and sensory quality of processed sweet potatoes as affected by low temperature blanching. Journal of Food Science 63, 739–743.

Tseng, Y.C., Xiong, Y.L., 2009. Effect of inulin on the rheological properties of silken tofu coagulated with glucono-δ-lactone. Journal of Food Engineering 90, 511–516.

Urrutia-Benet, G., Balogh, T., Schneider, J., Knorr, D., 2007. Metastable phases during high-pressure-low-temperature processing of potatoes and their impact on quality-related parameters. Journal of Food Engineering 78, 375–389.

Van Dijk, C., Fisher, M., Beekhuizen, J.G., Boeriu, C., Stolle-Smiths, T., 2002. Texture of cooked potatoes (*Solanum tuberosum*). 3. Preheating and the consequences for the texture and cell wall chemistry. Journal of Agricultural and Food Chemistry 50, 5098–5106.

Van Loey, A., Fransis, A., Hendrickx, M., Maesmans, G., Tobback, P., 1995. Kinetics of quality changes of green peas and white beans during thermal processing. Journal of Food Engineering 24, 361–377.

Verlinden, B.E., Nicola, B.M., De Baerdemaeker, J., 1995. The starch gelatinization in potatoes during cooking in relation to the modelling of texture kinetics. Journal of Food Engineering 24, 165–179.

Villegas, B., Carbonell, I., Costell, E., 2007. Inulin milk beverages: sensory differences in thickness and creaminess using r-index analysis of the ranking data. Journal of Sensory Studies 22, 377–393.

Wang, L., Xie, B.J., Xiong, G.Q., Wu, W.J., Wang, J., Qiao, Y., Liao, L., 2013. The effect of freeze-thaw cycles on microstructure and physicochemical properties of four starch gels. Food Hydrocolloids 31, 61–67.

Welti-Chanes, J., Lopez-Malo, A., Palou, E., Bermudez, D., Guerrero-Beltran, J.A., Barbosa-Canovas, G.V., 2005. Fundamentals and applications of high pressure processing to foods. In: Barbosa-Canovas, G.V., Tapia, M.S., Cano, P.M. (Eds.), Novel Food Processing Technologies. CRC Press LLC, Boca Raton, FL, USA, pp. 157–181 (Chapter 8).

Zhang, W.J., Chen, H.X., Wang, J., Wang, Y.W., Xing, L.S., Zhang, H., 2014. Physicochemical properties of three starches derived from potato, chestnut, and yam as affected by freeze-thaw treatment. Starch/Stärke 66, 353–360.

CHAPTER 15

Fried and Dehydrated Potato Products

Franco Pedreschi, María Salomé Mariotti, Pablo Cortés

Department of Chemical Engineering and Bioprocesses, Pontificia Universidad Católica de Chile, Santiago, Chile

1. Introduction

Global consumption of potatoes as food is shifting from fresh potatoes to added-value, processed food products. In this sense, potatoes are grown in over 110 countries throughout the world and represent one of the most important staples of the human diet (Bradshaw and Ramsay, 2005). Potatoes are processed into a great variety of products, including cooked potatoes, par-fried potato strips, french fries, potato chips, potato starch, potato granules, potato flakes, and dehydrated diced potatoes, among others (Pedreschi, 2012).

One of the most important potato-derived products is frozen potatoes, which includes most french fries served in restaurants and fast-food chains worldwide. Another processed potato product, potato chips, is the long-standing king of snack foods in many developed countries (Pedreschi, 2012). On the other hand, dehydrated potato flakes are used in retail mashed potato products, as ingredients in snacks, and even as food aid. Potato flour, another dehydrated product, is used by the food industry to bind meat mixtures and thicken gravies and soups. Potato starch provides higher viscosity than wheat and maize starches and delivers a tastier product (Bouchon and Pyle, 2005).

Par-fried potatoes and french fries are traditionally produced by cutting potato strips from fresh potatoes (parallelepiped 1×1 cm cross-section 4–7 cm in length), which are then deep fat fried. Three major kinds of chip are produced on a commercial scale: (1) deep-frozen completely fried strips, which just require oven heating; (2) deep-frozen partially fried chips, which require additional frying before eating; and (3) refrigerated partially fried chips, which have a short shelf life and need additional frying (Bouchon and Pyle, 2004; Lisińska and Leszczyński, 1989).

On the other hand, snack food manufacturers produce two types of potato chips: traditional chips made by thinly slicing fresh potatoes and chips processed from potato dough and formed into potato chip shapes. Restructured potato chips may not have flavor and textural characteristics similar to fresh potato chips; however, they have the advantages of uniformity and an absence of defects because they have been made from potato dough normally based on

Advances in Potato Chemistry and Technology. http://dx.doi.org/10.1016/B978-0-12-800002-1.00015-7

starch-containing ingredients such as potato flakes, flours, starches, ground slices, meal, granules, or mixtures (Bouchon and Pyle, 2004).

Potato heat processing generates products with unique sensorial attributes, which make them attractive to consumers and help increase the shelf life of the potato products. In this sense, chemical, physical, and microstructural changes of potato tissue during processing are responsible for the final attractive sensorial attributes of derived potato products such as fried potatoes (chips and french fries), par-fried potatoes, and cooked potatoes (Pedreschi, 2012). Heat processing also provokes enzyme inactivation and microbial destruction, which extends the shelf life of the thermally treated potato products. Unfortunately, as a result of heating, some undesirable toxicant contaminants such as acrylamide could be generated in potato products cooked at temperatures higher than 120 °C. The challenge in the potato processing industry is to generate attractive end products for the consumer, reducing or avoiding if possible the formation of some compounds which can negatively affect human health. This chapter provides an overview of some of the most important potato-derived products considering the importance of changes in their microstructure and chemical composition during processing using common unit operations such as cooking, baking, frying, blanching, and freezing, among others.

2. Importance of Chemical Composition in Potato Processing

Chemical potato components such as starch, nonstarch polysaccharides, and minerals affect the quality of potatoes and their derived products to a significant extent, and an understanding of the chemical structure of the potato component is a major issue for the development of unique or novel potato-derived products (Pedreschi, 2012). The chemical composition of potato varies with the cultivar, location of growth, agricultural practices, maturity at harvest, and subsequent storage history, among others (Andersson et al., 1994). The potato chemical composition and microstructural features of the potato vary considerably not only between potato tubers but also within, in such a way that some authors have found that the content of some components such as reducing sugars and the total dry matter increase gradually from the center of the tuber toward the external parenchyma region (Lamberg et al., 1990). Table 1 shows the average chemical composition of potato tuber tissue reported by Miranda and Aguilera (2006).

Potatoes are an important source of carbohydrates (reducing sugars, starch, and pectin among others) and other nutrients, including vitamins and minerals consumed widely in developing as well the developed world (Bordoloi et al., 2012). Potato tubers contain 70–80% water, 16–24% starch, and trace quantities (<4%) of proteins, lipids, anthocyanins, and minerals (Huang et al., 2005). Potato starch, which comprises 65–80% of the dry matter content of the potato tuber, is the most important nutritional component and its composition is about 21% amylose, 75% amylopectin, 0.1% protein, and 0.08% phosphorus (Lisińska and Leszczyński, 1989).

Table 1: Chemical Composition of Potato Tubers.

Substances	Content (%)	
	Range	Mean
Water	77.5	63.2–86.9
Total solids	22.5	13.1–36.8
Dry matter	21.5	20–23
Total carbohydrate	19.4	13.3–30.53
Crude fiber	0.6	0.17–3.48
Protein	2.0	0.7–4.6
Fat	0.1	0.02–0.96
Ash	1.0	0.44–1.9

Reprinted from Miranda and Aguilera (2006) with kind permission from Taylor and Francis.

The fat content of potato tubers is low (about 0.1% on a fresh weight basis) but still enough to cause oxidation and lower the quality of dried potato products. On the other hand, the mineral content of potato tubers is high because they contain iron, phosphorous, potassium, calcium, magnesium, and sulfur (Lamberg et al., 1990). The pectic substances, which are polymers of galacturonic acid, in which the carboxyl groups are more or less methylated, constitute about 1% of the dry matter content. In potato tissue, pectic substances are located in the cell wall and are related to cell–cell interactions.

The content of reducing sugars in raw potatoes is closely related to the final color of baked and fried potatoes (Pedreschi, 2012). The sugar content of the tuber varies over a wide range up to 10% of the dry weight and is highly influenced by postharvest factors such as the storage temperature. Reducing sugars influence the quality of heat-processed potato products through the Maillard reaction (Becalski et al., 2003), which has strong influence not only on the final color of high temperature–heated potato products but also on the presence and content of some toxic components such as acrylamide and furan (Mariotti et al., 2015; Pedreschi et al., 2014).

Finally, the potato is a commodity that sometimes needs to be stored up to 10 months under low temperatures to guarantee year-round potato chip availability (McCann et al., 2010). Low-temperature storage (2–5 °C) minimizes pathogen activity, water loss, and sprouting (Blenkinsop et al., 2002; McCann et al., 2010). However, temperatures below 10 °C stimulate a phenomenon known as cold-induced sweetening, characterized by the accumulation of reducing sugars in the tubers (Blenkinsop et al., 2002; Sowokinos, 2001). The free amino acids present in the tubers react with the reducing sugars when they are being processed into fries and chips through the Maillard reaction. This reaction produces dark color and bitter taste compounds, which render the product unacceptable to consumers. Acrylamide, a known neurotoxin and potential carcinogen, is also produced during this reaction (Friedman, 2003; Rommens et al., 2008). In this regard, there is need for developing early-stage predictors of quality that would allow decisions to be made about the fate of a specific batch and thus reduce food losses. Because the physiological state of the potato tubers immediately after harvest will be reflected in the gene expression pattern, a black box transcriptomics approach can help in finding gene predictors of susceptibility to cold-induced sweetening.

3. *Importance of Microstructure in Potato Processing*

Morphologically, a potato tuber is usually oval to round in shape with white flesh and a pale brown skin, although variations in size, shape, and flesh color are frequently encountered (Bordoloi et al., 2012). The microstructure of potato tissue and the properties of cell wall polymers (parenchyma cell size and nature of pectic materials) are two important factors influencing the deformation that occurs during mastication (Kaur et al., 2002; Waldron et al., 1997) and the texture of cooked potatoes, mealy and waxy (Kaur et al., 2002; van Marle et al., 1997). In this sense, potato cell walls can be presented as a cellulose–xyloglucan network embedded in a pectin matrix (Tucker and Mitchell, 1993). Cell wall strength is established by the cellulose microfibrils and the pectin matrix acts as glue that holds together the microfibrils around one cell and between cells.

Potato starch is the major component of the dry matter in potato tissue and it is thought that phenomena associated with gelatinization are involved in the changes of mechanical properties of potato during cooking (Pravisani and Calvelo, 1986; Salisbury and Ross, 1992). Granules of potato starch are oval or spherical units composed of amorphous and crystalline regions that tend to possess radial symmetry and often exhibit well-defined rings or lamellar structures (Andersson et al., 1994; van Marle et al., 1997). In raw and heat-processed potatoes, the abundance of starch in the cells and the shape and size of the starch granules have been reported to affect strongly the texture of the fresh potato and potato-derived products (Martens and Thybo, 2000; McComber et al., 1994).

Depending on the middle lamella characteristics of potato tissue, the cells can either separate or burst at the point of minimum resistance (Kaur et al., 2002). Raw potatoes generally show cell rupturing whereas cooked ones show cell separation as a result of the destabilization of pectic materials during thermal processing (Aguilera and Stanley, 1999). Therefore, a good understanding of potato texture requires sufficient knowledge of microstructural features such as cell shape, composition, and size, and cell wall characteristics. Advanced microscopic techniques such as confocal scanning laser microscopy (CSLM) and scanning electron microscopy have been used by various researchers to study the microstructural characteristics of foods including raw and processed potatoes (Kaur et al., 2002; McComber et al., 1994; Pedreschi and Aguilera, 2002; van de Velde et al., 2002). As shown in Figure 1, the three-dimensional surface microstructure reconstruction obtained from CSLM in the surface of a raw potato slice reveals the high heterogeneity of potato cell sizes and shapes clearly defined by their cell walls.

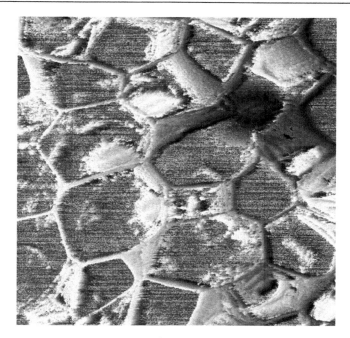

Figure 1
Reconstruction of surface topography of a raw potato slice obtained by confocal laser scanning
microscope (10x; 1 × 1 × 0.1 mm) in its reflective mode.

4. Potato Processing: Important Derived Products

Potato processing is highly industrialized, technologically advanced, and market-driven. For instance, some processed potato products such as potato flour are highly versatile in manufacturing convenience foods whose functional characteristics could depend on the type of tuber as well as the processing methods, among other factors (Singh et al., 2005).

Deep fat frying is one of the oldest and most common unit operations used for cooking foods by immersing them in an edible oil or fat heated above the boiling point of water (Farkas et al., 1996). This complex unit operation involves significant microstructural changes; in fact, most of the desirable characteristics of fried foods are derived from the formation of a composite structure: a dry, porous, crispy, and oily outer layer or crust, and a moist cooked interior or core whose microstructures form during the process (Bouchon et al., 2001). The high temperatures (around 160 and 180 °C) cause water evaporation, which is transferred from the food toward the surrounding oil, whereas oil is absorbed by the food replacing part of the released water. This process results in products with a unique flavor–texture combination (Mellema, 2003). On the other hand, drying is a widely used method of fruit and vegetable preservation. Water is removed to a final concentration, which ensures the microbial stability of the product and minimizes chemical and physical changes (Lewicki and Pawlak, 2005).

Nowadays, drying is regarded not only as a preservation process but also as a method for increasing added value of foods (Ramos et al., 2003). Processing of plant raw materials causes irreversible changes in the tissues of fruit and vegetables. These changes are particularly visible after heat treatment (Lisińska and Gołubowska, 2005).

Similarly, one main item in this category is frozen potatoes, which includes most of the french fries served in restaurants and fast-food chains worldwide (Bouchon and Pyle, 2004). Another processed product, the potato chip, is the long-standing king of snack foods in many developed countries. On the other hand, dehydrated potato products such as flakes are used in retail mashed potato products, as ingredients in snacks.

In potato processing, during harvesting, transportation, washing, sorting, and packing, tubers are subjected to mechanical stress that may lead to crushing of surface cell layers (Galindo et al., 2007). Handling of potato tubers during all steps of the production chain is of major importance because it could have a strong effect over the final quality of the final potato-derived products. The mechanical stress imposed on plant cells by a variety of physical stimuli during harvesting and handling of fresh potato induces a wide range of cellular responses such as increased respiration rate, ethylene production, and higher susceptibility to pathogen attack (Charron and Cantliffe, 1995; Galindo et al., 2007).

Potatoes are particularly susceptible to mechanical stress because physically stressed tuber tissue produces melanin-based pigments, leading to the blue-black discoloration of subdermal tissues known such as black spot bruising, which is a serious agronomic problem in the quality of potato products during harvesting, handling, and storage, and leading to significant levels of rejection of potato harvests (Johnson et al., 2008). Finally, mechanical impact, demonstrated by the use of varying peeling potato methods, also shows changes in the cell wall area, because bricklike cells and curly parenchyma cell walls develop in hard tissue (Kaaber et al., 2007; Kaack et al., 2002). The thickness of these cell layers corresponds to the depth of hard tissue and to an increased amount of cellulose and suberin (Kaack et al., 2002). Thus, as a result of peeling, grading, or shredding, a relatively stable agricultural product with a shelf life of several weeks or months will change into one that deteriorates rapidly from a food quality perspective (Galindo et al., 2007).

4.1 Blanched Potatoes

Blanching is a unit operation used in the potato processing industry to remove air from the potato tissue. This increases leaching of accumulated sugars to control the Maillard reaction during subsequent frying to avoid excessive darkening of potato chips or of french fries, to gelatinize the starch and inactivate enzymes present in the plant tissue (Pedreschi et al., 2004). Blanching consists of immersing potato pieces in hot water before final deep fat frying; starch gelatinization is the crucial microstructural event in the potato tissue that strongly affects the leaching out of some important solutes such as reducing sugars, ascorbic acid, and asparagine before final frying (Aguilera et al., 2001; Pedreschi et al., 2004).

The usual water temperature range for the blanching of potato slices is from 60 to 80 °C and the residence time could vary between 5 and 20 min (Califano and Calvelo, 1988). Previous investigations suggest that it is necessary to optimize blanching processing conditions to diminish the formation of some heat-toxic contaminants such as acrylamide and furan, and to control the oil uptake of potato pieces during final deep fat frying (Mariotti et al., 2015). In this sense, Figure 2 shows the effect of different temperature–time blanching treatments over the extraction of some furan precursors (reducing sugars and ascorbic acid) and over the final oil and furan content of these blanched potato slices fried at 170 °C until reaching a final moisture content of 2%.

There is a clear relationship between the amount of reducing sugars and ascorbic acid of the slices before frying and the furan content generated during frying. It would be interesting in future studies to elucidate which is the crucial precursor (reducing sugars or asparagine) in furan formation during frying of potato slices. On the other hand, there was no clear relationship between furan formed and oil absorbed under the same frying conditions for the different blanching treatments tested.

4.2 Cooked Potatoes

The texture of cooked potatoes is one of the most important quality attributes that is mostly determined by the degradation of cell walls and middle lamellae as a result of heating (temperature around 100 °C), and starch gelatinization (Andersson et al., 1994; Jarvis et al., 1992). Desirable qualities of cooked potatoes include a creamy white color, a moderately dry and mealy texture, and a good and natural potato flavor. Abnormal coloration and excessive softening or firming cause cooked potato products to be rejected by the consumer on many occasions (Andersson et al., 1994).

During potato cooking, a number of microstructural and chemical changes occur in the potato tuber whose magnitudes influence the quality of cooked potatoes. For instance, some cooking conditions may cause excessive quality loss as a result of the breakdown of cellular material such as pectin (Ibarz et al., 1999; Lozano and Ibarz, 1997). Simultaneously, starch present in potato tubers can absorb hot water and swell and therefore can create internal pressures that cause cell separation, reducing cohesiveness and softening the tissue (Binner et al., 2000; Jarvis et al., 1992).

Finally, during industrial production of frozen french fries, potato chips, blanched cubes, mashed potatoes, and many other cooked products, the original material is subjected to a variety of thermal treatments. Sometimes, undesirable hardness may develop in stored prepeeled potatoes after cooking. This problem is especially pronounced in tubers within the first 1–2 months after harvest and appears as subsurface hardness or hard lumps in the cooked tuber tissue, leading to impossible complete cooking and softening of the tubers (Kaaber et al., 2001; Sapers et al., 1995).

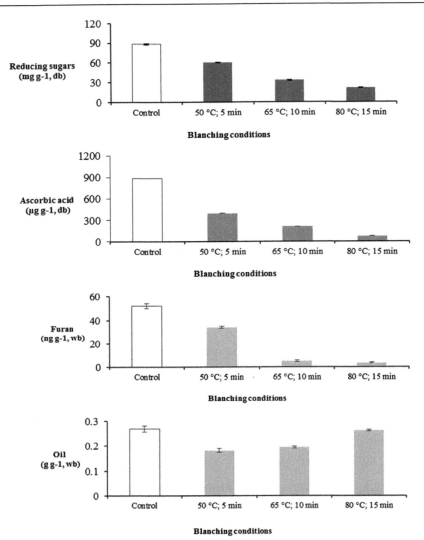

Figure 2
Effect of blanching conditions over furan precursors (reducing sugars and ascorbic acid) in raw potato slices and their corresponding furan and oil content after frying at 170 °C until reaching a final moisture content of 2%.

4.3 Fried Potatoes

Frying of potatoes is an unsteady-state process involving heat and mass transfer in which exposure to hot oil induces major microstructural changes in the raw potato (Aguilera and Stanley, 1999). Macroscopically, a dynamic state sets in which water evaporates in the form of steam bubbles and an outer dry crust is progressively formed by dehydration and

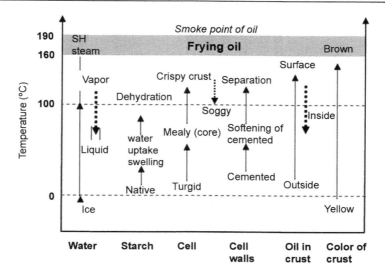

Figure 3

Physical, chemical, and structural changes occurring during deep frying of potatoes. *Reprinted from Miranda and Aguilera (2006) with kind permission from Taylor and Francis.*

infiltration with oil. In the presence of water, potato starch gelatinizes above 65 °C, resulting in hydration, swelling, and filling of the interior of the cell (Aguilera et al., 2001). The most important physical, chemical, and structural changes in potato tissue that take place during frying are shown in Figure 3.

During frying of potato pieces at temperatures well above the boiling point of water, the outer cells become dehydrated as water is released from the intercellular spaces in the form of steam. The main changes that potato tissue undergoes are associated with fast dehydration, starch gelatinization, softening of cell walls, and crust formation (Bouchon and Aguilera, 2001). Oil does not seem to penetrate into the potato piece to a large extent during frying; most of it remains on the surface until the piece is removed from the fryer and cools down (Aguilera and Gloria-Hernandez, 2000; Pedreschi et al., 1999; Ufheil and Escher, 1996).

Bouchon et al. (2003) explained that three different oil fractions can be identified as a consequence of the different absorption mechanisms in fried potato cylinders: (1) structural oil, which represents oil absorbed during frying; (2) penetrated surface oil, which represents oil suctioned into the food during cooling after removal from the fryer; and (3) surface oil, which is the oil that remains on the surface (Figure 4). These authors showed that a small amount of oil penetrates during frying because most oil was picked up at the end of the process, which suggests that oil uptake and water removal are not synchronous phenomena. These findings were confirmed by Pedreschi et al. (2008) and Durán et al. (2007) using glass micromodels by Cortés et al. (2014).

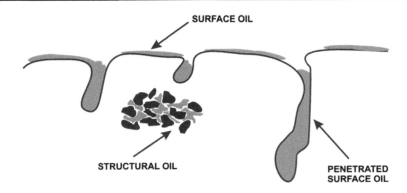

Figure 4
Types of oil fractions that can be identified in a fried product. *Adapted from Bouchon et al. (2003).*

Deep fat frying of potatoes leads to the appreciated surface color and mechanical characteristics of fried potatoes; heating of reducing sugars also affects a complex group of reactions termed caramelization, leading to browning that defines the color of the final product (Arabhosseini et al., 2011). In addition, some heat-induced toxic compounds such as acrylamide and furan are formed during this process and their formation has been mainly linked to the Maillard reaction (Gökmen and Palazoğlu, 2008; Mariotti et al., 2013; Pedreschi et al., 2004).

Other heat-induced toxicants that can be formed during frying are ethylcarbamate, heterocyclic amines, 5-hydroxymethylfurfural (HMF), polycyclic aromatic hydrocarbons, and nitrosoamines. Toxic compounds formed during frying of potatoes, such as HMF, acrylamide, furan, and a variety of Maillard reactants and lipid oxidation products, may constitute an increased cancer risk for consumers (Pedreschi, 2012).

Some undesirable effects derived from long-lasting high temperatures involved in the atmospheric deep fat frying process and exposure to oxygen are the degradation of important nutritional compounds and the generation of toxic molecules in the foodstuff or the frying oil itself. In this sense, vacuum frying is an excellent alternative to conventional frying, because when frying below atmospheric pressure, significant advantages such as healthier and higher quality products are reached (Troncoso et al., 2009). Owing to the lowering of pressure, the boiling points of both the oil and the moisture in the foods are lowered, which can reduce both oil and acrylamide content in the fried product, it may preserve natural color and flavors of the product as a result of the low temperature and oxygen content during the process, and it has less adverse effects on oil quality (Dueik et al., 2010; Garayo and Moreira, 2002; Granda et al., 2004; Yagua and Moreira, 2011).

French fries are among the highest saleable potato products all around the world and are traditionally produced by cutting potato strips from fresh potatoes which are then deep fat fried. Three major kinds of are produced at a commercial scale: (1) deep-frozen completely fried strips, which just require oven heating; (2) deep-frozen partially fried strips, which require additional frying before eating; and (3) refrigerated partially fried strips, which have a short shelf life and need additional frying (Lisińska and Leszczyński, 1989).

To prepare commercial french fries, raw potato strips are first blanched until they reach a generally translucent condition throughout. Once the strips are blanched, they are dehydrated by subjecting them to heated air currents to reduce the moisture content, losing 20–30% of their initial weight. Then, the partially dried strips are immersed in a deep fat fryer for a short period of time, at an oil temperature of about 190 °C, during which time the moisture content is still further reduced and the strips are partially fried. The strips are then frozen by placement directly into a freezer or by first cooling the strips and then freezing them to about −18 °C. They may be maintained in their frozen state for many months for use when necessary. They are packaged as desired. When consumers are ready to use the potatoes, they merely immerse the frozen strips into a deep fat or oil bath at a temperature of about 190 °C for a short time, depending on the bath size, the quantity and size of the strips, and their initial temperature, among other factors that will be apparent to those skilled in the art (Somsen et al., 2004).

Potato chips made from potato dough are normally based on starch-containing ingredients such as potato flakes, flour, starches, ground slices, meal, granules, or mixtures. Dry ingredients are mixed, normally followed by the addition of liquid ingredients. Thereafter, the mixture is introduced to the sheeting line, where it is formed into discrete pieces. Subsequently, the product is fried in a continuous fryer. Finally, restructured potato chips may not have flavor and textural characteristics similar to fresh potato chips; however, they have the advantages of uniformity and the absence of defects (Bouchon and Pyle, 2004).

4.4 Dehydrated Potatoes

Drying notoriously causes physical and structural modifications to potato tissue (Troncoso and Pedreschi, 2009). The most pronounced macroscopic modification is the shrinkage and deformation of food pieces. Transient thermal and moisture gradients develop tensional and compressional stresses that can cause tissue breakage and fracturing during drying (Lewicki and Pawlak, 2005). Wang and Brennan (1995) showed that surface layers of potato slabs dried by air convection are severely damaged at short times, whereas the inner structure appears intact. Further drying induces the formation of cracks; the inner tissue is pulled apart and numerous holes are produced (Lewicki and Pawlak, 2005). Loss of water and segregation of components occurring during drying could cause rigidity, damage, and disruption of cell walls, and even collapse of cellular tissue, leading to a reduction in the volume of the final product (Troncoso and Pedreschi, 2009).

Dehydrated potato products, including potato flakes and potato granules, are processed products made from whole raw potatoes that are cleaned, peeled, sliced, precooked,

| Standard Flakes | Granular & Fine Flour | Slices/Dices/ Shreds |

Figure 5
Main dehydrated products (http://www.potatoesusa.com/).

cooled, cooked, mashed, and either drum-dried into flake form or air-dried into granule form. The main dehydrated products are shown in Figure 5. Additives may be incorporated to improve texture, color, and shelf life (Troncoso and Pedreschi, 2009). The process of converting potatoes into flakes involves many stages culminating in drum-drying of potato slurry, which is spread onto the surface of a heated drum with steam condensing inside the drum. Heat is transferred through the metal of the drum and moisture is evaporated, usually with an appropriate scraping device, from the dried slurry adhering to the drum. Potato flakes generally have about 5–10% moisture and their potato cells are substantially intact with a minimum of free starch (Kakade et al., 2011). Low-leach potato flakes are chosen as the major ingredient in the restructured potato chip formulation because it is frequently used to manufacture pellets or die-cut sheet snacks because of its high stickiness. In addition, it is a desirable ingredient because of the expanded texture and rapid palate clearance it confers on finished products, mainly because the starch in potato flake is fully gelatinized (Bouchon and Pyle, 2004).

5. Conclusions

In the developed regions of the globe, a considerable part of potatoes, one- to two-thirds, is no longer consumed as fresh or table potatoes but in a processed form such as french fries or potato chips. Potato processing is highly industrialized, technologically advanced, and market-driven. However, the quality of its products and the economic success of this industry highly depend on the available raw materials: namely, the potatoes. This means that the availability of suitable cultivars, potato yield, quality of its various aspects during the growing season, and good postharvest (storage) performance are of utmost importance for the potato processing industry. Factors related to potato tubers such as microstructural aspects, chemical composition, and degree of maturity strongly affect processing and quality attributes of final potato products. Potatoes processed at high temperatures, such as those used in frying, cooking, and baking, lead to appreciated surface color and mechanical characteristics and to the undesirable formation of some toxic compounds such as acrylamide, whose generation is strongly linked to the Maillard reaction. Deep fat frying is a complex unit operation that induces significant microstructural changes in potato tissue; in fact, most of the desirable characteristics of fried foods are derived from the formation of a composite structure: (1) a dry, porous, crispy, and oily outer layer or crust; and (2) a moist cooked interior or core whose microstructures have formed during the process. On the other hand, when potatoes are cooked immersed in hot water (temperature ≤ 100 °C), no crust is formed and principally microstructural changes in potato tissue such as starch gelatinization and lamella media solubilization occur and end with a final microstructure similar to that present in the core of a french fry (moist cooked interior). Finally, blanching is an important unit operation because it could be useful to remove acrylamide precursors such as reducing sugars and asparagine, and in this way avoid too-dark final coloration in fried potatoes and excessive amounts of acrylamide generated during deep fat frying.

Acknowledgment

The authors acknowledge financial support from FONDECYT Project No. 1150146, entitled "Mitigation of neo-formed contaminants in Chilean starchy foods and its effect on the consumer acceptance."

References

Aguilera, J.M., Cadoche, L., Lopez, C., Gutierrez, G., 2001. Microstructural changes of potato cells and starch granules heated in oil. Food Research International 34, 939–947.

Aguilera, J.M., Gloria-Hernandez, H., 2000. Oil absorption during frying of frozen parfried potatoes. Journal of Food Science 65, 476–479.

Aguilera, J.M., Stanley, D.W., 1999. Microstructural Principles of Food Processing and Engineering, second ed. Aspen Publishers, Inc., Gaithersburg, Maryland.

Andersson, A., Gekas, V., Lind, I., Oliveira, F., Öste, R., Aguilfra, J.M., 1994. Effect of preheating on potato texture. Critical Reviews in Food Science and Nutrition 34, 229–251.

Arabhosseini, A., Padhye, S., Huisman, W., van Boxtel, A., Müller, J., 2011. Effect of drying on the color of Tarragon (*Artemisia dracunculus* L.) leaves. Food Bioprocess Technology 4, 1281–1287.

Becalski, A., Lau, B.P.Y., Lewis, D., Seaman, S.W., 2003. Acrylamide in foods: occurrence, sources, and modeling. Journal of Agricultural and Food Chemistry 51, 802–808.

Binner, S., Jardine, W.G., Renard, C.M.C.G., Jarvis, M.C., 2000. Cell wall modifications during cooking of potatoes and sweet potatoes. Journal of the Science of Food and Agriculture 80, 216–218.

Blenkinsop, R.W., Copp, L.J., Yada, R.Y., Marangoni, A.G., 2002. Changes in compositional parameters of tubers of potato (*Solanum tuberosum*) during low-temperature storage and their relationship to chip processing quality. Journal of Agricultural and Food Chemistry 50, 4545–4553.

Bordoloi, A., Kaur, L., Singh, J., 2012. Parenchyma cell microstructure and textural characteristics of raw and cooked potatoes. Food Chemistry 133, 1092–1100.

Bouchon, P., Aguilera, J.A., 2001. Microstructural analysis of frying potatoes. International Journal of Food Science and Technology 36, 669–676.

Bouchon, P., Aguilera, J.M., Pyle, D.L., 2003. Structure oil-absorption relationships during deep-fat frying. Journal of Food Science 68, 2711–2716.

Bouchon, P., Hollins, P., Pearson, M., Pyle, D.L., Tobin, M.J., 2001. Oil distribution in fried potatoes monitored by infrared microspectroscopy. Journal of Food Science 66, 918–923.

Bouchon, P., Pyle, D.L., 2004. Studying oil absorption in restructured potato chips. Journal of Food Science 69, E115–E122.

Bouchon, P., Pyle, D.L., 2005. Modelling oil absorption during post-frying cooling: I: model development. Food and Bioproducts Processing 83, 253–260.

Bradshaw, J., Ramsay, G., 2005. Utilisation of the commonwealth potato collection in potato breeding. Euphytica 146, 9–19.

Califano, A.N., Calvelo, A., 1988. Adjustment of surface concentration of reducing sugars before frying of potato strips. Journal of Food Processing and Preservation 12, 1–9.

Cortés, P., Badillo, G., Segura, L.A., Bouchon, P., 2014. Experimental evidence of water loss and oil-uptake during simulated deep-frying process using glass micromodels. Journal of Food Engineering 140, 19–27.

Charron, C.S., Cantliffe, D.J., 1995. Volatile emissions from plants. Horticultural Reviews 43–71.

Dueik, V., Robert, P., Bouchon, P., 2010. Vacuum frying reduces oil uptake and improves the quality parameters of carrot crisps. Food Chemistry 119, 1143–1149.

Durán, M., Pedreschi, F., Moyano, P., Troncoso, E., 2007. Oil partition in pre-treated potato slices during frying and cooling. Journal of Food Engineering 81, 257–265.

Farkas, B.E., Singh, R.P., Rumsey, T.R., 1996. Modeling heat and mass transfer in immersion frying. 1. Model development. Journal of Food Engineering 29, 211–226.

Friedman, M., 2003. Chemistry, biochemistry, and safety of acrylamide. A review. Journal of Agricultural and Food Chemistry 51, 4504–4526.

Galindo, F.G., Sjöholm, I., Rasmusson, A.G., Widell, S., Kaack, K., 2007. Plant stress physiology: opportunities and challenges for the food industry. Critical Reviews in Food Science and Nutrition 47, 749–763.

Garayo, J., Moreira, R., 2002. Vacuum frying of potato chips. Journal of Food Engineering 55, 181–191.

Gökmen, V., Palazoğlu, T., 2008. Acrylamide formation in foods during thermal processing with a focus on frying. Food and Bioprocess Technology 1, 35–42.

Granda, C., Moreira, R.G., Tichy, S.E., 2004. Reduction of acrylamide formation in potato chips by low-temperature vacuum frying. Journal of Food Science 69, E405–E411.

Huang, D., Ou, B., Prior, R.L., 2005. The chemistry behind antioxidant capacity assays. Journal of Agricultural and Food Chemistry 53, 1841–1856.

Ibarz, A., Pagán, J., Garza, S., 1999. Kinetic models for colour changes in pear puree during heating at relatively high temperatures. Journal of Food Engineering 39, 415–422.

Jarvis, M.C., Mackenzie, E., Duncan, H.J., 1992. The textural analysis of cooked potato. 2. Swelling pressure of starch during gelatinisation. Potato Research 35, 93–102.

Johnson, S.N., Anderson, E.A., Dawson, G., Griffiths, D.W., 2008. Varietal susceptibility of potatoes to wireworm herbivory. Agricultural and Forest Entomology 10, 167–174.

Kaaber, L., Bråthen, E., Martinsen, B., Shomer, I., 2001. The effect of storage conditions on chemical content of raw potatoes and texture of cooked potatoes. Potato Research 44, 153–163.

Kaaber, L., Kaack, K., Kriznik, T., Bråthen, E., Knutsen, S.H., 2007. Structure of pectin in relation to abnormal hardness after cooking in pre-peeled, cool-stored potatoes. LWT—Food Science and Technology 40, 921–929.

Kaack, K., Larsen, E., Thybo, A.K., 2002. The influence of mechanical impact and storage conditions on subsurface hardening in pre-peeled potatoes (*Solanum tuberosum* L.). Potato Research 45, 1–8.

Kakade, R.H., Das, H., Ali, S., 2011. Performance evaluation of a double drum dryer for potato flake production. Journal of Food Science and Technology 48, 432–439.

Kaur, L., Singh, N., Sodhi, N.S., 2002. Some properties of potatoes and their starches II. Morphological, thermal and rheological properties of starches. Food Chemistry 79, 183–192.

Lamberg, I., Hallström, B., Olsson, H., 1990. Fat uptake in a potato drying/frying process. Lebensmittel-Wissenschaft und -Technologie 23, 295–300.

Lewicki, P.P., Pawlak, G., 2005. Effect of mode of drying on microstructure of potato. Drying Technology 23, 847–869.

Lisińska, G., Gołubowska, G., 2005. Structural changes of potato tissue during french fries production. Food Chemistry 93, 681–687.

Lisińska, G., Leszczyński, W., 1989. Potato Science and Technology. Elsevier Applied Science, London.

Lozano, J.E., Ibarz, A., 1997. Colour changes in concentrated fruit pulp during heating at high temperatures. Journal of Food Engineering 31, 365–373.

Mariotti, M., Cortés, P., Fromberg, A., Bysted, A., Pedreschi, F., Granby, K., 2015. Heat toxicant contaminant mitigation in potato chips. LWT—Food Science and Technology 60, 860–866.

Mariotti, M.S., Granby, K., Rozowski, J., Pedreschi, F., 2013. Furan: a critical heat induced dietary contaminant. Food and Function 4, 1001–1015.

Martens, H.J., Thybo, A.K., 2000. An integrated microstructural, sensory and instrumental approach to describe potato texture. LWT—Food Science and Technology 33, 471–482.

McCann, L.C., Bethke, P.C., Simon, P.W., 2010. Extensive variation in fried chip color and tuber composition in cold-stored tubers of wild potato (Solanum) germplasm. Journal of Agricultural and Food Chemistry 58, 2368–2376.

McComber, D.R., Horner, H.T., Chamberlin, M.A., Cox, D.F., 1994. Potato cultivar differences associated with mealiness. Journal of Agricultural and Food Chemistry 42, 2433–2439.

Mellema, M., 2003. Mechanism and reduction of fat uptake in deep-fat fried foods. Trends in Food Science and Technology 14, 364–373.

Miranda, M.L., Aguilera, J.M., 2006. Structure and texture properties of fried potato products. Food Reviews International 22, 173–201.

Pedreschi, F., 2012. Frying of potatoes: physical, chemical, and microstructural changes. Drying Technology 30, 707–725.

Pedreschi, F., Aguilera, J., 2002. Some changes in potato chips during frying observed by confocal laser scanning microscopy (CLSM). Food Science and Technology International 8, 197–201.

Pedreschi, F., Aguilera, J.M., Arbildua, J.J., 1999. CLSM study of oil location in fried potato slices. Microscopy and Analysis 21–22.

Pedreschi, F., Kaack, K., Granby, K., 2004. Reduction of acrylamide formation in potato slices during frying. LWT—Food Science and Technology 37, 679–685.

Pedreschi, F., Kaack, K., Granby, K., 2008. The effect of asparaginase on acrylamide formation in french fries. Food Chemistry 109, 386–392.

Pedreschi, F., Mariotti, M.S., Granby, K., 2014. Current issues in dietary acrylamide: formation, mitigation and risk assessment. Journal of the Science of Food and Agriculture 94, 9–20.

Pravisani, C.I., Calvelo, A., 1986. Minimum cooking time for potato strip frying. Journal of Food Science 51, 614–617.

Ramos, I., Brandao, T., Silva, C., 2003. Structural changes during air drying of fruits and vegetables. Food Science and Technology International 9, 201–206.

Rommens, C.M., Yan, H., Swords, K., Richael, C., Ye, J., 2008. Low-acrylamide french fries and potato chips. Plant Biotechnology Journal 6, 843–853.

Salisbury, F.B., Ross, C.W., 1992. Plant Physiology, fourth ed. Wadsworth Publishing Co., Belmont, CA.

Sapers, G.M., Miller, R.L., Choi, S., 1995. Prevention of enzymatic browning in pre-peeled potatoes and minimally processed mushrooms. In: Lee, C.Y., Whitaker, J.R. (Eds.), Enzymatic Browning and Its Prevention. American Chemical Society, Washington, DC.

Singh, N., Kaur, L., Ezekiel, R., Singh Guraya, H., 2005. Microstructural, cooking and textural characteristics of potato (*Solanum tuberosum* L.) tubers in relation to physicochemical and functional properties of their flours. Journal of the Science of Food and Agriculture 85, 1275–1284.

Somsen, R.J., Jennings, J.R., Van der Molen, M.W., 2004. The cardiac cycle time effect revisited: temporal dynamics of the central-vagal modulation of heart rate in human reaction time tasks. Psychophysiology 41, 941–953.

Sowokinos, J.R., 2001. Pyrophosphorylase in *Solanum tuberosum* L.: allelic and isozyme patterns of UDP-Glucose pyrophosphorylase as a marker for cold-sweetening resistance in potatoes. American Journal of Potato Research 57–64.

Troncoso, E., Pedreschi, F., 2009. Modeling water loss and oil uptake during vacuum frying of pre-treated potato slices. LWT—Food Science and Technology 42, 1164–1173.

Troncoso, E., Pedreschi, F., Zúñiga, R.N., 2009. Comparative study of physical and sensory properties of pre-treated potato slices during vacuum and atmospheric frying. LWT—Food Science and Technology 42, 187–195.

Tucker, G.A., Mitchell, J., 1993. Cell walls, structure, utilization and manipulation. In: Grierson, D. (Ed.), Biosynthesis and Manipulation of Plant Products. Blackie Academic and Professional, London, UK.

Ufheil, G., Escher, F., 1996. Dynamics of oil uptake during deep-fat frying of potato slices. LWT—Food Science and Technology 29, 640–644.

van de Velde, F., van Riel, J., Tromp, R.H., 2002. Visualisation of starch granule morphologies using confocal scanning laser microscopy (CSLM). Journal of the Science of Food and Agriculture 82, 1528–1536.

van Marle, J.T., de Vries, R., Wilkinson, E.C., Yuksel, D., 1997. Sensory evaluation of the texture of steam-cooked table potatoes. Potato Research 40, 79–90.

Waldron, K.W., Smith, A.C., Parr, A.J., Ng, A., Parker, M.L., 1997. New approaches to understanding and controlling cell separation in relation to fruit and vegetable texture. Trends in Food Science and Technology 8, 213–221.

Wang, N., Brennan, J.G., 1995. A mathematical model of simultaneous heat and moisture transfer during drying of potato. Journal of Food Engineering 24, 47–60.

Yagua, C.V., Moreira, R.G., 2011. Physical and thermal properties of potato chips during vacuum frying. Journal of Food Engineering 104, 272–283.

CHAPTER 16

Textural Characteristics of Raw and Cooked Potatoes

Jaspreet Singh[1], Lovedeep Kaur[1], M.A. Rao[2]

[1]*Riddet Institute and Massey Institute of Food Science and Technology, Massey University, Palmerston North, New Zealand;* [2]*Cornell University, Geneva, NY, USA*

1. Introduction

Potatoes (*Solanum tuberosum* L.) are an important source of carbohydrates and are consumed widely in the developing as well as developed world (Bordoloi et al., 2012). Tuber crops contain 70–80% water, 16–24% starch, and trace quantities (<4%) of proteins, lipids, anthocyanins, minerals, etc. (Huang et al., 2006). Starch is the major component of the dry matter, accounting for about 70% of the total solids. Morphologically, a potato tuber is usually oval to round with white flesh and a pale brown skin, although variations in size, shape, and flesh color are frequently encountered. The color, size, and cooked potato texture are the main quality attributes assessed by the consumer for the acceptability of potatoes on a domestic scale. However, quality screening for the industrial processing of potatoes include several parameters such as dry matter, starch content and characteristics, and postharvest and post-processing shelf stability. Cultivar type, physicochemical composition, and postharvest storage are important factors that can influence the cooking characteristics of potato and potato products (Burton, 1989; Singh et al., 2002, 2003, 2008a,b; Kaur et al., 2007; Liu et al., 2007; Bordoloi et al., 2012).

Texture is related to the potato tuber's resistance to an applied force (Kaur et al., 2007; Nourian and Ramaswamy, 2003); this essential attribute of potatoes (raw and processed) is a function of the potato structure (Abu-Ghannam and Crowley, 2006). When a force is applied to the potato structure, accepted to be a network of interconnected cells, a failure or rupture occurs at the point of minimum resistance. In this system, either the cells can separate through rupture of the middle lamella or the cells can burst, depending on the strength of the middle lamella. Usually, raw potatoes undergo cell rupture, whereas cooked ones undergo cell separation as a result of thermal destabilization of pectic materials in the middle lamella (Aguilera and Stanley, 1990; Kaur et al., 2004). The microstructure of potato flesh, properties of cell wall polymers, abundance of starch in the potato cells, and shape and size of the starch granules are important for the texture of potatoes (Linehan and Hughes, 1969; Andersson

Advances in Potato Chemistry and Technology. http://dx.doi.org/10.1016/B978-0-12-800002-1.00016-9

et al., 1994; McComber et al., 1994; van Marle et al., 1997; Martens and Thybo, 2000). These factors can also influence deformation occurring during mastication or mechanical processing (Bordoloi et al., 2012; Waldron et al., 1997; Kaur, 2004).

During industrial processing, the potatoes are subjected to a variety of thermal treatments. Textural changes occurring during thermal processing and cooking of potato tubers have been associated with the gelatinization and retrogradation behavior of starch (Linehan and Hughes, 1969; Alvarez et al., 2001; Kaur et al., 2002; Shomer, 1995; Shomer et al., 1995; Ormerod et al., 2002) and cell wall and middle lamellae structural components (Ormerod et al., 2002; Alvarez and Canet, 1998; van Marle et al., 1997). Therefore, the optimization of industrial processes requires a good understanding of starch gelatinization in potato and microstructural features such as cell shape, size, and cell wall characteristics. Advanced microscopic techniques such as confocal laser scanning microscopy (CLSM) and electron microscopy have been used by various researchers to study the microstructural characteristics of foods including raw and cooked potatoes (McComber et al., 1994; Pedreschi and Aguilera, 2002; van Velde et al., 2002; Kaur, 2004; Bordoloi et al., 2012).

Postharvest storage also affects the cooking and textural quality of potato and potato products (Herrman et al., 1996; Cottrell et al., 1995; Spychalla and Desborough, 1990; Kazami et al., 2000; Nourian et al., 2003; Singh et al., 2008a). Low-temperature storage results in a change in the starch-to-sugar ratio of potatoes. Interrelationships between physical and biochemical quality characteristics of potato tubers have been reported between texture, starch, and reducing sugars during storage (Barichello et al., 1990; Kaur et al., 2002). Various methods of measuring texture-related and rheological properties can provide information on the behavior of foods when deformed; the best method depends on the type of food and the purpose of the measurement. On the basis of relaxation and creep experiments, several researchers have shown that fruit and vegetables exhibit viscoelastic behavior (Canet,1980; Alvarez and Canet, 1998; Kaur et al., 2002).

This chapter provides a detailed overview of factors that influence the cooking, textural, and rheological properties of potatoes.

2. Cooking and Sensory Characteristics

Cooking and sensory characteristics are important both to the domestic and industrial processing of potatoes. Cooking characteristics of potatoes such as the cooking time, cooked potato weight, and cooking loss have been used to select suitable potato cultivars with desirable processing characteristics. The cooking time of raw tubers can be determined approximately by cooking unpeeled tubers in boiling water until a kitchen fork can easily penetrate the tubers. Nguyen Do Trong et al. (2011) used a combination of hyperspectral imaging, chemometrics, and image processing for the contactless detection of the cooking front in potatoes (Figure 1). They reported that by modeling the evolution of the cooking front over time, the optimal cooking time could be predicted with less than 10% relative error. An estimation of

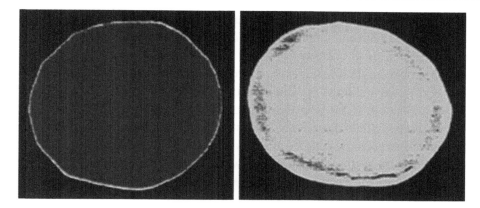

Figure 1

Partial Least Squares Discriminant Analysis model validation results for a raw (0 min; left) and fully cooked (30 min; right) potato. Brown (dark gray in print versions) color denotes raw, green (gray in print versions) color represents fully cooked. *Reproduced from Nguyen Do Trong et al. (2011), with permission from Elsevier.*

the cooking time of a potato using a texture analyzer or an Instron universal testing machine can also be an accurate and precise method. Sample preparation has an important role during the estimation of cooking time by texture analysis. The tubers are washed, and the periderm is peeled with a knife. Each tuber is cut into two equal halves and cylindrical pieces are taken from each half, excluding the core region, with the help of a cork borer, and then trimmed to a small height. The cooking time is determined by cooking several cylindrical pieces of a tuber in boiling water. After every 2–3 min of cooking, each cylindrical piece is tested for compression on a texture analyzer using a flat probe and a suitable load cell. The cooking time is determined as the time required to reach the lowest compression force during 50% compression of the sample (Kaur et al., 2002). The cooking time of potatoes in water may vary from 20 to 30 min whereas the compression force of cooked potatoes normally ranges between 0.15 and 0.25 N at their optimum cooking times (Kaur et al., 2002; Singh et al., 2008b). The effect of cooking time on the compressive force of cooked potatoes is shown in Figure 2.

The cooking time of tubers has been observed to depend mainly on dry solids, the starch content, and specific gravity. Although these three properties are interrelated, the starch characteristics of the tuber have a major effect on cooking time. Shorter cooking times have been reported for tubers that have starch with a lower gelatinization temperature. Tubers of different potato cultivars differ in tuber dry solids, starch content, and specific gravity (Table 1). The dry solids and starch contents of potato tubers are known to vary considerably between cultivars (Bu-Contreras and Rao, 2001; Kaur et al., 2002; Bordoloi et al., 2012). Factors that can affect the dry solids content and specific gravity of potatoes are the planting date, soil type, fertilizer use, harvesting period, and postharvest storage conditions (Toolangi, 1995). Singh et al. (2008b) reported a shorter cooking time for tubers of cultivars with high dry matter content. Also, a high degree of tuber mealiness is associated with a lower cooking time. To determine water uptake and total solids loss during cooking, preweighed raw cylindrical pieces are cooked in a small quantity of boiling water for approximately 30 min. After draining and rinsing, the cooked cylindrical pieces are weighed to determine water uptake (%). The rinse water is collected and dried to estimate total solids loss (percent weight of the cylindrical piece of potato). Potatoes with higher starch and dry matter content have higher water uptake and loss during cooking than those with lower starch content. Mealy/waxy characteristics of the potatoes are determined by sensory analysis after cutting a cooked tuber in half and subjectively scoring the texture of the tuber as waxy, slightly mealy, moderately mealy, or mealy. A positive correlation between dry matter content and the mealiness of potatoes was reported previously (O'Beirne et al., 1985). McComber et al. (1988) reported that starch granules from mealier potatoes gelatinized at a lower temperature.

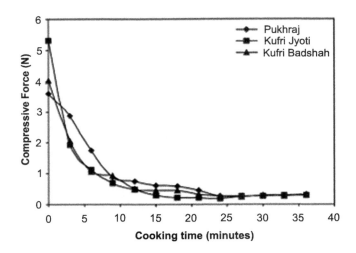

Figure 2

Effect of cooking time on compressive force of cooked potatoes. *Reproduced from Kaur et al. (2002), with permission from Elsevier.*

Table 1: Dry Solids Content (%), Starch Content (%), Specific Gravity, and Cooked Texture of Raw Tubers.

Cultivar	Dry Matter Content (%)	Starch Content (%)	Specific Gravity	Cooked Texture
Nadine	14.63 ± 0.29	6.40 ± 0.26	1.057 ± 0.002	Waxy
Karuparera	19.03 ± 0.44	10.32 ± 0.39	1.059 ± 0.001	Waxy
Tutaekuri	21.57 ± 0.38	12.58 ± 0.33	1.074 ± 0.003	Mealy
Huakaroro	20.97 ± 0.26	12.05 ± 0.23	1.062 ± 0.003	Slightly mealy
Moemoe	21.97 ± 0.27	12.94 ± 0.24	1.069 ± 0.004	Moderately mealy

Reproduced with permission from Singh et al. (2008b). © Wiley-VCH Verlag Gmbh & Co.

3. Textural Characteristics

Microstructural characteristics of raw potatoes, such as the parenchyma cell size and cell wall composition and thickness, have been reported to have a considerable effect on the final texture of potato (mealy, waxy, etc.) (van Marle et al., 1997). The rigid structure of the raw potato mainly results from the pectic substances, celluloses, and hemicelluloses. The pectic substances, which are the main constituents of the middle lamella, have a major role in intercellular adhesion and contribute to the mechanical strength of the cell wall. During cooking, changes in the potato microstructure and texture occur owing to continuous heat treatment over a certain period. These changes have been associated mainly with the gelatinization behavior of starch (Alvarez et al., 2001; Singh et al., 2002, 2008a; Ormerod et al., 2002), although cell wall and middle lamellae structural components have a significant part (van Marle et al., 1997; Alvarez and Canet, 1998; Ormerod et al., 2002; Bordoloi et al., 2012). Rupture of membranes and cell walls during processing leads to loss of turgor pressure. Pectic substances are brought more easily into solution than other cell wall polymers and subsequently contribute to the degradation of the potato's texture (Abu-Ghannam and Crowley, 2006).

Bordoloi et al. (2012) studied the microstructure of raw and cooked potato from different cultivars using CLSM and scanning electron microscopy (SEM). Confocal laser scanning microscopy allows direct visualization of the potato tissue. Micrographs of raw and cooked potato tuber cells (represented data from a single cultivar) are presented in Figure 3. Cell size and shape and the size and number of mature starch granules per cell were observed to vary among cultivars. Figure 4 shows a scanning electron micrograph of a fracture surface of raw potato parenchyma from Agria. Raw potatoes (Figure 4(a) and (b)) show cell rupture revealing

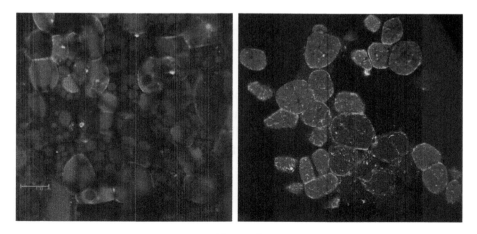

Figure 3

Confocal laser scanning micrographs of raw (left) and cooked (right) tuber parenchyma from Agria potato cultivar. *Reproduced from Bordoloi et al. (2012), with permission from Elsevier.*

Immature starch
granules

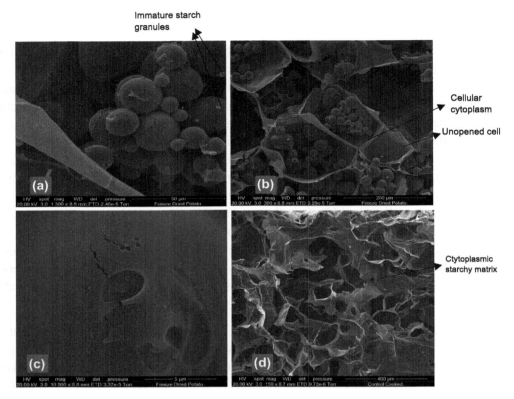

Cellular
cytoplasm

Unopened cell

Ctytoplasmic
starchy matrix

Figure 4
Scanning electron micrographs of raw (a–c) and cooked (d) tuber parenchyma from Agria potato
cultivar. In raw potato parenchyma micrographs, starch granules are seen embedded in the cellular
cytoplasm. Indentations and pores on the surface of starch granules are clearly seen, along with
other cellular remains and tiny granular structures resembling immature starch granules, as reported
by Singh et al. (2005). Cracks on the starch granules might have formed during microscopy.
Reproduced from Bordoloi et al. (2012), with permission from Elsevier.

starch granules, in contrast to the starchy matrix observed in cooked potato (Figure 4(d)).
Cavities, tiny holes, and remnants of the cell wall material were observed on the surface of
many raw starch granules present inside the cells (Figure 4(c)). Cooked potato parenchyma
showed separated cells containing gelled mass formed by the gelatinization of starch, as
observed using CLSM (Figure 3).

3.1 Texture Profile Analysis

A detailed method to perform texture profile analysis (TPA) on raw and cooked potatoes was
described by Kaur et al. (2007). The raw or cooked cylindrical pieces are prepared as
described in Section 2. Each sample is compressed with a flat platen using a suitable load
cell. The sample is preloaded with 0.5% of the total load. A slow cross-head speed is

Table 2: Texture Profile Analysis Parameters of Raw Potatoes of Different Cultivars.

Cultivar	Fracturability (Newton)	Hardness (Newton)	Cohesiveness	Springiness (meter)	Gumminess (Newton)	Chewiness (joules)
Kufri Chipsona-II	91 ± 2	132 ± 3	0.073 ± 0.002	0.0028 ± 0.0001	9.64 ± 0.50	0.027 ± 0.002
Kufri Chipsona-I	76 ± 2	105 ± 4	0.063 ± 0.001	0.0027 ± 0.0001	6.62 ± 0.35	0.018 ± 0.002
Kufri Lalima	91 ± 1	124 ± 2	0.072 ± 0.004	0.0033 ± 0.0003	8.93 ± 0.64	0.029 ± 0.004
Kufri Sindhuri	89 ± 2	115 ± 2	0.068 ± 0.002	0.0038 ± 0.0003	7.82 ± 0.37	0.030 ± 0.003
Kufri Anand	73 ± 1	99 ± 3	0.056 ± 0.001	0.0031 ± 0.0001	5.54 ± 0.27	0.017 ± 0.001

Reproduced with permission from Kaur (2004) and Kaur et al. (2007). © Wiley-VCH Verlag Gmbh & Co.

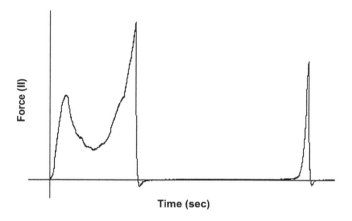

Figure 5

Typical texture profile analysis force–time curve for cooked potato. *Reproduced with permission from Singh et al. (2008b). © Wiley-VCH Verlag Gmbh & Co.*

preferred and the maximum extent of deformation is generally 75% of the original length of the sample. Textural parameters (fracturability, hardness, adhesiveness, cohesiveness, gumminess, and springiness), as defined by Singh et al. (2008b), are determined from the curve obtained through computer software. The parameters derived from textural profile analysis curves for raw potatoes vary considerably among different cultivars (Table 2). Raw tubers with high dry matter content generally show higher fracturability and hardness values. Potatoes with closely packed small and irregular parenchymatous cells and higher dry matter and starch contents have been observed to be relatively hard and cohesive. In contrast, potatoes with large, loosely packed cells are generally less hard (Kaur, 2004; Singh et al., 2005). Similar results were reported by Bordoloi et al. (2012).

A typical TPA curve for cooked potato is illustrated in Figure 5, whereas the TPA parameters of cooked potatoes from different cultivars are summarized in Table 3. The texture of the cooked potatoes depends on the texture of raw potatoes. Fracturability, hardness, and

Table 3: Texture Profile Analysis Parameters of Cooked Potatoes from Different Cultivars.

Cultivar	Fracturability (Newton)	Hardness (Newton)	Cohesiveness	Springiness (meter)	Adhesiveness ×10⁻⁴ (joules)	Gumminess (Newton)	Chewiness (joules)
Kufri Chipsona-II	3.2±0.08	3.9±0.22	0.088±0.002	0.0011	2.14±0.02	0.34±0.02	0.0004
Kufri Chipsona-I	2.2±0.12	3.3±0.09	0.088±0.003	0.0014	1.55±0.06	0.29±0.02	0.0004
Kufri Lalima	3.2±0.15	4.7±0.21	0.18±0.003	0.002	1.98±0.02	0.84±0.05	0.0017
Kufri Sindhuri	3.4±0.08	4.4±0.25	0.12±0.001	0.0019	2.22±0.02	0.52±0.04	0.001
Kufri Anand	2.7±0.06	3.9±0.20	0.065±0.002	0.0014	1.76±0.01	0.25±0.02	0.0003

Reproduced from Kaur (2004) and Kaur et al. (2007). © Wiley-VCH Verlag Gmbh & Co.

Table 4: Pearson Correlation Coefficients for Rheological and Selected Textural Properties.

	FrRaw	HdRaw	G'_{heat}	G'_{cool}	G'_{freq}
FrRaw					
HdRaw	0.968**				
G'_{heat}	0.955**	0.973**			
G'_{cool}	0.971**	0.958**	0.993**		
G'_{freq}	0.962**	0.970**	0.988**	0.986**	
Starch	0.903*	0.977**	0.915*	0.879*	0.912*
Dry matter	0.901*	0.978**	0.922*	0.885*	0.916*

$*p \le 0.05$; $**p \le 0.01$.
FrRaw, fracturability of raw potatoes; HdRaw, hardness of raw potatoes; G'_{heat}, storage modulus during heating cycle at Tgel; G'_{cool}, storage modulus during cooling cycle at 25 °C; G'_{freq}, storage modulus during frequency sweep at 20 Hz; Starch, starch content.
Reproduced with permission from Singh et al. (2008b). © Wiley-VCH Verlag Gmbh & Co.

cohesiveness are observed to be highest and chewiness and springiness to be lowest for potatoes with higher dry matter and starch content. Different textural properties of cooked potatoes, ranging from firmness to cohesiveness, in relation to mealiness, sogginess, stickiness, or gumminess, have been reported to be controlled by a number of factors such as starch, pectic substances, and cell size (Table 4). Cooked potato texture has also been associated with the dry solids content, specific gravity, amylose, sugars, proteins, and the total nitrogen content of the potato tuber. Many attempts have been made to show a relationship between the texture of the cooked potato and the physical or chemical properties of potato starch, which represents the predominant substance in the tuber. A highly significant correlation between the starch content of the raw tuber and various textural attributes of cooked potato, such as mealiness, consistency, sloughing, and sogginess, has been found by researchers (Kirkpatrick et al., 1951; Unrau and Nylund, 1957). The properties of starch and the changes in these properties during cooking are generally considered to explain the variations in texture. The internal pressures developed inside the cells owing to swelling of the gelled starch may result in slight distension in the tuber cells, causing them to become round and thus separate from each other. Correlations between starch granule size and the texture of cooked potatoes have been reported in the literature (Briant et al., 1945; Bu-Contreras and Rao, 2002). Mealy potatoes have higher specific gravities and starch and amylose content as well as a higher percentage of large starch granules than waxy cultivars (Barrios et al., 1963; Kaur et al., 2002). Bordoloi et al. (2012) reported that mealy cooked potatoes with higher starch and dry matter content resulted in higher values for textural parameters (such as fracturability, hardness, and cohesiveness) and also maintained the cellular structure well during cooking. On the contrary, waxy potatoes with lower starch content resulted in a disintegrated cellular structure upon cooking and had lower values for the textural parameters.

McComber et al. (1987) reported higher cohesiveness values for mealy American potatoes. Potatoes with a higher amylose content in their starches exhibit higher adhesiveness (Kaur et al., 2002). This may be because of the higher amylose content in the starches of these potatoes. Linehan and Hughes (1969) observed a correlation between intercellular adhesion and amylose content during studies on the texture of cooked potatoes. During cooking, amylose leaches out through weakened cell walls and acts as a cementing material between the cell walls of tuber cells, leading to an increase in intercellular adhesion (Linehan and Hughes, 1969). The fracturability of cooked tubers increases up to few days of refrigerated storage and then decreases slightly. However, this was not true for hardness, which had an irregular trend throughout the storage period in a study carried out on New Zealand potatoes (Singh et al., 2008b). Kaaber et al. (2007) reported an undesirable increase in hardness of prepeeled potatoes after cooking and attributed that partly to the removal of galactose and arabinose in the side chains of pectin. Possible reduction in steric hindrance by debranching the pectin molecules has been suggested to ease the formation of Ca^{2+} bridges. In this way, arabinan- and galactan-degrading enzymes favor hardening of the potato tissue (Kaaber et al., 2007; Van Buggenhout et al., 2009).

Sadowska et al. (2008) reported that mechanical resistance of the morphological elements of potatoes varied according to the type of tissue used to analyze the textural properties. They reported a highest puncture force for the cortex region followed by the pith and perimedullary zone. Moreover, they reported that the puncture force was higher near the stem end of the tuber and also that puncture forces measured perpendicularly to the length of the tuber axis were always lower than those measured parallel to the long axis.

Softening (%) calculated on the basis of changes in fracturability on cooking is another important textural parameter; it differs significantly among potato cultivars (Bordoloi et al., 2012). Mealy cultivars had lower softening than waxy or less mealy tubers. Starch characteristics such as the amylose to amylopectin ratio, cell separation, and cell wall softening during cooking of potatoes have been reported to influence the degree of softening (Singh et al., 2008b; Jarvis and Duncan, 1992).

The effect of different heating temperatures on the textural properties and microstructure of potatoes was reported by Fuentes et al. (2014). No microstructural or textural changes in the potato tissue were observed after cooking for 30 min at 30 °C and 40 °C. However, a significant loss in tissue firmness was reported when potato tissue was cooked at 70 °C, which was attributed to starch gelatinization and changes in the pectic substances (solubilization/dissolution of calcium-pectic gel) at higher temperatures (García-Segovia et al., 2008; Alvarez et al., 2001) (Figure 6). These changes in potato microstructure and texture with processing temperature have also been detected using real-time impedance spectroscopy (Fuentes et al., 2014). The researchers reported that samples processed at 60 and 70 °C had significantly different modules of impedance and phase values than those processed at about 50 °C (Figure 7).

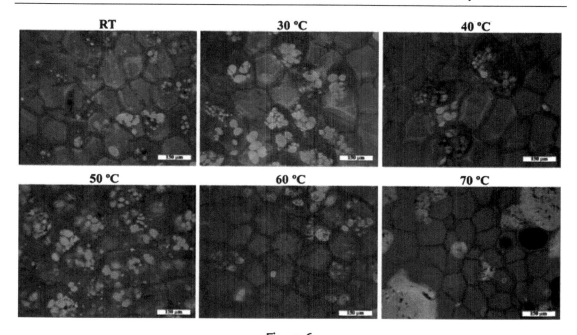

Figure 6
Confocal laser scanning microscopy micrographs of potato samples subjected to different heat treatments. *Reproduced from Fuentes et al. (2014), with permission from Elsevier.*

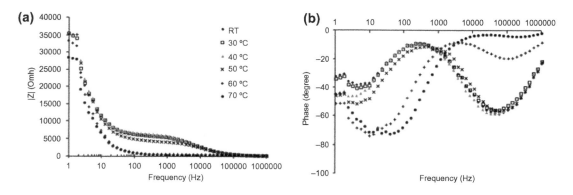

Figure 7
Module (a) and phase (b) of the impedance of potato samples subjected to different heat treatments. *Reproduced from Fuentes et al. (2014), with permission from Elsevier.*

Abu-Ghannam and Crowley (2006) reported that low-temperature treatment at 65 °C combined with a typical high-processing temperature range such as 95–100 °C resulted in an increase in the force required to shear whole new potatoes, compared with processing treatments without a preblanching step. They also reported increased activity of pectin methyl esterase (PME) when blanching was carried out at 65 °C, which may have resulted in subsequent strengthening in the potato structure upon further processing. Improved texture retention owing to low-temperature preheating has also been correlated with strengthened intercellular adhesion (Canet et al., 2005) and with significant modification in matrix-bonding, transforming the water-soluble pectin fraction into insoluble pectin (Sila et al., 2006; Van Buggenhout et al., 2009).

High-pressure pretreatment can be used as an alternative mode of preserving the textural properties of thermally processed fruits and vegetables because it significantly retards the rate of thermal softening (Sila et al., 2004; Shahidul Islam et al., 2007). The main reason for the increased positive effect of preheating when combined with pressures up to 400 MPa is increased catalytic activity of PME and, consequently, the extensive modification of pectins in terms of their dry matter (Sila et al., 2004; Van Buggenhout et al., 2009).

Raw potatoes are usually stored under controlled conditions until required for the market. Postharvest storage conditions can promote extensive changes in the chemical composition of potato tubers, thereby altering the quality characteristics of the final product (Spychalla and Desborough, 1990; McCay et al., 1987). Sugar and starch are the main components affected by postharvest metabolism in potato tubers, which may ultimately affect their textural, sensory, and cooking properties. It has been shown that potatoes undergo considerable changes in texture depending on the cultivar, storage temperature, and storage period (Sowokinos et al., 1987; Kazami et al., 2000; Bentini et al., 2009). Mechanisms of changes occurring during storage at higher temperatures are different from those at lower temperatures (Bourne, 1982; Nourian et al., 2003). Whereas cold storage may provide the necessary environment to prevent loss of weight, spoilage, and sprouting, the quality of potatoes continues to change as a result of physiological activity owing to accumulation of reducing sugars and depletion of starch (Bourne, 1982; Nourian et al., 2003). Extended storage at low temperatures has also been reported to change the mechanical properties of potato tubers (Bentini et al., 2009). The rate of sugar accumulation depends largely on the cultivar and temperature of storage and occurs most rapidly at cold temperatures. Hardness, cohesiveness, and gumminess of raw potatoes decrease progressively with an increase in the postharvest storage temperature. However, the decrease was observed to be significant only for potatoes stored above 12 °C (Kaur et al., 2007). The textural parameters of potato tubers stored at 16 °C and 20 °C did not show significant differences (Kaur et al., 2007). Elevated storage temperatures have an adverse effect on the texture of raw tubers because they develop wrinkles and become slightly softer than those stored at lower temperatures as a result of the loss of both solids and water. Postharvest low-dose γ-irradiation (0.05–0.15 kGy) has been reported to preserve textural quality and control storage losses and accumulation of sugars in potatoes during storage at 12 °C (Mahto and Das, 2015).

Springiness and chewiness of raw potatoes also decrease with an increase in storage temperature. Cooked potatoes from produce stored at low postharvest temperatures (4 °C) have a considerably different texture. The greater changes in textural properties of these cooked tubers may be attributed to changes in their starch content. The starch content of potatoes has been reported to decrease with a decrease in storage temperature through the process of starch conversion to sugars at lower temperatures (Smith, 1987). Nourian et al. (2003) also reported that starch content decreases considerably during prolonged storage of tubers at 4 °C. The change of starch to sugars at lower temperatures and the synthesis of starch from sugars at higher temperatures may be responsible for the change in textural properties of cooked potatoes stored at different temperatures.

3.2 Stress Relaxation Model

Davis et al. (1983) demonstrated that a generalized Maxwell model consisting of three elastic and two viscous elements could represent a stress relaxation response for cooked potatoes. Leung et al. (1983) studied the textural and rheological properties of cooked potatoes and correlated the hardness of cooked potatoes, by sensory evaluation, with fracturability, hardness, and elastic elements of the stress relaxation model. The reported effects of heat treatment on the rheological properties of potatoes include tests on the effects of cultivar and maturity on potato texture (Madsen and Christensen, 1996; Taguchi et al., 1991) and on starch properties (Briant et al., 1945). For the stress relaxation test, cylindrical potato pieces are cooked as described above in the case of cooking characteristics and TPA. The stress relaxation test can be performed on cooked potato pieces using an Instron universal testing machine or a texture analyzer. The unevenness of two ends of the sample is compensated by preloading with 0.5% of the load. The stress relaxation curve of cooked potato samples obtained using a texture analyzer (Figure 8) can be fitted with an equation consisting of three components: an equilibrium stress and two exponential decay

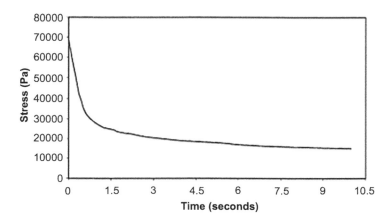

Figure 8
Typical stress relaxation curve of cooked potatoes. *Reproduced from Kaur et al. (2002), with permission from Elsevier.*

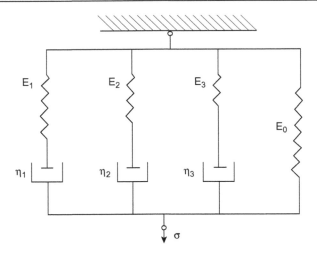

Figure 9

Seven-element Maxwell model representation for cooked potatoes. *Reproduced from Kaur et al. (2002), with permission from Elsevier.*

terms (Leung et al., 1983). The following equation represents the stress relaxation behavior of the cooked potato samples:

$$\sigma(t) = \varepsilon_0 \left(E_0 + E_1 e^{-t/T_1} + E_2 e^{-t/T_2} + E_3 e^{-t/T_3} \right) \tag{1}$$

where σ = stress; t = time; E_0 = constant strain; $T_1 = \eta_1/E_1$; $T_2 = \eta_2/E_2$; and $T_3 = \eta_3/E_3$; η_1, η_2, and η_3 = viscosities; and E_0, E_1, E_2, and E_3 = elastic moduli.

Equation (1) can be represented by the Maxwell model (Mohsenin, 1970) consisting of four elastic elements (dashpots), as illustrated in Figure 9. The 3 Mx elements, each consisting of a spring and a dashpot in series, correspond to the three exponential terms in Eqn (1). The lone elastic element corresponds to the equilibrium modulus (E_0) at infinite time. Kaur et al. (2002) reported the stress relaxation behavior of three different potato cultivars (Table 5), which agrees with the earlier findings of Davis et al. (1983) and Leung et al. (1983). Tubers with higher elastic moduli obtained by stress relaxation data represent higher degrees of hardness.

4. Rheological and Thermal Characteristics

4.1 Rheology of Potatoes

Rheological properties of foods can be studied using fundamental tests; empirical and imitative tests are used to determine textural characteristics. One advantage of fundamental tests is that they determine the properties of a material independent of the physical dimensions of the sample, the measurement procedure, or the equipment (Rao, 2007). Dynamic mechanical analysis is an effective analytical method because it applies small strains on the samples (less than 0.1%) over a

Table 5: Stress Relaxation Test Parameters of Cooked Potatoes from Different Cultivars.

Cultivar	E_0 (Pa)	E_1 (Pa)	E_2 (Pa)	E_3 (Pa)	η_1 (Pas)	η_2 (Pas)	η_3 (Pas)
Pukhraj	1.612×10^4	1.715×10^5	1.476×10^5	1.185×10^5	1.011×10^6	0.0785×10^6	0.00978×10^6
Kufri Jyoti	6.481×10^5	1.858×10^5	2.247×10^5	2.138×10^5	8.835×10^5	2.967×10^5	0.2605×10^5
Kufri Badshah	5.172×10^4	2.013×10^5	0.94×10^5	2.013×10^5	1.263×10^6	0.078×10^6	0.05×10^6

Reproduced from Kaur et al. (2002), with permission from Elsevier.

short time; as a result, the properties of solid foods can be studied with minimal physical and chemical changes. In addition, dynamic tests provide the ability to study the material's rheological properties at different frequencies. Another advantage that dynamic mechanical analysis provides is that one can detect changes resulting from to cross-linking, phase separation, and molecular aggregation of polymer chains (Bu-Contreras, 2001; Rao, 2007).

Dynamic mechanical analysis is also called small-amplitude oscillatory testing because an oscillatory (sinusoidal) stimulus in the form of stress or strain is applied to the sample and a sinusoidal response (stress or strain) is measured. Materials have different responses to the deformation applied. The response could be elastic (Hookean solid) when the angle between the stimulus and the response is in phase. If the phase angle of the response is 90°, the material has a purely viscous response (also considered a Newtonian liquid). Most materials have a response with a phase angle higher than 0° but lower than 90°; such a response is considered to be viscoelastic because it contains both viscous and elastic characteristics. For compressional/elongational deformation, the symbol E is used for the moduli. The generated stress within the linear viscoelastic range is expressed in terms of the storage modulus E' (Pa) and a loss modulus E'' (Pa); E' is a measure of the magnitude of the energy that is stored in the material and E'' is a measure of the energy that is lost as viscous dissipation per cycle of deformation, respectively (Bu-Contreras and Rao, 2002). Bu-Contreras and Rao (2001) studied the starch content and the rheological properties of American Russet Burbank (RB) and Yukon Gold (YG) potatoes. Based on the solids content, the starch content was estimated to be 7.0% for the RB and 11.7% for the YG potatoes. Potato discs (13.3 mm in diameter × 4.5 mm in height) were cut from cylinders extracted from the center of a potato using a cork borer. In all experiments, values of E' were much higher than those of E''; therefore, only the trends of E' are discussed here. The storage modulus of the raw sample, designated as E'_o, was measured at room temperature just before water was added to the submersion cup and was used to normalize the E' data of the heated potatoes:

$$Normalized Storage Modulus = \left(E' / E'_o \right) \qquad (2)$$

The value of E'_o of the RB sample was 1.87 MPa and that of the YG sample was 2.01 MPa. Swelling of starch granules affected the storage modulus of the heated potato samples, which

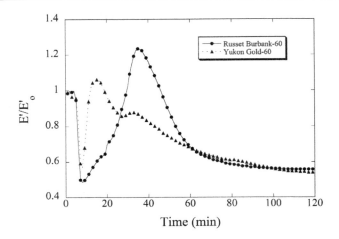

Figure 10

Profile of normalized storage modulus, E'/E'_o, of Russet Burbank potato disc as a function of heating time at 60 °C; E'_o is the storage modulus of the raw sample = 1.87 MPa. *Bu-Contreras (2001).*

in turn depended on the heating temperature and time. At all temperatures, immediately after the sample was immersed in water, the normalized storage modulus decreased for the first 5 min followed by an increase that depended on the heating temperature and time. The decrease in storage modulus after sample immersion resulted from absorption of water by the potato sample, causing the cells to separate, which resulted in the loss of rigidity (Bu-Contreras and Rao, 2001).

When samples were heated at low temperatures, 40 °C and 50 °C, at which starch granules did not swell, the storage modulus of the RB and YG samples did not increase (not shown here). At 60 °C, the normalized storage modulus of YG samples increased and reached a peak value of 1.1 in the first 10 min, whereas that of the RB samples reached a peak value of 1.25 after 35 min (Figure 10). After 60 min of heating, the normalized storage modulus values of both varieties were nearly equal and the modulus–temperature profiles were similar. At 70 °C and 80 °C, the storage modulus of RB potato samples increased but that of the YG samples decreased significantly (not shown here). For example, at 80 °C, the (E'/E'_o) values of RB were 30% higher than those of YG (Bu-Contreras and Rao, 2001).

The height of a typical DMA peak in heated potato tubers has been reported to depend on potato density and cultivar. A more waxy cultivar Nicola (frequently used as a salad variety) had a lower DMA peak and peak temperature than more a mealy cultivar 'Saturna' (cultivated for industrial use), which was attributed to more intensive swelling in the latter (Blahovec and Lahodová, 2011).

Continuous monitoring of the dynamic rheological changes in potato samples during heating and cooling cycles on the rheometer may provide a crucial basis for understanding changes that occur during *in-tuber* gelatinization and retrogradation of starch in potatoes. The structural transitions associated with phase change in starch-containing food systems are reflected by changes in rheological profiles and are described by parameters such as G', G'', and tan δ (= G''/G'). Small-amplitude oscillatory, three-step, rheological measurements (temperature sweeps during heating and subsequent cooling, and a frequency sweep on the cooled sample) on the raw potato slices using a controlled stress rheometer were reported by Singh et al. (2008b). They determined dynamic rheological parameters such as storage modulus (G'), loss modulus (G''), and loss tangent (tan δ) for potato tissue as functions of temperature and frequency. Softening of tissue and swelling of starch granules in the potato tubers affect rheological parameters, which in turn depended on the time and temperature of heating. Rheological profiles of the tubers from different potato cultivars of New Zealand are presented in Figures 11–14. During initial heating on a rheometer, the G' of the potato tuber tissue decreases up to a certain time/temperature. With further heating, the G' increases to a maximum and then drops. At early stages of heating, the tuber tissue loses its rigidity/strength owing to thermal softening, resulting in a decrease in the storage modulus (Bu-Contreras and Rao, 2001). With a further increase in temperature, G' increases steeply and reaches a maximum owing to swelling of starch granules and the formation of a network of swollen starch granules (gelatinization) in the tuber tissue (Figure 11(a)). A decrease in G' with a further increase in temperature indicates the destruction of this gel structure with prolonged heating (Hsu et al., 2000). The rheological profile of the tissue from waxy potatoes does not show a distinct and clear gelatinization curve as observed for mealy cultivars, probably owing to their significantly lower dry matter and starch content. The temperature at which G' was maximal (T_{gel}) shows a considerable variation depending on the starch characteristics of the cultivars. The mealy cultivars have a higher peak G' than the waxy cultivars. Also, the mealy potatoes require less time to complete gelatinization during heating than waxy potatoes (Bu-Contreras and Rao, 2001). A comparison of mealy cultivar Tutaekuri and waxy cultivar Karuparera is shown in Figure 11(a). In this figure, the gelatinization curve observed for a highly waxy cultivar Nadine is not sharp and distinct in the temperature/time range studied. This may have been the result of this cultivar's very low starch content failing to change the elastic character of the tissue significantly on gelatinization. The other waxy cultivar, Karuparera, also exhibited a relatively small peak, which, as for Nadine, may also be attributed to its relatively low dry matter and starch content. Generally, peak G' was directly related to dry matter and starch content (Figure 13). During holding at 95 °C, the G' of tubers showed a tendency to stabilize at a constant value. The G' of the heated tuber tissue increases during cooling from 90 °C to 25 °C (Figure 11(b)). The mealy tubers exhibit higher G' values than waxy tubers. The continuous increase in G' during cooling reflects gelation of the starch in the tuber tissue. Furthermore, retrogradation of starch leached from granules, and interaction between starch molecules, remaining inside granules, reinforce the gel structure during cooling (Hsu et al., 2000).

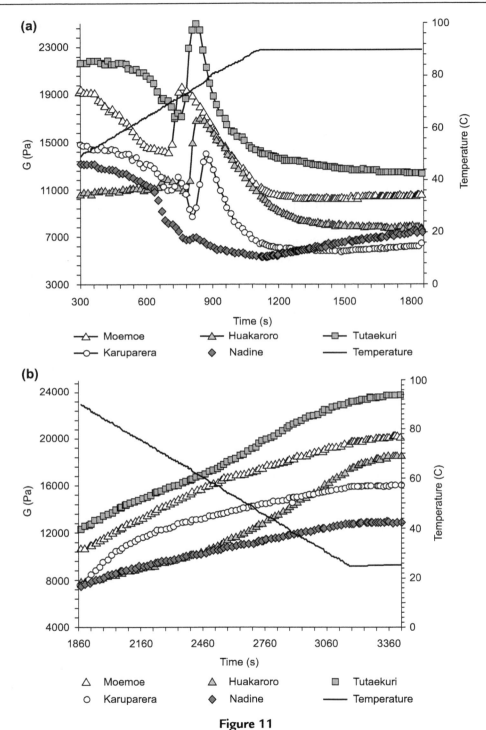

Figure 11

Storage modulus (*G'*) of raw tuber flesh during temperature sweeps ((a) heating and (b) cooling).
Reproduced with permission from Singh et al. (2008b). © Wiley-VCH Verlag Gmbh & Co.

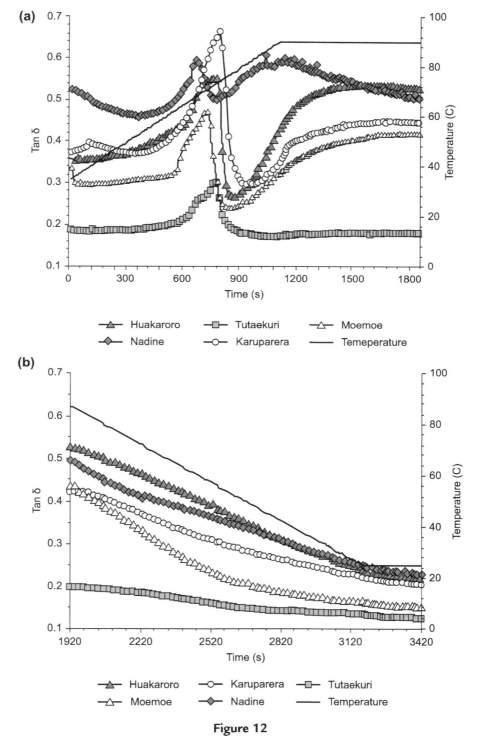

Figure 12
Loss tangent (tan δ) of raw tuber flesh during temperature sweeps ((a) heating and (b) cooling).
Reproduced with permission from Singh et al. (2008b). © Wiley-VCH Verlag Gmbh & Co.

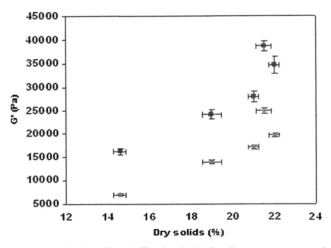

◆ G' during heating (at Tgel) ■ G' during frequency sweep (at 20 Hz)

Figure 13
Effect of dry solids (%) on the storage modulus (*G'*) of raw tuber flesh during temperature (heating) and frequency sweeps. *Reproduced with permission from Singh et al. (2008b). © Wiley-VCH Verlag Gmbh & Co.*

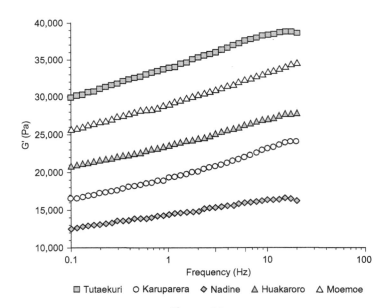

Figure 14
Storage modulus (*G'*) of raw tuber flesh during frequency sweeps. *Reproduced with permission from Singh et al. (2008b). © Wiley-VCH Verlag Gmbh & Co.*

The frequency dependence of G' can give valuable information about potato structure. A material that is frequency-independent over a large frequency range is solid-like; a true gel system is such a material. In contrast, strong frequency dependence suggests a material structure with molecular entanglements that behaves more like a solid at higher frequencies and more like a liquid at lower frequencies (Ross-Murphy, 1984). The G' of the tuber tissues increases with increasing frequency, but the frequency dependence is not marked, indicating that the cooled gelatinized tuber tissues are predominantly solid-like in character (Figure 14). The storage modulus of potatoes is affected by starch swelling, which has a crucial role in determining the final firmness characteristics of heated potatoes (Bu-Contreras and Rao, 2002). During heating and in the presence of water, the starch granules swell and apply pressure to the cells of the potato samples, resulting in an increase in storage modulus. Rheological parameters such as storage moduli during heating, cooling, and frequency sweep show significant positive correlations with starch, dry matter content, fracturability, and hardness (Table 4). These correlations imply that the dynamic rheology of potatoes can be a useful tool to predict the texture of raw and cooked potatoes.

The pulsed electric field (PEF) is becoming an increasingly popular nonthermal technology to process plant materials because it causes less degradation of nutritional compounds (Knorr et al., 2008; Vorobiev and Lebovka, 2010; Lebovka et al., 2014). Pulsed electric field treatment on cell material has also been reported to lead to pore formation in cell membrane and thus modify diffusion of intra- and extracellular media (Janositz et al., 2011) (Figure 15). Several studies have demonstrated a decrease in the stiffness of potato tissue to levels similar to those achieved by hyperosmotic treatment (Fincan and Dejmek, 2003). Pereira et al. (2009) reported that PEF causes a rapid change in the viscoelastic properties of the potato tissue and related that to partial loss in turgor pressure. They also reported that recovery of tan δ to values similar to those before pulsation strongly suggests recovery of cell membrane properties and turgor in response to stress, pointing to reversible permeabilization of the cells.

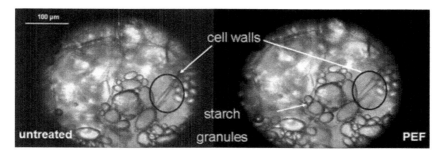

Figure 15
Untreated and PEF-treated ($E = 5$ kV/cm, $n = 20$, 5 min after treatment) potato tissue stained with ruthenium red (dark gray in print versions). *Reproduced from Janositz et al. (2011), with permission from Elsevier.*

Table 6: Gelatinization Temperatures of Starch from RB and YG Potato Samples Heated in Water for 120 min at Different Temperatures.

Sample	T_o (°C)	T_p (°C)	T_m (°C)
	RB		
Raw	68.0	70.4	73.9
40 °C	67.3	69.8	73.1
50 °C	69.2	71.7	75.0
60 °C	71.5	73.0	76.1
70 °C	ND	ND	ND
80 °C	ND	ND	ND
	YG		
Raw	70.1	72.3	75.1
40 °C	70.2	72.2	74.5
50 °C	72.8	75.2	77.7
60 °C	73.8	75.4	77.8
70 °C	76.5	78.6	80.9
80 °C	ND	ND	ND

T_o, T_p, T_m, gelatinization initiation, peak, and maximum temperatures, respectively.
Reproduced with permission from Bu-Contreras and Rao (2001). © Wiley-VCH Verlag Gmbh & Co.

This suggests a need to explore further complex cell stress physiology involving both cell membrane functional properties and cell wall structure that would influence the physical properties of the tissue (Pereira et al., 2009).

4.2 Differential Scanning Calorimetry of Potatoes

Differential scanning calorimetry provides information on the thermal transitions taking place in a food. However, rheological changes are strongly influenced by the physical structure of a food; in a food containing starch, such as a potato, rheological changes are strongly influenced by the state of the starch granules and the starch content. The DSC thermograms of unheated American RB potato samples (not shown here) had endothermic transition owing to starch gelatinization over the range 68–74 °C with a peak at 70.4 °C, whereas those of the YG samples had an endothermic transition over 70.1–75.1 °C with a peak at 72.3 °C (Table 6) (Bu-Contreras, 2001).

4.3 Kinetics of Firmness Changes in Potato

Modeling is an efficient approach to predict the texture of foods. Empirical, semiempirical, and statistical models have induced a tremendous impetus on food texture research (Chen and Opara, 2013). Because the firmness of potatoes seems to decrease rapidly at temperatures greater than 70 °C, in most studies, its values are modeled using first-order kinetics. Defining P as a texture property, the first-order kinetics for texture degradation can be written as (Rao and Lund, 1986):

$$\frac{dP}{dt} = -k_f P \tag{3}$$

The temperature dependence of the rate constant can be expressed by the Arrhenius equation:

$$k_f = k_o \exp\left(E_a/RT\right) \tag{4}$$

When a simple kinetics model is not enough, two separate first-order kinetics models for fast and slow degradation, respectively, are used. It is difficult to define the time when the fast mechanism ends and the slow mechanism begins during cooking (Lee et al., 2007). However, fractional conversion provides improved accuracy and reliability in determining the texture degradation kinetics of vegetables (Rizvi and Tong, 1997). The fractional conversion model is derived from the first-order model. For Young's modulus as the property, the fractional conversion, f, is defined as:

$$f = \frac{E_o - E}{E_o - E_\infty} \tag{5}$$

where E_o is the initial modulus, E is the modulus at time t, and E_∞ is the nonzero value of the modulus obtained after long times. The nonzero retainable modulus, E_∞, is a function of temperature.

A first-order equation is written using this fraction as:

$$\frac{d(1-f)}{dt} = -k_f(1-f) \tag{6}$$

$$E_\infty(T) = c_4 + \frac{c_1}{1 + \left(\dfrac{T}{c_3}\right)^{c_2}} \tag{7}$$

where c_1–c_4 are parameters are to be determined by fitting the equation to experimental data. Table 7 contains values of the reaction rate constant and activation energy for the first-order kinetics and the fractional conversion models (Lee et al., 2007).

Table 7: Fractional and First-Order Kinetics Parameters for Softening of Potato.

Variable	Value	Source
Fractional Conversion Kinetics		
k_o (per min)	1.18×10^8	Basis: Alvarez et al. (2001)
E_a (kJ/mol)	64.18	
E_∞ (parameters c_1 to c_4 from Eqn (6))	0.53, 21.97, 63.78, 0.47	Basis: Nourian and Ramaswamy (2003)
k_o (per min)	1.29×10^{10}	
E_a (kJ/mol)	77.86	
E_∞ (parameters c_1 to c_4 from Eqn (6))	0.80, 8.35, 73.76, 0.20	
First-Order Kinetics		
k_f at 100 °C (per min)	0.387	Basis: Harada et al. (1985) for maximum force
E_a (kJ/mol)	145	
k_f at 100 °C (per min)	0.069	Basis: Solomon and Jindal (2003) for tangent modulus
E_a (kJ/mole)	51.1	

At this stage, solution represents the kinetics effect and not diffusion. The diffusivity value used was $D_{eff} = 10^{-10}\,\mathrm{m^2/s}$.
Reproduced with permission from Lee et al. (2007). © Wiley-VCH Verlag Gmbh & Co.

5. Conclusions

The textural and rheological characteristics of raw and cooked tubers depend mainly on their starch and dry matter content. Other factors such as postharvest storage temperature, the microstructure of the potato tissue, the nature of the cell wall materials, the amylose content, and the processing temperature have also a significant effect. High-pressure or low-temperature pretreatment has been reported to preserve or improve the textural properties of thermally processed potatoes because it leads to modification of pectin and activation of PME. However, the role of endogenous enzymes in influencing textural attributes is still not clear. Nonthermal processing technologies such as PEF may also be used as an alternative method of processing potatoes because they cause less degradation of nutritional components, but there is a need to understand the effects these treatments may have on complex cell physiology, which in turn can affect the physical properties of the tissue.

References

Abu-Ghannam, N., Crowley, H., 2006. The effect of low temperature blanching on the texture of whole processed new potatoes. Journal of Food Engineering 74, 335–344.

Aguilera, J.M., Stanley, D.W., 1990. Microstructural Principles of Food Processing and Engineering, second ed. Elsevier Applied Science, New York, pp. 175–329.

Alvarez, M.D., Canet, W., Tortosa, M.E., 2001. Kinetics of thermal softening of potato tissue (cv. Monalisa) by water heating. European Food Research and Technology 212, 588–596.

Alvarez, M.D., Canet, W., 1998. Rheological characterization of fresh and cooked potato tissues (cv. Monalisa). Zeitschrift für Lebensmittel-Untersuchung und -Forschung 207, 55–65.

Andersson, A., Gekas, V., Lind, I., Oliveira, F., Oste, R., 1994. Effect of preheating on potato texture. Critical Reviews in Food Science and Nutrition 34, 229–251.

Barichello, V., Yada, R.Y., Coffin, R.H., Stanley, D.W., 1990. Low temperature sweetening in susceptible and resistant potatoes: starch structure and composition. Journal of Food Science 54, 1054–1059.

Barrios, E.P., Newson, D.W., Miller, J.C., 1963. Some factors influencing the culinary quality of Irish potatoes. II. Physical characters. American Potato Journal 40, 200–206.

Bentini, M., Caprara, C., Martelli, R., 2009. Physico-mechanical properties of potato tubers during cold storage. Biosystems Engineering 104, 25–32.

Blahovec, J., Lahodová, M., 2011. DMA peaks in potato cork tissue of different mealiness. Journal of Food Engineering 103, 273–278.

Bordoloi, A., Kaur, L., Singh, J., 2012. Parenchyma cell microstructure and textural characteristics of raw and cooked potatoes. Food Chemistry 133, 1092–1110.

Bourne, M.C., 1982. Effect of temperature on firmness of raw fruits and vegetables. Journal of Food Science 47, 440–444.

Briant, A.M., Personius, C.J., Grawemeyer- Cassel, E., 1945. Physical properties of starches from potatoes of different culinary quality. Food Research 10, 437–444.

Bu-Contreras, R., 2001. Influence of Physico-chemical Factors on the Firmness of Potatoes and Apples (Ph.D. thesis). Cornell University, Ithaca, New York, USA.

Bu-Contreras, R., Rao, M.A., 2001. Influence of heating conditions and starch on the storage modulus of Russet Burbank and Yukon Gold potatoes. Journal of the Science of Food and Agriculture 81, 1504–1511.

Bu-Contreras, R., Rao, M.A., 2002. Review: dynamic rheological behavior of heated potatoes. Food Science and Technology International 8, 3–10.

Burton, W.G., 1989. The Potato, third ed. Longman Scientific & Technical, Essex, UK.

Canet, W., 1980. Estudio de la influencia de los tratamientos termicos de escaldado, congelacion y descongelacion en la textura y estructura de patata (*Solanum tubersoum*, L.) (Ph.D. thesis). Univsidad Politecnica de Madrid, Madrid.

Canet, W., Alvarez, M.D., Fernandez, C., 2005. Optimization of low-temperature blanching for retention of potato firmness: effect of previous storage time on compression properties. European Food Research and Technology 221, 423–433.

Chen, L., Opara, U.L., 2013. Approaches to analysis and modeling texture in fresh and processed foods – a review. Journal of Food Engineering 119, 497–507.

Cottrell, J.E., Duffus, C.M., Paterson, L., George, R.M., 1995. Properties of potato starch: effects of genotype and growing conditions. Phytochemistry 40, 1057–1064.

Davis, D.C., McMahan, P.F., Leung, H.K., 1983. Rheological modeling of cooked potatoes. Transactions of the ASABE 26, 630–634.

Fincan, M., Dejmek, P., 2003. Effect of osmotic pre-treatment and pulsed electric field on the viscoelastic properties of potato tissue. Journal of Food Engineering 59, 169–175.

Fuentes, A., Vázquez-Gutiérrez, J.L., Pérez-Gago, M.B., Vonasek, E., Nitin, N., Barrett, D.M., 2014. Application of nondestructive impedance spectroscopy to determination of the effect of temperature on potato microstructure and texture. Journal of Food Engineering 133, 16–22.

García-Segovia, P., Andrés-Bello, A., Martínez-Monzo, J., 2008. Textural properties of potatoes (*Solanum tuberosum* L., cv. Monalisa) as affected by different cooking processes. Journal of Food Engineering 88, 28–35.

Harada, T., Tirtohusodo, H., Paulus, K., 1985. Influence of temperature and time on cooking kinetics of potatoes. Journal of Food Engineering 50, 459–462 472.

Herrman, T.J., Love, S.L., Shafaii, B., Dwelle, R.B., 1996. Chipping performance of three processing potato cultivars during long-term storage at two temperature regimes. American Potato Journal 73, 411–425.

Hsu, S., Lu, S., Huang, C., 2000. Viscoelastic changes of rice starch suspensions during gelatinization. Journal of Food Science 65, 215–220.

Huang, C.C., Lin, M.C., Wang, C.R., 2006. Changes in morphological, thermal and pasting properties of yam (*Dioscorea alata*) starch during growth. Carbohydrate Polymers 64, 524–531.

Janositz, A., Noack, A.-K., Knorr, D., 2011. Pulsed electric fields and their impact on the diffusion characteristics of potato slices. LWT – Food Science and Technology 44, 1939–1945.

Jarvis, M.C., Duncan, H.J., 1992. The textural analysis of cooked potato. 1. Physical principles of the separate measurements of softness and dryness. Potato Research 35, 83–91.

Kaaber, L., Kaack, K., Kriznik, T., Bråthen, E., Knutsen, S.H., 2007. Structure of pectin in relation to abnormal hardness after cooking in pre-peeled, cool-stored potatoes. LWT – Food Science and Technology 40, 921–929.

Kaur, L., 2004. Physico-chemical Properties of Potatoes in Relation to Thermal and Functional Properties of Their Starches (Ph.D. dissertation). Guru Nanak Dev University, Amritsar.

Kaur, L., Singh, J., Singh, N., Ezekiel, R., 2007. Textural and pasting properties of potatoes (*Solanum tuberosum* L.) as affected by storage temperature. Journal of the Science of Food and Agriculture 87, 520–526.

Kaur, L., Singh, N., Sodhi, N.S., Gujral, H.S., 2002. Some properties of potatoes and their starches I. Cooking, textural and rheological properties of potatoes. Food Chemistry 79, 177–181.

Kazami, D., Tsuchiya, T., Kobayashi, Y., Ogura, N., 2000. Effect of storage temperature on quality of potato tubers. Journal of the Japanese Society for Food Science and Technology – Nippon Shokuhin Kagaku Kagaku Kaishi 47, 851–856.

Kirkpatrick, M.E., Mountjoy, B.M., Albright, B.C., Heinze, P.H., 1951. Cooking quality, specific gravity and reducing sugar content of early-cropped potatoes. USDA Circular 872.

Knorr, D., Engel, K.-H., Vogel, R., Kochte-Clemens, B., Eisenbrand, G., 2008. Statement on the treatment of food using a pulsed electric field. Molecular Nutrition & Food Research 52, 1539–1542.

Lebovka, N.I., Mhemdi, H., Grimi, N., Bals, O., Vorobiev, E., 2014. Treatment of potato tissue by pulsed electric fields with time-variable strength: theoretical and experimental analysis. Journal of Food Engineering 137, 23–31.

Lee, S.H., Datta, A.K., Rao, M.A., 2007. 4. How does cooking time scale with size? A numerical modeling approach. Journal of Food Science 72, E001–E010.

Leung, H.K., Barron, F.M., Davis, D.C., 1983. Textural and rheological properties of cooked potatoes. Journal of Food Science 48, 1470–1474.

Linehan, D.J., Hughes, J.C., 1969. Texture of cooked potatoes. I. Introduction. Journal of the Science of Food and Agriculture 20, 110–112.

Liu, Q., Tarn, R., Lynch, D., Skjodt, N.M., 2007. Physicochemical properties of dry matter and starch from potatoes grown in Canada. Food Chemistry 105, 897–907.

Madsen, M.H., Christensen, D.H., 1996. Changes in viscosity properties of potato starch during growth. Starch 48, 245–249.

Mahto, R., Das, M., 2015. Effect of γ irradiation on the physico-mechanical and chemical properties of potato (*Solanum tuberosum* L), cv. 'Kufri Chandramukhi' and 'Kufri Jyoti', during storage at 12 °C. Radiation Physics and Chemistry 107, 12–18.

Martens, H.J., Thybo, A.K., 2000. An integrated microstructural, sensory and instrumental approach to describe potato texture. Lebensmittel-Wissenschaft & Technologie 33, 471–482.

McCay, C.M., McCay, J.B., Smith, O., 1987. The nutritive value of potatoes. In: Talburt, W.F., Smith, O. (Eds.), Potato Processing. Van Nostrand Reinhold Company Inc., New York, pp. 287–332.

McComber, D.R., Horner, H.T., Chamberlin, M.A., Cox, D.F., 1994. Potato cultivar differences associated with mealiness. Journal of Agricultural and Food Chemistry 42, 2433–2439.

McComber, D.R., Lohnes, R.A., Osman, E.M., 1987. Double direct shear test for potato texture. Journal of Food Science 52, 1302–1304.

McComber, D.R., Osman, E.M., Lohnes, R.A., 1988. Factors related to potato mealiness. Journal of Food Science 53, 1423–1426.

Mohsenin, N.N., 1970. Rheology and texture of food materials. In: Mohsenin, N.N. (Ed.), Physical Properties of Plant and Animal Materials. Gordon and Breach Science Publishers, New York, p. 309.

van Marle, J.T., Stollesmits, T., Donkers, J., VanDijk, C., Voragen, A.G.J., Recourt, K., 1997. Chemical and microscopic characterization of potato (*Solanum tuberosum* L.) cell walls during cooking. Journal of Agricultural and Food Chemistry 45, 50–58.

Nguyen Do Trong, N., Tsuta, M., Nicolaï, B.M., De Baerdemaeker, J., Saeys, W., 2011. Prediction of optimal cooking time for boiled potatoes by hyperspectral imaging. Journal of Food Engineering 105, 617–624.

Nourian, F., Ramaswamy, H.S., 2003. Kinetics of quality changes during cooking and frying of potatoes: Part I. Texture. Journal of Food Process Engineering 26, 377–394.

Nourian, F., Ramaswamy, H.S., Kushalappa, A.C., 2003. Kinetics changes in cooking quality of potatoes stored at different temperatures. Journal of Food Engineering 60, 257–266.

O'Beirne, D., Walshe, D., Egan, S., 1985. Searching for quality in cooked and processed potatoes. Farm & Food Research 16, 71–73.

Ormerod, A., Ralfs, J., Jobling, S., Gidley, M.J., 2002. The influence of starch swelling on the material properties of cooked potatoes. Journal of Materials Science 37, 1667–1673.

Pedreschi, F., Aguilera, J.M., 2002. Some changes in potato chips during frying observed by confocal laser scanning microscopy (CLSM). Food Science and Technology International 8, 197–201.

Pereira, R.N., Galindo, F.G., Vicente, A.A., Dejmek, P., 2009. Effects of pulsed electric field on the viscoelastic properties of potato tissue. Food Biophysics 4, 229–239.

Rao, M.A., 2007. Rheology of Fluid and Semisolid Foods: Principles and Applications, second ed. Springer, New York, p. 483.

Rao, M.A., Lund, D.B., 1986. Kinetics of thermal softening of foods – a review. Journal of Food Processing and Preservation 10, 311–329.

Rizvi, A.F., Tong, C.H., 1997. Fractional conversion for determining texture degradation kinetics of vegetables. Journal of Food Science 62, 1–7.

Ross-Murphy, S.B., 1984. Rheological methods. In: Chan, H.W.-S. (Ed.), Biophysical Methods in Food Research. Blackwell Publishers, pp. 138–199.

Sadowska, J., Fornal, J., Zgorska, K., 2008. The distribution of mechanical resistance in potato tuber tissues. Postharvest Biology and Technology 48, 70–76.

Shahidul Islam, M., Igura, N., Shimoda, M., Hayakawa, I., 2007. The synergistic effect of moderate heat and pressure on the physical properties and pectic substances of potato tissue. International Journal of Food Science & Technology 42, 434–440.

Shomer, I., 1995. Swelling behaviour of cell-wall and starch in potato (*Solanum tuberosum* L.) tuber cells. 1. Starch leakage and structure of single cells. Carbohydrate Polymers 26, 47–54.

Shomer, I., Vasiliver, R., Lindner, P., 1995. Swelling behaviour of cell-wall and starch in potato (*Solanum tuberosum* L.) tuber cells. 2. Permeability and swelling in macerates. Carbohydrate Polymers 26, 55–59.

Sila, D.N., Smout, C., Vu, T.S., Hendrickx, M., 2004. Effects of high-pressure pretreatment and calcium soaking on the texture degradation kinetics of carrots during thermal processing. Journal of Food Science 69, 205–211.

Sila, D., Doungla, E., Smout, C., Van Loey, A., Hendrickx, M., 2006. Pectin fractions interconversions: insight into understanding texture evolution of thermally processed carrot. Journal of Agricultural and Food Chemistry 54, 8471–8479.

Singh, J., Kaur, L., Mccarthy, O.J., Moughan, P.J., Singh, H., 2008b. Rheological and textural characteristics of raw and par-cooked Taewa (Maori potatoes) of New Zealand. Journal of Texture Studies 39, 210–230.

Singh, J., McCarthy, O.J., Singh, H., Moughan, P.J., 2008a. Low temperature post-harvest storage of New Zealand Taewa (Maori potato): effects on starch physico-chemical and functional characteristics. Food Chemistry 106 (583), 596.

Singh, J., Singh, N., Sharma, T.R., Saxena, S.K., 2003. Physicochemical, rheological and cookie making properties of corn and potato flours. Food Chemistry 83, 387–393.

Singh, N., Kaur, L., Ezekiel, R., Gurraya, H.S., 2005. Microstructural, cooking and textural characteristics of potato (*Solanum tuberosum* L.) tubers in relation to physico-chemical and functional properties of their flours. Journal of the Science of Food and Agriculture 85, 1275–1284.

Singh, N., Singh, J., Sodhi, N.S., 2002. Morphological, thermal, rheological and noodle-making properties of potato and corn starch. Journal of the Science of Food and Agriculture 82, 1376–1383.

Smith, O., 1987. Effect of cultural and environmental conditions on potatoes for processing. In: Talburt, W.F., Smith, O. (Eds.), Potato Processing. Van Nostrand Reinhold Company Inc., New York, pp. 73–148.

Solomon, W.K., Jindal, V.K., 2003. Modeling thermal softening kinetics of potatoes using fractional conversion of rheological parameters. Journal of Texture Studies 34, 231–247.

Sowokinos, J.R., Orr, P.H., Knoper, J.A., Varns, J.L., 1987. Influence of potato storage and handling stress, sugars, chip quality and integrity of the starch (amyloplast) membrane. American Potato Journal 64, 213–225.

Spychalla, N.P., Desborough, S.L., 1990. Fatty acids, membrane permeability, and sugars of stored potato tubers. Plant Physiology 94, 1207–1213.

Taguchi, M., Schafer, H.W., Breene, W.M., 1991. Influence of cultivar and prewarming on texture retention of thermally processed potatoes. Potato Research 34, 29–39.

Toolangi, T.K., 1995. Potatoes: factors affecting dry matter. Agriculture Notes 1329-8062 State of Victoria, Department of Primary Industries.

Unrau, A., Nylund, R., 1957. The relation of physical and chemical composition to mealiness in the potato I. Physical properties. American Potato Journal 34, 245–249.

Van Buggenhout, S., Sila, D.N., Duvetter, T., Van Loey, A., Hendrickx, M., 2009. Pectins in processed fruits and vegetables: Part III—Texture Engineering. Comprehensive Reviews in Food Science and Food Safety 8, 105–117.

van Velde, F., Riel, J.V., Tromp, R.H., 2002. Visualisation of starch granule morphologies using confocal scanning laser microscopy (CSLM). Journal of the Science of Food and Agriculture 82, 1528–1536.

Vorobiev, E., Lebovka, N., 2010. Enhanced extraction from solid foods and biosuspensions by pulsed electrical energy. Food Engineering Reviews 2, 95–108.

Waldron, K.W., Smith, A.C., Parr, A.J., Ng, A., Parker, M.L., 1997. New approaches to understanding and controlling cell separation in relation to fruit and vegetable texture. Trends in Food Science and Technology 8, 213–220.

CHAPTER 17

Mechanisms of Oil Uptake in French Fries

Anna Patsioura[1,2], Jean-Michaël Vauvre[1,2,3], Régis Kesteloot[4], Paul Smith[5], Gilles Trystram[6], Olivier Vitrac[1,2]

[1]INRA, UMR 1145 Ingénierie Procédés Alimentaires, Group Interaction between Materials and Media in Contact, Massy, France; [2]AgroParisTech, UMR 1145 Ingénierie Procédés Alimentaires, Massy, France; [3]McCain Alimentaire S.A.S., Parc d'entreprises de la Motte du Bois, Harnes, France; [4]Régis Kesteloot Conseil, Lambersart, France; [5]Cargill R&D Centre Europe, Vilvoorde, Belgium; [6]AgroParisTech, UMR 1145 Ingénierie Procédés Alimentaires, Group Interaction between Materials and Media in Contact, Massy, France

1. Introduction

Deep fat frying is considered an important food process unit operation (Achir et al., 2009) that has a poor image because most fried products are grouped as junk food. The current obesity epidemic, which has spreads in both developed (Guallar-Castillon et al., 2007; Sassi, 2010) and developing countries (Shi et al., 2008; Sichieri, 2002), has multiple sources including a lack of physical activity (Fair and Montgomery, 2009), a high-caloric diet (Swinburn et al., 2004), and oversized portions compared with the nutritional needs of children (Ello-Martin et al., 2005). As an example, french fry–type products are a significant source of starch (Zaheer and Akhtar, 2014) but they are also a source of fat with an oil content varying between a 0.25 and 0.52 kg/kg solid basis corresponding to 0.10 and 0.23 kg oil/kg consumed product (Aguilera and Gloria-Hernandez, 2000; Al-Khusaibi et al., 2012). Besides its caloric value, frying oil conveys various oxidation products such as polar and oligomer compounds (Mendonca Freire et al., 2013), heterocyclic amines (Burenjargal and Totani, 2009), and polycyclic aromatic hydrocarbons (Gertz and Klostermann, 2002; Mottram et al., 2002), as well as other products that may not be desirable, such as trans-fatty acids (Dalainas and Ioannou, 2008).

From a process point of view, deep fat frying consists of immersing a wet product into a liquid with a high boiling point. Surface boiling and the high renewal rate of the oil close to the product surface enable the fastest convective heat transfer among conventional cooking and drying food operation units (Vitrac and Trystram, 2005). Heat flux densities have been measured up to 150 and 230 kW/m^2 on cassava chips fried at 160 °C (Vitrac et al., 2003) and potato strips at 180 °C (Lioumbas et al., 2012), respectively, generated by a convective heat transfer coefficient up to 2 kW/m^2 K (Costa et al., 1999; Hubbard and Farkas, 2000;

Vitrac and Trystram, 2005). Heat and mass transfer coupled during frying and cooling have been reviewed in Achir et al. (2009) and Moreira et al. (1999). Beyond early phenomenological descriptions (Gamble and Rice, 1987), numerous models have successfully predicted coupled heat and mass transfer during frying. Such models describe the frying process as a moving vaporization front (Farkas et al., 1996a,b; Vitrac et al., 2002) or as a coupled set of heat and mass transfer equations in porous media with a phase change (Achir et al., 2008; Halder et al., 2007a,b; 2011; Ni and Datta, 1999). They appear to simulate reactions (Achir et al., 2008; Bassama et al., 2011) and texture modifications accurately in deep fat–fried products (Thussu and Datta, 2012). By contrast, oil uptake mechanisms are less extensively described and reviews regularly update mechanistic findings (Dana and Saguy, 2006; Mehta and Swinburn, 2001; Saguy et al., 1998; Ziaiifar et al., 2008). The generally accepted mechanisms come from the enlightening experiments of Ufheil and Escher (1996) involving successive frying in non-dyed and dyed oils, and the work of Moreira et al. (1997) using surface oil washing. Oil uptake was shown to be a superficial phenomenon (Achir et al., 2010; Bouchon et al., 2001; Gamble and Rice, 1987) occurring mainly during cooling (Aguilera and Gloria-Hernandez, 2000), when the product is removed from the oil bath. Indeed, internal overpressures measured at the maximum vaporization rate with values ranging from 5 to 50 kPa (Vitrac et al., 2000) are thought to strongly prevent oil from penetrating deeply inside the product during the immersion stage. The detailed mechanisms of oil uptake during cooling remain only partly understood. Oil invasion during cooling has been described variously either as spontaneous imbibition controlled mainly by capillary forces (Bouchon et al., 2003; Pedreschi and Aguilera, 2002) or as forced drainage driven by steam condensation (He et al., 2012; Vitrac et al., 2000).

Although consistent with mass balance considerations, most descriptions of oil uptake remain mechanistically unrealistic and provide little help to reduce oil uptake in fried products. Indeed, the multiphasic character of oil uptake is usually neglected, or oil and gas (steam or air) are considered miscible. In addition, the routes followed by oil in penetrating the fried product and their connection with cellular defects remain poorly described. This chapter presents advances in an understanding of oil uptake based on microscopic observations of oil penetration routes (Achir et al., 2010; Patsioura et al., 2015; Vauvre et al., 2014) and on the displacement of the gas phase by oil (Vauvre et al., 2015), as well as consistent thermodynamics for superheated steam in the crust of the fried product (Achir et al., 2008). The new results are highly counterintuitive and offer good opportunities for innovative strategies to reduce oil uptake: Cell walls are good barriers to oil. Without cellular damage, oil can pass from one cell to the next only through the small communication channels between cells, known as plasmodesmata. Air can hamper this, increasing the time from several seconds to hours of oil uptake, as shown independently by various research groups (Cortés et al., 2014; Kalogianni and Papastergiadis, 2014; Patsioura et al., 2015).

The chapter is organized as follows. Section 2 introduces the thermodynamics of steam in equilibrium with hygroscopic materials and its relationship to oil uptake during cooling. These concepts are particularly necessary to understand the dynamics of steam condensation during cooling and to control the equivalent thermodynamical conditions mimicking both immersion frying and cooling. Section 3 details the potato tuber parenchyma materials, especially the oil paths enabling oil flow from one cell to another one beneath. Section 4 presents a new oil uptake multiscale model that integrates a micromodel and its generalization at the product scale. Deviations of mechanisms from generally accepted descriptions are finally discussed in Section 5.

2. Physics of Cooling and Oil Uptake During Cooling

2.1 Oil Uptake is a Problem of Pressure

2.1.1 A First Description of Drying at a Product Scale

The iso-temperature T_{sat} separates the product inside the hygroscopic crust (above T_{sat}), where only residual water remains, from the inner crumb (below T_{sat}), where capillary water exists. As a first approximation, the internal vaporization can be described as a drying front moving towards the center of the product. This approximation as a Stefan problem was first suggested by Mittelman et al. (1984) and subsequently developed by Farkas et al. (1996a). It was relatively well-verified by more detailed models using a continuous mechanics approach (Halder et al., 2007a; Yamsaengsung and Moreira, 2002). The thermodynamical formulation of water vaporization and desorption in the hygroscopic domain of Achir et al. (2008) confirmed this approximation. The main characteristics of water, pressure, and temperature profiles inside a french fry–type product are summarized in Figure 1. Additional internal liquid water transport (Achir et al., 2008, 2009; Vitrac et al., 2003) caused by significant internal overpressure (Vitrac et al., 2000) was experimentally assessed in starchy materials in the crumb but also in the crust owing to non-equilibrium effects (liquid water can rapidly cross the crust without being vaporized).

2.1.2 Oil Capillary Pressure

During frying, positive internal pressures relatively to atmospheric pressure result from friction losses when steam crosses the dense crust. It is the main phenomenon that prevents oil from penetrating the crust during immersion frying by opposing capillary pressure, ΔP_C, which depends on the contact angle, θ_C and diverging angle, ϕ, shown in Figure 2 (see Eqn (2.3.1) of Dullien (1979) or Eqn (8.148) of Kaviany (1997)):

$$\Delta P_C = \frac{2\sigma}{r}\cos\left(\theta_C + \phi\right)\xrightarrow{\theta_C - \phi \to 0}\frac{2\sigma}{r}\left(1 - \frac{\left(\theta_C - \phi\right)^2}{2}\right) \tag{1}$$

where r is the pore radius at the contact line. A tube is obtained when $\phi = 0$. By considering a tube commensurable to the size of one potato parenchyma cell, capillary pressure averages 500 Pa (Vauvre et al., 2015) and oil uptake occurs only when $P < \Delta P_C$.

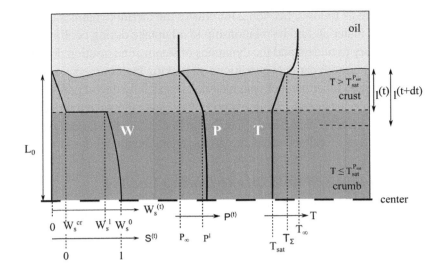

Figure 1
Simulation of water concentration profile in fried product from detailed modeling, after Achir et al., (2008, 2009) and Vitrac et al. (2000).

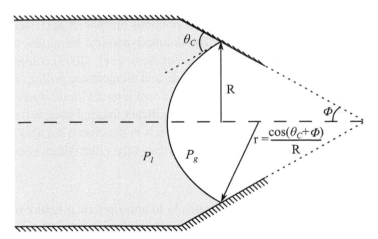

Figure 2
Sketch of tube and convergent filled by a wetting liquid, where P_l and P_g are the pressure of the liquid and gas phase, respectively.

2.2 Variations in Total Pressure During Frying and Postfrying Cooling

2.2.1 Pressure Gauge during Frying and Cooling

Total pressure is defined as the sum of pressures in the product and the gas phase. Few studies have reported the pressure within the core region during and after frying. Direct pressure measurement has been performed by casting a starchy alginate-gel material around a wireless autonomous sensor and by frying the whole system (Vitrac et al., 2000). Pressure varied with the solid content

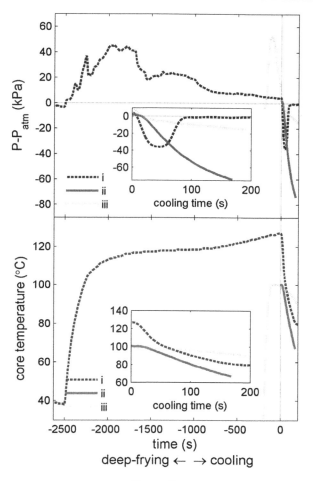

Figure 3

Pressure and temperature variations in starchy products during frying and cooling. (i) starchy alginate cylinders 34 mm in diameter and 150 mm long (frying temperature 160 °C), after Vitrac et al. (2000); (ii) potato cylinder 10 mm in diameter and 50 mm long (frying temperature 170 °C), after He et al. (2012); (iii) 8-mm-thick potato disk 48 mm in diameter (frying temperature 175 °C), after Sandhu et al. (2013).

of the gel (with or without starch) but followed a similar scheme reproduced in Figure 3. A positive pressure gauge develops during frying as a result of steam friction loss. During cooling, pressure decreases sharply below atmospheric pressure whereas core temperature decreases (although it is still above the saturation temperature, T_{sat}). A similar pressure drop during cooling was found by He et al. (2012) and Sandhu et al. (2013) (Figure 3). These results have important consequences in understanding oil uptake, especially its dynamics and the extent of oil penetration. A positive pressure gauge with values up to several tens of kilopascals during frying would overpass oil capillary pressure and therefore prevent oil from penetrating inside the product during frying, except perhaps at low-pressure edges. Conversely, the negative pressure gauge during cooling would be by far the main driving force responsible for oil uptake.

2.2.2 Positive Pressure Gauge Prevents Oil from Entering the Product during Immersion Frying

From Eqn (1), capillary pressure is maximal when $\theta_C = \phi$. The oil–gas surface tension, $\gamma = 2.35 \times 10^{-2}$ N/m, and contact angle between gas, oil, and potato matrix, $\theta_C = 0.66$ rad, at 120 °C, found in Bouchon and Pyle (2005) and in Pinthus and Saguy (1994), respectively, make it possible to calculate the pore geometry (d and ϕ), which theoretically counterbalances internal overpressure during frying. Based on p_g values reported in Vitrac et al. (2000) during frying, after 1500 and 200 s in starchy and rich-water gels, respectively, oil pressures, p_{oil}, higher than 10 and 2 kPa would be required to trigger oil uptake. For cylindrical capillaries ($\phi = 0$), Eqn (1) shows that such pressures are reached for diameters smaller than 17 and 87 µm, correspondingly. For a converging tube verifying $\theta_C = \phi$, the initial opening is increased up to 22 and 111 µm, respectively. In comparison, parenchyma cells of potatoes offer diameters of about 200 µm (Reeve et al., 1971), larger than any previously calculated critical diameters. Oil percolation through empty cells (without starch) therefore appears to be unlikely beyond the first cell layer during frying. On the contrary, oil uptake on french fry edges, where no significant overpressure exists, is thought to be likely during the immersion stage. Because of the absence of cellular structure, tortilla chips with pores smaller than parenchyma cells represent the worst-case scenario. The experimental determination of Moreira et al. (1997), showing that 20% of oil in tortilla chips was absorbed during the frying stage, confirms this interpretation.

2.3 Oil Transport Resulting from Pressure Gradient

2.3.1 Continuous Medium Approximations of Oil Transport: Considering Oil and Gas as Miscible Phases

Early descriptions of oil uptake found an apparent correlation between water loss during frying and oil uptake (Moyano and Berna, 2002; Rice and Gamble, 1989; Vitrac et al., 2001, 2002). They may be interpreted as the consequence of the creation of voids, which are subsequently filled with oil during cooling. However, there is a paradox: Drying the product before frying reduces the final oil uptake (Vitrac et al., 2002). The discrepancy between both modes of water removal on oil uptake could be justified by the modifications of the connectivity of cells after convective drying. Indeed, Achir et al. (2010) demonstrated that only superficial cells can be filled with oil, with three-quarters of oil located within the first cell layer damaged by cutting and with the last quarter down to the third cell layer.

The first consistent description of oil uptake continuously covering frying and postfrying stages was proposed by Halder et al. (2007a). The model was proposed as a significant improvement on earlier ones ignoring heat and mass transfer during cooling and consequently leading to unrealistic predictions of oil uptake. For example, simulations of Bouchon and Pyle (2005) predicted a fast capillary migration of oil in french fry–type products of distances up to 0.15 m (i.e., larger than the product itself) in less than 30 s. Oil mass transfer was best captured via the concept of local oil saturation, denoted S_o, and defined as the extent to which oil fills the pores. For proper initial and boundary conditions, the authors calculated S_o by solving a conservative formulation of the oil mass balance:

$$\frac{\partial}{\partial t}\left(\varepsilon \rho_o^{(T)} S_o\right) + \nabla \cdot \left(u_o \rho_o^{(T)}\right) = \nabla \cdot \left[D_{o,cap}^{(T,S_o)} \nabla \left(\varepsilon \rho_o^{(T)} S_o\right)\right] \tag{2}$$

where ρ_o is the intrinsic oil density. The volume averaged oil velocity u_o was derived from the generalized Darcy's law:

$$u_o = -\frac{k_{in,o}^p k_{r,o}^p}{\mu_o} \nabla P \tag{3}$$

whereas the effective diffusion coefficient was inferred from the relationship between the gradient of saturation and the gradient of capillary forces:

$$D_{o,cap}^{(T,S_o)} = -\frac{k_{in,o}^p k_{r,o}^{p\,(T,\,S_o)}}{\varepsilon \mu_o^{(T)}} \frac{\partial p_{cap}\left(S_o\right)}{\partial S_o}\bigg|_T \tag{4}$$

$k_{in,o}^p$ and $k_{r,o}^p$ appearing in Eqns (3) and (4) are the intrinsic permeability of the porous structure and relative permeability to oil, respectively.

The concept of relative permeability was first proposed by Muskat and Meres (1936) and is interpreted as the effect of phase distribution on viscous drag. As discussed in Kaviany (1997), no general treatment exists to predict these coefficients under drainage or imbibition conditions. In addition, because air can be incorporated into the porous structure, similar transport equations should be written for both air and steam. Owing to the different inherent complications, an empirical model independent of temperature was applied by the authors, based on an early suggestion of Ni and Datta (1999):

$$D_{o,cap}^{(S_o)} = D_o \exp\left(-2.8 + 2.0\frac{\varepsilon}{1-\varepsilon}\frac{\rho_o}{\rho_s}S_o\right) \tag{5}$$

where ρ_s is the intrinsic density of the solid matrix.

Reported values of the pre-exponential factor D_o ranging from 1.22×10^{-9} to $1.22 \times 10^{-7}\,m^2/s$ (Halder et al., 2007a; Warning et al., 2012) led to oil penetrating depths from a few millimeters to several centimeters, respectively, during the frying and cooling stages. Neglecting the specific contribution of the total pressure-driven oil flow on final oil uptake (Halder et al., 2007b) makes the penetration of oil continuous during frying and cooling. The discontinuity behavior has been shown in experiments with variable pressure conditions during cooling (He et al., 2012).

2.3.2 Oil and Gas as Immiscible Phases

The formulation of Halder et al. (2007a) incorporates advection terms (see Eqns (2) and (3)) but makes explicit displacements of the oil–air interface only in a mean-field sense. Further understanding can be gained by adopting the simplified form of continuity and momentum equations proposed by Bucklet and Leverett (1942), with more details in Dullien (1979) and Kaviany (1997). By assuming that relative permeabilities to gas (mixture of water vapor and air), $k_{r,g}^p$, and to oil, $k_{r,o}^p$, do not vary significantly with S_o, the Bucklet–Leverett equation gives the variation of the position, z, of the iso-saturation lines:

$$\frac{dz}{dt}\bigg|_{S_o} = \frac{\langle u_t \rangle}{\varepsilon} \frac{df_o}{dS_o} \tag{6}$$

where f_o is the relative volumetric oil flow rate and $\langle u_t \rangle$ is the total local superficial (Darcy) velocity averaged on a representative elementary volume of the crust. By neglecting the weight of the air column and buoyancy forces, one gets:

$$\langle u_t \rangle = \langle u_o \rangle + \langle u_g \rangle = -\frac{k_{in}^p k_{r,o}^p}{\mu_o}\left(\frac{\partial p_o}{\partial z} + \rho_o g \sin(\theta_u)\right) - \frac{k_{in}^p k_{r,g}^p}{\mu_g}\left(\frac{\partial p_g}{\partial z}\right) \tag{7}$$

where $\langle u_o \rangle$ and $\langle u_g \rangle$ are the local velocities of oil and gas, respectively, and θ_u is the polar angle measured from the horizontal plane, with $\theta_u = \pi/2$ corresponding to positive velocities pointing against the gravity vector. By considering that oil is the wetting phase in oil–gas systems, capillary pressure is defined as the excess pressure seen by the non-wetting phase (gas) $p_c = p_g - p_o$. In fried products, superficial velocity is controlled by the superposition of two driving forces: the gradient of capillary forces and the gradient of gas pressure as measured by Vitrac et al. (2000) and Sandhu et al. (2013):

$$\langle u_t \rangle = \frac{k_{in}^p k_{r,o}^p}{\mu_o}\frac{\partial p_c}{\partial z} - \left(\frac{k_{in}^p k_{r,o}^p}{\mu_o} + \frac{k_{in}^p k_{r,g}^p}{\mu_g}\right)\frac{\partial p_g}{\partial z} - k_{in}^p g \sin(\theta_u)\left(\frac{k_{r,o}^p}{\mu_o}\rho_0\right) \tag{8}$$

Three limiting cases are particularly meaningful (Vauvre et al., 2014):

- spontaneous imbibition of oil when $\dfrac{\partial p_g}{\partial z} = 0$ (e.g., no steam condensation);

- gas–oil co-current flow in direction z, when $\dfrac{\partial p_g}{\partial z} < 0$ when air is pushed or pulled in the same direction (e.g., case of steam condensation);

- gas–oil countercurrent flow, when $\dfrac{\partial p_g}{\partial z} > 0$ (e.g., air is displaced by oil and compressed inside the sample).

2.3.3 Variation of the Pressure Inside One Single Potato Cell: Effect of an Incondensable Phase (e.g., air)

The lowest oil invasion is obtained with the oil–gas countercurrent and when an internal over-pressure is achieved. These conditions are met naturally during the frying stage and during cooling only when the initial internal steam generation can be sustained. This possibility was investigated by He et al. (2012) by cooling the fried product under vacuum. On the other hand, the highest penetration rate is achieved when a positive pressure gradient exists in the material.

It happens particularly when fast cooling rates cause steam condensation. An intermediate penetration rate is expected when air can compensate partly or fully for the pressure drop owing to condensation. In the absence of a pressure drop, the wicking experiments of Kalogianni and Papastergiadis (2014) showed a surprisingly slow penetration rate in the tangential direction in the crust. This study provides an explanation by showing that the trapping-release of air bubbles can dramatically slow down spontaneous oil imbibition. Patsioura et al. (2015) and Vauvre et al. (2015) demonstrated in particular that oil uptake is mainly governed by microscopic details and fails to be captured correctly through continuous mechanics descriptions.

Figure 4 plots the theoretical gas pressure during cooling according to the composition of the partial pressure of water vapor, by considering that residual steam was not escaping the product during cooling. As a result, the presented pressure drops represent upper bounds of real values. The equilibrium curves demonstrate that pressure drop can occur far above T_{sat} during postfrying.

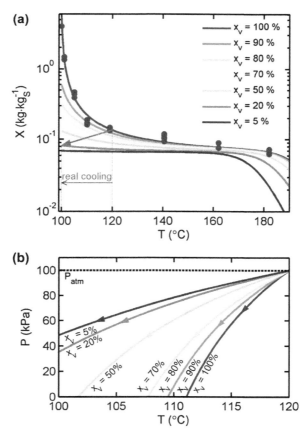

Figure 4

Thermodynamical equilibrium curves of cell walls in frying and cooling conditions according to the volume fraction of water vapor x_v: (a) residual water content (X, solid basis) versus T at thermal equilibrium; (b) equivalent total gas pressure (P).

Predicted negative pressure gauges are in the range of values experimentally assessed by Vitrac et al. (2000), He et al. (2012), and Sandhu et al. (2013) for an air fraction of about 10–20%.

3.   *Typology of Defects Provoking Oil Uptake*

The observations of Achir et al. (2010) showed that the percolation of oil in fried potato strips prepared from fresh potato tubers is highly heterogeneous in space. The honeycomb arrangement of cells hinders oil percolation by imposing specific routes governed essentially by the connectivity of parenchyma cells between layers and the distribution of cell damage. This section reports the size of capillaries (i.e., reservoirs and connections) enabling oil uptake in french fries, as shown in Figure 5.

3.1 *Natural Defects in Cellular Tissues*

In undamaged tissue, cells or cells filled with starch are thought to behave as capillary dead ends. Thus, oil could flow either through intercellular spaces at cell junctions (see Chapter 3

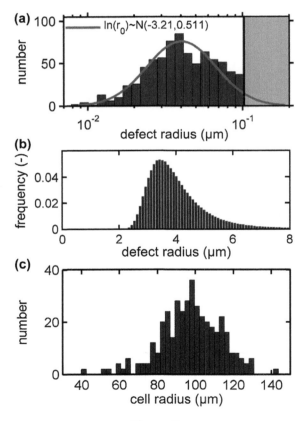

Figure 5

Determination of oil pathway radius: (a) natural cell connections; (b) small defects at cell walls; (c) empty potato parenchyma cells.

of Mauseth (2008)) with radii of a few to tens of micrometers, or through sparingly distributed channels connecting cells called plasmodesmata (Michalski and Saramago, 2000) with diameters of about 100 nm. Both passages were investigated using a breakthrough oil permeation test involving undamaged potato parenchyma tissue and lasting 2 h at 120 °C (Patsioura et al., 2015). The experimental time was much longer than frying and cooling, to observe oil permeation through the smallest defects. Surprisingly, droplets covered only from 2.6% to 13.5% of the surface, but such a result was in accordance with the extremely low connectivity of undamaged parenchyma cells reported in Achir et al. (2010). Half of the spills had a circular shape and could be reasonably approximated as spherical caps. With the theory that oil permeated through defects to form these caps, they were analyzed individually to determine the distribution of defects. Such inverse modeling detailed in Patsioura et al. (2015) led to the determination of apparent defects, which cylindrical approximation gives radii reported in Figure 5(a) and ranges from 7.8 to 100 nm with a median value of 40 nm. Their small sizes were consistent with an assumption of oil flow through plasmodesmata. According to Robards (1975), their numbers in cells vary between 10^3 and 10^5. The junctions connect adjacent cells and enable mass transfer of molecules in living cells. Nevertheless, little is known about their structure once they are subjected to superheated steam.

3.2 Distribution of Cell Defects' Radii and Typical Oil Impregnation Profiles

Larger defects between cells were determined from oil uptake kinetics obtained on the scale of cells. The methodology of Patsioura et al. (2015) combined a deep ultraviolet synchrotron source and a photo-amplified charge-coupled device camera to image simultaneously the epifluorescence of cell walls and of dyed oil. Specific image analysis software was developed to deconvolve the dense phases and their three-dimensional (3D) arrangements. Filling times were determined and used along the inverted oil uptake model of Vauvre et al. (2015) to infer the likely distribution of defect radii. It led to an almost symmetric distribution of radii centered on 3.8 μm, as reported in Figure 5(b).

Empty potato cells are larger capillaries that could be filled by oil. Cells were digitized from the observations of Patsioura et al. (2015). Their apparent radii were calculated from disks of similar cross-sectioned areas by assuming the conservation of flow rates. In this configuration, cell radii were found normally distributed with an average of about 100 μm (Vauvre et al., 2015), as illustrated in Figure 5(c), which was consistent with the mean radius value of 100 μm reported in the review of Reeve (1967). Under deep-frying conditions, cell damage can lead to disruptions and leads to the possibility of cavities larger than cells with a size of 1 mm or more, as observed in MacMillan et al. (2008) and Vauvre et al. (2014).

3.3 Summary of Defects Creating Pathways to Oil Percolation

Whereas natural oil paths consist of cells and their connections (i.e., plasmodesmata smaller than 1 μm), larger defects leading to cell disruptions and even cavities are

Table 1: List of Oil Pathways and Relative Oil Capillary Pressure.

Type of Oil Path	Plasmodesmata	Defect owing to Process	Path in Partially Starch-Filled Parenchyma Cell	Empty Parenchyma Cell	Cavity
Average radius (μm)	4.0×10^{-2}	3.8	20	10^2	10^3
Capillary pressure (Pa)[a]	1.2×10^6	1.2×10^4	2.4×10^3	4.7×10^2	47

[a]Calculated by considering oil temperature $T = 120\,°C$, i.e., oil surface tension $\gamma = 0.03\,N/m$ (Bouchon and Pyle, 2005) and contact angle between oil, air, and restructured potato product $\theta_C = 0.663\,rad$ (Pinthus and Saguy, 1994).

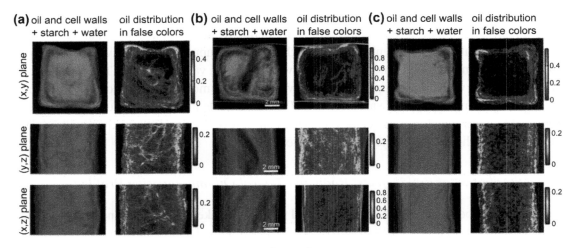

Figure 6

Two-dimensional profiles of dense fractions in industrial french fries finish-fried for 300 s. Dense volume fractions are presented in false colors (green (gray in print versions) for solid and water, and red (dark gray in print versions) for oil), on projections in (x,y), (y,z), and (x,z) planes: (a) a lot, (b) a little, and (c) no oil in the core region (in the section observed).

induced by the process (Patsioura et al., 2015; Vauvre et al., 2014, 2015). The variation in hydraulic radius from the cellular connection or defect to cells or cavities induces large variations in capillary pressure. Table 1 lists the size and defect and the corresponding capillary pressure enabling spontaneous oil uptake. The microscopic observations of Vauvre et al. (2014) revealed that oil was percolating in industrial french fries through a network with increasing pore size from the external surface to the center. Although it is much easier to stop oil uptake in larger pores such as cavities, they are also filled with oil, as shown in Figure 6.

4. Multiscale Modeling of Oil Uptake

A general and flexible framework was designed to simulate oil uptake with an arbitrary cell arrangement (i.e., based on 3D real digitized structures) and underlying physics (with or without air, and with constant pressure or not). This section reports the principles in a simple case: (1) all cells have similar geometry and are arranged as a regular honeycomb structure; (2) oil percolates through faces parallel to the surface, and (3) oil uptake inside individual cells is described at a coarse-grained level. The main goal of this section is to propose a consistent framework for the time-resolved 3D description of oil penetration in a potato parenchyma structure at all scales of time (from 10^{-4} to 10^4 s) and space (from cell content up to 10^6 cells). In particular, a dual-scale modeling approach is applied to simulate oil penetration with or without total pressure gradients. It couples an efficient kinetic Monte-Carlo (KMC) method to represent oil mass transfer between cells (macroscale) with a capillary-driven flow description (microscale).

4.1 Air Transport Mechanisms on the Cellular Scale

A model integrating cellular defects and the cellular scale was published in Vauvre et al. (2015) simulating oil–air biphasic flow on the scale of an entire potato tissue using a KMC formulation parameterized on the microscopic observations of Patsioura et al. (2015). Figure 7 presents typical oil flow configurations from one cell layer to the next, with subsequent gas-phase displacements on the cellular scale and related oil uptake kinetics on the cellular scale. The kinetics of Figure 7(c) represent oil flow between two cells connected through one constriction. Without a gas phase, the meniscus velocity increases significantly in thin constrictions as thick as the cell walls, but it drops dramatically when the meniscus enters the next cell. Similar effects have been described in a convergent-divergent configuration in Erickson et al. (2002) and Liou et al. (2009). The reduction in oil flow rate in the second cell originates from the oil balance conservation principle, causing that cell to be supplied by a thin tube (i.e., constriction) while the first cell was initially supplied by a tube as large as the cell (the cell was cut). Figure 7(d) describes the presence of an air bubble obstructing 75% of the section that is released when 2% of the next cell has been filled. Entrapped bubbles reduce the effective hydraulic diameter and decrease the oil flow rate, which seems to freeze at the entrance of the second cell. Similar kinetics can be found in Figures 4 and 5 of Dawson et al. (2013). One can notice that the kinetics in Figure 7(d) can be recovered from those of Figure 7(c) by adding a lag-time corresponding to bubble-trapping time after filling the first cell. Such kinetics are highly consistent with the experimental observations of Patsioura et al. (2015), in which samples comprised two cell layers, denoted L_1 and L_2, for a total thickness averaging 250, and in which cells were initially filled by air.

4.2 Experimental Penetration Kinetics

Typical kinetics are reported in Figure 8, with related distributions of filling and lag-times depicted after Vauvre et al. (2015). All cells of L_1 are in contact with oil and filled within a few seconds, but only a few cells of L_2 are filled with oil, with a filling ratio that

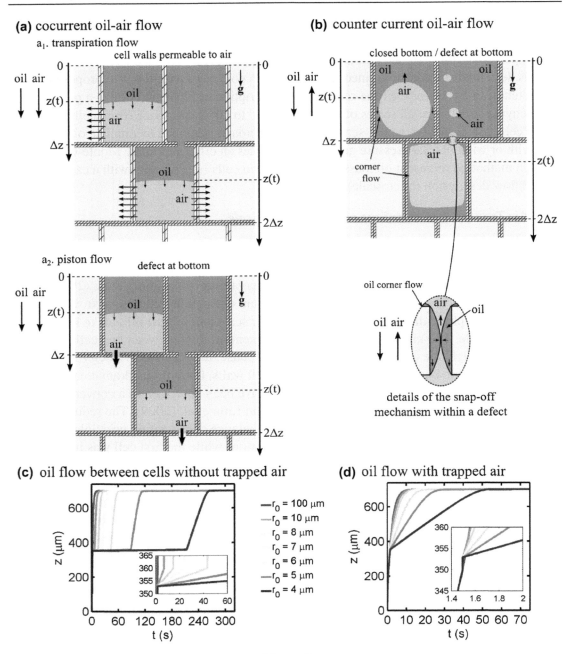

(a) cocurrent oil-air flow

a₁. transpiration flow

(b) counter current oil-air flow

Figure 7
Oil–air displacements in the crust of a fried product (with no starch). (a) Co-current air and (b) countercurrent air escape mechanisms in realistic parenchyma structure (transversal cross-section of hexagonal-shaped cells; $z/\Delta z$ represents the average volume filling ratio of the cell). (c, d) Oil–air front displacements between two circular-shaped cells ($r = 100\,\mu m$; $\Delta z = 350\,\mu m$) connected by a defect of radius r_0 (c) without and (d) with trapped bubble. The trapped bubble is assumed to reduce the oil passage cross-section by 75% and to be released when 2% of the next cell is filled with oil.

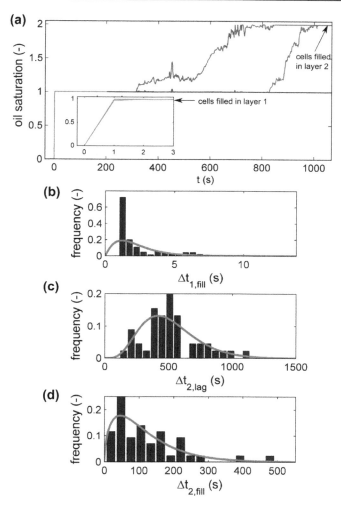

Figure 8
Dynamics of oil percolation of cells in a region of interest of the potato cell layers. (a) Oil saturation of individual cells of L_1 (top cell layer of the potato crust) and L_2 (second cell layer, below L_1) corresponding to the specific region of interest. Distributions of $\Delta t_{1,fill}$, $\Delta t_{2,lag}$, and $\Delta t_{2,fill}$ are depicted in (b), (c) and (d), respectively.

remains incomplete at the end of the experiment. Such a result is consistent with observations on a larger scale on 1 cm^2 potato tissue, as depicted in Patsioura et al. (2015). The distributions of filling time and lag-time in the two first cell layers are key parameters to elucidate the presence of incondensable gas phase, determine the size of defects (Figure 5(b)), and model oil uptake on a larger scale under the conditions of experimental observations.

4.3 Extensions of the Proposed KMC Approach

The combination of the KMC framework proposed in Vauvre et al. (2015) with the oil momentum equations integrated on a microscopic scale enabled extrapolation to more general configurations. Percolation routes were implemented as a Bayesian formulation following the 3D honeycomb arrangement of cells. In this context, three major applications are presented. First, KMC simulations homogenize sparse dynamic observations on the scale of isolated cells up to the scale of a full tissue. Second, the effect of cell damage on oil uptake can be investigated. Finally, our general KMC formulation was successfully compared with the diffusive model of oil uptake of Halder et al. (2007b). Comprehensive rules to set the distribution parameters of all quantities (kinetic and structure parameters) from scarce observations or general assumptions are discussed.

5. Deviations to Generally Accepted Oil Uptake Mechanisms

Contrary to these oil uptake methods, this section details simple phenomena that deviate from conventional descriptions. It is generally recognized that oil uptake occurs during postfrying cooling, but only when french fries prepared from fresh potatoes are studied. Results revealed that oil penetration may also occur during the immersion stage; the latter conclusion was extracted after experiments on industrial par-fried french fries. Potatoes show more damage and ruptures after frying, as demonstrated by Vauvre et al. (2014). In particular, in frozen par-fried products, inner vaporization and inner overpressure are delayed enough to enable oil penetration at the early stages of immersion frying. However, the mechanisms observed during par-frying have not yet been validated for finish-frying. Such a question has been investigated in Vauvre (2014) and is reported here.

5.1 Experimental Setup

Oil uptake was compared for two different frying steps: par-frying (i.e., from fresh potatoes purchased from a local market and with a 9×9-mm^2 cross-section area) and finish-frying (i.e., from industrial par-fried frozen french fries supplied by an industrial partner and with the same initial cross-section area). Several periods of frying times were tested using a modified version of the staining method of Ufheil and Escher (1996). Three successive frying baths were combined with no cooling time between them. Frying baths were either not stained or stained with Sudan blue (blue-stained bath) or Sudan red (red-stained bath). Because oil is stained blue or red at different periods of frying, this experimental setup facilitated discrimination of oil uptake periods during par-frying or finish-frying. The configuration of the reported frying experiment is presented in Figure 9. Fresh or par-fried frozen potato strips were fried first in unstained oil for t_0, then in blue-stained bath for t_1, and finally in red-stained bath for t_2. After the frying experiment for a total duration $t_0 + t_1 + t_2$ of 5 min, the product was immersed in liquid nitrogen, stored at $-80\,°C$, and cut in

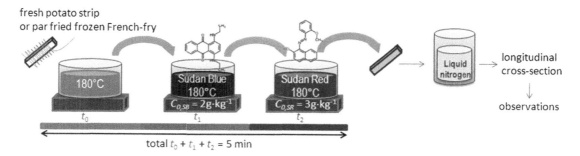

Figure 9

Experimental configuration of frying to identify the periods of oil uptake. Frying experiment involved (1) unstained frying bath for t_0, (2) blue-stained (light gray in print versions) frying bath stained with Sudan blue for t_1, and (3) red-stained (dark gray in print versions) frying bath stained with Sudan red for t_2. Experimental tests were carried out using high oleic sunflower oil at 180 °C for a constant total duration $t_0 + t_1 + t_2 = 5$ min for each frying experiment. Fresh and frozen par-fried potato strips were tested and fried independently for each frying experiment.

a frozen state for subsequent observations. The freezing step (liquid nitrogen) caused a strong contraction of oil. Thus, the volume of oil in successive observations is lower than the volume occupied at frying temperatures. Preliminary tests verified that reversing the order between the red-stained and blue-stained baths did not change the conclusion. However, the sequence of the baths was not random; the blue-stained bath was selected before the red-stained one because it facilitated oil uptake identification and quantification by image analysis (the product and oil are naturally yellowish with a significant red contribution).

5.2 Typical Results with $t_0 = 0$ and $t_1 = 1$ min

The most surprising results were observed during the first minute of frying, when the product was immersed in the blue-stained bath (Figure 9). Results obtained on fresh par-fried frozen potatoes are illustrated in detail in Figure 10. On the one hand, red oil was visible on the external surface of the products prepared from fresh and industrial par-fried potato strips. This external surface corresponded to a thickness of approximately two cell layers in par-fried products. When red oil penetrated deeper in the product during finish-frying (between 0.5 and 1 mm), no significant amount of red oil was identified in the core region. Such a description is consistent with the generally recognized mechanisms describing surface oil uptake occurring during cooling. On the other hand, blue oil belonging to the first minute of frying was not identified in the potato strips prepared from fresh potatoes, but it was visible in all potato strips prepared from industrial french fries. In particular, blue oil penetrated throughout the product as it was detected in the core region.

(a) from fresh potato   **(b)** from par-fried frozen French-fries

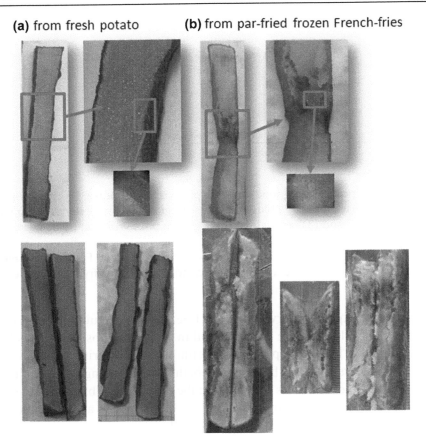

Figure 10

Comparison of oil uptake for $t_0 = 0$ and $t_1 = 1$ min: (a) during par-frying step (red (gray in print versions) oil at surface), and (b) during finish-frying step (red (gray in print versions) oil at surface and blue (dark gray in print versions) oil spots in central regions).

Additional microscopic details obtained by assembling 30 to 55 images at a resolution of 2048×1532 pixels (3.2 µm/pixel) are presented in Figure 11. Such illustrations demonstrated that oil followed the routes from the surface and deeper to inner regions, where cells lose cohesion. Oil was even able to cross the plane of symmetry. These observations match the descriptions of oil and cavities presented in Figure 6. The reported frying experiment constitutes the first clear proof confirming the difference of oil uptake mechanisms between par-frying and finish-frying. Indeed, in finish-frying conditions, oil penetrated french fries in core regions but also from the first minute of frying.

5.3 Generalization for $t_0 > 0$ and $t_1 = 1$ min

This section presents a generalization of previous experiments by shifting the period of immersion frying in the blue-stained bath by increasing t_0 by a step of 1 min. Consistent conclusions are obtained as summarized in Figure 12. Blue-stained oil penetrated inside the

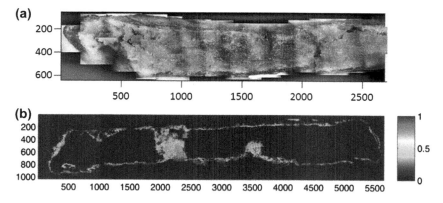

Figure 11

Micrographs of french fry cross-section with stained oil detected in the external and internal region of the potato product: (a) microscopic details of oil distribution in false colors obtained by merging 45 partly overlapping images. Potato structure is depicted in gray with oil appearing superimposed in red (light gray in print versions) and blue (dark gray in print versions) owing to subsequent frying steps in red- (light gray in print versions) and blue-stained baths (dark gray in print versions), respectively; (b) intensity image of blue (dark gray in print versions) oil penetrating the potato product. The lateral resolution in images was 3.2 μm.

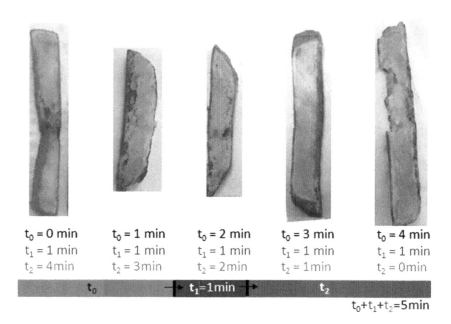

Figure 12

Identification of oil uptake during finish-frying (in par-fried frozen french fries) at different periods. Frying time in undyed oil (for t_0) increases to shift frying time in blue (dark gray in print versions) oil (for $t_1 = 1$ min). $t_0 = 0$ min: blue (dark gray in print versions) oil spots at center and edges, red (gray in print versions) oil at surface; $t_0 = 1–3$ min: red (gray in print versions) oil at surface; $t_0 = 4$ min: blue (dark gray in print versions) oil at surface.

industrial par-fried french fry only during the first minute of frying. Beyond that, no blue oil was detected. Only the condition of $t_0 = 4\,min$ and $t_1 = 1\,min$ confirmed the conclusions obtained with red oil, showing that penetration of blue-stained oil was limited to the superficial crust of the potato. Such an experiment verified that oil uptake occurs during two frying periods: during the beginning of frying (the first minute according to our conditions) as well as during the end of frying or the cooling stage. This description confirms the deviation from the generally accepted mechanism and observations on par-fried french fries, which are not easily transposable to finish-fried french fries.

6. Conclusions

Updated oil uptake mechanisms presented in this chapter provide ground for new experimental routes to understand the role of cellular structure on oil uptake in fried products for postfrying conditions, and also provide initial evidence regarding the role of the incondensable phase in the slowdown of oil penetration. Microscopic observations captured the main events controlling the penetration of oil. The dynamics of oil uptake evidenced by experiments involving two-layer-thick parenchyma tissues (free of starch and with cells open at only one end) were counterintuitive. Cell walls appeared as high barriers to oil. The filling kinetics of the second cell layer was surprisingly slow, from minutes to 1 h, whereas cells of the first cell layer were filled in less than 1 s in most cases.

A more extended mechanistic interpretation of the current results and their generalization to tissues with a larger number of cells and arbitrary cellular damage is given by the developed oil uptake multiscale modeling based on Kinetic Monte-Carlo simulation. This general framework was devised to describe oil uptake at cellular scale in the presence (or not) of an incondensable phase, using filling and lag-times instead of conventional effective transport properties to describe oil uptake. The KMC method developed in the referred study is used to simulate dynamically the displacements of all involved phases on 3D digitized tissues. It was found that the final oil pickup was not a direct function of porosity but depended in a complex manner on the amount of oil available at the surface, on the connectivity of damaged cells, and on the size of passage sections. Percolation from one cell to another is controlled by the distribution of time constants arising from the dynamics of air penetration or escape and steam condensation. It was concluded that surface oil uptake is a fast process owing to the large openings and the relatively short lengths of cells, but filling times increase dramatically with depth as a result of percolation fed through a series of small defects connecting cells. In addition, the proposed analytical solutions on a cellular scale make the simulations of oil adhesion and percolation tractable at the scale of hundreds of thousands of cells.

Beyond generally accepted mechanisms considering that oil uptake occurs during the postfrying cooling, it was demonstrated that finish-frying differed significantly from par-frying. The difference is two-fold: First, finish-fried potato strips are deep-fried in the frozen state and have large internal cavities after frying; then, oil uptake occurs during the beginning of finish-frying, as well. Such a new description envisions new applications to optimize the current process. A simple recommendation to limit oil uptake during the beginning of deep-frying consists of thawing french fries to limit boiling retardation. However, such a precaution does not solve the presence of oil paths from the surface to the core region. Forces opposed to oil uptake such as overpressure should be preserved when the product is removed from the oil bath, by heating or by reducing ambient pressure. Such an approach is possible only if domestic or catering frying tools are adapted to drip oil before oil uptake occurs. This project was launched according to the framework of the French research program "Fry'in," in collaboration with fryer manufacturer SEB and Joint Research Unit 1145.

Acknowledgment

The authors gratefully acknowledge the DISCO beamline team of synchrotron SOLEIL for supporting this study (Proposal 20110137) and "Association Nationale de la Recherche et de la Technologie" (ANRT) agency for co-funding the study. The authors also thank the group McCain, who made possible updates on oil uptake mechanisms.

References

Achir, N., Vitrac, O., Trystram, G., 2008. Simulation and ability to control the surface thermal history and reactions during deep fat frying. Chemical Engineering and Processing 47, 1953–1967.

Achir, N., Vitrac, O., Trystram, G., 2009. Heat and mass transfer during frying. In: Sahin, S., Sumunu, S.G. (Eds.), Advances in Deep-fat Frying. CRC Press, Boca-Raton, pp. 5–32.

Achir, N., Vitrac, O., Trystram, G., 2010. Direct observation of the surface structure of french fries by UV-VIS confocal laser scanning microscopy. Food Research International 43, 307–314.

Aguilera, J.M., Gloria-Hernandez, H., 2000. Oil absorption during frying of frozen par-fried potatoes. Journal of Food Science 65, 476–479.

Al-Khusaibi, M., Gordon, M.H., Lovegrove, J.A., Niranjan, K., 2012. Provenance of the oil in par-fried french fries after finish frying. Journal of Food Science 77, E32–E36.

Bassama, J., Brat, P., Bohuon, P., Hocine, B., Boulanger, R., Guenata, Z., 2011. Acrylamide kinetic in plantain during heating process: precursors and effect of water activity. Food Research International 44, 1452–1458.

Bouchon, P., Pyle, D.L., 2005. Modelling oil absorption during post-frying cooling - II: solution of the mathematical model, model testing and simulations. Food and Bioproducts Processing 83, 261–272.

Bouchon, P., Hollins, P., Pearson, M., Pyle, D.L., Tobin, M.J., 2001. Oil distribution in fried potatoes monitored by infrared microspectroscopy. Journal of Food Science 66, 918–923.

Bouchon, P., Aguilera, J.M., Pyle, D.L., 2003. Structure oil-absorption relationships during deep-fat frying. Journal of Food Science 68, 2711–2716.

Bucklet, S.E., Leverett, M.C., 1942. Mechanism of fluid displacement in sands. Transactions of the AIME 146, 107–116.

Burenjargal, M., Totani, N., 2009. Cytotoxic compounds generated in heated oil and assimilation of oil in Wistar rats. Journal of Oleo Science 58, 1–7.

Cortés, P., Badillo, G., Segura, L., Bouchon, P., 2014. Experimental evidence of water loss and oil uptake during simulated deep-fat frying using glass micromodels. Journal of Food Engineering 140, 19–27.

Costa, R.M., Oliveira, F.A.R., Delaney, O., Gekas, V., 1999. Analysis of the heat transfer coefficient during potato frying. Journal of Food Engineering 39, 293–299.

Dalainas, I., Ioannou, H.P., 2008. The role of trans fatty acids in atherosclerosis, cardiovascular disease and infant development. International Angiology 27, 146–156.

Dana, D., Saguy, I.S., 2006. Review: mechanism of oil uptake during deep-fat frying and the surfactant effect-theory and myth. Advances in Colloid and Interface Science 128, 267–272.

Dawson, G., Lee, S., Juel, A., 2013. The trapping and release of bubbles from a linear pore. Journal of Fluid Mechanics 722, 437–460.

Dullien, F.A.L., 1979. Porous Media: Fluid Transport and Pore Structure. Academic Press, New-York.

Ello-Martin, J.A., Ledikwe, J.H., Rolls, B.J., 2005. The influence of food portion size and energy density on energy intake: implications for weight management. American Journal of Clinical Nutrition 82, S236–S241.

Erickson, D., Li, D., Park, C.B., 2002. Numerical simulations of capillary-driven flows in nonuniform cross-sectional capillaries. Journal of Colloid and Interface Science 250, 422–430.

Fair, A.M., Montgomery, K., 2009. Energy balance, physical activity, and cancer risk. Methods in Molecular Biology 472, 57–88.

Farkas, B.E., Singh, R.P., Rumsey, T.R., 1996a. Modeling heat and mass transfer in immersion frying. 1. Model development. Journal of Food Engineering 29, 211–226.

Farkas, B.E., Singh, R.P., Rumsey, T.R., 1996b. Modeling heat and mass transfer in immersion frying. II. Model solution and verification. Journal of Food Engineering 29, 227–248.

Gamble, M.H., Rice, P., 1987. Effect of pre-fry drying of oil uptake and distribution in potato crisp manufacture. International Journal of Food Science and Technology 22, 535–548.

Gertz, C., Klostermann, S., 2002. Analysis of acrylamide and mechanisms of its formation in deep-fried products. European Journal of Lipid Science and Technology 104, 762–771.

Guallar-Castillon, P., Rodriguez-Artalejo, F., Fornes, N.S., Banegas, J.R., Etxezarreta, P.A., Ardanaz, E., Barricarte, A., Chirlaque, M.-D., Iraeta, M.D., Larranaga, N.L., Losada, A., Mendez, M., Martinez, C., Quiros, J.R., Navarro, C., Jakszyn, P., Sanchez, M.J., Tormo, M.J., Gonzalez, C.A., 2007. Intake of fried foods is associated with obesity in the cohort of Spanish adults from the European Prospective Investigation into Cancer and Nutrition. American Journal of Clinical Nutrition 86, 198–205.

Halder, A., Dhall, A., Datta, A.K., 2007a. An improved, easily implementable, porous media based model for deep-fat frying - Part I: Model development and input parameters. Food and Bioproducts Processing 85, 209–219.

Halder, A., Dhall, A., Datta, A.K., 2007b. An improved, easily implementable, porous media based model for deep-fat frying - Part II: Results, validation and sensitivity analysis. Food and Bioproducts Processing 85, 220–230.

Halder, A., Dhall, A., Datta, A.K., 2011. Modeling transport in porous media with phase change: applications to food processing. Journal of Heat Transfer-Transactions of the ASME 133, 031010 (031013 pp.).

He, D.-B., Xu, F., Hua, T.-C., Song, X.-Y., 2013. Oil absorption mechanism of fried food during cooling process. Journal of Food Process Engineering 36(4), 412–417.

Hubbard, L.J., Farkas, B.E., 2000. Influence of oil temperature on convective heat transfer during immersion frying. Journal of Food Processing and Preservation 24, 143–162.

Kalogianni, E.P., Papastergiadis, E., 2014. Crust pore characteristics and their development during frying of french-fries. Journal of Food Engineering 120, 175–182.

Kaviany, M., 1997. Principles of Heat Transfer in Porous Media, second ed. Springer-Verlag, Berlin, Germany.

Liou, W.W., Peng, Y., Parker, P.E., 2009. Analytical modeling of capillary flow in tubes of nonuniform cross section. Journal of Colloid and Interface Science 333, 389–399.

Lioumbas, J.S., Kostoglou, M., Karapantsios, T.D., 2012. Surface water evaporation and energy components analysis during potato deep fat frying. Food Research International 48, 307–315.

MacMillan, B., Hickey, H., Newling, B., Ramesh, M., Balcom, B., 2008. Magnetic resonance measurements of french fries to determine spatially resolved oil and water content. Food Research International 41, 676–681.

Mauseth, J.D., 2008. Plant Anatomy. Blackburn Press, Caldwell, NJ.

Mehta, U., Swinburn, B., 2001. A review of factors affecting fat absorption in hot chips. Critical Reviews in Food Science and Nutrition 41, 133–154.

Mendonca Freire, P.C., Mancini-Filho, J., Pinto de Castro Ferreira, T.A., 2013. Major physical and chemical changes in oils and fats used for deep frying: regulation and effects on health. Revista De Nutricao-Brazilian Journal of Nutrition 26.

Michalski, M.C., Saramago, B.J.V., 2000. Static and dynamic wetting behavior of triglycerides on solid surfaces. Journal of Colloid and Interface Science 227, 380–389.

Mittelman, N., Mizrahi, S., Berk, Z., 1984. Heat and mass transfer in frying. In: McKenna, B.M. (Ed.), Engineering and Food. Elsevier Applied Science Publisher, London, pp. 109–116.

Moreira, R.G., Sun, X.Z., Chen, Y.H., 1997. Factors affecting oil uptake in tortilla chips in deep-fat frying. Journal of Food Engineering 31, 485–498.

Moreira, R.G., Castell Perez, M.E., Barrufet, M.A., 1999. Deep-Fat Frying: Fundamentals and Applications. Aspen Publishers Inc., MD 20878, USA.

Mottram, D.S., Wedzicha, B.L., Dodson, A.T., 2002. Acrylamide is formed in the Maillard reaction. Nature 419, 448–449.

Moyano, P.C., Berna, A.Z., 2002. Modeling water loss during frying of potato strips: effect of solute impregnation. Drying Technology 20, 1303–1318.

Muskat, M., Meres, M.W., 1936. The flow of heterogeneous fluids through porous media. Physics 7, 346–363.

Ni, H., Datta, A.K., 1999. Moisture, oil and energy transport during deep-fat frying of food materials. Food and Bioproducts Processing 77, 194–204.

Patsioura, A., Vauvre, J.M., Kesteloot, R., Jamme, F., Hume, P., Vitrac, O., 2015. Microscopic imaging of biphasic air-oil flow in french-fries using synchrotron radiation. AIChE Journal 61(4), 1427–1446.

Pedreschi, F., Aguilera, J.M., 2002. Some changes in potato chips during frying observed by confocal laser scanning microscopy (CLSM). Food Science and Technology International 8, 197–201.

Pinthus, E.J., Saguy, I.S., 1994. Initial interfacial tension and oil uptake by deep-fat fried foods. Journal of Food Science 59, 804–807.

Reeve, R.M., Timm, H., Weaver, M.L., 1971. Cell size in Russet Burbank potato tubers with various levels of nitrogen and soil moisture tensions. American Potato Journal 48, 450–456.

Reeve, R.M., 1967. A review of cellular structure, starch, and texture qualities of processed potatoes. Economic Botany 21, 294–308.

Rice, P., Gamble, M.H., 1989. Modeling moisture loss during potato slice frying. International Journal of Food Science and Technology 24, 183–187.

Robards, A.W., 1975. Plasmodesmata. Annual Review of Plant Physiology and Plant Molecular Biology 26, 13–29.

Saguy, I.S., Ufheil, G., Livings, S., 1998. Oil uptake in deep-fat frying: review. OCL-Oleagineux Corps Gras Lipides 5, 30–35.

Sandhu, J., Bansal, H., Takhar, P.S., 2013. Experimental measurement of physical pressure in foods during frying. Journal of Food Engineering 115, 272–277.

Sassi, F., 2010. Obesity and the Economics of Prevention: Fit Not Fat. OECD Publishing, Paris.

Shi, Z., Hu, X., Yuan, B., Hu, G., Pan, X., Dai, Y., Byles, J.E., Holmboe-Ottesen, G., 2008. Vegetable-rich food pattern is related to obesity in China. International Journal of Obesity (London) 32, 975–984.

Sichieri, R., 2002. Dietary patterns and their associations with obesity in the Brazilian city of Rio de Janeiro. Obesity Research 10, 42–48.

Swinburn, B.A., Caterson, I., Seidell, J.C., James, W.P.T., 2004. Diet, nutrition and the prevention of excess weight gain and obesity. Public Health Nutrition 7, 123–146.

Thussu, S., Datta, A.K., 2012. Texture prediction during deep frying: a mechanistic approach. Journal of Food Engineering 108, 111–121.

Ufheil, G., Escher, F., 1996. Dynamics of oil uptake during deep-fat frying of potato slices. Food Science and Technology-Lebensmittel-Wissenschaft & Technologie 29, 640–644.

Vauvre, J.M., Kesteloot, R., Patsioura, A., Vitrac, O., 2014. Microscopic oil uptake mechanisms in fried products. European Journal of Lipid Science and Technology 116, 741–755.

Vauvre, J.M., Patsioura, A., Kesteloot, R., Vitrac, O., 2015. Multiscale modeling of oil uptake in fried products. AIChE Journal 61(7), 2329–2353.

Vauvre, J.M., 2014. Oil Uptake Mechanisms in French-Fry Type Products. UMR 1145 Food Process and Engineering, AgroParisTech, Massy, p. 389.

Vitrac, O., Trystram, G., 2005. A method for time and spatially resolved measurement of convective heat transfer coefficient (h) in complex flows. Chemical Engineering Science 60, 1219–1236.

Vitrac, O., Trystram, G., Raoult-Wack, A.L., 2000. Deep-fat frying of food: heat and mass transfer, transformations and reactions inside the frying material. European Journal of Lipid Science and Technology 102, 529–538.

Vitrac, O., Dufour, D., Trystram, G., Raoult-Wack, A.L., 2001. Deep-fat frying of cassava: influence of raw material properties on chip quality. Journal of the Science of Food and Agriculture 81, 227–236.

Vitrac, O., Dufour, D., Trystram, G., Raoult-Wack, A.L., 2002. Characterization of heat and mass transfer during deep-fat frying and its effect on cassava chip quality. Journal of Food Engineering 53, 161–176.

Vitrac, O., Trystram, G., Raoult-Wack, A.L., 2003. Continuous measurement of convective heat flux during deep-frying: validation and application to inverse modeling. Journal of Food Engineering 60, 111–124.

Warning, A., Dhall, A., Mitrea, D., Datta, A.K., 2012. Porous media based model for deep-fat vacuum frying potato chips. Journal of Food Engineering 110, 428–440.

Yamsaengsung, R., Moreira, R.G., 2002. Modeling the transport phenomena and structural changes during deep fat frying - Part 1: Model development. Journal of Food Engineering 53, 1–10.

Zaheer, K., Akhtar, M.H., 2014. Recent advances in potato production, usage, nutrition-a Review. Critical Reviews in Food Science and Nutrition. http://dx.doi.org/10.1080/10408398.2012.724479.

Ziaiifar, A.M., Achir, N., Courtois, F., Trezzani, I., Trystram, G., 2008. Review of mechanisms, conditions, and factors involved in the oil uptake phenomenon during the deep-fat frying process. International Journal of Food Science and Technology 43, 1410–1423.

CHAPTER 18

Acrylamide in Potato Products

Bruno De Meulenaer, Raquel Medeiros, Frédéric Mestdagh
NutriFOODchem Unit, Department of Food Safety and Food Quality, Faculty of Bioscience Engineering, Ghent University, Ghent, Belgium

1. Introduction

Acrylamide has been used in the chemical industry since the 1950s as an intermediate in the production of polyacrylamide polymers and copolymers in order to improve adhesion and cross-linking of polymers. Polymerized acrylamide is widely used as a flocculant in wastewater treatment and paper and textile manufacturing. Its application also extends to grouting agents for the construction of dam foundations, sewers, and tunnels; cosmetics; and electrophoresis gels. Incomplete polymerization may result in residual amounts of the acrylamide monomer in final products (Shipp et al., 2006), with a maximum concentration permitted of up to 5 mg/kg (Cosmetic Ingredient Review, 2009). In addition, acrylamide is found in filtered mainstream cigarette smoke (Smith et al., 2000).

After the leakage of acrylamide that occurred at the Hallandsås tunnel construction in Sweden (Reynolds, 2002), unexpectedly high levels of acrylamide were found in the blood of a control group of subjects (nonsmokers) without occupational exposure. These findings led to the hypothesis that acrylamide might be present via dietary exposure (Tareke et al., 2000). The Swedish National Food Administration reported in 2002 the presence of relevant amounts of acrylamide in several carbohydrate-rich foods baked at high temperatures (Swedish National Food Administration, 2002). These findings were soon confirmed by other research groups and, together with stakeholders, efforts were carried out to build greater understanding of acrylamide concerning the mechanism of its formation in foods, the risks associated for consumers, and possible strategies to lower acrylamide levels in foodstuffs.

This chapter is largely based upon a review paper earlier published by Medeiros et al. (2012). This review has been extended, however, on the basis of additional material published from 2012 onward.

1.1 Pathways for Acrylamide Formation

Soon after its discovery in heat-processed foods, scientists reported that acrylamide was formed during the Maillard reaction (Mottram et al., 2002; Stadler et al., 2002) (Figure 1).

Advances in Potato Chemistry and Technology. http://dx.doi.org/10.1016/B978-0-12-800002-1.00018-2

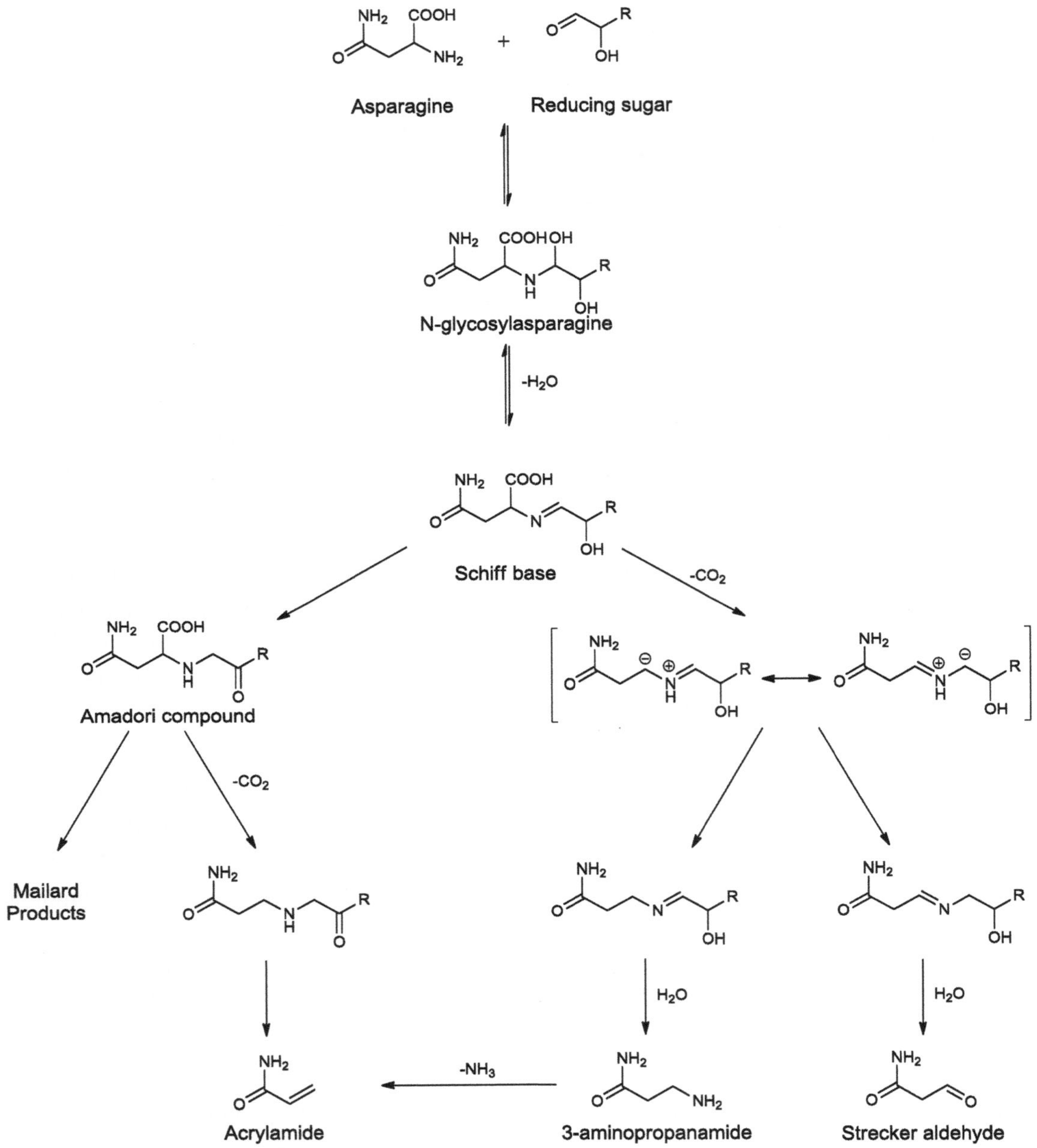

Figure 1

Proposed mechanism for acrylamide formation as a side reaction of the Maillard reaction. Based on Mottram et al. (2002), Stadler et al. (2004), and Granvogl and Schieberle (2006).

This nonenzymatic browning reaction influences several aspects of food quality, such as flavor, color, and aroma formation. Mass spectral studies using ^{15}N-labeled asparagine and ^{13}C-labeled glucose confirmed that the three carbon atoms and the nitrogen of the amide group are derived from asparagine. Although asparagine alone may release acrylamide by thermally initiated decarboxylation and deamination, in the presence of reducing sugars acrylamide formation from asparagine is significantly increased (Yaylayan et al., 2003). The major mechanism of acrylamide formation therefore involves the reaction of a carbonyl compound (preferably an α-hydroxycarbonyl) with asparagine, resulting in the corresponding *N*-glycosyl conjugation and the formation of a decarboxylated Schiff base (after dehydration under high temperatures) (Zyzak et al., 2003; Stadler et al., 2004). This reaction involves a cascade of reactions with various highly reactive intermediates resulting in acrylamide formation in food. The following intermediates have been proposed (Figure 1): (a) the decarboxylation of the Schiff base, which may lead after decomposition directly to acrylamide and an imine or be followed by hydrolysis to 3-aminopropamide (3-APA) and carbonyl compounds; in this respect, it should be noted that 3-APA may also occur in potatoes as such (Granvogl and Schieberle, 2006); (b) subsequent elimination of ammonia from 3-APA can yield acrylamide (Granvogl and Schieberle, 2006); (c) alternatively, the hydrolysis of the imine that furnishes the Strecker aldehyde of asparagine (3-oxopropanamide) may also yield acrylamide, although to a limited extent (Stadler and Scholz, 2004; Blank et al., 2005). On the basis of this generally accepted reaction mechanism, Parker et al. (Parker et al., 2012) presented a mathematical model that, on basis of compositional data of the potato, can predict accurately the acrylamide concentration in french fries after frying for a particular time at a particular temperature.

Other minor reaction routes have been proposed for acrylamide formation, such as from ammonia and acrolein in the absence of asparagine, a pathway that was suggested to play a role in lipid-rich foods (Yasuhara et al., 2003). Acrolein can be formed by oxidative lipid degradation or from glycerol, leading to acrylic acid. Acrylic acid can react with ammonia to form acrylamide (Yaylayan et al., 2005). However, this pathway does not seem to be involved in acrylamide formation in fried potatoes (Becalski et al., 2003). It was also reported that oil degradation products, such as glycerol and mono and diacylglycerols, had no significant impact on acrylamide formation in a potato model system or in french fries (Mestdagh et al., 2007a). Nonetheless, studies have demonstrated that some oxidized lipids may potentially promote the decarboxylation of asparaginase and the later deamination of the produced 3-APA (Zamora and Hidalgo, 2008; Zamora et al., 2009; Hidalgo et al., 2010). Similarly, in 2014 Urbančič et al. univocally showed in a consecutive frying experiment using the same oil that upon increased oil oxidation, the acrylamide content of the finished french fries doubled when the frying oil reached its critical polar content. Addition of several antioxidants to the oil not only slowed down the production of polar lipids, but also led to a reduced increase in acrylamide formation upon progressive oil degradation. The role of antioxidants in acrylamide formation can, however, be considerably more complex and thus also quite confusing, as outlined later (Jin et al., 2013). Acrylamide formation has also been proposed from protein pyrolysis in dry heated wheat gluten (Claus et al., 2006).

Nevertheless, a general consensus has been achieved that the major route for acrylamide formation in potato products remains the route via asparagine and reducing sugars.

1.2 Occurrence of Acrylamide in Foods and Dietary Exposure Assessment

Acrylamide is present in several carbohydrate-rich foods when cooked at high temperatures (>120 °C) as in frying, baking, and roasting. The European Union recommended in 2007 that member states perform annually the monitoring of acrylamide in certain foodstuffs (EC, 2007). The results for 2007 and 2008 have been published (EFSA, 2010) and are summarized in Table 1. Also, the Joint Food and Agriculture Organization/World Health Organization (FAO/WHO) Expert Committee has reported data regarding acrylamide levels in foodstuffs analyzed between 2004 and 2009 from 31 countries (12,582 analytical results from single or composite samples). Mean concentrations of acrylamide in major foods were found to range from 399 to 1202 μg/kg for potato chips, from 159 to 963 μg/kg for french fries, from 169 to 518 μg/kg for cookies, from 87 to 459 μg/kg for crispbread and crackers, and from 3 to 68 μg/l for coffee (ready to drink) (FAO/WHO, 2011).

A great variability in acrylamide levels can be observed within each food category, as shown in Table 1. Moreover, such variability has also been reported within the same batch (Roach et al., 2003; Sanny et al., 2010). Factors such as variability in acrylamide precursors in the raw material, difference in food composition, difference in process parameters, and final baking conditions could be sources of fluctuations. Furthermore, concentrations of acrylamide present in heated foods are the result of simultaneously occurring formation and elimination mechanisms (Claeys et al., 2005b; Gökmen and Senyuva, 2006a). This big variability of acrylamide levels in foods becomes important when considering dietary exposure to acrylamide and moreover when evaluating possible mitigation strategies, as will be discussed later.

In 2011, the Joint FAO/WHO Expert Committee on Food Additives (JECFA) estimated the mean dietary acrylamide intake for the general population, including children, to be between 1 and 4 μg/kg bw/day. It was noted that children have dietary acrylamide exposures at least twice as high as adult consumers when expressed on a body weight basis (FAO/WHO, 2011). Foods contributing the most to dietary intake will differ from country to country, according to different dietary patterns and the way in which foodstuffs are processed and prepared. However, several acrylamide intake studies indicate that fried potato products (french fries and potato crisps), bread and bakery products, coffee, and breakfast cereals are the food commodities that contribute the most to dietary acrylamide exposure. Other food items contribute less than 10% of the total dietary intake (Konings et al., 2003; Svensson et al., 2003; Boon et al., 2005; Dybing et al., 2005; Wilson et al., 2006; Mestdagh et al., 2007b; Claeys et al., 2010; Mojska et al., 2010; FAO/WHO, 2011). Acrylamide concentrations obtained in 2007 and 2008 reported by the

Table 1: Acrylamide Levels (μg/kg)[a] of Foodstuffs Monitored in 2007 and 2008 Reported by the European Food Safety Authority.

	2007				2008			
	N[b]	Mean	SD[c]	Maximum	N	Mean	SD	Maximum
Biscuits								
Crackers	66	284	315	1526	131	204	178	1042
Infant biscuit	97	204	352	2300	88	110	147	1200
Not specified	291	303	433	4200	260	209	247	1940
Wafers	38	210	256	1378	48	252	416	2353
Bread								
Bread, crisp	153	228	328	2430	90	235	273	1538
Bread, soft	123	70	116	910	191	49	56	528
Not specified	54	190	424	2565	17	23	19	86
Coffee								
Instant	51	357	327	1047	58	502	285	1373
Not specified	41	261	268	1158	10	241	215	720
Roasted	151	253	203	958	253	208	182	1524
Other products								
Gingerbread	357	425	494	3615	246	437	545	3307
Muesli and porridge	47	215	183	805	18	43	27	112
Not specified	378	271	355	2529	445	198	309	2592
Substitute coffee	59	800	1062	4700	73	1124	1138	7095
Breakfast cereals	132	152	184	1600	120	170	247	2072
Cereal-based baby food	92	69	72	353	96	45	81	660
Jarred baby food	87	44	35	162	128	35	39	297
Home-cooked potato products								
Deep fried	54	354	413	1661	39	228	253	1220
Not specified	82	277	392	2175	100	192	402	3025
Oven fried	8	385	342	941	94	235	268	1439
French fries	647	357	382	2668	521	280	279	2466
Potato crisps	273	565	259	4180	435	616	634	4382

[a]Values based on an upper bound scenario (values below LOD and values between LOD and LOQ were set to the LOD or the LOQ value, respectively).
[b]Number of individual samples analyzed for each food category.
[c]Standard deviation of the upper bound scenario.
Adapted from EFSA (2010).

European Food Safety Authority (EFSA) in 2010 have not yet been used for an exposure assessment. The assessment will be performed later in 2015 when acrylamide levels of the three monitoring years are gathered. However, when comparing 2008 to 2007 data, acrylamide contents of potato crisps, instant coffee, and substitute coffee were significantly higher in 2008. On the other hand, for other foods, including french fries and fried potato products for home cooking, significantly lower acrylamide levels in 2008 compared to 2007 were obtained. Whether these differences are coincidence or are merely a result of the growing awareness of the food industry of the acrylamide issue needs to be confirmed from results obtained in the coming years.

1.3 Health Risks and Risk Assessment

Acrylamide contains an electrophilic double bond, which can react with nucleophilic groups, hence covalently interacting in vivo with cellular nucleophiles such as the sulfydryl groups in reduced glutathione and in proteins and to a lesser extent protein amino groups (Dybing et al., 2005). Acrylamide is known to be neurotoxic, and several toxicological studies have demonstrated its genotoxic carcinogenicity in animals, thus indicating potential human health risks (Johnson et al., 1986; Friedman et al., 1995; Rice, 2005). In 1994 the International Agency for Research on Cancer classified acrylamide as a possible carcinogen for humans (Group 2A), based on its carcinogenicity in rodents (IARC, 1994).

According to data derived from animal studies, acrylamide is quickly absorbed by the skin and by the mucosa if inhaled. If taken by the oral route, acrylamide is well absorbed and widely distributed to the tissues as well as to the fetus, in pregnant women (Friedman, 2003). It has been postulated that acrylamide is carcinogenic through a genotoxic pathway (Dybing et al., 2005), after conversion to glycidamide, a DNA-reactive epoxide. However, studies indicate that acrylamide-induced carcinogenicity might also be modulated by hormonal systems (Besaratinia and Pfeifer, 2007; Wilson et al., 2010; Pedersen et al., 2010). The relative risks of cancers of certain organs regulated by the human hormonal systems turned out to be greater (Hogervorst et al., 2007; Wilson et al., 2010) than those predicted by linear extrapolation from animal models (Dybing and Sanner, 2003). This and other relative risks determined in epidemiological studies turned out to be 10- to 100-fold higher than expected from animal studies and therefore suggest that rodents might not be good models for all types of cancer (Hogervorst et al., 2010). Numerous case–control and prospective cohort studies have investigated possible associations between dietary acrylamide intake and risk for several types of human cancers. Yet, most breast cancer research done to date has not found an association with dietary acrylamide exposure (Mucci et al., 2005; Pelucchi et al., 2006; Hogervorst et al., 2007; Wilson et al., 2009b; Larsson et al., 2009b; Pedersen et al., 2010; Wilson et al., 2010; Burley et al., 2010). Burley et al. (2010), however, reported a weak correlation for premenopausal breast cancer and acrylamide intake. A cell culture study published in 2011 demonstrated the effects of acrylamide on early molecular changes known to be linked to the development of breast cancer (Lyn-Cook et al., 2011); therefore, further investigation is still warranted. So far, there is no evidence either that acrylamide intake could influence brain, bladder, prostate, lung, thyroid, and gastrointestinal cancers (Pelucchi et al., 2006; Hogervorst et al., 2008a,b, 2009a,b; Schouten et al., 2009; Larsson et al., 2009a; Wilson et al., 2009a). On the other hand, increased risks were reported for ovarian and endometrial cancers (Hogervorst et al., 2007; Wilson et al., 2010), although other authors found no association between these cancers and acrylamide intake (Pelucchi et al., 2006; Larsson et al., 2009c,d). Other positive associations were reported for renal (Hogervorst et al., 2008a), oral cavity (Schouten et al., 2009), and esophageal (Lin et al., 2011) cancers. A 2010 review paper summarizes animal and epidemiological studies up to 2010 on the carcinogenicity of dietary acrylamide intake (Hogervorst et al., 2010). Many of the human cancer studies present certain uncertainties related to imprecise estimates of acrylamide intake based on calculations from food consumption databases or from the use of food frequency questionnaires. Moreover, the large variability between individual foods is a further limitation in these studies, leading to a relatively small difference between high- and low-level intake and therefore lacking sufficient power to detect effects on cancer incidence (Mills et al., 2009).

Genotoxic carcinogens are often considered to have no threshold limit of exposure, meaning that a single exposure to one molecule of carcinogen could trigger the biological process leading to cancer (FAO/WHO, 2005). Since acrylamide is present at quite high levels in many food products consumed daily, the EFSA and the Joint FAO/WHO Committee recommend the MOE (margin of exposure) approach as the preferred method of risk assessment for this contaminant (EFSA, 2005; FAO/WHO, 2005). The value of the MOE is represented by the ratio between a particular point on the dose–response curve leading to tumors in experimental animals and the human dietary intake. Thus, this value indicates the level of concern to assist risk managers in setting priorities for implementing measures to protect public health. The higher the MOE, the lower the risk of exposure to the component concerned. The MOE values reported for acrylamide range from 200 to 50, for the general population and for consumers with high exposure, respectively. These low MOE values not only reveal that this contaminant, being genotoxic and carcinogenic, indeed presents a human health concern (FAO/WHO, 2011), but also indicate that acrylamide is a much more "severe" process contaminant compared to others (e.g., benzo(*a*)pyrene, with MOE values ranging from 10,000 to 25,000). Thus, authorities and industry urge that a solution be found for acrylamide formation in foods. This chapter focuses on acrylamide formation and mitigation in fried potato products, since these are so important to the total intake of dietary acrylamide.

2. Aspects Affecting Acrylamide Formation in Fried Potato Products and Possible Mitigation Strategies

Potato products are strongly susceptible to acrylamide formation. On one hand, this food commodity contains the acrylamide precursors (asparagine and reducing sugars) and, on the other hand, the traditional applied baking conditions (temperatures >120 °C), such as frying and roasting. favor the Maillard reaction, which is linked to acrylamide formation (Figure 1).

Thus, all the potential strategies to prevent acrylamide formation may be covered in two major approaches, removal of the acrylamide precursors or interfering with the Maillard reaction (Table 2). However, since the Maillard reaction is essential for the desired and characteristic flavor and color formation of potato products, this constitutes the first challenge for food scientists: how to reduce acrylamide formation without affecting final product specifications and quality. Research performed as of this writing, and described in this section, demonstrates the necessity of a farm-to-fork approach to reduce acrylamide in fried potato products. Several relevant aspects regarding acrylamide formation at various stages of commercial production of french fries and other potato products are presented in Table 2.

2.1 Potato Cultivar

Potato composition varies greatly between cultivars but generally comprises 63–87% water, 13–37% dry matter, 13–30% carbohydrates, 0.7–4.6% proteins, 0.02–0.96% lipids,

Table 2: Industrial Production Process of French Fries and Potato Crisps and Possible Variables Influencing Acrylamide Formation in the Final Product, with Reference to the Main Mechanism of Action on Acrylamide Formation (Effect on Precursors or the Maillard Reaction).

Industrial Production Process for French Fries and Potato Crisps		Precursors	Maillard Reaction
Raw material production	• Preharvest		
	• cultivar	X	
	• fertilization	X	
	• climate	X	
	• maturity	X	
	• Postharvest		
	• storage (time and temperature)	X	
Processing	• Quality control of raw material	X	
	• Sorting (size)	Related to color	
	• Peeling/washing	X	
	• Cutting		
	• Blanching (time and temperature)		X
	• Dip (dextrose and SAPP and other additives)	X	Surface/volume ratio
	• Drying		X
	• Par-frying[a]		X
	• Freezing[a]		X
	• Packaging		
	• Final frying		X color formation

SAPP, sodium acid pyrophosphate.
[a]Not applicable to potato crisps.

0.2–3.5% fiber, and 0.4–2% ash (Torres and Parreño, 2009). In potatoes, asparagine concentrations are relatively in excess compared to reducing sugar contents. Thus, it is the last that represents the limiting factor in acrylamide formation and therefore will largely determine acrylamide formation in potato products (Amrein et al., 2003, 2004; Becalski et al., 2004; De Wilde et al., 2005).

Some varieties are therefore more suitable than others for french fry production. These generally consist of cultivars with large, long, oval tubers containing moderately high dry matter and low reducing sugar contents. Regarding production of crisps, cultivars with even lower reducing sugar contents, higher dry matter, and moderate-sized oval tubers are preferred (Torres and Parreño, 2009). De Wilde et al. (2006a) reported that acrylamide levels in fried potatoes derived from 16 different varieties correlated to reducing sugar content of the potatoes ($R^2 = 0.82$, $n = 96$). The correlation between reducing sugars and acrylamide formation has been demonstrated in several other studies, whereas sucrose and asparagine presented no correlation (Biedermann et al., 2002; Amrein et al., 2003, 2004; De Wilde et al., 2005; Mestdagh et al., 2008a; Medeiros et al., 2010).

2.2 Soil Properties and Fertilization

Soil type may influence the specific gravity of tubers because of its water-holding capacity, drainage, aeration, structure, temperature, and fertility (Torres and Parreño, 2009). However, only minor differences in acrylamide formation from 16 different varieties grown in a sandy loam soil vs a clay soil were reported by De Wilde et al. (2006a). A 2010 study examined the effect of tuber mineral composition (influenced by both the mineral content and the pH of the soil) on the expression of asparagine and reducing sugars in tubers (Whittaker et al., 2010). The authors concluded that the cultivation location had a significant effect on the mineral composition of the tubers, and subsequently reducing sugar contents were negatively correlated with potassium and calcium and positively correlated with zinc and copper contents.

Nitrogen fertilization has been reported in the literature to have an impact on asparagine and reducing sugar concentrations. A decrease in nitrogen fertilization enhanced reducing sugar expression, resulting in an increase in acrylamide formation (De Wilde et al., 2006b). Moderate nitrogen fertilization combined with a good provision of potassium may result in low levels of free asparagine and reducing sugars in tubers (Heuser et al., 2005). The combination of high phosphorus and low potassium supply has been reported to increase asparagine and reducing sugars in potatoes (Gerendas et al., 2007). These studies clearly indicate that the mineral composition involved in potato tubers development, due to either the soil composition or fertilization, may have an impact on the concentrations of acrylamide precursors in potato products. An appropriate balance between the levels of fertilizers therefore should be considered in order to obtain tubers less prone to acrylamide formation, while taking into account possible environmental impacts and legal fertilization limits (De Wilde et al., 2006b).

2.3 Climatological Conditions and Maturity of the Tuber

Climatological conditions during tuber development and near the harvest period may affect the susceptibility to acrylamide formation in potatoes. The effects of seasonal variation on the chemical composition of four potato varieties (used in french fry and crisp production) over a period of 9-month storage have been studied (De Meulenaer et al., 2008). Results showed a significant impact of variable climatological conditions on reducing sugar, dry matter, total free amino acid, and free asparagine contents of tubers. Lower reducing sugar contents were observed in exceptionally warm summers, which resulted in lower acrylamide generation in fried potatoes.

Throughout tuber maturation, nutrients are transported from the leaves to the tuber, and during vine senescence prior to harvesting a drop in sugar levels in the tuber is normally observed. This drop in sugar concentration is therefore an indication of chemical maturity and suggests that harvesting can proceed with the maximum likelihood of quality being maintained in storage (Torres and Parreño, 2009). The harvest of immature and thus smaller tubers is related to higher reducing sugar contents in the tuber and therefore increased acrylamide formation upon final frying (De Wilde et al., 2006a). However, the maturity of the tuber upon harvest influences not only the reducing sugar levels in the tubers but also the enzymatic system responsible for cold-induced sweetening (Hertog et al., 1997). Accordingly, this agronomical aspect is strongly related to the tuber storage behavior, which is discussed below.

2.4 Potato Storage

Generally after harvest, tubers are stored up to several months to maintain supplies of potatoes throughout the year. Nonetheless, certain storage conditions can cause potatoes to accumulate unacceptable quantities of sugars, even though levels were acceptable at harvest. Senescent sweetening and cold temperatures are the main causes of sugar accumulation during storage. Senescent sweetening results from an enzymatic process that occurs more rapidly at higher storage temperature (>8 °C) and is related to the start of sprout growth (Amrein et al., 2004). Chemical sprout suppressing agents may be used to avoid tuber sprouting, although this is not always an option owing to customer demands. An alternative solution for inhibiting tubers from sprouting in addition to making them less susceptible to diseases is cold temperature storage (<8 °C) (Burton and Wilson, 1978; Blenkinsop et al., 2002). This alternative, however, has a major impact on reducing sugar accumulation (Figure 2) and subsequently on acrylamide formation in fried potato products (Biedermann et al., 2002; Noti et al., 2003; Olsson et al., 2004; De Wilde et al., 2005; Ohara-Takada et al., 2005; Matsuura-Endo et al., 2006; Viklund et al., 2008; Knutsen et al., 2009; Viklund et al., 2010). Thus, cold-induced sweetening is presumably a defense mechanism of tubers to protect themselves from frost, and therefore they start mobilizing sugars from starch at temperatures <8 °C (Blenkinsop et al., 2002). Asparagine contents appear not to be susceptible to various storage temperatures and long storage time (De Wilde et al., 2005). Reducing sugar content is not significantly influenced when potatoes are stored at 8 °C (Biedermann et al., 2002; Noti et al., 2003; De Wilde et al., 2005, 2006a) (Figure 2). Therefore, ideally potato tubers should be stored at an intermediate temperature of approximately 8 °C (Kumar et al., 2004). On the other hand, cold conditions are sometimes unavoidable during winter periods. While senescence sweetening is a permanent condition and only gets worse with time, cold-induced sweetening may be partially reversible (Burton, 1989). Submission of cold-stored tubers at 15 °C for a period of 3 weeks may result in a reversible reduction of the reducing sugar levels (Blenkinsop et al., 2002; Biedermann et al., 2002; De Wilde et al., 2005). This solution, however, may affect the production yield, because of changes in dry matter content. In the case of sugar accumulation due to senescence sweetening, it is critical that no attempt is made to recondition the potatoes, since warm temperatures will only speed up the aging process, therefore incrementing the problem (Torres and Parreño, 2009).

2.5 Quality Control of Incoming Potatoes

All the above-mentioned agronomical aspects unavoidably lead to great variability of the raw material between seasons and even within the same storage season. Reducing sugars are an important parameter in the quality control of potatoes for processing given their influence not only on acrylamide formation (Biedermann et al., 2002; Amrein et al., 2004; De Wilde et al., 2005) but also on the final color of fried potato products (Márquez and Añón, 1986; Pedreschi et al., 2005; Pedreschi et al., 2007b; Mestdagh et al., 2008a). Current quality control criteria

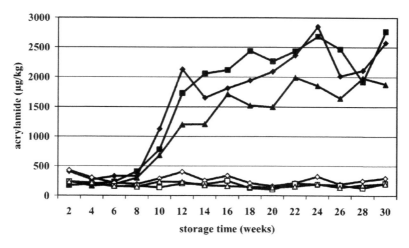

Figure 2

Influence of storage time and temperature on acrylamide formation during frying of three varieties (Bintje, Ramos, Saturna) stored at 4 and 8 °C over 24 weeks, expressed in µg/kg (♦ = Bintje, 4 °C; ■ = Ramos, 4 °C; ▲ = Saturna, 4 °C; ◊ = Bintje, 8 °C; □ = Ramos, 8 °C; Δ = Saturna, 8 °C). *Reprinted from De Wilde et al. (2005) with kind permission from the American Chemical Society (© 2005).*

before accepting potato lots for production comprise size grading (as mentioned above, smaller tubers tend to accumulate more reducing sugars), dry matter determination, and color evaluation with a USDA (U.S. Department of Agriculture)/Munsell color chart (after a short frying test, typically 180 °C for 3 min) (EUPPA, 2007). The last relates raw material to color specifications of the final product (customer demand) and, accordingly, either the raw material is rejected for processing or appropriate adjustments are taken (e.g., optimized blanching conditions). Thus, an effective entrance control of the raw potato tubers could identify batches of potatoes prone to acrylamide formation and therefore prevent the problem of acrylamide formation at the start of the production process. A 2010 study evaluated whether alternative color measurements (Agtron process analyzer and with CIE L*a*b* color parameters) and/or reducing sugar content determination would correlate better with acrylamide levels of french fries upon final frying (Medeiros et al., 2010). To investigate whether seasonal variation would influence the data, the authors repeated the study in the next potato storage season (Medeiros et al., 2012). It was demonstrated that a more effective entrance control for incoming potatoes was indeed possible and differences in the raw material due to seasonal variability did affect overall conclusions. Color determination by means of the Agtron methodology and reducing sugar contents in the raw material allows a better identification of potatoes prone to acrylamide formation compared to the current entrance control. This potentially enables the industry to apply preventive measures to minimize the acrylamide risk in their products. From an implementation point of view, measuring color with an Agtron process analyzer as an acrylamide predictor in raw material is probably more practical than analyzing sugar content. Moreover, color is one of the most important quality specifications demanded by customers.

2.6 Cutting

Acrylamide is formed in the surface layer of the potato product and therefore, size and cut shape of the product (surface-to-volume ratio) will also influence final acrylamide contents. Accordingly, thinner and smaller cut sizes result in increased acrylamide formation upon final frying (Matthäus et al., 2004). Moreover, since the peripheral region of tubers has higher reducing sugar content, these fine cuts from the outer sphere of tubers tend to overheat upon frying. Therefore, removing these fines may contribute to an acrylamide reduction (Foot et al., 2007).

2.7 Blanching Process

Blanching is an important unit operation in the industrial process of french fry production and its complexity may differ between production lines (e.g., the use of one, two, or three blanchers). During this step enzymes are inactivated and a layer of gelatinized starch is formed, which limits oil absorption and improves texture (Moreira et al., 1999). In addition, blanching also contributes to a uniform color of the product after final frying. Moreover, during this step, acrylamide precursors are leached out, resulting in the reduction of acrylamide content in the final product (Pedreschi et al., 2004; Pedreschi et al., 2007c; Mestdagh et al., 2008b; Pedreschi et al., 2009; Medeiros et al., 2010; Viklund et al., 2010). Blanching conditions (time and temperature) therefore can be manipulated until an optimized reducing sugar extraction is reached. To keep the final product specifications constant, potato processors typically increase the intensity of blanching conditions toward the end of the potato season owing to senescence sweetening. However, extreme blanching conditions result in textural and nutrient loss issues and therefore blanching can be adapted only within certain limitations. Mestdagh et al. (2008b) reported an acrylamide reduction of 65% and 96% for french fries and potato crisps, respectively, after blanching (70 °C, 10–15 min). In addition to time and temperature, the concentration of soluble components extracted from the potato cuts during this continuous process will also influence the efficiency of sugar extractability and therefore affect acrylamide formation in potato products. A decrease in sugar extraction of 10% was ascribed to the concentration of soluble components in the blanching water compared to extraction rates in fresh water (Mestdagh et al., 2008b). On the other hand, the continuous replacement of the blanching water with fresh water is not feasible, from both environmental and economical points of view. Moreover, since in general dextrose is added in the subsequent step (as discussed below), this aspect becomes less relevant on the acrylamide content of the final product.

2.8 Use of Additives or Processing Aids

Two common substances used in the potato processing industry (e.g., french fries) are sodium acid pyrophosphate (SAPP, E450) and dextrose (glucose). SAPP is added (pH level ≈4.7) to reduce the darkening of the blanched potato cuts (caused by the ferridichlorogenic acid complex formation during cooking and air exposure) and dextrose contributes to a uniform and standardized color of the final product (according to customer demand). Dextrose dip treatments may affect acrylamide formation in potato products. In North America, the use of color additives such as caramel and annatto is permitted in potato processing instead of dextrose. Two U.S. patent applications claim that these color agents in replacement of dextrose prevent acrylamide formation in potato products (up to 86% and 93% reduction with caramel and annatto, respectively) (U.S. 2005/0214411 A1, 2005; U.S. 2010/0080872 A1, 2010). However, the use of these additives is still restricted in Europe for this particular food commodity. In addition to these, other food additives have been described in the literature to influence acrylamide formation in model systems (Table 3) and in potato products (Table 4). Organic acids are known for their mitigating effect due to the protonation of asparagine amino groups at low pH. This would block the nucleophilic addition of asparagine with a carbonyl compound, preventing the formation of the corresponding Schiff base, a key intermediate in the Maillard reaction and in the formation of acrylamide (Jung et al., 2003; Rydberg et al., 2003; Kita et al., 2004; Gama-Baumgartner et al., 2004; Pedreschi et al., 2004, 2007c; Low et al., 2006; Mestdagh et al., 2008c,d).

Table 3: Summary of Studies and Findings on Mitigation Strategies of Acrylamide Formation Tested in Potato Model Systems.

	Matrix	Additives	Control Sample Conditions	Results	Sensorial Aspects	Reference
Potato model systems	Potato cake	Glutamine, glycine, 4-aminobutyric acid, lysine, sodium glutamate, alanine, and sodium ascorbate (pH 5.72–6.16)	• pH 5.7–5.9 • Acrylamide concentration 439–488 µg/kg	92%, 91%, 86%, 88%, 54%, 50%, and 46% acrylamide reduction for the different additives tested, respectively	N/A	Rydberg et al. (2003)
	Potato cake	Citric acid and glycine	–	20% and 70% acrylamide reduction for citric acid and glycine	N/A	Low et al. (2006)
	Potato cake	Asparaginase	–	Up to 97% acrylamide reduction	N/A	Ciesarova et al. (2006)
	Potato powder (fructose and glucose (0.03 g/100 g powder) and asparagine (0.9 g/100 g powder))	Citric, acetic, and L-lactic acids (pH 4.5–3.7)	• pH 5.4	78%, 46%, and 62% acrylamide reduction, respectively	N/A	Mestdagh et al. (2008c)
		L-cysteine, L-glycine, L-lysine, and glutamine		92%, 24%, and 39% acrylamide reduction for cysteine, glycine, and lysine, respectively; 50% acrylamide increase for glutamine		
		NaCl, SAPP, MgCl$_2$, and CaCl$_2$		5%, 10%, 25%, 58%, and 82% acrylamide reduction, respectively		
	Potato cake	Virgin olive oil and oregano phenolic compounds	–	Up to 45% acrylamide increase for samples treated with virgin olive oil and up to 30% acrylamide reduction when treated with oregano phenolic extracts	N/A	Kotsiou et al. (2010)

Table 4: Summary of Studies and Findings on Mitigation Strategies of Acrylamide Formation Tested in Potato Products in the Lab.

	Matrix	Additives	Control Sample Conditions	Results	Sensorial Aspects	Reference
Potato products—lab experiments	French fries	Citric acid	• Dipped in distilled water for 1 h (pH 6.2) • Acrylamide concentration 645 µg/kg	Up to 75% acrylamide reduction	No differences between control and citric acid solution 1%; solution at 2% resulted in sour and harder in texture fries	Jung et al. (2003)
	Crisps	Antioxidants	• Acrylamide concentration 2160 µg/kg	~50% acrylamide reduction	-	
	French fries	Citric acid (pH 4.54)	• pH 5.8 • Acrylamide concentration 275 µg/kg	46% acrylamide reduction	-	Rydberg et al. (2003)
	French fries	Citric acid	• Immersed in tap water at 60 °C for 15 min • Acrylamide concentration 290 µg/kg	14% acrylamide reduction	Treatment affects sensorial properties	Gama-Baumgartner et al. (2004)
	Crisps	Citric and acetic acid	• Blanched in water at 70 °C for 3 min • Acrylamide concentration 428 µg/kg	32% acrylamide reduction for both acidic treatments	Sour taste detected for citric acid treatment	Kita et al. (2004)
	Crisps	Citric acid	• Rinsed with distilled water • Acrylamide concentration 356 µg/kg	70% acrylamide reduction	–	Pedreschi et al. (2004)
	Crisps	Lysine and glycine	• Soaked in water at 65 °C for 5 min • Acrylamide concentration ~1048 µg/kg	Up to 94% acrylamide reduction	–	Kim et al. (2005)
	Crisps	Glycine and glutamine	• Blanched in water at 80 °C for 2 min • Acrylamide concentration 4127 µg/kg	55% and 25% acrylamide reduction, respectively	–	Brathen et al. (2005)
	French fries	Glycine and glutamine	• Blanched in water at 70–90 °C for 10–15 min • Acrylamide concentration 211–404 µg/kg	No effect on acrylamide contents	–	

Continued

Table 4: Summary of Studies and Findings on Mitigation Strategies of Acrylamide Formation Tested in Potato Products in the Lab.—cont'd

Matrix	Additives	Control Sample Conditions	Results	Sensorial Aspects	Reference
French fries	NaCl and SAPP	• Cut, washed and fried immediately • Acrylamide concentration 489 µg/kg	71% and 95% acrylamide reduction when samples were blanched at 80 °C for 4 min and 70 °C for 30 min, respectively, followed by a soaking treatment in NaCl and SAPP	–	Lindsay and Jang (2005)
French fries	Lactic acid fermentation	• Blanched in water at 80 °C for 5 min • pH 5.7 • Acrylamide concentration ~1000 µg/kg	Up to 86% acrylamide reduction when fermentation lasted 120 min	No difference from control sample	Baardseth et al. (2006)
French fries	NaCl and CaCl$_2$	• Dipped in distilled water at room temperature for 60 min • Acrylamide concentration 589 ± 41 µg/kg	49% and 93% acrylamide reduction, respectively	No effect on final product	Gökmen and Senyuva (2007a)
Crisps	Antioxidant of bamboo leaves	• Acrylamide concentration ~3743 µg/kg • Commercial french fries, washed to remove fat coating. • Acrylamide concentration ~580 µg/kg	~74% acrylamide reduction ~76% acrylamide reduction	No significant differences with control sample	Zhang et al. (2007)
French fries	Citric acid and SAPP	• Rinsed with water • Acrylamide concentrations 1180, 3427, and 4028 µg/kg for the three frying conditions, respectively	53% and 17% acrylamide reduction (average values of three frying conditions tested) for citric acid and SAPP treatments, respectively	–	Pedreschi et al. (2007c)
Crisps	NaCl	• Blanched in distilled water at 85 °C for 3.5 min • Acrylamide concentrations, ~800, 930, and 3860 µg/kg for the three frying conditions, respectively	97%, 92%, and 82% acrylamide reduction for the three frying conditions, respectively	–	Pedreschi et al. (2007a)

Product	Agent	Conditions	Reduction	Effects	Reference
French fries	Asparaginase	• Blanched in distilled water at 75 °C for 10 min • Acrylamide concentration 1264 µg/kg.	Up to 62% acrylamide reduction	—	Pedreschi et al. (2008)
Crisps	Citric, acetic, and L-lactic acid	• Blanched in distilled water (65 °C for 5 min)	Up to 100% acrylamide reduction with all acids	Crisps treated with citric acid were sour, while for other acidic treatments crisps were not significantly different from control	Mestdagh et al. (2008d)
	Glycine and L-lysine		Up to 68% and 85% acrylamide reduction, respectively	Crisps treated with glycine were not significantly different from control, while L-lysine resulted in popcorn-like taste	
	NaCl, SAPP, CaCl$_2$, and Ca-lactate		Up to 43%, 90%, 93%, and 82% acrylamide reduction, respectively	No significant effects for treated samples regarding taste, but increased "snap" and "crispiness"	
Crisps	NaHSO$_3$ (5 g/l) CaCl$_2$ (5 g/l) Cysteine (5 g/l)	• Rinsed in distilled water (1 min) • Acrylamide concentration 5433 µg/kg	47% acrylamide reduction 100% acrylamide reduction 100% acrylamide reduction	Unpleasant odor Increased brittleness Unpleasant odor	Ou et al. (2008)
Potato cubes	Glycine, lactic acid fermentation, and combination of both (pH 5.32–5.83)	• Immersed in distilled water at 37 °C for 75 min (pH 5.9) • Acrylamide concentration ~5000 µg/kg	35%, 50%, and 70% acrylamide reduction, respectively	No effect of treatments on taste, lactic acid treatment resulted in lighter in color final product	Anese et al. (2009)
French fries	Asparaginase	• Blanched by pouring water at 90–95 °C and holding in water at 70 °C for 20 min (pH 6.5–5.5) • Acrylamide concentration 600–1900 µg/kg	60–85% acrylamide reduction	—	Hendriksen et al. (2009)
Crisps		• Blanched at 80 °C for 1 min • Acrylamide concentration 1750 µg/kg	Up to 60% acrylamide reduction	—	

Continued

Table 4: Summary of Studies and Findings on Mitigation Strategies of Acrylamide Formation Tested in Potato Products in the Lab.—cont'd

Matrix	Additives	Control Sample Conditions	Results	Sensorial Aspects	Reference
Crisps	Fruit extracts	–	Up to 26% acrylamide reduction	–	Cheng et al. (2010)
French fries	Hydrocolloids	• Immersed in water at room temperature for 1 and 5 h	Up to 60% acrylamide reduction	–	Zeng et al. (2010)
Crisps	NaCl	• Blanched at 90 °C for 5 min in water • Acrylamide concentration ~2600 µg/kg	~62% acrylamide reduction	–	Pedreschi et al. (2010)
Frozen par-fried french fries	Acetic, citric, lactic, L-ascorbic, and succinic acid (pH 3.4–2.6)	• Potato strips previously blanched, SAPP treated, and stored at 4 °C • Dipped in distilled water at 60 °C for 1 min	45%, 64%, 30%, 25%, and 7% acrylamide reduction, respectively	Acceptable	In supporting information (Medeiros et al., 2011)
	CaCl₂, Ca lactate, MgCl₂, Mg lactate, KCl, and NaCl		44%, 12%, 32%, 18%, 23%, and 49% acrylamide reduction, respectively		
	Glycine		7% acrylamide reduction		
	Asparaginase	• Potato strips previously blanched, SAPP treated, and stored at 4 °C • Dipped in distilled water at 60 °C for 5 min	Up to 65% acrylamide reduction		
Crisps	Asparaginase	• Blanched in distilled water at 85 °C for 3.5 min followed by soaking in water at 50 °C for 20 min • Acrylamide concentration ~800 µg/kg	Up to 80% acrylamide reduction	–	Pedreschi et al. (2011)

Mono- and divalent cations (e.g., Na^+ and Ca^{2+}) were indicated to efficiently reduce acrylamide formation. It was postulated that these ions could interact with asparagine so that the Schiff base formation was again prevented (Park et al., 2005; Lindsay and Jang, 2005; Pedreschi et al., 2007a; Gökmen and Senyuva, 2007b; Mestdagh et al., 2008c,d; Ou et al., 2008; Pedreschi et al., 2010). The addition of calcium ions has also been related to a pH decrease in the food matrices and accordingly associated with acrylamide reductions (Mestdagh et al., 2008c; Levine and Ryan, 2009). The pH decrease may be explained by the competitive displacement of protons from ionizable functional groups, containing oxygen, nitrogen, or sulfur atoms that share electrons with hydrogen atoms (Clydesdale, 1988). NaCl has also been proposed to accelerate acrylamide elimination via polymerization in a model food matrix (Kolek et al., 2006).

Free amino acids such as glycine, cysteine, and lysine have also been suggested for reducing acrylamide formation, either by promoting competitive reactions or by covalently binding acrylamide, which is formed through Michael-type addition reactions (Rydberg et al., 2003; Brathen et al., 2005; Kim et al., 2005; Hanley et al., 2005; Claeys et al., 2005a; Low et al., 2006; Mestdagh et al., 2008c,d; Ou et al., 2008; Anese et al., 2009) resulting in adduct formation (Friedman, 2003). Thus, the presence of amino acids other than asparagine can have an effect on the formation and/or elimination kinetics of acrylamide.

Hydrocolloids are described to efficiently reduce fat uptake when used as coating agents for fried potato products and snacks (Khalil, 1999; Garcia et al., 2004; Kowalczyk and Gustaw, 2009). More recently, Zeng et al. (2010) studied the effect of several hydrocolloid coatings on acrylamide formation in potato strips. Coating treatments with alginic acid and pectin reduced acrylamide contents of the final product, whereas carob gum, carrageenan, hydroxypropyl distarch phosphate, and xanthan gum enhanced acrylamide formation (Zeng et al., 2010).

Antioxidants have been reported to influence the Maillard reaction and thereby affect acrylamide formation positively and negatively (Becalski et al., 2003; Zhang et al., 2007; Zhu et al., 2009, 2010; Bassama et al., 2010; Ou et al., 2010; Cheng et al., 2010). The mechanism of antioxidants on acrylamide formation is, however, not yet fully understood. In a review (Jin et al., 2013), seven different mechanisms were presented, which comprehensively summarized the sometimes conflicting outcome of mitigation studies using antioxidants. Indeed, antioxidants can have various impacts on the complex chemistry of acrylamide formation. Some antioxidants (e.g., chlorogenic acid) may facilitate sucrose degradation, thus increasing the pool of reactive carbonyl compounds, which will, upon reaction with asparagine, generate acrylamide. On the other hand, antioxidants may trap Maillard intermediates (e.g., epicatechin). Given the reactivity of acrylamide itself, some antioxidants may react with acrylamide too. Obviously, antioxidants may delay lipid oxidation, which as discussed already seems to affect acrylamide formation as well. Acrylamide reductions were reported in potato cuts when treated with extract of bamboo leaves (Zhang et al., 2007) and oregano phenolic extract (Kotsiou et al., 2010). Kotsiou et al. (2010) concluded that phenolic compounds without aldehydic groups in their structures are more effective at acrylamide reduction. Moreover, higher levels of phenolic compounds were related to lower acrylamide contents in potato powders from 16 commercial potato varieties when submitted to 185 °C for 25 min (Zhu et al., 2010). These experimental conditions, however, do not represent realistic baking conditions applied to potato products. In contrast, other studies either described positive correlations between acrylamide and antioxidants or found no correlations (Rydberg et al., 2003; Ehling and Shibamoto, 2005; Serpen and Gokmen, 2009; Bassama et al., 2010; Becalski et al., 2010). Supplementary factors associated with the addition of antioxidants, such as pH decrease or the presence of amino acids in the extracts, could influence acrylamide contents and therefore hinder the comparison of results between studies (Mestdagh et al., 2009).

Asparaginase, an enzyme that hydrolyzes asparagine to aspartic acid and ammonia, can reduce acrylamide formation in foods by removal of the precursor asparagine (Ciesarova et al., 2006; Pedreschi et al., 2008; Hendriksen et al., 2009; Pedreschi et al., 2011). Thus, acrylamide formation is successfully reduced upon addition of the enzyme to dough-based products, given that asparagine is readily available for the enzyme to be converted to aspartic acid. On the other hand, asparaginase application on potato products that consist of solid cut pieces is more complex because of the less optimal contact between enzyme and substrate (Hendriksen et al., 2009). For that reason, a blanching step prior to enzyme application is required for better asparaginase–asparagine contact (Pedreschi et al., 2008).

Studies on lactic acid fermentation with *Lactobacillus plantarum* have shown it to reduce acrylamide formation in fried potato products upon final frying (Baardseth et al., 2006; Anese et al., 2009). Accordingly, glucose is metabolized in addition to an acidifying effect of lactic acid production (Baardseth et al., 2006).

2.9 Drying

Drying of blanched or soaked potato cuts prior to final frying reduces fat absorption and oil hydrolysis (Krokida et al., 2001; Choe and Min, 2007). In addition, it also reduces acrylamide formation in french fries owing to the need of shorter finish frying times to obtain the same product quality in terms of color and crispiness (Franke et al., 2005; Gökmen et al., 2006).

2.10 Frying

Finally, but not less important, frying conditions dramatically affect acrylamide levels of the products as they are eaten, since it is during this last step that acrylamide is actually formed. Simultaneous with acrylamide formation, browning, texture, and flavor development caused by the Maillard reaction equally occur during frying. And therefore, acrylamide formation is correlated with color development, given that both are linked to the Maillard reaction and, moreover, the applied frying conditions (time and temperature) affect both in similar manners. Intense frying conditions (time and temperature) lead to darker fries and higher acrylamide contents (Pedreschi et al., 2005, 2006, 2007b; Gökmen and Senyuva, 2006b; Viklund et al., 2007; Mestdagh et al., 2008a; Gökmen and Mogol, 2010). On the other hand, frying at lower temperatures (below 140 °C) results in increased frying time and enhances fat uptake (Foot et al., 2007). Thus, frying time and oil temperature should be controlled to avoid high acrylamide levels, meaning that temperature should not exceed 170–175 °C, and lower temperatures toward the end of the process may reduce acrylamide formation. Other factors, such as the product/oil ratio, may influence a drop in the initial frying temperature and therefore longer frying periods would be needed, resulting in higher acrylamide contents.

The type of oil used for frying was investigated for acrylamide formation in a potato model system and french fries. Some studies indicated that palm oil relative to rapeseed and sunflower oils generated higher acrylamide contents (Gertz and Klostermann, 2002) and olive oil in comparison to corn oil (Becalski et al., 2003), while other authors reported that oil type did not influence acrylamide in the final product (Matthäus et al., 2004; Mestdagh et al., 2005). Moreover, oil oxidation and hydrolysis products, which have been proposed as possible acrylamide precursors, seem to be negligible in acrylamide formation in fried potato products (Becalski et al., 2003; Mestdagh et al., 2007b).

Frying under reduced pressure by means of a vacuum fryer may result in great acrylamide reductions (up to 94%) (Granda and Moreira, 2005). This decrease in acrylamide content can be attributed to the lower temperature used during vacuum frying. However, this frying method is practically applicable only in the production of crisps.

A 2010 study compared the effects of frying versus baking on acrylamide contents of potato chips (Palazoglu et al., 2010). The authors concluded that baking at 170 °C resulted in more than double the acrylamide contents compared to frying at the same temperature. However, at 180 and 190 °C, acrylamide levels of chips prepared by baking were lower in than their fried counterparts.

3. Additives or Processing Aids—from Lab Tests to Industrial Scale

Various mitigation recipes have been proposed in the literature (Tables 3 and 4). However, most of them have been tested only on a lab scale, and therefore it is of crucial importance to perform further in-depth studies before transposing these possible treatments to the factory environment. From a food safety and industrial feasibility perspective, the main criteria regarding the successfulness of any acrylamide mitigation strategy ought to be the efficient reduction of acrylamide formation without affecting overall quality of the final product. As of this writing, only two studies are described in the literature regarding the application of additives or processing aids on an actual industrial setting of potato processing (Table 5). One of these studies investigated the effects of various additives or processing aids on the industrial production of french fries (Medeiros et al., 2011). The application of acetic and citric acid, calcium lactate, and asparaginase was evaluated on the production of frozen par-fried

Table 5: Summary of Studies and Findings on Mitigation Strategies of Acrylamide Formation Tested in Potato Products in Industrial Settings.

Product	Additive	Control Sample Conditions	Results	Sensorial Aspects	Reference
Potato products— industrial settings					
Crisps	CaCl$_2$ (5 g/l)	–	85–96% acrylamide reduction	–	Ou et al. (2008)
Frozen par-fried french fries	Citric acid (pH 4.0–2.0) Acetic acid (pH 3.6–3.3)	• SAPP treated, pH 4.7 • Acrylamide concentration 111 ± 33 to 289 ± 36 µg/kg	Up to 39% acrylamide reduction (pH 3.0) No consistent results in three trials (not significantly different from control increased and decreased acrylamide formation)	Negative impact on sensorial properties	Medeiros et al. (2011)
	Ca lactate (6–36 g/l) Asparaginase (5000–20,000 ASNU/l)	• SAPP treated, pH 4.7 • Acrylamide concentration 247 ± 40 to 298 ± 11 µg/kg	Up to 36% acrylamide reduction (pH 3.0) Up to 66% acrylamide increase		
Chilled (not par-fried) french fries	Asparaginase (625–2500 ASNU/l)	• SAPP treated, pH 4.7 • Acrylamide concentration 90 ± 9.1 to 124 ± 21.5 µg/kg	~100% acrylamide reduction	No effect detected	

french fries regarding their effects on acrylamide reduction and sensorial properties of the final product in addition to other parameters. Even though the authors obtained significant acrylamide reductions in preliminary lab experiments, these treatments did not translate into consistent acrylamide reductions in the industrial production of par-fried french fries during four different trials. pH reduction treatments were effective only at reducing acrylamide formation (up to 39%) when applied at extremely exaggerated low pH values (e.g., pH 3 and 2 vs pH 4.7 of the standard production process). This obviously affected negatively the sensorial properties of the fries. An acidic treatment below pH 4.7 (standard process conditions) is therefore not realistic or feasible in terms of acrylamide mitigation in the industrial production of french fries. Calcium lactate treatment provided inconsistent results when tested throughout three trials and reduced acrylamide formation (up to 36%) only when exaggerated concentrations were used, but also with negative consequences on the sensorial properties of the fries. In addition, this additive precipitates in the presence of SAPP, which could represent problems in production (e.g., cleaning of the line). Regarding the use of asparaginase, this treatment was not successful in reducing acrylamide formation when tested on the production of frozen french fries even at high enzyme concentrations. Crucial factors that might have affected the enzyme performance in the production of frozen french fries are the pH of the standard process not being optimal for enzyme activity, temperature of the dip treatment not optimal either, and time allowed for the enzyme to react before reaching the par-frying step. Medeiros et al. (2011) additionally tested asparaginase in the production of chilled french fries (not par-fried), which accounts only for a small segment of the french fry business. In this type of product, asparaginase reduced up to 90% of acrylamide formation after 3 days of product storage at 4 °C. Since, for this product, there is no par-frying step or any other process step that would deactivate the enzyme, asparaginase could remain active throughout the product shelf life.

The second study reports the addition of an additive for acrylamide mitigation in an industrial line of potato crisp production (Ou et al., 2008) (Table 5). These authors describe an acrylamide reduction of 85–96% in potato crisps when $CaCl_2$ (5 g/l) is added in the blanching process. This, however, is the only result Ou et al. (2008) described and no reference to acrylamide content in control or treated sample was made. The sensorial properties of the final product were not documented either. Moreover, $CaCl_2$ presents other disadvantages such as incompatibility with SAPP (as mentioned above), limitations related to the chloride content of the effluent, and the possible formation of 3-monoclhloropropanediole esters upon frying. This last point is merely speculative since it has not yet been investigated.

Although there are many promising acrylamide-reducing agents, special care should be given to the interpretation of the results while considering the possibility of implementation in an industrial setting. In most of the reported studies (Tables 3 and 4), the use of additives in potato products does not consider the sensorial properties of the final product; therefore unrealistic concentrations of various additives might have been tested and consequently resulted in significant acrylamide reductions. In addition, the pH employed by potato processors at a certain stage of the production process is generally already acidic ($\approx$4.7) owing to the addition of pyrophosphate. This important aspect is still too frequently overlooked and not considered in the reported lab experiments. Higher acrylamide levels of control samples simply immersed in distilled water therefore correspond to higher reduction yields obtained in lab studies. Moreover, laboratory experiments generally reflect standardized "process conditions" and the variability of experimental conditions and of raw material is controlled. However, in industry raw material variability is a daily challenge and may represent an obstacle when accessing implementation of a mitigation strategy on an industrial scale. Model systems based on potato flakes or powder to test the effects of additives on acrylamide formation deserve special interpretation of the results. In model systems, additives are distributed homogeneously throughout the potato cake, while standard potato crisps and french fries, being solid-cut pieces, have restricted contact with the surface layers only. The time that the surface of a potato slice or stick is exposed to the additive is limited by the production process, representing another constriction in acrylamide reduction in industrial settings that is not covered by model systems.

This leads to the conclusion that, as of this writing, although many additives have been demonstrated to prevent acrylamide formation in potato products on the lab scale, these are still not proven on the industrial scale, and therefore further research is warranted. Recently, the Atlantic Canada Opportunities Agency (ACOA) released $2.5 million to Functional Technologies Corp. and Phyterra Yeast, Inc., to support the development and commercialize the application of a yeast, Acryleast™, in the processed potato industry. This yeast is capable of preventing acrylamide formation by reducing asparagine contents in dough-based products, and this investment is intended to extend its application to potato products (Functional Technologies Corp., 2011).

4. Evolution of Risk Management

To date, there are no legal regulations (worldwide) regarding acrylamide levels in food. Many authorities have therefore adopted minimization strategies by working closely with industry, with focus on effective mitigation strategies and following up whether these strategies are applied. Recommendations given by the Joint FAO/WHO Committee highlight the need for further efforts on developing and implementing mitigation methods for acrylamide in foods of major importance for dietary exposure (FAO/WHO, 2011). The main initiatives taken so far by governmental agencies and industry are summarized below.

In 2002, the German Federal State authorities, together with the Federal Office of Consumer Protection and Food Safety (BVF), industry, and the Federal Ministry of Food Agriculture and Consumer Protection, developed a risk management tool for acrylamide in foods (Göbel and Kliemant, 2007). This tool, also known as the German signal values, was based on values that represent the lowest 10% of products of each food commodity that have the highest acrylamide content. Thus, the German Federal State prompts food producers to take adequate actions to lower the acrylamide contents whenever these exceed the signal value. These signal values were evaluated from 2002 until 2005, and the latest update included the values of 530 and 1333 μg/kg for french fries (chips) and potato crisps, respectively (Göbel and Kliemant, 2007). However, since 2007 no other German signal values have been published. Recently, the European Commission adopted a somewhat alternative approach and published a Recommendation on the Web site of the European Commission's Directorate General for Health and Consumer Policy (DG SANCO) on investigations into the levels of acrylamide in food (EC, 2011). Member states are therefore recommended to investigate cases in which acrylamide contents of foodstuffs after final preparation exceed prescribed acrylamide indicative values. For french fries and potato crisps, these indicative values are 600 and 1000 μg/kg, respectively. The member states are recommended to report the results back to the Commission for further assessment by December 2012 and decision-making for additional measures if necessary.

Other governments have taken action toward consumer information. The state of California, for example, requires that producers or businesses that knowingly expose individuals to significant amounts of acrylamide must provide clear and reasonable warning to those individuals. Based on this, California state government settled a court agreement in 2008 with a number of french fry and crisp producers and some fast-food chains to reduce the acrylamide content of their products significantly and to place warning labels regarding the presence of this contaminant on the packaging, in addition to the payment of monetary fines (Mills et al., 2009). Health Canada informs consumers by making public on their Web site acrylamide monitoring results from 2002 to 2008, including the brand names of products (Health Canada, 2009).

The Swiss federal authorities chose to take direct action against certain food products present in the market that contained high acrylamide levels. In 2005, an infant biscuit exceeding 1000 μg/kg of acrylamide was removed from the market (Grob, 2007), and approximately 2 years later, "hard measures" were undertaken regarding a brand of potato chips in the Swiss market that contained up to 7000 μg/kg of acrylamide (Biedermann et al., 2010).

A Code of Practice for the reduction of acrylamide in foods has been adopted by the Codex Alimentarius Commission (2009). This code intends to provide authorities, manufacturers, and other relevant bodies with guidance to prevent and reduce acrylamide formation in potato and cereal products.

Since the discovery of acrylamide in food, the industry has voluntarily taken action to understand the mechanism of acrylamide formation and possible mitigation strategies in food by supporting and cofunding several research initiatives. The European Food and Drink Federation (Food Drink Europe) established a Technical Acrylamide Expert Group in 2003 and created the

"Acrylamide Toolbox," which is available online (http://www.fooddrinkeurope.eu/uploads/publications_documents/Toolboxfinal260911.pdf). This toolbox presents a regularly updated description of the possible mitigation strategies that could be employed to reduce acrylamide levels in foods contributing the most to acrylamide intake. The mitigation strategies covered in the toolbox comprise aspects related to agronomy, recipe, processing, and final preparation that contribute to acrylamide formation. Brochures summarizing the most important information of the toolbox relating to french fries and crisps are also published on the European Commission's Web site in 22 different languages (EC, 2015). The toolbox allows an open communication, and consequently joint efforts are being taken to reduce the acrylamide problem in most relevant food commodities. Accordingly, small and medium enterprises, which normally have limited R&D resources, have access to updated information. In addition to the toolbox, Food Drink Europe made available also very practical pamphlets in the 23 European languages for each of the critical food groups with respect to the acrylamide issue (http://www.fooddrinkeurope.eu/S=0/publication/Download-FoodDrinkEurope-Acrylamide-Pamphlets-in-23-languages/).

5. Future Outlook

Acrylamide is formed in potato products during industrial processing, retail, catering, and home preparation. In the literature, several possibilities have been proposed for acrylamide reduction in potato products. These have been described throughout this chapter and most of them are summarized in the acrylamide toolbox developed by Food Drink Europe (2011).

Current industrial practices in the potato processing sector already consider the selection of potato varieties with low reducing sugar contents, and storage temperatures of about 8 °C are normally respected. However, retail businesses and consumers might still be unaware of the importance of cultivar selection and optimal storage conditions for acrylamide formation. Therefore food safety authorities and the media may play a key role in this by ensuring that potatoes available in the retail market for frying and roasting contain low reducing sugar contents and by implementing campaigns regarding ideal storage conditions and suitable baking conditions. Nevertheless, from an agronomical point of view, substantial improvements can still be made in reducing acrylamide formation in potato products by developing varieties that are resistant to low-temperature sweetening and with lower asparagine contents either by breeding or by genetic engineering. Rommens et al. (2006, 2008) has demonstrated that with genetic engineering it is indeed possible to improve potato storage conditions and nutritional properties of potatoes and consequently lower levels of acrylamide. Owing to legal constraints and public acceptability, it is, however, currently not possible to use genetically engineered products within the European Union.

From the process point of view, blanching conditions and the acidifying effect of added pyrophosphate reduces considerably the acrylamide formation in french fries. Further acrylamide reductions could probably be obtained by avoiding the use of dextrose as a browning agent and using caramel and annatto colors as replacement. Important aspects in this respect are the significantly lower acceptable daily intake for annatto compared to caramel, and the fact that these food additives are not yet regulated in Europe for use in potato processing.

Regarding the use of processing aids in the industrial production process, although asparaginase treatment was not successful in reducing acrylamide formation in par-fried french fries, this treatment reduced acrylamide to levels less than the limit of detection (LOD) in "chilled" french fries (not par-fried), thus indicating its potential as a mitigation strategy. However, certain limitations still need to be evaluated before possible implementation. An optimal pH and temperature for enzyme activity, modifications to the process line, and clarification of regulatory status regarding residual enzyme activity in the product still need to be addressed. Further work is therefore warranted in possibly finding an enzyme with improved activity (faster action) and with increased resistance to higher temperatures and lower pH. Or alternatively, process treatments other than dipping, such as spraying in an enclosed system, should be explored.

So far, none of the acrylamide-mitigating agents studied on the lab scale is ready to be implemented in the industrial processing of potato products; therefore, further work is necessary in exploring the possibilities studied in the lab in an industrial setting. Any concepts established to minimize acrylamide formation must certainly ensure that the sensorial properties of the final product are not negatively affected. In addition, factors such as seasonal variability of the raw material, raw material characteristics (before and after blanching), the complexity of the blanching step (one, two, or three blanchers), dip tank parameters (temperature, pH, and duration), and other variables should be considered when implementing an acrylamide mitigation strategy on the industrial level while maintaining the expected product quality for the consumer. Other considerations such as feasibility, food legislation, cost, effluent treatment, safety and comfort of the employees, ability to control dosage, etc., are equally relevant when considering the implementation of any change to an industrial process. Ultimately, with regard to french fries, the success of any acrylamide mitigation strategy implemented by industry is strongly dependent upon the final baking conditions adopted by the household consumers or caterers. For lower acrylamide contents, frying temperature of 160–170 °C and to fry until the french fries have a light yellow color is suggested. To enable better control over frying temperatures, it is necessary to improve the reliability and accuracy of temperature controls on frying equipment. This is partially possible by changing technical specifications of the fryers as suggested by Grob et al. (2007); however, it is of crucial importance that consumers/caterers are well informed that darker fries correspond to higher acrylamide contents.

Finally, any promising mitigation strategy and regulatory resolutions should be assessed with regard to the possible impact on consumer exposures as exemplified by several authors (Boon et al., 2005; Claeys et al., 2010; Seal et al., 2008). In this respect, if acrylamide formation in french fries would be reduced to levels below the LOD, as referred to in Section 3, this would reduce the P50 and P95 acrylamide total intake for the Belgian population up to 12% and 25%, respectively (Claeys et al., 2010). This example demonstrates that in addition to the efforts of the potato industry to reduce acrylamide contents in potato products, it still remains necessary to mitigate acrylamide in other food commodities to effectively reduce acrylamide intake.

Mitigating acrylamide in potato products while safeguarding other quality aspects still remains a challenge.

References

Amrein, T.M., Bachmann, S., Noti, A., Biedermann, M., Barbosa, M.F., Biedermann-Brem, S., et al., 2003. Potential of acrylamide formation, sugars, and free asparagine in potatoes: a comparison of cultivars and farming systems. Journal of Agricultural and Food Chemistry 51, 5556–5560.

Amrein, T.M., Schönbächler, B., Rohner, F., Lukac, H., Schneider, H., Keiser, A., et al., 2004. Potential for acrylamide formation in potatoes: data from the 2003 harvest. European Food Research and Technology 219, 572–578.

Anese, M., Bortolomeazzi, R., Manzocco, L., Manzano, M., Giusto, C., Nicoli, M.C., 2009. Effect of chemical and biological dipping on acrylamide formation and sensory properties in deep-fried potatoes. Food Research International 42, 142–147.

Baardseth, P., Blom, H., Skrede, G., Mydland, L.T., Skrede, A., Slinde, E., 2006. Lactic acid fermentation reduces acrylamide formation and other Maillard reactions in french fries. Journal of Food Science 71, C28–C33.

Bassama, J., Brat, P., Bohuon, P., Boulanger, R., Gunata, Z., 2010. Study of acrylamide mitigation in model system: effect of pure phenolic compounds. Food Chemistry 123, 558–562.

Becalski, A., Lau, B.P.Y., Lewis, D., Seaman, S.W., 2003. Acrylamide in foods: occurrence, sources, and modeling. Journal of Agricultural and Food Chemistry 51, 802–808.

Becalski, A., Lau, B.P.Y., Lewis, D., Seaman, S.W., Hayward, S., Sahagian, M., et al., 2004. Acrylamide in french fries: influence of free amino acids and sugars. Journal of Agricultural and Food Chemistry 52, 3801–3806.

Becalski, A., Stadler, R., Hayward, S., Kotello, S., Krakalovich, T., Lau, B.P.Y., et al., 2010. Antioxidant capacity of potato chips and snapshot trends in acrylamide content in potato chips and cereals on the Canadian market. Food Additives and Contaminants Part A-Chemistry Analysis Control Exposure & Risk Assessment 27, 1193–1198.

Besaratinia, A., Pfeifer, G.P., 2007. A review of mechanisms of acrylamide carcinogenicity. Carcinogenesis 28, 519–528.

Biedermann, M., Grundbock, F., Fiselier, K., Biedermann, S., Burgi, C., Grob, K., 2010. Acrylamide monitoring in Switzerland, 2007–2009: results and conclusions. Food Additives and Contaminants Part A-Chemistry Analysis Control Exposure & Risk Assessment 27, 1352–1362.

Biedermann, M., Noti, A., Beidermann-Brem, S., Mozzetti, V., Grob, K., 2002. Experiments on acrylamide formation and possibilities to decrease the potential of acrylamide formation in potatoes. Mitteilungen aus Lebensmitteluntersuchung und Hygiene 93, 668–687.

Blank, I., Robert, F., Goldmann, T., Pollien, P., Varga, N., Devaud, S., et al., 2005. Mechanisms of acrylamide formation–Maillard-induced transformation of asparagine. In: Friedman, M., Mottram, D. (Eds.), Chemistry and Safety of Acrylamide in Food, 561 ed. Springer, New York, pp. 171–189.

Blenkinsop, R.W., Copp, L.J., Yada, R.Y., Marangoni, A.G., 2002. Changes in compositional parameters of tubers of potato (*Solanum tuberosum*) during low-temperature storage and their relationship to chip processing quality. Journal of Agricultural and Food Chemistry 50, 4545–4553.

Boon, P.E., de Mul, A., van der Voet, H., van Donkersgoed, G., Brette, M., van Klaveren, J.D., 2005. Calculations of dietary exposure to acrylamide. Mutation Research-Genetic Toxicology and Environmental Mutagenesis 580, 143–155.

Brathen, E.B., Kita, A., Knutsen, S.H., Wicklund, T., 2005. Addition of glycine reduces the content of acrylamide in cereal and potato products. Journal of Agricultural and Food Chemistry 53, 3259–3264.

Burley, V.J., Greenwood, D.C., Hepworth, S.J., Fraser, L.K., de Kok, T.M., van Breda, S.G., et al., 2010. Dietary acrylamide intake and risk of breast cancer in the UK women's cohort. British Journal of Cancer 103, 1749–1754.

Burton, W.G., 1989. Post-harvest physiology. In: Burton, W.G. (Ed.), The Potato, third ed. Longman Scientific & Technical, Harlow, UK, pp. 423–522.

Burton, W.G., Wilson, A.R., 1978. Sugar content and sprout growth of tubers of potato cultivar record, grown in different localities, when stored at 10, 2 and 20 °C. Potato Research 21, 145–162.

Cheng, K.W., Shi, J.J., Ou, S.Y., Wang, M.F., Jiang, Y., 2010. Effects of fruit extracts on the formation of acrylamide in model reactions and fried potato crisps. Journal of Agricultural and Food Chemistry 58, 309–312.

Choe, E., Min, D.B., 2007. Chemistry of deep-fat frying oils. Journal of Food Science 72, R77–R86.

Ciesarova, Z., Kiss, E., Boegl, P., 2006. Impact of L-asparaginase on acrylamide content in potato products. Journal of Food and Nutrition Research 45, 141–146.

Claeys, W., Baert, K., Mestdagh, F., Vercammen, J., Daenens, P., De Meulenaer, B., et al., 2010. Assessment of the acrylamide intake of the Belgian population and the effect of mitigation strategies. Food Additives and Contaminants Part A-Chemistry Analysis Control Exposure & Risk Assessment 27, 1199–1207.

Claeys, W., De Vleeschouwer, K., Hendrickx, M.E., 2005a. Effect of amino acids on acrylamide formation and elimination kinetics. Biotechnology Progress 21, 1525–1530.

Claeys, W., De Vleeschouwer, K., Hendrickx, M.E., 2005b. Kinetics of acrylamide formation and elimination during heating of an asparagine-sugar model system. Journal of Agricultural and Food Chemistry 53, 9999–10005.

Claus, A., Weisz, G.M., Schieber, A., Carle, R., 2006. Pyrolytic acrylamide formation from purified wheat gluten and gluten-supplemented wheat bread rolls. Molecular Nutrition & Food Research 50, 87–93.

Clydesdale, F.M., 1988. Minerals: their chemistry and fate in food. In: Smith, K. (Ed.), Trace Minerals in Food. Marcel Dekker, New York, pp. 57–94.

Codex Alimentarius Commission, 2009. Code of Practice for the Reduction of Acrylamide in Foods (CAC/RCP 67-2009)., pp. 1–8.

Cosmetic Ingredient Review, 2009. Cosmetic Ingredients Review Reports through June 2010-Reference Table.

De Meulenaer, B., De Wilde, T., Mestdagh, F., Govaert, Y., Ooghe, W., Fraselle, S., et al., 2008. Comparison of potato varieties between seasons and their potential for acrylamide formation. Journal of the Science of Food and Agriculture 88, 313–318.

De Wilde, T., De Meulenaer, B., Mestdagh, F., Govaert, Y., Ooghe, W., Fraselle, S., et al., 2006a. Selection criteria for potato tubers to minimize acrylamide formation during frying. Journal of Agricultural and Food Chemistry 54, 2199–2205.

De Wilde, T., De Meulenaer, B., Mestdagh, F., Govaert, Y., Vandeburie, S., Ooghe, W., et al., 2005. Influence of storage practices on acrylamide formation during potato frying. Journal of Agricultural and Food Chemistry 53, 6550–6557.

De Wilde, T., De Meulenaer, B., Mestdagh, F., Govaert, Y., Vandeburie, S., Ooghe, W., et al., 2006b. Influence of fertilization on acrylamide formation during frying of potatoes harvested in 2003. Journal of Agricultural and Food Chemistry 54, 404–408.

Dybing, E., Farmer, P.B., Andersen, M., Fennell, T.R., Lalljle, S.P.D., Muller, D.J.G., et al., 2005. Human exposure and internal dose assessments of acrylamide in food. Food and Chemical Toxicology 43, 365–410.

Dybing, E., Sanner, T., 2003. Risk assessment of acrylamide in foods. Toxicological Sciences 75, 7–15.

EC, 2007. European Commission Recommendation of 3 May 2007 on the monitoring of acrylamide levels in food 2007/331/EC. Official Journal of the European Union L123, 33–40.

EC, 2011. European Commission Recommendation of 10 January 2011 on investigations into the levels of acrylamide in food. http://ec.europa.eu/food/food/chemicalsafety/contaminants/acrylamide_en.htm.

EC, 2015. http://ec.europa.eu/food/food/chemicalsafety/contaminants/acrylamide_en.htm.

European Food Safety Authority (EFSA), 2005. Opinion of the scientific committee on a request from EFSA related to a harmonised approach for risk assessment of substances which are both genotoxic and carcinogenic (Request No. EFSA-Q-2004-020, adopted on 18.10.05.). The EFSA Journal 282, 1–31.

EFSA, 2010. Results on acrylamide levels in food from monitoring year. Scientific Report of EFSA 8 (5), 1599.

Ehling, S., Shibamoto, T., 2005. Correlation of acrylamide generation in thermally processed model systems of asparagine and glucose with color formation, amounts of pyrazines formed, and antioxidative properties of extracts. Journal of Agricultural and Food Chemistry 53, 4813–4819.

EUPPA, 2007. Entrance control for raw material (potatoes) currently used in the french fries industry. www.euppa.eu.

FAO/WHO, 2005. Joint FAO/WHO Expert Committee on Food Additives: Summary and Conclusions Report from Sixty-fourth Meeting, Rome, 8–17 February 2005. (Rep. No. JECFA/64/SC).

FAO/WHO, 2011. Joint FAO/WHO Expert Committee on Food Additives: Evaluation of Certain Contaminants in Food Report from Sevety-second Meeting. (Rep. No. WHO technical report series; no. 959).

Food Drink Europe, 2011. Food Drink Europe Acrylamide Toolbox 2011.

Foot, R.J., Haase, N.U., Grob, K., Gondé, P., 2007. Acrylamide in fried and roasted potato products: a review on progress in mitigation. Food Additives and Contaminants 24, 37–46.

Franke, K., Sell, M., Reimerdes, E.H., 2005. Quality related minimization of acrylamide formation–an integrated approach. In: Friedman, M., Mottram, D. (Eds.), Chemistry and Safety of Acrylamide in Food, 561 ed. Springer, New York, pp. 357–369.

Friedman, M., 2003. Chemistry, biochemistry, and safety of acrylamide. A review. Journal of Agricultural and Food Chemistry 51, 4504–4526.

Friedman, M.A., Dulak, L.H., Stedham, M.A., 1995. A lifetime oncogenicity study in rats with acrylamide. Fundamental and Applied Toxicology 27, 95–105.

Functional Technologies Corp, 2011. http://www.functionaltechcorp.com/s/Phyterra_Yeast.asp.

Gama-Baumgartner, F., Grob, K., Biedermann, M., 2004. Citric acid to reduce acrylamide formation in french fries and roasted potatoes. Mitteilungen aus Lebensmitteluntersuchung und Hygiene 95, 110–117.

Garcia, M.A., Ferrero, C., Campana, A., Bertola, N., Martino, M., Zaritzky, N., 2004. Methylcellulose coatings applied to reduce oil uptake in fried products. Food Science and Technology International 10, 339–346.

Gerendas, J., Heuser, F., Sattelmacher, B., 2007. Influence of nitrogen and potassium supply on contents of acrylamide precursors in potato trubers and on acrylamide accumulation in french fries. Journal of Plant Nutrition 30, 1499–1516.

Gertz, C., Klostermann, S., 2002. Analysis of acrylamide and mechanisms of its formation in deep-fried products. European Journal of Lipid Science and Technology 104, 762–771.

Göbel, A., Kliemant, A., 2007. The German minimization concept for acrylamide. Food Additives and Contaminants 24, 82–90.

Gökmen, V., Mogol, B.A., 2010. Computer vision-based image analysis for rapid detection of acrylamide in heated foods. Quality Assurance and Safety of Crops & Foods 2, 203–207.

Gökmen, V., Palazoglu, T.K., Senyuva, H.Z., 2006. Relation between the acrylamide formation and time-temperature history of surface and core regions of french fries. Journal of Food Engineering 77, 972–976.

Gökmen, V., Senyuva, H.Z., 2006a. A simplified approach for the kinetic characterization of acrylamide formation in fructose-asparagine model system. Food Additives and Contaminants 23, 348–354.

Gökmen, V., Senyuva, H.Z., 2006b. Study of colour and acrylamide formation in coffee, wheat flour and potato chips during heating. Food Chemistry 99, 238–243.

Gökmen, V., Senyuva, H.Z., 2007a. Acrylamide formation is prevented by divalent cations during the Maillard reaction. Food Chemistry 103, 196–203.

Gökmen, V., Senyuva, H.Z., 2007b. Effects of some cations on the formation of acrylamide and furfurals in glucose-asparagine model system. European Food Research and Technology 225, 815–820.

Granda, C., Moreira, R.G., 2005. Kinetics of acrylamide formation during traditional and vacuum frying of potato chips. Journal of Food Process Engineering 28, 478–493.

Granvogl, M., Schieberle, P., 2006. Thermally generated 3-aminopropionamide as a transient intermediate in the formation of acrylamide. Journal of Agricultural and Food Chemistry 54, 5933–5938.

Grob, K., 2007. Options for legal measures to reduce acrylamide contents in the most relevant foods. Food Additives and Contaminants 24, 71–81.

Hanley, A.B., Offen, C., Clarke, M., Ing, B., Roberts, M., Burch, R., 2005. Acrylamide reduction in processed foods. In: Friedman, M., Mottram, D. (Eds.), Chemistry and Safety of Acrylamide in Food, 561 ed. Springer, New York, pp. 387–392.

Health Canada, 2009. Acrylamide Levels in Selected Canadian Foods. http://www.hc-sc.gc.ca/fn-an/securit/ chem-chim/food-aliment/acrylamide/acrylamide_level-acrylamide_niveau_table-eng.php (On-line).

Hendriksen, H.V., Kornbrust, B.A., Ostergaard, P.R., Stringer, M.A., 2009. Evaluating the potential for enzymatic acrylamide mitigation in a range of food products using an asparaginase from *Aspergillus oryzae*. Journal of Agricultural and Food Chemistry 57, 4168–4176.

Hertog, M.L.A.T., Putz, B., Tijskens, L.M.M., 1997. The effect of harvest time on the accumulation of reducing sugars during storage of potato (*Solanum tuberosum* L.) tubers: experimental data described, using a physiological based, mathematical model. Potato Research 40, 69–78.

Heuser, H., Gerendás, J., Sattelmacher, B., 2005. Einfluss der N- und K-Dünung auf die Gehalte an reduzierenden Zuckern und freien Aminosäuren. Kartoffelbau 56, 308–313.

Hidalgo, F.J., Delgado, R.M., Navarro, J.L., Zamora, R., 2010. Asparagine decarboxylation by lipid oxidation products in model systems. Journal of Agricultural and Food Chemistry 58, 10512–10517.

Hogervorst, J.G., Schouten, L.J., Konings, E.J., Goldbohm, R.A., van den Brandt, P.A., 2007. A prospective study of dietary acrylamide intake and the risk of endometrial ovarian, and breast cancer. Cancer Epidemiology Biomarkers & Prevention 16, 2304–2313.

Hogervorst, J.G., Schouten, L.J., Konings, E.J., Goldbohm, R.A., van den Brandt, P.A., 2008a. Dietary acrylamide intake and the risk of renal cell, bladder, and prostate cancer. American Journal of Clinical Nutrition 87, 1428–1438.

Hogervorst, J.G.F., Baars, B.J., Schouten, L.J., Konings, E.J.M., Goldbohm, R.A., van den Brandt, P.A., 2010. The carcinogenicity of dietary acrylamide intake: a comparative discussion of epidemiological and experimental animal research. Critical Reviews in Toxicology 40, 485–512.

Hogervorst, J.G.F., Schouten, L., Konings, E.J.M., Goldbohm, R.A., van den Brandt, P.A., 2008b. Dietary acrylamide intake is not associated with gastrointestinal cancer risk. Journal of Nutrition 138, 2229–2236.

Hogervorst, J.G.F., Schouten, L.J., Konings, E.J.M., Goldbohm, R.A., van den Brandt, P.A., 2009a. Dietary acrylamide intake and brain cancer risk. Cancer Epidemiology Biomarkers & Prevention 18, 1663–1666.

Hogervorst, J.G.F., Schouten, L.J., Konings, E.J.M., Goldbohm, R.A., van den Brandt, P.A., 2009b. Lung cancer risk in relation to dietary acrylamide intake. Journal of the National Cancer Institute 101, 651–662.

IARC, 1994. Acrylamide. International Agency for Research on Cancer, Lyon, France.

Jin, C., Wu, X., Zhang, Y., 2013. Relationship between antioxidants and acrylamide formation: a review. Food Research International 51, 611–620.

Johnson, K.A., Gorzinski, S.J., Bodner, K.M., Campbell, R.A., Wolf, C.H., Friedman, M.A., et al., 1986. Chronic toxicity and oncogenicity study on acrylamide incorporated in the drinking water of Fischer 344 rats. Toxicology and Applied Pharmacology 85, 154–168.

Jung, M.Y., Choi, D.S., Ju, J.W., 2003. A novel technique for limitation of acrylamide formation in fried and baked corn chips and in french fries. Journal of Food Science 68, 1287–1290.

Khalil, A.H., 1999. Quality of french fried potatoes as influenced by coating with hydrocolloids. Food Chemistry 66, 201–208.

Kim, C.T., Hwang, E.S., Lee, H.J., 2005. Reducing acrylamide in fried snack products by adding amino acids. Journal of Food Science 70, C354–C358.

Kita, A., Brathen, E., Knutsen, S.H., Wicklund, T., 2004. Effective ways of decreasing acrylamide content in potato crisps during processing. Journal of Agricultural and Food Chemistry 52, 7011–7016.

Knutsen, S.H., Dimitrijevic, S., Molteberg, E.L., Segtnan, V.H., Kaaber, L., Wicklund, T., 2009. The influence of variety, agronomical factors and storage on the potential for acrylamide formation in potatoes grown in Norway. Lwt-Food Science and Technology 42, 550–556.

Kolek, E., Simko, P., Simon, P., 2006. Effect of NaCl on the decrease of acrylamide content in a heat-treated model food matrix. Journal of Food and Nutrition Research 45, 17–20.

Konings, E.J.M., Baars, A.J., van Klaveren, J.D., Spanjer, M.C., Rensen, P.M., Hiemstra, M., et al., 2003. Acrylamide exposure from foods of the Dutch population and an assessment of the consequent risks. Food and Chemical Toxicology 41, 1569–1579.

Kotsiou, K., Tasioula-Margari, M., Kukurova, K., Ciesarova, Z., 2010. Impact of oregano and virgin olive oil phenolic compounds on acrylamide content in a model system and fresh potatoes. Food Chemistry 123, 1149–1155.

Kowalczyk, D., GustKaw, W., 2009. Effect of hydrocolloid coatings on quality parameters of french fries. Zywnosc-Nauka Technologia Jakosc 16, 72–80.

Krokida, M.K., Oreopoulou, V., Maroulis, Z.B., Marinos-Kouris, D., 2001. Effect of pre-drying on quality of french fries. Journal of Food Engineering 49, 347–354.

Kumar, D., Singh, B.P., Kumar, P., 2004. An overview of the factors affecting sugar content of potatoes. Annals of Applied Biology 145, 247–256.

Larsson, S.C., Akesson, A., Wolk, A., 2009a. Dietary acrylamide intake and prostate cancer risk in a prospective cohort of Swedish men. Cancer Epidemiology Biomarkers & Prevention 18, 1939–1941.

Larsson, S.C., Akesson, A., Wolk, A., 2009b. Long-term dietary acrylamide intake and breast cancer risk in a prospective cohort of Swedish women. American Journal of Epidemiology 169, 376–381.

Larsson, S.C., Akesson, A., Wolk, A., 2009c. Long-Term dietary acrylamide intake and risk of epithelial ovarian cancer in a prospective cohort of Swedish women. Cancer Epidemiology Biomarkers & Prevention 18, 994–997.

Larsson, S.C., Hakansson, N., Akesson, A., Wolk, A., 2009d. Long-term dietary acrylamide intake and risk of endometrial cancer in a prospective cohort of Swedish women. International Journal of Cancer 124, 1196–1199.

Levine, R.A., Ryan, S.M., 2009. Determining the effect of calcium cations on acrylamide formation in cooked wheat products using a model system. Journal of Agricultural and Food Chemistry 57, 6823–6829.

Lin, Y.L., Lagergren, J., Lu, Y.X., 2011. Dietary acrylamide intake and risk of esophageal cancer in a population-based case-control study in Sweden. International Journal of Cancer 128, 676–681.

Lindsay, R.C., Jang, S.J., 2005. Chemical intervention strategies for substantial suppression of acrylamide formation in fried potato products. In: Friedman, M., Mottram, D. (Eds.), Chemistry and Safety of Acrylamide in Food, 561 ed. Springer, New York, pp. 393–404.

Low, M.Y., Koutsidis, G., Parker, J.K., Elmore, J.S., Dodson, A.T., Mottram, D.S., 2006. Effect of citric acid and glycine addition on acrylamide and flavor in a potato model system. Journal of Agricultural and Food Chemistry 54, 5976–5983.

Lyn-Cook, L.E., Tareke, E., Word, B., Starlard-Davenport, A., Lyn-Cook, B.D., Hammons, G.J., 2011. Food contaminant acrylamide increases expression of Cox-2 and nitric oxide synthase in breast epithelial cells. Toxicology and Industrial Health 27, 11–18.

Márquez, G., Añón, M.C., 1986. Influence of reducing sugars and amino acids in the color development of fried potatoes. Journal of Food Science 51, 157–160.

Matsuura-Endo, C., Ohara-Takada, A., Chuda, Y., Ono, H., Yada, H., Yoshida, M., et al., 2006. Effects of storage temperature on the contents of sugars and free amino acids in tubers from different potato cultivars and acrylamide in chips. Bioscience Biotechnology and Biochemistry 70, 1173–1180.

Matthäus, B., Haase, N.U., Vosmann, K., 2004. Factors affecting the concentration of acrylamide during deep-fat frying of potatoes. European Journal of Lipid Science and Technology 106, 793–801.

Medeiros, R., Mestdagh, F., Van Poucke, C., Van Peteghem, C., De Meulenaer, B., 2012. A two-year investigation towards an effective quality control of incoming potatoes and acrylamide mitigation strategy in French fries. Food Additives and Contaminants: Part A 29 (3), 362–370.

Medeiros, R., Mestdagh, F., De Muer, N., Van Peteghem, C., De Meulenaer, B., 2010. Effective quality control of incoming potatoes as an acrylamide mitigation strategy for the french fries industry. Food Additives and Contaminants Part A-Chemistry Analysis Control Exposure & Risk Assessment 27, 417–425.

Medeiros, R., Mestdagh, F., Van Poucke, C., Kerkaert, B., De Muer, N., Denon, Q., et al., 2011. Implementation of acrylamide mitigation strategies on industrial production of french fries: challenges and pitfalls. Journal of Agricultural and Food Chemistry 59, 898–906.

Medeiros, R., Mestdagh, F., De Meulenaer, B., 2012. Acrylamide formation in fried potato products–present and future, a critical review on mitigation strategies. Food Chemistry 133, 1138–1154.

Mestdagh, F., De Meulenaer, B., Van Peteghem, C., 2007a. Influence of oil degradation on the amounts of acrylamide generated in a model system and in french fries. Food Chemistry 100, 1153–1159.

Mestdagh, F., De Wilde, T., Castelein, P., Németh, O., Van Peteghem, C., De Meulenaer, B., 2008a. Impact of the reducing sugars on the relationship between acrylamide and Maillard browning in french fries. European Food Research and Technology 227, 69–76.

Mestdagh, F., De Wilde, T., Fraselle, S., Govaert, Y., Ooghe, W., Degroodt, J.M., et al., 2008b. Optimization of the blanching process to reduce acrylamide in fried potatoes. Lwt-Food Science and Technology 41, 1648–1654.

Mestdagh, F., Lachat, C., Baert, K., Moons, E., Kolsteren, P., Van Peteghem, C., et al., 2007b. Importance of a canteen lunch on the dietary intake of acrylamide. Molecular Nutrition & Food Research 51, 509–516.

Mestdagh, F., Maertens, J., Cucu, T., Delporte, K., Van Peteghem, C., De Meulenaer, B., 2008c. Impact of additives to lower the formation of acrylamide in a potato model system through pH reduction and other mechanisms. Food Chemistry 107, 26–31.

Mestdagh, F., De Wilde, T., Delporte, K., Van Peteghem, C., De Meulenaer, B., 2008d. Impact of chemical pre-treatments on the acrylamide formation and sensorial quality of potato crisps. Food Chemistry 106, 914–922.

Mestdagh, F., Van Peteghem, C., De Meulenaer, B., 2009. A farm-to-fork approach to lower acrylamide in fried potatoes. In: Yee, N., Bussel, W.T. (Eds.), IV Potatoes. Global Science Books (Food 3, Special issue Second ed., pp. 66–75).

Mestdagh, F.J., De Meulenaer, B., Van Poucke, C., Detavernier, C., Cromphout, C., Van Peteghem, C., 2005. Influence of oil type on the amounts of acrylamide generated in a model system and in french fries. Journal of Agricultural and Food Chemistry 53, 6170–6174.

Mills, C., Mottram, D.S., Wedzicha, B.L., 2009. Acrylamide. In: Stadler, R.H., Lineback, D.R. (Eds.), Process-Induced Food Contaminants–Occurrence, Formation, Mitigation, and Health Risks. Wiley, pp. 23–50.

Mojska, H., Gielecinska, I., Szponar, L., Oltarzewski, M., 2010. Estimation of the dietary acrylamide exposure of the Polish population. Food and Chemical Toxicology 48, 2090–2096.

Moreira, R.G., Castell-Perez, M., Barrufet, M.A., 1999. Deep-Fat Frying: Fundamentals and Applications. Aspen Publishers, Gaithersburg, MD.

Mottram, D.S., Wedzicha, B.L., Dodson, A.T., 2002. Acrylamide is formed in the Maillard reaction. Nature 419, 448–449.

Mucci, L.A., Sandin, S., Balter, K., Adami, H.O., Magnusson, C., Weiderpass, E., 2005. Acrylamide intake and breast cancer risk in Swedish women. Jama-Journal of the American Medical Association 293, 1326–1327.

Noti, A., Biedermann-Brem, S., Biedermann, M., Grob, K., Albisser, P., Realini, P., 2003. Storage of potatoes at low temperature should be avoided to prevent increased acrylamide formation during frying and roasting. Mitteilungen aus Lebensmitteluntersuchung und Hygiene 94, 167–180.

Ohara-Takada, A., Matsuura-Endo, C., Chuda, Y., Ono, H., Yada, H., Yoshida, M., et al., 2005. Change in content of sugars and free amino acids in potato tubers under short-term storage at low temperature and the effect on acrylamide level after frying. Bioscience Biotechnology and Biochemistry 69, 1232–1238.

Olsson, K., Svensson, R., Roslund, C.A., 2004. Tuber components affecting acrylamide formation and colour in fried potato: variation by variety, year, storage temperature and storage time. Journal of the Science of Food and Agriculture 84, 447–458.

Ou, S.Y., Lin, Q., Zhang, Y.P., Huang, C.H., Sun, X., Fu, L., 2008. Reduction of acrylamide formation by selected agents in fried potato crisps on industrial scale. Innovative Food Science & Emerging Technologies 9, 116–121.

Ou, S.Y., Shi, J.J., Huang, C.H., Zhang, G.W., Teng, J.W., Jiang, Y., et al., 2010. Effect of antioxidants on elimination and formation of acrylamide in model reaction systems. Journal of Hazardous Materials 182, 863–868.

Palazoglu, T.K., Savran, D., Gokmen, V., 2010. Effect of cooking method (Baking compared with frying) on acrylamide level of potato chips. Journal of Food Science 75, E25–E29.

Park, Y.W., Yang, H.W., Storkson, J.M., Albright, K.J., Liu, W., Lindsay, R.C., et al., 2005. Controlling acrylamide in french fry and potato chip models and a mathematical model of acrylamide formation–acrylamide: acidulants, phytate and calcium. In: Friedman, M., Mottram, D. (Eds.), Chemistry and Safety of Acrylamide in Food, 561 ed. Springer, New York, pp. 343–356.

Parker, J.K., Balagiannis, D.P., Higley, J., Smith, G., Wedzicha, B.L., Mottram, D.S., 2012. Kinetic model for the formation of acrylamide during the finish-frying of commercial french fries. Journal of Agricultural and Food Chemistry 60, 9321–9331.

Pedersen, G.S., Hogervorst, J.G.F., Schouten, L.J., Konings, E.J.M., Goldbohm, R.A., van den Brandt, P.A., 2010. Dietary acrylamide intake and estrogen and progesterone receptor-defined postmenopausal breast cancer risk. Breast Cancer Research and Treatment 122, 199–210.

Pedreschi, F., Bustos, O., Mery, D., Moyano, P., Kaack, K., Granby, K., 2007a. Color kinetics and acrylamide formation in NaCl soaked potato chips. Journal of Food Engineering 79, 989–997.

Pedreschi, F., Granby, K., Risum, J., 2010. Acrylamide mitigation in potato chips by using NaCl. Food and Bioprocess Technology 3, 917–921.

Pedreschi, F., Kaack, K., Granby, K., 2004. Reduction of acrylamide formation in potato slices during frying. LWT–Food Science and Technology 37, 679–685.

Pedreschi, F., Kaack, K., Granby, K., 2006. Acrylamide content and color development in fried potato strips. Food Research International 39, 40–46.

Pedreschi, F., Kaack, K., Granby, K., 2008. The effect of asparaginase on acrylamide formation in french fries. Food Chemistry 109, 386–392.

Pedreschi, F., Leon, J., Mery, D., Moyano, P., Pedreschi, R., Kaack, K., et al., 2007b. Color development and acrylamide content of pre-dried potato chips. Journal of Food Engineering 79, 786–793.

Pedreschi, F., Kaack, K., Granby, K., Troncoso, E., 2007c. Acrylamide reduction under different pre-treatments in french fries. Journal of Food Engineering 79, 1287–1294.

Pedreschi, F., Mariotti, S., Granby, K., Risum, J., July 2011. Acrylamide reduction in potato chips by using commercial asparaginase in combination with conventional blanching. LWT - Food Science and Technology 44 (6), 1473–1476.

Pedreschi, F., Moyano, P., Kaack, K., Granby, K., 2005. Color changes and acrylamide formation in fried potato slices. Food Research International 38, 1–9.

Pedreschi, F., Travisany, X., Reyes, C., Troncoso, E., Pedreschi, R., 2009. Kinetics of extraction of reducing sugar during blanching of potato slices. Journal of Food Engineering 91, 443–447.

Pelucchi, C., Galeone, C., Levi, F., Negri, E., Franceschi, S., Talamini, R., et al., 2006. Dietary acrylamide and human cancer. International Journal of Cancer 118, 467–471.

Reynolds, T., 2002. Acrylamide and cancer: tunnel leak in Sweden prompted studies. Journal of the National Cancer Institute 94, 876–878.

Rice, J.M., 2005. The carcinogenicity of acrylamide. Mutation Research-Genetic Toxicology and Environmental Mutagenesis 580, 3–20.

Roach, J.A.G., Andrzejewski, D., Gay, M.L., Nortrup, D., Musser, S.M., 2003. Rugged LC-MS/MS survey analysis for acrylamide in foods. Journal of Agricultural and Food Chemistry 51, 7547–7554.

Rommens, C.M., Yan, H., Swords, K., Richael, C., Ye, J.S., 2008. Low-acrylamide french fries and potato chips. Plant Biotechnology Journal 6, 843–853.

Rommens, C.M., Ye, J., Richael, C., Swords, K., 2006. Improving potato storage and processing characteristics through all-native DNA transformation. Journal of Agricultural and Food Chemistry 54, 9882–9887.

Rydberg, P., Eriksson, S., Tareke, E., Karlsson, P., Ehrenberg, L., Törnqvist, M., 2003. Investigations of factors that influence the acrylamide content of heated foodstuffs. Journal of Agricultural and Food Chemistry 51, 7012–7018.

Sanny, M., Luning, P.A., Marcelis, W.J., Jinap, S., van Boekel, M.A.J.S., 2010. Impact of control behaviour on unacceptable variation in acrylamide in french fries. Trends in Food Science & Technology 21, 256–267.

Schouten, L.J., Hogervorst, J.G.F., Konings, E.J.M., Goldbohm, R.A., van den Brandt, P.A., 2009. Dietary acrylamide intake and the risk of head-neck and thyroid cancers: results from the Netherlands cohort study. American Journal of Epidemiology 170, 873–884.

Seal, C.J., de Mul, A., Eisenbrand, G., Haverkort, A.J., Franke, K., Lalljie, S.P.D., et al., 2008. Risk-benefit considerations of mitigation measures on acrylamide content of foods–a case study on potatoes, cereals and coffee. British Journal of Nutrition 99, S1–+.

Serpen, A., Gokmen, V., 2009. Evaluation of the Maillard reaction in potato crisps by acrylamide, antioxidant capacity and color. Journal of Food Composition and Analysis 22, 589–595.

Shipp, A., Lawrence, G., Gentry, R., McDonald, T., Bartow, H., Bounds, J., et al., 2006. Acrylamide: review of toxicity data and dose-response analyses for cancer and noncancer effects. Critical Reviews in Toxicology 36, 481–608.

Smith, C.J., Perfetti, T.R., Rumple, M.A., Rodgman, A., Doolittle, D.J., 2000. "IARC group 2A Carcinogens" reported in cigarette mainstream smoke. Food and Chemical Toxicology 38, 371–383.

Stadler, R.H., Blank, I., Varga, N., Robert, F., Hau, J., Guy, P.A., et al., 2002. Acrylamide from Maillard reaction products. Nature 419, 449–450.

Stadler, R.H., Robert, F., Riediker, S., Varga, N., Davidek, T., Devaud, S., et al., 2004. In-depth mechanistic study on the formation of acrylamide and other vinylogous compounds by the Maillard reaction. Journal of Agricultural and Food Chemistry 52, 5550–5558.

Stadler, R.H., Scholz, G., 2004. Acrylamide: an update on current knowledge in analysis, levels in food, mechanisms of formation, and potential strategies of control. Nutrition Reviews 62, 449–467.

Svensson, K., Abramsson, L., Becker, W., Glynn, A., Hellenäs, K.E., Lind, Y., et al., 2003. Dietary intake of acrylamide in Sweden. Food and Chemical Toxicology 41, 1581–1586.

Swedish National Food Administration, 2002. Swedish National Food Administration: Analysis of Acrylamide in Food. http://www.slv.se/acrylamide.

Tareke, E., Rydberg, P., Karlsson, P., Eriksson, S., Törnqvist, M., 2000. Acrylamide: a cooking carcinogen? Chemical Research in Toxicology 13, 517–522.

Torres, M.D.A., Parreño, W.C., 2009. Thermal processing and quality optimization. In: Singh, J., Kaur, L. (Eds.), Advances in Potato Chemistry and Technology, first ed. Elsevier, pp. 163–219.

Urbancic, S., Kolar, M.H., Dimitrijevic, D., Demsar, L., Vidrih, R., 2014. Stabilisation of sunflower oil and reduction of acrylamide formation of potato with rosemary extract during deep-fat frying. LWT–Food Science and Technology 57, 671–678.

US 2005/0214411 A1, 2005. Methods for Surpressing Acrylamide Formation and Restoring Browned Color and Flavor (United States).

US 2010/0080872 A1, 2010. Low Acrylamide French Fry and Preparation Process (United States).

Viklund, G., Mendoza, F., Sjöholm, I., Skog, K., 2007. An experimental set-up for studying acrylamide formation in potato crisps. LWT–Food Science and Technology 40, 1066–1071.

Viklund, G., Olsson, K., Sjoholm, I., Skog, K., 2008. Variety and storage conditions affect the precursor content and amount of acrylamide in potato crisps. Journal of the Science of Food and Agriculture 88, 305–312.

Viklund, G.A.I., Olsson, K.M., Sjoholm, I.M., Skog, K.I., 2010. Acrylamide in crisps: effect of blanching studied on long-term stored potato clones. Journal of Food Composition and Analysis 23, 194–198.

Whittaker, A., Marotti, I., Dinelli, G., Calamai, L., Romagnoli, S., Manzelli, M., et al., 2010. The influence of tuber mineral element composition as a function of geographical location on acrylamide formation in different Italian potato genotypes. Journal of the Science of Food and Agriculture 90, 1968–1976.

Wilson, K., Rimm, E., Thompson, K., Mucci, L., 2006. Dietary acrylamide and cancer risk in humans: a review. Journal für Verbraucherschutz und Lebensmittelsicherheit 1, 19–27.

Wilson, K.M., Balter, K., Adami, H.O., Gronberg, H., Vikstrom, A.C., Paulsson, B., et al., 2009a. Acrylamide exposure measured by food frequency questionnaire and hemoglobin adduct levels and prostate cancer risk in the Cancer of the Prostate in Sweden Study. International Journal of Cancer 124, 2384–2390.

Wilson, K.M., Mucci, L.A., Cho, E., Hunter, D.J., Chen, W.Y., Willett, W.C., 2009b. Dietary acrylamide intake and risk of premenopausal breast cancer. American Journal of Epidemiology 169, 954–961.

Wilson, K.M., Mucci, L.A., Rosner, B.A., Willett, W.C., 2010. A prospective study on dietary acrylamide intake and the risk for breast, endometrial, and ovarian cancers. Cancer Epidemiology Biomarkers & Prevention 19, 2503–2515.

Yasuhara, A., Tanaka, Y., Hengel, M., Shibamoto, T., 2003. Gas chromatographic investigation of acrylamide formation in browning model systems. Journal of Agricultural and Food Chemistry 51, 3999–4003.

Yaylayan, V.A., Locas, C.P., Wnorowski, A., O'Brien, J., 2005. Mechanistic pathways of formation of acrylamide from different amino acids. In: Friedman, M., Mottram, D. (Eds.), Chemistry and Safety of Acrylamide in Food, 561 ed. Springer, New York, pp. 191–203.

Yaylayan, V.A., Wnorowski, A., Locas, C.P., 2003. Why asparagine needs carbohydrates to generate acrylamide. Journal of Agricultural and Food Chemistry 51, 1753–1757.

Zamora, R., Delgado, R.M., Hidalgo, F.J., 2009. Conversion of 3-aminopropionamide and 3-alkylaminopropionamides into acrylamide in model systems. Molecular Nutrition & Food Research 53, 1512–1520.

Zamora, R., Hidalgo, F.J., 2008. Contribution of lipid oxidation products to acrylamide formation in model systems. Journal of Agricultural and Food Chemistry 56, 6075–6080.

Zeng, X.H., Cheng, K.W., Du, Y.G., Kong, R., Lo, C., Chu, I.K., et al., 2010. Activities of hydrocolloids as inhibitors of acrylamide formation in model systems and fried potato strips. Food Chemistry 121, 424–428.

Zhang, Y., Chen, J., Zhang, X., Wu, X., Zhang, Y., 2007. Addition of antioxidant of bamboo leaves (AOB) effectively reduces acrylamide formation in potato crisps and french fries. Journal of Agricultural and Food Chemistry 55, 523–528.

Zhu, F., Cai, Y.Z., Ke, J., Corke, H., 2010. Compositions of phenolic compounds, amino acids and reducing sugars in commercial potato varieties and their effects on acrylamide formation. Journal of the Science of Food and Agriculture 90, 2254–2262.

Zhu, F., Cai, Y.Z., Ke, J.X., Corke, H., 2009. Evaluation of the effect of plant extracts and phenolic compounds on reduction of acrylamide in an asparagine/glucose model system by RP-HPLC-DAD. Journal of the Science of Food and Agriculture 89, 1674–1681.

Zyzak, D.V., Sanders, R.A., Stojanovic, M., Tallmadge, D.H., Eberhart, B.L., Ewald, D.K., et al., 2003. Acrylamide formation mechanism in heated foods. Journal of Agricultural and Food Chemistry 51, 4782–4787.

CHAPTER 19

Advanced Analytical Techniques for Quality Evaluation of Potato and Its Products

Carmen Jarén, Ainara López, Silvia Arazuri

Department of Agricultural Projects and Engineering, Universidad Pública de Navarra, Pamplona, Navarra, Spain

1. Introduction

Potato (*Solanum tuberosum* L.) is considered one of the main food products worldwide. In 2013, its production exceeded 368 million tonnes according to the Food and Agriculture Organization of the United Nations (FAOSTAT, 2013). In addition, these tubers represent a staple crop in many countries around the world today (Alva et al., 2011).

Freshly harvested raw tubers contain about 80% water and 20% dry matter. About 60% to 80% of the dry matter is starch, a carbohydrate able to supply high levels of energy. On a dry weight basis, the protein content of potato is similar to that of cereals and is very high in comparison with other roots and tubers. These tubers are also rich in calcium, potassium, and vitamin C and have an especially good amino acid balance.

Potato is a versatile food highly popular worldwide, prepared and served in a variety of ways: cooked, fried, dehydrated, or as an ingredient in industrial prepared foods.

Because potato products are highly consumed globally, the food industry is challenged to meet the current demand of quality products by consumers and worldwide standards. To meet this challenge, the potato manufacturing industry is relying on the adoption of increasingly advanced and sustainable techniques to determine the quality of its products. Development in related technology areas such as computing, optical devices, and miniaturization combined with a considerable decrease in prices have made possible actual production improvements.

In this chapter, a detailed review of the advanced analytical techniques to evaluate the quality of potatoes and potato products is given.

Advances in Potato Chemistry and Technology. http://dx.doi.org/10.1016/B978-0-12-800002-1.00019-4

2. Analytical Techniques

2.1 Dry Matter and Specific Gravity

Dry matter (DM) represents approximately 20% of raw potato tubers. According to some authors, DM is a suitable index for determining potato quality (Subedi and Walsh, 2009).

The most commonly used method to estimate DM content of potato tubers is indirectly, by calculating the specific gravity (SG) (Storey and Davies, 1992). SG is a practical method for determining DM, since these two parameters are known to be highly correlated (Wilson and Lindsay, 1969). According to many authors SG could be used to directly measure quality characteristics of potatoes because of its direct relationship to cooking quality of tubers (Bewell, 1937; Clark et al., 1940; Smith and Nash, 1940). SG can be calculated using Eqn (1):

$$SG = \frac{\text{weight in air}}{\text{weight in air-weight in water}} \tag{1}$$

In this method, samples are suspended from a load cell and are first weighed in air and then under water. After these measurements, the DM can be calculated using Eqn (2) (Lunden, 1956):

$$DM = 215.73 \, (SG - 0.9825) \tag{2}$$

A hydrometer is also a useful method to determine DM in potatoes. However, this method has an inconvenience given that the tubers need to be trimmed to a specific weight. Then, tubers of known weight are placed in a basket with an attached hydrometer and suspended in water. The graduated stem of the hydrometer gives the measure of DM at the point where the water level crosses the scale.

Other authors describe a computer-based method to determine potato SG (Tai et al., 1985). This method is called "grav-o-tater" and presents an advantage to the use of the hydrometer since it is capable of assessing whole tubers.

Another method to estimate DM in potatoes consists of analyzing the difference between the weight of potato samples before freeze-drying and after. To accomplish freeze-drying of samples, tubers should be cut lengthwise to obtain representative samples of the different tissues. Then, samples must be introduced into a freeze-dryer, for example, an Alpha d1-4 (Christ, Osterode, Germany), at −50 °C and 0 atm until 250 g of fresh weight is achieved. Finally, when the samples have lost all their water content, they should be ground with liquid nitrogen to fine dust and stored at −20 °C before analyses (López et al., 2013b).

The need for a nondestructive technique for determining DM in potatoes has led to numerous studies in past years employing near-infrared (NIR) spectroscopy (NIRS). Some authors have estimated DM by using NIR techniques in potato tubers (Haase, 2003; Hartmann and Büning-Pfaue, 1998; Helgerud et al., 2012) and potato chips (Pedreschi et al., 2010).

2.2 Starch

Starch is the major carbohydrate present in potato, representing about 60–80% of the DM of the tubers (Anon, 2008). Starch plays a major role in human nutrition since it supplies high levels of energy (Woolfe and Poats, 1987). It is principally composed of two polymers of glucose: amylose, an essentially linear molecule, and amylopectin, a highly branched molecule (Ellis et al., 1998). Starch can be classified into three types according to its in vitro digestion: rapidly digestible starch (RDS), slowly digestible starch (SDS), and resistant starch (RS) (Rosin et al., 2002). RDS increases blood glucose levels after ingestion, SDS is completely digested in the small intestine, while RS cannot be digested in the small intestine but is fermented in the large intestine. According to some authors, SDS and RS have significant implications for human health (Englyst et al., 1992; Lehmann and Robin, 2007). Therefore, it becomes crucial to study the physicochemical and structural properties of potato starch and the amylose/amylopectin ratio that influences the starch retrogradation (Sievert and Pomeranz, 1989), for quality control of potatoes and its products.

To study its structure and properties, starch must be extracted from potato. Potato starch can be isolated from potatoes by soaking in distilled water, followed by disintegrating in a centrifugal juice extractor and finishing by either centrifugation or filtration (Liu et al., 2002). The amylose content of starch can be determined by various methods. Some authors have opted to determine amylose content by a colorimetric procedure according to the blue value (Morrison and Laignelet, 1983), defined as the absorbance of 10 mg of anhydrous starch in 100 ml of diluted I_2–KI solution at 635 nm. The amylose percentage is calculated according to Eqn (3):

$$\text{Amylose } (\%) = (28.414 \times \text{Blue value}) - 6.218 \tag{3}$$

Other methods are based on differential scanning calorimetry (DSC) (see Section 4.2) and size-exclusion chromatography (SEC) (Colonna and Mercier, 1984). In the case of amylopectin, two techniques are generally used to estimate the chain length and distribution of this polysaccharide: SEC and high-performance anion-exchange chromatography with pulsed amperometric detection (Hizukuri, 1986; Wong and Jane, 1997).

In 2012, Lu et al. studied the synergistic effects of amylose and phosphorus on the gel properties and in vitro starch digestibility of potato starch. These authors selected potato starches with variable contents of amylose and phosphorus. Potato tubers were soaked in distilled water containing 20 mM sodium bisulfite and 10 mM citric acid for 2 h and disintegrated using a centrifugal juice extractor, and the starch milk was collected. The milk was allowed to sediment for a minimum of 30 min and after the suspended solids were removed by decantation, the starch sediment was resuspended in water. The starch granules were recovered by vacuum filtration and dried at ambient temperature. The starch was then packed in air-tight plastic bags and stored at room temperature. The amylose content of the starch was determined by iodine colorimetry according to Williams et al. (1970), and the total phosphorus content was determined using a Varian Vista Pro ICP-OES (inductively coupled plasma-optical emission spectrometer; Mississauga, ON, Canada). Thermal analyses for gelatinization and retrogradation of starches were performed using a differential scanning calorimeter (2920 Modulated DSC, TA Instruments, New Castle, DE, USA). Infrared spectra of the potato starch and dry gel powder were recorded on a Digilab FTS 7000 spectrometer (Digilab USA, Randolph, MA, USA) equipped with a thermoelectrically cooled deuterated triglycine sulfate (DTGS) detector at a resolution of $4\,cm^{-1}$ by 128 scans.

The authors found that phosphorus and amylose contents in potato starch were highly correlated to some rheological, thermal, structural, and nutritional properties of potato starch and its concentrated starch gel. At a high level of amylose content, higher phosphorus content in potato starch enhanced recrystallization of amylopectin during retrogradation of the starch gel. Phosphorus content was positively correlated to the RS content in raw starch, as well as the SDS in starch gel. On the other hand, increases in amylose content showed less effect than phosphorus on the gelation and RS, RDS, and SDS fractions of potato starch.

More recently, Schimer et al. (2013) carried out a study to characterize and compare the morphological and physicochemical properties of starches of cereal and tuber origin. The authors studied the influence of the amylose/amylopectin (AM/AP) content on the characteristics of various starches. According to Tester et al. (2004) regular starches contain approximately 70–80% AP and 20–30% AM, waxy starches less than 10% AM, and high-AM starches more than 40% AM. In this study, the AM and AP content was determined using a Megazyme kit (Megazyme International Ireland Ltd). This method is based on the specific formation and precipitation of AP–concanavalin A complexes, which occur after a sample is treated to solubilize RS and remove lipids and free D-glucose. The pasting properties of starch samples were determined by a Rapid Visco analyzer (Newport Scientific Pty Ltd, Warriewood, Australia). Finally, starch thermal characteristics were analyzed by DSC. The results obtained showed significant differences in the morphological and physicochemical characteristics of starches with varying AM/AP contents. The waxy starches were found to have a higher gelatinization temperature and enthalpy, which indicates a higher crystalline and molecular order.

2.3 Protein

The protein content of potatoes is considerably smaller than that of DM and starch, ranging between 0.5% and 2% of the fresh weight (López et al., 2013a). Protein is commonly determined using the Kjeldahl method, which is the official reference method to determine crude protein (CP). The Kjeldahl method first estimates total nitrogen content of the samples by following three basic steps: (1) Digestion: this step consists of digesting the sample in sulfuric acid with a catalyst, which helps in the conversion of the amine nitrogen to ammonium ions. (2) Distillation: separating the ammonia (nitrogen) from the digestion mixture. In this step, the ammonium ions are converted into ammonia gas, heated, and distilled. The ammonia gas is directed into a trapping solution where it dissolves and becomes an ammonium ion again. (3) Titration: the last step consists of quantifying the ammonia by titration with a standard solution. Once the total nitrogen of samples is calculated, the CP content is determined using Eqn (4):

$$CP\ (\%) = Nitrogen\ (\%) \times 6.25 \tag{4}$$

A conversion factor is used to convert percentage of nitrogen into percentage of CP. Since most of the proteins typically contain 16% nitrogen, a conversion factor of 6.25 is used ($100/16 = 6.5$).

Some authors have used a nitrogen analyzer such as the Thermoquest CE Instrument (NA 2100 Protein, Thermo-Quest Italia S.P.A., Ann Arbor, MI, USA) to estimate protein content of the DM of potatoes (Chung et al., 2014; Lu et al., 2011).

Other authors have studied the estimation of protein content of potatoes by NIRS. They realized that that was a challenging mission owing to the low content of this compound in potatoes and the high error of the reference method (Fernández-Ahumada et al., 2006). Despite this, good predictions of coagulating protein were achieved (Brunt et al., 2010). Lower correlations were found for crude and recoverable protein (Brunt et al., 2010; Haase, 2006).

2.4 Sugars

The major sugars in potato are sucrose, fructose, and glucose. They are present in small concentrations of around 1% in potato tubers (López et al., 2013a). However, sugars play a main role in potato processing, since high levels of reducing sugars (fructose and glucose) will lead to browning during frying of tubers, an undesirable attribute for consumers (Mehrubeoglu and Cote, 1997). Moreover, the reaction of reducing sugars with asparagines via the Maillard reaction at high temperatures generates acrylamide (Chen et al., 2010) (see Section 2.5). Storage and lower temperatures influence in an increase in the accumulation of sugars in the tubers.

An upper limit of reducing sugars for potato chip production has not been specified; however, varieties with less than 3 g/kg of fresh weight are recommended by CIAA (2009). Despite this, in some countries the preferred limit of reducing sugars for potato chip producers has been decreased to 0.5 g/kg (Foot et al., 2007).

Traditional analyses for determining sugar content of potatoes are generally accomplished by high-performance liquid chromatography (HPLC). To determine sugars by HPLC, the flesh of the potato tuber samples is cut and mashed by a mixer. The mixture is then centrifuged at 3000 rpm for 30 min, diluted 1:5 by adding distilled water, and subsequently passed through a 0.45-μm pore-size filter. This juice is later used for sugar analysis. The sugar composition (fructose, glucose, sorbitol, sucrose) of the potato juice is analyzed using an HPLC (Shimadzu, SCL-10A VP) apparatus with a refractive index detector (Shimadzu, RID10A). Sample mixtures (10 μl) are separated on a Shodex Asahipak NH2P-50 4A column at 40 °C using an acetonitrile/water (3/1 v/v) solution as the eluate (Chen et al., 2010).

As previously described for other potato components, some authors have studied the estimation of sugars in potatoes by NIRS as a nondestructive alternative over traditional techniques. In 1997, Mehrubeoglu and Cote were able to accurately predict the total reducing sugars in slices of a single potato type by NIR. Later, in a similar study carried out by Scanlon et al. (1999), it was not possible to determine the sugars of potato coins by NIRS. More recently, in an investigation completed by Chen et al. (2010), good correlation coefficients were obtained for fructose and glucose of intact potatoes estimated by NIRS in the wavelength range 400–1100 nm.

2.5 Acrylamide

Acrylamide is a chemical compound classified as a probable human carcinogen by the International Agency for Research on Cancer (IARC, 1994). In 2002, public awareness rose importantly regarding this compound since its presence was reported in many fried and oven-cooked products (Rufián-Henares and Morales, 2006). High concentrations of acrylamide were found in potato chips (U.S. terminology) and french fries (Amrein et al., 2003). Acrylamide formation occurs in the Maillard reaction of reducing sugars with asparagine at temperatures higher than 120 °C (Gökmen et al., 2005).

According to many authors, the acrylamide content of foods is not easy to detect (Jezussek and Schieberle, 2003), mainly because of its high water solubility and reactivity (Mottram et al., 2002). As of this writing, the analytical methods available for determining acrylamide in foods are based on HPLC or gas chromatography (GC) techniques coupled with mass spectrometry (MS) (Ahn et al., 2002; Gertz and Klostermann, 2002; Roach et al., 2003; Rosén and Hellenäs, 2002).

Some authors reported a few drawbacks of those techniques in terms of time consumption and difficulty of performance (Fernandes and Soares, 2007), and therefore, they searched for alternative methods to determine acrylamide in potato-based products. In 2007, Fernandes and Soares proposed an alternative sample preparation technique based on matrix solid-phase dispersion (MSPD) for determining acrylamide in potato chips. For this purpose, an aliquot of 0.5 g of a previously ground potato chip sample, 2 g of C_{18} sorbent, and 0.25 μg of working standard solution ($[^{13}C_3]$acrylamide, 99% purity), were blended until complete disruption and dispersion of the sample in the solid support was achieved. Then, the sample was packed into an MSPD column between two polyethylene frits, placed in a vacuum manifold, and defatted with 20 ml of *n*-hexane. Vacuum aspiration was then applied for drying and after, the crylamide was extracted with 4 ml of water applied twice. The obtained extract was brominated and analyzed by GC–MS. The authors compared this method with liquid extraction–GC–MS and obtained similar results. They stated that this new method was simpler and easier to implement.

Other authors (Gökmen et al., 2005), investigated a method using LC coupled to diode array detection (DAD) for determination of acrylamide in potato chips and french fries at low levels. In this method, potato chip samples were ground, suspended in 10 ml of methanol, and homogenized. The suspension was then centrifuged and the clear supernatant treated with Carrez I and II solutions for the coextractives to precipitate. The clear supernatant was placed in a water bath at 40 °C and allowed to dry. The remaining residue was dissolved in water. The acrylamide residue was completely transferred into water by changing the solvent. Then, a solid-phase extraction cleanup was done and 1 ml of the extract was passed through a cartridge (1 drop/s). Finally, 20 µl of the final solution was used for the quantification of acrylamide. This quantification was performed by an HPLC system equipped with a DAD and by LC–MS analysis for confirmation. The authors reported good results by using this new and low-cost technique.

Other authors have focused in reducing the acrylamide content in potato chips and french fries by adding amino acids (Kim et al., 2005), antioxidants of bamboo leaves (Zhang et al., 2007), or glycine (Bråthen et al., 2005). With the same objective, Granda and Moreira investigated the formation of acrylamide during vacuum frying of potato chips compared to traditional frying. They achieved a 94% decrease in acrylamide content when samples were fried under vacuum (Granda and Moreira, 2005).

2.6 Phenolic Compounds and Carotenoids

Carotenoids are lipophilic compounds synthesized in plastids from isoprenoids (DellaPenna and Pogson, 2006). These compounds are of considerable interest because of their antioxidant activity in humans (Fraser and Bramley, 2004). Major carotenoids present in potatoes are lutein, zeaxanthin, violaxanthin, and neoxanthin, β-Carotene can be found in trace amounts (Brown, 2008). Potatoes, especially intensely colored ones, are an important source of carotenoids for the human diet (Lu et al., 2001).

Phenolic compounds are also gaining interest mainly because of their antioxidant nature, which is said to have health benefits (Manach et al., 2004). High concentrations of these compounds are contained in potatoes (Al Saikhan et al., 1995). Phenolic content varies considerably among different types of potatoes. Thus, red- and purple-fleshed varieties present around double the concentration of phenolic acids compared to white-fleshed ones (Ezekiel et al., 2013).

Current analytical methods to determine polyphenols in potatoes include gas–liquid chromatography, thin-layer chromatography, HPLC, and ultraviolet (UV)–visible (VIS) spectrophotometry (Friedman, 1997).

Before analysis by those methods, phenolic compounds have to be extracted and purified. According to Voigt and Noske (1964) an optimized extraction of chlorogenic acid from potatoes is achieved with the aid of methanol, ethanol, and acetone, in that order. Tisza et al. (1996) quantified chlorogenic, citric, malic, and caffeic acids in freeze-dried potato samples by GC–MS. They found that chlorogenic acid content correlated with values obtained by UV spectroscopy. More recently, Kvasnička et al. (2008) studied the separation and determination of 12 phenolic acids of potatoes by HPLC and capillary zone electrophoresis (CZE). A mixture of 50 g of potato and 150 ml of methanol with 0.02% butylhydroxytoluene was placed in an ultrasonic bath for 15 min and cooled at 20 °C, and then methanol was added up to the volume. The filtrate was analyzed by HPLC and CZE. CZE was performed in a hydrodynamically open system using a 365-mm fused silica capillary. Chromatographic separation by HPLC was performed on an Eclipse XBD C8 column (150×4.6 mm, 3.5 μm, Agilent Technologies, Santa Clara, CA, USA) at 30 °C. Both methods gave similar results. The authors stated that CZE presented advantages over HPLC since lower consumption of a water-based electrolyte and lower running costs were achieved.

Im et al. (2008) analyzed the phenolic content of plant flowers, stems, and tubers in home-processed potatoes by HPLC and LC–MS. HPLC analysis was accomplished in a Hitachi liquid chromatograph Model 665-II equipped with a UV–Vis detector (SPD-10Avp, Kyoto, Japan) set at 280 and 340 nm. LC–MS analysis was carried out in an ion trap mass spectrometer (LCQ, Thermo Fisher Scientific, Waltham, MA, USA) equipped with an HPLC system connected to a DAD. The authors were able to determine the content and distribution of phenolic compounds in potato flowers, leaves, and tubers.

In 2008, Barba et al. investigated the influence of various baking treatments (boiling and microwaving) on the amount of phenolic marker compounds and tryptophan left over in potatoes of the Agria cultivar. HPLC with DAD was used to identify and quantify phenolic compounds and tryptophan. Baking treatments included boiling in water of peeled and unpeeled tubers and microwaving at various powers between 300 and 1000 W. The dielectric behavior of the irradiated samples was also studied to investigate the feasible application of microwave treatments in terms of nutritional factor preservation. According to the results obtained by this study, optimal microwave baking was achieved at 500 W in terms of short time, reduced water, and phenolic losses. The study of the dielectric behavior of potatoes showed that water plays a crucial role in keeping lossy features, avoiding thermal damage, and promoting starch gelatinization.

Determination of phenolic compounds of fruit and vegetables by NIRS has been broadly studied, whereas for potatoes few studies can be found. López et al. (2014) investigated the prediction of total phenolic content (TPHEN) in potatoes by NIRS. TPHEN was determined by an adapted microscale protocol for the Fast Blue BB spectrophotometric method (Lester et al., 2012; Medina, 2011) using gallic acid as standard. The range of the samples covered by the study comprised yellow-, red-, and purple-fleshed potatoes. High correlations were found, indicating that NIRS was a technology capable of predicting the TPHEN content of that set of potatoes with good accuracy.

The carotenoid content of potatoes has also been investigated generally, in addition to phenolic content. Murniece et al. (2014) carried out a study to determine TPHEN and carotenoid content in relationship to the color of organically and conventionally grown potatoes with different flesh colors. Carotenoids were analyzed by the spectrophotometric method with a UV–Vis spectrophotometer (Jenway 6705) at 440 nm. TPHEN was determined by the Folin–Ciocalteu spectrophotometric method (Singleton et al., 1999) with some modifications. The color of the samples was measured by a Color Tec-PCM device (USA) in at least seven locations on the samples. The authors found a correlation between TPHEN and carotenoid content with color, L^* ($R > 0.8$). Variety was found to be the most significant factor in the changes in TPHEN and carotenoid content. The same authors investigated TPHEN and carotenoid content of organically and conventionally cultivated varieties before and after a period of storage. They found a significant influence on carotenoid content between organic and conventional tubers and among varieties. TPHEN content varied significantly among varieties and by storage condition.

3. Quality Evaluation

Quality is the totality of characteristics of an entity (product, service, process, activity, system, organization, physical person) that bear on its ability to satisfy stated and implied needs (ISO 9000:2000 Quality Management Systems).

Quality evaluation of potatoes can be accomplished either by texture analyses or by sensory tests. Textural properties of potatoes are some of the most important sensorial attributes and also have an influence on others, such as appearance and taste. Color is also considered an essential attribute since is the first characteristic that consumers take into account to evaluate the quality of many products. Undoubtedly, an optimal texture and color of potatoes will lead to a better consumer acceptance of the product. However, a lot of time, expense, and resources are employed in training taste panels to assess the food as objectively as possible. On the other hand, sensory evaluation is considered to be an inefficient means of texture evaluation for quality assurance purposes (Essex, 1969; Ross and Scanlon, 2004; Ross and Porter, 1969). This situation has motivated the development and widespread use of empirical mechanical tests that correlate with sensory analysis of potato texture. Over the years, a wide range of instrumental tests have been used in both research and industry to assess potato texture. Often the choice of any particular instrument or analytical procedure depends on cost and availability of expertise within the organization.

3.1 Texture

Texture is one of the four principal quality factors in foods (Bourne, 2002; Chen and Opara, 2013), along with appearance, flavor, and nutrition. Food texture is defined by Bourne as the properties of a food that are that group of physical characteristics arising from the structural elements of the food, primarily sensed by the feeling of touch, related to the deformation, disintegration, and flow of the food under a force, and measured objectively by functions of mass, time, and distance. Given that the potato is one of the most important foods in the human diet, we can assume this definition of Bourne as valid for the potato texture and its products. In general, texture is an essential factor in the consumer's perception of the quality of potatoes (García-Segovia et al., 2008).

The texture of potatoes is determined by several mutually dependent factors. The genetically determined characteristics of the fresh product (Ducreux et al., 2008), the structural and biochemical properties of tuber tissue, the chemical composition (dry solids content, SG, starch, AM, sugars, proteins, and total nitrogen), and both the agronomic and the storage conditions (Kaur et al., 2002; Liu et al., 2009).

Potato processing generates products such as table stock (31%), frozen french fries (30%), chips (12%), and dehydrated items (12%) (Miranda and Aguilera, 2006). Therefore, the texture of potatoes and their products (cooked potatoes, chips, dehydrated, dry, starch) has been the subject of study of many research papers with various objectives. Potato texture was studied to differentiate varieties (Kaur et al., 2002) and cultivars (Sadowska et al., 2008), compare cultivation techniques, assess mechanical damage (Singh et al., 2013), manage bruising (Baritelle and Hyde, 2001), and research adaptation to mechanical harvest (Geyer et al., 2009), harvesting damage (Bentini et al., 2006), storage techniques (Amaral et al., 2014; Kumar et al., 2007; Mahto and Das, 2014, 2015), time of storage (Bentini et al., 2009), storage of plant tissue with several high-pressure treatments (Luscher et al., 2005), high-pressure–low-temperature processes (Nath et al., 2007; Urrutia-Benet et al., 2007), etc.

French fries are one of the most studied products regarding various aspects of their texture. Firmness (Sundara et al., 2001), fracture properties of the crust with different fry times (Ross and Scanlon, 2004), oil absorption and its kinetics (Pedreschi and Moyano, 2005), the influence of various pretreatments and drying temperature on low-pressure superheated steam drying (Pimpaporn et al., 2007), drying kinetics (Leeratanarak et al., 2006), and frying medium and temperature effects on fat content (Kita et al., 2007) are some of the studies carried out.

Many food scientists, engineers, and technologists have evaluated potato chip texture to know the effects of vacuum microwaving (Song et al., 2007), the effects of type of commercial potato chip on different characteristics (Salvador et al., 2009), the crispness (Srisawas and Jindal, 2003; Taniwaki and Kohyama, 2012; Taniwaki et al., 2010), the effects of various processing conditions (Troncoso et al., 2009), and how frying of chips from red- and purple-fleshed potatoes influences the stability of phenolics, especially anthocyanins, and antioxidant activity of ready products (Kita et al., 2015).

Regarding cooked potatoes, Garcia-Segovia et al. (2008) studied the effects of various heat and pressure cooking treatments on the texture and cooking losses of potatoes (cv. Monalisa). Alvarez and Canet (2002) modeled the kinetics of thermal softening of potato tissue (cv. Monalisa) by water cooking and pressure steaming. Abu-Ghannam and Crowley (2006) investigated the effect of blanching temperature of whole new potatoes before and after processing at temperatures of 95 and 100 °C. Chiavaro et al. (2006) compared the effects of various cooking treatments on selected physicochemical properties of cooked potatoes to evaluate the importance of steam introduction at various steps of the cooking cycle and to consequently identify the best cooking conditions for home and food service operations.

Other researchers have studied rehydration methods (Rojo and Vincent, 2008; Setiady et al., 2009), such as the recovery of various treatments with high pressure, drying, and rehydrating on fresh samples (Al-Khuseibi et al., 2005) and mashed potatoes (Alvarez et al., 2012, 2011; Fernández et al., 2009a,b, 2006) and dry conditions (Bondaruk et al., 2007), among other topics.

The objective methods to evaluate texture have allowed objective fundamental engineering definitions of the mechanical attributes associated with texture in food products (Mohsenin, 1986). These fundamental mechanical parameters include fracture stress (Rojo and Vincent, 2008) and strain (Nath and Chattopadhyay, 2007), yield stress (Fernández et al., 2008), elastic modulus (Thussu and Datta, 2012), the Poisson ratio (Baritelle and Hyde, 2001), stress relaxation (Lebovka et al., 2005), coefficient of friction (Bishop et al., 2012), and fracture toughness. Various methods have been used, such as uniaxial compression (Alvarez and Canet, 2002; Bentini et al., 2009; Bondaruk et al., 2007; Luscher et al., 2005; Thybo and Van den Berg, 2002; Urrutia-Benet et al., 2007), tensile test (Alvarez and Canet, 1998; Scanlon and Long, 1995), a penetration or puncture test (Anzaldúa-Morales et al., 1992; Kumar et al., 2007; Sadowska et al., 2008; Song et al., 2007), three-point bend rig (Moyano et al., 2007; Pedreschi and Moyano, 2005; Pedreschi et al., 2007), small deformation test (Scanlon et al., 1998), impact (Geyer et al., 2009), cutting probes (Figure 1) (Jarén and Sanmartín, 2004; Sanmartín, 2003), back extrusion (Wischmann et al., 2007), and rupture test (Salvador et al., 2009). Back extrusion has been proposed for evaluation of textural parameters of raw and treated potato tubers. Among them, a puncture test is commonly used to describe the textural behavior of various plant-derived food products because the puncture can be done using a small punch directed exactly to the required point of the tuber (Solomon and Jindal, 2003). These advantages favor the puncture method (Sadowska et al., 2008). Of these, texture profile analysis stands out and has been extensively studied by Alvarez et al. (2002). Figure 2 shows a texture analyzer performing a compressing test on a potato (left) and the graphical result obtained (right).

The use of mechanical instruments still holds the dominant position in food texture measurement (Chen and Opara, 2013). Literature evidence shows that there are three main instruments used in texture research of solid and semisolid foods: the texture analyzer (TA)

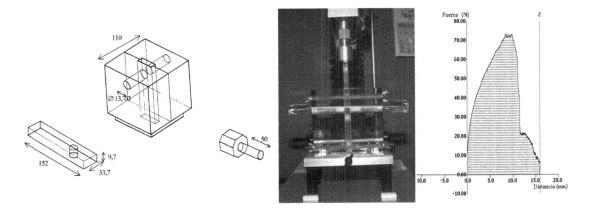

Figure 1
A prototype of the methacrylate cutting probe (left). A texture analyzer (Stable Micro Systems Ltd.) performing a cutting test (center). Graphical result from the cutting test (right).

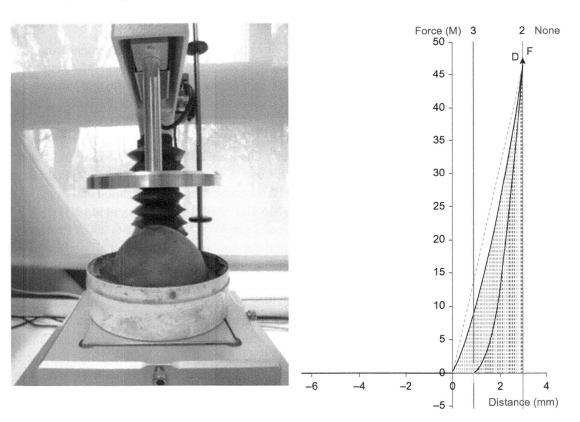

Figure 2
A texture analyzer on the left executing a compression test on a raw tuber. Graphical result from the compression test (right).

by Stable Micro Systems Ltd (Abu-Ghannam and Crowley, 2006; Amaral et al., 2014; García-Segovia et al., 2008; Luscher et al., 2005; Pedreschi and Moyano, 2005; Pimpaporn et al., 2007; Troncoso et al., 2009; Urrutia-Benet et al., 2007), the Lloyd Instruments analyzer (Liu and Scanlon, 2007; Ross and Scanlon, 2004; Zdunek and Bednarczyk, 2006), and the Instron Ltd testing machine (Alvarez et al., 2002; Kita et al., 2015, 2007; Leeratanarak et al., 2006; Mahto and Das, 2014; Sadowska et al., 2008; Singh et al., 2014). Among the models covered by TA, the TA-XT2i, TA-XT2, and TA.XT Plus appear to be the most popular in food texture research. Compared with the Instron, the TA focuses more on food texture measurement and is convenient for both academic and industrial use. However, the Instron is a general and professional instrument for studying mechanical properties of various materials. Nowadays, most researchers combine one of these instruments with other measuring techniques for obtaining more information during experiments. For instance, researchers have combined the TA with an acoustic envelope device (Salvador et al., 2009; Sanz et al., 2007; Taniwaki and Kohyama, 2012; Taniwaki et al., 2010).

3.2 Sensory

Sensory evaluation of cooked potato is normally completed by a panel of trained or untrained evaluators and is based on appearance, smell, taste, and texture. Different attributes are measured according to each experiment, including yellowness, reflection, surface hardness, firmness, mealiness, adhesiveness, moistness, potato flavor, off-flavor, and rancidness (Thybo et al., 2006). These attributes are generally evaluated by quantitative descriptive analysis (Thybo and Martens, 1998).

Some authors have studied the sensory properties of peeled cooked potatoes, dehydrated potato flakes (Neilson et al., 2006), deep-fried potatoes (Anese et al., 2009), and potato chips (Mestdagh et al., 2008). Thybo et al. studied the effect of cultivar and storage on sensory quality of cooked potatoes. Samples were boiled in water for 20–25 min and analyzed by the sensory panel after 0, 1.5, and 6 months of storage. Results obtained indicated that the choice of cultivar was a fundamental factor since some of the cultivars analyzed developed surface hardening and negative flavor.

In a different type of study, Neilson et al. investigated the sensory quality of dehydrated potato flakes packaged in cans and held at ambient temperatures for up to 30 years. Samples were collected from various cans, opened and repacked, and stored at −18 °C until sensory analyses. The authors found that 80% of the initial sensory quality of flakes was maintained up to 16 years and then decreased significantly.

3.3 Color

According to Kays (1999), the appearance of foods directly affects final purchase decisions of customers. Color is a crucial sensory attribute since it is associated with freshness, desirability, food safety, and other quality factors by consumers (McCaig, 2002). Therefore, it becomes important to measure the color information of food products in an effective and objective manner before these products reach the market (Wu and Sun, 2013).

For visual inspection of potato color, there are reference charts available, such as the USDA Color Standards for Frozen French Fried Potatoes (Munsell Color Co., Baltimore, MD, USA) and Color Standards Reference Chart for Potato Chips (Snack Food Association).

Nevertheless, colorimeters and spectrophotometers have also been used for color measurements of potatoes, such as the Minolta chromameter (Minolta Corp., Ramsey, NJ, USA) and the HunterLab colorimeter (HunterLab, Inc., Reston, VA, USA) (Krokida et al., 2001). These measurements are easy and fast to perform. Colorimeters must be calibrated before measurements against a standard tile (Oliveira and Balaban, 2006). Even though these systems offer a simple solution for color measurements, they have a few drawbacks (Balaban and Odabasi, 2006), since only a small surface of the product is measured, and for heterogeneous surfaces, multiple measurements have to be accomplished. Consequently, many authors have chosen computer vision systems for potato color measurements as a more reliable technique than traditional methods.

Scanlon et al. (1994) used computer vision for characterization of potato chip color. They used video image analysis and a software to analyze mean gray level values from specific regions of potato chips. They studied the correlation between color assessed by mean gray level and color measured by two colorimeters (Agtron and HunterLab), obtaining a good value. Segnini et al. (1999a) developed a computer-based video image analysis system to quantify the color of potato chips in the $L*$ $a*$ $b*$ color space. They found a good correlation between the capability of the system to separate $L*$, $a*$, or $b*$ and the ability of the human eye. In another study, the same authors investigated the capability of a previously developed computer-based video image analysis system to quantify the color of commercial potato chips compared to sensory analysis (Segnini et al., 1999b). A good correlation was found between $L*$ and most of the sensory color attributes; also, $a*$ correlated highly with "burnt aspect," while $b*$ did not show any correlation with any of the sensory parameters. In a more recent study by Pedreschi et al. (2011) the automatic classification of commercial potato chips by computer vision was studied. Sensory evaluations were performed by 11 assessors using a standard color chart (Belgapom, Brussels, Belgium) that classifies potato chips into seven color categories and, simultaneously, the color of the same samples was determined objectively by a computer vision system in $L*$, $a*$, $b*$ units. A high correlation was observed between the color measured by the sensory panel and the color determined by the computer vision system.

3.4 Glycemic Index

The glycemic index (GI) is a term first introduced by Jenkins et al. (1981), as an alternative classification of foods containing carbohydrates. The GI establishes a ranking of carbohydrate-containing foods according to their potential to increase blood glucose

(Hätönen et al., 2006). It is defined as the quotient between the area under the blood glucose curve after the consumption of 50 g of carbohydrates from a test food and the area under the curve after consumption of 50 g of carbohydrate from a reference food (white bread or glucose), multiplied by 100 (Ludwig, 2000). The Food and Agriculture Organization/World Health Organization provides recommendations on GI measurements (Anon, 1998). Glucose response can be measured by either capillary or venous blood samples at 0, 15, 30, 45, 60, 90, and 120 min after consumption. It is also recommended that at least six subjects be included in the test and the standard food response be measured a minimum of three times per subject.

A wide range of GI values of potatoes have been reported in literature; however, it is difficult to make comparisons since GI values vary from one variety to another and among cooking methods. In general potatoes have one of the highest GI values compared to other foods (Henry et al., 2005).

Many authors have studied the GI of potatoes subjected to various cooking methods. In all the studies subjects underwent an overnight fast before the experiments. Leeman et al. (2007) investigated the glycemic and satiating properties of potatoes in two studies. Study 1 was composed of a random sample of 13 healthy subjects (nine men and four women), while 14 subjects (six men and eight women) took part in the second study. Study 1 consisted of four energy-equivalent (1000 kJ) meals of cooked potatoes, french fries, and mashed potatoes. Study 2 included four carbohydrate-equivalent meals (50 g available carbohydrates) of french fries, potatoes cooked with and without oil, and white bread as reference. Satiety was measured according to a subjective rating scale. In both studies capillary blood samples were collected before meal consumption and at 20, 45, 70, 180, and 240 min (Study 1 only) after. Results showed greater satiating properties of cooked potatoes compared to french fries on an energy-equivalent basis, while no such differences were observed on a carbohydrate-equivalent basis.

A similar study was conducted to determine the GI of potatoes frequently consumed in North America. The investigation was divided into Study 1 and Study 2. A random sample of 4 men and 6 women participated in Study 1 and 11 men and 1 woman in Study 2. In Study 1 the subjects consumed 200 g of Russet potatoes that were precooked, refrigerated, and cooked again or just cooked and consumed. In Study 2 subjects consumed 50 g of white bread or potatoes from six different varieties cooked by two different methods. All subjects were subjected to capillary blood glucose measurements before the meal and at 2-h intervals after consumption. The authors stated that GI of potatoes varied according to variety and cooking method. The lowest GI values were found in red cooked potatoes consumed cold. The Russet variety had a moderately high GI (Fernandes et al., 2005).

3.5 Damage

Damage to tubers during harvesting is one of the most important causes of lower potato quality and value and increases the incidence of losses and diseases during storage

(Bentini et al., 2006). From harvest to packing, potatoes move at different velocities along the chains of the harvester and processing machines (Van Kempen et al., 1999). Potatoes can impact against one another, against clods, or against the mechanical components of the machines. During this process, potential and kinetic energies join together and, depending on the total energy absorbed or dissipated, potatoes will suffer more or less damage (Van Kempen et al., 1999). There are two categories of damage, internal and external, and often both occur during harvesting and subsequent handling of tubers (Baritelle et al., 2000). Reducing such defects can also increase food safety by decreasing the potential for microbial infestation. The aesthetic, trimming, and disposal problems associated with these defects can also be reduced by reducing damage (Baritelle and Hyde, 2001).

Concerning potato variety characteristics or management quality such as harvesting and processing, there are several variables that one can manage in order to reduce the damages. Potato sensitivity to damage is related to factors like DM, low temperatures during manipulation, or low content of potassium.

Cárdenas and Lema (2006) indicated that DM content above 21–22% increased the risk of internal discoloration due to the impacts during processing. Van Canneyt et al. (2004) evaluated the potatoes' response to impacts as discoloration index (I_D), concluding that DM content was positively correlated with I_D as well as temperature of the tuber during processing. Capurro et al. (2004) found differences among four cultivars due to storage temperature.

Potassium fertilization has a great influence on the DM content, reducing its quantity and, as consequence, reducing the susceptibility to damage. This is in accordance with Baritelle and Hyde (2003), who showed that tubers with higher SG were more sensitive to impacts, obtaining lower bruise resistance and lower bruise thresholds with increasing SG. DM content is proportional to SG (Van Canneyt et al., 2004).

Apart from cultivar or variety characteristics, control of harvesters and the postharvest process lines may be considered an important factor in reducing damage by reducing the level of impacts produced during these phases of tuber management. Although it is not easy work, there are several researchers focused on it at the time of this writing. The electronic fruit is the most useful equipment developed with the objective of evaluating critical points at which impacts produce damage. An electronic fruit records loads and impacts as it passes through the handling system, and several types have been developed worldwide. The most widely used is the American instrumented sphere (IS) (Techmark, Inc., Lansing, MI, USA), which consists of a triaxial accelerometer mounted inside a molded plastic ball a few centimeters in diameter. A microprocessor device designed for data collection, an internal real-time clock, and a rechargeable Ni–Cd battery provide the possibility of collecting and analyzing the maximum acceleration, the velocity change (i.e., the area under the acceleration–time curve), and the time of each event. The output (acceleration and velocity change) has to be interpreted in terms of damage done to real fruit (Studman, 2001). The latest version of this product is called an impact-recording device (IRD) (Opara and Pathare, 2014).

Arazuri et al. (2001, 2010) established a test method to evaluate tomato harvesters: the IRD was attached to a plant, which was then collected by the cutting and lifting system and placed inside the harvester, so the IRD followed the whole route with the tomatoes and was dropped into the trailer. Following this method, Salar (2009) analyzed four different potato harvesters and two processing lines for potato packaging in Navarra (Spain). Although impacts were registered in all the elements of the harvester, measurements of potato samples were collected at three zones: control (hand-harvested potatoes), zone 1 (after separator), and zone 2 (at bunker). Maximum acceleration values were those obtained in the elements included in zone 1, reaching, in one of the harvester, 607.6 m/s^2. The registered impacts produced around 50% of tuber damage after harvesting and a progressive loss of firmness from control to zone 2. With respect to processing lines, the recorded impacts showed similar data. The most critical points were the screen graders with 605.6 m/s^2 of maximum impact level. Similar tests were performed by Bentini et al. (2006), obtaining higher values of acceleration peaks. In this case, the research was focused on the influence of the forward speed of the potato harvester and soil moisture content on impact intensity and the damage suffered by the tubers. Under conditions of wet soil, the increased flow of soil in the machine, deriving from the higher forward speed, reduced impact intensity and extent of the damage. Harvesting in dry soil led to greater impact intensity and damage to the tubers.

In addition to IRD, there are other electronic fruits like the PTR-200 and Smart Spud. The PTR-200, or digital potato, manufactured by SM Engineering (Denmark), consists of a measuring and transmitting unit (the sensor) and a receiver. The sensor has an elongated semiellipsoidal shape measuring 53 by 53 by 83 mm, comparable to two equal hemispheres joined by a cylinder. Compared with the spherical shape of the IS, the elongated semiellipsoidal shape of the PTR-200 may be considered an improvement in the dynamic behavior of the sensor under practical measuring conditions (Van Canneyt et al., 2003). In the Smart Spud (Sensor Wireless, Canada), the manufacture provides the possibility of choosing different urethane casings into which the sensor is embedded for the measurement of impact acceleration. This electronic fruit is powered by a user-replaceable lithium battery. Data measurements can be monitored on a handheld wireless computer (Jarén et al., 2008; Praeger et al., 2013).

Salar (2009) compared IRD, PTR-200, and Smart Spud registered impact data and concluded that IRD and Smart Spud were more sensitive to low-intensity impacts than PTR-200. But, PTR-200-recorded data were better correlated to real damage than those of the others. Figure 3 shows the three types of electronic fruits described.

From the point of view of finding devices with a shape similar to that of potatoes, TuberLog (ESYS GmbH, Berlin, Germany) can be an option. It consists of a triaxial impact acceleration calculator embedded in a synthetic potato-shaped device with a length of 90 mm. Communication with the logger, for starting a measurement or data readout, is accomplished wirelessly over a Bluetooth interface via USB.

Nowadays, there are studies focused on the development of new types of electronic fruits. This is the case of Roa et al. (2013), who contributed a new solution, the wireless instrument sphere

Figure 3
A potato tuber compared with three electronic fruits. From left to right: IRD, PTR-200, and Smart Spud.

(WIS). This device is linked to a real-time analysis software that is able to acquire, process, and visualize the three-axis acceleration of the WIS, allowing the identification and measurements of rotations, vibrations, and impacts in real time. The WIS includes acceleration sensors, a microcontroller, a radio-frequency module, a battery, and a connector to charge the battery. It is packaged using a spherical molded housing, 63 mm in diameter. Like in other fruits, the obtained results are acceleration data and number of impacts. The added value of this equipment is the possibility of installing and synchronizing a video camera controlled by the software to visualize the element and the mode in which the device records the impacts. This provides the possibility of a rapid evaluation of a packaging line.

With the new types of electronic fruits developed to predict damages in potatoes, the Acceleration Measuring Unit (AMU) and Mikras can be cited. These are devices that work when embedded into real fruit. So, the expected recorded impacts will be more realistic that those obtained by dummies.

Mikras is a miniaturized triaxial acceleration measuring unit that is implanted in a real agricultural product of adequate size. The device contains a rechargeable NiMH battery, a data acquisition system, and a triaxial acceleration sensor. The components are encased in epoxy resin and a ceramic coating. After measurements, stored data (acceleration courses and peak values of impact) are transferred by a PC-USB adapter (Praeger et al., 2013). In a comparison carried out by Praeger et al. (2013), the Mikras unit in real potatoes measured peak acceleration values much lower than those measured by IRD and TuberLog, owing to the texture of the material. The AMU is the other miniaturized triaxial acceleration measuring unit. The dimensions and mass of the AMU are $42 \times 13 \times 13$ mm and 15 g, respectively. The device contains a rechargeable battery as power supply, a triaxial acceleration sensor, a data processor, and a wireless data transmitter. All parts are cast in epoxy resin, which ensures resistance against water and fruit acid. Studies with two potato varieties showed that the AMU was able to detect specific differences in tissue properties, closely reflecting the results of impact force measurements with a drop simulator (Geyer et al., 2009).

Finally, it is necessary to briefly speak about another source of damage, the compression forces that tubers suffer at the harvester's bulk or during storage. To detect them, a second type of instrument has been developed. The advantage of them is that they can measure both impacts and compressions. The PMS-60 is one of these devices. PMS-60 is a pressure- or force-measuring device constructed by ATB Bornim (Potsdam, Germany) and is capable of measuring static and dynamic loads above a preset threshold. The outer layer is a 4-mm-thick rubber skin. An inner 42-mm diameter electronic unit is centered by means of 16 conical steel springs. The space between the inner and the outer ball is filled with silicon oil. The inner ball contains all the electronic parts. The oil transmits external pressure loads to the built-in electronic pressure sensor (Shahbazi et al., 2011). Müller et al. (2009, 2012) developed a low-cost IS based on piezoelectric transducers. It was able to quantify all the compression forces, including those produced by misalignment in bulk fruit transportation. This device consisted of six sensors, three around each hemisphere of the pseudo-fruit. The sensors were completely decoupled, concerning each Cartesian axis. This arrangement permitted individual monitoring of each pair of sensors and thus the detection of misalignment in some degree.

There are many alternatives to evaluate all the processes that potatoes suffer from field to market (harvesters, packaging lines, storage). Only the good use of them can lead to reducing great economic losses and improving the quality of potatoes.

4. Advanced Techniques

4.1 Microscopy

Microscopy techniques are commonly used to study the structure and morphology of potato starch. Potato starch granules are generally large, between 30 and 50 μm (Šimková et al., 2013), with a uniform circular shape compared to other starches (Park et al., 2011). Advanced microscopy techniques like scanning electron microscopy (SEM), atomic force microscopy (AFM), and transmission electron microscopy (TEM) have been used for starch characterization during the past years. Surface information can be acquired by using both SEM and AFM, while internal information at high resolution requires the use of TEM (Gallant et al., 1997).

Park et al. (2011) investigated the application of AFM for the study and comparison of corn and potato starch subjected to various humidity conditions. AFM imaging was accomplished in an Agilent 5500 AFM using the tapping mode. Ambient conditions were change in humidification by placing warm water at the bottom of the chamber; iodine exposure by placing a tray with iodine crystals (2 g) at the bottom, next to the dish with water; dehydration by locating a dish of Drierite (10 g) at the bottom while removing the water and iodine dishes; and finally rehumidification achieved by replacing the Drierite plate with the water dish. Studies showed that potato starch granules were more uniformly distributed than corn starch. Additionally, when starches were exposed to iodine vapor under humid environments, it was possible to observe the hair-like structures on the surface of granular starches. The authors concluded that this study demonstrated the potential of AFM to identify the morphology of starch granules. In the same year, Alvani et al. (2011) published a study of the physicochemical properties of potato starch measured by various methods, including SEM, nuclear magnetic resonance (NMR), and high-performance SEC, among others. Starches were extracted from raw potatoes of 10 cultivars and purified before analyses. SEM (Zeiss EVO 50, Peabody, MA, USA) was used to characterize the surface and structure of the starch samples. The authors found that the physicochemical properties of starches did not vary as was expected between different cultivars grown under the same conditions.

4.2 X-Ray Analysis and DSC

X-ray diffractometry (XRD) is an X-ray analysis that has been used to investigate the physicochemical properties of starch (Dündar et al., 2009). The AP molecules of potato starch form crystalline regions that can be analyzed through various techniques including X-rays. The semicrystalline starch granules are insoluble in cold water (Steeneken and Woortman, 2009). However; a heating process with water provokes a disruption of that granule structure, known as gelatinization or melting (Parker and Ring, 2001). Melting occurs with low water content (Steeneken and Woortman, 2009). At temperatures above 130 °C entire molecular dissolution is completed, which results in a significant molecular degradation (Hanselmann et al., 1996). The temperature range of the melting transition is generally studied by DSC.

Many authors have investigated the changes in potato starch subjected to various temperatures and pressures by X-rays, either by wide-angle X-ray scattering or by small-angle X-ray scattering, as well as by DSC.

Liu et al. (2008) investigated the effects of high pressure on some physicochemical properties of potato starch using XRD and DSC. Starch samples were subjected to three types of high pressure: a 0.5-g starch sample was pressed using 10–12 tons force for periods between 5 min and 2 h for the first type. In the second, 13–15 tons force was used for 24 h and in the third, 2 g starch under vacuum underwent 1500 MPa for 24 h. DSC measurements were performed on a differential scanning calorimeter (Mettler-Toledo DSC822^e, Switzerland). A 5-mg starch sample was weighed, mixed with 20 µl of distilled water, and left in a pan for 2 h. After that, the sample pan was heated from 30 to 110 °C until complete gelatinization of the starch. XRD measurements were accomplished with a diffractometer (Siemens D5005). Samples were scanned at steps of 0.02 °C from 5 to 40 °C. According to the DSC results, the gelatinization temperature of the starch was lowered by 3.3–6.6 °C after high pressure. On the other hand, X-ray diffraction patterns were similar for native and high-pressure-treated starches.

Later, Han et al. (2009) studied the effects of pulsed electric field (PEF) treatments on the physicochemical properties of starch using XRD and DSC. For the PEF treatments, potato starch suspensions were subjected to 30, 40, and 50 kV/cm. XRD analysis was performed on a D/Max 2200 X-ray diffractometer (Tokyo, Japan) and for DSC a PerkinElmer DSC-7 (PerkinElmer Corp., Norwalk, CT, USA). X-ray diffraction patterns showed a loss of crystalline structure after the PEF at 50 kV/cm. DSC analysis showed a decrease in gelatinization temperature with increasing PEF treatment.

More recently, Carlstedt et al. (2014) investigated the hydration of acid-hydrolyzed potato starch by X-ray scattering and DSC. They were able to construct the first-phase diagram of acid-hydrolyzed starch water.

4.3 Nuclear Magnetic Resonance (NMR)

NMR is a spectroscopic technique based on the magnetic properties of some nuclei. NMR has made its way into the food industry as a valuable tool in the study of the structure and composition of many foods (Spyros and Dais, 2009). Advances in high-field magnet and probe design have contributed to the rise in the use of this technique in food science. There are three nuclei mostly used in food investigations, namely ^{1}H, ^{13}C, and ^{31}P. Low-field ^{1}H NMR relaxation and high-field NMR imaging (MRI) have shown their potential to estimate quality attributes of foods (Thybo et al., 2000). NMR relaxation is a method capable of noninvasively determining water distribution in foods in a fast way (Cornillon, 1998; McCarthy, 1994). Since water distribution in foods is directly related to textural properties, low-field NMR relaxation has been proven to be highly correlated with texture in many foods such as rice (Ruan et al., 1997), bread (Seow and Teo, 1996), and fish (Steen and Lambelet, 1997), among others. In addition, MRI has shown its potential in measuring water properties and has been successfully applied in the study of the internal structure of fruit and vegetables (Clark et al., 1997; Létal et al., 2003).

Regarding potatoes, Thybo et al. (2000) studied the prediction of sensory texture of 24 cooked potato samples by different spectroscopic techniques including low-field pulsed ^{1}H NMR (LF-NMR). NMR measurements were performed on a Maran benchtop pulsed ^{1}H NMR analyzer (Resonance Instruments, Witney, UK) measuring the transverse relaxation time (T_2) of the water protons. The potato samples were randomly measured at 35 °C (magnet temperature). LF-NMR was found to give the best predictions of the sensory texture profile of cooked potatoes among the techniques studied. Later, in a similar study, Thygesen et al. (2001) investigated the correlation between LF-NMR, chemical constituents, and sensory quality of cooked potatoes. They found that the attributes adhesiveness and springiness could be slightly better described by LF-NMR than by chemical descriptors. In 2003, Thybo et al. studied the prediction of DM content of potatoes by LF-NMR and MRI. Four potato varieties were analyzed. Potato images were acquired by a 7.0 T SISCO scanner (SIS-300/183, Varian).

LF-NMR measurements were performed with the previously described analyzer. DM content was determined as the difference in weight before and after freeze-drying. LF-NMR was found to be correlated with DM content, whereas no correlation was achieved between DM content of whole potatoes and MRI. Nevertheless, in 2004, the same authors (Thybo et al., 2004b) investigated the application of MRI for the prediction of sensory texture quality of cooked potatoes, obtaining promising results. They found that hardness and adhesiveness could be predicted with high accuracy and moistness only to a certain extent. However, they were not able to estimate the attributes of mealiness and graininess.

MRI has also been investigated for internal defects in potatoes. In a study by Thybo et al. (2004a) MRI was applied for the first time in the identification of internal bruising in potatoes of the cultivar Saturna and spraing symptoms in the cultivar Estima. According to the result obtained, MRI was capable of accurately detecting nonvisible internal bruising and spraing symptoms in potatoes of those cultivars.

4.4 Near-Infrared Spectroscopy (NIRS)

NIRS is a noninvasive technique widely used for quality assessment of food products (Osborne et al., 1993). NIR radiation covers the range of the electromagnetic spectrum between 780 and 2500 nm, the region between the visible and the midinfrared (MIR). It is used to study the interaction between electromagnetic radiation and various materials. NIRS consists in irradiating a sample with NIR radiation and measuring its reflected or transmitted radiation. When a radiation in the NIR region of the spectrum is directed to a sample, the incident radiation can be reflected, absorbed, or transmitted, resulting in an NIR spectrum depending on the chemical constitution and physical parameters of the sample (Nicolaï et al., 2007), which are unique for each sample analyzed. NIR spectra are characterized for overlapping absorption bands mainly corresponding to overtones and combinations of vibrational modes involving C–H, O–H, N–H, and S–H chemical bonds (Osborne, 2000). The NIR spectrum of fruits and vegetables is dominated by water absorption bands (Palmer and Williams, 1974) mainly because water constitutes about 80% to 90% of these products (Gómez et al., 2006). Figure 4 shows typical reflectance spectra of raw potato samples. To extract information from the NIR spectra, advanced multivariate statistical techniques are required (Nicolaï et al., 2007) to develop calibration equations that relate spectral data with the quality parameters measured through standard destructive laboratory methods (Peiris et al., 1999).

The first application of NIRS in agriculture was accomplished in the 1960s (Norris, 1964) to measure moisture in grain. Since then the application of NIRS for the estimation of compounds, mainly moisture, protein, and fat, in a wide variety of agricultural and food products has increased substantially (Davies and Grant, 1987; Gunasekaran and Irudayaraj, 2000; Nicolaï et al., 2007). Other NIRS applications include internal damage determination

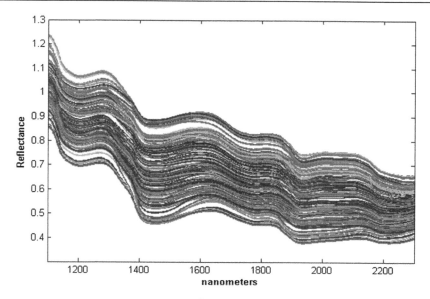

Figure 4
Reflectance spectra of a set of raw potato samples.

(Clark et al., 2003), estimation of the optimal harvest date of products (Jarén et al., 2006), identification of their origin (Vitale et al., 2013), and sensory attributes (Mehinagic et al., 2004). NIRS applications were initiated for postharvest analyses; however, this technology soon showed its potential for in-line applications and it is consequently being implemented by many manufacturers to measure various quality attributes (Nicolaï et al., 2007).

NIRS analysis offers the potential to analyze samples in a rapid, clean, nondestructive, and environmentally friendly manner (Jarén et al., 2001) with little or no sample preparation (Niemöller and Behmer, 2008).

The potential of NIRS has extended further, including the use of multi- and hyperspectral imaging systems (see Section 4.6) that have the advantage of including spatial information along with spectral data (Lu, 2003; Martinsen and Schaare, 1998).

NIRS applications in potatoes started in the 1980s and have been extensively used for major and minor components estimation (see Section 2) and damage evaluation among other characteristics (López et al., 2013a). One of the first successful applications was performed by McDermott (1988), who obtained good accuracy while predicting the moisture content of chips. Similar investigations followed that study. Shiroma and Rodriguez-Saona (2009) investigated the prediction of fat and moisture content of commercially available potato chips using of NIR and MIR spectroscopy. The moisture of samples was determined by loss of weight, while fat was quantified by the Soxhlet method. Samples were ground before analyses. NIR measurements were performed on an Excalibur 3500 Fourier-transform IR spectrometer (Varian) with a quartz beam splitter and lead selenide detector operating at 8 cm^{-1} resolution for all the readings. The spectra were collected in the range 1000–2500 nm in absorbance mode. For MIR measurements, a Varian 3100 FT-IR Excalibur series spectrometer (Varian) with a potassium bromide beam splitter and DTGS detector operating at 4 cm^{-1} resolution was used. Spectra were collected over the frequency region from 2500 to 14,000 nm. The spectral data were analyzed by partial least-squares regression and calibration models generated. The correlation coefficient (R) obtained for moisture was >0.97 with a standard error of cross validation (SECV) of <0.3% for both NIR and MIR techniques. Prediction models for fat had $R > 0.96$ and SECV < 1.60 for both NIR and MIR. Despite the good performance of both methods, the authors stated that NIR allowed faster preparation of sample and was easier to use than MIR.

Although most of the NIR applications in potatoes have been in the form of quantitative analysis, other authors have focused on different applications like studying the correlation between NIRS and texture profiling of cooked potatoes (Boeriu et al., 1998; Van Dijk et al., 2002). Boeriu et al. investigated the correlation between NIRS and texture profiling of steam-cooked potatoes. The texture of 87 potato samples was sensorily evaluated by 16 trained panelists at 1, 3, and 6 months after storage, and at each period, samples were scanned using a Bran + Luebbe InfraAlyzer500 in the 1100–2500 nm spectral range. A quantitative model based on partial least squares was developed. High correlation coefficients were obtained for moistness ($R = 0.91$) and mealiness ($R = 0.89$) attributes. According to the results obtained, the authors determined that NIRS was capable of accurately evaluating the texture of cooked potatoes with good accuracy. A similar study was developed later by Van Dijk et al. to analyze the relationship between DM content, sensory-perceived texture, and NIRS in steam-cooked potatoes. Eighty-one potato tuber samples representing various types of cooking behavior were used for this assessment. Sensory texture analysis was accomplished by a panel of 16 trained judges. The authors obtained correlations above 0.82 for all the attributes studied: hardness, firmness, graininess, mealiness, moistness, and waxiness.

As explained before, NIRS covers a wide range of applications nowadays; an example is the more recent study by Jeong et al. (2008) in which they investigated the correlation between sprouting capacity in potato tubers and NIRS. A set of 380 potato tubers was used, divided into four groups: two groups of the same variety (Superior) harvested at two consecutive years, a group of another variety (Atlantic), and the last group containing the total number of samples. NIR spectra were measured in reflectance mode in the 400–2500 nm wavelength range (including the visible region). First derivative, standard normal variate, and detrend pretreatments were applied to the data. Modified partial least squares was used as a regression method to correlate spectral data and sprouting capacity. The coefficients of determination (R^2) obtained for cross and external validation ranged from 0.69 to 0.93 with SECV and Standard Error of Prediction (SEP) values between 0.40 and 0.68. On the basis of the results obtained, Jeong et al. resolved that it was possible to predict the sprouting capacity of potato tubers by NIRS with a reliable accuracy.

4.5 Microwave Heating

Microwave heating is a method commonly used to cook food products. Microwave heating depends on the dielectric properties of the material. Dielectric properties are a measure of the interaction of a food with electromagnetic fields (Kuang and Nelson, 1998; Seyhun et al., 2009). Dielectric properties are represented by the dielectric constant (ε') and the dielectric loss factor (ε'') (Wang et al., 2011). The ability of a material to store electromagnetic energy is reflected by the dielectric constant, while the ability of a material to dissipate electric energy as heat is measured by the dielectric loss factor. According to some authors, dielectric properties and the resultant dielectric heating can be affected by many factors such as the frequency of the electromagnetic waves and temperature or the water and salt content of food materials (Engelder and Buffler, 1991; Galema, 1997).

Some authors have studied the dielectric properties of potatoes and potato products. Regier et al. (2001) carried out a study to measure the dielectric properties of mashed potatoes as a function of sample temperature and preparation procedure. Samples were prepared using a dried semiproduct (producer: Naturamyl, joint stock company, Pohledsti Dvoraci, local producer Bohdalov, Czech Republic). Three hundred grams of dried mashed potatoes were added to a mixture of 800 g of cold water and 1200 g of boiling water and stirred. Two different preparation methods were used before dielectric characterization. In the first method, samples were dielectrically measured directly after mixing, while in the second method, samples were previously evacuated for 5 min at an absolute pressure of 0.1 bar. The temperature range for the first method was between 10 and 80 °C and for the second between 17.7 and 20.7 °C. Dielectric properties were measured by an open-ended coaxial-line reflectory probe, a device that uses the complex reflection coefficient of the system consisting of the probe and the contacted sample, to determine its complex dielectric constant. A frequency of 2.45 GHz was used. The authors found that both measured values ε' and ε'' were influenced by the sample preparation and temperature. Later, Guan et al. (2004) studied how various factors like frequency, moisture, and salt content influenced the dielectric properties of mashed potatoes. Dielectric properties were measured using a probe similar to that previously described because it did not require particular sample shapes and offered broadbrand measurement. The measurement system mainly consisted of a radio-frequency analyzer equipped with a calibration kit (4291B, Agilent Technologies), a dielectric probe kit (85070B, Hewlett Packard Corp., Palo Alto, CA, USA), and an oil bath. This system was able to measure sample dielectric properties with a frequency ranging from 1 to 1800 MHz and a temperature between 10 and 130 °C. After calibration, each sample was measured at 201 discrete frequencies between 1 and 1800 MHz at 20, 40, 60, 80, 100, and 120 °C. Regression equations were developed for temperature, moisture, and salt content factors at 27, 40, 433, and 915 MHz. The authors found that both ε' and ε'' decreased with increasing frequency. Additionally, according to the result obtained, moisture content did not significantly affect the dielectric properties of mashed potatoes. The addition of salt to mashed potatoes was found to increase the dielectric loss factor. The authors also investigated the penetration depth of electromagnetic energy in mashed potatoes, discovering that it decreased with the temperature and frequency, while it was less dependent on moisture content.

More recently, Wang et al. (2011) investigated the effect of temperature and water content on the dielectric properties of potato purees and the effects of salt and sucrose content on the dielectric properties and microwave freeze-drying (MFD). Experiments were carried out with a lab-scale microwave freeze-dryer (YT2S-01, Nanjing Yatai Microwave Power Technology Research Institute, China). A dielectric probe kit (Agilent 85070E) was used to measure the dielectric properties of potato samples connected to a network analyzer (E5062A, Agilent Technologies). Results similar to those in previous studies were obtained. Dielectric properties of potato purees were influenced by water content, salt content, sucrose content, and temperature. The dielectric properties of frozen and nonfrozen potato purees showed large differences. The ε' and ε'' values decreased with increasing water content in nonfrozen samples, while small variations were found in frozen samples. The addition of salt and sucrose at low temperatures increased ε' and ε'' values. MFD experiments showed that an increase in the salt and sucrose content resulted in a reduction in drying time.

4.6 Imaging Systems

Machine vision techniques have gained much interest within the food industry, translating into an increase in their applications (Sun, 2011) in many agricultural and fruit products (Cubero et al., 2011), among others. These systems are able to provide spatial information useful for quality evaluation and sorting of agricultural products (Zheng et al., 2006) based on shape, size, texture, and external defects. However, since such vision systems are generally based on red/green/blue (RGB) color cameras, they are usually unable to provide information about the internal composition of products (Lorente et al., 2012). To overcome this lack of information, hyperspectral imaging is becoming an emerging technology in the field of food quality assessment (Sun, 2010).

Hyperspectral imaging provides both spatial and spectral information, offering more information and a more reliable characterization of a product than vision techniques (Elmasry et al., 2012b). A hyperspectral image is a set of monochromatic images that are combined to form a three-dimensional data cube, which can provide physical and chemical information on the product examined, such as size, color, and texture, along with water, fat, and other constituents (Lawrence et al., 2003). Hyperspectral systems typically consist of a CCD (charge-coupled device) or complementary metal oxide semiconductor camera and a spectrograph, along with an illumination unit and image acquisition software (Elmasry et al., 2012b). A spectrograph is an optical device that disperses incident broadband light into various wavelengths by using a prism on a detector, such as a camera. This optimal device normally operates by scanning each line of an object until acquiring the whole field of view

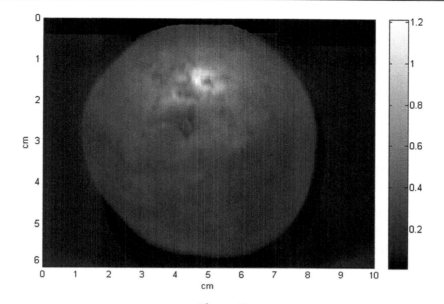

Figure 5
An image of a potato sample from a Vis–NIR hyperspectral camera (TXG14, Baumer, Germany)
with optical sensitivity from 400 to 1000 nm.

(Lorente et al., 2012). Depending on the detector type, hyperspectral images can be acquired in the UV, NIR, or infrared regions of the electromagnetic spectrum (Elmasry et al., 2012b). Figure 5 shows an image of a potato obtained from a CCD camera (TXG14, Baumer, Germany) with optical sensitivity from 400 to 1000 nm.

Imaging systems based on either RGB cameras or hyperspectrum have been successfully applied in many studies regarding potatoes in the past years.

Defects such as blemishes in raw potatoes were detected with a 89.6% accuracy by the use of a Sony DSRL-A350K color camera with a resolution of 1536×1024 (Barnes et al., 2010). In a similar study, Jin et al. investigated the detection of defects in three potato cultivars using a CCD Olympus camera (C5060WZ). They achieved better classification rates than the previous study, an above 90% correct classification rate of defects (Jin et al., 2009). More recently, Razmjooy et al. (2012) achieved a 95% correct classification rate of external defects by using a machine vision system and support vector machines (SVMs) as the classification technique. Other authors (Al-Mallahi et al., 2008) developed a machine vision system to discriminate raw tubers from soil clods, achieving very good results above a 92% discrimination rate. The same authors were able to develop a system to detect potato tubers on the harvester. Results showed that above 98% of both tubers and clods were successfully detected (Al-Mallahi et al., 2010). Other authors have focused on the in-line implementations of such a vision system for the detection of irregular potatoes (Elmasry et al., 2012a). With the development of an image processing algorithm they were able to detect misshapen potatoes with an overall accuracy of 96%. Noordamm et al. (2005) carried out a study to compare multispectral imaging and RGB color imaging for the identification of defects and diseases on raw french fries. The system consisted of a Sony 3-CCD color camera for acquiring color images and a monochrome PMI-1400EC camera with a mounted ImSpector V9 spectrograph for multispectral images. These authors also investigated the effects of various preprocesses on the spectral data. The best classification performance was given by an SVM classifier for multispectral images and *k*-nearest neighbors for RGB images. Detection of latent greening was successfully achieved only by multispectral images, since this type of defect is not visible on the color images. Other authors have employed computer vision to characterize the color of potatoes (see Section 3.3).

Hyperspectral imaging for potato defect identification has also been studied by many authors. Dacal-Nieto et al. (2011b) investigated the ability of a hyperspectral imaging technology in the infrared region to detect the presence of an internal disorder (hollow heart) in potatoes. A Xenics Xeva 1.7-320 camera coupled with a short-wavelength infrared–NIR spectrograph (Specim ImSpector N17E) was used to obtain images of raw tubers in the spectral range from 900 to 1700 nm. A sample set of 234 potato tubers belonging to the Agria variety was scanned with the hyperspectral imaging system and after the tubers were cut to check for the presence of hollow heart. Four classification algorithms were tried, giving similar results with a percentage of correctly classified samples above 86%. The best classification performance was achieved by using SVM (89.1%). In a similar study by the same authors, the detection of common scab by hyperspectral imaging was investigated (Dacal-Nieto et al., 2011a). The same hyperspectral system as the previous study was used to scan 234 raw tubers with different degrees of common scab. SVM and Random Forest classifiers were used, both achieving very high accuracy of correctly classified samples (>96%). However, in this study, once more, SVM performed slightly better, with a 97.1% of accuracy. In 2013, a comparative study of transmission and reflection hyperspectral imaging technology for potato damage detection was performed (Gao et al., 2013). Potato images of three directions (damaged part facing front, back, and side to the camera) were acquired by transmission and reflection with a hyperspectral system. The best classification performance was achieved by transmittance information with the damaged part facing front and back (100%), and very high accuracy was obtained when tubers were faced side to the camera (99.53%). Other authors studied the prediction of optimal cooking time for cooked potatoes by hyperspectral imaging (Nguyen Do Trong et al., 2011). The hyperspectral imaging setup mainly consisted of an ImSpector V10 spectrograph (Spectral Imaging Ltd, Oulu, Finland) coupled with a monochrome CCD camera KP-F120 (Hitachi Denshi Ltd, UK) covering the wavelength range from 400 to 1000 nm. Samples were boiled for 0 (raw), 3, 6, 9, 12, 15, 18, 21, 24, 27, and 30 min and one-half of each potato was scanned. Partial least-squares discriminant analysis was performed to discriminate between cooked regions and raw regions of each potato. Image processing techniques were applied to detect the cooking front and then, by modeling the evolution of the cooking front over time, the authors were able to estimate the optimal cooking time for boiled potatoes.

5. Final Remarks

This chapter highlights the various techniques currently being used for quality control of potato tubers and products. Potatoes in their many forms, cooked, mashed, or fried, have always represented a major source of food in many countries worldwide. Therefore, quality control at every step of the potato industry line becomes essential to ensure a supply of quality products for consumers. A better understanding of the structural and physical changes that occur during potato handling and processing would benefit potato suppliers and manufacturers for providing products with superior quality. Moreover, increasing consumer awareness of the value of food products forces the industry to hunt for technologies capable of guaranteeing the quality of the products being offered.

It is now possible to automate several production control analyses, permitting the improvement of processes, reduction of analytical costs, and increase in product quality to comply with quality standards and regulation in addition to providing customer satisfaction. Studies suggest that the future of potato control is moving toward the in-line applications of nondestructive, relatively cheap, and easy to use techniques for real-time monitoring of potato quality parameters. Such techniques have shown their potential to accurately meet potato quality requirements, confirmed by the studies performed during the past years.

References

Abu-Ghannam, N., Crowley, H., 2006. The effect of low temperature blanching on the texture of whole processed new potatoes. Journal of Food Engineering 74, 335–344.

Ahn, J., Castle, L., Clarke, D., Lloyd, A., Philo, M., Speck, D., 2002. Verification of the findings of acrylamide in heated foods. Food Additives & Contaminants 19, 1116–1124.

Al-Khuseibi, M.K., Sablani, S.S., Perera, C.O., 2005. Comparison of water blanching and high hydrostatic pressure effects on drying kinetics and quality of potato. Drying Technology 23, 2449–2461.

Al-Mallahi, A., Kataoka, T., Okamoto, H., 2008. Discrimination between potato tubers and clods by detecting the significant wavebands. Biosystems Engineering 100, 329–337.

Al-Mallahi, A., Kataoka, T., Okamoto, H., Shibata, Y., 2010. Detection of potato tubers using an ultraviolet imaging-based machine vision system. Biosystems Engineering 105, 257–265.

Al Saikhan, M., Howard, L., Miller, J., 1995. Antioxidant activity and total phenolics in different genotypes of potato (*Solanum tuberosum* L.). Journal of Food Science 60, 341–343.

Alva, A., Fan, M., Qing, C., Rosen, C., Ren, H., 2011. Improving nutrient-use efficiency in Chinese potato production: experiences from the United States. Journal of Crop Improvement 25, 46–85.

Alvani, K., Qi, X., Tester, R.F., Snape, C.E., 2011. Physico-chemical properties of potato starches. Food chemistry 125, 958–965.

Alvarez, M., Canet, W., López, M., 2002. Influence of deformation rate and degree of compression on textural parameters of potato and apple tissues in texture profile analysis. European Food Research and Technology 215, 13–20.

Alvarez, M.D., Canet, W., 1998. Rheological characterization of fresh and cooked potato tissues (cv. Monalisa). Zeitschrift für Lebensmitteluntersuchung und-Forschung A 207, 55–65.

Alvarez, M.D., Canet, W., 2002. A comparison of various rheological properties for modelling the kinetics of thermal softening of potato tissue (cv Monalisa) by water cooking and pressure steaming. International Journal of Food Science & Technology 37, 41–55.

Alvarez, M.D., Fernández, C., Olivares, M.D., Canet, W., 2012. A rheological characterisation of mashed potatoes enriched with soy protein isolate. Food Chemistry 133, 1274–1282.

Alvarez, M.D., Fernández, C., Solas, M.T., Canet, W., 2011. Viscoelasticity and microstructure of inulin-enriched mashed potatoes: influence of freezing and cryoprotectants. Journal of Food Engineering 102, 66–76.

Amaral, R.D.A., Benedetti, B.C., Pujola, M., Achaerandio, I., Bachelli, M.L.B., 2014. Effect of ultrasound on quality of fresh-cut potatoes during refrigerated storage. Food Engineering Reviews 1–9.

Amrein, T.M., Bachmann, S., Noti, A., Biedermann, M., Barbosa, M.F., Biedermann-Brem, S., Grob, K., Keiser, A., Realini, P., Escher, F., 2003. Potential of acrylamide formation, sugars, and free asparagine in potatoes: a comparison of cultivars and farming systems. Journal of Agricultural and Food Chemistry 51, 5556–5560.

Anese, M., Bortolomeazzi, R., Manzocco, L., Manzano, M., Giusto, C., Nicoli, M.C., 2009. Effect of chemical and biological dipping on acrylamide formation and sensory properties in deep-fried potatoes. Food Research International 42, 142–147.

Anon, 1998. Carbohydrates in Human Nutrition: Report of a Joint FAO/WHO Expert Consultation. Food and Agriculture Organization/World Health Organization, Rome.

Anon, 2008. Potatoes, Nutrition and Diet, International Year of the Potato. FAO.

Anzaldúa-Morales, A., Bourne, M., Shomer, I., 1992. Cultivar, specific gravity and location in tuber affect puncture force of raw potatoes. Journal of Food Science 57, 1353–1356.

Arazuri, S., Arana, I., Jaren, C., 2010. Evaluation of mechanical tomato harvesting using wireless sensors. Sensors 10, 11126–11143.

Arazuri, S., Jaren, C., Juanera, A., Martínez, F., 2001. Daños producidos por las cosechadoras de tomate. Horticultura 19, 528–535.

Balaban, M.O., Odabasi, A.Z., 2006. Measuring color with machine vision. Food Technology 60 (12), 32–36.

Barba, A.A., Calabretti, A., d'Amore, M., Piccinelli, A.L., Rastrelli, L., 2008. Phenolic constituents levels in cv. Agria potato under microwave processing. LWT-Food Science and Technology 41, 1919–1926.

Baritelle, A., Hyde, G., 2003. Specific gravity and cultivar effects on potato tuber impact sensitivity. Postharvest Biology and Technology 29, 279–286.

Baritelle, A., Hyde, G., Thornton, R., Bajema, R., 2000. A classification system for impact-related defects in potato tubers. American Journal of Potato Research 77, 143–148.

Baritelle, A.L., Hyde, G.M., 2001. Commodity conditioning to reduce impact bruising. Postharvest Biology and Technology 21, 331–339.

Barnes, M., Duckett, T., Cielniak, G., Stroud, G., Harper, G., 2010. Visual detection of blemishes in potatoes using minimalist boosted classifiers. Journal of Food Engineering 98, 339–346.

Bentini, M., Caprara, C., Martelli, R., 2006. Harvesting damage to potato tubers by analysis of impacts recorded with an instrumented sphere. Biosystems engineering 94, 75–85.

Bentini, M., Caprara, C., Martelli, R., 2009. Physico-mechanical properties of potato tubers during cold storage. Biosystems Engineering 104, 25–32.

Bewell, E., 1937. The determination of the cooking quality of potatoes. American Potato Journal 14, 235–242.

Bishop, C., Gash, A., Heslim, C., Hanney, S., 2012. The Coefficient of Friction of Individual Potatoes and Various Handling Materials–Short Communication. Research in Agricultural Engineering-UZEI (Czech Republic).

Boeriu, C., van der Vuurst de Vries, Y., Stolle-Smits, T., van Dijk, C., 1998. Correlation between near infrared spectra and texture profiling of steam cooked potatoes. Journal of Near Infrared Spectroscopy 6, 291–297.

Bondaruk, J., Markowski, M., Błaszczak, W., 2007. Effect of drying conditions on the quality of vacuum-microwave dried potato cubes. Journal of Food Engineering 81, 306–312.

Bourne, M., 2002. Food Texture and Viscosity: Concept and Measurement. Academic Press.

Bråthen, E., Kita, A., Knutsen, S.H., Wicklund, T., 2005. Addition of glycine reduces the content of acrylamide in cereal and potato products. Journal of Agricultural and Food Chemistry 53, 3259–3264.

Brown, C., 2008. Breeding for phytonutrient enhancement of potato. American Journal of Potato Research 85, 298–307.

Brunt, K., Smits, B., Holthuis, H., 2010. Design, construction, and testing of an automated NIR in-line analysis system for potatoes. Part II. Development and testing of the automated semi-industrial system with in-line NIR for the characterization of potatoes. Potato Research 53, 41–60.

Capurro, J.A., Cuenca, I., Exilart, J.P., Nolasco, M.E., 2004. Daño mecánico de cuatro cultivares de papa (*Solanum tuberosum*) a tres temperaturas de conservación. RIA 33, 41–53.

Cárdenas, J.E., Lema, J.B., 2006. Mecanización del cultivo de la patata. I parte. Horticultura: Revista de industria, distribución y socioeconomía hortícola: frutas, hortalizas, flores, plantas, árboles ornamentales y viveros 20–25.

Carlstedt, J., Wojtasz, J., Fyhr, P., Kocherbitov, V., 2014. Hydration and the phase diagram of acid hydrolyzed potato starch. Carbohydrate Polymers 112, 569–577.

CIAA, 2009. A "Toolbox" for the Reduction of Acrylamide in Fried Potato Products/French Fries. Confederation of the Food and Drink Indrustries of the EU, Brussels.

Clark, C., Hockings, P., Joyce, D., Mazucco, R., 1997. Application of magnetic resonance imaging to pre-and post-harvest studies of fruits and vegetables. Postharvest Biology and Technology 11, 1–21.

Clark, C., Lombard, P., Whiteman, E.F., 1940. Cooking quality of the potato as measured by specific gravity. American Journal of Potato Research 17, 38–45.

Clark, C., McGlone, V., Jordan, R., 2003. Detection of Brownheart in 'Braeburn'apple by transmission NIR spectroscopy. Postharvest Biology and Technology 28, 87–96.

Colonna, P., Mercier, C., 1984. Macromolecular structure of wrinkled-and smooth-pea starch components. Carbohydrate Research 126, 233–247.

Cornillon, P., 1998. Applications of NMR relaxometry to food products. Seminars in Food Analysis (Chapman and Hall) 235–250.

Cubero, S., Aleixos, N., Moltó, E., Gómez-Sanchis, J., Blasco, J., 2011. Advances in machine vision applications for automatic inspection and quality evaluation of fruits and vegetables. Food and Bioprocess Technology 4, 487–504.

Chen, J.Y., Zhang, H., Yelian, M., Asakura, M., 2010. Nondestructive determination of sugar content in potato tubers using visible and near infrared spectroscopy. Japan Journal of Food Engineering 11, 59–64.

Chen, L., Opara, U.L., 2013. Texture measurement approaches in fresh and processed foods—a review. Food Research International 51, 823–835.

Chiavaro, E., Barbanti, D., Vittadini, E., Massini, R., 2006. The effect of different cooking methods on the instrumental quality of potatoes (cv. Agata). Journal of Food Engineering 77, 169–178.

Chung, H.-J., Li, X.-Q., Kalinga, D., Lim, S.-T., Yada, R., Liu, Q., 2014. Physicochemical properties of dry matter and isolated starch from potatoes grown in different locations in Canada. Food Research International 57, 89–94.

Dacal-Nieto, A., Formella, A., Carrión, P., Vazquez-Fernandez, E., Fernández-Delgado, M., 2011a. Common scab detection on potatoes using an infrared hyperspectral imaging system. In: Image Analysis and Processing–ICIAP 2011. Springer, pp. 303–312.

Dacal-Nieto, A., Formella, A., Carrión, P., Vazquez-Fernandez, E., Fernández-Delgado, M., 2011b. Non–destructive detection of hollow heart in potatoes using hyperspectral imaging. In: Computer Analysis of Images and Patterns. Springer, pp. 180–187.

Davies, A.M.C., Grant, A., 1987. Review: near infra-red analysis of food. International Journal of Food Science and Technology 22, 191.

DellaPenna, D., Pogson, B.J., 2006. Vitamin synthesis in plants: tocopherols and carotenoids. Annual Review of Plant Biology 57, 711–738.

Ducreux, L.J., Morris, W.L., Prosser, I.M., Morris, J.A., Beale, M.H., Wright, F., Shepherd, T., Bryan, G.J., Hedley, P.E., Taylor, M.A., 2008. Expression profiling of potato germplasm differentiated in quality traits leads to the identification of candidate flavour and texture genes. Journal of Experimental Botany 59, 4219–4231.

Dündar, E., Turan, Y., Blaurock, A.E., 2009. Large scale structure of wheat, rice and potato starch revealed by ultra small angle X-ray diffraction. International Journal of Biological Macromolecules 45, 206–212.

Elmasry, G., Cubero, S., Moltó, E., Blasco, J., 2012a. In-line sorting of irregular potatoes by using automated computer-based machine vision system. Journal of Food Engineering 112, 60–68.

Elmasry, G., Kamruzzaman, M., Sun, D.-W., Allen, P., 2012b. Principles and applications of hyperspectral imaging in quality evaluation of agro-food products: a review. Critical Reviews in Food Science and Nutrition 52, 999–1023.

Ellis, R.P., Cochrane, M.P., Dale, M.F.B., Duffus, C.M., Lynn, A., Morrison, I.M., Prentice, R.D.M., Swanston, J.S., Tiller, S.A., 1998. Starch production and industrial use. Journal of the Science of Food and Agriculture 77, 289–311.

Engelder, D.S., Buffler, C.R., 1991. Measuring dielectric properties of food products at microwave frequencies. Microwave World 12, 6–15.

Englyst, H.N., Kingman, S., Cummings, J., 1992. Classification and measurement of nutritionally important starch fractions. European Journal of Clinical Nutrition 46, S33–S50.

Essex, E., 1969. Objective measurements for texture in foods. Journal of Texture Studies 1, 19–37.

Ezekiel, R., Singh, N., Sharma, S., Kaur, A., 2013. Beneficial phytochemicals in potato—a review. Food Research International 50, 487–496.

FAOSTAT, 2013. Production quantities by country, Statistics Division. Food and Agriculture Organization of the United Nations. Available at http://faostat3.fao.org/browse/Q/QC/E.

Fernandes, G., Velangi, A., Wolever, T., 2005. Glycemic index of potatoes commonly consumed in North America. Journal of the American Dietetic Association 105, 557–562.

Fernandes, J.O., Soares, C., 2007. Application of matrix solid-phase dispersion in the determination of acrylamide in potato chips. Journal of Chromatography A 1175, 1–6.

Fernández-Ahumada, E., Garrido-Varo, A., Guerrero-Ginel, J.E., Wubbels, A., Van der Sluis, C., Van der Meer, J.M., 2006. Understanding factors affecting near infrared analysis of potato constituents. Journal of Near Infrared Spectroscopy 14, 27–35.

Fernández, C., Alvarez, M.D., Canet, W., 2008. Steady shear and yield stress data of fresh and frozen/thawed mashed potatoes: effect of biopolymers addition. Food Hydrocolloids 22, 1381–1395.

Fernández, C., Canet, W., Alvarez, M.D., 2009a. The effect of long-term frozen storage on the quality of frozen and thawed mashed potatoes with added cryoprotectant mixtures. International Journal of Food Science & Technology 44, 1373–1387.

Fernández, C., Canet, W., Alvarez, M.D., 2009b. Quality of mashed potatoes: effect of adding blends of kappa-carrageenan and xanthan gum. European Food Research and Technology 229, 205–222.

Fernández, C., Dolores Alvarez, M., Canet, W., 2006. The effect of low-temperature blanching on the quality of fresh and frozen/thawed mashed potatoes. International Journal of Food Science & Technology 41, 577–595.

Foot, R., Haase, N., Grob, K., Gonde, P., 2007. Acrylamide in fried and roasted potato products: a review on progress in mitigation. Food Additives and Contaminants 24, 37–46.

Fraser, P.D., Bramley, P.M., 2004. The biosynthesis and nutritional uses of carotenoids. Progress in Lipid Research 43, 228–265.

Friedman, M., 1997. Chemistry, biochemistry, and dietary role of potato polyphenols. A review. Journal of Agricultural and Food Chemistry 45, 1523–1540.

Galema, S.A., 1997. Microwave chemistry. Chemical Society Reviews 26, 233–238.

Gallant, D.J., Bouchet, B., Baldwin, P.M., 1997. Microscopy of starch: evidence of a new level of granule organization. Carbohydrate Polymers 32, 177–191.

Gao, H., Li, X., Xu, S., Tao, H., Li, X., Sun, J., 2013. Comparative study of transmission and reflection hyperspectral imaging technology for potato damage detection. Guang Pu Xue Yu Guang Pu Fen Xi/Spectroscopy and Spectral Analysis 33, 3366–3371.

García-Segovia, P., Andrés-Bello, A., Martínez-Monzó, J., 2008. Textural properties of potatoes (*Solanum tuberosum* L., cv. Monalisa) as affected by different cooking processes. Journal of Food Engineering 88, 28–35.

Gertz, C., Klostermann, S., 2002. Analysis of acrylamide and mechanisms of its formation in deep-fried products. European Journal of Lipid Science and Technology 104, 762–771.

Geyer, M., Praeger, U., Konig, C., Graf, A., Truppel, I., Schluter, O., Herold, B., 2009. Measuring behavior of an acceleration measuring unit implanted in potatoes. Transactions of the ASAE (American Society of Agricultural Engineers) 52, 1267.

Gökmen, V., Şenyuva, H.Z., Acar, J., Sarıoğlu, K., 2005. Determination of acrylamide in potato chips and crisps by high-performance liquid chromatography. Journal of Chromatography A 1088, 193–199.

Gómez, A.H., He, Y., Pereira, A.G., 2006. Non-destructive measurement of acidity, soluble solids and firmness of Satsuma mandarin using Vis/NIR-spectroscopy techniques. Journal of Food Engineering 77, 313–319.

Granda, C., Moreira, R.G., 2005. Kinetics of acrylamide formation during traditional and vacuum frying of potato chips. Journal of Food Process Engineering 28, 478–493.

Guan, D., Cheng, M., Wang, Y., Tang, J., 2004. Dielectric properties of mashed potatoes relevant to microwave and Radio-frequency pasteurization and sterilization processes. Journal of Food Science 69, 30–37.

Gunasekaran, S., Irudayaraj, J., 2000. Optical methods: visible NIR and FTIR spectroscopy. In: Nondestructive Food Evaluation: Techniques to Analyse Properties and Quality. Marcel Dekker Inc., New York.

Haase, N., 2003. Estimation of dry matter and starch concentration in potatoes by determination of under-water weight and near infrared spectroscopy. Potato Research 46, 117–127.

Haase, N.U., 2006. Rapid estimation of potato tuber quality by near-infrared spectroscopy. Starch-Stärke 58, 268–273.

Han, Z., Zeng, X.A., Yu, S.J., Zhang, B.S., Chen, X.D., 2009. Effects of pulsed electric fields (PEF) treatment on physicochemical properties of potato starch. Innovative Food Science & Emerging Technologies 10, 481–485.

Hanselmann, R., Burchard, W., Ehrat, M., Widmer, H., 1996. Structural properties of fractionated starch polymers and their dependence on the dissolution process. Macromolecules 29, 3277–3282.

Hartmann, R., Büning-Pfaue, H., 1998. NIR determination of potato constituents. Potato Research 41, 327–334.

Hätönen, K.A., Similä, M.E., Virtamo, J.R., Eriksson, J.G., Hannila, M.-L., Sinkko, H.K., Sundvall, J.E., Mykkänen, H.M., Valsta, L.M., 2006. Methodologic considerations in the measurement of glycemic index: glycemic response to rye bread, oatmeal porridge, and mashed potato. The American Journal of Clinical Nutrition 84, 1055–1061.

Helgerud, T., Segtnan, V.H., Wold, J.P., Ballance, S., Knutsen, S.H., Rukke, E.O., Afseth, N.K., 2012. Near-infrared spectroscopy for rapid estimation of dry matter content in whole unpeeled potato tubers. Journal of Food Research 1, p. 55.

Henry, C.J.K., Lightowler, H.J., Strik, C.M., Storey, M., 2005. Glycaemic index values for commercially available potatoes in Great Britain. British Journal of Nutrition 94, 917–921.

Hizukuri, S., 1986. Polymodal distribution of the chain lengths of amylopectins, and its significance. Carbohydrate Research 147, 342–347.

IARC, 1994. IARC monographs on the evaluation of carcinogenic risks to humans. In: Monographs on the Evaluation of Carcinogenic Risks to Humans: Some Industrial Chemicals. International Agency for Research on Cancer, Lyon, France, pp. 389–433.

Im, H.W., Suh, B.-S., Lee, S.-U., Kozukue, N., Ohnisi-Kameyama, M., Levin, C.E., Friedman, M., 2008. Analysis of phenolic compounds by high-performance liquid chromatography and liquid chromatography/mass spectrometry in potato plant flowers, leaves, stems, and tubers and in home-processed potatoes. Journal of Agricultural and Food Chemistry 56, 3341–3349.

Jarén, C., Arazuri, S., Arana, I., 2008. Electronic fruits and other sensors. Chronica Horticulturae 48, 4–6.

Jarén, C., Arazuri, S., García, M., Arnal, P., Arana, J., 2006. White asparagus harvest date discrimination using NIRS technology. International Journal of Infrared and Millimeter Waves 27, 391–401.

Jarén, C., Sanmartín, M., 2004. Study the main parameters for the shear stress determination by cutting probes. In: EurAgeng 2004, Engineering the Future, Belgium.

Jarén, C., Ortuño, J.C., Arazuri, S., Arana, J.I., Salvadores, M.C., 2001. Sugar determination in grapes using NIR Technology. International Journal of Infrared and Millimeter Waves 22, 1521–1530.

Jenkins, D., Wolever, T., Taylor, R.H., Barker, H., Fielden, H., Baldwin, J.M., Bowling, A.C., Newman, H.C., Jenkins, A.L., Goff, D.V., 1981. Glycemic index of foods: a physiological basis for carbohydrate exchange. The American Journal of Clinical Nutrition 34, 362–366.

Jeong, J.-C., Ok, H.-C., Hur, O.-S., Kim, C.-G., 2008. Prediction of sprouting capacity using near-infrared spectroscopy in potato tubers. American Journal of Potato Research 85, 309–314.

Jezussek, M., Schieberle, P., 2003. A new LC/MS-method for the quantitation of acrylamide based on a stable isotope dilution assay and derivatization with 2-mercaptobenzoic acid. Comparison with two GC/MS methods. Journal of Agricultural and Food Chemistry 51, 7866–7871.

Jin, J., Li, J., Liao, G., Yu, X., Viray, L.C.C., 2009. Methodology for potatoes defects detection with computer vision. In: International Symposium on Information Processing, Huangshan, pp. 346–351.

Kaur, L., Singh, N., Singh Sodhi, N., Singh Gujral, H., 2002. Some properties of potatoes and their starches I. Cooking, textural and rheological properties of potatoes. Food Chemistry 79, 177–181.

Kays, S.J., 1999. Preharvest factors affecting appearance. Postharvest Biology and Technology 15, 233–247.

Kim, C.T., Hwang, E.S., Lee, H.J., 2005. Reducing acrylamide in fried snack products by adding amino acids. Journal of Food Science 70, C354–C358.

Kita, A., Bąkowska-Barczak, A., Lisińska, G., Hamouz, K., Kułakowska, K., 2015. Antioxidant activity and quality of red and purple flesh potato chips. LWT-Food Science and Technology 62(1), 525–531.

Kita, A., Lisińska, G., Gołubowska, G., 2007. The effects of oils and frying temperatures on the texture and fat content of potato crisps. Food Chemistry 102, 1–5.

Krokida, M.K., Maroulis, Z.B., Saravacos, G.D., 2001. The effect of the method of drying on the colour of dehydrated products. International Journal of Food Science & Technology 36, 53–59.

Kuang, W., Nelson, S., 1998. Low-frequency dielectric properties of biological tissues: a review with some new insights. Transactions of the ASAE 41, 173–184.

Kumar, S., Khade, H., Dhokane, V., Behere, A., Sharma, A., 2007. Irradiation in combination with higher storage temperatures maintains chip-making quality of potato. Journal of Food Science 72, S402–S406.

Kvasnička, F., Čopíková, J., Ševčík, R., Krátká, J., Syntytsia, A., Voldřich, M., 2008. Determination of phenolic acids by capillary zone electrophoresis and HPLC. Central European Journal of Chemistry 6, 410–418.

Lawrence, K., Park, B., Windham, W., Mao, C., 2003. Calibration of a pushbroom hyperspectral imaging system for agricultural inspection. Transactions of the ASAE 46, 513–521.

Lebovka, N.I., Praporscic, I., Ghnimi, S., Vorobiev, E., 2005. Does electroporation occur during the ohmic heating of food? Journal of Food Science 70, E308–E311.

Leeman, M., Östman, E., Björck, I., 2007. Glycaemic and satiating properties of potato products. European Journal of Clinical Nutrition 62, 87–95.

Leeratanarak, N., Devahastin, S., Chiewchan, N., 2006. Drying kinetics and quality of potato chips undergoing different drying techniques. Journal of Food Engineering 77, 635–643.

Lehmann, U., Robin, F., 2007. Slowly digestible starch–its structure and health implications: a review. Trends in Food Science & Technology 18, 346–355.

Lester, G.E., Lewers, K.S., Medina, M.B., Saftner, R.A., 2012. Comparative analysis of strawberry total phenolics via Fast Blue BB vs. Folin–Ciocalteu: assay interference by ascorbic acid. Journal of Food Composition and Analysis 27, 102–107.

Létal, J., Jirak, D., Šuderlová, L., Hajek, M., 2003. MRI 'texture' analysis of MR images of apples during ripening and storage. LWT-Food Science and Technology 36, 719–727.

Liu, E.Z., Scanlon, M.G., 2007. Modeling the effect of blanching conditions on the texture of potato strips. Journal of Food Engineering 81, 292–297.

Liu, Q., Charlet, G., Yelle, S., Arul, J., 2002. Phase transition in potato starch–water system I. Starch gelatinization at high moisture level. Food Research International 35, 397–407.

Liu, Q., Donner, E., Tarn, R., Singh, J., Chung, H.-J., 2009. Advanced analytical techniques to evaluate the quality of potato and potato starch. In: Advances in Potato Chemistry and Technology, pp. 221–248.

Liu, Y., Selomulyo, V.O., Zhou, W., 2008. Effect of high pressure on some physicochemical properties of several native starches. Journal of Food Engineering 88, 126–136.

López, A., Arazuri, S., García, I., Mangado, J., Jarén, C., 2013a. A review of the application of near-infrared spectroscopy for the analysis of potatoes. Journal of Agricultural and Food Chemistry 61, 5413–5424.

López, A., Arazuri, S., Jarén, C., Mangado, J., Arnal, P., Ruiz de Galarreta, J.I., Riga, P., López, R., 2013b. Crude Protein content determination of potatoes by NIRS technology. Procedia Technology 8, 488–492.

López, A., Jarén, C., Arazuri, S., Mangado, J., Tierno, R., Ruiz de Galarreta, J.I., Riga, P., 2014. Estimation of the total phenolic content in potatoes by NIRS. In: International Conference of Agricultural Engineering, Zurich.

Lorente, D., Aleixos, N., Gómez-Sanchis, J., Cubero, S., García-Navarrete, O.L., Blasco, J., 2012. Recent advances and applications of hyperspectral imaging for fruit and vegetable quality assessment. Food and Bioprocess Technology 5, 1121–1142.

Lu, R., 2003. Detection of bruises on apples using near-infrared hyperspectral imaging. Transactions-American Society of Agricultural Engineers 46, 523–530.

Lu, W., Haynes, K., Wiley, E., Clevidence, B., 2001. Carotenoid content and color in diploid potatoes. Journal of the American Society for Horticultural Science 126, 722–726.

Lu, Z.-H., Donner, E., Yada, R.Y., Liu, Q., 2012. The synergistic effects of amylose and phosphorus on rheological, thermal and nutritional properties of potato starch and gel. Food Chemistry 133, 1214–1221.

Lu, Z.-H., Yada, R.Y., Liu, Q., Bizimungu, B., Murphy, A., De Koeyer, D., Li, X.-Q., Pinhero, R.G., 2011. Correlation of physicochemical and nutritional properties of dry matter and starch in potatoes grown in different locations. Food Chemistry 126, 1246–1253.

Ludwig, D.S., 2000. Dietary glycemic index and obesity. The Journal of Nutrition 130, 280S–283S.

Lunden, A., 1956. Undersokelser over forholdet mellom potetenes spesifikke vekt og deres torrstoff-og stivelsesinnhold. Forskning og Forsøk i Landbruket 7, 81–107.

Luscher, C., Schlüter, O., Knorr, D., 2005. High pressure–low temperature processing of foods: impact on cell membranes, texture, color and visual appearance of potato tissue. Innovative Food Science & Emerging Technologies 6, 59–71.

Mahto, R., Das, M., 2014. Effect of gamma irradiation on the physico-mechanical and chemical properties of potato (*Solanum tuberosum* L.), cv.'Kufri Sindhuri', in non-refrigerated storage conditions. Postharvest Biology and Technology 92, 37–45.

Mahto, R., Das, M., 2015. Effect of γ irradiation on the physico-mechanical and chemical properties of potato (*Solanum tuberosum* L.), cv.'Kufri Chandramukhi'and 'Kufri Jyoti', during storage at 12 °C. Radiation Physics and Chemistry 107, 12–18.

Manach, C., Scalbert, A., Morand, C., Rémésy, C., Jiménez, L., 2004. Polyphenols: food sources and bioavailability. The American Journal of Clinical Nutrition 79, 727–747.

Martinsen, P., Schaare, P., 1998. Measuring soluble solids distribution in kiwifruit using near-infrared imaging spectroscopy. Postharvest Biology and Technology 14, 271–281.

McCaig, T., 2002. Extending the use of visible/near-infrared reflectance spectrophotometers to measure colour of food and agricultural products. Food Research International 35, 731–736.

McCarthy, M.J., 1994. Magnetic Resonance Imaging in Foods. Chapman and Hall, New York.

McDermott, L., 1988. Nar-infrared reflectance analysis of processed foods. Cereal Foods World 33, 498–502.

Medina, M.B., 2011. Determination of the total phenolics in juices and superfruits by a novel chemical method. Journal of Functional Foods 3, 79–87.

Mehinagic, E., Royer, G., Symoneaux, R., Bertrand, D., Jourjon, F., 2004. Prediction of the sensory quality of apples by physical measurements. Postharvest Biology and Technology 34, 257–269.

Mehrubeoglu, M., Cote, G., 1997. Determination of total reducing sugars in potato samples using near-infrared spectroscopy. Cereal Foods World (USA) 42 (5), 409–413.

Mestdagh, F., De Wilde, T., Delporte, K., Van Peteghem, C., De Meulenaer, B., 2008. Impact of chemical pre-treatments on the acrylamide formation and sensorial quality of potato crisps. Food Chemistry 106, 914–922.

Miranda, M.L., Aguilera, J.M., 2006. Structure and texture properties of fried potato products. Food Reviews International 22, 173–201.

Mohsenin, N.N., 1986. Physical Properties of Plant and Animal Materials. Gordon & Breach Science, New York.

Morrison, W.R., Laignelet, B., 1983. An improved colorimetric procedure for determining apparent and total amylose in cereal and other starches. Journal of Cereal Science 1, 9–20.

Mottram, D.S., Wedzicha, B.L., Dodson, A.T., 2002. Food chemistry: acrylamide is formed in the Maillard reaction. Nature 419, 448–449.

Moyano, P., Troncoso, E., Pedreschi, F., 2007. Modeling texture kinetics during thermal processing of potato products. Journal of Food Science 72, E102–E107.

Muller, I., Basso, D., Brusamarello, V., Pereira, C.E., 2012. Three-independent axis instrumented sphere for compression measurement based on piezoelectric transducers. In: Instrumentation and Measurement Technology Conference (I2MTC), 2012 IEEE International. IEEE, pp. 628–632.

Müller, I., de Brito, R.M., Pereira, C.E., Bender, R.J., 2009. Wireless instrumented sphere for three-dimensional force sensing. In: IEEE Sensors Applications Symposium (February 17–19, 2009: New Orleans, LA). IEEE SAS Proceedings. IEEE, Piscataway, NJ.

Murniece, I., Kruma, Z., Skrabule, I., Vaivode, A., 2014. Carotenoids and phenols of organically and conventionally cultivated potato varieties. In: 9th Baltic Conference on Food Science and Technology "Food for Consumer Well-being", Latvia.

Nath, A., Chattopadhyay, P., 2007. Optimization of oven toasting for improving crispness and other quality attributes of ready to eat potato-soy snack using response surface methodology. Journal of Food Engineering 80, 1282–1292.

Nath, A., Chattopadhyay, P., Majumdar, G., 2007. High temperature short time air puffed ready-to-eat (RTE) potato snacks: process parameter optimization. Journal of Food Engineering 80, 770–780.

Neilson, A.P., Pahulu, H.F., Ogden, L.V., Pike, O.A., 2006. Sensory and nutritional quality of dehydrated potato flakes in long-term storage. Journal of Food Science 71, S461–S466.

Nguyen Do Trong, N., Tsuta, M., Nicolaï, B., De Baerdemaeker, J., Saeys, W., 2011. Prediction of optimal cooking time for boiled potatoes by hyperspectral imaging. Journal of Food Engineering 105, 617–624.

Nicolaï, B.M., Beullens, K., Bobelyn, E., Peirs, A., Saeys, W., Theron, K.I., Lammertyn, J., 2007. Nondestructive measurement of fruit and vegetable quality by means of NIR spectroscopy: a review. Postharvest Biology and Technology 46, 99–118.

Niemöller, A., Behmer, D., 2008. Use of near infrared spectroscopy in the food industry. In: Irudayaraj, J., Reh, C. (Eds.), Nondestructive Testing of Food Quality. John Wiley & Sons.

Noordam, J.C., van den Broek, W.H., Buydens, L., 2005. Detection and classification of latent defects and diseases on raw French fries with multispectral imaging. Journal of the Science of Food and Agriculture 85, 2249–2259.

Norris, K.H., 1964. Design and development of a new moisture meter. Agricultural Engineering 45, 370–372.

Oliveira, A., Balaban, M., 2006. Comparison of a colorimeter with a machine vision system in measuring color of Gulf of Mexico sturgeon fillets. Applied Engineering in Agriculture 22 (4), 583–587.

Opara, U.L., Pathare, P.B., 2014. Bruise damage measurement and analysis of fresh horticultural produce—a review. Postharvest Biology and Technology 91, 9–24.

Osborne, B., Fearn, T., Hindle, P.H., 1993. Introduction. In: Brownin, D. (Ed.), Practical NIR Spectroscopy with Applications in Food and Beverage Analysis, second ed. Longman Scientific and Technical, Harlow, p. 227.

Osborne, B.G., 2000. Near-infrared spectroscopy in food analysis. Encyclopedia of Analytical Chemistry. BRI Australia Ltd, North Ryde, Australia.

Palmer, K.F., Williams, D., 1974. Optical properties of water in the near infrared. JOSA 64, 1107–1110.

Park, H., Xu, S., Seetharaman, K., 2011. A novel in situ atomic force microscopy imaging technique to probe surface morphology features of starch granules. Carbohydrate Research 346, 847–853.

Parker, R., Ring, S., 2001. Aspects of the physical chemistry of starch. Journal of Cereal Science 34, 1–17.

Pedreschi, F., Mery, D., Bunger, A., Yanez, V., 2011. Computer vision classification of potato chips by color. Journal of Food Process Engineering 34, 1714–1728.

Pedreschi, F., Moyano, P., 2005. Oil uptake and texture development in fried potato slices. Journal of Food Engineering 70, 557–563.

Pedreschi, F., Moyano, P., Santis, N., Pedreschi, R., 2007. Physical properties of pre-treated potato chips. Journal of Food Engineering 79, 1474–1482.

Pedreschi, F., Segtnan, V., Knutsen, S., 2010. On-line monitoring of fat, dry matter and acrylamide contents in potato chips using near infrared interactance and visual reflectance imaging. Food Chemistry 121, 616–620.

Peiris, K., Dull, G., Leffler, R., Kays, S., 1999. Spatial variability of soluble solids or dry-matter content within individual fruits, bulbs, or tubers: implications for the development and use of NIR spectrometric techniques. HortScience 34, 114–118.

Pimpaporn, P., Devahastin, S., Chiewchan, N., 2007. Effects of combined pretreatments on drying kinetics and quality of potato chips undergoing low-pressure superheated steam drying. Journal of Food Engineering 81, 318–329.

Praeger, U., Surdilovic, J., Truppel, I., Herold, B., Geyer, M., 2013. Comparison of electronic fruits for impact detection on a laboratory scale. Sensors 13, 7140–7155.

Razmjooy, N., Mousavi, B.S., Soleymani, F., 2012. A real-time mathematical computer method for potato inspection using machine vision. Computers & Mathematics with Applications 63, 268–279.

Regier, M., Housova, J., Hoke, K., 2001. Dielectric properties of mashed potatoes. International Journal of Food Properties 4, 431–439.

Roa, Y.H.H., Fruett, F., Ferreira, M.D., 2013. Real time measurement system based on wireless instrumented sphere. SpringerPlus 2, 582.

Roach, J.A., Andrzejewski, D., Gay, M.L., Nortrup, D., Musser, S.M., 2003. Rugged LC-MS/MS survey analysis for acrylamide in foods. Journal of Agricultural and Food Chemistry 51, 7547–7554.

Rojo, F., Vincent, J., 2008. Fracture properties of potato crisps. International Journal of Food Science & Technology 43, 752–760.

Rosén, J., Hellenäs, K.-E., 2002. Analysis of acrylamide in cooked foods by liquid chromatography tandem mass spectrometry. Analyst 127, 880–882.

Rosin, P.M., Lajolo, F.M., Menezes, E.W., 2002. Measurement and characterization of dietary starches. Journal of Food Composition and Analysis 15, 367–377.

Ross, K.A., Scanlon, M.G., 2004. A fracture mechanics analysis of the texture of fried potato crust. Journal of Food Engineering 62, 417–423.

Ross, L., Porter, W., 1969. Objective measurements of French fried potato quality. Laboratory techniques for research use. American Potato Journal 46, 192–200.

Ruan, R.R., Zou, C., Wadhawan, C., Martinez, B., Chen, P., Addis, P., 1997. Studies of hardness and water mobility of cooked wild rice using nuclear magnetic resonance. Journal of food processing and preservation 21, 91–104.

Rufián-Henares, J.A., Morales, F.J., 2006. Determination of acrylamide in potato chips by a reversed-phase LC–MS method based on a stable isotope dilution assay. Food Chemistry 97, 555–562.

Sadowska, J., Fornal, J., Zgórska, K., 2008. The distribution of mechanical resistance in potato tuber tissues. Postharvest Biology and Technology 48, 70–76.

Salar, M., 2009. Determinación de daños durante la recolección mecanizada y la manipulación de patata. Agricultural Projects and Engineering, Universidad Pública de Navarra, Pamplona.

Salvador, A., Varela, P., Sanz, T., Fiszman, S., 2009. Understanding potato chips crispy texture by simultaneous fracture and acoustic measurements, and sensory analysis. LWT-Food Science and Technology 42, 763–767.

Sanmartín, M., 2003. Influencia de los principales parámetros de ensayo de esfuerzo cortante por corte de probetas de material vegetal (Study the main parameters for the shear stress test by cutting probes on vegetal products). Agricultural Projects and Engineering, Universidad Pública de Navarra, Pamplona.

Sanz, T., Primo-Martín, C., Van Vliet, T., 2007. Characterization of crispness of French fries by fracture and acoustic measurements, effect of pre-frying and final frying times. Food Research International 40, 63–70.

Scanlon, M., Roller, R., Mazza, G., Pritchard, M., 1994. Computerized video image analysis to quantify color of potato chips. American Potato Journal 71, 717–733.

Scanlon, M.G., Day, A.J., Povey, M.J., 1998. Shear stiffness and density in potato parenchyma. International Journal of Food Science & Technology 33, 461–464.

Scanlon, M.G., Long, A.E., 1995. Fracture strengths of potato tissue under compression and tension at two rates of loading. Food Research International 28, 397–402.

Scanlon, M.G., Pritchard, M.K., Adam, L.R., 1999. Quality evaluation of processing potatoes by near infrared reflectance. Journal of the Science of Food and Agriculture 79, 763–771.

Schirmer, M., Höchstötter, A., Jekle, M., Arendt, E., Becker, T., 2013. Physicochemical and morphological characterization of different starches with variable amylose/amylopectin ratio. Food Hydrocolloids 32, 52–63.

Segnini, S., Dejmek, P., Öste, R., 1999a. A low cost video technique for colour measurement of potato chips. LWT-Food Science and Technology 32, 216–222.

Segnini, S., Dejmek, P., ÖSTE, R., 1999b. Relationship between instrumental and sensory analysis of texture and color of potato chips. Journal of Texture Studies 30, 677–690.

Seow, C., Teo, C., 1996. Staling of Starch-based products: a comparative study by firmness and pulsed NMR measurements. Starch-Stärke 48, 90–93.

Setiady, D., Rasco, B., Younce, F., Clary, C., 2009. Rehydration and sensory properties of dehydrated russet potatoes (*Solanum tuberosum*) using microwave vacuum, heated air, or freeze dehydration. Drying Technology 27, 1116–1122.

Seyhun, N., Ramaswamy, H., Sumnu, G., Sahin, S., Ahmed, J., 2009. Comparison and modeling of microwave tempering and infrared assisted microwave tempering of frozen potato puree. Journal of Food Engineering 92, 339–344.

Shahbazi, F., Geyer, M., Praeger, U., König, C., Herold, B., 2011. Comparison of two impact detecting devices to measure mechanical load on potatoes. Agricultural Engineering International: CIGR Journal 13.

Shiroma, C., Rodriguez-Saona, L., 2009. Application of NIR and MIR spectroscopy in quality control of potato chips. Journal of Food Composition and Analysis 22, 596–605.

Sievert, D., Pomeranz, Y., 1989. Enzyme-resistant starch. I. Characterization and evaluation by enzymatic, thermoanalytical, and microscopic methods. Cereal Chemistry 66, 342–347.

Šimková, D., Lachman, J., Hamouz, K., Vokál, B., 2013. Effect of cultivar, location and year on total starch, amylose, phosphorus content and starch grain size of high starch potato cultivars for food and industrial processing. Food Chemistry 141, 3872–3880.

Singh, F., Katiyar, V., Singh, B., 2013. A new strain energy function to characterize apple and potato tissues. Journal of Food Engineering 118, 178–187.

Singh, F., Katiyar, V., Singh, B., 2014. Analytical study of turgor pressure in apple and potato tissues. Postharvest Biology and Technology 89, 44–48.

Singleton, V.L., Orthofer, R., Lamuela-Raventos, R.M., 1999. Analysis of total phenols and other oxidation substrates and antioxidants by means of Folin-Ciocalteu reagent. Methods in Enzymology 299, 152–178.

Smith, O., Nash, L., 1940. Potato quality. I. Relation of fertilizers and rotation systems to specific gravity and cooking quality. American Journal of Potato Research 17, 163–169.

Solomon, W., Jindal, V., 2003. Modeling thermal softening kinetics of potatoes using fractional conversion of rheological parameters. Journal of Texture Studies 34, 231–247.

Song, X-j., Zhang, M., Mujumdar, A.S., 2007. Effect of vacuum-microwave predrying on quality of vacuum-fried potato chips. Drying Technology 25, 2021–2026.

Spyros, A., Dais, P., 2009. ^{31}P NMR spectroscopy in food analysis. Progress in Nuclear Magnetic Resonance Spectroscopy 54, 195–207.

Srisawas, W., Jindal, V., 2003. Acoustic testing of snack food crispness using neural networks. Journal of Texture Studies 34, 401–420.

Steen, C., Lambelet, P., 1997. Texture changes in frozen cod mince measured by low-field nuclear magnetic resonance spectroscopy. Journal of the Science of Food and Agriculture 75, 268–272.

Steeneken, P.A., Woortman, A.J., 2009. Identification of the thermal transitions in potato starch at a low water content as studied by preparative DSC. Carbohydrate Polymers 77, 288–292.

Storey, R., Davies, H., 1992. Tuber Quality, the Potato Crop. Springer, pp. 507–569.

Studman, C., 2001. Computers and electronics in postharvest technology—a review. Computers and Electronics in Agriculture 30, 109–124.

Subedi, P., Walsh, K.B., 2009. Assessment of potato dry matter concentration using short-wave near-infrared spectroscopy. Potato Research 52, 67–77.

Sun, D.-W., 2010. Hyperspectral Imaging for Food Quality Analysis and Control. Elsevier.

Sun, D.-W., 2011. Computer Vision Technology for Food Quality Evaluation. Academic Press, London.

Sundara, V., Pourzand, F., Berberat, A., 2001. A technique to measure firmness in potato tissue during frying. Food Science and Technology International 7, 141–144.

Tai, G., Misener, G., Allaby, E., McMillan, L., 1985. Grav-o-tater: a computer apparatus for measuring specific gravity. American Potato Journal 62, 403–408.

Taniwaki, M., Kohyama, K., 2012. Mechanical and acoustic evaluation of potato chip crispness using a versatile texture analyzer. Journal of Food Engineering 112, 268–273.

Taniwaki, M., Sakurai, N., Kato, H., 2010. Texture measurement of potato chips using a novel analysis technique for acoustic vibration measurements. Food Research International 43, 814–818.

Tester, R.F., Karkalas, J., Qi, X., 2004. Starch—composition, fine structure and architecture. Journal of Cereal Science 39, 151–165.

Thussu, S., Datta, A.K., 2012. Texture prediction during deep frying: a mechanistic approach. Journal of Food Engineering 108, 111–121.

Thybo, A.K., Andersen, H.J., Karlsson, A.H., Dønstrup, S., Stødkilde-Jørgensen, H., 2003. Low-field NMR relaxation and NMR-imaging as tools in differentiation between potato sample and determination of dry matter content in potatoes. LWT-Food Science and Technology 36, 315–322.

Thybo, A.K., Bechmann, I., Martens, M., Engelsen, S., 2000. Prediction of sensory texture of cooked potatoes using uniaxial compression, near infrared spectroscopy and low ^{1}H NMR spectroscopy. LWT-Food Science and Technology 33, 103–111.

Thybo, A.K., Christiansen, J., Kaack, K., Petersen, M.A., 2006. Effect of cultivars, wound healing and storage on sensory quality and chemical components in pre-peeled potatoes. LWT-Food Science and Technology 39, 166–176.

Thybo, A.K., Jespersen, S.N., Lærke, P.E., Stødkilde-Jørgensen, H.J., 2004a. Nondestructive detection of internal bruise and spraing disease symptoms in potatoes using magnetic resonance imaging. Magnetic Resonance Imaging 22, 1311–1317.

Thybo, A.K., Martens, M., 1998. Development of a sensory texture profile of cooked potatoes by multivariate data analysis. Journal of Texture Studies 29, 453–468.

Thybo, A.K., Szczypiński, P.M., Karlsson, A.H., Dønstrup, S., Stødkilde-Jørgensen, H.S., Andersen, H.J., 2004b. Prediction of sensory texture quality attributes of cooked potatoes by NMR-imaging (MRI) of raw potatoes in combination with different image analysis methods. Journal of Food Engineering 61, 91–100.

Thybo, A.K., Van den Berg, F., 2002. Full uniaxial compression curves for predicting sensory texture quality of cooked potatoes. Journal of Texture Studies 33, 119–134.

Thygesen, L., Thybo, A.K., Engelsen, S., 2001. Prediction of sensory texture quality of boiled potatoes from low-field ^{1}H NMR of raw potatoes. The role of chemical constituents. LWT-Food Science and Technology 34, 469–477.

Tisza, S., Molnár-Perl, I., Friedman, M., Sass, P., 1996. Simultaneous capillary GC of acids and sugars as their silyl (oxime) derivatives: quantitation of chlorogenic acid, raffinose, and pectin substances. Journal of High Resolution Chromatography 19, 54–58.

Troncoso, E., Pedreschi, F., Zúñiga, R., 2009. Comparative study of physical and sensory properties of pre-treated potato slices during vacuum and atmospheric frying. LWT-Food Science and Technology 42, 187–195.

Urrutia-Benet, G., Balogh, T., Schneider, J., Knorr, D., 2007. Metastable phases during high-pressure–low-temperature processing of potatoes and their impact on quality-related parameters. Journal of Food Engineering 78, 375–389.

Van Canneyt, T., Tijskens, E., Ramon, H., Verschoore, R., Sonck, B., 2003. Characterisation of a potato-shaped instrumented device. Biosystems Engineering 86, 275–285.

Van Canneyt, T., Tijskens, E., Ramon, H., Verschoore, R., Sonck, B., 2004. Development of a predictive tissue discolouration model based on electronic potato impacts. Biosystems Engineering 88, 81–93.

Van Dijk, C., Fischer, M., Holm, J., Beekhuizen, J.-G., Stolle-Smits, T., Boeriu, C., 2002. Texture of cooked potatoes (*Solanum tuberosum*). 1. Relationships between dry matter content, sensory-perceived texture, and near-infrared spectroscopy. Journal of Agricultural and Food Chemistry 50, 5082–5088.

Van Kempen, P., Le Corre, P., Bedin, P., 1999. Fitotecnia. In: Rousselle, P., Robert, Y., Crosnier, J.C. (Eds.), La Patata. Mundi-Prensa, Madrid.

Vitale, R., Bevilacqua, M., Bucci, R., Magrì, A.D., Magrì, A.L., Marini, F., 2013. A rapid and non-invasive method for authenticating the origin of pistachio samples by NIR spectroscopy and chemometrics. Chemometrics and Intelligent Laboratory Systems 121, 90–99.

Voigt, J., Noske, R., 1964. On the question of chlorogenic acid content of raw and stewed potatoes. Nahrung 8, 19–26.

Wang, R., Zhang, M., Mujumdar, A.S., Jiang, H., 2011. Effect of salt and sucrose content on dielectric properties and microwave freeze drying behavior of re-structured potato slices. Journal of Food Engineering 106, 290–297.

Wilson, J., Lindsay, A.M., 1969. The relation between specific gravity and dry matter content of potato tubers. American Potato Journal 46, 323–328.

Williams, P., Kuzina, F., Hlynka, I., 1970. Rapid colorimetric procedure for estimating the amylose content of starches and flours. Cereal Chemistry 47, 411–421.

Wischmann, B., Ahmt, T., Bandsholm, O., Blennow, A., Young, N., Jeppesen, L., Thomsen, L., 2007. Testing properties of potato starch from different scales of isolations—a ringtest. Journal of Food Engineering 79, 970–978.

Wong, K., Jane, J., 1997. Quantitative analysis of debranched amylopectin by HPAEC-PAD with a postcolumn enzyme reactor. Journal of Liquid Chromatography & Related Technologies 20, 297–310.

Woolfe, J., Poats, S., 1987. The nutritional value of the components of the tuber. The Potato in the Human Diet 19–51.

Wu, D., Sun, D.-W., 2013. Colour measurements by computer vision for food quality control–a review. Trends in Food Science & Technology 29, 5–20.

Zdunek, A., Bednarczyk, J., 2006. Effect of mannitol treatment on ultrasound emission during texture profile analysis of potato and apple tissue. Journal of Texture Studies 37, 339–359.

Zhang, Y., Chen, J., Zhang, X., Wu, X., Zhang, Y., 2007. Addition of antioxidant of bamboo leaves (AOB) effectively reduces acrylamide formation in potato crisps and French fries. Journal of Agricultural and Food Chemistry 55, 523–528.

Zheng, C., Sun, D.-W., Zheng, L., 2006. Recent applications of image texture for evaluation of food qualities—a review. Trends in Food Science & Technology 17, 113–128.

CHAPTER 20

The Role of Potatoes in Biomedical/Pharmaceutical and Fermentation Applications

Shrikant A. Survase[1], Jaspreet Singh[2], Rekha S. Singhal[1]
[1]*Food Engineering and Technology Department, Institute of Chemical Technology, Mumbai, Maharashtra, India;* [2] *Riddet Institute and Massey Institute of Food Science and Technology, Massey University, Palmerston North, New Zealand*

1. Introduction

The increasing human population has greatly influenced the demand for food products. Potato (*Solanum tuberosum* of the Solanaceae family) is a major world crop of which more than 300 million tons are produced worldwide annually. It is the world's most widely grown tuber crop and the fourth largest food crop in terms of fresh produce after rice, wheat, and maize (corn). It is the most important vegetable in European countries today. Potatoes are used for human consumption, industrial processing (potato starch, alcohol, etc.), and recultivation. Nutritionally, potatoes are best known for their carbohydrate content, with starch being the predominant form of carbohydrate. A small but significant portion of the starch in potatoes called "resistant starch" is considered to have physiological effects and health benefits similar to those of fiber (e.g., provides bulk, offers protection against colon cancer, improves glucose tolerance and insulin sensitivity, lowers plasma cholesterol and triglyceride concentrations, increases satiety, and possibly even reduces fat storage) (Cummings et al., 1996; Hylla et al., 1998; Raban et al., 1994). Potatoes contain toxic glycoalkaloids of which the most prevalent are solanine and chaconine (Talburt, 1987; Feustel, 1987).

Potato starch was first produced in Germany at the end of the seventeenth century. Industrial starch production includes steps as cleaning and soil removal on rotating bar screens; opening of the cells by high-speed raspers; separation of the starch granules and juice from the cell walls (the pulp) on rotating conical screens; concentrating the crude starch milk (starch+juice) on hydrocyclones, which is washed on multistage hydrocyclones in a countercurrent with water; and drying of the dewatered cake. The starch yield in a modern process is close to 96% of the granular starch in raw potatoes. Potato starch has been reported to have applications in the biomedical/pharmaceutical industry as well as in fermentation for production of various biomolecules.

Advances in Potato Chemistry and Technology. http://dx.doi.org/10.1016/B978-0-12-800002-1.00020-0

Table 1: Components of Conventional Wet Potato Pulp (Mayer, 1998).

Components	% (w/w) of Wet Pulp	% (w/w) of Dry Matter
Dry matter	13.0	–
Ashes	0.5	4.0
Starch	4.9	37.0
Cellulose	2.2	17.0
Hemicellulose	1.8	14.0
Pectin	2.2	17.0
Fiber (unidentified)	0.9	7.0
Protein/amino acids	0.5	4.0

During starch production from potatoes, a huge amount of a mass comprising fruit liquid (up to 90% of the mass), cell debris, intact starch cells, and cells or cell aggregates of potato skin is also produced. The potato fruit liquid can be separated from the particulate fraction. It is characterized by a high content of proteins, free amino acids, and salts. This protein-rich juice could be exploited for food, biotechnological, and pharmaceutical applications. It is primarily used to enrich for proteins and amino acids and as a fertilizer because of its high nitrogen content.

The particulate fraction, called potato pulp, contains starch, cellulose, hemicelluloses, pectin, proteins, and some amount of fruit liquid and water in intact cells. On a dry matter basis, the pulp contains 1.5–2.5% protein, which is around 74% of that found in potato tubers (Kempf, 1980). The components of conventional wet potato pulp are listed in Table 1.

In countries where a strong environmental regulation for industrial wastewater exists, purification of waste streams (both the fruit water and the pulp) from potato factories is required. Several attempts have been made to dehydrate the by-products and to utilize them for various purposes. Its high moisture content (80%) requires expensive drying owing to the problem of spoilage if left untreated. The starch industry tries to sell as much pulp as possible as wet or partially dried cattle feed. However, the need for potato pulp by farmers is limited. Potato pulp is being used as cattle feed as well as a medium for solid-state fermentation for the production of various biomolecules. Conventional applications of potato pulp are listed in Table 2.

Potato peels are a waste by-product of the potato processing industry. They are a good source of vitamin C, vitamin B_6, copper, potassium, manganese, and dietary fiber. They also contain a variety of phytonutrients, which are a natural source of antioxidants that help to prevent cellular deterioration of the body. The phytonutrients found in potato skins as well as the flesh include polyphenols, carotenoids, flavonoids, and caffeic acid.

2. Biomedical Applications

Cells are often implanted into artificial structures typically called "scaffolds" that are capable of supporting three-dimensional (3D) tissue formation. These structures are often critical,

Table 2: Conventional Applications of Potato Pulp (Mayer, 1998).

Treatment/Product	Application
Pulp supplemented by potato proteins or other nitrogen-containing components (wet or partially dried and pelleted)	Cattle feed
Preparation of pectin or pectin–starch mixtures	Nutritional and technical applications
Conversion into sugars and extraction of a syrup	Treatment of potato chips and pommes fries
Hydrolysis for substrates used in fermentation	Alcohol production
Extraction of nitrogen-containing components from the liquid phase	Fertilizer
Dilution with water	Stabilizing factor in deep drilling (lubricant)
Untreated, substrate for growth of yeast	Vitamin B_{12} production
Untreated, component of growth substrate	Biogas production

both ex vivo and in vivo, to recapitulating the in vivo milieu and allowing cells to influence their own microenvironments. Scaffolds usually serve at least one of the following purposes: (1) allow cell attachment and migration, (2) deliver and retain cells and biochemical factors, (3) enable diffusion of vital cell nutrients and expressed products, and (4) exert certain mechanical and biological influences to modify the behavior of the cell phase. Biodegradability is often an essential factor since scaffolds should preferably be absorbed by the surrounding tissues without the necessity of surgical removal. The rate of degradation should coincide as much as possible with the rate of tissue formation. Injectability is also important for clinical uses. Many different materials (natural and synthetic, biodegradable and permanent) have been investigated. New biomaterials have been engineered to have ideal properties and functional customization such as injectability, synthetic manufacture, biocompatibility, nonimmunogenicity, transparency, nanoscale fibers, and low concentration and resorption rates. A commonly used synthetic material is polylactic acid (PLA). This is a polyester that is degraded within the human body to form lactic acid, a naturally occurring chemical that is easily removed from the body. Similar materials are polyglycolic acid and polycaprolactone. Repair or regeneration of bone is one of the most challenging areas in tissue engineering because of the specific characteristics of the skeleton. A wide array of properties in materials is desirable for bone tissue engineering applications. A very promising approach for the same combines the use of adequate scaffold materials and site-specific cells to create a hybrid material that can enhance repair.

The use of starch-based materials can be justified by the well-known acidification phenomenon when using systems constituted by PLA. Using materials of natural origin has an advantage because starch can be degraded within the body by several enzymes (Gomes et al., 2001), resulting in degradation products that can be readily metabolized and excreted. Its natural origin, together with its mechanical properties (Gomes et al., 2002; Reis et al., 1996; Sousa et al., 2002) and biocompatibility (Marques et al., 2001; Mendes et al., 2001), supports its potential in the biomedical field. Starch-based polymers are being studied for a wide range of bone-related therapy applications such as tissue engineering scaffolds (Reis and Cunha, 2001) and bone cements (Boesel et al., 2003; Elvira et al., 2002; Pereira et al., 1998).

Silva et al. (2004) synthesized and evaluated a novel bioactive composite of starch/bioactive glass microparticles. The biodegradable character, good controlled-release properties, and natural origin of starch-based biomaterials were combined with the bioactive and bone-bonding properties of bioactive glass (BG). The bioactive composite starch–BG microparticles were synthesized from a blend of starch and PLA (50/50 w/w) with BG 45S5 powder using a simple emulsion method. The short-term cytotoxicity of these materials was also tested by placing 24 h leachables of the materials extracted in culture medium in contact with a fibroblastic cell line (L929) for up to 72 h. The results showed these materials to be free of any cytotoxicity. Silva et al. (2005a) reported on the synthesis and bioactivity of newly developed polymer-soluble potato starch and composite (with BG 45S5) micrometer-size particles. They found that both polymer and composite particles were able to form a calcium phosphate layer at their surface. They also found that both types of trials allowed rat bone marrow cells to attach and proliferate on their surface and to express osteogenic markers such as alkaline phosphatase and osteopontin. Their results indicated that these carriers might be used as substrates for cell culture in vitro to form constructs that might be used as part of a tissue engineering strategy. In addition to developing starch-based microparticles and evaluating them for bioactivity, cytotoxicity, and ability to serve as substrates for cell adhesion, these authors (Silva et al., 2006) also checked their potential to be used as delivery systems for either anti-inflammatory agents or growth factors. Two starch-based materials were used for the development of starch-based particulate systems: (1) a blend of starch and PLA (SPLA) (50:50 w/w) and (2) a chemically modified potato starch, Paselli II (PaII). Both materials enabled the synthesis of particulate systems, both polymer and composite (with BG 45S5). A simple solvent extraction method was employed for the synthesis of SPLA and SPLA/BG microparticles, while for PaII and PaII/BG microparticles an emulsion cross-linking method using trisodium trimetaphosphate as a cross-linker was developed.

Cellulose nanocrystals derived from potato peel were studied for strength reinforcement and water permeability of biocomposites. The average fiber length of these nanocrystals was 410 nm with an aspect ratio of 41, which is considerably larger than cotton-derived nanocrystals. Cellulose nanocrystals filled polyvinyl alcohol and thermoplastic starch films prepared by a solution casting method showed significant improvement in tensile modulus. The water permeability in the case of starch composites was unchanged (Chen et al., 2012).

Torres et al. (2007) reported a novel microwave processing technique to produce biodegradable scaffolds for tissue engineering from various types of starch-based polymers. Potato, sweet potato, corn starch, and amaranth and quinoa starch were used along with water and glycerol as plasticizers to produce porous structures. Figure 1 shows the manufacturing procedure of microwaved starch scaffolds.

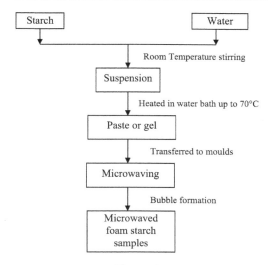

Figure 1

Manufacturing of microwaved starch scaffolds (Torres et al., 2007).

Polyvinyl alcohol (PVA) and starch were chemically cross-linked to form a 3D scaffold with characteristics such as a stable structure, effective water absorption, and supporting cell growth. The PVA-g-starch 3D scaffold polymer showed an absorbency of up to 800%, a strength of 4×10^{-2} MPa, and a high porosity. The 3D scaffold was degraded by various enzymes at a rate of up to approximately 30–60% in 28 days. Cell proliferation and growth were checked in vitro and biocompatibility was confirmed by energy-dispersive spectrometer analysis (Hsieh and Liau, 2013). Similarly, starch/cellulose nanofiber composites prepared by combining film casting, salt leaching, and freeze-drying methods have been reported by Nasri-Nasrabadi et al. (2014). The diameter of 70% of the nanofibers was in the range of 40–90 nm, with scaffolds having adequate mechanical properties for cartilage tissue engineering applications. The addition of 10% cellulose nanofibers enhanced the water uptake ratio of nanocomposites remarkably. The scaffolds were partially destroyed after more than 20 weeks. The incorporation of nanofibers into starch structure improved cell attachment and proliferation.

Vigneshwaran et al. (2006) synthesized stable silver nanoparticles by using soluble starch as both reducing and stabilizing agents. The use of environmentally benign and renewable materials like soluble starch offers numerous benefits of eco-friendliness and compatibility for pharmaceutical and biomedical applications. Shape-memory polymers such as natural starch can now be considered for manufacturing innovative biodegradable devices without chemical grafting for less-invasive surgery. Shape-memory extruded starch can recover its original shape in water at 37 °C. Samples implanted in a rat model exhibited normal tissue integration with a low inflammatory response (Beilvert et al., 2014).

Torres et al. (2011) reported the use of starch films as substrate for 3T3 fibroblast cells. They studied 17 novel starch-based films from various Andean crops. After a culture period of 3 days, films made from starch of "tunta," "muro-huayro" potato, and white carrot showed the highest levels of living cells and cell viability. These results indicate that native starches from Andean crops can be used for biomedical applications.

3. Pharmaceutical Applications

Synthetic and natural biodegradable polymers have been a major focus of interest in pharmaceutical research. They are used to control the rate of drug release from parenteral controlled-delivery systems (Asano et al., 1991). Furthermore, drugs encapsulated within injectable biodegradable micro- or nanospheres can be targeted directly to the site of action (Laakso et al., 1987). In addition, they have enormous potential in the delivery of peptides and proteins by protecting them from premature inactivation (Woo et al., 2001). Because of the considerable advantage of their clearance from the body after the release of therapeutic agents, biodegradable polymers are among the most widely used materials for controlled drug delivery applications. Various starches and their derivatives have been widely studied because of their good molding and film-forming properties, high oxygen barrier, and good mechanical strength (Forssell et al., 2002; Gilleland et al., 2001; Lawton, 1996; Lee and Rhim, 2000; Mali et al., 2002).

Native starches are used as disintegrants, diluents, and wet binders. However, their poor flow and high lubricant sensitivity make them less favorable in direct compression. Various chemical, mechanical, and physical modifications of native starches have been used to improve both their direct compression and their controlled-release properties (van Aerde and Remon, 1988; Sanghvi et al., 1993). Schinzinger and Schmidt (2005) used potato starch as an excipient and compared its granulating behavior with that of α-lactose monohydrate and anhydrous dicalcium phosphate in a laboratory fluidized bed granulator using statistical methods.

Yoshizawa and Koishi (1990) reported on the coating of an interactive powder mixture for sustained-release preparations. The interactive powder mixture consisted of drug particles coated on the surface of potato starch that were then encapsulated in magnesium stearate to make a sustained-release preparation. The nanoparticles of indomethacin were deposited directly on potato starch as the excipient powder by fluidizing the powder by vibration (Nagare and Senna, 2004) and by pulsed laser deposition using a rotation method (Nagare et al., 2007).

Korhonen et al. (2000) evaluated starch acetates and the effects of substitution of the starch acetates on physical and tablet properties and compared them to commercially available direct compression excipients. Tuovinen et al. (2003) compared the drug-release rates from native and acetylated starches. Their results showed that the acetylation of potato starch could substantially retard drug release and that the release profiles could be controlled by the degree of substitution. They further studied drug release from starch acetate microparticles and films with and without incorporation of α-amylase (Tuovinen et al., 2004). The study concluded that the release of model drugs of various molecular weights from the starch acetate microparticles and films was slow compared to that from the native starch preparations. Figure 2 shows the preparation of starch acetate microparticles by the water-in-oil-in-water double-emulsion technique. Szepes et al. (2008) characterized and studied the drug-delivery behavior of starch-based (potato and maize starch) hydrogels prepared via isostatic ultrahigh pressure. They found that potato starch as a gel-forming polymer exhibited faster drug dissolution compared to an aqueous theophylline suspension, used as a reference, while the pressurization of maize starch resulted in a gel exhibiting sustained drug release.

Tarvainen et al. (2002) studied the film-forming ability of starch acetate (DS 2.8) and the effects of commonly used plasticizers on the physical properties of starch acetate films. The properties were compared with ethyl cellulose films. Mechanical studies, water vapor and drug permeability tests, and thermal analysis by differential scanning calorimetry (DSC) were used to characterize the film-forming ability of starch acetate and efficiency of tested plasticizers. Starch acetate films were found to be tougher and stronger than ethyl cellulose films at

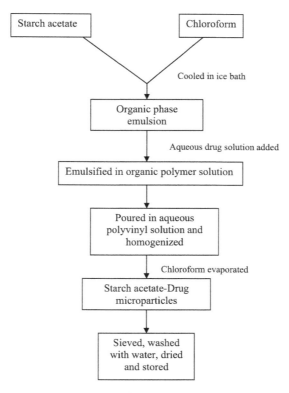

Figure 2
Preparation of starch acetate microparticles by water-in-oil-in-water double-emulsion technique.

the same plasticizer concentration. Also, in most cases, the water vapor permeability of starch acetate films was lower than that of ethyl cellulose films. DSC thermograms supported the findings of the tensile test: plasticizers with several small ester groups (e.g., triacetin and triethyl citrate) were the most compatible with starch acetate. Owing to the good mechanical properties, low water vapor, and drug permeability of the films, starch acetate seems to be a promising film former for pharmaceutical coatings. Tarvainen et al. (2004) evaluated the film-forming properties of a novel, organic, solvent-free aqueous dispersion of potato starch acetate (degree of substitution 2.8) and its ability to control drug release from a coated tablet. The starch acetate dispersion was found to be suitable for the fluid-bed coating process, forming strong films with complete coalescent polymeric spheres. These results clearly demonstrated starch acetate dispersion to have high utility as a novel aqueous coating material for controlled-release products.

Oral controlled drug-release systems are increasingly used for short half-life drugs to reduce peak blood levels and side effects, to maintain optimum drug concentration, and to stimulate patient compliance. To maintain a constant level of the drug in the blood during an extended period, a constant in vitro drug release rate is desired. The most popular controlled-release system is the matrix tablet (Desai et al., 1965). Te Wierik et al. (1996) reported on preparation and binding properties of high-surface-area potato starch products. Te Wierik et al. (1997a,b) reported retrograded modified starch to be a suitable candidate as an excipient in controlled-release matrix tablets. In addition to the ease of tablet manufacturing, it is able to sustain the release of many active ingredients with different physicochemical properties. Acceleration or retardation to a desired profile can be achieved easily by changing tablet geometries, drug load, and incorporation of water-soluble excipients or weak acid or basic compounds or by a combination of two or more of these factors.

Grafting of polyacrylic acid into starch has been considered as an alternative procedure to produce nonirritant delivery systems in tablet form with good bioadhesion and controlled-release properties for buccal application. Geresh et al. (2004) studied grafted starch copolymers as platforms for peroral drug delivery. Grafting onto starch was performed with acrylic acid salts of mono- and divalent cations. The release kinetics of theophylline from graft copolymer tablets was studied as a function of variables such as radiation time, ratio of starch to acrylic acid, type of cation, type of starch (corn, rice, or potato), and pH of the dissolution medium. The suitability of the tablets for buccal application was assessed by determination of the extent of tablet erosion and swelling.

Stepto (1997) focused on the injection molding of potato starch including the basis of the process. In addition, the rheological behavior of starch/water melts during the refill part of the injection molding cycle was analyzed quantitatively to give apparent melt viscosities. Finally, the mechanical properties of molded starch materials and the drug-delivery behavior of starch capsules were also discussed.

Although curcumin has been proved to be highly cytotoxic, its poor solubility in aqueous medium and rapid degradation under physiological pH are the common drawbacks preventing its efficient use. Hence, Pereira et al. (2013) reported the development of starch microspheres cross-linked with N,N-methylenebisacrylamide as an efficient delivering system. The new system showed high loading efficiency with various curcumin concentrations. Because of slow and sustained release of curcumin, starch microspheres could improve the cytotoxicity of curcumin toward Caco-2 and HCT-116 tumor cell lines 40-fold compared to pure curcumin.

Starch microparticles have been shown to be excellent for the controlled release of meclofenamic acid, an anti-inflammatory agent (Malafaya et al., 2001), and for the release of glucocorticoid agents such as dexamethasone (Silva et al., 2005b). When combined with the advantageous bone-bonding properties of BG 45S5, there is a distinct potential for these particles to be used as controlled-release systems of either bone-acting drugs or growth factors. In principle, these systems would be able to bond to the bone and at the same time act as drug-release systems. In theory, the presence of a bone-bonding material (BG) would enhance small-defect bone repair, whereas the biodegradable material would simultaneously act as a scaffold for cell growth by releasing the growth factors incorporated therein. This release would stimulate cell proliferation and differentiation, thus achieving a faster repair. Friciu et al. (2013) proposed a novel excipient for colon delivery for drugs expected to act locally in the colon, to limit the systemic absorption, which decreases effectiveness and causes undesirable side effects. This new preparation consists of the complexation between carboxymethyl starch and lecithin, which is a pH-independent delivery system for drugs such as mesalamine used in the treatment of inflammatory bowel disease.

Inhibitory RNA therapeutics is a powerful tool for treating diseases by sequence-specific targeting of genes using small interfering RNA (siRNA). Modified potato starch, i.e., Q-starch, with cationic properties was developed as a delivery vector that was able to bind siRNA. The resulting complexes with a small diameter (~30 nm) and positive surface charge were able to protect siRNA from enzymatic degradation. Additionally, Q-starch-based complexes showed high cellular uptake during a 24-h study (Amar-Lewis et al., 2014). Janaswamy (2014) reported on the opportunity for the design and development of novel starch-based stable carriers that not only have altered digestion rates but could also deliver health-promoting and disease-preventing compounds. Water channels native to root starches, e.g., potato starch, could embed drugs, nutraceuticals, flavor compounds, and vitamins.

4. Fermentation Applications

Bioethanol production by yeasts is widely used for biodegradation of potato. However, yeasts cannot ferment starch directly, and a two-step enzymatic reaction to glucose is necessary. Various potato wastes such as industrial residues, low-grade potatoes, and spoiled potatoes can be used for acetone/ethanol production. Nimcevic et al. (1998) used whole potato media to study the solvent production by *Clostridium beijerinckii* NRRL B592 and showed that potato can serve as an excellent substrate for acetone/butanol fermentation. Their investigations were also directed toward the development of a continuous culture with online product removal, whereby productivity and yield could be significantly improved, effluent problems could be reduced, and complete utilization of the substrate could be achieved.

The global market for lactic acid and lactate (polymers excluded) production ranges at about 100,000 tons per year with an annual growth rate of 15% (Akerberg and Zacchi, 2000). Lactic acid production is currently attracting a great deal of research and development interest since it is one of the most important organic acids owing to its potential for becoming a very large volume commodity–chemical intermediate that can be produced from renewable carbohydrates. Lactic acid can be used as feedstock for biodegradable polymers, oxygenated chemicals, plant growth regulators, environmentally friendly "green" solvents, and special chemical intermediates (Datta et al., 1995; Huang et al., 2005a). One of its most promising applications is its use in biodegradable and biocompatible polylactic acid polymers, an environmentally friendly alternative to nonbiodegradable plastics derived from petrochemicals. Jin et al. (2003) reported a new strain of *Rhizopus arrhizus* DAR 36017, for simultaneous saccharification and fermentation of starch waste effluents for lactic acid production. They found supplementation of the nitrogen source to be unnecessary if potato or corn starch waste effluent was used as the production medium. Similar reports were obtained by Jin et al. (2005), quoting a single-stage simultaneous saccharification and fermentation process using potato, corn, wheat, and pineapple waste streams as production media. Direct fermentation of lactic acid production from starch wastewater with respect to growth pH, temperature, and substrate was also reported by Huang et al. (2005b). The performance of direct fermentation was characterized by starch hydrolysis, accumulation of reducing sugar, production of lactic acid, and fungal biomass.

The practical technique for lactic acid fermentation of potato pulp has been developed (Oda et al., 2002). They screened 38 strains of the fungus *Rhizopus oryzae*; either lactic acid or fumaric acid and ethanol were formed, and the ratio differed among the strains tested. Saito et al. (2003) studied the effects of pectinolytic enzymes on lactic acid fermentation of potato pulp by different *R. oryzae* NRRL 395 and NBRC 4707 strains. When a commercial preparation of pectinase was added to potato pulp inoculated with fungal spores and incubated for 7 days, both strains effectively produced larger amounts of lactic acid and ethanol. These data indicated that the fermentation of potato pulp depends on the degradation of pectic substances by NRRL 395 and NBRC 4707. Saito et al. (2006) found the pectin content in potato pulp to be dependent on the season.

Potato pulp and potato pulp residue (after acidic treatment) and nutrient salt solution and potato protein liquor, respectively, were used for the production of cellulases and hemicellulases by *Trichoderma reesei* Rut C30 in a continuously operated bioreactor (Klingspohn and Schügerl, 1993). Trojanowski et al. (1995) reported on the utilization of potato pulp and potato liquor for the production of laccase by basidiomycetes. Transglutaminase is an enzyme used as a food additive to improve the textural properties of proteinaceous foods. Vázquez and Rodríguez (2011) simulated transglutaminase production from potatoes and evaluated its feasibility using the software "SuperPro Designer." The industrial production of transglutaminase by *Streptomyces mobaraensis* from potato can be a feasible alternative with a payback time of 5 years.

Meyer et al. (2009) have reported the release of high-molecular-weight pectinaceous dietary fiber polysaccharides from starch-free potato pulp by using a multicomponent pectinase preparation from *Aspergillus aculeatus* (Viscozyme® L), suggesting that enzymatic upgrading of potato pulp is a possible route for obtaining new nutri-functional fibers for food supplementation (Figure 3).

Potato peel waste contains starch, cellulose, hemicellulose, and fermentable sugars to warrant use as a feedstock. Shukla and Kar (2006) studied potato peel as a solid-state substrate for

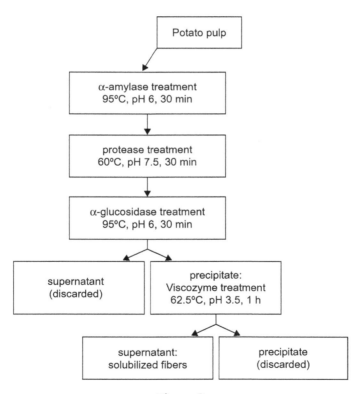

Figure 3
Flow sheet for enzyme-assisted release of a soluble potato fiber fraction from raw potato pulp. Each separation step resulting in a supernatant and a precipitate involves centrifugation for 10 min at 30,000 × g, washing of the pellet with doubly distilled water, and recentrifugation for 10 min at 30,000 × g. *Reproduced from Meyer et al. (2009), with permission from Elsevier.*

thermostable α-amylase production by thermophilic *Bacillus* isolates. Under optimal conditions, *Bacillus licheniformis* produced 270 and 175 units/ml of α-amylase on potato peel and wheat bran, respectively, while the corresponding values for *Bacillus subtilis* were 600 and 265 units/ml. Mukherjee et al. (2008) used potato peel and *Imperata cylindrica* grass mixed in a ratio of 1:1 (w/w) for the maximum production of alkaline protease. Arotupin (2007) studied growth and production of polygalacturonase of *Aspergillus flavus* isolated from cropped soils on various raw and commercial carbon substrates. Ripe banana peel supported maximum growth, followed by orange bagasse, unripe plantain peel, and potato peel, whereas the order for production of polygalacturonase production was potato > ripe banana peel > orange bagasse > ripe plantain peel > unripe plantain peel > soluble starch > unripe banana > cassava peel. Mabrouk and El Ahwany (2008) studied the production of β-mannanase by *Bacillus amyloliquefaciens* 10A1 cultured on potato peels. They reported that potato peels at 14 g/l as carbon source and ammonium nitrate as a nitrogen source produced maximum enzyme activity (61.5 units/mg protein). The optimum incubation temperature and pH for enzyme production were 35 °C and 7, respectively.

Enzymatic- and acid-hydrolyzed potato peel waste was fermented by *Saccharomyces cerevisiae* var. *bayanus* for ethanol production. Enzymatic hydrolysis with a combination of three enzymes released 18.5 g/l reducing sugar and produced 7.6 g/l of ethanol after fermentation, which demonstrated its potential for ethanol production (Arapoglou et al., 2010). Khawla et al. (2014) demonstrated the possibility of substituting commercial amylase by an onsite-produced enzyme using potato peel waste and its potential as a bioethanol feedstock. First, an economical medium for amylase production from *Bacillus* sp. UEB-S (amylase UEB-S) was formulated using potato peel as a carbon source. Amylase production reached 15 units/ml at flask scale and increased to 87 units/ml at fermenter scale. A comparative study between acidic and enzymatic hydrolysis of peel waste was investigated and found that a combined enzymatic treatment using amylase UEB-S and amyloglucosidase under optimal conditions gave significant yields of reducing sugars (69 g/l) and ethanol (21 g/l).

Biosurfactants (surface-active molecules produced by microorganisms) are commonly used in specialty markets such as food and textiles and have potential for improved oil recovery. Currently, biosurfactants are not widely utilized in the petroleum industry because of high production costs associated with the use of expensive substrates and inefficient product-recovery methods. The economics of biosurfactant production could be significantly affected through the use of media optimization and application of inexpensive carbon substrates such as agricultural process residuals. Utilization of biosurfactants produced from agricultural residuals may result in an economic advantage for surfactant production and technology application and convert a substantial agricultural waste stream into a value-added product. Bala et al. (2002) discussed the production of surfactants from microbiological growth media based on simple sugars, chemically pure starch medium, simulated liquid and solid potato-process effluent media, a commercially prepared potato starch in mineral salts, and process effluent from a potato processor. Thompson et al. (2001) studied the effects of pretreatments such as heat, removal of starch particulates, and acid hydrolysis on surfactin production from potato process effluent by *B. subtilis*. Thompson et al. (2000) tested high-solid and low-solid potato process effluents as substrates for surfactin production. They concluded that low solids could potentially be used without sterilization for surfactin production for low-value applications such as environmental remediation or oil recovery. In continuation, Noah et al. (2002) described continuous surfactin production from low-solid potato process effluent by *B. subtilis* in an airlift reactor. Das and Mukherjee (2007) used potato peels as a cheap carbon source for the production of crude lipopeptide biosurfactants from *B. subtilis* DM-03 and DM-04 strains through fermentation. They reported that these thermostable biosurfactants exhibited strong emulsifying properties and could release an appreciable amount of oil from a saturated sand-pack column (Das and Mukherjee, 2007).

Govindsamy et al. (2013) attempted to utilize potato powder as a low-cost carbon source and statistically evaluated the medium for production of red pigment by *Monascus purpureus* (MTCC 369). The potato-processing wastewater containing high concentrations of solids, starch, and nutrients was used for microbial lipid production by the oleaginous filamentous fungus *Aspergillus oryzae* with the purpose of recycling potato-processing wastewater for biodiesel production. Microbial lipids of *A. oryzae* contained major fatty acids such as palmitic acid (11.6%), palmitolic acid (15.6%), stearic acid (19.3%), oleic acid (30.3%), linolenic acid (5.5%), and linoleic acid (6.5%), suggesting that the lipids were suitable for the production of second-generation biodiesel (Muniraj et al., 2013). Table 3 compiles the use of potato starch as a carbon source for the production of various biomolecules such as enzymes and organic acids.

Laurinavichene et al. (2010) studied an integrated process utilizing potato homogenate for dark, fermentative production of hydrogen followed by photoproduction of hydrogen using purple nonsulfur bacteria. The utilization of potato juice for production of biogas was

Table 3: Fermentation Products Obtained by Using Potato Starch as a Carbon Source.

Product	Microorganism	References
Thermostable pullulanase	*Clostridium thermosulfurogenes* SV2	Rama Mohan Reddy et al. (1999)
Glucoamylase	*S. cerevisiae* strain C468	Pavezzi et al. (2008)
Lactic acid	*R. arrhizus* WEBL 0501	Zhang et al. (2007)
	Lactobacillus amylophilus GV6	Vishnu et al. (2002)
Citric acid	*A. niger* GCB-47 (parental strain) and GCMC-7 (mutant strain)	Haq et al. (2003)
Cholesterol oxidase	*Streptomyces lavendulae* NCIM 2421	Varma and Nene (2003)
Thermostable β-amylase	*C. thermosulfurogenes* SV2	Rama Mohan Reddy et al. (2003)
Amylase inhibitor	*Streptomyces nigrifaciens* NTU-3314	Su et al. (1984)
Ethanol	*A. niger* and *S. cerevisiae*	Jeon et al. (2007)

investigated in batch mode and in high-rate anaerobic reactors. They also proposed that the reactor effluent could be concentrated by evaporation to minimize its volume and later be utilized as fertilizer (Fang et al., 2011). An economically sustainable process was developed for propionic acid production by fermentation of glycerol using *Propionibacterium acidipropionici* and potato juice as a nitrogen/vitamin source. Succinic acid was 13 g% of propionic acid and could represent a potential coproduct covering the cost of the nitrogen/vitamin source (Dishisha et al., 2013).

5. Other Applications

5.1 Animal Feed

Silage produced to preserve forage with high moisture content by controlled fermentation is an important winter feed for cattle. Efforts toward an increased use of potato pulp were primarily directed to a broader application as animal feed (Lisinska and Leszczynski, 1989). Okine et al. (2005) studied the effect of addition of two bacteria inoculants as *Lactobacillus rhamnosus* and *R. oryzae* at ensiling on the fermentation quality, change in nutrient composition, and nutritive value of potato pulp silage. They concluded that the potato pulp can ensile well with or without bacterial inoculants.

A study by Liu et al. (2014) has reported the conversion of potato starch processing waste into single-cell protein (SCP) using a two-step fermentation process. They selected the mutant strain *Aspergillus niger* H3 after UV irradiation and ethylmethyl sulfone mutagenesis for more cellulase production. Treatment of the potato waste with strain H3 followed by liquid fermentation using *B. licheniformis* was reported to produce SCP that could be used as animal feed.

5.2 Technical Applications

Only small fractions of pulp are used for technical applications such as production of glue (Mayer, 1998). Mayer and Hillebrandt (1997) reported the microbiological characterization, physical modification, and application of potato pulp. They identified autochthonic microbial flora (bacterial, fungi) and studied them with a view toward the degradative potential of the microorganisms and ways of conserving the pulp for subsequent technical applications such as animal feed and production of glue.

Gupta et al. (2011) have reported the removal of dye pollutants from the textile, paper, and leather industries using agricultural waste from the potato plant. Potato stem powder (PSP) and potato leaves powder (PLP) were used as adsorbents for removal of cationic dyes (methylene blue (MB) and malachite green (MG)) from aqueous solution (Figure 4). They reported that rough surface and the presence of functional groups in PSP and PLP made them beneficial for this application.

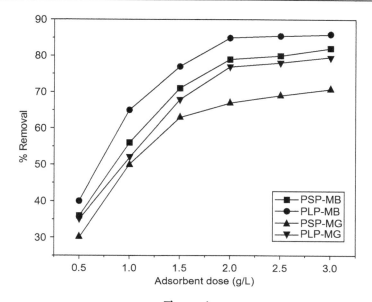

Figure 4

Effect of adsorbent dose on adsorption of MB/MG onto PSP/PLP (initial dye concentration 10 mg/l, pH 7, temperature 303 K, contact time 25 min). *Reproduced from Gupta et al. (2011).*

5.3 Functional Food

Tapé is a popular Indonesian delicacy with a sweet–acid taste and mild alcoholic flavor. It is prepared by fermenting glutinous rice or cassava tuber using *ragi tapé* and consumed as a sweet dessert or snack. Among the microorganisms present in *ragi tapé*, the fungus *Amylomyces rouxii* plays a crucial role in *tapé* fermentation. *A. rouxii* degrades starch to glucose, which sustains its growth and that of some yeasts and bacteria and simultaneously synthesizes both lactic acid and ethanol. Abe et al. (2004) reported that potato pulp fermented by *ragi tapé* may be consumed as other *tapé* products. They converted potato pulp to palatable foodstuff by microbial fermentation and compared microflora of *tapé* made with potato pulp to *tapé ketan*, the conventional *tapé* made with glutinous rice.

In recent years, there has been an increasing interest in finding natural antioxidants, since they can protect the human body from free radicals and retard the progress of many chronic diseases. Potato peel could be considered as a new source of natural antioxidant. Potato peel contains phenolic acids (Lisinska and Leszczynski, 1987). Al-Weshahy and Rao (2009) analyzed the polyphenolic content and antioxidant properties of peel samples from potatoes grown in Ontario, Canada (Figure 5). They reported that red-colored potato varieties had the highest antioxidant potential compared to other varieties.

The high contents of chlorogenic acid and caffeic acid in the extracts of the downstream wastes from potato processing make it a promising candidate for functional foods and

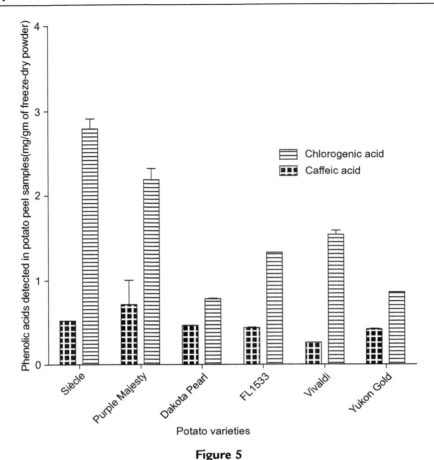

Figure 5
HPLC analysis of potato peel sample extracts from six varieties. *Reproduced from Al-Weshahy and Rao (2009); with permission from Elsevier.*

nutraceutical ingredients (Wu et al., 2012). The antioxidant activity of potato peel extract has been studied in food systems (Rodriguez de Sotillo et al., 1994a,b). Singh et al. (2008) reported potato peel extract to have the potential to offer protection against acute liver injury in rats because of its antioxidant propensity. Singh and Rajini (2008) investigated the ability of potato peel extract to protect erythrocytes against oxidative damage in vitro. The protection rendered by extract in erythrocytes was studied in terms of resistance to oxidative damage, morphological alterations, and membrane structural alterations.

Fats and oils undergo pronounced oxidative changes at elevated temperature during storage, thereby decreasing the nutritional value and the consumer acceptability. However, the addition of some suitable antioxidants in fats and oils retards the oxidation process. Synthetic antioxidants, especially butylated hydroxytoluene (BHT) and butylated hydroxyanisole (BHA), are commonly used to prevent the oxidation process (Sobedio et al., 1991). These synthetic antioxidants are known to have toxic and carcinogenic effects on humans (Ito et al., 1986). Zia-ur-Rehman et al. (2004) evaluated potato peel extract as a natural antioxidant during 60 days storage of refined soybean oil at 25 and 45 °C. Free fatty acids, peroxide values, and iodine values were used as criteria to assess the antioxidant activity. They concluded that potato peel extract exhibits an antioxidant activity, which was almost equal to that of BHA and BHT. Therefore potato peel extract in oils, fats, and other food products can safely be used as a natural antioxidant to suppress lipid oxidation.

Potato peel waste is a potential source of steroidal alkaloids such as α-solanine, α-chaconine, solanidine, and demissidine, which are biologically active secondary metabolites and which could serve as precursors to agents with apoptotic, chemopreventive, and anti-inflammatory properties (Hossain et al., 2014).

Dietary fiber comes from the portion of plants that is not digested by enzymes in the intestinal tract. Various types of plants have varying amounts and kinds of fiber, including pectin, gum, mucilage, cellulose, hemicellulose, and lignin. Insoluble fiber is helpful in the treatment and prevention of constipation, hemorrhoids, and diverticulosis. Water-soluble fiber binds bile acids, suggesting that a high-fiber diet may result in an increased excretion of cholesterol. Dietary fiber is known to reduce the risk of some cancers, especially colon cancer. High-fiber diets are also useful in weight loss regimens. Kaack et al. (2006) compared the functional properties of five fiber fractions by substitution of wheat flour using dry potato pulp, a commercial potato fiber, two fibers prepared from potato pulp by enzymatic hydrolysis, and one solubilized fiber in bread making. The effects of the chemical composition of fiber on texture, color, specific weight, and volume of wheat bread were studied. They concluded that the enzymatically solubilized fiber with a high concentration of soluble fiber, and a low concentration of cellulose and lignin could be used for substitution of at least 12% wheat flour for baking of bread, with an attractive color and delicious texture and flavor.

References

Abe, A., Sujaya, I.-N., Sone, T., Asano, K., Oda, Y., 2004. Microflora and selected metabolites of potato pulp fermented with an Indonesian starter *ragi tapé*. Food Technology and Biotechnology 42 (3), 169–173.

Akerberg, C., Zacchi, G., 2000. An economic evaluation of the fermentative production of lactic acid from wheat flour. Bioresource Technology 75, 119–126.

Al-Weshahy, A., Rao, A.V., 2009. Isolation and characterization of functional components from peel samples of six potatoes varieties growing in Ontario. Food Research International 42, 1062–1066.

Amar-Lewis, E., Azagury, A., Chintakunta, R., Goldbart, R., Traitel, T., Prestwood, J., Landesman-Milo, D., Peer, D., Kost, J., 2014. Quaternized starch-based carrier for siRNA delivery: from cellular uptake to gene silencing. Journal of Controlled Release 185, 109–120.

Arapoglou, D., Varzakas, Th, Vlyssides, A., Israilides, C., 2010. Ethanol production from potato peel waste (PPW). Waste Management 30 (10), 1898–1902.

Arotupin, D.J., 2007. Effect of different carbon sources on the growth and polygalacturonase activity of *Aspergillus flavus* isolated from cropped soils. Research Journal of Microbiology 2 (4), 362–368.

Asano, M., Fukuzaki, H., Yoshida, M., Kumakura, M., Mashimo, T., Yuasa, H., Imai, K., Yamanaka, H., Kawaharada, U., Suzuki, K., 1991. In vivo controlled release of a luteinizing hormone-releasing hormone agonist from poly(DL-lactic acid) formulation of varying degradation pattern. International Journal of Pharmaceutics 67, 67–77.

van Aerde, P., Remon, J.P., 1988. In vitro evaluation of modified starches as matrices for sustained release dosage form. International Journal of Pharmaceutics 45, 145–152.

Bala, G.A., Bruhn, D.F., Fox, S.L., Noah, K.S., Thompson, D.N., 2002. Microbiological production of surfactant from agricultural residuals for IOR application. In: Proceedings – SPE Symposium on Improved Oil Recovery, pp. 1003–1011.

Beilvert, A., Chaubet, F., Chaunier, L., Guilois, S., Pavon-Djavid, G., Letourneur, D., Meddahi-Pellé, A., Lourdin, D., 2014. Shape-memory starch for resorbable biomedical devices. Carbohydrate Polymers 99 (2), 242–248.

Boesel, L.F., Mano, J.F., Elvira, C., San Roman, J., Reis, R.L., 2003. Hydrogels and hydrophilic partially degradable bone cements based on biodegradable blends incorporating starch. In: Chiellini, E. (Ed.), Biodegradable Polymers and Plastics. Kluwer Academic, Dordrecht.

Chen, D., Lawton, D., Thompson, M.R., Liu, Q., 2012. Biocomposites reinforced with cellulose nanocrystals derived from potato peel waste. Carbohydrate Polymers 90 (1), 709–716.

Cummings, J.H., Beatty, E.R., Kingman, S.M., Bingham, S.A., Englyst, H.N., 1996. Digestion and physiological properties of resistant starch in the human large bowel. British Journal of Nutrition 75, 733–747.

Das, K., Mukherjee, A.K., 2007. Comparison of lipopeptide biosurfactants production by *Bacillus subtilis* strains in submerged and solid state fermentation systems using a cheap carbon source: some industrial applications of biosurfactants. Process Biochemistry 42, 1191–1199.

Datta, R., Tsai, S.P., Bonsignor, P., Moon, S., Frank, J., 1995. Technological and economic potential of poly(lactic acid) and lactic acid derivatives. FEMS Microbiology Reviews 16, 221–231.

Desai, S.J., Simonelli, A.P., Higuchi, W.I., 1965. Investigation of factors influencing release of solid drug dispersed in inert matrices. Journal of Pharmaceutical Sciences 54, 1459–1464.

Dishisha, T., Ståhl, Å., Lundmark, S., Hatti-Kaul, R., 2013. An economical biorefinery process for propionic acid production from glycerol and potato juice using high cell density fermentation. Bioresource Technology 135, 504–512.

Elvira, C., Mano, J.F., San Roman, J., Reis, R.L., 2002. Starch-based biodegradable hydrogels with potential biomedical applications as drug delivery systems. Biomaterials 23 (9), 1955–1966.

Fang, C., Boe, K., Angelidaki, I., 2011. Biogas production from potato-juice, a by-product from potato-starch processing, in upflow anaerobic sludge blanket (UASB) and expanded granular sludge bed (EGSB) reactors. Bioresource Technology 102 (10), 5734–5741.

Feustel, I.C., 1987. Miscellaneous product from potato. In: Talburt, W.F., Smith, O. (Eds.), Potato Processing. Van Nostrand, New York, pp. 727–746.

Friciu, M.M., Le, T.C., Ispas-Szabo, P., Mateescu, M.A., 2013. Carboxymethyl starch and lecithin complex as matrix for targeted drug delivery: I. Monolithic mesalamine forms for colon delivery. European Journal of Pharmaceutics and Biopharmaceutics 85 (3 Part A), 521–530.

Forssell, P., Lahtinen, R., Lahelin, M., Myllarinen, P., 2002. Oxygen permeability of amylose and amylopectin films. Carbohydrate Polymers 47, 125–129.

Geresh, S., Gdalevsky, G.Y., Gilboa, I., Voorspoels, J., Remon, J.P., Kost, J., 2004. Bioadhesive grafted starch copolymers as platforms for peroral drug delivery: a study of theophylline release. Journal of Controlled Release 94 (2–3), 391–399.

Gilleland, G.M., Turner, J.L., Patton, P.A., Harrison, M.D., 2001. Modified starch as a replacement for gelatin in soft gel films and capsules. International Patent Number, WO 01/91721 A2.

Gomes, M.E., Ribeiro, A.S., Malafaya, P.B., Reis, R.L., Cunha, A.M., 2001. A new approach based on injection moulding to produce biodegradable starch based polymeric scaffolds: morphology, mechanical, and degradation behaviour. Biomaterials 22, 883–889.

Gomes, M.E., Godinho, J.S., Tchalamov, D., Cunha, A.M., Reis, R.L., 2002. Alternative tissue engineering scaffolds based on starch: processing methodologies, morphology, degradation, mechanical properties and biological response. Materials Science and Engineering C 20 (1–2), 19–26.

Govindsamy, S., Bhagat, N., Chandrasekaran, M., 2013. Sequential statistical optimization of red pigment production by *Monascus purpureus* (MTCC 369) using potato powder. Industrial Crops and Products 44, 158–164.

Gupta, N., Kushwaha, A.K., Chattopadhyaya, M.C., 2011. Application of potato (*Solanum tuberosum*) plant wastes for the removal of methylene blue and malachite green dye from aqueous solution. Arabian Journal of Chemistry Available online July 30, 2011.

Haq, I., Ali, S., Iqbal, J., 2003. Direct production of citric acid from raw starch by *Aspergillus niger*. Process Biochemistry 38, 921–924.

Hossain, M.B., Tiwari, B.K., Gangopadhyay, N., O'Donnell, C.P., Brunton, N.P., Rai, D.K., 2014. Ultrasonic extraction of steroidal alkaloids from potato peel waste. Ultrasonics Sonochemistry 21 (4), 1470–1476.

Hsieh, W.-C., Liau, J.-J., 2013. Cell culture and characterization of cross-linked poly(vinyl alcohol)-g-starch 3D scaffold for tissue engineering. Carbohydrate Polymers 98 (1), 574–580.

Huang, L.P., Jin, B., Lant, P., 2005a. Direct fermentation of potato starch wastewater to lactic acid by *Rhizopus oryzae* and *Rhizopus arrhizus*. Bioprocess and Biosystems Engineering 27, 229–238.

Huang, L.P., Jin, B., Lant, P., Zhou, J., 2005b. Simultaneous saccharification and fermentation of potato starch wastewater to lactic acid by *Rhizopus oryzae* and *Rhizopus arrhizus*. Biochemical Engineering Journal 23, 265–276.

Hylla, S., Gostner, A., Dusel, G., Anger, H., Bartram, H.P., Christl, S.U., Kasper, H., Scheppach, W., 1998. Effects of resistant starch on the colon in healthy volunteers: possible implications for cancer prevention. The American Journal of Clinical Nutrition 67, 136–142.

Ito, N., Hiroze, M., Fukushima, G., Tauda, H., Shira, T., Tatematsu, M., 1986. Studies on antioxidant; their carcinogenic and modifying effects on chemical carcinogensis. Food and Chemical Toxicology 24, 1071–1081.

Janaswamy, S., 2014. Encapsulation altered starch digestion: toward developing starch-based delivery systems. Carbohydrate Polymers 101, 600–605.

Jeon, B.Y., Kim, S.J., Kim, D.H., Na, B.K., Park, D.H., Tran, H.T., Zhang, R., Ahn, D.H., 2007. Development of a serial bioreactor system for direct ethanol production from starch using *Aspergillus niger* and *Saccharomyces cerevisiae*. Biotechnology and Bioprocess Engineering 12 (5), 566–573.

Jin, B., Huang, L.P., Lant, P., 2003. *Rhizopus arrhizus* – a producer for simultaneous saccharification and fermentation of starch waste materials to L(+)-lactic acid. Biotechnology Letters 25, 1983–1987.

Jin, B., Yin, P., Ma, Y., Zhao, L., 2005. Production of lactic acid and fungal biomass by *Rhizopus* fungi from food processing waste streams. Journal of Industrial Microbiology & Biotechnology 32, 678–686.

Kaack, K., Pedersen, L., Laerke, H.N., Meyer, A., 2006. New potato fibre for improvement of texture and colour of wheat bread. European Food Research and Technology 224, 199–207.

Kempf, W., 1980. Beitrag zur mechanischen Entwasserung von Kartoffelpulpe. Starch/Starke 32, 14–20.

Khawla, B.J., Sameh, M., Imen, G., Donyes, F., Dhouha, G., Raoudha, E.G., Oumèma, N.-E., 2014. Potato peel as feedstock for bioethanol production: a comparison of acidic and enzymatic hydrolysis. Industrial Crops and Products 52, 144–149.

Klingspohn, U., Schügerl, K., 1993. Integrated enzyme production in continuous operation by utilization of potato pulp. Journal of Biotechnology 29 (1–2), 109–119.

Korhonen, O., Raatikainen, P., Harjunen, P., Nakari, J., Suihko, E., Peltonen, S., Vidgren, M., Paronen, P., 2000. Starch acetates—multifunctional direct compression excipients. Pharmaceutical Research 17 (9), 1138–1143.

Laakso, T., Stjärnkvist, P., Sjöholm, I., 1987. Biodegradable microspheres VI: lysosomal release of covalently bound antiparasitic drugs from starch microparticles. Journal of Pharmaceutical Sciences 76 (2), 134–140.

Laurinavichene, T.V., Belokopytov, B.F., Laurinavichius, K.S., Tekucheva, D.N., Seibert, M., Tsygankov, A.A., 2010. Towards the integration of dark- and photo-fermentative waste treatment. 3. Potato as substrate for sequential dark fermentation and light-driven H_2 production. International Journal of Hydrogen Energy 35 (16), 8536–8543.

Lawton, J.W., 1996. Effect of starch type on the properties of starch containing films. Carbohydrate Polymers 29, 203–208.

Lee, J.J., Rhim, J.W., 2000. Characteristics of edible films based with various cultivars of sweet potato starch. Korean Journal of Food Science and Technology 32 (4), 834–842.

Lisinska, G., Leszczynski, W., 1989. In: Lisinska, G., Leszczynski, W. (Eds.), Potato Science and Technology. Elsevier Applied Science, London, UK.

Lisinska, G., Leszczynski, W., 1987. Potato tubers as raw materials for processing and nutrition. In: Lisinska, G., Leszczynski, W. (Eds.), Potato Science and Technology, Chapter 2. Elsevier Applied Science, London, England, pp. 34–38.

Liu, B., Li, Y., Song, J., Zhang, L., Dong, J., Yang, Q., 2014. Production of single-cell protein with two-step fermentation for treatment of potato starch processing waste. Cellulose 21, 3637–3645.

Mabrouk, M.E.M., El Ahwany, A.M.D., 2008. Production of β-mannanase by *Bacillus amylolequifaciens* 10A1 cultured on potato peels. African Journal of Biotechnology 7 (8), 1123–1128.

Malafaya, P.B., Elvira, C., Gallardo, A., San Roman, J., Reis, R.L., 2001. Porous starch-based drug delivery systems processed by a microwave treatment. Journal of Biomaterials Science Polymer Edition 12 (11), 1227–1241.

Mali, S., Grossmann, M.V.E., Garcia, M.A., Martino, M.N., Zaritzky, N.E., 2002. Microstructural characterization of yam starch films. Carbohydrate Polymers 50, 379–386.

Marques, A.P., Reis, R.L., Hunt, J.A., 2001. In vitro evaluation of the biocompatibility of novel starch based polymeric and composite material. Biomaterials 21, 1471–1478.

Mayer, F., 1998. Potato pulp: properties, physical modification and application. Polymer Degradation and Stability 59, 231–235.

Mayer, F., Hillebrandt, J.O., 1997. Potato pulp: microbiological characterization, physical modification, and application of this agricultural waste product. Applied Microbiology and Biotechnology 48, 435–440.

Mendes, S.C., Bovell, Y.P., Reis, R.L., Cunha, A.M., de Bruijn, J.D., van Blitterswijk, C.A., 2001. Biocompatibility testing of novel starch-based materials with potential application in orthopaedic surgery. Biomaterials 22, 2057–2064.

Meyer, A.S., Dam, B.P., Lærke, H.N., 2009. Enzymatic solubilization of a pectinaceous dietary fiber fraction from potato pulp: optimization of the fiber extraction process. Biochemical Engineering Journal 43, 106–112.

Mukherjee, A.K., Adhikari, H., Rai, S.K., 2008. Production of alkaline protease by a thermophilic *Bacillus subtilis* under solid-state fermentation (SSF) condition using *Imperata cylindrical* grass and potato peel as low-cost medium: characterization and application of enzyme in detergent formulation. Biochemical Engineering Journal 39, 353–361.

Muniraj, I.K., Xiao, L., Hu, Z., Zhan, X., Shi, J., 2013. Microbial lipid production from potato processing wastewater using oleaginous filamentous fungi *Aspergillus oryzae*. Water Research 47 (10), 3477–3483.

Nagare, S., Senna, M., 2004. Indomethacin nanoparticles directly deposited on the fluidized particulate excipient by pulsed laser deposition. Journal of Nanoparticle Research 6 (6), 589–593.

Nagare, S., Sagawa, J., Senna, M., 2007. Investigation and control of the uniformity of drug nanoparticles directly deposited on the particulate surfaces of excipient by PLD. Journal of Physics: Conference Series 59, 88–91.

Nasri-Nasrabadi, B., Mehrasa, M., Rafienia, M., Bonakdar, S., Behzad, T., Gavanji, S., 2014. Porous starch/cellulose nanofibers composite prepared by salt leaching technique for tissue engineering. Carbohydrate Polymers 108 (8), 232–238.

Nimcevic, D., Schuster, M., Gapes, J.R., 1998. Solvent production by *Clostridium beijerinckii* NRRL B592 growing on different potato media. Applied Microbiology and Biotechnology 50, 426–428.

Noah, K.S., Fox, S.L., Bruhn, D.F., Thompson, D.N., Bala, G.A., 2002. Development of continuous surfactin production from potato process effluent by *Bacillus subtilis* in an airlift reactor. Applied Biochemistry and Biotechnology 99 (1–3), 803–814.

Oda, Y., Saito, K., Yamauchi, H., Mori, M., 2002. Lactic acid fermentation of potato pulp by the fungus *Rhizopus oryzae*. Current Microbiology 45, 1–4.

Okine, A., Hanada, M., Aibibula, Y., Okamoto, M., 2005. Ensiling of potato pulp with or without bacterial inoculants and its effect on fermentation quality, nutrient composition and nutritive value. Animal Feed Science and Technology 121 (3–4), 329–343.

Pereira, A.G.B., Fajardo, A.R., Nocchi, S., Nakamura, C.V., Rubira, A.F., Muniz, E.C., 2013. Starch-based microspheres for sustained-release of curcumin: preparation and cytotoxic effect on tumor cells. Carbohydrate Polymers 98 (1), 711–720.

Pavezzi, F.C., Gomes, E., da Silva, R., 2008. Production and characterization of glucoamylase from fungus *Aspergillus awamori* expressed in yeast *Saccharomyces cerevisiae* using different carbon sources. Brazilian Journal of Microbiology 39, 108–114.

Pereira, C.S., Cunha, A.M., Reis, R.L., Vazquez, B., San Roman, J., 1998. New starch-based thermoplastic hydrogels for use as bone cements or drug-delivery carriers. Journal of Materials Science: Materials in Medicine 9 (12), 825–833.

Raban, A., Tagliabue, A., Christensen, N.J., Madsen, J., Host, J.J., Astrup, A., 1994. Resistant starch: the effect on postprandial glycemia, hormonal response, and satiety. The American Journal of Clinical Nutrition 60, 544–551.

Rama Mohan Reddy, P., Ramesh, B., Mrudula, S., Reddy, G., Seenayya, G., 2003. Production of thermostable β-amylase by *Clostridium thermosulfurogenes* SV2 in solid-state fermentation: optimization of nutrient levels using response surface methodology. Process Biochemistry 39, 267–277.

Rama Mohan Reddy, P., Reddy, G., Seenayya, G., 1999. Production of thermostable pullulanase by *Clostridium thermosulfurogenes* SV2 in solid-state fermentation: optimization of nutrients levels using response surface methodology. Bioprocess Engineering 21, 497–503.

Reis, R.L., Cunha, A.M., 2001. Starch and starch based thermoplastics. Biological and biomimetic materials. In: Jurgen Buschow, K.H., Cahn, R.W., Flemings, M.C., Ilschner, B., Kramer, E.J., Mahajan, S. (Eds.). Jurgen Buschow, K.H., Cahn, R.W., Flemings, M.C., Ilschner, B., Kramer, E.J., Mahajan, S. (Eds.), Encyclopedia of Materials Science and Technology, vol. 11. Pergamon–Elsevier Science, Amsterdam, pp. 8810–8816.

Reis, R.L., Cunha, A.M., Allan, P.S., Bevis, M.J., 1996. Mechanical behavior of injection moulded starch based polymers. Journal of Polymers for Advanced Technologies 7, 784–790.

Rodriguez de Sotillo, D., Hadley, M., Holm, E.T., 1994a. Phenolics in aqueous potato peel extract: extraction, identification and degradation. Journal of Food Science 59, 649–651.

Rodriguez de Sotillo, D., Hadley, M., Holm, E.T., 1994b. Potato peel waste: stability and antioxidant activity of a freeze-dried extract. Journal of Food Science 59, 1031–1033.

Saito, K., Kawamura, Y., Oda, Y., 2003. Role of the pectinolytic enzyme in the lactic acid fermentation of potato pulp by *Rhizopus oryzae*. Journal of Industrial Microbiology & Biotechnology 30, 440–444.

Saito, K., Noda, T., Tsuda, S., Mori, M., Hasa, Y., Kito, H., Oda, Y., 2006. Effect of the dates of extraction on the quality of potato pulp. Bioresource Technology 97, 2470–2473.

Sanghvi, P.P., Collins, C.C., Shukla, A.J., 1993. Evaluation of Preflo® modified starches as new direct compression excipients I. Tabletting characteristics. Pharmaceutical Research 10, 1597–1603.

Schinzinger, O., Schmidt, P.C., 2005. Comparison of the granulation behavior of three different excipients in a laboratory fluidized bed granulator using statistical methods. Pharmaceutical Development and Technology 10 (2), 175–188.

Shukla, J., Kar, R., 2006. Potato peel as a solid state substrate for thermostable α-amylase production by thermophilic *Bacillus* isolates. World Journal of Microbiology & Biotechnology 22, 417–422.

Silva, G.A., Costa, F.J., Coutinho, O.P., Radin, S., Ducheyne, P., Reis, R.L., 2004. Synthesis and evaluation of novel bioactive composite starch/bioactive glass microparticles. Journal of Biomedical Materials Research Part A 70A (3), 442–449.

Silva, G.A., Coutinho, O.P., Ducheyne, P., Shapiro, I.M., Reis, R.L., 2006. Starch-based microparticles as a novel strategy for tissue engineering applications. Key Engineering Materials 309–311, 907–910.

Silva, G.A., Dias, A.C.P., Coutinho, O.P., Reis, R.L., 2005b. Entrapment ability and release profile of corticosteroids using starch-based microparticles. Journal of Biomedical Materials Research 73A, 234–243.

Silva, G.A., Pedro, T.A., Costa, F.J., Neves, N.M., Coutinho, O.P., Reis, R.L., 2005a. Soluble starch and composite starch Bioactive Glass 45S5 particles: synthesis, bioactivity, and interaction with rat bone marrow cells. Materials Science and Engineering: C 25, 237–246.

Singh, N., Rajini, P.S., 2008. Antioxidant-mediated protective effect of potato peel extract in erythrocytes against oxidative damage. Chemico-Biological Interactions 173 (2), 97–104.

Singh, N., Kamath, V., Narasimhamurthy, K., Rajini, P.S., 2008. Protective effect of potato peel extracts against carbon tetrachloride-induced liver injury in rats. Environmental Toxicology and Pharmacology 26 (2), 241–246.

Sobedio, J.L., Kaitaramita, J., Grandgiral, A., Malkkl, Y., 1991. Quality assessment of industrial pre-fried french fries. Journal of the American Oil Chemists Society 68, 299–302.

Sousa, R.A., Mano, J.F., Reis, R.L., Cunha, A.M., Bevis, M.J., 2002. Mechanical performance of starch based bioactive composite biomaterials molded with preferred orientation for potential medical applications. Polymer Engineering & Science 42 (5), 1032–1045.

Stepto, R.F.T., 1997. Thermoplastic starch and drug delivery capsules. Polymer International 43, 155–158.

Su, Y.C., Chiu, R.J., Yu, N., Chang, W.R., 1984. The microbial production of amylase inhibitor and its application. I. Isolation and cultivation of *Streptomyces nigrifaciens* NTU-3314. Proceedings of the National Science Council, Republic of China. Part B, Life Sciences 8 (4), 292–301.

Szepes, A., Makai, Z., Blümer, C., Mäder, K., Kása Jr., P., Szabó-Révész, P., 2008. Characterization and drug delivery behaviour of starch-based hydrogels prepared via isostatic ultrahigh pressure. Carbohydrate Polymers 72 (4), 571–578.

Talburt, W.F., 1987. History of potato processing. In: Talburt, W.F., Smith, O. (Eds.), Potato Processing. Van Nostrand, New York, pp. 1–10.

Tarvainen, M., Peltonen, S., Mikkonen, H., Elovaara, M., Tuunainen, M., Paronen, P., Ketolainen, J., Sutinen, R., 2004. Aqueous starch acetate dispersion as a novel coating material for controlled release products. Journal of Controlled Release 96 (1), 179–191.

Tarvainen, M., Sutinen, R., Peltonen, S., Tiihonen, P., Paronen, P., 2002. Starch acetate-a novel film-forming polymer for pharmaceutical coatings. Journal of Pharmaceutical Sciences 91 (1), 282–289.

Te Wierik, G.H.P., Bergsma, J., Arends, A.W., Boersma, T., Eissens, A.C., Lerk, C.F., 1996. A new generation of starch products as excipient in pharmaceutical tablets. I. Preparation and binding properties of high surface area potato starch products. International Journal of Pharmaceutics 134, 27–36.

Te Wierik, G.H.P., Eissens, A.C., Bergsma, J., Arends, A.W., Lerk, C.F., 1997b. A new generation of starch products as excipient in pharmaceutical tablets. II. High surface area retrograded pregelatinized potato starch products in sustained-release tablets. Journal of Controlled Release 45, 25–33.

Te Wierik, G.H.P., Eissens, A.C., Bergsma, J., Arends-Scholte, A.W., Bolhuis, G.K., 1997a. A new generation starch product as excipient in pharmaceutical tablets. III. Parameters affecting controlled drug release from tablets based on high surface area retrograded pregelatinized potato starch. International Journal of Pharmaceutics 157 (2), 181–187.

Thompson, D.N., Fox, S.L., Bala, G.A., 2001. The effect of pretreatments on surfactin production from potato process effluent by *Bacillus subtilis*. Applied Biochemistry and Biotechnology 93 (1–3), 487–501.

Thompson, D.N., Fox, S.L., Bala, G.A., 2000. Biosurfactants from potato process effluents. Applied Biochemistry and Biotechnology 84–86 (1–9), 917–930.

Torres, F.G., Boccaccini, A.R., Troncoso, O.P., 2007. Microwave processing of starch-based porous structures for tissue engineering scaffolds. Journal of Applied Polymer Science 103 (2), 1332–1339.

Torres, F.G., Troncoso, O.P., Grande, C.G., Díaz, D.A., 2011. Biocompatibility of starch-based films from starch of Andean crops for biomedical applications. Materials Science and Engineering: C 31 (8), 1737–1740.

Trojanowski, J., Mayer, F., Hutterman, A., 1995. Utilization of potato pulp and potato liquor for the production of laccase by selected basidiomycetes. Starch/Starke 47, 116–118.

Tuovinen, L., Peltonen, S., Järvinen, K., 2003. Drug release from starch-acetate films. Journal of Controlled Release 91 (3, 4), 345–354.

Tuovinen, L., Peltonen, S., Liikola, M., Hotakainen, M., Lahtela-Kakkonen, M., Poso, A., Järvinen, K., 2004. Drug release from starch-acetate microparticles and films with and without incorporated α-amylase. Biomaterials 25 (18), 4355–4362.

Varma, R., Nene, S., 2003. Biosynthesis of cholesterol oxidase by *Streptomyces lavendulae* NCIM 2421. Enzyme and Microbial Technology 33, 286–291.

Vázquez, M.V., Rodríguez, M.E.G., 2011. Computer simulation of microbial transglutaminase production from potato wastes. Procedia Food Science 1, 736–741 11th International Congress on Engineering and Food (ICEF11).

Vigneshwaran, N., Nachane, R.P., Balasubramanya, R.H., Varadarajan, P.V., 2006. A novel one-pot 'green' synthesis of stable silver nanoparticles using soluble starch. Carbohydrate Research 341 (12), 2012–2018.

Vishnu, C., Seenayya, G., Reddy, G., 2002. Direct fermentation of various pure and crude starchy substrates to L(+) lactic acid using *Lactobacillus amylophilus* GV6. World Journal of Microbiology & Biotechnology 18, 429–433.

Woo, B.H., Jiang, G., Jo, Y.W., De Luca, P.P., 2001. Preparation and characterization of a composite PLGA and poly(acryloyl hydroxyethyl starch) microsphere system for protein delivery. Pharmaceutical Research 18 (11), 1600–1606.

Wu, T., Yan, J., Liu, R., Marcone, M.F., Aisa, H.A., Tsao, R., 2012. Optimization of microwave-assisted extraction of phenolics from potato and its downstream waste using orthogonal array design. Food Chemistry 133 (4), 1292–1298.

Yoshizawa, H., Koishi, M., 1990. The coating and the encapsulation of an interactive powder mixture and its application to sustained release preparations. Journal of Pharmacy and Pharmacology 42 (10), 673–678.

Zhang, Z.Y., Jin, B., Kelly, J.M., 2007. Production of lactic acid and byproducts from waste potato starch by *Rhizopus arrhizus*: role of nitrogen sources. World Journal of Microbiology & Biotechnology 23, 229–236.

Zia-ur-Rehman, Habib, F., Shah, W.H., 2004. Utilization of potato peels extract as a natural antioxidant in soy bean oil. Food Chemistry 85 (2), 215–220.

CHAPTER 21

Novel Applications of Potatoes

Lovedeep Kaur, Jaspreet Singh

Riddet Institute and Massey Institute of Food Science and Technology, Massey University, Palmerston North, New Zealand

1. Introduction

Nanoscience is the fabrication, study, and modeling of principles of devices and structures for which at least one dimension is several 100 nm or smaller (Ladisch, 2004). The aim of nanotechnology is to understand and apply atomic- or molecular-level manipulation to design novel molecular goods with improved properties. The potential advantages of nanotechnology have been identified in many areas, such as microelectronics, aerospace, and pharmaceutical industries. These achievements and discoveries have begun to have an impact on many aspects of the food and agriculture systems, such as food security, disease-treatment delivery methods, protection of the environment, and the molecular synthesis of new food products and ingredients (Weiss et al., 2006). The development of novel coatings, barriers, release devices, and packaging materials are some of the areas where food nano-technology can play its part. Novel barriers are being developed in the synthetic polymer field through the use of composite structures produced from successive molecular layers of various polymers; this approach can be applied to the food area (Morris, 2003). The development of biocompatible surfaces for medical or pharmaceutical use may lead to novel surfaces or coatings that repel or combat bacterial adhesion and biofilm formation. There are new opportunities for colloid scientists in the design and production of nanocrystals and nanoparticles, which are finding new applications as nonviral gene vectors and as molecular delivery systems (Morris, 2003).

Polymers from renewable sources have received great attention over the past many years predominantly because of environmental concerns. Potato starch is a promising biopolymer for various food, pharmaceutical, and biomedical applications because of its higher water solubility, which raises its degradability and speed of degradation; nontoxicity; easy availability; and abundance. The role of starch in tissue engineering of bone and bone fixation, as a carrier for the controlled release of drugs and hormones, and as hydrogels has already been documented (Mano and Reis, 2004; Won et al., 1997; Lenaerts et al., 1998; Pal et al., 2006; Pereira et al., 1998; Chakraborty et al., 2005). Starch-based biodegradable bone cements

627

may provide for immediate structural support and, in addition, as they degrade from the site of application they allow the ingrowth of new bone for complete healing of bone fractures (Domb et al., 1996; Pereira et al., 1998). Starch nanoparticles, nanospheres, and nanogels have also been used as base materials for nanoscale construction of sensors, tissues, mechanical devices, and drug delivery systems (Chakraborty et al., 2005). Rajisha et al. (2014) have reported on the use of potato starch nanocrystals as a reinforcing agent for natural rubber. Similarly, cellulose nanocrystals (CNCs) from potato peel waste are being explored as a high-value-adding reinforcing aid for polymer matrix composite manufacturing (Chen et al., 2012).

Biodegradable plastics are seen as one of many strategies to minimize the environmental impact of plastics and to develop sustainable plastics (McGlashan and Halley, 2003). Osés et al. (2009) reported that potato starch films act as oxygen barriers and delay the development of rancidity in vegetable oil. They also predicted that this result could be extended to all foods with a high content of polyunsaturated fatty acids. The use of starch-based protective coatings and biodegradable packaging materials may extend the shelf life of food products along with solving global environmental problem (Sorrentino et al., 2007). However, because of poor barrier and mechanical properties of starches, their use in biodegradable food packaging has been limited. These limitations can be overcome by either blending of the starches with synthetic polymers or by chemical modification of the starches. Starch-based biodegradable plastics materials can be prepared by various methods: embedding starch in synthetic polymeric matrices such as polyethylene, polypropylene, polystyrene, polyvinyl chloride; blending with hydrophilic polymers such as polyvinyl alcohol (PVA); using the modified starch by substitution, copolymerization, oxidation, and hydrolysis; foaming of starch within the extruder; and preparing thermoplastic starch (TPS) by melting under high pressure and temperature (Kim et al., 2003). Properties of biodegradable starch films are dependent on the starch type, starch concentration, type of starch modification and modification conditions, type and concentration of the plasticizer used, and drying conditions. Composites made from genetically modified potato starch along with wheat gluten have been reported to possess excellent gas barrier properties and may find applications in multilayer film packaging (Muneer et al., 2015). Also, the use of hybrid starch–nanoclay composites for the preparation of packaging films and other packaging materials has been reported (Park et al., 2002, 2003). The incorporation of clay nanoparticles markedly increases the physical properties of the polymers such as shear resistance, heat resistance, hydrophobicity, and strength of the packaging materials. However, many features of starch–nanoparticle systems are still to be explored.

In this chapter, we discuss the potential role of the potato and its starch in nanotechnology, with an emphasis on some of their novel applications in food and nonfood areas.

2. Biodegradable Packaging

Potato producers are currently aiming at three major areas of biodegradable polymers: food and nonfood packaging, personal and health care items, and other disposables. The use of these biodegradable materials may slow down the emission of fossil-fuel-derived carbon dioxide into the air (Stearns et al., 1994). Potato starch-based polymers have been used to make packaging for peanuts, candle cups for churches, and golf trees and are currently exploited for food packaging and many other nonfood uses. These products decompose in sewage treatment plants or in soil composts.

Polymer–clay nanocomposites (PCN) are a class of hybrid materials composed of organic polymer matrices and organophilic clay fillers, introduced in late 1980s by the researchers of Toyota (Kawasumi, 2004). They observed an increase in the mechanical and thermal properties of nylons with the addition of a small amount of nanosized clays. This new and emerging class of polymers has found several applications in the food and nonfood sector, such as in construction, automobiles, aerospace, military, electronics, food packaging, and coatings, because of their superior mechanical strength, heat and flame resistance, and improved barrier properties (Ray et al., 2006).

The structure and properties of PCN can be controlled depending on their end use by manipulating the polymer–clay interactions during their manufacturing. The most commonly used nanoscale clay particles in the preparation of PCN is montmorillonite, a hydrated alumina-silicate-layered clay with layer thickness in the nanometer dimension (Weiss et al., 2006). The layered silicates are of great interest because of their low cost, abundance, and high aspect ratio, which give greater possibility of energy transfer from one phase to another (Pandey et al., 2005). Solid layered dispersion in polymers involves two major steps: intercalation and exfoliation (Figure 1). In intercalation, the spacing between the clay layers increases as the extended polymer chains diffuse into the clay galleries, but the inorganic layers remain parallel to one another. On the other hand, in exfoliation, the silicate layers are completely separated from one another and dispersed in a continuous polymer matrix with no interclay particle interactions. The level of intercalation and exfoliation is generally confirmed through techniques such as X-ray diffraction (XRD) and transmission electron microscopy (Figure 2). The intercalation or exfoliation of a clay–polymer mixture depends on the characteristics of the polymer matrix and the organic modifiers. These characteristics include the nature of the polymer and the type, packing density, and size of the organic modifiers on the inorganic surface (Sorrentino et al., 2007; Alexandre and Dubois, 2000; Pantoustier et al., 2001).

The manufacturing process of PCN involves various methods: (1) *Solution method*: clay and polymer are dissolved in a polar organic solvent to avoid coiling of the polymer in the inter-gallery space. After solvent evaporation, generally intercalated nanocomposites may result. (2) *In situ method or interlamellar polymerization*: a nano-filler is swollen within the liquid

Figure 1
Schematic picture of a PCN material with completely exfoliated (molecular dispersed) clay sheets within the polymer matrix material. *Reprinted from Fischer (2003) with permission from Elsevier.*

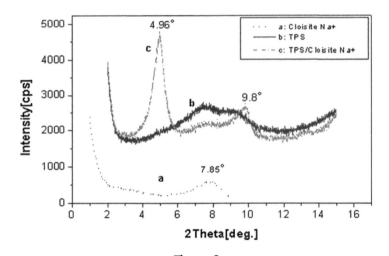

Figure 2
XRD patterns of Cloisite Na$^+$ (a), TPS (b), and the TPS/Cloisite Na$^+$ hybrid (trace c). *Reproduced with permission from Park et al. (2002). © Wiley-VCH Verlag GmbH & Co. KGaA.*

monomer and after complete dispersion in the monomer solvent, a curing agent is added, and this generally results in exfoliated nanocomposites. (3) *Melt intercalation method*: the layered inorganic is mixed with the polymer in a molten state; this may result in either exfoliated or intercalated nanocomposites. However, this process is quite temperature sensitive, as the temperature should not exceed the decomposition temperature of the clay modifier (Pandey et al., 2005). A new method for the preparation of nanocomposites has been reported, which involves solid-state mixing at room temperature using milling tools (ball milling; Mangiacapra et al., 2005; Sorrentino et al., 2005). This method has the advantage of not requiring the use of high temperatures or solvent treatments (Sorrentino et al., 2007).

Halloysite, a type of nanoclay, is a natural aluminum-silicate hollow cylinder with a length of 0.5–1 μm and an outer diameter and lumen of about 50 and 15 nm, respectively (Ismail et al., 2008). Halloysite is able to entrap a range of molecules with specific sizes and can be used as a viable nanocage for active molecules, owing to its empty space inside the nanotubes (He et al., 2012). He et al. (2012) prepared halloysite/potato starch composites by adding modified natural halloysite nanotubes into potato starch matrices that reinforced the mechanical properties of potato starch films. These biodegradable, low-cost, and high-strength films possess a good potential as green packaging materials (He et al., 2012).

Biodegradable polymers have been referred to as polymeric materials for which at least one step in the degradation process is through metabolism in the presence of naturally occurring organisms (Sorrentino et al., 2007). These can be classified into categories according to their source: (1) Polymers such as polysaccharides (starch, cellulose), proteins, polypeptides, or polynucleotides, which are directly extracted from biomass. (2) Polymers such as polylactic acid or biopolyester, which are produced by chemical synthesis using renewable sources. (3) Polymers such as bacterial cellulose, xanthan, or pullan, which are produced by microorganisms (Sorrentino et al., 2007). Starch-based biodegradable materials come under the first category. Potato starch increases the biodegradability of a nonbiodegradable plastic and also can be used together with a completely biodegradable synthetic plastic producing biodegradable plastic of low cost (Park et al., 2003). Starch-based absorbent pads may also be a potential alternative to conventional absorbent pads for meat exudation (Smith et al., 1995). And also it can be used to make packaging films for fruits and vegetables, snacks, and other low-moisture food/pharmaceutical products. However, the major drawback of granular potato starch is its limited processability, due to the big granule sizes (5–100 μm). Therefore, it is difficult to make blown thin films for package applications that require efficient mechanical and barrier properties. Chemical modification of starch or plasticization of starch may help in improving the mechanical strength of the final product.

TPS is the most widely used bioplastic, accounting for 50% of the bioplastics market, and is commonly derived from potatoes or corn (Robertson, 2006). TPS has been developed by gelatinizing starch with 6–10% moisture with heat and pressure (along with a plasticizer), which gives superior product properties (George et al., 1994; Usuki et al., 1993). Common plasticizers for the preparation of TPS are glycerol and other low-molecular-weight polyhydroxy compounds, polyethers, urea, and water, which are added to reduce intramolecular hydrogen bonds and to provide stability to product properties (Sorrentino et al., 2007). However, TPS still needs blending with other polymers, such as poly(ethylene-co-vinyl

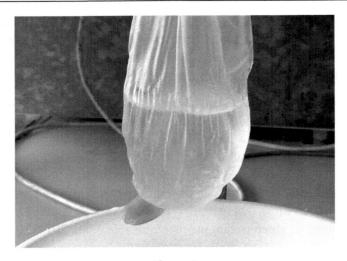

Figure 3
Starch–clay nanocomposite bag made by film-blowing technology. This bag was stored filled with water and did not show any release nor failure after 3 weeks. *Reprinted from Fischer (2003) with permission from Elsevier.*

alcohol), because of the low strength and poor water resistance of the final product (Park et al., 2003). The addition of clay, as a potential filler, has also been reported to improve the properties of TPS in various applications (Sorrentino et al., 2007; Park et al., 2003, 2002; Chen and Evans, 2005; De Carvalho et al., 2001; McGlashan and Halley, 2003; Wilhelm et al., 2003; Yoon and Deng, 2006) (Figure 3). However, the extent of improvement in the mechanical and thermal properties depends on the type of starch and nanoclay used along with the extent of dispersion of the filler in the polymer matrix (as discussed above). Park et al. (2003) investigated the effects of two different clays—one organically modified, Cloisite 30B, with ammonium cations located in the silicon gallery and one unmodified, Cloisite Na^+—on the tensile, mechanical, and thermal properties of potato TPS–clay nanocomposites using a melt intercalation method. They found that TPS–Cloisite Na^+ nanocomposites had higher tensile strength, higher mechanical stability, and better barrier properties compared to water vapor than TPS–Cloisite 30B nanocomposites (Figures 4 and 5; Table 1). Also, the addition of a small amount of nanoclay (less than 5%) increased the decomposition temperature and decreased the relative water vapor diffusion coefficient of TPS.

The light weight and bulky nature of conventional foam plastic packaging present a major disposal problem for companies because of its high handling and transportation costs (Nabar et al., 2006). Novel TPS–clay nanocomposite foams with potentially enhanced mechanical and barrier properties have been prepared by melt processing using urea as a plasticizer (Chen et al., 2005). The use of urea as a plasticizer has been reported to avoid the cracking of TPS during storage and enhance the dispersion of ammonium-treated clay in TPS. Some

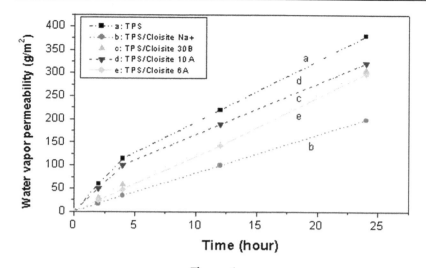

Figure 4
Water vapor permeability behavior of TPS–clay hybrids of different kinds of clay at 24 °C. *Reproduced with permission from Park et al. (2002). © Wiley-VCH Verlag GmbH & Co. KGaA.*

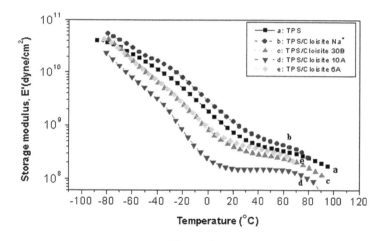

Figure 5
Typical storage modulus behavior of TPS–clay hybrids of different kinds of clay. *Reproduced with permission from Park et al. (2002). © Wiley-VCH Verlag GmbH & Co. KGaA.*

researchers have prepared extruded starch foams with PVA that perform similar to conventional packaging foams and can be safely disposed of in soil or in composting operations (Lacourse and Alteiri, 1991; Roesser et al., 2000). Chemically modified starch, such as hydroxypropylated and acetylated starch, has also been used in the preparation of these foam products for use in cushioning and insulation applications. Altieri and Tessler (1996) prepared water-resistant foams from blends of starch with starch esters. Polymers such as

Table 1: Tensile Properties of TPS–Clay Hybrids of 5 wt% Clay Contents.

Clay Type	Tensile Strength (MPa)	Elongation at Break (%)
Cloisite Na$^+$	3.32	57.2
Cloisite 30B	2.80	44.5
Cloisite 10A	2.14	34.9
Cloisite 6A	2.51	38.0
Blank (0%)	2.61	47.0

Reproduced with permission from Park et al. (2002). © Wiley-VCH Verlag GmbH & Co. KGaA.

poly(caprolactone), cellulose acetate, poly(ethylene vinyl alcohol), and poly(ethylene-co-acrylic acid) are some of the other polymers used to increase the hydrophobicity of starch-based foam packaging (Nabar et al., 2006). Addition of fiber (>30%) has been reported to increase the mechanical strength of starch-based foam trays (Yu et al., 2006). In 2005, an application of potato starch as filler for a composite material with high-density polyethylene was reported (Szymanowski et al., 2005). Starch surface properties were altered using methane radio-frequency glow discharge and then composite materials were prepared using modified potato starch with polyethylene. Polyethylene samples filled with modified starch presented an improvement in mechanical properties compared to those filled with unmodified starch.

Muneer et al. (2015) reported the use of potato starch genetically modified to yield increased lengths of the glucan chains in the amylopectin fraction and higher amylose content, along with wheat gluten, to produce composites with attractive mechanical and gas barrier properties. The nanomorphology of the composites produced through extrusion showed the modified potato starch had the semicrystalline structure of a characteristic lamellar arrangement with an approximately 100-Å period and a B-type crystal structure as observed by small-angle X-ray scattering and wide-angle X-ray scattering analysis, respectively (Figure 6). Under ambient conditions these wheat gluten–potato starch composites (70/30) with 45% glycerol showed excellent gas barrier properties to be further explored in multilayer film packaging applications.

Problems of applications of bioplastics arise not only from (price) competition with the highly developed synthetic polymers but also from their insufficient property levels. Possible ways to decrease the hydrophilicity and increase the values of the mechanical properties so far are: (1) application of a coating(s) of the processed bioplastic material with hydrophobic materials; (2) blending with different, hydrophobic, biodegradable synthetic polymers (polyesters); and (3) new methods of reactive extrusion of natural polymers (graft- and copolymerization, esterification during extrusion process). A new possibility in this direction is seen in the creation of composite materials of thermoplastic organic biopolymer and nanoscopic inorganic particles incorporated on a molecular scale (Fischer et al., 2001; Fischer, 2003).

Starch-based biodegradable plastics are the future of the global packaging industry; however, not enough attention has been provided to studying the durability and degradability of these

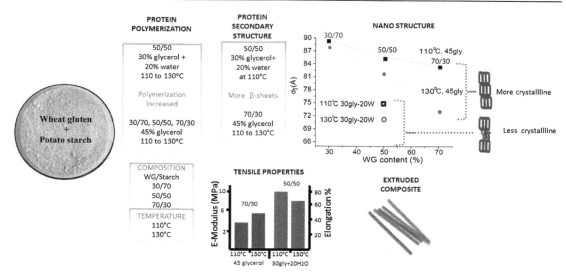

Figure 6

Schematic diagram of the summarized effects of temperature and composition of the blend on the wheat gluten protein polymerization and secondary structure, as well as nanomorphology and mechanical properties of extruded wheat gluten-modified potato starch composites. *Reprinted with permission from Muneer et al. (2015). Copyright (2015) American Chemical Society.*

plastics. Degradation is a process in which deterioration occurs in polymer properties owing to factors such as light, heat, mechanical disruption, etc. And as a consequence the life of a material becomes limited; thus, the study of degradation and stabilization of these polymers should be carried out to better understand this phenomenon, which will ensure the long life of the product (Pandey et al., 2005). Synthetic polymeric components used with starch remain in an undegraded state even after the starch is fully biodegraded. Also, when starch is blended with some polymeric materials, there is a problem of phase separation that has an undesirable effect on the mechanical properties of the product (Kim et al., 2003). Reinforcement of starch-based plastic films with short pulp fiber has been reported to improve mechanical properties (Kim et al., 2003; Tsiapouris and Dresden, 2000). One major advantage of using natural fibers to reinforce natural polymers is their compatibility (Yu et al., 2006).

3. Fiber-Reinforced Biodegradable Composites for Constructive Parts in Aerospace, Automotive, and Other Areas

Fiber-reinforced biodegradable plastics are becoming increasingly popular for use in various applications because of their excellent mechanical properties, e.g., high strength and stiffness, light weight, and low cost. Various plant fibers studied for this purpose are jute, flax, ramie, oil palm fibers, and fibers made from regenerated cellulose. Natural fibers have excellent mechanical properties of breaking length and elastic moduli characteristics such as E-glass, and they enhance the mechanical strength of biodegradable polymers such as potato starch (Riedel, 1999). These biocomposites are produced by embedding these fibers into a biopolymeric matrix made of derivatives of starch, lactic acid, cellulose, etc. (Yu et al., 2006).

These have been formulated to meet the processing requirements for commonly used manufacturing techniques, such as hot press molding, extrusion cooking, injection molding, etc. An essential requirement for a good fiber matrix adhesion system is optimization for impregnation of the reinforcing system (Riedel, 1999). Potato starch (usually TPS) reinforced by cellulose is a typical example of a natural polymer composite. Starch needs to be modified physically or chemically to be suitable for processing as thermoplastic resins. Another commonly used option is to add copolymers, which can even be of petrochemical origin (such as Mater-Bi; Bastioli, 1998). However, esterification of potato starch by adding plasticizers or by acetylation can give the same effect.

Apart from anisotropic and specially tailored structural parts with continuous fiber reinforcements, biocomposites have been reported to be well suited to paneling elements in cars, railways, and airplanes, using various kinds of nonwovens from single fibers (Riedel, 1999). In 2001, a fiber-reinforced molded plastic part manufacturing process using cellulose propionate, thermoplastic potato starch, and flax fibers was reported (Foelster et al., 2001). These molded plastic parts have excellent stability and strength and can be used as molded parts for trucks or passenger cars or rail vehicles or aircraft, particularly their body and/or their paneling. It has been suggested that biocomposites might be suitable for interior applications, such as furniture and packaging for electronics (Riedel and Nickel, 2003). However, further research should be conducted to develop biocomposites with flame-retardant properties for use in some applications, such as interior cladding of railway carriages and aircraft bodies. Further applications of starch-based biocomposites needs to be explored in automobile and railway design, in the furniture industry, and in the leisure industry (Riedel, 1999).

4. Edible Films

Edible film and coating are defined as a thin, continuous layer of edible material used as a coating or as a film placed between food components to provide a barrier to mass transfer (Balasubramanian et al., 1997; Guilbert et al., 1997). These films/coatings have the potential to replace conventional packaging in some applications. Starches such as potato starch, cellulose derivatives, and plant gums are some of the materials used as edible coatings and films in food packaging and preservation (Bertuzzi et al., 2007). Use of these edible films could extend shelf life and improve quality and handling properties of food and pharmaceutical products (Talja et al., 2008a). Edible films can be a possible means to incorporate substances, such as antimicrobial agents, flavors, and coloring agents, into foods (Han, 2002). These can also be used for the controlled release of drugs (Tuovinen et al., 2003) or active components from foods and packaging materials (Talja et al., 2008a).

An edible film should have good water vapor barrier properties (low or no water permeation and diffusion through film), which should not increase or increase very little with increasing relative vapor pressure (Lawton, 1996). Films should stand mechanical stress and strain to such an extent that they do not break easily under a decent mechanical force (Talja et al., 2008a). Thus, the composition of starch-based films is an important factor influencing its barrier and mechanical properties. Also, starch-based edible films may have an impact on the sensory and textural characteristics of the food.

Edible coatings can be applied and formed directly on the food product, by the addition of a liquid film-forming solution or by molten compounds, using a brush or by spraying, dipping, or fluidizing, and are an integral part of the food product (Sorrentino et al., 2007). On the other hand, edible films are first formed using casting or extrusion processing and then applied to foods. Potato starch-based films can provide an effective barrier to oxygen and carbon dioxide transport but they are a poor barrier to water vapor. The hydrophilic nature of starch is the major limitation for its use in edible films, which can be improved by the addition of lipids.

The functional, organoleptic, nutritional, and mechanical properties of an edible film have been reported to be altered by the addition of plasticizers such as glycerol or surfactants or both in small amounts (Rodriguez et al., 2006). Glycerol is a plasticizer compatible with starch, which is often used to modify the mechanical characteristics of the film, such as film flexibility, facilitating its handling and preventing cracks. The presence of glycerol in the film slows down starch digestion, which is a feature of potential dietetic use (Hernández et al., 2008).

Talja et al. (2008b) prepared potato starch-based films plasticized by a binary mixture of xylitol and sorbitol using various potato starches extracted from different potato cultivars. The amylose content of the potato starches was found to vary from 11.9% to 20.1%. They reported that the mechanical properties and the water vapor permeability of the films were not influenced by the amylose content of the starch.

Surfactants such as Tween 20 improve wettability properties of film solutions by decreasing surface tension (Hiemenz and Rajagopalan, 1997; Rodriguez et al., 2006). The presence of a surfactant has been reported to enhance plasticity and increase water vapor permeability of the films with glycerol, which could be due to some interaction between glycerol and surfactants allowing more molecular mobility (Rodriguez et al., 2006).

Zavareze et al. (2012) made potato starch films using oxidized and heat–moisture-treated starches; however, they used only one modification degree to produce the films. Fonseca et al. (2015) studied the effects of sodium hypochlorite concentration on film-forming capacity, water solubility, mechanical properties, and water vapor permeability of the films. They reported that an increase in active chlorine concentration increased the intensity of oxidation of the potato starches. Also, the films produced with oxidized starches had different properties depending on the degree of oxidation. The films made with oxidized starch with the highest level of active

Table 2: Thickness, Solubility, Color, Tensile Strength, and Water Vapor Permeability (WVP) of Films of Oxidized Potato Starches with Various Concentrations of Active Chlorine.

Active Chlorine (g/100 g)[a]	Starch (g/100 ml)	Thickness (mm)	Solubility (g/100 g)	L*	Tensile Strength (MPa)	WVP (g mm/m² day kPa)
Native	3	0.15c	14.0d	89.6a	3.8d	9.3a
	4	0.18c	23.9b	89.2a	4.3d	9.4a
	5	0.23b	24.8ab	87.7b	6.7a	9.9a
0.5	3	0.16c	12.7e	89.3a	3.5d	8.5a
	4	0.23b	18.2c	88.3ab	3.9d	9.3a
	5	0.28a	21.7bc	87.6b	5.3c	9.5a
1.0	3	0.17c	15.3d	89.4a	4.1d	7.4ab
	4	0.23b	21.9bc	89.1a	5.8bc	7.7ab
	5	0.29a	26.0a	87.7b	6.4ab	7.6ab
1.5	3	0.16c	12.9e	89.7a	1.8e	5.8bc
	4	0.20bc	12.7e	89.4a	1.9e	4.4c
	5	0.25ab	18.3c	88.6b	2.3e	4.5c

[a]Different letters in the same column differ statistically ($P < 0.05$).
Reproduced from Fonseca et al. (2015) with permission from Elsevier.

chlorine (1.5 g/100 g) showed lower tensile strength, regardless of the amount of starch used, compared to native starch films. However, these films exhibited lower water solubility and water vapor permeability, which enables the use of the oxidized starch films in products with higher water activity compared to the native starch films (Fonseca et al., 2015; Table 2).

5. Textiles and Paper

Potato starch is used in the textile industry for sizing cotton, worsted, and spun rayon warps (Stearns et al., 1994). In the sizing process, a film of the sizing agent is applied to the textile yarn in order to bind the loose fibers tightly to the surface of the thread. This strengthens and protects the warp from abrasion during weaving. Potato starch films have been reported to have a higher degree of toughness and flexibility compared with other starches, which allows potato starch-sized warps to be woven at lower humidity than those sized with corn starch (Stearns et al., 1994). Chemically modified (carboxymethylated) amylopectin potato starch has been reported to be used as sizing agents for natural and/or synthetic textile yarns. This modified starch has very favorable sizing properties and imparts excellent weaving properties to the sized yarn and washing-out properties to the woven cloth (Huizenga et al., 1998). The amylose component of potato starch can be used to prepare zinc oxide-soluble starch nanocomposites (Nano-ZnO) to impart antibacterial and UV-protection functions to cotton fabrics (Vigneshwaran et al., 2006). Cotton fabrics with Nano-ZnO impregnation show better protection against UV radiation in comparison with the untreated cotton fabrics.

Native and chemically modified potato starch is the second most consumed starch in the paper industry, primarily for use as a furnish additive, only surpassed by corn starch. Starch is used in the paper industry during furnish preparation (as a flocculent, retention aid, drainage aid, and carrier for alkaline size), for surface sizing (as an adhesive to enhance paper strength and stiffness), for coating (as a binder for fine pigments to impart smoothness and gloss), as an effluent treatment (as a cationic polymer to control discharge of cellulose fiber and pigments), and during conversion of paperboard to packaging grades (as an adhesive in the manufacture of multi-ply board and as a glue for corrugation and laminating applications) (Maurer and Kearney, 1998).

Chemical modification of potato starch via grafting of vinyl monomers can be used to incorporate desirable properties into starch without sacrificing its biodegradable nature (Athawale and Lele, 2000). Potato starch–*graft*–polyacrylonitrile has been used as a superabsorbent in diapers and sanitary napkins, for example (Athawale and Lele, 1998; Fang et al., 2005). Fu et al. (2002) synthesized *p*-acryloazoanilide derivatives grafted onto potato starch copolymers, which may have potential utilization in heavy metal ions removal and paper or textile manufacturing industries. Graft copolymerization of potato starch with zwitterionic monomers may also hold great promise for the future, as the resulting copolymers may be useful in many commercial applications concerned with textiles, dispersion agents, coatings, personal care formulation, and water remediation. Of particular interest is the possibility of viscosity maintenance or increase of such copolymers in the presence of electrolytes and surfactants due to shielding of coloumbic attractions (Zhang and Hu, 2002).

Wastewater and dyeing effluent generated by textile and other industries are generally discharged to the surrounding environment without any further treatment. These pollutants, apart from adding color to water, also are toxic to aquatic and other forms of life (Khan and Husain, 2007). Immobilized potato enzyme polyphenol oxidase (celite bound) has been reported to be a cheaper option for decolorization of a number of reactive textile dyes, such as Reactive Blue 4 and Reactive Orange 86, present in polluted water.

6. Starch Spherulites and Nanocrystals

Spherulites are semicrystalline entities with some degree of radial symmetry displaying a "Maltese cross" extinction pattern when viewed between crossed polarizers (Ziegler et al., 2003). The Maltese cross is oriented with its arms parallel to the polarizer and analyzer vibration directions. These small particles or spherulites are birefringent, which is generally defined as the difference between the radial and the tangential refractive indices. In synthetic polymer melts, they form at a high degree of undercooling (rapid quenches) in media of high viscosity and in the presence of some impurity that can be rejected at the growing crystal front (Ziegler et al., 2003). Starch's ability to bind volatile flavor compounds can be exploited in the food industry by using nanosized starch spherulites for this purpose, which can bind more flavors because of their very exceptionally small size (Conde-Petit et al., 2006). Potato starch is the cheapest available biopolymer and therefore can be used as a raw material to produce spherulites (Zeigler et al., 2003). Starch spherulites may also find applications as encapsulants for the controlled release of drugs and may be used as a fat mimetic. Starch spherulites can be produced using differential scanning calorimetry (DSC) or steam-jet cooking. Steam-jet cooking has been used commercially for decades to prepare aqueous starch solutions for industrial application. Dispersions for steam-jet cooking are prepared by dispersing starch (and fatty acid) in water at temperatures sufficient to cause starch granule swelling. DSC is the most common technique used for detecting both first-order (melting) and second-order (glass) thermal transitions. DSC is an efficient tool for the study of starches and the production of spherulites because of its high sensitivity and ease of use. Furthermore, analyses are carried out using a sealed pan, which prevents the loss of water from the suspension during heating. Native starch granules generally exhibit positive birefringence, since the refractive index is largest along the chain axis, which is oriented radially (Nordmark et al., 2002a,b). But for starch spherulites, studies (Nordmark et al., 2002a,b) revealed the presence of predominantly negative birefringence. Spherulites appear to consist of radially oriented "fibers" that, if composed of chain-folded lamellae, double helices, or single helices with inclusion complexes with their main chain axis perpendicular to the lamella surface, would result in negative birefringence.

The addition of various fatty acids to the starch may enhance the process of spherulite formation. The formation of spherulites from helical inclusion complexes of amylose with various lipids or fatty acids has been reported in the literature. Many studies on amylose–guest complexes carried out by electron microscopy and XRD have demonstrated that these complexes form lamellar single crystals approximately 100 Å thick, with the amylose chains perpendicular to the crystal surface in a chain-folded configuration (Shogren et al., 2006). Fanta et al. (2006) reported the formation of large spherical/lobed and small torus-shaped spherulites using starch and various fatty acids. These spherulites consisted of layers of chain-folded amylose–lipid helices (i.e., crystalline lamellae) about 80 Å in thickness with the helices oriented radially. The morphology of spherulites has been reported to depend on the branching of lamellae and also on imperfect stacking of individual lamellae during the growth process because of repulsive forces caused by portions of uncrystallized polymer chains that are not chain-folded and thus extend from lamella surfaces (Shogren et al., 2006). It was also suggested that spherulite morphology is influenced by the rate at which the spherulites are formed, which, in turn, would depend upon factors such as water solubility of the fatty acid, rate of cooling, and the amount of fatty acid used relative to starch (Fanta et al., 2008).

Singh et al. (2010) subjected superheated gels of various starches treated using DSC (heated from 10 to 180 °C at 5 °C/min, with a holding time of 1 min at maximal temperature, and then quenched to 10 °C at 150 °C/min) to polarized microscopy, to confirm the formation of spherulites. They observed that the high-amylose corn starch gels exhibited birefringence and a weak crystalline pattern, showing the presence of spherulites. No birefringence was observed for waxy corn starch gel, while potato starch gels had some birefringence (Figure 7). Their results suggested that the amylose content of the starch and melting of crystallites during heating play important roles during recrystallization of amylose.

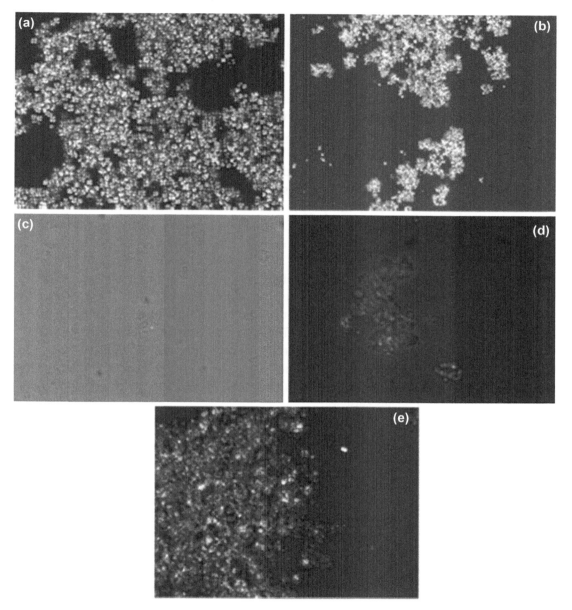

Figure 7
Polarized microscopy images of starch spherulites. (a) High-amylose maize starch I, (b) high-amylose maize starch II, (c) waxy maize starch, (d) potato starch I, and (e) potato starch II. *Reprinted from Singh et al. (2010), with permission from Elsevier.*

The formation of starch nanocrystals by acid hydrolysis of native starch has also been reported. By subjecting native starch to acid hydrolysis below its gelatinization temperature, the amorphous regions are hydrolyzed, allowing the separation of crystalline residues (Song et al., 2008; Angellier et al., 2004). During the hydrolysis process the amorphous region of starch was hydrolyzed and removed, and nanocrystals around 5–7 nm thick with a length of 20–50 nm and a width of 15–30 nm were obtained. Nanocrystals have the shape of parallelepiped nanoplatelets. Even if the parallelepipedic shape is the general shape observed, a lot of varying organizations can be distinguished. A few stacks of nanoplatelets oriented edge-on may be observed in a very small proportion. Such nanocrystals can be obtained from potato starch granules by acid hydrolysis, using H_2SO_4. It is used as a reinforcing phase in a polymeric matrix and displays substantially improved mechanical properties. Starch disruption by acid hydrolysis depends on many factors such as the botanical origin, namely the crystalline type, the relative proportions of amylose and amylopectin, and the granular morphology. The conditions during acid hydrolysis, such as acid type, acid concentration, temperature, and hydrolysis duration, also have a significant effect (Angellier et al., 2004). Rajisha et al. (2014) reported the use of potato starch nanocrystals as a reinforcing agent for natural rubber. They observed, through X-ray diffractograms, that the addition of starch nanocrystals into natural rubber resulted in an increase in the global crystallinity and strength of the composite material (Figures 8 and 9). Their results indicated that these environmentally friendly potato starch nanocrystal-reinforced natural rubber nanocomposites have the potential for replacing conventional polymer composites (e.g., carbon black) and nanocomposites. The overall improvement in properties was attributed to the possible formation of a three-dimensional network of starch nanocrystals within the nanocomposite network.

7. Potato Waste Utilization and Other Miscellaneous Uses

Genetically modified (GM) potatoes have been introduced that produce several times more fructose than the normal ones (Somasekhar, 2001). Fructose is generally produced by enzymatic conversion of corn starch through industrial processes. However, scientists have applied gene fusion technology to convert starch (40–60%) stored in potato to fructose in the plant itself. Fructose is released upon heating and mashing of the potatoes. Potato has been modified by inserting two genes coding enzymes, α-amylase and glucose isomerase. Apart from use in the food industry, fructose from these GM potatoes can also be used for low-cost ethanol production to fuel automobiles. Thanavala et al. (2005) conducted a double-blind placebo-controlled trial to evaluate the immunogenicity of hepatitis B surface antigen expressed in transgenic potatoes and delivered orally to previously vaccinated individuals. They concluded that the potato tuber-derived orally delivered vaccine for prevention of hepatitis B virus should be considered as a viable component of a global immunization program.

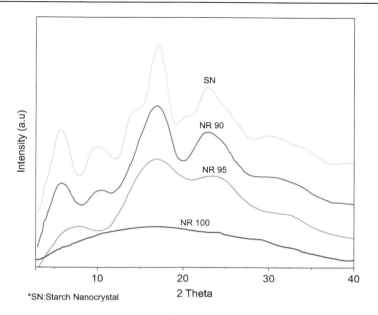

Figure 8

XRD patterns of potato starch nanocrystals/natural rubber nanocomposite films. NR 100, NR 95, and NR 90 are films containing 0, 5%, and 10% potato starch, respectively; SN is potato starch nanocrystals. *Reprinted from Rajisha et al. (2014), with permission from Elsevier.*

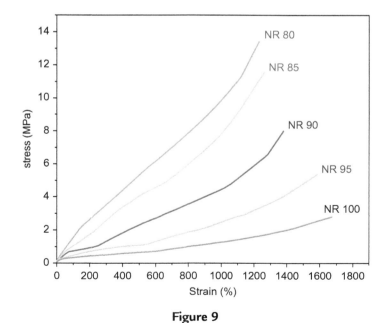

Figure 9

Stress–strain curves of starch nanocrystal/NR nanocomposites. NR 100, NR 95, NR 90, NR 85, and NR 80 are films containing 0, 5%, 10%, 15%, and 20% potato starch, respectively. *Reprinted from Rajisha et al. (2014), with permission from Elsevier.*

Various research projects in Europe are trying to find a less expensive manufacturing process for polyhydroxy alkanoates (PHA), which can be used for the production of biodegradable packaging materials, by identifying the individual genes, and gene sets, in the bacteria that are responsible for the production of PHA (e.g., 3HB, 3HV, and 4HV-PHA) and then expressing them in plants like tobacco, rape, pea, and potato. This new manufacturing process has come to be known as molecular farming (Canadian Agri-Food Research Council, 2003).

Waste from the potato processing industry has been reported to be turned into biodegradable packaging trays that offer major environmental advantages and give the same performance as those made from conventional plastics (Storey, 2003; Murphy et al., 2004). Around 17,000 tonnes a year of the waste has been estimated to be generated by the British potato product producers, which can be converted into biodegradable plastics to meet foreseeable packaging needs. The starch from processed waste material has been reported to perform even better than the pure potato starch and the packaging has been shown to be well suited to single-use applications with fruit and vegetables that have a relatively short shelf life (Figure 10).

Kang and Min (2010) applied high-pressure homogenization (138 MPa) to potato peel homogenate to break down biopolymer particles in solution small enough to allow for bio-polymer film formation. They reported that the produced potato peel films have a good potential for application to low- or intermediate-moisture foods. Also, the concentrations of glycerol and soy lecithin were identified as important variables in producing biopolymer films from potato peel.

Cellulose microfibrils extracted from potato pulp have been reported as an effective reinforcing additive (Dufresne et al., 2000). The microfibrils exhibited diameters on the order of 10 nm and lengths of 10–100 μm, prepared by alkaline washing, bleaching, and finally mechanical attrition. These long fibers retained amorphous regions that weakened their stiffness (Chen et al., 2012). Chen et al. (2012) prepared CNCs from potato peel waste to remove these weaker domains, for the purpose of being a high-value-adding reinforcing aid for polymer matrix composite manufacturing. They reported that the mechanical and barrier properties of CNC-filled PVA and TPS films were improved by the incorporation of these CNCs, even at low loadings of 1–2%. Despite the improvements made by the inclusion of these CNCs in the tested polymer films, the changes observed were small when factoring in the cost to prepare them relative to other comparable commercial fillers like nanoclays or carbon nanotubes (Chen et al., 2012). CNCs are already being used in structural plastics, smart coatings, cosmetics, pharmaceuticals, and solar energy collection (Chen et al., 2012).

Some researchers are converting potato waste to some high-value products, such as xanthan and polylactic acid, which is otherwise an expensive waste management challenge (Robertson, 2006). Ansari et al. (2014) used potato peel as a cheap nutrient source for the production of biosurfactants from bacterial isolates. They reported that potato peel enhanced the growth

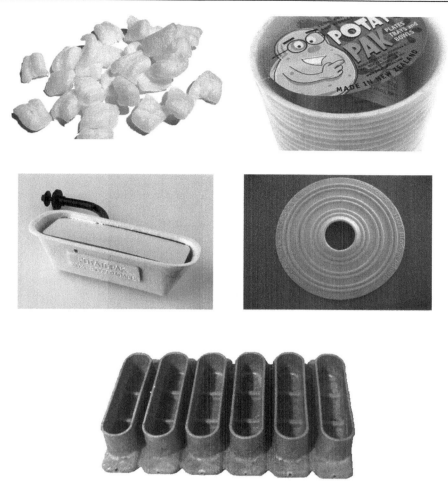

Figure 10
Various potato starch-based packaging products. *Potatopak NZ Ltd.*

rate of culture and positively affected the biosurfactant production. Potato starch is converted to glucose through enzymatic hydrolysis, which is a raw material for the production of lactic acid. Then the lactic acid is polymerized to produce polylactic acid (PLA). PLA is becoming increasingly popular in the production of a wide range of biodegradable materials (board, sheet, films, fiber, paint, etc.) because of low energy requirements during its production compared to other plastics of petroleum origin. Fiber-reinforced PLA composite materials have been used in the production of interior components for automobiles (Kawamoto, 2007). However, the cost of production of these plastics is significantly high compared to those of petroleum origin and it is essential to develop a technology that uses energy more efficiently and is cost-effective.

References

Alexandre, M., Dubois, P., 2000. Polymer-layered silicate nanocomposites preparation, properties and uses of a new class of materials. Material Science and Engineering 28, 1–11.

Altieri, P.A., Tessler, M.M., 1996. Water Humidity Resistant Starch Foams, U.S. Patent 5,554,660.

Angellier, H., Choisnard, L., Molina-Boisseau, S., Ozil, P., Dufresne, A., 2004. Optimization of the preparation of aqueous suspensions of waxy maize starch nanocrystals using a response surface methodology. Biomacromolecules 5, 1545–1551.

Ansari, F.A., Hussain, S., Ahmed, B., Akhter, J., Shoeb, E., 2014. Use of potato peel as cheap carbon source for the bacterial production of biosurfactants. International Journal of Biological Research 2, 27–31 2014.

Athawale, V.D., Lele, V., 1998. Graft copolymerization onto starch. II. Grafting of acrylic acid and preparation of its hydrogels. Carbohydrate Polymers 35, 21–27.

Athawale, V.D., Lele, V., 2000. Syntheses and characterisation of graft copolymers of maize starch and methacrylonitrile. Carbohydrate Polymers 41, 407–416.

Balasubramaniam, V.M., Chinnan, M.S., Mallikarjunan, P., Philips, R.D., 1997. The effect of edible film on oil uptake and moisture retention of deep-fat fried poultry product. Journal of Food Process Engineering 20 (1), 17–29.

Bastioli, C., 1998. Starch based bioplastics: properties, applications and future perspectives. In: Proceedings of the Symposium, Wurzburg, 11/12. "Biologisch Abbaubare Werkstoffe (BAW)".

Bertuzzi, M.A., Castro Vidaurrea, E.F., Armadaa, M., Gottifredia, J.C., 2007. Water vapor permeability of edible starch based films. Journal of Food Engineering 80, 972–978.

Canadian Agri-Food Research Council, 2003. An Assessment of the Opportunities and Challenges of a Bio-based Economy for Agriculture and Food Research in Canada. http://www.agwest.sk.ca/bioproducts/documents/OpportunitiesandChallengesofaBio-BasedEconomy.pdf.

Chakraborty, S., Sahoo, B., Teraka, I., Miller, L.M., Gross, R.A., 2005. Enzyme-catalyzed regioselective modification of starch nanoparticles. Macromolecules 38, 61–68.

Chen, B., Evans, J.R.G., 2005. Thermoplastic starch-clay nanocomposites and their characteristics. Carbohydrate Polymers 61, 455–463.

Chen, D., Lawton, D., Thompson, M.R., Liu, Q., 2012. Biocomposites reinforced with cellulose nanocrystals derived from potato peel waste. Carbohydrate Polymers 90, 709–716.

Chen, M., Chen, B., Evans, J.R.G., 2005. Novel thermoplastic starch-clay nanocomposite foams. Nanotechnology 16, 2334–2337.

Conde-Petit, B., Escher, F., Nuessli, J., 2006. Structural features of starch-flavor com-plexation in food model systems. Trends in Food Science and Technology 17, 227–235.

De Carvalho, A.J.F., Curvelo, A.A.S., Agnelli, J.A.M., 2001. A first insight on composites of thermoplastic starch and kaolin. Carbohydrate Polymers 45, 189–194.

Domb, A.J., Manor, N., Elmalak, O., 1996. Biodegradable bone cement compositions based on acrylate and epoxide terminated poly (propylene fumarate) oligomers and calcium salt compositions. Biomaterials 17, 411–417.

Dufresne, A., Dupeyre, D., Vignon, M.R., 2000. Cellulose microfibrils from potato tuber cells: processing and characterization of starch–cellulose microfibril composites. Journal of Applied Polymer Science 76, 2080–2092.

Fang, J.M., Fowler, P.A., Hill, C.A.S., 2005. Studies on the grafting of acryloylated potato starch with styrene. Journal of Applied Polymer Science 96, 452–459.

Fanta, G.F., Felker, F.C., Shogren, R.L., Salch, J.H., 2006. Effect of fatty acid structure on the morphology of spherulites formed from jet cooked mixtures of fatty acids and defatted cornstarch. Carbohydrate Polymers 66, 60–70.

Fanta, G.F., Felker, F.C., Shogren, R.L., Salch, J.H., 2008. Preparation of spherulites from jet cooked mixtures of high amylose starch and fatty acids. Effect of preparative conditions on spherulite morphology and yield. Carbohydrate polymers 71, 253–262.

Fischer, H., 2003. Polymer nanocomposites: from fundamental research to specific applications. Materials Science and Engineering C 23, 763–772.

Fischer, S., de Vlieger, J., Kock, T., Batenburgand, L., Fischer, H.R., 2001. "Green" nano-composite materials–new possibilities for bioplastics. MRS Symposium Series 2001, 628.

Foelster, T., Greiner, R., Schaefer, D., 2001. Fiber-reinforced moulded plastic part and process for its manufacture. U.S. Patent 6184272.

Fonseca, L.M., Gonçalves, J.R., El Halal, S.L.M., Pinto, V.Z., Dias, A.R.G., Jacques, A.C., da Rosa Zavareze, E., 2015. Oxidation of potato starch with different sodium hypochlorite concentrations and its effect on biodegradable films. LWT–Food Science and Technology 60, 714–720.

Fu, Z.S., Liang, W.D., Yang, A.M., Wang, Z.G., 2002. Role and relevance of polarity and hindrance of vinyl monomers in graft copolymerization onto potato starch. Journal of Applied Polymer Science 85, 896–899.

George, E.R., Sullivan, T.M., Park, E.H., 1994. Thermoplastic starch blends with a poly(ethylene-co-vinyl alcohol): processability and physical properties. Polymer Engineering and Science 34, 17–23.

Guilbert, S., Cuq, B., Gontard, N., 1997. Recent innovations in edible and/or biodegradable packaging materials. Food Additives and Contaminants 14, 741–751.

Han, J.H., 2002. Protein-based edible films and coatings carrying antimicrobial agents. In: Gennadios, A. (Ed.), Protein-Based Edible Films and Coatings. CRC Press LCC, Boca Raton, FL.

He, Y., Kong, W., Wang, W., Liu, T., Liu, Y., Gong, Q., Gao, J., 2012. Modified natural halloysite/potato starch composite films. Carbohydrate Polymers 87, 2706–2711.

Hernández, O., Emaldi, U., Tovar, J., 2008. In vitro digestibility of edible films from various starch sources. Carbohydrate Polymers 71, 648–655.

Hiemenz, P.C., Rajagopalan, R., 1997. Surface tension and contact angle. In: Hiemenz, P.C. (Ed.), Principles of Colloid and Surface Chemistry. Marcel Dekker, New York, pp. 248–255.

Huizenga, R., Mantingh, J., Pomp-De Wit, F., 1998. Amylopectin potato starch products as sizing agents for textile yarns. World Intellectual Property Organization, International Patent Application PCT/NL98/00063.

Ismail, H., Pasbakhsh, P., Ahmad Fauzi, M.N., Abu Bakar, A., 2008. Morphological, thermal and tensile properties of halloysite nanotubes filled ethylene propylene diene monomer (EPDM) nanocomposites. Polymer Testing 27, 841–850.

Kang, H.J., Min, S.C., 2010. Potato peel-based biopolymer film development using high-pressure homogenization, irradiation, and ultrasound. LWT–Food Science and Technology 43, 903–909.

Kawamoto, H., 2007. Trends in research and development of plastics of plant origin–from the perspective of nanocomposite polylactic acid for automobile use. Science and Technology Trends 22, 62–75.

Kawasumi, M., 2004. The discovery of polymereclay hybrids. Journal of Polymer Science, Part A: Polymer Chemistry 42, 819–824.

Khan, A.A., Husain, Q., 2007. Decolorization and removal of textile and non-textile dyes from polluted wastewater and dyeing effluent by using potato (*Solanum tuberosum*) soluble and immobilized polyphenol oxidase. Bioresource Technology 98, 1012–1019.

Kim, D.-H., Na, S.-K., Park, J.-S., 2003. Preparation and characterization of modified starch-based plastic film reinforced with short pulp fiber. I. Structural properties. Journal of Applied Polymer Science 88, 2100–2107.

Lacourse, N.L., Altieri, P.A., 1991. Biodegradable packaging material and the method of preparation Thereof, U.S. Patent 4,863,655(1989); biodegradable shaped products and the method of preparation Thereof, U.S. Patent 5,043,196.

Ladisch, M.R., 2004. Fundamentals of nanotechnology: relationship to food science and technology. In: IFT Annual Meeting, July 12-16-Las Vegas, NV. http://ift.confex.com/ift/2004/techprogram/paper_21360.htm.

Lawton, J.W., 1996. Effect of starch type on the properties of starch containing films. Carbohydrate Polymers 29, 203–208.

Lenaerts, V., Moussa, I., Dumoulin, Y., Mebsout, F., Chouinard, F., Szabo, P., Mateescu, M.A., Cartilier, L., Marchessault, R., 1998. Cross-linked high amylose starch for controlled release of drugs: recent advances. Journal of Controlled Release 53, 225–234.

Mangiacapra, P., Gorrasi, G., Sorrentino, A., Vittoria, V., 2005. Biodegradable nanocomposites obtained by ball milling of pectin and montmorillonites. Carbohydrate Polymers 64, 516–523.

Mano, J.F., Reis, R.L., 2004. Viscoelastic monitoring of starch-based biomaterials in simulated physiological conditions. Materials Science and Engineering A 370, 321–325.

Maurer, H.W., Kearney, R.L., 1998. Opportunities and challenges for starch in the paper industry. Starch 50, 396–402.

McGlashan, S.A., Halley, P.J., 2003. Preparation and characterization of biodegradable starch-based nanocomposite materials. Polymer International 52, 1767–1773.

Morris, V.J., 2003. Nanotechnology in Food. http://www.nanotec.org.uk/evidence/26aVicMorris.htm.

Muneer, F., Andersson, M., Koch, K., Menzel, C., Hedenqvist, M.K.S., Gällstedt, M., Plivelic, T.S., Kuktaite, R., 2015. Nanostructural morphology of plasticized wheat gluten and modified potato starch composites: relationship to mechanical and barrier properties. Biomacromolecule 16, 695–705.

Murphy, R., Bonin, M., Hillier, B., 2004. Life Cycle Assessment (LCA) of Potato Starch Based Packaging- Strategic Industry Report STI/3/011. British Potato Council, UK.

Nabar, Y., Narayan, R., Schindler, M., 2006. Twin-screw extrusion production and characterizatin of starch foam products for use in cushioning and insulation applications. Polymer Engineering and Science 46, 438–451.

Nordmark, T.S., Ziegler, G.R., 2002a. Spherulitic crystallisation of gelatinized maize starch and its fractions. Carbohydrate Polymers 49, 439–448.

Nordmark, T.S., Ziegler, G.R., 2002b. Structural features of non-granular spherulitic maize starch. Carbohydrate Research 337, 1467–1475.

Osés, J., Niza, S., Ziani, K., Maté, J.I., 2009. Potato starch edible films to control oxidative rancidity of polyunsaturated lipids: effects of film composition, thickness and water activity. International Journal of Food Science and Technology 44, 1360–1366.

Pal, K., Banthia, A.K., Majumdar, D.K., 2006. Starch based hydrogel with potential biomedical application as artificial skin. African Journal of Biomedical Research 9, 23–29.

Pandey, J.K., Reddy, K.R., Kumar, A.P., Singh, R.P., 2005. An overview on the degradability of polymer nanocomposites. Polymer Degradation and Stability 88, 234–250.

Pantoustier, N., Alexandre, M., Degee, P., Calberg, C., Jerome, R., Henrist, C., et al., 2001. Poly(3-caprolactone) layered silicate nanocomposites: effect of clay surface modifiers on the melt intercalation process. e-Polymer 9, 1–9.

Park, H.W., Lee, W.K., Park, C.Y., Cho, W.J., Ha, C.S., 2003. Environmentally friendly polymer hybrids: part I. Mechanical, thermal, and barrier properties of thermoplastic starch/clay nanocomposites. Journal of Materials Science 38, 909–915.

Park, H.-M., Li, X., Jin, C.-Z., Park, C.-Y., Cho, W.-J., Ha, C.-S., 2002. .Preparation and properties of biodegradable thermoplastic starch/clay hybrids. Macromolecular Materials and Engineering 287, 553–558.

Pereira, C.S., Cunha, A.M., Reis, R.L., Vazquez, B., San Roman, S.J., 1998. New starch-based thermoplastic hydrogels for use as bone cements or drug-delivery carriers. Journal of Materials Science: Materials in Medicine 9, 825–833.

Rajisha, K.R., Maria, H.J., Pothan, L.A., Ahmad, Z., Thomas, S., 2014. Preparation and characterization of potato starch nanocrystal reinforced natural rubber nanocomposites. International Journal of Biological Macromolecules 67, 147–153.

Ray, S., Quek, S.Y., Easteal, A., Chen, X.D., 2006. The potential use of polymer-clay nanocomposites in food packaging. International Journal of Food Engineering 2, 1–11.

Riedel, U., 1999. Natural fibre-reinforced biopolymers as construction materials-new discoveries. In: 2nd International Wood and Natural Fibre Composites Symposium, June 28–29, Kassel/Germany, pp. 1–10.

Riedel, U., Nickel, J., 2003. Applications of natural fiber composites for constructive parts in aerospace, automobiles, and other areas. In: Steinbüchel, A. (Ed.). Steinbüchel, A. (Ed.), Biopolymers, vol. 10. Wiley-VCH, pp. 1–11.

Robertson, T., 2006. Study Seeks New Uses for Potato Waste. Minnesota Public Radio. http://news.minnesota.publicradio.org/features/2005/01/27_robertsont_potatowaste/.

Rodríguez, M., Osés, J., Ziani, K., Maté, J.I., 2006. Combined effect of plasticizers and surfactants on the physical properties of starch based edible films. Food Research International 39, 840–846.

Roesser, D.S., Nevling, J., Rawlins, D.C., Billmers, R.J., 2000. Biodegradable expanded starch products and the method of preparation, U.S. Patent 6,107,371.

Shogren, R.L., Fanta, G.F., Felker, F.C., 2006. X-ray diffraction study of crystal transformations in spherulitic amylose/lipid complexes from jet-cooked starch. Carbohydrate Polymers 64, 444–451.

Singh, J., Lelane, C., Stewart, R.B., Singh, H., 2010. Formation of starch spherulites: role of amylose content and thermal events. Food Chemistry 121, 980–989.

Smith, J.P., Hoshino, J., Abe, Y., 1995. Interactive packaging involving sachet technology. In: Rooney, M.L. (Ed.), Active Food Packaging. Blackie Academic and Professional, Glasgow, pp. 143–173.

Somasekhar, M., 2001. Potatoes to Power Cars, Make Soft Drinks. The Hindu. July 30 http://www.hinduonnet.com/businessline/2001/07/30/stories/073067b1.htm.

Song, S., Wang, C., Pan, Z., Wang, X., 2008. Preparation and characterization of amphiphilic starch nanocrystals. Journal of Applied Polymer Science 107, 418–422.

Sorrentino, A., Gorrasi, G., Tortora, M., Vittoria, V., Costantino, U., Marmottini, F., et al., 2005. Incorporation of Mg-Al hydrotalcite into a biodegradable poly(3-caprolactone) by high energy ball milling. Polymer 46, 1601–1608.

Sorrentino, A., Gorrasia, G., Vittoriaa, V., 2007. Potential perspectives of bio-nanocomposites for food packaging applications. Trends in Food Science and Technology 18, 84–95.

Stearns, L.D., Petry, T.A., Krause, M.A., 1994. Potential Food and Nonfood Utilization of Potatoes and Related Byproducts in North Dakota. Agricultural Experiment Station, Fargo. Agricultural Economics Report Number 322.

Storey, M., 2003. Biodegradable Packaging: Spuds Get Packing. Sustainable Technologies Initiative. Oakdene Hollins. http://www.ohlsti.co.uk/ohl/stipdfs/ohl_sti56.pdf.

Szymanowski, H., Kaczmarek, M., Gazicki-Lipman, M., Klimek, L., Woźniak, B., 2005. New biodegradable material based on RF plasma modified starch. Surface and Coatings Technology 200, 539–543.

Talja, R.A., Helén, H., Roos, Y.H., Jouppila, K., 2008a. Effect of type and content of binary polyol mixtures on physical and mechanical properties of starch-based edible films. Carbohydrate Polymers 71, 269–276.

Talja, R.A., Peura, M., Serimaa, R., Jouppila, K., 2008b. Effect of amylose content on physical and mechanical properties of potato-starch-based edible films. Biomacromolecules 9, 658–663.

Thanavala, Y., Mahoney, M., Pal, S., Scott, A., Richter, L., Natarajan, N., Goodwin, P., Arntzen, C.J., Mason, H.S., 2005. Immunogenicity in humans of an edible vaccine for hepatitis B. PNAS 102, 3378–3382.

Tsiapouris, A., Dresden, L.L., 2000. Water vapor sorption determination of starch based porous packaging materials. Starch 52, 53–57.

Tuovinen, L., Peltonen, S., Järvinen, K., 2003. Drug release from starch-acetate films. Journal of Controlled Release 91 (3), 345–354.

Usuki, A., Kojima, Y., Kawasumi, M., Okada, A., Fukushima, Y., Kurauchi, T., Kamigaito, O., 1993. Synthesis of nylon6-clay hybrid. Journal of Materials Research 8, 1179–1184.

Vigneshwaran, N., Kumar, S., Kathe, A.A., Varadarajan, P.V., Prasad, V., 2006. Functional finishing of cotton fabricsusing zinc oxide-soluble starch nanocomposites. Nanotechnology 17, 5087–5095.

Weiss, J., Takhistov, P., McClements, J., 2006. Functional materials in food nanotechnology. Journal of Food Science 71, R106–R107.

Wilhelm, H.M., Sierakowski, M.R., Souza, G.P., Wypych, F., 2003. Starch film reinforced with mineral clay. Carbohydrate Polymers 52, 101–110.

Won, C.Y., Chu, C.C., Yu, T.J., 1997. Synthesis of starch-based drug carrier for the control/release of estrone hormone. Carbohydrate Polymers 32, 239–244.

Yoon, S.-Y., Deng, Y., 2006. Claystarch composites and their application in papermaking. Journal of Applied Polymer Science 100, 1032–1038.

Yu, L., Dean, K., Li, L., 2006. Polymr blends and composites from renewable sources. Progress in Polymer Science 31, 576–602.

Zhang, L.-M., Hu, Z.-H., 2002. Synthesis and thermal degradation of a novel starch graft copolymer incorporating a sulfobetaine zwitterionic monomer. Starch 54, 290–295.

Ziegler, G.R., Nordmark, T.S., Woodling, S.E., 2003. Spherulitic crystallisation of starch: influence of botanical origin and extent of thermal treatment. Food Hydrocolloids 17, 487–494.

Zavareze, E.D.R., Pinto, V.Z., Klein, B., El Halal, S.L.M., Elias, M.C., Prentice-Hernández, C., Dias, A.R.G., 2012. Development of oxidised and heat–moisture treated potato starch film. Food Chemistry 132, 344–350.

CHAPTER 22

Potato Proteomics: A New Approach for the Potato Processing Industry

Cristina Barsan

Université de Toulouse, INP-ENSA Toulouse, Génomique et Biotechnologie des Fruits, Castanet-Tolosan, France; INRA, Génomique et Biotechnologie des Fruits, Chemin de Borde Rouge, Castanet-Tolosan, France; Led Academy, Toulon, France

1. Introduction

Food quality has been a priority for humans all through history. People empirically selected foods with higher nutritional qualities, more appealing aspect, and richer aromas. Potato is the third most important food crop worldwide, behind wheat and rice (http://faostat.fao.org). According to the International Potato Center (http://cipotato.org) there are over 4000 edible varieties of potato. Potato is unique among the major world food crops in producing tubers that are an important dietary source of starch, protein, antioxidants, and vitamins (Potato Genome Sequencing Consortium, 2011). It can be used as food, by itself, or can provide ingredients for the food industry such as starch or proteins like patatin galactose, galactooligosaccharides, and galactan conjugates with high nutritional quality and multiple health benefits (Seo et al., 2014). Today an array of new methods allow the improvement of potato quality in shorter periods of time with targeted results. The cell is a complex mechanism and all of its pieces work together. To select plant varieties with specific qualities required by the consumer or the food industry, as well as to find novel ingredients, it is essential to understand how these pieces interact and how various components of interest are biosynthesized and degraded. The potato genome was sequenced in 2011 by the Potato Genome Sequencing Consortium, providing a powerful tool in potato selection and improvement. However, as powerful as it is, genomics has several limitations: it cannot give us the set of all protein isoforms and modifications, the interactions between them, the structural description of proteins and their higher-order complexes, or almost anything "postgenomic." Proteomics complements functional genomics approaches and is by definition expected to yield direct biological insights (Tyers and Mann, 2003). It identifies the crucial roles of proteins in cell structure and diverse biological processes, including signal transduction and nutrient utilization (Wang et al., 2006). Protein expression is the functional outcome of gene transcription and translation. Even from the early stages of biological research the fundamental role of proteins in supporting life was recognized. Approximately

9000 specialized protein types per nucleated cell regulate multivariate functions from movement to mitosis (Cristea et al., 2004). Proteomic approaches can be used for several purposes: proteome profiling, comparative expression analysis of two or more protein samples, the localization and identification of posttranslational modifications, the study of protein–protein interactions (Chandramouli and Qian, 2009), and quantitative comparison of various proteomes. The use of proteomics is a powerful tool for the food industry in terms of process optimization, quality, safety, and nutritional assessment (Pedreschi et al., 2010). In food science, increasing effort is being put toward gaining more insight in the underlying produce physiology through proteomics-related approaches (Gašo-Sokac et al., 2010). Understanding the physiological processes behind postharvest disorders of fruits and selecting markers associated with harvested fruit maturity would decrease costs related to produce degradation.

2. Proteomics Techniques—An Overview

The proteome could be defined as the array of proteins found within a cell at a specific time. High-throughput proteomics was born in the late 1980s with the invention of soft ionization techniques, matrix-assisted laser desorption ionization (MALDI), and electrospray ionization (ESI) to vaporize peptides and proteins to make possible their analysis by mass spectrometry (MS) (Wang et al., 2006). The first step in any proteomics analysis (Figure 1) is cell lysis, followed by protein purification and separation. The protein preparation is a critical step in the analysis of the proteome, since both the quality and the quantity of the protein isolate determine the reliability and reproducibility of the findings (Fuchs et al., 2005). Various protein extraction protocols are available depending on species, tissue type, protein solubility, and subcellular localization. The most popular are the trichloroacetic acid/acetone precipitation first described by Damerval et al. (1986) and the phenol extraction method first described by Hurkman and Tanaka (1986). The next step is the protein separation by electrophoresis. The most commonly practiced is sodium dodecyl sulfate–polyacrylamide gel electrophoresis. It allows the estimation of the molecular weight of the proteins and the number of polypeptides in a sample (Garfin, 2003). Another electrophoresis technique, first described by O'Farrell (1975), is two-dimensional polyacrylamide gel electrophoresis (2-DE). The proteins are separated in two steps according to two independent properties (isoelectric point and molecular weight), and this technique has evolved as one of the core technologies for the analysis of complex protein mixtures. 2-DE is being used less and less, researchers arguing that it is cumbersome, time-consuming, and lacking in automation (Cristea et al., 2004). However, its days are not over yet: 2-DE is the only technique that can be routinely applied for parallel quantitative expression profiling of complex protein mixtures and is the most widely used method for efficiently separating proteins, their variants, and their modifications (Chandramouli and Qian, 2009). Gel electrophoresis analysis is typically followed by protein digestion, a key procedure prior to the identification of proteins by MS. Before protein digestion the proteins in most samples need to be denatured, reduced,

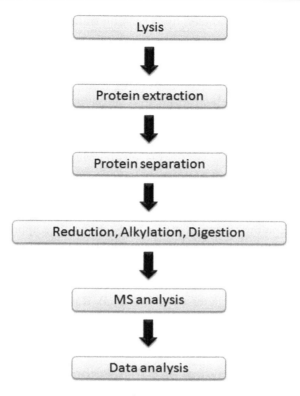

Figure 1
A simplified course of action for a proteomics analysis. The main steps are cell lysis; protein extraction; protein separation; reduction, alkylation and digestion; MS analysis; and data analysis.

and alkylated, using various reagents to allow the proteolytic enzyme to efficiently cleave the peptide chains of the proteins (Hustoft et al., 2012). Different proteolytic agents are used for protein digestion, including enzymes such as trypsin, the most popular, various endoproteases (Lys-C, Arg-C, Asp-N, Glu-C); or chymotrypsin, as well as chemical reagents such as hydroxylamine or cyanogen bromide. The specificity of the bonds cleaved by these reagents allows one to obtain specific peptides that facilitate the interpretation of their mass spectra and database search (Martínez-Maqueda et al., 2013). Regardless of the choice of a given proteomic separation technique, gel-based or gel-free, the mass spectrometer is always the next step in protein identification (Chandramouli and Qian, 2009). In general, a mass spectrometer has several components: an ion source, one or several mass analyzers, and a detector (Wang et al., 2006). Several types of commercially available mass spectrometers that combine ESI or MALDI with a variety of mass analyzers are available. Gas-phase ions, produced in the ion source, are introduced into the mass analyzer and differentiated according to their mass/charge ratio on the basis of their motion in a vacuum under the influence of electric or magnetic fields (Cristea et al., 2004). If we were to summarize all in one phrase, one would say that peptides of predicted open reading frames are generated in silico and used to match the MS-derived data.

Several possibilities of MS analysis are available today depending on the desired outcome, such as targeted MS or quantitative techniques using isotope-mediated approaches; 10,000 proteins can be identified in a single experiment (Low et al., 2013). MS quantitative proteomics methods can be categorized into "label-free" approaches, or peptide counting, and "stable isotope labeling" methods by which proteins and/or peptides are metabolically or chemically encoded by heavy stable isotopes (Ong and Mann, 2005). Commonly used metabolic labeling methods in plant proteomics include stable isotope labeling by ^{15}N (Nelson et al., 2007) and stable isotope labeling by amino acids in cell culture (Ong et al., 2002). Chemical methods for stable isotope labeling include iTRAQ (Ross et al., 2004) and isotope-coded protein labeling (Schmidt et al., 2005). Label-free approaches are rather straightforward to implement even if their robustness and accuracy rely on multiple replicate runs and rather complex comparative data analysis (Vowinckel et al., 2014). The main advantages of stable isotope labeling techniques are their accuracy of quantification and the ability to perform multiple experiments (Petersen et al., 2013). Raw data from MS/MS are then searched using database search engines and software such as MASCOT, ProteinPilot, the ProteinLynx Global Server, PEAKS, and X!tandeem (Noordin and Othman, 2013).

3. Proteomics in Food Nutrition and Processing

Every cellular process relies on the interaction of a very large number of proteins expressed at the level of the cell, the organ, or the whole organism. The flux of a nutrient or metabolite through a biochemical pathway is controlled by alterations in levels of mRNA and, in turn, of the corresponding proteins (Fuchs et al., 2005). In addition to in vivo posttranslational modifications, food processing induces further modifications of the proteins, such as side-chain oxidation, cross-link formation, and backbone cleavage, critically influencing key food properties (Clerens et al., 2012). Proteomics has been used in a variety of purposes: (1) in pointing out differences in food proteomes relevant to nutrition and to the identification of food supplements (Carbonaro, 2004); (2) in following changes in the protein composition of various tissues during growth, during maturation, postmortem or postharvest, as well as after downstream treatments such as cooking (Clerens et al., 2012); (3) in the identification of biomarkers that could allow the prediction of the processed products' quality; and (4) in the characterization and standardization of raw materials (Gašo-Sokac et al., 2010). Some of the most consumed products today, meat, milk, cereal, fruits, and vegetables, could benefit from proteomics in terms of nutrient content, shelf life, digestibility, health effects, and processed product quality.

Tomatoes, for example, are widely treasured especially for their carotenoids. The most famous tomato carotenoid is lycopene, a compound associated with major health benefits

(Dahan et al., 2008; Riccioni et al., 2008; Wood et al., 2008). A proteomic analysis revealed that the chromoplast-specific carotenoid-associated protein probably regulates carotenoid sequestration and storage, which seems to be important for the enhanced accumulation of carotenoids in high-pigmented transgenic tomato fruits (Kilambi et al., 2013). The production site for carotenoids and other essential metabolites for nutritional and sensory quality of agricultural products, the chromoplast, was the target of proteomic studies in several species, sweet orange (*Citrus sinensis*) (Zeng et al., 2011), red bell pepper (*Capsicum annuum*) (Siddique et al., 2006), and tomato (*Solanum lycopersicum*) (Barsan et al., 2010; Zeng et al., 2011), and a proteomic comparative characterization of chromoplasts from six carotenoid-rich crops, watermelon, tomato, carrot, orange cauliflower, red papaya, and red bell pepper, was made (Wang et al., 2013).

Texture is another trait of major importance on the fruit market and is mainly related to the cell wall. The cell wall proteome of three ripe tomato fruit lines with contrasting texture traits was analyzed by Konozy et al. (2013). Weakly bound and soluble proteins were isolated from the cell wall. Some variations in the protein repertoires were observed among the lines, opening the possibility of characterizing cell wall protein genetic variability.

Another major food group is represented by cereal grains (wheat, maize, rice, and barley) (Cordain, 1999). Komatsu and Tanaka (2004) provided a catalog of rice proteins. A more recent rice proteome database is OryzaPG-DB, based on shotgun proteogenomics (Helmy et al., 2011).

A proteomic study was performed on transgenic rice to detect the effects of genetic modification on crops (Xue et al., 2010). Safety assessments were conducted for MON810 (Bt maize, Monsanto), a variety of transgenic maize resistant to the European corn borer (*Ostrinia nubilalis*), using grains from plants grown in fields (Giuliano Albo et al., 2007) and growth chambers (Zolla et al., 2008). Fourteen novel allergenic proteins were detected in both the transgenic and the nontransgenic varieties (Fonseca et al., 2012). The first maize allergen, a 9-kDa lipid-transfer protein (LTP), was discovered by Pastorello et al. (2000). The much stronger allergenic potential of maize seeds was revealed using MS techniques that led to the identification of six additional allergens (Fasoli et al., 2009). In wheat, a comprehensive characterization of allergens led to the identification of nine subunits of low-molecular-weight glutenins. 2-DE maps were generated to be used for the diagnosis of wheat-allergic patients and the identification of wheat allergens in food (Akagawa et al., 2007).

The barley cultivar has a major impact on the protein profiles of beers, the malting process, and the brewing yeast. 2-D gel maps of various beer proteomes representing different cultivars and malting types have been created as a tool in the quality control of the beers and to detect and identify beer type-specific proteins or protein isoforms that might represent taste, flavor, or texture. In the long term this will potentially enable the manipulation of flavor proteins (Fasoli et al., 2010; Iimure et al., 2010).

4. Potato Proteomics and Quality Improvement

Potato (*Solanum tuberosum*), as the third most important food crop worldwide, is used not only for human and animal consumption but also as a raw material for starch and alcohol production. There has been a significant increase in the demand for potatoes by the food industry. For example, in 2005 in the United States and in several European countries more than 50% of the tuber yield was purchased by the processing industry (Carputo et al., 2005). This calls for potatoes with specific shapes, less disease on the skin, and minimal external damage. Size, shape, and shallow eyes are important with regard to the appearance of the product and the influence on wastage during peeling. Other important traits are dry matter (both the processing efficiency and the quality of the finished product benefit from a high dry matter content); the reducing sugar, with great influence on the color of fried products; color distribution (unevenness in color distribution results in french fries with a brown color at one end); and darkening (Netherlands Potato Consultative Foundation—http://www.nivaa.nl). A major problem for the industry is represented by black spot, a condition that is the result of tissue injury due to transportation, handling, and storage conditions.

One of the most important aspects of potato production is tuber quality (Figure 2)

Tubers contain significant concentrations of vitamin C and essential amino acids. They are also a source of at least 12 essential vitamins and minerals. Several factors affect tuber quality

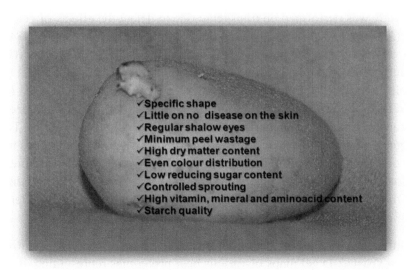

Figure 2
Main potato tuber qualities demanded by consumers and the food industry: shape, shallow eye distribution, absence of skin disease, minimum peel wastage, dry matter content, color, reducing sugar content, sprouting, and nutrient and starch content.

(Figure 3): the genetic makeup of the cultivar, crop maturity, agronomic practices, environmental conditions, storage temperature, and the presence of pests and diseases (Carputo et al., 2005). They have a double function, as a storage organ and as a vegetative propagation system. Potato tubers are developed from underground stems that originate from thickened stolons (Sarkar, 2008) and go through a series of morphological changes of stolon formation, initial tuber development, and tuber maturation (Yu et al., 2012). To understand the molecular basis of tuberization in potato, comparative proteomic approaches have been applied to monitor differentially expressed proteins at various developmental stages. A study following changes in the tuber proteome from tuberization, through tuber development and storage, into the sprouting phase found that the development process was characterized by the accumulation of isoforms of the major storage protein patatin and enzymes involved in disease and defense reactions. Furthermore, enzymes involved in carbohydrate and energy metabolism and protein processing were associated with development, but decreased during tuber maturation. Two hundred proteins were found to express differential abundance (Lehesranta et al., 2006). A comparative proteomic analysis of tuber induction, development, and maturation revealed nearly 100 proteins. Among them, some were presumably associated with tuber cell differentiation or regulation of diverse functions like protein biogenesis and storage, bioenergy and metabolism, and cell defense and rescue. Reactive oxygen species-catabolizing enzymes were activated at the initial tuber stage with functionally diverse proteins (Agrawal et al., 2008). An alternative in vitro tuberization model system was applied to identify the

Figure 3
Factors that play a role in potato tuber quality: they can be both internal (the genetic makeup) and external (environmental conditions, agricultural practices, etc.).

large-scale proteome using a shotgun proteomic approach. Two hundred fifty-one stage-specific identified proteins showed dynamic changes in their expression at different growth stages. Stolon stage-specific proteins were primarily involved in the photosynthetic machinery. Proteins specific to each tuber stage were found: patatin for the initial tuber stage, and 6-fructokinase, phytoalexin-deficient 4-1, metallothionein II-like protein, and malate dehydrogenase for the developing tuber stage. Novel stage-specific proteins identified during in vitro tuberization were ferredoxin–NADP reductase, 34-kDa porin, aquaporin, calmodulin, ripening-regulated protein, and starch synthase. Superoxide dismutase, dehydroascorbate reductase, and catalase I were most abundantly expressed in the stolon; however, the enzyme activities of these proteins were stronger in the initial tuber (Yu et al., 2012).

4.1 Transgenic Potatoes—Meeting Specific Demands

Basic studies in genetics, biochemistry, and physiology served as starting points for improving potato quality (Figure 4). The introduction of recombinant DNA technology led to the development of transgenic plants. Traits such as starch composition, antioxidant content, and pest resistance were targets for the creation of new potato varieties that answered specific demands. And more, the pesticidal proteins ectopically expressed in transgenic crops were followed by an increase in the protein content (Munger et al., 2012). The production of starches with modified amylose-to-amylopectin ratio was achieved after altering various enzymes of the starch biosynthetic pathway. The reduction of the activity of ADP-glucose

STARCH

✓ Smaller starch granues – reduced amylose to amylopectin ratio (Lloyd *et al.*, 1999)
✓ High amylose, low amylopectin ,high phosphorus levels (Schwall *et al.*, 2000)
✓ Freeze-thaw-stable potato starch - amylose-free with short-chain amylopectin (Jobling *et al.*, 2002)
✓ Starch suitable for industrial applications: low in amylopectin, Amflora - BASF Plant Science

AMINOACIDS

✓ 1.5- to 8-fold increase in all essential amino acids (Chakaborty *et al.*, 2000)

CAROTENOIDS

✓ Increase of zeaxanthin content 4 to 130 fold (Römer *et al.*, 2002)
✓ Increase in total carotenoids 7 fold (Ducreux *et al.*,2005)
✓ Increase in β-carotenoid levels 14 fold and total carotenoids 2.5 fold (Diretto *et al.*, 2006)
✓ Increase in β-carotenoid levels 3600 fold and total carotenoids 20 fold (Diretto *et al.*, 2007)

Figure 4

Examples of improved nutritional qualities and better response to the demands of the industry in transgenic potatoes: alterations in starch composition, improved amino acid concentration, and higher carotenoid content.

pyrophosphorylase significantly reduced the amylose content and shortened the amylopectin chains, thus reducing the size of the starch granules (Lloyd et al., 1999). The simultaneous antisense inhibition of two isoforms of starch-branching enzymes (SBE A and B) gave transgenic lines with increased amount of amylose, low amylopectin, and high phosphorus levels (Schwall et al., 2000). An amylose-free starch with short-chain amylopectin was produced by simultaneous antisense downregulation of three starch synthase genes: SSII, SSIII, and GBSS. The result was a freeze–thaw-stable potato starch (Jobling et al., 2002). In 1996 BASF Plant Science developed Amflora, a genetically modified starch potato that produces only one particular type of starch, low in amylopectin, for industrial applications, thus optimizing the further processing of starch and saving energy and water—and that was the first transgenic crop approved by the European Commission in 2010. Potatoes transformed with the gene *AmA1* from *Amaranthus hypochondriacus*, encoding a protein with a nutritionally balanced amino acid composition, showed a 1.5- to 8-fold increase in all essential amino acids compared to the wild type (Chakaborty et al., 2000).

Another target of choice was some of the most health-promoting ingredients of the human diet: the carotenoids. They have antioxidant properties and are supposedly beneficial in preventing cancer, cardiac disease, and eye diseases (Krinsky et al., 2004). β-Carotene, also known as provitamin A, is a nutrient of major importance since vitamin A deficiency affects several hundred million people worldwide (Giuliano et al., 2000). The main carotenoids present in cultivated potato are lutein, violaxanthin, zeaxanthin, and antheraxanthin. β-Carotene is present only in trace amounts (Nesterenko and Sink, 2003). The synthesis of zeaxanthin epoxidase was downregulated specifically in tubers through antisense technology and cosuppression approaches. Both strategies achieved decreased conversion of zeaxanthin to violaxanthin in transgenic tubers with a corresponding increase in zeaxanthin content of 4- to 130-fold. (Römer et al., 2002). Potatoes with a 7-fold increase in total carotenoids, mainly β-carotene and lutein, were obtained by the overexpression of the bacterial *CrtB* gene, encoding phytoene synthase (Ducreux et al., 2005). The silencing of lycopene ε-cyclase, the first dedicated step in the β–ε branch of carotenoid biosynthesis, led to tubers with up to 14-fold increase in β-carotenoid levels and up to 2.5-fold increase in total carotenoids (Diretto et al., 2006). The transformation of potato with a minipathway of bacterial origin driving the synthesis of β-carotene from geranylgeranyl diphosphate led to an impressive increase of approximately 20-fold in carotenoids and 3600-fold in β-carotene (Diretto et al., 2007).

Hectares of transgenic crops have increased exponentially since 1996, when such crops began to be commercialized, making transgenic biotechnology, together with conventional breeding, the main approach to improving agronomic crop traits (Gong and Wang, 2013). This came with public concerns about transgenic crop safety for consumption and their influence on the environment, namely a possible impact on insect populations and biodiversity in agricultural fields. In the case of transgenic potatoes one study in particular created a media sensation.

The study was conducted in 1998 by Dr Árpád Pusztai and it claimed that consumption of transgenic potatoes in rats has significant effects on organ development, body metabolism, and immune function. His results were never published and the reviewers stated that the experiment was poorly conducted. However, the harm was done: it negatively influenced public opinion against genetically modified (GM) foods in general and GM potatoes in particular (Smyth and Phillips, 2000). Today, according to the http://www.gmo-compass.org web site, transgenic potatoes can be cultivated only in the United States (four cultivars), Canada (four cultivars), and Europe (one cultivar), but can be used as food or livestock feed in Australia, Mexico, Japan, Korea, and Philippines.

There is an increasing demand for time- and cost-effective tools suitable to assess the characteristics of transgenic crops. Comparative proteomics was used to evaluate variation in tuber proteomes of a large collection of potato genotypes, including different varieties, landraces, and transgenic lines. The results revealed clear separations, indicating extensive genotypic variation, but no significant differences between transgenic and nontransgenic lines (Lehesranta et al., 2005). Other studies found inconsistent proteome changes in various tissues of transgenic potato. Transgenic potato expressing cathepsin D inhibitor showed no quantitative proteome difference in tubers compared to the nontransgenic counterpart (Khalf et al., 2010). Transgenic crops are full of surprises. Unintended pleiotropic effects were reported for a number of pesticidal proteins ectopically expressed in transgenic crops, but the nature and significance of such effects in plants remain poorly understood (Munger et al., 2012). The so-called pleiotropic effects of recombinant protease inhibitors might simply reflect a lack of knowledge on stress-related proteolysis in plants but represent a source of potentially useful traits for crop improvement (Schlüter et al., 2010). It is thus important to take into account not only the characteristics of expressed recombinant proteins, but also their possible effects on protein accumulation in the host plant. The effects of corn cystatin II, a potent inhibitor of C1A cysteine protease considered for insect and pathogen control and resistance, were assessed by 2-DE in potato lines constitutively expressing it. Most upregulated proteins were abiotic or biotic stress-responsive proteins and pathogenesis-related proteins (Munger et al., 2012). The tomato cathepsin D inhibitor ectopically expressed in potato was shown to have a significant impact on protein levels (Goulet et al., 2010). A clinically useful protease inhibitor, the bovine aprotinin used to reduce bleeding during complex surgery, was expressed in transgenic potato lines. No visible phenotypic effects of aprotinin were observed in the transgenic lines, but aprotinin retention in the endoplasmic reticulum was associated with a significant decrease in leaf soluble protein content. A 2-D gel assessment of control and transgenic lines revealed a possible link between this altered protein content and the downregulation of proteins involved in protein synthesis and maturation. These observations, supported by complementary 2-DE analyses with potato lines targeting aprotinin to the apoplast, suggest an aprotinin-mediated feedback in plants negatively altering protein anabolism (Badri et al., 2009).

4.2 Potato Pathogens and Tuber Quality

One of the purposes of transgenic crops is to reduce the utilization of pesticides to prevent disease, thus increasing the yield and providing produce with fewer defects. In potato, disease can have an impact on tuber quality in many ways: indirectly by pesticide and fungicide use (Burra et al., 2014) and directly by defects, such as rotting (in late blight), or low crop yield. Understanding the defense mechanisms and finding new ways of fighting disease would lead to increased potato production. Major drawbacks of many resistant clones are late maturity and nonmarketable tuber characteristics preventing them from achieving commercial acceptance—this should make selection for tuber quality a priority for inbreeding disease-resistant cultivars (Bisognin and Douches, 2002). Understanding the defense mechanisms could lead to the discovery of new methods to fight against pathogens that would have a direct impact on agricultural methods. Natural defense systems that could be potentially enhanced do exist. The potato tuber periderm is a protective tissue containing an array of plant defense components (Kolattukudy, 1977). A proteomic analysis of the potato skin revealed that the major functional category of its proteins (63%) was involved in plant defense responses to biotic and abiotic stresses (Barel and Ginzberg, 2008).

Potato suffers from many pests and diseases, among which late blight, caused by the oomycete *Phytophthora infestans*, is the worst. Despite over a century of resistance breeding, fungicide use, and other control measures, it is still a major threat to sustainable food production worldwide (Haverkort et al., 2009). The compatible and incompatible interactions between *P. infestans* and potato were investigated using quantitative label-free proteomics. Four potato clones, Desiree, Sarpo, Mira, and SW93-1015, were tested, the last being a highly resistant potato clone. Over 17,000 transcripts and 1000 secreted proteins changed in abundance at at least one time point. Certain apoplastic proteins were found to be suppressed by the pathogen during the compatible interaction. Major proteins that accumulated in the susceptible interaction were glucanases and glucosidases. The study provided an ample number of transcripts and proteins that could be further used in understanding the plant–pathogen interactions (Ali et al., 2014).

Extensive efforts to find more efficient weapons against late blight are ongoing worldwide. One alternative could be integration of induced resistance in current management strategies, or the equivalent of vaccines in plants. β-Aminobutyric acid (BABA), thiamine (vitamin B1), thiadiazole-7-carbothioc acid *S*-methyl ester, phosphite ($H_2PO_3^-$) (Burra et al., 2014), and curdlan oligosaccharide (CurdO) (Li et al., 2014) can induce resistance in a wide range of plants against several types of pathogens including *P. infestans*. However, their mechanisms of action are poorly understood. To elucidate the defense responses activated by BABA in potato, a genome-wide transcript microarray analysis in combination with label-free quantitative proteomics analysis of the apoplast secretome was performed 2 days after treatment of the leaf canopy with 1 and 10 mM BABA. At 10 mM BABA affected several processes related to plant hormones and amino acid metabolism. A global effect was observed on the transcriptome and there was an increase in the abundance of a large number of secreted proteins (Bengtsson et al., 2014). The tuber yield of potatoes grown in the field and under greenhouse conditions was assessed after weekly applications of BABA and it was found to be unaffected (Liljeroth et al., 2010). The phosphite-mediated induced resistance was also studied. Phosphite enters the cell via phosphate transporters and interferes with phosphate signaling mechanisms because of its close steric resemblance, which potentially could lead to indirect induction of resistance (Berkowitz et al., 2013). Complex and multiple processes are involved in phosphite-induced resistance. A broad proteomic and microscopy study performed by Lim et al. (2013) suggested that phosphite triggers a hypersensitive response that is responsible for induced resistance of potato leaves against *P. infestans*. The team analyzed soluble proteins from potato leaf treated with the phosphite product Confine and found that the proteins involved in salicylic acid (SA)-dependent defense responses, reactive oxygen species (ROS), and calcium-dependent pathways were upregulated, while proteins involved in amino acid and starch metabolism were downregulated. An induction of hypersensitive response and callose formation after pathogen attack in phosphite-treated leaves was also shown.

The transcriptome of *S. tuberosum* (cv. Desiree) was investigated and in parallel its secretome was analyzed by quantitative proteomics after foliar application of the protective agent phosphite. The distribution of phosphite in plants after application and in tested transgenic potato lines with impaired salicylic and jasmonic acid signaling was also studied. The results revealed that multiple defense pathways are rapidly induced by phosphite treatment, possibly causing heightened defense leading to enhanced resistance after pathogen infection in local tissues and also an influence on primary metabolism and cell wall-associated metabolic processes (Burra et al., 2014). The tuber yield of potatoes grown in the field and under greenhouse conditions after the application of potassium phosphites (KPhi) to seed tubers of two potato cultivars on crop emergence and early growth was studied. The application of KPhi reduced the period between planting and emergence and increased leaf area and dry matter. The ratio between dry matter of aerials and underground organs was unaffected (Tambascio et al., 2014). Foliar applications of KPhi to field-grown crops resulted in postharvest tubers with a reduced susceptibility to *P. infestans*, *Fusarium solani*, and *Erwinia carotovora* infections. No negative effects were observed in potato yield at harvest, measured as total tuber weight and dry matter (Lobato et al., 2011).

Curdlan, a linear water-insoluble β-1,3-glucan produced by fermentation of *Agrobacterium* sp., is approved as a safe food additive (Spicer et al., 1999). An effective process to produce curdlan β-1,3-oligosaccharides was developed (Li et al., 2013). This could be seen as an alternative to the toxic compounds used in disease control today. The defense responses after the application of CurdO on foliar tissues of potato (*S. tuberosum* L. cv. McCain G1) were investigated at the biochemical and proteomic levels. It had a boosting effect on the early and late defense responses in potato leaves. A 2-D polyacrylamide gel electrophoresis analysis revealed that after the treatment a number of proteins involved in disease/defense, metabolism, transcription, and cell structure were upregulated. This molecule could also be used in prevention; plants treated with CurdO 1 day before infection with *P. infestans* had fewer lesions on their leaves. Furthermore, the results on potato yield and induction reactions indicated that the defense responses induced by CurdO lasted for a short period of time but disappeared gradually (Li et al., 2014).

Aspergillus terreus is one of the most harmful filamentous fungal pathogen of humans, animals, and plants (De Lucca, 2007) and can cause foliar blight disease in potato by an unelucidated infection process (Louis et al., 2013). To understand the mechanism by which *A. terreus* interacts with the potato host, the changes in leaf and tuber proteins of potato cv. Kufri Jyoti were studied. A significant proteome change hallmarked by differential expression of class I patatin, lipoxygenase, catalase–peroxidase complex, and cysteine proteinase inhibitor was observed during tuber colonization. These proteins are often involved in signal transduction pathways and cross talk in pathogenic responses (Louis et al., 2014).

Plant–insect interactions are another source of concern. It appears that plants modulate their responses also in functions of the wounding patterns during herbivory (Bricchi et al., 2010). The response of cultivated potato to either mechanical wounding or herbivory by two specialized insect herbivores, the defoliating pest Colorado potato beetle *Leptinotarsa decemlineata* Say and the potato aphid *Macrosiphum euphorbiae* Thomas, was studied by a 2-DE/MS approach. Of approximately 500 leaf proteins monitored, 31 were up- or downregulated by at least one stress treatment compared to healthy control plants. Twenty-nine were regulated by beetle chewing, eight by wounding, and eight by aphid feeding. The modulated proteins were mainly typical defense proteins and proteins involved in photosynthesis (modulated by potato beetle feeding). These observations illustrate the differential impacts of wounding and insect herbivory on defense- and photosynthesis-related components of the potato leaf proteome, probably associated with the perception of distinct physical and chemical cues in plants (Duceppe et al., 2012).

4.3 Proteomics and Stress Adaptation in Potatoes

Stress can be defined as any condition that is above or below the optimum for plant growth and tuber production. Stressed potatoes often exhibit high levels of internal and external defects, as well as low specific gravity (Thornton, 2002). High temperatures, above 30 °C, for example, decrease cell division and lower the supply of carbohydrates available to the tuber (Geigenberger et al., 1998). Numerous major external defects (growth cracks, knobs, malformations, greening, heat sprouts, enlarged lenticels, elephant hide) and internal problems (brown center/hollow heart, internal brown spot, heat necrosis, vascular discolorations, translucent end, blackheart) (Hiller and Thornton, 1993) make stress management a top priority in potato research.

Low specific gravity in the stem end of the tuber is often associated with high levels of reducing sugars, making sugar-end fries a common problem in the food industry (Thornton, 2002). The disorder brown fleck, a major cause of potato consumer dissatisfaction that reduces potato growing profitability, is also a stress-related condition due to elevated night temperature (Harper, 2004). Water stress can result in yield reductions and loss of tuber grade (Eldredge et al., 1992). The creation of stress-resistant cultivars could lead to major improvements in product quality and crop yield and diminished losses with visible economical impact.

Cryopreservation, the storage of biological material in liquid nitrogen ($-196\,°C$), complements classical conservation methods carried out in the field or in vitro, allowing the safe storage of vegetative propagated plants over an unlimited period of time. An important application of this technique is obtaining virus-free potatoes (Wang et al., 2006). The recovery after cryopreservation is improved by drought acclimation. Better knowledge on responses to dehydration stress could help to improve the existing cryopreservation protocols for potato, since plant tissues processed for cryopreservation are often submitted to similar in vitro stress conditions (Folgado et al., 2013). A proteomic analysis was performed on potato to elucidate the mechanism behind this improvement. Plants were treated both before and after cryopreservation, with sorbitol up to 0.11 M for 21 days and 0.22 M as a pretreatment. In the first case sorbitol carbohydrate and polyol concentrations increased. When sorbitol was applied as a pretreatment, most sugar concentrations decreased. Since high intracellular osmolyte concentrations are needed to allow successful recovery after cryopreservation, the results from carbohydrate and polyol analyses may explain the higher recovery rate after cryopreservation when sorbitol pretreatment is used (Criel et al., 2006). The response to osmotic stress and chilling temperature was investigated in two potato species, *S. tuberosum* and *S. commersonii*, the latter being resistant to frost, by 2-D difference gel electrophoresis (DIGE) experiments followed by high-performance anion-exchange chromatography with pulsed amperometric detection analysis. Both species responded with an inhibition of carbon metabolism and oxidative homeostasis to all the used stress treatments. The differences between *S. commersonii* and the cv. Desiree were reflected in the protein metabolism. The sucrose and sorbitol treatments induced more changes in *S. commersonii*, whereas cold affected Desiree more. Patatin (a lipid metabolism-related protein) increased in *S. commersonii* only when treated with osmotic stress, suggesting a more important role for this protein in *S. commersonii* than in Desiree. Sorbitol and sucrose induce similar changes for both species; however, more stress-related proteins were identified in Desiree under sorbitol treatment than under sucrose, suggesting that both sugars induce different pathways in this cultivar (Folgado et al., 2013).

Plants have evolved various strategies to adapt to cold temperatures. One of them is the accumulation of amino acids, glucosides, or sugars (Fischer et al., 2013). In mature potato tubers, the accumulation of soluble sugars during cold adaptation is referred to as cold-induced sweetening (CIS) (Guy et al., 1992), an important issue for the potato processing industry. Long-term storage of potato tubers at low temperatures is a strategy currently used to reduce sprouting and extend potato marketability. However, high concentrations of the reducing sugars glucose and fructose, either inherently or caused by CIS, negatively affect the quality of potato chips and french fries (Dale and Bradshaw, 2003). The accumulated reducing sugars react with free amino acids in the potato cell forming unacceptably brown to black chips or fries and, more problematically, acrylamide, a neurotoxin and a potential carcinogen (Sowokinos, 2001; Mottram et al., 2002). 2-DE was used to compare tuber proteomes of 40 tetraploid potato cultivars highly diverse for CIS. Significant differences in protein expression were detected between CIS-tolerant and CIS-sensitive cultivars before the onset as well as during cold storage. Identified proteins were classified into protease inhibitors, patatins, heat-shock proteins, lipoxygenase, phospholipase A1, and leucine aminopeptidase (Lap). Association mapping based on single-nucleotide polymorphisms (SNPs) supported a role for Lap in the natural variation of the quantitative traits tuber starch and sugar content. Association of DNA polymorphisms in loci encoding differentially expressed proteins further validated a role for Lap in chip quality. Novel SNPs diagnostic for increased tuber starch content, starch yield, and chip quality were identified, traits useful for selecting improved potato processing cultivars (Fischer et al., 2013). Another frequently encountered problem that touches the potato industry is after-cooking darkening (ACD). It appears in cooked potatoes exposed to air and it presents itself in the form of colored areas ranging from gray though blue and purple to black. Even if it does not influence the nutritional quality of the product, it has a great visual impact, making it undesirable to the consumer. The mechanism behind it relies on the formation during cooking of a complex between chlorogenic acid and ferrous ions that is subsequently oxidized in air in a nonenzymatic reaction to the colored ferric form (Muneta and Kaisaki, 1985). Potatoes with higher concentrations of chlorogenic acid are more prone to this phenomenon and it appears that citric acid reduces the degree of darkening: the severity of the darkening is dependent on the ratio of chlorogenic acid to citric acid concentrations in the potato tubers (Wang-Pruski and Nowak, 2004). Proteomic profiles of potato tubers were analyzed in an attempt to identify proteins involved in ACD. A set of 30 proteins was found to be correlated with ACD, including protease inhibitors. Three aspartase proteases in this group implicated enhanced wound-response mechanisms involving protease inhibitors in potatoes with high ACD. There were also increased abundances of patatins and lipoxygenases. The conclusions were that wound response takes place in conjunction with an increase in polyphenol (chlorogenic acid) synthesis, leading to darkening (Murphy et al., 2010).

Agricultural practices have a great influence on tuber quality and crop yield. A proteomic approach to the effects of agricultural production systems on the protein profiles of potato tubers (Lehesranta et al., 2007) showed that the fertilization practice (organic matter vs mineral-based fertilizer) had a significant effect on protein composition, organic fertilization leading to an increased stress response in potato tubers. However, replacing mineral with organic fertilizers (animal and green manures) and restricting the use of chemosynthetic crop protection may significantly reduce crop yields. The effects of replacing mineral fertilizers with composted cattle manure and omitting pesticides on potato tuber yield, leaf and tuber mineral nutrient content, and leaf protein profiles were investigated by quantitative proteomics. Switching to organic fertilizer had a greater effect on yield and protein profiles than the omission of chemosynthetic crop protection. Leaf N and P compositions were significant drivers to changes in protein expression, particularly proteins involved in photosynthesis, which were more abundant in potato leaves grown under mineral fertilizer regimes. Stress-response proteins were also upregulated under mineral fertilization, whereas two proteins known to be involved in biotic stress (1,3-β-D-glucan glucanohydrolase and putative Kunitz-type tuber invertase inhibitor) were more abundant under compost fertilization. Switching from mineral to organic fertilizers reduced the N availability, had a significant change in leaf protein expression, and resulted in a lower tuber yield. In contrast, omission of chemosynthetic crop protection inputs had limited effects on protein expression and no significant effect on tuber yield (Rempelos et al., 2013).

4.4 Postharvest Proteomics

The postharvest system should be thought of as encompassing the delivery of a crop from the time and place of harvest to the time and place of consumption, with minimum loss, maximum efficiency, and maximum return for all involved (Spurgeon, 1976). In potato, losses due to physical wounding, such as cuts and bruises, can be as high as 40%, with damage to tuber quality such as increased moisture loss and the apparition of easy entry points for postharvest rots (Jobling, 2000). Another problem is the physiological aging of the potato tuber during storage that can further affect its quality by means of loss of mass due to respiration and growth of sprout tissue, decreased turgor, and increased sugar concentrations (Burton et al., 1992; Davies, 1990; van Es and Hartmans, 1987). Improper storage at low temperatures, even if it reduces the sprouting and prolongs the dormancy, leads to low-temperature sweetening (Burton et al., 1992).

Potato tubers were harvested and sampled after 0, 30, 90, 150, 210, or 270 days of storage at 4 °C (95% relative humidity, darkness). A proteomic approach was performed based on fluorescence DIGE coupled with tandem MS. Fifty-two and 41 spots were up- or downregulated, respectively, and four spots exhibited a transient abundance. The differentially expressed proteins were mainly involved in starch catabolism, control of protein conformation, protein recycling, and stress response. Moreover, 14 breakdown products of patatin increased during aging, indicating enhanced patatin proteolysis. The postharvest development of potato tubers actually seems to induce an oxidative challenge that is efficiently taken up by the proteomic and metabolic responses of the tubers, as no significant accumulation of oxidative damage on polyunsaturated fatty acids or proteins was noticed, even when a decrease in sprouting vigor was observed. Some of the results of this study were, however, correlated with the sprouting phenotypes and could also be assessed as potential aging biomarkers (Delaplace et al., 2009).

Another trait of high agronomic interest is the tuber susceptibility to bruising upon mechanical impact involving an enzymatic browning reaction. A proteomic comparison of 20 potato varieties revealed that seven genes or gene families both were differentially expressed at the protein level between groups and had DNA polymorphisms associated with the investigated traits. A novel factor contributing to the natural variation in bruising, a putative class III lipase, was identified. Additionally, tuber proteome changes triggered by mechanical impact, within and between groups, were monitored over time, showing time-dependent protein variation. Differentially expressed proteins were lipases, patatins, and annexins (Urbany et al., 2012). Wounding can occur during handling and transportation of potato tubers. The ability to rapidly produce a wound periderm, upon damage, is vital for the maintenance of tuber quality (Stark et al., 1994). A proteomic experiment on dormant potato tubers, cv. Asterix, 3 months from harvesting revealed that slicing triggered differentiation processes that led to changes in metabolism, activation of defense, and cell-wall reinforcement. Proteins related to storage, cell growth and division, cell structure, signal transduction, energy production, disease/defense mechanisms, and secondary metabolism were detected. There was a succession of proteomic events leading to wound-periderm reconstruction, suggesting that the processes of wound-periderm formation are extended in time. The slice metabolism in the later stages tended to acquire similarities to the native periderm. Numerous patatins and protease inhibitors were found to form a group of proteins in the slices, possibly involving the vacuole, the main reserve of proteins in the cell, in the healing process. Various classes of protease inhibitors were detected with distinct expression patterns along the sampling dates, suggesting their possible complex and marked roles during the wound-healing process (Chaves et al., 2009).

4.5 Proteomics of Potato Subcellular Compartments

The plant cell is a complex system composed of several parts—called organelles—that play specific roles in the well-being of the whole. In the potato cell they all have distinct functions (Figure 5). Organelles all work together in the cell, linked by the cytoskeleton and directed by the nucleus. They are interdependent and can send and receive information. Their cooperation is what determines the properties of a tissue, an organ, and eventually the entire organism. Determining their functions and the way they communicate is essential to obtaining better products that respond to specific demands of quality, nutrition, and shelf life. The complete sequencing of the potato genome (Potato Genome Sequencing Consortium, 2011) opened a door to endless possibilities, bringing, among other advantages, the accurate annotation of protein-coding genes. According to the Potato Genome Sequencing Consortium, the potato has 39,031 predicted protein-coding genes (http://potatogenome.net). Even if these data do not inform us about posttranscriptional modifications, we can have a general idea about their distribution inside the potato cell, by using predictive software that aims at identifying

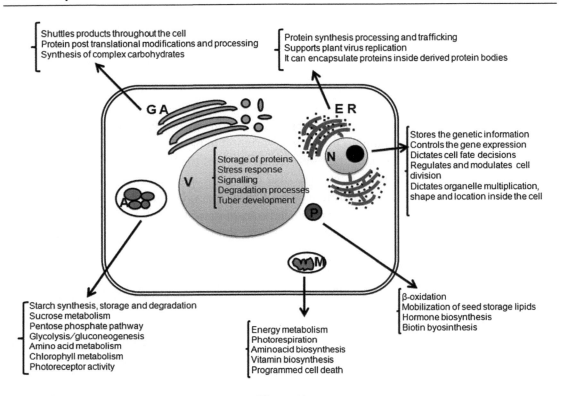

Shuttles products throughout the cell
Protein post translational modifications and processing
Synthesis of complex carbohydrates

Protein synthesis processing and trafficking
Supports plant virus replication
It can encapsulate proteins inside derived protein bodies

GA

ER

Storage of proteins
Stress response
Signalling
Degradation processes
Tuber development

Stores the genetic information
Controls the gene expression
Dictates cell fate decisions
Regulates and modulates cell division
Dictates organelle multiplication, shape and location inside the cell

V

N

P

A

M

Starch synthesis, storage and degradation
Sucrose metabolism
Pentose phosphate pathway
Glycolysis/gluconeogenesis
Amino acid metabolism
Chlorophyll metabolism
Photoreceptor activity

Energy metabolism
Photorespiration
Aminoacid biosynthesis
Vitamin biosynthesis
Programmed cell death

β-oxidation
Mobilization of seed storage lipids
Hormone biosynthesis
Biotin byosinthesis

Figure 5
The main roles of the organelles inside the potato tuber cell.

putative N-terminal targeting sequences. Proper subcellular localization is a prerequisite for the accurate determination of protein function (Gakh et al., 2002). Many bioinformatic tools, such as TargetP (Emanuelsson et al., 2000), MitoProt (Claros and Vincens, 1996), iPSORT (Bannai et al., 2002), Predotar (Small et al., 2004), and ngLOC (King and Guda, 2007) have been developed for predicting protein subcellular locations within cells. The knowledge we have today allows us to link specific metabolites of interest to the organelles that produce them. For this chapter the list of putative potato proteins available on the http://potatogenome. net Web site was retrieved and analyzed with several prediction softwares for protein localization, ngLOC (King and Guda, 2007), Predotar (Small et al., 2004), and PredPlantPTS1 (Reumann et al., 2012), to get a rough estimate of the size of the proteomes of specific organelles. However, these data need to be confirmed experimentally.

4.5.1 Biomass and Vitamin Production

Plastids and mitochondria are two organelles of vital importance for the cell. To oversimplify, one is the engine of the cell and the other the nutrient producer. Both organelles possess their own genome but approximately 95% of their constitutive proteins are imported and only approximately 5% synthesized within their membrane boundaries (Soll, 2002; Millar et al., 2006). They work closely together, as in the case of plant respiration (Kromer, 1995; Raghavendra and Padmasree, 2003). Plant biomass production is ultimately determined by the ratio between photosynthetic CO_2 assimilation and respiratory CO_2 (Millenaar and Lambers, 2003). There is experimental evidence suggesting the existence of an effective signaling network between chloroplasts and mitochondria that involves ROS and antioxidants (Foyer and Noctor, 2003; Noctor et al., 2007). The mitochondrion is the powerhouse of the cell, but its functions go way beyond that and vary from energy metabolism, photorespiration, amino acid biosynthesis, and coenzyme (vitamin) biosynthesis to programmed cell death (Salvato et al., 2014). If we use Predotar, particularly good at distinguishing mitochondrial and plastid targeting sequences (Small et al., 2004), of the 39,031 presumed potato proteins 2906 are estimated to be mitochondrial, the same as in the case of *Arabidopsis* (2000 to 3000 gene products) (Cui et al., 2011). Salvato et al. (2014) identified 1060 nonredundant proteins in the proteome of highly purified mitochondria isolated from dormant potato tubers. The team found evidence of extensive posttranslational modifications (PTMs) throughout the dormant potato tuber proteome, with proteins involved in or associated with the tricarboxylic acid cycle showing the highest levels. The widespread PTMs found throughout the dormant potato tuber proteome may suggest that this could represent an important method for functional regulation. Ascorbic acid is a major antioxidant and a highly valued vitamin. The last step in the biosynthesis of vitamin C involves the conversion of L-galactono-γ-lactone to ascorbic acid, a reaction catalyzed by the membrane-bound enzyme L-galactono-γ-lactone dehydrogenase(GLDH) (Ostergaard et al., 1997). GLDH activity was localized in the mitochondria of potato tubers (Oba et al., 1994; Salvato et al., 2014). Other enzymes involved in the biosynthesis of coenzymes of major interest (pyrimidine, biotin (vitamin B8), folates) were identified in the tuber mitochondria (Salvato et al., 2014). Evidence suggests that higher mitochondrial activity is linked to higher respiratory degradation of starch and less extended shelf life (Shukla et al., 2005).

4.5.2 Starch

Starch mainly comes from plastids in plants. They are typical organelles unique to lower and higher plants that are thought to originate from the endosymbiotic integration of a photosynthetic prokaryote, cyanobacterium, into a eukaryotic ancestor of algae (Pyke, 2007). The central role of plastids derives primarily from the chloroplast's ability to supply the cell with fixed carbon and energy as a result of photosynthesis. They are also a site of several other major plant metabolic pathways, such as lipid and vitamin biosynthesis and amino acid metabolism (Galili, 1995; Ohlrogge and Browse, 1995) and can be found in both photosynthetic and nonphotosynthetic tissues. There are a variety of types of plastids, which fulfill different roles in different types of plant cells: proplastids, chloroplasts, chromoplasts, and leukoplasts (amyloplasts, elaioplasts, proteinoplasts, etioplasts) (Pyke, 2009). Predotar predicted the existence of 2683 plastidial proteins of the 39,031 putative *S. tuberosum* proteins, but there is no tool available, other than experimentation, that would allow us to discriminate the exact proteins belonging to a specific type of plastid.

In potato, only one plastid type was studied using a proteomics approach, by Stensballe et al. (2008): the amyloplast—responsible for starch synthesis, storage, and degradation. Ninety soluble proteins were identified. As expected, the majority of identified proteins were involved in carbohydrate metabolism—starch and sucrose metabolism, the pentose phosphate pathway, and glycolysis/gluconeogenesis—and in amino acid metabolism. Several starch metabolic enzymes were partitioned to starch granules (α-glucan water dikinase, starch synthase II, granule-bound starch synthase I). Subunit variants of chaperonin 60 and Hsp70 were also identified. Monodehydroascorbate reductase variants appear to play an important role in amyloplasts as well as chloroplasts. Other enzymes thought to participate in chlorophyll metabolism and photoreceptor activity were also prominent in young amyloplasts. There are no proteomic studies available on the *S. tuberosum* chloroplasts; however, there is proof that the plastid DNA of potato tuber amyloplasts and of potato leaf chloroplasts is essentially the same (Steele Scott et al., 1984). Starch degradation is an important issue since several models have been postulated. One model involves membrane breakdown, thus giving access to all enzymes of the cell to starch granules, with a dominating role for α-amylase (Ziegler, 1995; Asatsuma et al., 2005). In the case of regreening the plastidial membranes of turions of *Spirodela polyrhiza* remain intact, giving access to the starch only to the plastidial enzymes (Appenroth et al., 2011); we could assume that this is also true for the potato tuber. Another aspect of starch is the elevated glycemic index (GI) of cooked potatoes (Atkinson et al., 2008), a feature that is undesirable. Health professionals advise that potatoes should be substituted by a low-GI carbohydrate to reduce the risk of chronic disease (Brand-Miller et al., 2009). Since potatoes are an important part of the diet worldwide, the discovery and development of low-GI potatoes would bring benefits to consumers, agriculture, and food industry. There already exist cultivars with low and medium GIs, like the Carisma cultivar (GI ¼ 53—low GI) and the Nicola cultivar (GI ¼ 69—medium GI), and it was shown that under identical cooking conditions, potato cultivars with similar amylose contents differ significantly in both in vitro digestion and in vivo blood glucose responses (Ek et al., 2013). The fine structure of amylose and amylopectin has also been shown to be important in the determination of the digestibility of starch (Syahariza et al., 2013).

4.5.3 Protein Content

The vacuole is another organelle of interest because it contains the majority of potato tuber proteins, especially patatin and protease inhibitor variants (Pouvreau et al., 2001; Bauw et al., 2006), presenting considerable interest to knowledge-based breeding for increased protein content or production of protein products for industrial and medical use. According to the ngLOC software (King and Guda, 2007) there are 296 predicted vacuolar proteins. Four hundred fifty putative vacuolar proteins were identified by mass spectrometric sequencing by Jørgensen et al. (2011). The difference between the predicted number of proteins and the number given by the experiment could be explained by the fact that some proteins might be imported by an autophagic pathway or be nonvacuolar impurities. The soluble and insoluble fractions were analyzed. The tuber vacuole is a typical protein storage vacuole, so predictably, no proteolytic or glycolytic enzymes were identified. In the soluble fraction were identified storage and stress response proteins: Kunitz protease inhibitors, protease inhibitors 1 and 2, carboxypeptidase inhibitors, patatins, and lipoxygenases. In the insoluble vacuole fraction, two variants of phytepsin (aspartate protease) were identified.

5. Getting the Best of Potato—Perspectives of Proteomics Applied to Product Quality

With all the experimental work done so far we still have a long way to go before mapping the potato proteome (Figure 6). Organelles such as peroxisomes, Golgi apparatus, endoplasmic reticulum, chloroplasts, proplastids, and nucleus lay still unexplored. Even if mitochondria and plastids have their own genomes, the rest of the organelles are completely dependent on the nucleus; thus understanding its genetic information and structure is an essential step in applying novel methods of quality improvement.

5.1 Crop and Genotype-Specific Adaptation

The nucleus houses the nuclear genome and provides a functional environment for the information that will maintain and reproduce that cell and potentially the whole organism (Martins et al., 2012). Nuclear proteins (NPs) are predicted to comprise about 10–20% of the total cellular proteins, suggesting the involvement of the nucleus in a number of diverse functions (Narula et al., 2013). The nucleus keeps the genetic material separate from other activities of the cell (Roix and Misteli, 2002), directly controls the gene expression (Wilson and Dawson, 2011), and plays an important morphoregulatory role in chloroplast division (Cavalier-Smith, 2006). The organization of NPs controls the shape, size, and composition of the nucleus: plasticity in the nucleus allows cell differentiation, while rigidity in the nucleus determines its mechanical stiffness (Jiang et al., 2006). The nucleus serves a multifunctional role, as a regulator and modulator during cell division and a controller and integrator for fertilization and inheritance—modulating the cellular phenotype (Franklin et al., 2011). Understanding protein diversity and shared features can open a door to the most fundamental aspects of nuclear structure and protein organization. The nuclear proteome could hold the key to the understanding of crop and genotype-specific adaptation and the biochemical and physiological processes that come with phenotypic variation. ngLOC (King and Guda, 2007) predicts the existence of 11,927 putative potato NPs—approximately 30% of the total cellular proteins.

	Number of Predicted proteins	Number of Proteins described by proteomic analysis
Nucleus	11927 - ngLoc	0
Endoplasmic reticulum	4806 - predotar 450 - ngLoc	0
Golgi apparatus	169 - ngLoc	0
Mitochondria	2906 - predotar	1,060 (Salvato *et al.*, 2014)
Amyloplast	2683 - predotar	90 (Stensballe *et al.*, 2008)
Peroxizome	365 - PredPlantPTS1 66 – ng Loc	0
Vacuole	296 – ngLoc	450 (Jørgensen *et al.*, 2011)

Figure 6

An evaluation of the state of the art of potato subcellular proteomics. The Potato Genome Sequencing Consortium potato has 39,031 predicted protein-coding genes and the presumed protein sequences are available on their Web site (http://potatogenome.net). The list was analyzed using several softwares for the prediction of the subcellular localization, ngLOC (King and Guda, 2007), Predotar (Small et al., 2004), and PredPlantPTS1 (Reumann et al., 2012), to have an idea about the protein distribution inside the cell. Proteomic analyses of the potato mitochondria (Salvato et al., 2014), amyloplast (Stensballe et al., 2008), and vacuole (Jørgensen et al., 2011) are available.

5.2 *Protein Synthesis and Transport*

The secretory pathway of the cell is composed of the Golgi apparatus and the endoplasmic reticulum (ER). The fundamental question of how it transports biosynthetic material remains still unanswered. There are no proteomic studies in potato addressing this essential question even if isolation protocols currently exist for other species (Walter and Blobel, 1983; Croze and Morre, 1984; Joseph et al., 2012). The Golgi apparatus shuttles products between a variety of destinations throughout the cell. It is a major site for the posttranslational modification and processing of proteins, has a significant role in the synthesis of complex carbohydrates, and exists as a contiguous component of the endomembrane system, which encompasses the ER, plasma membrane, vacuoles, endosomes, and lysosomes (Morre and Mollenhauer, 2009). One hundred sixty-nine putative potato proteins are predicted in the Golgi apparatus by ngLOC (King and Guda, 2007). The ER is a multifunctional intracellular organelle responsible for the synthesis, processing, and trafficking of a wide variety of proteins essential for development and life support (Chen et al., 2010). Comprehensive characterization of the ER proteome is of great importance to the understanding of its functions. By using prediction softwares we can estimate its size at 4806 proteins according to Predotar (Small et al., 2004). The ER is also of particular interest because plant viruses use the cellular endomembrane system to support virus replication, and in the case of RNA viruses in particular, the viral replicase is typically anchored to membranes associated with the ER (Rubino and Russo, 1998), thus becoming an important step in viral disease. Knowledge of the ER proteome could serve either to provide markers for selection in the purpose of obtaining plants resistant to viruses (potyvirus, comovirus, diantho-virus, pecluvirus, and bromovirus, the mosaic virus, the red, or the Brome mosaic virus or the X TGBp2 virus (Ju et al., 2005)) or to find new strategies of treatment and prevention. Another practical application of ER intimate knowledge is the use of ER-derived protein bodies (PBs). It was shown by Joseph et al. (2012) that the N-terminal proline-rich domain (Zera) of the maize storage protein γ-zein is able to induce the formation of ER-derived PBs when fused to proteins of interest in tobacco. This encapsulation enables a recombinant fused protein to escape from degradation and facilitates its recovery from plant biomass by gradient purification. This could lead further to optimized downstream purification of recombinant proteins in molecular farming applications. The understanding of the encapsulation process could also be useful to increase the recovery and production of already existing proteins of interest.

5.3 Peroxisomes—A Site of Hormone and Biotin Biosynthesis

In plants, peroxisomes are the organelles involved in various metabolic processes and physiological functions including β-oxidation (Gerhardt, 1983), mobilization of seed storage lipids (Mano and Nishimura, 2005), photorespiration (Nishimura et al., 1996), hormone biosynthesis—jasmonic acid (Weber, 2002) and auxin (Woodward and Bartel, 2005), and biotin biosynthesis (Maruyama et al., 2012). They change their functions in response to developmental stages and environmental conditions (Fransen, 2012). In potato, 365 proteins were predicted as peroxisomal according to PredPlantPTS1 (Reumann et al., 2012), a software specifically designed for the prediction of peroxisomal protein location.

5.4 How Are Tubers Made?

Of most importance is the tuberization process, not yet completely understood. Studies suggest the influence of Ca^{2+} in potato tuberization (Balamani et al., 1986; Chang et al., 2007) through Ca^{2+} modulator proteins, CaM, CDPK, CBLs, and channel proteins (Nookaraju et al., 2012). Two organelles stand out regarding calcium signaling and accumulation: mitochondria (Rizzuto et al., 2012) and the vacuole (Machlon, 1984). The vacuole is a reservoir for calcium (Machlon, 1984) and consequently plays an important role in calcium homeostasis (Miller et al., 1990; Gelli and Blumwald, 1993). Mitochondria, like a bandmaster, can propagate calcium-driven signals in two ways: acting as a calcium sink to prevent feedback inhibition or acting as a calcium reservoir, releasing more calcium to the cytosol to amplify signals (Whelan and Zuckerbraun, 2013). Mitochondria are able to encode and decode Ca^{2+} signals because the respiratory chain generates an electrochemical gradient for protons across the inner mitochondrial membrane. By taking up and subsequently releasing Ca^{2+} ions, it can alter both the spatial extent and the duration of cytosolic calcium signals (Poburko and Demaurex, 2012). Cytosolic Ca^{2+} elevations have a direct effect on mitochondrial pH, decreasing it. Changes in cytosolic pH are paralleled by changes in mitochondrial pH (Poburko et al., 2011).

The main external cue affecting the tuberization process is the photoperiod. The chloroplastic protein phytochrome B can perceive the photoperiod in the leaf, leading to the initiation of tuberization in the subapical tip of the stolon (Sarkar, 2008). Plastids are not the only light-sensing organelles in the cell. Mitochondria also have the ability to absorb light by the cytochrome c oxidase (COX) (Douce, 1985), and COX has been increasingly shown to be the photoacceptor and photosignal transducer in the red to near-infrared region of light (Pastore et al., 2000; Karu, 2010). COX can also reduce nitrite to nitric oxide (Poyton et al., 2009). Low-intensity light enhances nitric oxide synthesis by COX without altering its ability to reduce oxygen (Poyton and Ball, 2011), but no studies have explored this astonishing quality of potato mitochondria and its possible implications for tuber quality. Tubers exposed to light will turn green owing to chlorophyll formation (Salunkhe and Salunkhe, 1974), rendering the potato unacceptable on the marketplace. This phenomenon is influenced by the potato variety (Zhu et al., 1984). Along with chlorophyll formation there is an increase in glycoalkaloid levels, which pose a taste and toxicity concern (Percival, 1999). Potatoes are more sensitive to chlorophyll accumulation when exposed to wavelengths of light in the blue (475 nm) and red (675 nm) regions of the light spectrum (Petermann and Morris, 1985), so there could also be an effect on the mitochondrial COX. The regreening could hold a key to understanding one of the most important players in starch metabolism: the plastids. Reversible differentiation of the amyloplast to chloroplast in response to exposure to light leads to fast starch degradation, carotenoid accumulation (Zhu et al., 1984), and internal rearrangements of the organelle (Badenhuizen and Salema, 1976). Preberg et al. (2008) quote a number of situations in which regreening of tissues occurs as a consequence of redifferentiation of gerontoplasts, etioplasts, or chromoplasts into chloroplasts. The phenomenon is truly a redifferentiation process without any evidence of de novo generation of plastids or plastid division (Thomson et al., 1967; Preberg et al., 2008). Light is probably the most important factor of regreening via phytochromes; however, nutritional factors are also involved (Huff, 1983).

6. Conclusions

Potato is a complex food valued for its ease of cultivation, minimum transportation and storage requirements, and high carbohydrate, vitamin, and antioxidant content. Proteomic research in potato has been mainly oriented toward increasing crop yield, reducing fertilizers, understanding defense mechanisms against pests, and stress adaptation. The apparition of the transgenic potato varieties led to the need for tests that could assess the impact of these newcomers on the environment and health. This demand was met by proteomics. However, even if a great deal of work has been done on potato, we are still far from a comprehensive proteomics database, essential to understanding the various phenotypic aspects that could be used in agricultural practice. A proteomic potato map followed by metabolomics analysis is necessary in the quest of plants with increased nutritional quality, resistance to pests and injury, long storability, and higher synthesis of one or more molecules of interest.

Acknowledgments

I would like to thank Dr Jaspreet Singh for his help and guidance in writing this chapter; Dr Thomas Lingner for his help with the PredPlantPTS1 software, and Dr Jean-Claude Pech for his feedback.

References

Agrawal, L., Charkraborty, S., Jaiswal, D.K., Gupta, S., Datta, A., Chakraborty, N., 2008. Comparative proteomics of tuber induction, development and maturation reveal the complexity of tuberization process in potato (*Solanum tuberosum* L.). Journal of Proteome Research 7, 3803–3817.

Akagawa, M., Handoyo, T., Ishii, T., Kumazawa, S., Morita, N., Suyama, K., 2007. Proteomic analysis of wheat flour allergens. Journal of Agricultural and Food Chemistry 55, 6863–6870.

Ali, A., Alexandersson, E., Sandin, M., Resjö, S., Lenman, M., Hedley, P., Levander, F., Andreasson, E., 2014. Quantitative proteomics and transcriptomics of potato in response to *Phytophthora infestans* in compatible and incompatible interactions. BMC Genomics 15, 497.

Appenroth, K.J., Keresztes, A., Krzysztofowicz, A., Gabrys, H., 2011. Light-induced degradation of starch granules in turions of *Spirodela polyrhiza* studied by electron microscopy. Plant Cell Physiology 52, 384–391.

Asatsuma, S., Sawada, C., Itoh, K., Okito, M., Kitajima, A., Mitsui, T., 2005. Involvement of alpha-amylase I-1 in starch degradation in rice chloroplasts. Plant Cell Physiology 46, 858–869.

Atkinson, F.S., Foster-Powell, K., Brand-Miller, J.C., 2008. International tables of glycaemic index and glycaemic load values. Diabetes Care 31, 2281–2283.

Badenhuizen, N.P., Salema, R., 1976. Observations of the development of chloroamyloplasts. Revista de Biologia (Lisbon) 6, 139–155.

Badri, M.A., Rivard, D., Coenen, K., Michaud, D., 2009. Unintended molecular interactions in transgenic plants expressing clinically-useful proteins: the case of bovine aprotinin travelling the potato leaf cell secretory pathway. Proteomics 9, 746–756.

Balamani, V., Veluthambi, K., Poovaiah, B.W., 1986. Effect of calcium on tuberization in potato (*Solanum tuberosum* L.). Plant Physiology 80, 856–858.

Bannai, H., Tamada, Y., Maruyama, O., Nakai, K., Miyano, S., 2002. Extensive feature detection of N-terminal protein sorting signals. Bioinformatics 18, 298–305.

Barel, G., Ginzberg, I., 2008. Potato skin proteome is enriched with plant defence components. Journal of Experimental Botany 59 (12), 3347–3357.

Barsan, C., Sanchez-Bel, P., Rombaldi, C., Egea, I., Rossignol, M., Kuntz, M., Zouine, M., Latche, A., Bouzayen, M., Pech, J.C., 2010. Characteristics of the tomato chromoplast revealed by proteomic analysis. Journal of Experimental Botany 61, 2413–2431.

Bauw, G., Nielsen, H.V., Emmersen, J., Nielsen, K.L., Jørgensen, M., Welinder, K.G., 2006. Patatins, Kunitz protease inhibitors and other major proteins in tuber of potato cv. Kuras. FEBS Journal 273, 3569–3584.

Bengtsson, T., Weighill, D., Proux-Wéra, E., Levander, F., Resjö, S., Burra, D.D., Moushib, L.I., Hedley, P.E., Liljeroth, E., Jacobson, D., Alexandersson, E., Andreasson, E., 2014. Proteomics and transcriptomics of the BABA-induced resistance response in potato using a novel functional annotation approach. BMC Genomics 15, 315.

Berkowitz, O., Jost, R., Kollehn, D.O., Fenske, R., Finnegan, P.M., O'Brien, P.A., Hardy, G.E., Lambers, H., 2013. Acclimation responses of *Arabidopsis thaliana* to sustained phosphite treatments. Journal of Experimental Botany 64, 1731–1743.

Bisognin, D.A., Douches, D.S., 2002. Early generation selection for potato tuber quality in progenies of late blight resistant parents. Euphytica 127, 1–9.

Brand-Miller, J., McMillan-Price, J., Steinbeck, K., Caterson, I., 2009. Dietary glycaemic index: health implications. The Journal of the American College of Nutrition 28 (Suppl.), 446S–449S.

Bricchi, I., Leitner, M., Foti, M., Mithöfer, A., Boland, W., Maffei, M.E., 2010. Robotic mechanical wounding (MecWorm) versus herbivore-induced responses: early signaling and volatile emission in Lima bean (*Phaseolus lunatus* L.). Planta 232, 719–729.

Burra, D.D., Berkowitz, O., Hedley, P.E., Morris, J., Resjö, S., Levander, F., Liljeroth, E., Andreasson, E., Alexandersson, E., 2014. Phosphite-induced changes of the transcriptome and secretome in *Solanum tuberosum* leading to resistance against *Phytophthora infestans*. BMC Plant Biology 14, 254.

Burton, W.G., van Es, A., Hartmans, K.J., 1992. The physics and physiology of storage. In: Harris, P.M. (Ed.), The Potato Crop: The Scientific Basis for Improvement, second ed. Chapman and Hall, London, UK, pp. 608–727.

Carbonaro, M., 2004. Proteomics: present and future in food quality evaluation. Trends in Food Science and Technology 15, 209–216.

Carputo, D., Aversano, R., Frusciante, L., 2005. Breeding potato for quality traits. Acta Horticulturae 684, 55–64.

Cavalier-Smith, T., 2006. Cell evolution and earth history: stasis and revolution. Philosophical Transactions of the Royal Society B: Biological Sciences 361, 969–1006.

Chakaborty, S., Chakaborty, N., Datta, A., 2000. Increased nutritive value of transgenic potato by expressing a nonallergenic seed albumin gene from *Amaranthus hypochondriacus*. Proceedings of the National Academy of Sciences of the United States of America 97, 3724–3729.

Chandramouli, K., Qian, P.Y., 2009. Proteomics: challenges, techniques and possibilities to overcome biological sample complexity. Human Genomics and Proteomics 2009, 1–92.

Chang, D.C., Park, C.S., Lee, J.G., Jeong, J.C., Kim, S.Y., 2007. Effect of pre-plant calcium application on calcium concentration, yield, and incidence of internal brown spot in potatoes (cv. Atlantic) grown under greenhouse conditions. Acta Horticulturae (ISHS) 761, 449–455.

Chaves, I., Pinheiro, C., Paiva, J.A.P., Planchon, S., Sergeant, K., Renaut, J., Graça, J.A., Costa, G., Coelho, A.V., Pinto Ricardo, C.P., 2009. Proteomic evaluation of wound-healing processes in potato (*Solanum tuberosum* L.) tuber tissue. Proteomics 9, 4154–4175.

Chen, X., Karnovsky, A., Sans, M.D., Andrews, P.C., Williams, J.A., 2010. Molecular characterization of the endoplasmic reticulum: insights from proteomic studies. Proteomics 10, 4040–4052.

Claros, M.G., Vincens, P., 1996. Computational method to predict mitochondrially imported proteins and their targeting sequences. European Journal of Biochemistry 241, 779–786.

Clerens, S., Plowman, J.E., Dyer, J.M., 2012. In: Heazlewood, J.L., Petzold, C.J. (Eds.), Food Proteomics: Mapping Modifications, Proteomic Applications in Biology. InTech.

Cordain, L., 1999. Cereal grains: humanity's double-edged sword. World Review of Nutrition and Dietetics 84, 19–73.

Criel, B., Hausman, J.F., Oufir, M., Swennen, R., Panis, B., Renaut, J., 2006. Proteome and sugar analysis of abiotic stress underlying cryopreservation in potato. Communications in Agricultural and Applied Biological Sciences Ghent University 71 (1), 3–6.

Cristea, I.M., Gaskell, S.J., Whetton, A.D., 2004. Proteomics techniques and their application to hematology. Blood 103, 3624–3634.

Croze, E.M., Morre, D.J., 1984. Isolation of plasma membrane, golgi apparatus, and endoplasmic reticulum fractions from single homogenates of mouse liver. Journal of Cellular Physiology 119, 46–57.

Cui, J.A., Liu, J.H., Li, Y.H., Shi, T.L., 2011. Integrative identification of *Arabidopsis* mitochondrial proteome and its function exploitation through protein interaction network. PLoS One 6 (1), e16022.

Dahan, K., Fennal, M., Kumar, N.B., 2008. Lycopene in the prevention of prostate cancer. Journal of the Society for Integrative Oncology 6 (1), 29–36.

Dale, M.F., Bradshaw, J.E., 2003. Progress in improving processing attributes in potato. Trends in Plant Science 8 (7), 310–312.

Damerval, C., Vienne, D.D., Zivy, M., Thiellement, H., 1986. Technical improvements in two-dimensional electrophoresis increase the level of genetic variation detected in wheat-seedling proteins. Electrophoresis 7 (1), 52–54.

Davies, H.V., 1990. Carbohydrate metabolism during sprouting. American Potato Journal 67, 815–827.

De Lucca, A.J., 2007. Harmful fungi in both agriculture and medicine. Revista Iberoamericana de Micología 24, 3–13.

Delaplace, P., Fauconnier, M.L., Sergeant, K., Dierick, J.F., Oufir, M., van der Wal, F., America, A.H.P., Renaut, J., Hausman, J.F., du Jardin, P., 2009. Potato (*Solanum tuberosum* L.) tuber ageing induces changes in the proteome and antioxidants associated with the sprouting pattern. Journal of Experimental Botany 60 (4), 1273–1288.

Diretto, G., Al-Babili, S., Tavazza, R., Papacchioli, V., Beyer, P., Giuliano, G., 2007. Metabolic engineering of potato carotenoid content through tuber-specific overexpression of a bacterial minipathway. PLoS One 2, e350.

Diretto, G., Tavazza, R., Welsch, R., Pizzichini, D., Mourgues, F., Papacchioli, V., Beyer, P., Giuliano, G., 2006. Metabolic engineering of potato tuber carotenoids through tuber-specific silencing of lycopene epsilon cyclase. BMC Plant Biology 6, 13.

Douce, R., 1985. Mitochondria in Higher Plants: Structure, Function, and Biogenesis. Academic Press, Orlando, FL.

Duceppe, M.O., Cloutier, C., Michaud, D., 2012. Wounding, insect chewing and phloem sap feeding differentially alter the leaf proteome of potato, *Solanum tuberosum* L. Proteome Science 10, 73.

Ducreux, L.J., Morris, W.L., Hedley, P.E., Shepherd, T., Davies, H.V., Millam, S., Taylor, M.A., 2005. Metabolic engineering of high carotenoid potato tubers containing enhanced levels of beta-carotene and lutein. Journal of Experimental Botany 56 (409), 81–89.

Ek, K.L., Wang, S., Copeland, L., Brand-Miller, J.C., 2013. Discovery of a low-glycaemic index potato and relationship with starch digestion in vitro. British Journal of Nutrition 111 (4), 699–705.

Eldredge, E.P., Shock, C.C., Stieber, T.D., 1992. Plot sprinklers for irrigation research. Agronomy Journal 84, 1081–1084.

Emanuelsson, O., Nielsen, H., Brunak, S., von Heijne, G., 2000. Predicting subcellular localization of proteins based on their N-terminal amino acid sequence. Journal of Molecular Biology 300, 1005–1016.

Fasoli, E., Aldini, G., Regazzoni, L., Kravchuk, A.V., Citterio, A., Righetti, P.G., 2010. Les maitres de l'Orge: the proteome content of your beer mug. Journal of Proteome Research 9, 5262–5269.

Fasoli, E., Pastorello, E.A., Farioli, L., Scibilia, J., Aldini, G., Carini, M., Marocco, A., Boschetti, E., Righetti, P.G., 2009. Searching for allergens in maize kernels via proteomic tools. Journal of Proteomics 72, 501–510.

Fischer, M., Schreiber, L., Colby, T., Kuckenberg, M., Tacke, E., Hofferbert, H.R., Schmidt, R., Gebhardt, C., 2013. Novel candidate genes influencing natural variation in potato tuber cold sweetening identified by comparative proteomics and association mapping. BMC Plant Biology 13, 113.

Folgado, R., Panis, B., Sergeant, K., Renaut, J., Swennen, R., Hausman, J.F., 2013. Differential protein expression in response to abiotic stress in two potato species: *Solanum commersonii* Dun and *Solanum tuberosum* L. International Journal of Molecular Sciences 14, 4912–4933.

Fonseca, C., Planchon, S., Renaut, J., Oliveira, M.M., Batista, R., 2012. Characterization of maize allergens – MON810 vs. its nontransgenic counterpart. Journal of Proteomics 75, 2027–2037.

Foyer, C.H., Noctor, G., 2003. Redox sensing and signalling associated with reactive oxygen in chloroplasts, peroxisomes and mitochondria. Physiologia Plantarum 119, 355–364.

Franklin, S., Zhang, M.J., Chen, H., Paulsson, A.K., Jordan, S.S.M., Li, Y., Ping, P., Vondriska, T.M., 2011. Specialized compartments of cardiac nuclei exhibit distinct proteomic anatomy. Proteomics 10, 1015–1074.

Fransen, M., 2012. Peroxisome dynamics: molecular players, mechanisms, and (Dys) functions. ISRN Cell Biology 2012 Article ID: 714192.

Fuchs, D., Winkelmann, I., Johnson, I.T., Mariman, E., Wenzel, U., Daniel, H., 2005. Proteomics in nutrition research: principles, technologies and applications. British Journal of Nutrition 94 (3), 302–314.

Gakh, O., Cavadini, P., Isaya, G., 2002. Mitochondrial processing peptidases. Biochimica et Biophysica Acta 1592, 63–77.

Galili, G., 1995. Regulation of lysine and threonine biosynthesis. Plant Cell 7, 899–906.

Garfin, D.E., 2003. Gel electrophoresis of proteins. Essential Cell Biology 1, 197–268.

Gašo-Sokac, D., Kovac, S., Josic, D., 2010. Application of proteomics in food technology and biotechnology: process development, quality control and product safety. Food Technology and Biotechnology 48, 284–295.

Geigenberger, P., Geiger, M., Stitt, M., 1998. High-temperature perturbation of starch synthesis is attributable to inhibition of ADP-glucose pyrophosphorylase by decreased levels of glycerate-3-phosphate in growing potato tubers. Plant Physiology 117, 1307–1316.

Gelli, A., Blumwald, E., 1993. Calcium retrieval from vacuolar pools. (Characterization of a vacuolar calcium channel). Plant Physiology 102, 1139–1146.

Gerhardt, B., 1983. Localization of β-oxidation enzymes in peroxisomes isolated from nonfatty plant tissues. Planta 159 (3), 238–246.

Giuliano Albo, A., Mila, S., Digilio, G., Motto, M., Aime, S., Corpillo, D., 2007. Proteomic analysis of a genetically modified maize flour carrying cry1Ab gene and comparison to the corresponding wild-type. Maydica 52, 443–455.

Giuliano, G., Aquilani, R., Dharmapuri, S., 2000. Metabolic engineering of plant carotenoids. Trends in Plant Science 5 (10), 406–409.

Gong, C.Y., Wang, T., 2013. Proteomic evaluation of genetically modified crops: current status and challenges. Frontiers in Plant Science 4, 41.

Goulet, C., Benchabane, M., Anguenot, R., Brunelle, F., Khalf, M., Michaud, D., 2010. A companion protease inhibitor for the protection of cytosol-targeted recombinant proteins in plants. Plant Biotechnology Journal 8, 142–154.

Guy, C.L., Huber, J.L.A., Huber, S.C., 1992. Sucrose phosphate synthase and sucrose accumulation at low-temperature. Plant Physiology 100 (1), 502–508.

Harper, S., 2004. Potato Tuber Quality Management in Relation to Environmental and Nutritional Stress. HAL Project No. PT99052. Horticultural Australia Ltd.

Haverkort, A., Struik, P., Visser, R., Jacobsen, E., 2009. Applied biotechnology to combat late blight in potato caused by Phytophthora infestans. Potato Research 52 (3), 249–264.

Helmy, M., Tomita, M., Ishihama, Y., 2011. OryzaPG-DB: rice proteome database based on shotgun proteogenomics. BMC Plant Biology 11, 63.

Hiller, L.K., Thornton, R.E., 1993. Management of physiological disorders. In: Rowe, R.C. (Ed.), Potato Health Management. The American Phytopathological Society Press.

Huff, A., 1983. Nutritional control of regreening and degreening in citrus peel segments. Plant Physiology 73, 243–249.

Hurkman, W.J., Tanaka, C.K., 1986. Solubilization of plant membrane proteins for analysis by two-dimensional gel electrophoresis. Plant Physiology 81 (3), 802–806.

Hustoft, H.K., Malerod, H., Wilson, S.R., Reubsaet, L., Lundanes, E., Greibrokk, T., 2012. A critical review of trypsin digestion for LC-MS based proteomics. In: Leung, H.C.E. (Ed.), Biochemistry, Genetics and Molecular Biology. Integrative Proteomics, pp. 73–92.

Iimure, T., Nankaku, N., Hirota, N., Zhou, T.S., Hoki, T., Kihara, M., Katsuhiro, H., Kazutoshi, I., Kazuhiro, S., 2010. Construction of a novel beer proteome map and its use in beer quality control. Food Chemistry 118, 566–574.

Jiang, G., Huang, A.H., Cai, Y., Tanase, M., Sheetz, M.P., 2006. Rigidity sensing at the leading edge through $\alpha_v\beta_3$ integrins and RPTPα. Biophysical Journal 90, 1804–1809.

Jobling, J., 2000. Potatoes: handle with care. Good Fruit and Vegetables Magazine 11 (4), 34–35.

Jobling, S.A., Westcott, R.J., Tayal, A., Jeffcoat, R., Schwall, G.P., 2002. Production of a freeze-thaw-stable potato starch by antisense inhibition of three starch synthase genes. Nature Biotechnology 20, 295–299.

Jørgensen, M., Stensballe, A., Welinder, K.G., 2011. Extensive post-translational processing of potato tuber storage proteins and vacuolar targeting. FEBS Journal 278, 4070–4087.

Joseph, M., Ludevid, M.D., Torrent, M., Rofidal, V., Tauzin, M., Rossignol, M., Peltier, J.B., 2012. Proteomic characterisation of endoplasmic reticulum-derived protein bodies in tobacco leaves. BMC Plant Biology 12, 36.

Ju, H.J.C., Samuels, T.D., Wang, Y.S., Blancaflor, E., Payton, M., Mitra, R., Krishnamurthy, K., Nelson, R.S., Verchot-Lubicz, J., 2005. The potato virus X TGBp2 movement protein associates with endoplasmic reticulum-derived vesicles during virus infection. Plant Physiology 138, 1877–1895.

Karu, T.I., 2010. Multiple roles of cytochrome c oxidase in mammalian cells under action of red and IR-A radiation. IUBMB Life 62, 607–610.

Khalf, M., Goulet, C., Vorster, J., Brunelle, F., Anguenot, R., Fliss, I., Michaud, D., 2010. Tubers from potato lines expressing a tomato Kunitz protease inhibitor are substantially equivalent to parental and transgenic controls. Plant Biotechnology Journal 8, 155–169.

Kilambi, H.V., Kumar, R., Sharma, R., Sreelakshmi, Y., 2013. Chromoplast-specific carotenoid-associated protein appears to be important for enhanced accumulation of carotenoids in *hp1* tomato fruits. Plant Physiology 161, 2010–2085.

King, B.R., Guda, C., 2007. ngLOC: an n-gram-based Bayesian method for estimating the subcellular proteomes of eukaryotes. Genome Biology 8 (5), R68.

Kolattukudy, P.E., 1977. Lipid polymers and associated phenols, their chemistry, biosynthesis and role in pathogenesis. Recent Advances in Phytochemistry 77, 185–246.

Komatsu, S., Tanaka, N., 2004. Rice proteome analysis: a step toward functional analysis of the rice genome. Proteomics 5, 938–949.

Konozy, E.H., Rogniaux, H., Causse, M., Faurobert, M., 2013. Proteomic analysis of tomato (*Solanum lycopersicum*) secretome. Journal of Plant Research 126 (2), 251–266.

Krinsky, N.I., Mayne, S.T., Sies, H., 2004. Carotenoids in Health and Disease. Marcel-Dekker, New York.

Kromer, S., 1995. Respiration during photosynthesis. Annual Review of Plant Physiology and Plant Molecular Biology 46, 45–70.

Lehesranta, S., Koistinen, K.M., Massat, N., Davies, H.V., Shepherd, L., Mc-Nicol, J.W., Cakmak, I., Cooper, J., Luck, L., Karenlampi, S., Leifert, C., 2007. Effects of agricultural production systems and their components on protein profiles of potato tubers. Proteomics 7, 597–604.

Lehesranta, S.J., Davies, H.V., Shepherd, L.V.T., Koistinen, K.M., Massat, N., Nunan, N., McNicol, J.W., Karenlampi, S.O., 2006. Proteomic analysis of the potato tuber life cycle. Proteomics 6, 6042–6052.

Lehesranta, S.J., Davies, H.V., Shepherd, L.V.T., Nunan, N., McNicol, J.W., Auriola, S., Koistinen, K.M., Suomalainen, S., Kokko, H.I., Kärenlampi, S.O., 2005. Comparison of tuber proteomes of potato varieties, land races, and genetically modified lines. Plant Physiology 138, 1690–1699.

Li, J., Zhu, L., Lu, G., Zhan, X.B., Lin, C.C., Zheng, Z.Y., 2014. Curdlan b-1,3-glucooligosaccharides induce the defense responses against *Phytophthora infestans* infection of potato (*Solanum tuberosum* L. cv. McCain G1) leaf cells. PLoS One 9 (5), e97197.

Li, J., Zhu, L., Zheng, Z.Y., Zhan, X.B., Lin, C.C., Zong, Y., Li, W.J., 2013. A new effective process for production of curdlan oligosaccharides based on alkali-neutralization treatment and acid hydrolysis of curdlan particles in water suspension. Applied Microbiology and Biotechnology 97, 8495–8503.

Liljeroth, E., Bengtsson, T., Wiik, L., Andreasson, E., 2010. Induced resistance in potato to *Phytphthora infestans*-effects of BABA in greenhouse and field tests with different potato varieties. European Journal of Plant Pathology 127, 171–183.

Lim, S., Borza, T., Peters, R.D., Coffin, R.H., Al-Mughrabi, K.I., Pinto, D.M., Wang-Pruski, G., 2013. Proteomics analysis suggests broad functional changes in potato leaves triggered by phosphites and a complex indirect mode of action against *Phytophthora infestans*. Journal of Proteomics 93, 207–223.

Lloyd, J.R., Springer, F., Buléon, A., Müller-Rober, B., Willmitzer, L., Kossmann, J., 1999. The influence of alterations in ADP glucose pyrophosphorylase activities on starch structure and composition in potato tubers. Planta 209, 230–238.

Lobato, M.C., Machinandiarena, M.F., Tambascio, C., Dosio, G.A.A., Caldiz, D.O., Daleo, G.R., Andreu, A.B., Olivieri, F.P., 2011. Effect of foliar applications of phosphate on post-harvest potato tubers. European Journal of Plant Pathology 130 (2), 155–163.

Louis, B., Roy, P., Sayanika, D.W., Talukdar, N.C., 2013. *Aspergillus terreus* Thom a new pathogen that causes foliar blight of potato. Plant Pathology & Quarantine 3 (1), 29–33.

Louis, B., Waikhom, D., Roy, P., Bhardwaj, P.K., Singh, M.W., Chandrade, S.K., Talukdar, N.C., 2014. Invasion of *Solanum tuberosum* L. by *Aspergillus terreus*: a microscopic and proteomics insight on pathogenicity. BMC Research Notes 7, 350.

Low, T.Y., van Heesch, S., van den Toorn, H., Giansanti, P., Cristobal, A., Toonen, P., Schafer, S., Hübner, N., van Breukelen, B., Mohammed, S., Cuppen, E., Heck, A.J., Guryev, V., 2013. Quantitative and qualitative proteome characteristics extracted from in-depth integrated genomics and proteomics analysis. Cell Reports 5 (5), 1469–1478.

Machlon, A.E.S., 1984. Calcium fluxes at plasmalemma and tonoplast. Plant, Cell & Environment 7, 423–429.

Mano, S., Nishimura, M., 2005. Plant peroxisomes. Vitamins & Hormones 72, 111–154.

Martínez-Maqueda, D., Hernández Ledesma, B., Amigo, L., Miralles, B., Gómez Ruiz, J.Á., 2013. Extraction/Fractionation techniques for proteins and peptides and protein digestion. In: Toldrá, F., Nollet, L.M.L. (Eds.), Proteomics in Foods: Principles and Applications. Springer, US, pp. 21–51.

Martins, R.P., Finan, J.D., Guilak, F., Lee, D.A., 2012. Mechanical regulation of nuclear structure and function. Annual Review of Biomedical Engineering 14, 431–455.

Maruyama, J., Yamaoka, S., Matsuo, I., Tsutsumi, N., Kitamoto, K., 2012. A newly discovered function of peroxisomes: involvement in biotin biosynthesis. Plant Signaling & Behavior 7 (12), 1589–1593.

Millar, A.H., Whelan, J., Small, I., 2006. Recent surprises in protein targeting to mitochondria and plastids. Current Opinion in Plant Biology 9, 610–615.

Millenaar, F.F., Lambers, H., 2003. The alternative oxidase: in vivo regulation and function. Plant Biology 5, 2–15.

Miller, A.J., Vogg, G., Sanders, D., 1990. Cytosolic calcium homeostasis in fungi: roles of plasma membrane transport and intracellular sequestration of calcium. Proceedings of the National Academy of Sciences of the United States of America 87, 9348–9352.

Morre, J., Mollenhauer, H.H., 2009. The Golgi Apparatus: The First 100 Years. Springer, New York.

Mottram, D.S., Wedzicha, B.L., Dodson, A.T., 2002. Acrylamide is formed in the Maillard reaction. Nature 419, 448–449.

Muneta, P., Kaisaki, F., 1985. Ascorbic acid-ferrous iron (Fe^{++}) complexes and after cooking darkening of potatoes. American Potato Journal 62 (10), 531–536.

Munger, A., Coenen, K., Cantin, L., Goulet, C., Vaillancourt, L.-P., Goulet, M.C., Tweddell, R., Sainsbury, F., Michaud, D., 2012. Beneficial unintended effects of a cereal cystatin in transformed lines of potato, *Solanum tuberosum*. BMC Plant Biology 12, 198.

Murphy, J.P., Kong, F., Pinto, D.M., Wang-Pruski, G., 2010. Relative quantitative proteomic analysis reveals wound response proteins correlated with after-cooking darkening. Proteomics 10, 4258–4269.

Narula, K., Datta, A., Chakraborty, N., Chakraborty, S., 2013. Comparative analyses of nuclear proteome: extending its function. Frontiers in Plant Science 4, 100.

Nelson, C.J., Huttlin, E.L., Hegeman, A.D., Harms, A.C., Sussman, M.R., 2007. Implications of N-15-metabolic labeling for automated peptide identification in *Arabidopsis thaliana*. Proteomics 7, 1279–1292.

Nesterenko, S., Sink, K.C., 2003. Carotenoid profiles of potato breeding lines and selected cultivars. HortScience 38, 1173–1177.

Nishimura, M., Hayashi, M., Kato, A., Yamaguchi, K., Mano, S., 1996. Functional transformation of microbodies in higher plant cells. Cell Structure and Function 21, 387–393.

Noctor, G., De Paepe, R., Foyer, C.H., 2007. Mitochondrial redox biology and homeostasis in plants. Trends in Plant Science 12, 125–134.

Nookaraju, A., Pandey, S.K., Upadhyaya, C.P., Heung, J.J., Kim, H.S., Chun, S.C., Kim, D.H., Park, S.W., 2012. Role of Ca^{2+}-mediated signaling in potato tuberization: an overview. Botanical Studies 53, 177–189.

Noordin, R., Othman, N., 2013. Proteomics technology – a powerful tool for the biomedical scientists. Malaysian Journal of Medical Sciences 20 (2), 1–2.

O'Farrell, P.H., 1975. High resolution two-dimensional electrophoresis of proteins. The Journal of Biological Chemistry 250, 4007–4021.

Oba, K., Fukui, M., Imai, Y., Iriyama, S., Nogaru, K., 1994. L-Galactono-g-lactone dehydrogenase: partial characterization, induction of activity and role in synthesis of ascorbic acid in wounded white potato tuber tissue. Plant Cell Physiology 35, 473–478.

Ohlrogge, J., Browse, J., 1995. Lipid biosynthesis. Plant Cell 7, 957–970.

Ong, S.E., Blagoev, B., Kratchmarova, I., Kristensen, D.B., Steen, H., Pandey, A., Mann, M., 2002. Stable isotope labeling by amino acids in cell culture, SILAC, as a simple and accurate approach to expression proteomics. Molecular & Cellular Proteomics 1, 376–386.

Ong, S.E., Mann, M., 2005. Mass spectrometry-based proteomics turns quantitative. Nature Chemical Biology 1, 252–262.

Ostergaard, J., Persiau, G., Davey, M.W., Bauw, G., Van Montagu, M., 1997. Isolation of a cDNA coding for L-galactono-g-lactone dehydrogenase, an enzyme involved in the biosynthesis of ascorbic acid in plants. The Journal of Biological Chemistry 272, 30009–30016.

Pastore, D., Greco, M., Passarella, S., 2000. Specific helium-neon laser sensitivity of the purified cytochrome c oxidase. International Journal of Radiation Biology 76, 863–870.

Pastorello, E.A., Farioli, L., Pravettoni, V., Ispano, M., Scibola, E., Trambaioli, C., Giuffrida, M.G., Ansaloni, R., Godovac-Zimmermann, J., Conti, A., Fortunato, D., Ortolani, C., 2000. The maize major allergen, which is responsible for food induced allergic reactions, is a lipid transfer protein. Journal of Allergy and Clinical Immunology 106, 744–751.

Pedreschi, R., Hertog, M., Lilley, K.S., Nicolaï, B., 2010. Proteomics for the food industry: opportunities and challenges. Critical Reviews in Food Science and Nutrition 50 (7), 680–692.

Percival, G., 1999. Light-induced glycoalkaloid accumulation of potato tubers (*Solanum tuberosum* L.). Journal of the Science of Food and Agriculture 79, 1305–1310.

Petermann, J.B., Morris, S.C., 1985. The spectral responses of chlorophyll and glycoalkaloid synthesis in potato tubers (*Solanum tubersosum*). Plant Science 39, 105–110.

Petersen, J., Rogowska-Wrzesinska, A., Jensen, O.N., 2013. Functional proteomics of barley and barley chloroplasts – strategies, methods and perspectives. Frontiers in Plant Science 4, 52.

Poburko, D., Demaurex, N., 2012. Regulation of the mitochondrial proton gradient by cytosolic Ca^{2+} signals. European Journal of Physiology 464, 19–26.

Poburko, D., Santo-Domingo, J., Demaurex, N., 2011. Dynamic regulation of the mitochondrial proton gradient during cytosolic calcium elevations. The Journal of Biological Chemistry 286 (13), 11672–11684.

Potato Genome Sequencing Consortium, 2011. Genome sequence and analysis of the tuber crop potato. Nature 475, 189–195.

Pouvreau, L., Gruppen, H., Piersm, S.R., van den Broek, L.A.M., van Koningsveld, G.A., Voragen, A.G.J., 2001. Relative abundance and inhibitory distribution of protease inhibitors in potato juice from cv. Elkana. Journal of Agricultural and Food Chemistry 49, 2864–2874.

Poyton, R.O., Ball, K.A., 2011. Therapeutic photobiomodulation: nitric oxide and a novel function of mitochondrial cytochrome c oxidase. Discovery Medicine 57, 154–159.

Poyton, R.O., Ball, K.A., Castello, P.R., 2009. Mitochondrial generation of free radicals and hypoxic signaling. Trends in Endocrinology and Metabolism 20, 332–340.

Preberg, T., Wrisher, M., Fulgosi, H., Ljubesic, N., 2008. Ultrastructural characterization of the reversible differentiation of chloroplasts in cucumber fruit. Journal of Plant Physiology 51, 122–131.

Pyke, K.A., 2007. Plastid biogenesis and differentiation. In: Bock, R. (Ed.), Cell and Molecular Biology of Plastids, vol. 19. Springer-Verlag, Berlin, pp. 1–28.

Pyke, K.A., 2009. Different Types of Plastids and Their Structure, Plastid Biology. Cambridge University Press, pp. 9–30.

Raghavendra, A.S., Padmasree, K., 2003. Beneficial interactions of mitochondrial metabolism with photosynthetic carbon assimilation. Trends in Plant Science 8, 546–553.

Rempelos, R., Cooper, J., Wilcockson, S., Eyre, M., Shotton, P., Volakakis, N., Orr, C.H., Leifert, C., Gatehouse, A.M.R., Tétard-Jones, C., 2013. Quantitative proteomics to study the response of potato to contrasting fertilisation regimes. Molecular Breeding 31 (2), 363–378.

Reumann, S., Buchwald, D., Lingner, T., 2012. PredPlantPTS1: a web server for the prediction of plant peroxisomal proteins. Frontiers in Plant Science 3, 194.

Riccioni, G., Mancini, B., di Ilio, E., Bucciarelli, T., D'Orazio, N., 2008. Protective effect of lycopene in cardiovascular disease. European Review for Medical and Pharmacological Sciences 12 (3), 183–190.

Rizzuto, R., De Stefani, D., Raffaello, A., Mammucari, C., 2012. Mitochondria as sensors and regulators of calcium signalling. Nature Reviews Molecular Cell Biology 13, 566–578.

Roix, J., Misteli, T., 2002. Genomes, proteomes, and dynamic networks in the cell nucleus. Histochemistry and Cell Biology 118, 105–116.

Römer, S., Lübeck, J., Kauder, F., Steiger, S., Adomat, C., Sandman, G., 2002. Genetic engineering of a zeaxanthin-rich potato by antisense inactivation and cosuppression of carotenoid epoxidation. Metabolic Engineering 4, 263–272.

Ross, P.L., Huang, Y.L.N., Marchese, J.N., Williamson, B., Parker, K., Hattan, S., Khainovski, N., Pillai, S., Dey, S., Daniels, S., Purkayastha, S., Juhasz, P., Martin, S., Bartlet-Jones, M., He, F., Jacobson, A., Pappin, D.J., 2004. Multiplexed protein quantitation in *Saccharomyces cerevisiae* using amine-reactive isobaric tagging reagents. Molecular and Cellular Proteomics 3, 1154–1169.

Rubino, L., Russo, M., 1998. Membrane targeting sequences in tombusvirus infections. Virology 252, 431–437.

Salunkhe, D.K., Salunkhe, K., 1974. Chlorophyll and solanine in potato tubers: formation and control. Acta Horticulturae (ISHS) 38, 257–276.

Salvato, F., Havelund, J.F., Chen, M., Rao, R.S., Rogowska-Wrzesinska, A., Jensen, O.N., Gang, D.R., Thelen, J.J., Møller, I.M., 2014. The potato tuber mitochondrial proteome. Plant Physiology 164 (2), 637–653.

Sarkar, D., 2008. The signal transduction pathways controlling in planta tuberization in potato: an emerging synthesis. Plant Cell Reports 27, 1–8.

Schlüter, U., Benchabane, M., Munger, A., Kiggundu, A., Vorster, J., Goulet, M.C., Cloutier, C., Michaud, D., 2010. Recombinant protease inhibitors for herbivore pest control: a multitrophic perspective. Journal of Experimental Botany 61, 4169–4183.

Schmidt, A., Kellermann, J., Lottspeich, F., 2005. A novel strategy for quantitative proteomics using isotope-coded protein labels. Proteomics 5, 4–15.

Schwall, G.P., Safford, R., Westcott, R.J., Jeffcoat, R., Tayal, A., Shi, Y.C., Gidley, M.J., Jobling, S.A., 2000. Production of very-high-amylose potato starch by inhibition of SBE A and B. Nature Biotechnology 18, 551–554.

Seo, S., Karboune, S., Archelas, A., 2014. Production and characterisation of potato patatin-galactose, galactooligosaccharides, and galactan conjugates of great potential as functional ingredients. Food Chemistry 158, 480–489.

Shukla, P.K., Misra, P., Bains, G., Guru, S.K., Shukla, A., Pant, R.C., 2005. Mitochondria isolation, SDH assay and protein profiling in potato (*Solanum tuberosum* L.) tubers for improving shelf life. In: Keshavachandran, R., Nazeem, P., Girija, D., John, P.S., Peter, K.V. (Eds.), Recent Trends in Horticultural Biotechnology, ICAR National Symposium on Biotechnological Interventions for Improvement of Horticultural Crops: Issues and Strategies, Vellanikkara, Kerala, India, 10–12 January, 2005, vol. 2, pp. 941–946.

Siddique, M.A., Grossmann, J., Gruissem, W., Baginsky, S., 2006. Proteome analysis of bell pepper (*Capsicum annuum* L.) chromoplasts. Plant and Cell Physiology 47, 1663–1673.

Small, I., Peeters, N., Legeai, F., Lurin, C., 2004. Predotar: a tool for rapidly screening proteomes for N-terminal targeting sequences. Proteomics 4, 1581–1590.

Smyth, S., Phillips, P., 2000. Science and regulation: assessing the impacts of incomplete institutions and information in the global agricultural biotechnology industry. In: Proceedings of the ICABR Conference. University of Rome 'Tor Vergara'.

Soll, J., 2002. Protein import into chloroplasts. Current Opinion in Plant Biology 5, 529–535.

Sowokinos, J.R., 2001. Biochemical and molecular control of cold-induced sweetening in potatoes. American Journal of Potato Research 78, 221–236.

Spicer, E.J.F., Goldenthal, E.I., Ikeda, T., 1999. A toxicological assessment of curdlan. Food and Chemical Toxicology 37, 455–479.

Spurgeon, D., 1976. Hidden Harvest: A Systems Approach to Postharvest Technology. International Development Research Centre, Ottawa.

Stark, R.E., Sohn, W., Pacchiano Jr., R.A., Al-Bashir, M., Garbow, J.R., 1994. Following suberization in potato wound periderm by histochemical and solid-state 13C nuclear magnetic resonance methods. Plant Physiology 104, 527–533.

Steele Scott, N., Tymms, M.J., Possingham, J.V., 1984. Plastid-DNA levels in the different tissues of potato. Planta 161 (1), 12–19.

Stensballe, A., Hald, S., Bauw, G., Blennow, A., Welinder, K.G., 2008. The amyloplast proteome of potato tuber. FEBS Journal 275, 1723–1741.

Syahariza, Z.A., Sar, S., Hasjim, J., Tizzotti, M.J., Gilbert, R.G., 2013. The importance of amylose and amylopectin fine structures for starch digestibility in cooked rice grains. Food Chemistry 136, 742–749.

Tambascio, C., Covacevich, F., Lobato, M., Lasa, C., Caldiz, D., Dosio, G., Andreu, A., 2014. The Application of K Phosphites to seed tubers enhanced emergence, early growth and Mycorrhizal Colonization in potato (*Solanum tuberosum*). American Journal of Plant Sciences 5 (1), 132–137.

Thomson, W.W., Lewis, L.N., Coggins, C.W., 1967. The reversion of chromoplasts to chloroplasts in Valencia oranges. Cytologia 32, 117–124.

Thornton, M.K., 2002. Effects of heat and water stress on the physiology of potatoes. In: Presented at the Idaho Potato Conference.

Tyers, M., Mann, M., 2003. From genomics to proteomics. Nature 422, 193–197.

Urbany, C., Colby, T., Stich, B., Schmidt, L., Schmidt, J., Gebhardt, C., 2012. Analysis of natural variation of the potato tuber proteome reveals novel candidate genes for tuber bruising. Journal of Proteome Research 11 (2), 703–716.

van Es, A., Hartmans, K.J., 1987. Starch and sugars during tuberization, storage and sprouting. In: Rastovski, A., van Es, A. (Eds.), Storage of Potatoes. Wageningen, Netherlands, pp. 79–113.

Vowinckel, J., Capuano, F., Campbell, K., Deery, M.J., Lilley, K.S., Ralser, M., 2014. The beauty of being (label)-free: sample preparation methods for SWATH-MS and next-generation targeted proteomics. F1000 Research 2, 272.

Walter, P., Blobel, G., 1983. Preparation of microsomal membranes for cotranslational protein translocation. Methods in Enzymology 96, 84–93.

Wang, Q., Liu, Y., Xie, Y., You, M., 2006. Cryotherapy of potato shoot tips for efficient elimination of potato leafroll virus (PLRV) and potato virus Y (PVY). Potato Research 49, 119–129.

Wang, Y.Q., Yang, Y., Fei, Z., Yuan, H., Fish, T., Thannhauser, T.W., Mazourek, M., Kochian, L.V., Wang, X., Li, L., 2013. Proteomic analysis of chromoplasts from six crop species reveals insights into chromoplast function and development. Journal of Experimental Botany 64, 949–961.

Wang-Pruski, G., Nowak, J., 2004. Potato after-cooking darkening. American Journal of Potato Research 81, 7–16.

Weber, H., 2002. Fatty acid-derived signals in plants. Trends in Plant Science 7, 217–224.

Whelan, S.P., Zuckerbraun, B.S., 2013. Mitochondrial signaling: forwards, backwards, and in between. Oxidative Medicine and Cellular Longevity 2013, 351613.

Wilson, K.L., Dawson, S.C., 2011. Functional evolution of nuclear structure. Journal of Cell Biology 195, 171–181.

Wood, L.G., Garg, M.L., Powell, H., Gibson, P.G., 2008. Lycopene-rich treatments modify noneosinophilic airway inflammation in asthma: proof of concept. Free Radical Research 42 (1), 94–102.

Woodward, A.W., Bartel, B., 2005. Auxin: regulation, action, and interaction. Annals of Botany 95, 707–735.

Xue, K., Liu, B., Yang, J., Xue, D., 2010. The integrated risk assessment of transgenic rice: a comparative proteomics approach. Electronic Journal of Environmental, Agricultural and Food Chemistry 9, 1693–1700.

Yu, J.W., Choi, J.S., Upadhyaya, C.P., Kwon, S.O., Gururani, M.A., Nookaraju, A., Nam, J.H., Choi, C.W., Kim, S.I., Ajappala, H.A., Kim, H.S., Jeon, J.H., Park, S.W., 2012. Dynamic proteomic profile of potato tuber during its in vitro development. Plant Science 195, 1–9.

Zeng, Y., Pan, Z., Ding, Y., Zhu, A., Cao, H., Xu, Q., Deng, X., 2011. A proteomic analysis of the chromoplasts isolated from sweet orange fruits [*Citrus sinensis* (L.) Osbeck]. Journal of Experimental Botany 62, 5297–5309.

Zhu, Y.S., Merkle-Lehman, D.L., Kung, S.D., 1984. Light-induced transformation of amyloplasts into chloroplasts in potato tubers. Plant Physiology 75, 142–145.

Ziegler, P., 1995. Carbohydrate degradation during seed development. In: Kigel, J., Galili, G. (Eds.), Seed Development and Germination. Dekker, New York, pp. 447–474.

Zolla, L., Rinalducci, S., Antonioli, P., Righetti, P.G., 2008. Proteomics as a complementary tool for identifying unintended side effects curring in transgenic maize seeds as a result of genetic modifications. Journal of Proteome Research 7, 1850–1861.

CHAPTER 23

Potatoes and Human Health

Mary E. Camire

School of Food and Agriculture, University of Maine, Orono, ME, USA

1. Introduction

The potato (*Solanum tuberosum* L.) is cultivated in much of the world as a nutritious, convenient, shelf-stable food. Research findings have clashed, though, regarding the health benefits and possible hazards of potato consumption. Popular low-carbohydrate weight-loss diets have eschewed the potato for being "fattening." Controversies about the potato's value to human health have spurred debates on whether public subsidies for fresh potato purchase should be prohibited. On the other hand, potato offers a myriad of healthful compounds, which I once referred to as a "phytochemical jewel box" at a Potato Association of America conference. The potato is a satisfying, nutritious, and inexpensive vegetable. This chapter focuses on new findings on potato health benefits since the 2009 review on this subject (Camire et al., 2009). Other related reviews include King and Slavin (2013) and Zaheer and Akhta (2014). Please refer to other chapters in this book for detailed information on related information such as nutrient content and glycemic index (GI).

2. Nutrient Contributions

Potatoes are rarely consumed raw. Table 1 compares selected nutrient contents among potatoes prepared by various methods. Cooking increases the digestibility of macronutrients that, in turn, may improve the bioavailability of vitamins and minerals. The method of cooking and cooking time and temperature control the water loss from potatoes. As more water is removed, other constituents may become more concentrated. Removal of the peel tends to reduce the dietary fiber content. Frying simultaneously dehydrates and adds oil to potatoes, thereby raising the energy content significantly. Decker and Ferruzzi (2013) recommended that the potato industry explores new forms of food processing that protect nutrients without using immersion frying. Diverse types of phytochemicals are also found in potatoes (Liu, 2013) (Table 2), but distribution varies by cultivar and tuber maturity. Potato phytochemical reviews have summarized the types and quantities of these compounds (Burlingame et al., 2009; Ezekiel et al., 2013). Lovat et al. (2015) reviewed the biochemical pathways involved in antioxidant synthesis in potato; understanding these pathways could lead to new cultivars

Advances in Potato Chemistry and Technology. http://dx.doi.org/10.1016/B978-0-12-800002-1.00023-6

Table 1: Selected Nutrient Contents of Raw and Prepared Potatoes per 100 g.[a]

Potato Type	Water (g)	Energy (kcal)	Fat (g)	Dietary Fiber (g)	Potassium (mg)	Vitamin C (mg)	Type Size
Baked Russet with skin	74.45	97	0.13	2.3	550	8.30	Medium, 173 g
Plain chips with salt	1.86	532	33.98	3.1	1275	21.6	28-g (1-oz) bag
Fast food french fries	36.63	323	15.47	3.9	596	5.6	Medium, 117 g
Raw	78.58	79	0.08	1.3	417	5.7	Small, 170 g

[a]Adapted from U.S. Department of Agriculture, Agricultural Research Service (2013). USDA National Nutrient Database for Standard Reference, Release 26. Nutrient Data Laboratory Home Page, http://www.ars.usda.gov/ba/bhnrc/ndl.

Table 2: Phytochemicals in Potatoes.[a]

Phytochemical Class	All Potatoes	Yellow-Fleshed Potatoes	Potatoes with Red, Blue, or Purple Coloring
Phenolic acids	+		
Anthocyanins			+
Flavonols	+		
Flavanols	+		
Carotenoids	+	++	
Glycoalkaloids	+		
Alkaloids	+		
Protease inhibitors	+		

[a]+, indicates present; ++, indicates present in higher quantities than in other types of potato.

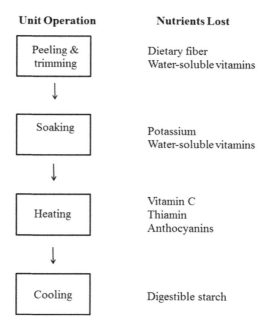

Figure 1

Nutrient and phytochemicals lost during various steps of potato processing.

with higher antioxidant activity. Nutrient and phytochemical losses from potatoes may be due to thermal degradation, oxidation, physical removal such as peeling, and soaking and cooking cut, peeled potatoes in water (Figure 1). Ascorbic acid, for example, is thermally labile, an oxygen scavenger, and water soluble.

Processed potatoes are not necessarily devoid of nutrients, however. Kondo et al. (2014) used a strain of mice that could not synthesize ascorbic acid (SMP30/GNL knockout mice) to assess the contribution of potato chip vitamin C to in vivo oxidative stress. Mice fed a diet

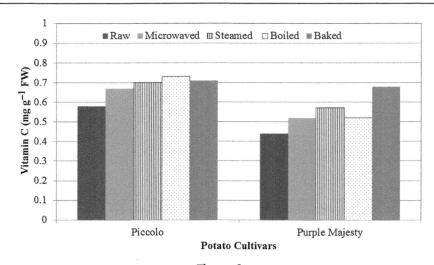

Figure 2
Vitamin C content of two potato cultivars for raw cooked tubers. *Data adapted from Navarre et al. (2010).*

with 20% or 10% potato chips had reduced oxidative stress in multiple tissues versus mice fed a diet devoid of ascorbic and dehydroascorbic acids. Clearly, a diet with such high potato chip consumption is not advisable for humans; however, the study demonstrated that nutritionally relevant amounts of ascorbic acid are retained during chip manufacture. A pilot study involving five healthy young Japanese men established that ascorbic acid was well absorbed from mashed potato and potato chips, but the food matrices slowed absorption relative to an aqueous solution of the vitamin (Kondo et al., 2012).

Home and food service cooking practices vary in their effects on nutrient and phytochemical retention and bioavailability. Immature, also referred to as new, potatoes had higher values of total and individual phenolics on a dry matter basis after being cooked by several methods (Navarre et al., 2010). Vitamin C showed a similar trend (Figure 2). Boiled deep-yellow-fleshed potatoes lost some lutein and zeaxanthin during in vitro digestion, but retained sufficient amounts to supply 14–50% of the recommended levels of those carotenoids for the Andean population (Burgos et al., 2013). Another study employing in vitro digestion assessed the release of polyphenols from boiled *S. tuberosum* cv. Nicola and cv. Vitelotte (purple-fleshed and a source of anthocyanins) (Miranda et al., 2013). The simulated digestions freed bound polyphenolic compounds in the gastric and intestinal phases, but extracts containing those compounds also inhibited ferritin synthesis, which in turn might compromise iron status in vivo.

Potato GI and resistant starch are addressed in detail in another chapter of this book, but starch digestibility has important health effects that will be summarized here. Raw potato starch

Table 3: Potential Health Effects of Resistant Starch.[a]

Primary Effect	Secondary Effect	Tertiary Effect
Prebiotic	↑ Production of short-chain fatty acids	↓ Serum cholesterol levels ↓ Serum triglycerides ↓ Risks for colorectal cancer ↓ Inflammatory bowel disease ↓ Appetite ↓ Diarrhea
Laxation	↑ Mineral absorption Improved immune function ↓ Starch digestibility	↓ Postprandial serum glucose Weight control

[a]Adapted from Birt et al. (2013) and Raigond et al. (2015).

granules are indigestible; potatoes must be cooked prior to consumption. Resistant starch in foods, including potato, has numerous potential health benefits (Table 3). Slavin (2013) advocated consumption of white vegetables, including potatoes, for their dietary fiber and resistant starch content. Some reviews have addressed food processing effects on potato GI (Nayak et al., 2014) and resistant starch health effects (Birt et al., 2013; Raigond et al., 2015).

The popularity of potato in the United States influences the relative dietary contributions of the vegetable. Freedman and Keast (2011) analyzed the 2003–2006 U.S. National Health and Nutrition Examination Survey (NHANES) to assess how many nutrients were provided by white potatoes, oven-baked fries, and french fries consumed by children ages 2–13 years and adolescents ages 14–18 years. Adolescents consumed considerably more potatoes than did younger children. Potato products, excluding chips and snacks, provided important levels of key nutrients such as fiber, vitamin C, niacin, and potassium (Table 4). Branum and Rossen (2014) analyzed the 2003–2008 NHANES data and concluded that mixed dishes supply additional vegetables to children's diets. When whole foods only were included in calculations, fried potatoes accounted for one-third of the vegetables consumed by children and adolescents, but if mixed dishes were included in estimates, then the fried potato share dropped to 18.7% of vegetable intake. Whole vegetables were 57%, and mixed dishes 43%, of overall total vegetable intake. Fried potatoes and potato chips were among the most frequently consumed whole vegetables (Branum and Rossen, 2014). Olsho and Fernandes (2013) questioned whether policies to reduce white potato, particularly fried forms, would result in dietary changes among children and adolescents.

Despite having relatively low total phenolic content relative to spinach and other vegetables, potato was estimated to provide 24.9% of phenolics and cellular antioxidant activity in American diets (Song et al., 2010). Zhang et al. (2010) surveyed residents of the northern China city of Harbin and calculated that potato supplied 8% of the daily dietary intake of

Table 4: Comparison of Selected Nutrients Supplied by Potato Products Consumed by American Adolescents Ages 14–18 years with Dietary Reference Intakes for those Nutrients.

Nutrient	Male		Female	
	Consumed[a]	Percentage of Recommended[b,c,d]	Consumed[a]	Percentage of Recommended[b,c,d,e]
Dietary fiber	3.7 g	9.7	2.5 g	9.6
Vitamin C	10.6 mg	14.1	5.7 mg	8.8
Niacin	2.6 mg	16.2	1.8 mg	12.9
Potassium	673.3 mg	14.3	483.5 mg	10.3

[a]Adapted from Freedman and Keast (2011).
[b]Calculated from values in National Research Council (2005a).
[c]Calculated from values in National Research Council (2000).
[d]Calculated from values in National Research Council (1998).
[e]Calculated from values in National Research Council (2005b).

flavonols and flavones (approximately 1.51 mg/day). Potato is a popular crop in that region because of the cold climate.

3. Dietary Guidance

King and Slavin (2013) pointed out that while potatoes provide dietary fiber and potassium, two nutrients highlighted in the 2010 Dietary Guidelines for Americans, nutritional guidance programs have struggled with the best classification for potatoes. Are they a vegetable or a starch? Classifications based on color cannot easily include the breadth of potato colors—from white to yellow to deep purple.

Changing lifestyles influence the ratio of meals consumed at home to foods prepared away from home. An evaluation of the 2003–2004 U.S. NHANES revealed that the Americans surveyed consumed 0.25 cups of potato per day versus 0.15 cups prepared away from home (Lin et al., 2013). The beneficial effects of potato nutrients may have been overshadowed by the added lipid and sodium in fried potatoes, which have dominated the potatoes eaten away from home food category. Portion size is especially difficult to assess for potatoes because of vast differences in tuber size (2 cm) to specially selected baking potatoes for premium steak restaurants (>12 cm, 453 g). Median recommended potato serving sizes in the United Kingdom ranged from 125 to 240 g (Lewis et al., 2012), creating additional challenges for consumers who dine out.

4. Obesity

Potatoes are often criticized as being a fattening food. A review of the Nurses' Health Studies and the Health Professionals Follow-up Study reported that increased consumption of potato chips by nurses and physicians was associated with a 1.5-kg weight gain, while higher consumption of boiled, mashed, and fried potatoes led to a gain of just 0.26 kg (Mozaffarian et al., 2011). Over a 12-week period, overweight adults provided with free potatoes had a small weight loss, whether or not they were advised to cut 500 kcal per day from their diets or were advised to follow a low- or high-GI diet (Randolph et al., 2014). Several possible mechanisms may contribute to potato's ability to curb obesity. An extract from Onaway and Russet Burbank potatoes that was high in polyphenols prevented weight gain in mice fed a high-fat diet (Kubow et al., 2014).

Curbing appetite by prolonging feelings of fullness or satiety is one strategy for reducing energy intake. Although consumption of boiled potatoes resulted in higher serum glucose levels in healthy young adult volunteers ($n = 13$) with normal body mass index (BMI), compared with a test meal of french fries, boiled and mashed potatoes were more satiating when meals were adjusted for equal energy (Leeman et al., 2008). A dozen young, healthy adults who were not overweight or obese were fed different test meals containing various carbohydrate sources and then were provided with an ad libitum lunch (Geliebter et al., 2013). Although no differences were found for hunger or fullness among treatments, after 120 min subjects reported being able to eat less after consuming a baked potato than after consuming brown rice ($p = 0.04$). Overweight and obese persons may have different reactions to satiating foods, so further testing with individuals with high BMI would be useful in determining whether potato products could help them reduce daily energy consumption.

Rising incidences of overweight and obese children in many nations have spurred evaluation of energy-dense foods in the diets of children and adolescents. A survey of American children ages 2–3 years revealed that french fries and other fried potato products were the vegetables most often consumed by the children (Fox et al., 2010). Mashed or whipped potatoes were also among the top five vegetables eaten by the young children on a regular basis. Fox et al. (2009) reviewed the 2004–2005 School Nutrition Dietary Assessment study and concluded that elementary school children were more likely to become obese if they attended a school that served desserts and french fries, potato puffs, and oven-baked fries more than once per week. The cross-sectional nature of that study does not necessarily prove cause and effect. Over a 20-year period, national nutrition surveys revealed that U.S. children were consuming smaller amounts of fried potato in 2009–2010 compared with diets in 1994–1998 and 2003–2004 (Slining et al., 2013). This change was probably due to changes in school lunch policy as a result of the Dietary Guidelines for Americans. Children who engaged in extra screen time, a risk factor for obesity, tended to eat more french fries and chips (crisps) (Bornhorst et al., 2015). The extra energy and fat from those products may interfere with sleep.

5. Diabetes

The global type 2 diabetes incidence was more than 8% of adults in 2014, and concerted efforts are needed to control this disease (Hu et al., 2015). The relatively high GI of potato dishes has made the vegetable a target for diabetes control. Overweight and obese adults with elevated blood pressure took part in a study that evaluated low and high carbohydrate consumption with a DASH diet for hypertension (Sacks et al., 2014). Low- and high-GI carbohydrates were also assessed for a total of four experimental diets; subjects consumed two different diets during two 5-week test periods. The researchers were surprised to find that after the high-carbohydrate diet, the low-GI diet reduced insulin sensitivity by 20% ($p = 0.002$) from that of the high-GI, high-carbohydrate diet, and for the low-carbohydrate diets, the low-GI version had higher fasting glucose levels ($p = 0.007$). Achieving normal glycemic balance may be more challenging than once thought. Hätönen et al. (2011) reported that consumption of mashed potatoes with high protein food such as chicken or a high-fat food such as rapeseed oil modulated GI. A test meal of mashed potato, chicken breast, rapeseed oil, and salad had a GI of 54, while a GI of 108 was found for mashed potatoes alone.

6. Immune Function and Inflammation

Although potatoes contain many different types of antioxidants, breeding programs had not selected for these compounds. Lovat et al. (2015) reviewed the biosynthetic pathways for all the main classes of antioxidants to facilitate the development of new antioxidant-rich cultivars. The diversity of antioxidants in potatoes also modulates inflammation. Potato consumption was associated with reduced incidence of current wheeze (adjusted OR (oddsratio) = 0.44, $p = 0.005$) and eczema (adjusted OR = 0.54, $p = 0.03$) in Colombian children ages 6–7 years (Cepeda et al., 2015), but the research team did not explain the observed effect. Men fed purple potatoes (experiment breeding variety PORO4PG82-1) daily for 6 weeks had lower levels of plasma high-sensitivity C-reactive protein ($p < 0.05$) and interleukin-6 ($p < 0.07$) than did men in a control group fed white potatoes (cv. Ranger Russet) (Kaspar et al., 2011). Other markers of oxidative stress and inflammation were not different among potato types. The purple potatoes had higher levels of anthocyanins and total antioxidant activity than did the white or yellow potatoes evaluated.

7. Cardiovascular Disease

Cardiovascular health is multifaceted. Blood pressure (too low or too high), arterial blockage by plaque, congestive heart failure, infarction, stroke, and other conditions are encompassed by the term cardiovascular disease (CVD). Potatoes contain potassium, dietary fiber, and antioxidants that may attenuate risks for CVD (McGill et al., 2013). Figure 3 shows compounds in potatoes and their possible mechanisms of action. Although potatoes naturally contain low amounts of sodium, potatoes served in restaurants, particularly fried potatoes, and processed potato products may contain added sodium chloride and

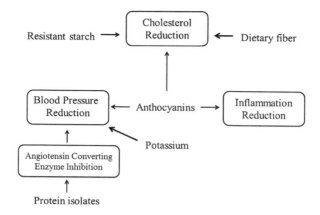

Figure 3

Potato constituents that reduce risks for CVD. Thicker arrows indicate stronger research support for inhibition.

other sodium salts. Concerns about excess sodium and hypertension have led to some changes in the food industry. Ahuja et al. (2015) compared sodium values in popular U.S. foods and found that sodium in restaurant french fries had been reduced by 28% since 2010. Potato chips had a 14–21% reduction depending upon flavoring, but mashed potatoes increased in sodium by 35%.

Consumption of six to eight small Purple Majesty potatoes for 4 weeks by 18 healthy young adults modestly reduced diastolic blood pressure by 4 mm Hg (Vinson et al., 2012). The small pool of subjects varied considerably in their blood pressure status at the start of the study, and their systolic blood pressure decreased, although the change was not statistically significant. No changes in serum lipids or weight were found in the subjects compared with the arm of the study in which potatoes were not eaten. A case–control study in Iran was one of the few human studies to assess separately fried versus boiled potato consumption (Khosravi-Boroujeni et al., 2013). One hundred ninety-five people who were hospitalized with a stroke were compared to patients who were hospitalized for other reasons. People in the third quartile for potato consumption were calculated to have a higher risk of stroke (OR = 2.2, 95% CI (confidence interval) = 1.0–4.7). Curiously, persons in the highest quartile for potato consumption were younger, had lower BMI, were more active, and ate fruits, vegetables, legumes, and grains—all factors that would be expected to lower one's risk.

High levels of serum cholesterol, particularly low-density lipoprotein (LDL) cholesterol, increase the risk for atherosclerosis. Rats fed supplementary potato pulp residue from a starch factory for 4 weeks had no difference in high-density lipoprotein compared with rats fed a control diet, but total cholesterol and LDL plus other forms of cholesterol were significantly lower in the potato pulp diet (Hashimoto et al., 2011). Potato prebiotic activity was thought to be responsible for the cholesterol reduction because cecal pH dropped, and counts of probiotic bacterial species rose, with the potato pulp diet. A 24-week study of rats fed a diet supplemented with potato pulp fiber or cellulose reported that individual and total short-chain fatty acids (SCFAs) were higher in the ceca of rats on the potato fiber diet (Pastuszewska et al., 2010). Although serum triglycerides and very low density lipoprotein cholesterol were significantly lower in the potato fiber-fed rats after 12 weeks, by 24 weeks the only significant differences between diet groups were lower glucose and testosterone in the potato fiber-fed animals. Concerns about the atherosclerotic effects of *trans* fatty acids (TFAs) has prompted many manufacturers to alter their products to exclude these fats that are usually associated with partially hydrogenated vegetable oils. In Poland, an average crisps (chips) and fries intake of about 20 g contributed 0.187 g of daily TFAs, which was thought to be within a safe level (Żbikowska et al., 2015).

8. Cancer

A "meat and potatoes" diet is a classic food pattern in the United States, Canada, and many other nations. Potato consumption was positively associated with consumption of red and processed meats ($p < 0.001$) by Finnish adults (Fogelholm et al., 2015), thereby complicating attempts to dissect the roles of certain foods and food groups on disease and obesity risks. Postmenopausal Canadian women who consumed a meat and potatoes diet had higher risks for breast cancer ($p = 0.043$) than did women who were younger or who consumed "healthy" or "ethnic" diets (Catsburg et al., 2015), leading the researchers to recommend avoiding red meats and consuming a plant-based diet. Consumption of starchy tubers such as taro, potato, and sweet potato at three times per week had a significant protective effect (trend $p = 0.03$) against death from renal cell carcinoma among subjects in the Japan Collaborative Cohort Study for the Evaluation of Cancer (Washio et al., 2014). Over 300 Iranian women who had recently been diagnosed with breast cancer completed food-frequency questionnaires that were then compared with the diets of healthy women (Tajaddini et al., 2015). Women in the highest tertile for consumption of baked or boiled potato, an excellent source of resistant starch, had an adjusted OR of 0.45 (95% CI = 0.28–0.71). Breast cancer risk reduction was not as clear for consumption of fried potatoes; women who ate 1.8–17.1 g of fried potatoes daily had a high adjusted OR (1.2; 95% CI = 0.77–2.00) compared to women in the lowest tertile, but women who consumed more than 17.1 g of fried potatoes had a lower OR of 0.64 ($p = 0.005$, 95% CI = 0.39–1.05). Dietary fiber consumption has long been linked with prevention of cancer, but Cheng et al. (2013) found that isolated potato rhamnogalacturonan I pectin was able to retard human HT-29 colorectal adenocarcinoma cells in vitro. Further study of potato cell wall constituents and their role in cancer prevention and mitigation is needed.

Several theories to explain how potato may reduce cancer risks have been proposed. Reddivari et al. (2010) reported that gallic acid and α-chaconine stimulated apoptosis and subsequent cell death of the prostate cancer cell lines LNCaP and PC-3 in vitro, with inhibitory concentrations of 50%. Although chlorogenic acid extracts from raw potatoes were nearly 10 times more effective in vitro for controlling colon human cancer (Caco-2) and liver cancer (HepG2) cell growth compared with anthocyanins (Wang et al., 2011), synergy among potato constituents may occur in vivo. Several research groups have focused on colored-flesh potatoes. Anthocyanin extracts from purple Vitelotte Noire potatoes reduced the viability of four cancer cell lines in a dose-dependent manner (Ombra et al., 2015). Anthocyanin-rich extracts from the same cultivar showed antiproliferative activity against several cancer cell lines (cervical, breast, prostate, and leukemia) in vitro (Bontempo et al., 2013). While extraction of anthocyanins from raw potato may have some commercial benefit, understanding how the anticancer activity is affected by cooking potatoes is undoubtedly more relevant to consumers. Zuber et al. (2015) evaluated 13 potato cultivars at varying stages of maturity and cooked at various temperatures. Colored-flesh cultivars generally showed higher activity against HT-29 cancer cells, and immature tubers were more effective than mature, stored samples. Although tuber and extract heating reduced effectiveness against cancer cells, some extraction techniques showed promise for pharmaceutical processing. Potato stored for 90 days resulted in slightly lower antiproliferative activity against HCT-116 human colon cancer cells, and purple-fleshed potatoes exhibited greater anticancer activity than did varieties with white or yellow flesh (Madiwale et al., 2012). Baking or chipping the tubers further reduced potential health benefits.

Potato glycoalkaloids may also offer some protection against cancer. Lu et al. (2010a) found that a 23 μM solution of α-solanine reduced the viability of melanoma (skin cancer) cell cultures and that a lower concentration of 18.4 μM was effective in slowing the melanoma cells' motility related to metastasis. The glycoalkaloid impaired matrix metalloproteinases MMP-2 and MMP-9 and blocked phosphorylation of c-Jun N-terminal kinase, protein kinase B, and phosphatidylinositide 3-kinase, which are involved in cell signaling related to cancer cell spread. In vitro assays have many limitations, thus Sun et al. (2014) compared the effects of α-solanine in the pancreatic cancer cell lines SW1990 and Panc-1 as well as using athymic *nu/nu* nude mice. Solanine increased cancer cell membrane permeability, and apoptosis was stimulated in vitro. Fifteen days after injection with the human pancreatic cancer cells, half of the mice were fed 5 mg solanine/kg of body weight daily for another 2 weeks. Tumors were smaller in the solanine-fed mice, and their tumors showed higher Bax and lower Bcl-2 levels. Similar results were reported by Lv et al. (2014), with xenograft pancreatic tumor volume decreased by 61% and tumor weight by 43% after administration of α-solanine ($p < 0.05$). This glycoalkaloid may also have value against prostate cancer. Human prostate cancer PC-3 cells lost invasive capacity when treated with α-solanine via mechanisms similar to those described previously for other cancer models (Shen et al., 2014). The other primary glycoalkaloid in potato, α-chaconine, suppressed angiogenesis in cultured bovine aortic endothelial cells by inhibiting MMP-2 (Lu et al., 2010a). Friedman (2015) has compiled a structure-focused review of glycoalkaloids produced by commonly eaten members of the Solanaceae family and the anticancer mechanisms underlying their utility as cancer treatments. The review includes a thought-provoking list of research needs.

Patatin isolated from potato starch manufacturing by-products inhibited proliferation of mouse melanoma B16 cells by spurring apoptosis, possibly related to the protein's antioxidant activity (Sun et al., 2013). The diversity of potential anticancer compounds in potato warrants further research. Translation of in vitro research to human clinical studies is needed since concentrations of potato chemicals used in the studies discussed here may exceed physiological levels obtained by ingestion of potatoes as food.

9. Gastrointestinal Health

Dietary fiber aids laxation and may reduce risks of diverticular disease, which is characterized by small pouches in the small intestine that can become painfully inflamed. The type of fiber consumed may influence gastrointestinal health. Dietary fiber intakes and sources were evaluated from food surveys sent to participants in the United Kingdom Million Women Study (Crowe et al., 2014). Potatoes supplied about 2 g of dietary fiber in all quintiles of fiber intake that represented 15% of total daily fiber consumption. Total fiber intakes reduced the relative risk for diverticular disease, but the relative risk rose to 1.08 in the highest potato fiber group. Women in the higher potato fiber quintiles also consumed red meats, possibly confounding the findings.

The "Paleo Diet" excludes potato on the basis that primitive humans lacked access to potatoes and that high-carbohydrate foods were not a natural part of our ancestors' diet. Vegan humans and gelada baboons donated feces that were in turn mixed with media (Frost et al., 2014). Medium containing shredded raw potato produced more bacterial flora, SCFAs, and gut hormones than that supplemented with grass clippings. A small ($n=32$) study of elderly persons in Spain revealed a strong association between potato consumption and colonic production of the SCFAs acetate, butyrate, and propionate (Cuervo et al., 2013). The leaders of that study postulated that the mixture of dietary fibers (cellulose, pectin, hemicellulose, and lignin) was responsible for the increased acid production, which is thought to confer a variety of health benefits. Interest in the antimicrobial benefits of anthocyanin-rich potatoes spurred Ombra et al. (2015) to evaluate the in vitro inhibition of *Escherichia coli*, *Bacillus cereus*, and *Pseudomonas aeruginosa* by the purple potato variety Vitelotte Noire. Cooked potato extracts subjected to an in vitro digestion produced a higher inhibition of *E. coli* than did the antibiotic drug chloramphenicol.

Gastroparesis, a condition characterized by delayed stomach emptying and discomfort, is frequently made worse by highly flavored and rough foods. Wytiaz et al. (2015) noted that the bland nature of potatoes made the tuber better tolerated by persons with gastroparesis. Researchers using two different mouse models of intestinal inflammation revealed that the glycoalkaloid content of fried potato skins increased colonic inflammation (Iablokov et al., 2010). Dutch adults who had undergone an ileal pouch–anal anastomosis for Crohn disease, ulcerative colitis, and other gastrointestinal conditions found that potato products increased stool consistency, thereby relieving diarrhea often associated with that surgical procedure (Steenhagen et al., 2006).

10. Chronic Kidney Disease

Potassium is a critical mineral with diverse functions in the human body, but persons with chronic kidney disease (CKD) who also have high serum levels of potassium must limit their consumption of the mineral to reduce further strain on the kidneys. Diabetes and hypertension exert strains on kidneys and genetic conditions, and race predisposes individuals to this life-threatening condition (Beto et al., 2014). Potatoes are often a favorite food that CKD patients miss. Particle size reduction (shredding, mincing, cubing) prior to boiling raw potato in water significantly reduces potassium and other mineral content (Bethke and Jansky, 2008). Leaching prior to boiling did not increase mineral losses. Soaking reduced potato potassium content by 16%, but boiling in an excess of water successfully extracted 73% of the mineral (Picq et al., 2014). The addition of the ion-exchange drug Kayexalate®, a sodium polystyrene sulfonate resin, did not increase removal of potassium from solid foods.

11. Consumer Benefits

Potatoes are an ideal vegetable for older adults because they are inexpensive, keep for long periods without refrigeration, and can be easily prepared. Boiled potatoes are a popular dish in Norway. A cross-sectional survey of more than 74,000 Norwegian women revealed that over half (56%) of the women ages 41–70 years ate two or more potatoes daily (Åsli et al., 2015). Potato consumption was associated with older age, smoking, living with others, and lower income and educational levels, leading researchers to conclude that socioeconomic factors must be considered as confounders when assessing the role of potato consumption in health. A mobile vegetable van that delivered fresh vegetables at wholesale cost to low-income seniors in New York increased potato consumption ($p=0.013$) and reduced the number of trips that the program participants made to grocery stores ($p=0.001$) (AbuSabha et al., 2011). Buying potatoes close to their homes enabled these seniors to purchase easily carried amounts. Supermarkets frequently sell fresh potatoes in 4- to 10-lb (1.8- to 4.5-kg) bags that may be too heavy for some elderly persons to carry along with other foods.

Immature or new potatoes have a higher concentration of phytochemicals than do larger tubers of the same cultivar (Payyavula et al., 2013) and thus offer an opportunity for marketing based on more healthful composition. Gourmet new and novel potatoes are sold in smaller (1.5-lb, 0.68-kg) bags at premium prices.

The global spread of potato production is due to many factors, but certainly low cost is an important criterion. School food service programs face higher costs and limited reimbursement from federal funding. Selection of tasty, nutritious, and well-liked foods is a priority for food service directors in these institutions. Drewnowski and Rehm (2013) analyzed the nutritional benefits of vegetables served in the U.S. National School Lunch Program. The cost of each vegetable that provided 10% of the daily value for six nutrients (vitamins A, C, and K; dietary fiber; potassium; and magnesium) was calculated for the Rank 6 Affordability Index. Only beans and peas scored higher than potatoes (Table 5). Drewnowski (2013) also compared the relative cost and nutritional value of vegetables in the American diet by the nutrient-rich foods (NRF) index. The NRF considers nine nutrients that are desirable in the diet (protein; dietary fiber; vitamins A, C, and E; and minerals calcium, iron, magnesium, and potassium) and saturated fat, added sugar, and sodium. The 2010 Dietary Guidelines for Americans (2011) identified the last three nutrients as ones that Americans should reduce in their diets. Potatoes had an NRF per reference amount customarily consumed of 40.0 ± 25.1 and an affordable nutrition index score of 233.7 ± 204.8; both values were below those for dark green and deep yellow vegetables. However, potatoes had a high NRF per dollar, indicating that potatoes provide substantial nutrients at a relatively low cost.

Sensitivity analyses indicated that women allowed to purchase white potatoes with vouchers from the Women, Infants, and Children (WIC) nutrition aid program could increase their and their children's intake of potassium and dietary fiber for less than $2.00 per month (IOM, 2015). The Institute of Medicine recommended that the U.S. Department of Agriculture Food and Nutrition Service allow the purchase of fresh potatoes with WIC vouchers for this reason. Although most Americans consume adequate amounts of potato, WIC program participants are nutritionally and often financially disadvantaged.

12. Conclusions

The potato is a crop easily adapted to a variety of agronomic conditions at relatively low cost that provides a variety of health benefits. Potatoes with colored flesh offer additional health benefits and their consumption should be encouraged. Access to fresh potatoes should not be denied to persons of limited means because these tubers supply numerous nutrients. The food processing and food service industries face the challenge of developing healthful processes that reduce added fat in potato products while preserving nutrient and phytochemical content.

Table 5: Median Dollar Cost per 10% Daily Value for School Lunch Vegetables.[a]

Vegetable Subgroup	Dietary Fiber	Potassium	Magnesium	Vitamin C	Vitamin A	Vitamin K	Rank 6 Score[b]	Rank 6/Cost[c]
Potatoes	0.19	0.14	0.29	0.10	–	6064	49.5	177.9
Beans/peas	0.05	0.10	0.10	0.58	–	3.70	48.5	257.5
Dark green	0.32	0.54	0.60	0.12	0.30	0.17	67.5	108.0

[a]Adapted from Drewnowski and Rehm (2013).
[b]The score is based on the median ranking of the nutrient density of the six nutrients shown. The range of possible scores is 1–98.
[c]Higher ratios indicate most nutritious at least cost.

References

AbuSabha, R., Namjoshi, D., Klein, A., 2011. Increasing access and affordability of produce improves perceived consumption of vegetables in low-income seniors. Journal of the American Dietetic Association 111, 1549–1555.

Ahuja, J.K., Pehrsson, P.R., Haytowitz, D.B., Wasswa-Kintu, S., Nickle, M., Showell, B., Thomas, R., Roseland, J., Williams, J., Khan, M., Nguyen, Q., Hoy, K., Martin, C., Rhodes, D., Moshfegh, A., Gillespie, C., Gunn, J., Merritt, R., Cogswell, M., 2015. Sodium monitoring in commercially processed and restaurant foods. American Journal of Clinical Nutrition 101, 622–631.

Åsli, L.A., Braaten, T., Olsen, A., Lund, E., Skeie, G., 2015. What characterises women who eat potatoes? A cross-sectional study among 74,208 women in the Norwegian Women and Cancer cohort. Food & Nutrition Research 59. http://dx.doi.org/10.3402/fnr.v3459.25703.

Bethke, P.C., Jansky, S.H., 2008. The effects of boiling and leaching on the content of potassium and other minerals in potatoes. Journal of Food Science 73, H80–H85.

Beto, J.A., Ramirez, W.E., Bansal, V.K., 2014. Medical nutrition therapy in adults with chronic kidney disease: integrating evidence and consensus into practice for the generalist registered dietitian nutritionist. Journal of the Academy of Nutrition and Dietetics 114, 1077–1087.

Birt, D.F., Boylston, T., Hendrich, S., Jane, J.L., Hollis, J., Li, L., McClelland, J., Moore, S., Phillips, G.J., Rowling, M., Schalinske, K., Scott, M.P., Whitley, E.M., 2013. Resistant starch: promise for improving human health. Advances in Nutrition 4, 587–601.

Bontempo, P., Carafa, V., Grassi, R., Basile, A., Tenore, G.C., Formisano, C., Rigano, D., Altucci, L., 2013. Antioxidant, antimicrobial and anti-proliferative activities of *Solanum tuberosum* L. var. Vitelotte. Food & Chemical Toxicology 55, 304–312.

Bornhorst, C., Wijnhoven, T.M., Kunesova, M., Yngve, A., Rito, A.I., Lissner, L., Duleva, V., Petrauskiene, A., Breda, J., 2015. WHO European Childhood Obesity Surveillance Initiative: associations between sleep duration, screen time and food consumption frequencies. BMC Public Health 15, 442.

Branum, A.M., Rossen, L.M., 2014. The contribution of mixed dishes to vegetable intake among US children and adolescents. Public Health Nutrition 17, 2053–2060.

Burgos, G., Munoa, L., Sosa, P., Bonierbale, M., zum Felde, T., Diaz, C., 2013. *In vitro* bioaccessibility of lutein and zeaxanthin of yellow fleshed boiled potatoes. Plant Foods for Human Nutrition 68, 385–390.

Burlingame, B., Mouillé, B., Charrondière, R., 2009. Nutrients, bioactive non-nutrients and anti-nutrients in potatoes. Journal of Food Composition and Analysis 22, 494–502.

Camire, M.E., Kubow, S., Donnelly, D.J., 2009. Potatoes and human health. Critical Reviews in Food Science & Nutrition 49, 823–840.

Catsburg, C., Kim, R.S., Kirsh, V.A., Soskolne, C.L., Kreiger, N., Rohan, T.E., 2015. Dietary patterns and breast cancer risk: a study in 2 cohorts. American Journal of Clinical Nutrition 101, 817–823.

Cepeda, A., Del Giacco, S., Villalba, S., Tapias, E., Jaller, R., Segura, A., Reyes, G., Potts, J., Garcia-Larsen, V., 2015. A traditional diet is associated with a reduced risk of eczema and wheeze in Colombian children. Nutrients 7, 5098.

Cheng, H., Zhang, Z., Leng, J., Liu, D., Hao, M., Gao, X., Tai, G., Zhou, Y., 2013. The inhibitory effects and mechanisms of rhamnogalacturonan I pectin from potato on HT-29 colon cancer cell proliferation and cell cycle progression. International Journal of Food Science & Nutrition 64, 36–43.

Crowe, F.L., Balkwill, A., Cairns, B.J., Appleby, P.N., Green, J., Reeves, G.K., Key, T.J., Beral, V., 2014. Source of dietary fibre and diverticular disease incidence: a prospective study of UK women. Gut 63, 1450–1456.

Cuervo, A., Salazar, N., Ruas-Madiedo, P., Gueimonde, M., González, S., 2013. Fiber from a regular diet is directly associated with fecal short-chain fatty acid concentrations in the elderly. Nutrition Research 33, 811–816.

Decker, E.A., Ferruzzi, M.G., 2013. Innovations in food chemistry and processing to enhance the nutrient profile of the white potato in all forms. Advances in Nutrition 4, 345S–350S.

Drewnowski, A., 2013. New metrics of affordable nutrition: which vegetables provide most nutrients for least cost? Journal of the Academy of Nutrition & Dietetics 113, 1182–1187.

Drewnowski, A., Rehm, C.D., 2013. Vegetable cost metrics show that potatoes and beans provide most nutrients per penny. PLoS One 8, e63277.

Ezekiel, R., Singh, N., Sharma, S., Kaur, A., 2013. Beneficial phytochemicals in potato — a review. Food Research International 50, 487–496.

Fogelholm, M., Kanerva, N., Mannisto, S., 2015. Association between red and processed meat consumption and chronic diseases: the confounding role of other dietary factors. European Journal of Clinical Nutrition 69. http://dx.doi.org/10.1038/ejcn.2015.63.

Fox, M.K., Condon, E., Briefel, R.R., Reidy, K.C., Deming, D.M., 2010. Food consumption patterns of young preschoolers: are they starting off on the right path? Journal of the American Dietetic Association 110, S52–S59.

Fox, M.K., Dodd, A.H., Wilson, A., Gleason, P.M., 2009. Association between school food environment and practices and body mass index of US public school children. Journal of the American Dietetic Association 109, S108–S117.

Freedman, M.R., Keast, D.R., 2011. White potatoes, including french fries, contribute shortfall nutrients to children's and adolescents' diets. Nutrition Research 31, 270–277.

Friedman, M., 2015. Chemistry and anticarcinogenic mechanisms of glycoalkaloids produced by eggplants, potatoes, and tomatoes. Journal of Agricultural & Food Chemistry 63, 3323–3337.

Frost, G.S., Walton, G.E., Swann, J.R., Psichas, A., Costabile, A., Johnson, L.P., Sponheimer, M., Gibson, G.R., Barraclough, T.G., 2014. Impacts of plant-based foods in ancestral hominin diets on the metabolism and function of gut microbiota *in vitro*. MBio 5, e00814–00853.

Geliebter, A., Lee, M.I., Abdillahi, M., Jones, J., 2013. Satiety following intake of potatoes and other carbohydrate test meals. Annals of Nutrition & Metabolism 62, 37–43.

Hashimoto, N., Nakamura, Y., Noda, T., Han, K.-H., Fukushima, M., 2011. Effects of feeding potato pulp on cholesterol metabolism and its association with cecal conditions in rats. Plant Foods for Human Nutrition 66, 401–407.

Hätönen, K.A., Virtamo, J., Eriksson, J.G., Sinkko, H.K., Sundvall, J.E., Valsta, L.M., 2011. Protein and fat modify the glycaemic and insulinaemic responses to a mashed potato-based meal. British Journal of Nutrition 106, 248–253.

Hu, F.B., Satija, A., Manson, J.E., 2015. Curbing the diabetes pandemic: the need for global policy solutions. JAMA 313, 2319–2320.

Iablokov, V., Sydora, B.C., Foshaug, R., Meddings, J., Driedger, D., Churchill, T., Fedorak, R.N., 2010. Naturally occurring glycoalkaloids in potatoes aggravate intestinal inflammation in two mouse models of inflammatory bowel disease. Digestive Diseases and Sciences 55, 3078–3085.

Institute of Medicine (IOM), 2015. Review of WIC Food Packages: An Evaluation of White Potatoes in the Cash Value Voucher: Letter Report. The National Academies Press, Washington, DC.

Kaspar, K.L., Park, J.S., Brown, C.R., Mathison, B.D., Navarre, D.A., Chew, B.P., 2011. Pigmented potato consumption alters oxidative stress and inflammatory damage in men. Journal of Nutrition 141, 108–111.

Khosravi-Boroujeni, H., Saadatnia, M., Shakeri, F., Keshteli, A.H., Esmaillzadeh, A., 2013. A case-control study on potato consumption and risk of stroke in central Iran. Archives of Iranian Medicine 16, 172–176.

King, J.C., Slavin, J.L., 2013. White potatoes, human health, and dietary guidance. Advances in Nutrition 4, 393S–401S.

Kondo, Y., Higashi, C., Iwama, M., Ishihara, K., Handa, S., Mugita, H., Maruyama, N., Koga, H., Ishigami, A., 2012. Bioavailability of vitamin C from mashed potatoes and potato chips after oral administration in healthy Japanese men. British Journal of Nutrition 107, 885–892.

Kondo, Y., Sakuma, R., Ichisawa, M., Ishihara, K., Kubo, M., Handa, S., Mugita, H., Maruyama, N., Koga, H., Ishigami, A., 2014. Potato chip intake increases ascorbic acid levels and decreases reactive oxygen species in SMP30/GNL knockout mouse tissues. Journal of Agricultural and Food Chemistry 62, 9286–9295.

Kubow, S., Hobson, L., Iskandar, M.M., Sabally, K., Donnelly, D.J., Agellon, L.B., 2014. Extract of Irish potatoes (*Solanum tuberosum* L.) decreases body weight gain and adiposity and improves glucose control in the mouse model of diet-induced obesity. Molecular Nutrition & Food Research 58, 2235–2238.

Leeman, M., Ostman, E., Bjorck, I., 2008. Glycaemic and satiating properties of potato products. European Journal of Clinical Nutrition 62, 87–95.

Lewis, H.B., Ahern, A.L., Jebb, S.A., 2012. How much should I eat? A comparison of suggested portion sizes in the UK. Public Health Nutrition 15, 2110–2117.

Lin, B.-H., Wendt, M., Guthrie, J.F., 2013. Impact on energy, sodium and dietary fibre intakes of vegetables prepared at home and away from home in the USA. Public Health Nutrition 16, 1937–1943.

Liu, R.H., 2013. Health-promoting components of fruits and vegetables in the diet. Advances in Nutrition 4, 384S–392S.

Lovat, C., Nassar, A.M., Kubow, S., Li, X.Q., Donnelly, D.J., 2015. Metabolic biosynthesis of potato (*Solanum tuberosum* L.) antioxidants and implications for human health. Critical Reviews in Food Science & Nutrition in press.

Lu, M.K., Chen, P.H., Shih, Y.W., Chang, Y.T., Huang, E.T., Liu, C.R., Chen, P.S., 2010a. alpha-Chaconine inhibits angiogenesis in vitro by reducing matrix metalloproteinase-2. Biological & Pharmaceutical Bulletin 33, 622–630.

Lu, M.K., Shih, Y.W., Chang Chien, T.T., Fang, L.H., Huang, H.C., Chen, P.S., 2010b. alpha-Solanine inhibits human melanoma cell migration and invasion by reducing matrix metalloproteinase-2/9 activities. Biological & Pharmaceutical Bulletin 33, 1685–1691.

Lv, C., Kong, H., Dong, G., Liu, L., Tong, K., Sun, H., Chen, B., Zhang, C., Zhou, M., 2014. Antitumor efficacy of alpha-solanine against pancreatic cancer in vitro and in vivo. PLoS One 9, e87868.

Madiwale, G.P., Reddivari, L., Stone, M., Holm, D.G., Vanamala, J., 2012. Combined effects of storage and processing on the bioactive compounds and pro-apoptotic properties of color-fleshed potatoes in human colon cancer cells. Journal of Agricultural & Food Chemistry 60, 11088–11096.

McGill, C.R., Kurilich, A.C., Davignon, J., 2013. The role of potatoes and potato components in cardiometabolic health: a review. Annals of Medicine 45, 467–473.

Miranda, L., Deuber, H., Evers, D., 2013. The impact of *in vitro* digestion on bioaccessibility of polyphenols from potatoes and sweet potatoes and their influence on iron absorption by human intestinal cells. Food & Function 4, 1595–1601.

Mozaffarian, D., Hao, T., Rimm, E.B., Willett, W.C., Hu, F.B., 2011. Changes in diet and lifestyle and long-term weight gain in women and men. New England Journal of Medicine 364, 2392–2404.

National Research Council, 1998. Dietary Reference Intakes for Thiamin, Riboflavin, Niacin, Vitamin B_6, Folate, Vitamin B_{12}, Pantothenic Acid, Biotin, and Choline. The National Academies Press, Washington, DC.

National Research Council, 2000. Dietary Reference Intakes for Vitamin C, Vitamin E, Selenium, and Carotenoids. The National Academies Press, Washington, DC.

National Research Council, 2005a. Dietary Reference Intakes for Energy, Carbohydrate, Fiber, Fat, Fatty Acids, Cholesterol, Protein, and Amino Acids (Macronutrients). The National Academies Press, Washington, DC.

National Research Council, 2005b. Dietary Reference Intakes for Water, Potassium, Sodium, Chloride, and Sulfate. The National Academies Press, Washington, DC.

Navarre, D., Shakya, R., Holden, J., Kumar, S., 2010. The effect of different cooking methods on phenolics and vitamin C in developmentally young potato tubers. American Journal of Potato Research 87, 350–359.

Nayak, B., Berrios, J.D.J., Tang, J., 2014. Impact of food processing on the glycemic index (GI) of potato products. Food Research International 56, 35–46.

Olsho, L.E., Fernandes, M.M., 2013. Relationship of white potato to other vegetable consumption by schoolchildren and adolescents in the USA: National Health and Nutrition Examination Survey, 2003–2008. Public Health Nutrition 16, 1933–1936.

Ombra, M.N., Fratianni, F., Granese, T., Cardinale, F., Cozzolino, A., Nazzaro, F., 2015. *In vitro* antioxidant, antimicrobial and anti-proliferative activities of purple potato extracts (*Solanum tuberosum* cv Vitelotte noire) following simulated gastro-intestinal digestion. Natural Products Research 29, 1087–1091.

Pastuszewska, B., Taciak, M., Tusnio, A., Misztal, T., Ochtabinska, A., 2010. Physiological effects of long-term feeding diets supplemented with potato fibre or cellulose to adult rats. Archives of Animal Nutrition 64, 155–169.

Payyavula, R.S., Navarre, D.A., Kuhl, J., Pantoja, A., 2013. Developmental effects on phenolic, flavonol, anthocyanin, and carotenoid metabolites and gene expression in potatoes. Journal of Agricultural and Food Chemistry 61, 7357–7365.

Picq, C., Asplanato, M., Bernillon, N., Fabre, C., Roubeix, M., Ricort, J.M., 2014. Effects of water soaking and/or sodium polystyrene sulfonate addition on potassium content of foods. International Journal of Food Science & Nutrition 65, 673–677.

Raigond, P., Ezekiel, R., Raigond, B., 2015. Resistant starch in food: a review. Journal of the Science of Food and Agriculture 95, 1968–1978.

Randolph, J.M., Edirisinghe, I., Masoni, A.M., Kappagoda, T., Burton-Freeman, B., 2014. Potatoes, glycemic index, and weight loss in free-living individuals: practical implications. Journal of the American College of Nutrition 33, 375–384.

Reddivari, L., Vanamala, J., Safe, S.H., Miller, J.C., 2010. The bioactive compounds α-chaconine and gallic acid in potato extracts decrease survival and induce apoptosis in LNCaP and PC3 prostate cancer cells. Nutrition and Cancer 62, 601–610.

Sacks, F.M., Carey, V.J., Anderson, C.M., Miller, E.R., Copeland, T., Charleston, J., Harshfield, B.J., Laranjo, N., McCarron, P., Swain, J., White, K., Yee, K., Appel, L.J., 2014. Effects of high vs low glycemic index of dietary carbohydrate on cardiovascular disease risk factors and insulin sensitivity: the OmniCarb randomized clinical trial. JAMA 312, 2531–2541.

Slavin, J.L., 2013. Carbohydrates, dietary fiber, and resistant starch in white vegetables: links to health outcomes. Advances in Nutrition 4, 351S–355S.

Slining, M.M., Mathias, K.C., Popkin, B.M., 2013. Trends in food and beverage sources among US children and adolescents: 1989–2010. Journal of the Academy of Nutrition and Dietetics 113, 1683–1694.

Song, W., Derito, C.M., Liu, M.K., He, X., Dong, M., Liu, R.H., 2010. Cellular antioxidant activity of common vegetables. Journal of Agricultural and Food Chemistry 58, 6621–6629.

Steenhagen, E., de Roos, N.M., Bouwman, C.A., van Laarhoven, C.J., van Staveren, W.A., 2006. Sources and severity of self-reported food intolerance after ileal pouch–anal anastomosis. Journal of the American Dietetic Association 106, 1459–1462.

Sun, H., Lv, C., Yang, L., Wang, Y., Zhang, Q., Yu, S., Kong, H., Wang, M., Xie, J., Zhang, C., Zhou, M., 2014. Solanine induces mitochondria-mediated apoptosis in human pancreatic cancer cells. BioMed Research International 805926, 1–9.

Sun, Y., Jiang, L., Wei, D., 2013. Partial characterization, *in vitro* antioxidant and antiproliferative activities of patatin purified from potato fruit juice. Food & Function 4, 1502–1511.

Tajaddini, A., Pourzand, A., Sanaat, Z., Pirouzpanah, S., 2015. Dietary resistant starch contained foods and breast cancer risk: a case-control study in northwest of Iran. Asian Pacific Journal of Cancer Prevention 16, 4185–4192.

U.S. Department of Agriculture, Agricultural Research Service, 2013. USDA National Nutrient Database for Standard Reference, Release 26. http://www.ars.usda.gov/ba/bhnrc/ndl (accessed 02.07.15.).

U.S. Department of Agriculture and U.S. Department of Health and Human Services, 2010. Dietary Guidelines for Americans, 2010, seventh ed. U.S. Government Printing Office, Washington, DC.

Vinson, J.A., Demkosky, C.A., Navarre, D.A., Smyda, M.A., 2012. High-antioxidant potatoes: acute *in vivo* antioxidant source and hypotensive agent in humans after supplementation to hypertensive subjects. Journal of Agricultural and Food Chemistry 60, 6749–6754.

Wang, Q., Chen, Q., He, M., Mir, P., Su, J., Yang, Q., 2011. Inhibitory effect of antioxidant extracts from various potatoes on the proliferation of human colon and liver cancer cells. Nutrition and Cancer 63, 1044–1052.

Washio, M., Mori, M., Mikami, K., Miki, T., Watanabe, Y., Nakao, M., Kubo, T., Suzuki, K., Ozasa, K., Wakai, K., Tamakoshi, A., 2014. Risk factors for renal cell carcinoma in a Japanese population. Asian Pacific Journal of Cancer Prevention 15, 9065–9070.

Wytiaz, V., Homko, C., Duffy, F., Schey, R., Parkman, H., 2015. Foods provoking and alleviating symptoms in gastroparesis: patient experiences. Digestive Diseases and Sciences 60, 1052–1058.

Zaheer, K., Akhtar, M.H., 2014. Recent advances in potato production, usage, nutrition-a review. Critical Reviews in Food Science & Nutrition in press.

Żbikowska, A., Rutkowska, J., Kowalska, M., 2015. Consumption safety of pastries, confectioneries, and potato products as related to fat content. Journal of the American College of Nutrition 34, 507–514.

Zhang, Y., Li, Y., Cao, C., Cao, J., Chen, W., Zhang, Y., Wang, C., Wang, J., Zhang, X., Zhao, X., 2010. Dietary flavonol and flavone intakes and their major food sources in Chinese adults. Nutrition and Cancer 62, 1120–1127.

Zuber, T., Holm, D., Byrne, P., Ducreux, L., Taylor, M., Kaiser, M., Stushnoff, C., 2015. Optimization of *in vitro* inhibition of HT-29 colon cancer cell cultures by *Solanum tuberosum* L. extracts. Food & Function 6, 72–83.

Index

Note: Page numbers followed by "f" indicate figures and "t" indicate tables.